Lepton and Baryon Number Violation in Particle Physics, Astrophysics and Cosmology

Lepton and Baryon Number Violation in Particle Physics, Astrophysics and Cosmology

Proceedings of the First International Symposium on Lepton and Baryon Number Violation (Lepton–Baryon 98), European Centre for Theoretical Physics, Trento, Italy, 20–25 April 1998

Edited by H V Klapdor-Kleingrothaus and I V Krivosheina

Institute of Physics Publishing
Bristol and Philadelphia

British Library Cataloguing in Publication Data

A catalogue record for this book is available from the British Library.

ISBN 0 7503 0616 5

Library of Congress Cataloging-in-Publication Data are available

Published by Institute of Physics Publishing, wholly owned by The Institute of Physics, London
Institute of Physics Publishing, Dirac House, Temple Back, Bristol BS1 6BE, UK
US Office: Institute of Physics Publishing, The Public Ledger Building, Suite 1035, 150 South Independence Mall West, Philadelphia, PA 19106, USA

Printed in the UK by Galliard (Printers) Ltd, Great Yarmouth, Norfolk

Contents

Preface

The baryon number (B) and lepton number (L) are additive quantum numbers playing an intriguing role in modern physics. No processes have yet been observed in which these quantum numbers have not been conserved, except perhaps for the atmospheric neutrino results. On the other hand there is no underlying symmetry known which would require this, and theories of grand unification generally allow for B and L violation. This naturally plays an important role in the early Universe.

These Proceedings contain the lectures presented at the First International Conference on Lepton and Baryon Number Violation (Lepton–Baryon 98), held at the ECT (European Centre for Theoretical Studies in Nuclear Physics and Related Areas), Trento, Italy, 20–25 April 1998. The lectures cover topics related to B and L violation in particle physics, nuclear physics, astrophysics and cosmology, both from the theoretical and the experimental side. They start with an overview of B and L violation in Grand Unified Theories, with lepton number and lepton flavour violation in left–right symmetric models and and the flavour problem in SUSY. They then discuss more specifically the potential of neutrino physics—including solar and atmospheric neutrino research and double beta decay, proton decay and neutron–antineutron oscillations to probe B and L violation. B and L violation in the early Universe, including electroweak phase transitions, leptogenesis scenarios, neutralino relic densities, cosmic antimatter search and R-parity violation from cosmic-ray antiprotons are covered in a subsequent chapter. In astrophysics, implications of neutrino oscillations for supernovae and dark matter (neutralinos, higgsinos) are discussed. There then follows a study of lepton family violation and R-parity violation in rare decay processes. Finally the search for B and L violation in collider physics is outlined, including inverse neutrinoless double beta decay, sneutrino physics and lepton number violation, LEP searches for R-parity violation, the search for new physics at HERA, and R-parity breaking in future linear colliders.

The organizers express their gratitude to the scientists from many countries who followed up their invitation and contributed to the success of the meeting. It is also a pleasure to thank the staff of the ECT for their efficient cooperation. The help of Mrs C Klehr and Mrs V Traeumer in designing the book cover is gratefully acknowledged. We thank Genevieve Belanger for her permission to use one of her figures for the cover. The organizers are grateful to Professor D Schwalm, director at the Max-Planck-Institut für Kernphysik, for continuous scientific interest and generous support.

H V Klapdor-Kleingrothaus
Conference Chairman

I Krivosheina
Scientific Secretary

Heidelberg, November 1998

1. L. Bento
2. S. Constantini
3. S. Davidson
4. H.V. Klapdor-Kleingrothaus
5. Z. Berezhiani
6. M.G. Giammarchi
7. D. Bryman
8. I.V. Krivosheina
9. R. Nicolaidou
10. K. Huitu
11. H.B. Nielsen
12. M. Raidal
13. R. Molzon
14. G. Belanger
15. Y. Kuno
16. T. Falk
17. L. Covi
18. M. Losada
19. S. Kolb
20. G.W. Buschhorn
21. U. Sarkar
22. M. Corradi
23. J. Maalampi
24. Z. Fodor
25. W. Wallraff
26. A. Guiliani
27. A. v. der Schaaf
28. P. Wintz
29. O. Miranda
30. M. Sato

WOLF

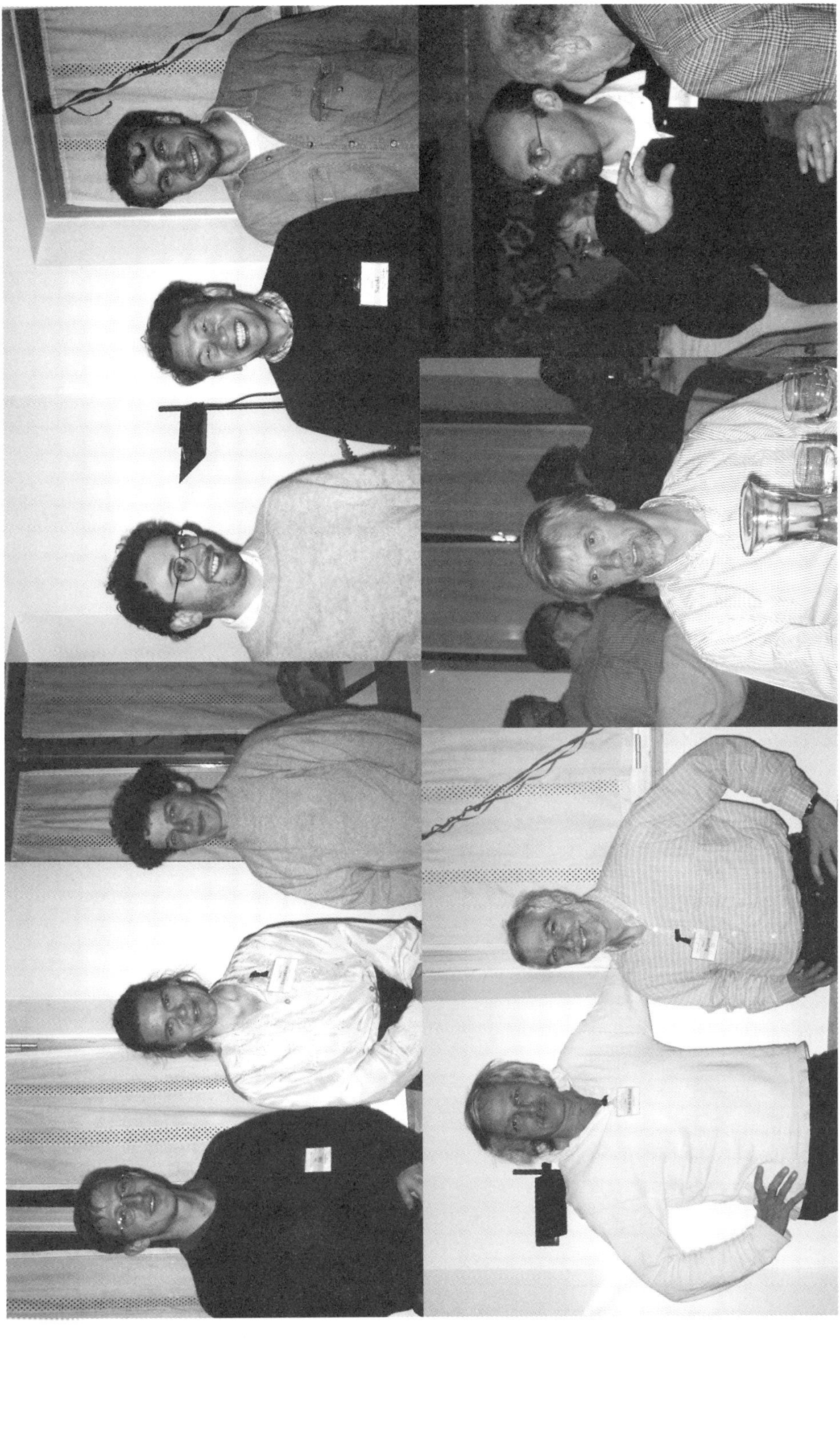

SÜDTIROL ALTO ADIGE
Bënudui Willkommen Benvenuti

SOUTH STREET SEAPORT

PART 1

B and L Violation, Grand Unified Theories and SUSY, Theory

Grand Unification and B & L Conservation

Pran Nath†[1] and R. Arnowitt‡

† Department of Physics, Northeastern University, Boston, MA, 02115, USA

‡ Center for Theoretical Physics, Texas A & M University, ollege Station, TX 77843, USA

Abstract. A review of baryon and lepton conservation in supersymmetric grand unified theories is given. Proton stability is discussed in the minimal SU(5) supergravity grand unification and in several non-minimal extensions such as the $SU(3)^3$, SO(10) and some string based models. Effects of dark matter on proton stability are also discussed and it is shown that the combined constraints of dark matter and proton stability constrain the sparticle spectrum. It is also shown that proton lifetime limits put severe constraints on the event rates in dark matter detectors. Future prospects for the observation of baryon and lepton number violation are also discussed.

1. Introduction

While string theory purports to unify all the fundamental interactions of physics including gravity, there is currently no string model which is fully viable. Pending the discovery of such a model, it is imperative that one look to the bottom up approach to unification. Grand unification is such an approach wherein one attempts to unify the electro-weak and the strong interactions in a single framwork. In this review we will focus on such an approach, specifically, the approach of supersymmetric and supergravity grand unification[1]. Our aim is to discuss the phenomena of baryon and lepton number conservation and their violation in such theories. The grand

[1] E-mail: nath@neu.edu

unification approach is expected to be valid up to the scale $M_G \sim 2 \times 10^{16}$ GeV but it could be valid even up to the string scale of

$$M_{str} \approx 5 \times g_{string} \times 10^{17}\ GeV \quad (1)$$

Beyond that scale one is in the domain of quantum gravity where the full framework of string unification is needed for consistency.

The outline of the paper is as follows: In Sec.2 we discuss the LEP data which lends support to the ideas of supersymmetry and gauge coupling unification. In Sec.3 we give a general discussion of the various sources of baryon number (B) and lepton number (L) violation in SUSY theories. In Sec.4 we discuss proton decay from B & L violating dimension five operators. In Sec.5 we discuss proton decay within the framework of supergravity unification. In Sec.6 we discuss B & L violation in non-minimal models. These include the $(SU(3))^3$ models, the unflipped and the flipped no-scale models, and SO(10) models. Effects of textures on proton lifeitme are discussed in Sec. 7 and effects of dark matter constraints on proton lifetime are discussed in Sec.8. A brief discussion of proton decay in gauge mediated breaking of supersymmetry is given in Sec.9, and a discussion of Planck effects on proton decay is given in Sec.10. Exotic decay modes of the proton are discussed in Sec.11. In Sec.12 we give a brief discussion of the relation between SUSY GUTS and strings. Conclusions and prospects are given in Sec.13.

2. Grand Unification and LEP Data

While the LEP data extrapolated to high scales exhibits unification[2] in the minimal SUSY SU(5), one finds that there could be a small discrepancy in the predicted value of α_s and experiment, i.e., that the predicted value of α_s lies $\sim (1-2)\ \sigma$ higher than the world average. Such deviations can be corrected by Planck scale corrections[3, 4] which can produce $\sim (1-2)$ σ effects on α_s. In fact, effects of such a size are expected because of the proximity of the grand unification scale to the Planck scale, and so effects of size $O(\frac{M_G}{M_{Planck}})$, i.e., O(few%) are quite natural[5]. A source of such corrections in supergravity grand unification is from the gauge kinetic energy function $f_{\alpha\beta}$ where

$$[A\delta_{\alpha\beta} + \frac{c}{2M_{Planck}} d_{\alpha\beta\gamma}\Sigma^{\gamma}]F^{\alpha}{\mu\nu}F^{\beta\mu\nu} \quad (2)$$

Analyses show that values of $c \sim O(1)$ give agreement with LEP data. One may also use the LEP data to put a limit on the range of c .One finds

$$-1 \leq c \leq 3, \quad LEP\ DATA \quad (3)$$

It is also possible to get 1-2 σ effects on α_s from extensions of SU(5), such as, for example, in the missing doublet model with Peccei-Quinn symmetry[6].

3. Sources of Baryon Number Violation in SUSY

We dicuss now the main sources of B & L violation in supersymmetric theories. These consist of baryon and lepton number violation from (i)lepto-quark exchange, (ii) dimension 4 operators, (iii) dimension dimension 5 operators, and (iv) higher dimensional operators. A summary of the important operators in given in Table 1.

Table 1: Baryon and lepton number violating operators

Dim	Operator	B & L violation	comment
dim 6	qqql	$\Delta B = 1$, $\Delta L = 1$	
dim 9	qqqqqq	$\Delta B = 2$, $\Delta L = 0$	$n - \bar{n}$ oscillation
SUSY			
dim 4	$(qqq)_F$	$\Delta B = 1$, $\Delta L = 0$	
dim 5	$(qqql)_F$	$\Delta B = 1$, $\Delta L = 1$	mass scale $\sim 10^{16}$ GeV
dim 7	$(qqqqqq)_F$	$\Delta B = 2$, $\Delta L = 0$	$n - \bar{n}$ oscillation

The experimental possibilites for testing the baryon and lepton number violations are (i) the proton decay decay experiments, (ii) experiments that look for double beta decay[7], and (iii) proposed experiments for the test of $n - \bar{n}$ oscillation[8, 9]. Of these the first two are the ones that have been most vigorously pursued. In this review we will mostly focus on the implications of B & L violations on proton stability.

3.1. Proton Decay via Lepto-quark Exchange

As mentioned already there are various sources of proton decay in grand unified theories. Most grand unified theories allow for baryon and lepton number violation because quarks and leptons belong to the same multiplets and proton decay in these theories can proceed via lepto-quark exchange. In SU(5) models the dominant mode via lepto-quark exchange is $p \to e^+\pi^0$ which gives a life time for this mode of[10]

$$\tau(p \to e^+\pi^0) \approx (\frac{M_V}{3.5 \times 10^{14} GeV})^4 10^{31\pm1}\ yr \tag{4}$$

For the non-supersymmetric SU(5) the lifetime is too small to be consistent with experiment. For the supersymmetric SU(5) one estimates $\tau(p \to e^+\pi^0)$ to be[10]

$$1 \times 10^{35\pm1}\ yr \tag{5}$$

The current experimental limit for this decay mode is[11]

$$\tau(p \to e^+\pi^0) > 2.1 \times 10^{33} yr, (90\% CL) \tag{6}$$

It is expected that in the future Super K will reach a sensitivity of[12]

$$\tau(p \to e^+\pi^0) > 1 \times 10^{34} yr, (90\% CL) \tag{7}$$

Thus the $e^+\pi^0$ mode in SUSY SU(5) may be on the edge of detection if Super-K and Icarus reach their maximum sensitivity. However, as mentioned already in suprsymmetric theories there are other sources of B & L violation, such as dimension 4 and dimension 5 operators.

3.2. p Decay via Dimension 4 Operators

Supersymmetric theories generically have dimension four operators which violate baryon and lepton number. Thus, for example, in the minimal supersymmetric standard model (MSSM) one has in general dimension four operators in the superpotential of the form

$$\begin{aligned} W = \lambda_u Q u^c H_2 + \lambda_d Q d^c H_1 + \lambda_e L e^c H_1 + \mu H_1 H_2 \\ +(\lambda'_B u^c d^c d^c + \lambda'_L Q d^c L + \lambda''_L L L e^c) \end{aligned} \tag{8}$$

Here the terms proportional to λ'_B, λ'_L, and λ''_L induce B & L violation. Suppression of fast p decay requires

$$\lambda'_B \lambda'_L < (\frac{m^2_{\tilde{d}}}{10^{16} GeV})^2 \sim O(10^{-26\pm 1}) \tag{9}$$

In the MSSM one eliminates fast p decay via a discrete R symmetry $R = (-1)^{3B+L+2S}$. However, in general R symmetry which is a global symmetry is not preserved by gravitational interactions. Thus for example, worm holes can generate dimension 4 operators and catalize p decay[13]. To protect against fast p decay of the above type one must promote the global R symmetry to a gauge symmetry, since gauge symmetries are protected against wormhole effects. Even if the local symmetry breaks down leaving behind a residual discrete symmetry, that residual discrete symmetry will be sufficient to protect against fast p decay induced by worm hole effects[14]. R parity is an interesting symmetry in that it appears that it is the only Z_2 symmetry which is free of anomalies with just the MSSM spectrum[15]. Thus there is a possibility that it could arise in an automatic fashion from the spontaneous breakdown of groups that contain a $U(1)_{B-L}$ such as $SU(4)_C$ and SO(10). It should be noted, however, that even if a theory is originally free of dangerous dimension four operators, such operators can be induced from higher dimensional operators via spontaneous symmetry breaking. For example, one may have a dimension five operator in SO(10) which contains an SU(5) singlet field ν^c. The following provide examples where spontaneous VEV formation of an SU(5) singlet generates dangerous dimension four operators

$$\frac{1}{M_P}(u^c d^c d^c \nu^c) \to (u^c d^c d^c), \quad \frac{1}{M_P}(Q L d^c \nu^c) \to (Q L d^c) \tag{10}$$

4. Dimension 5 B and L Violation and p Decay

In MSSM one can write many dimension 5 operators that violate B and L number such as $QQQL$, $u^cu^cd^ce^c$, $QQQH_1$, $Qu^ce^cH_1$ etc. These operators contribute to p decay at the loop level. The only operators that arise in the minimal SU(5) GUT model and generate observable p decay are the first two operators in the list above[16, 17], i.e., $QQQL$, and $u^cu^cd^ce^c$.

B and L violating dimension five operators occur in most supersymmetric theories and string theories and lead to p instability in these models. In SUSY grand unified models proton decay via dimension five operators is governed by the interaction

$$\bar{H}_1 J + \bar{K} H_1 + \bar{H}_i M_{ij} H_j \tag{11}$$

where $H_1, \bar{H}_1$ are the Higgs triplets, J and $\bar{K}$ are matter currents, and M_{ij} is the Higgs triplet mass matrix. The suppression of p decay in these theories can come about if

$$(M^{-1})_{11} = 0 \tag{12}$$

A suppression of this type can occur by discrete symmetries, by non-standard embeddings, or by the presence of additional Higgs triplets[18, 19]. However, in most SUSY/string models (except for the flipped $SU(5) \times U(1)$ models) one does not have a a natural suppression, and a suppression requires a doublet-triplet splitting in the Higgs multiplets. Many mechanisms for doublet-triplet splittings have been discussed in the literature, such as (i) the sliding singlet mechanism[20] which works for SU(n) for n$\geq$ 6, (ii) the missing partner mechanism[21], (iii) the mechanism of VEV alignment[22], (iv) the mechanism where the Higgs doublets are pseudo-Goldstones[23], and (v) the mechanism with more than one adjoint Higgs.

In the following we will discuss p decay via dimension 5 operators in several SUSY GUTS: SU(5), $(SU(3))^3$, SO(10), the flipped and the unflipped no-scale models, with the most emphasis on the simplest SUSY GUTs, i.e., the minimal SU(5) model. The p decay in the minimal SU(5) model is governed by

$$W_Y = -\frac{1}{8} f_{1ij} \epsilon_{uvwxy} H_1^u M_i^{vw} M_j^{xy} + f_{2ij} \bar{H}_{2u} \bar{M}_{iv} M_j^{uv} \tag{13}$$

where M_{ix} and M_i^{xy} (i=1,2,3) are the $\bar{5}$ and 10 of SU(5) which contain the three generations of quarks and leptons, and H_1, H_2 are the $\bar{5}$,5 of Higgs. After the breakdown of the GUT symmetry there is a splittiing of the Higgs multiplets where the Higgs triplets become superheavy and the Higgs doublets remain light by one of the mechanisms listed earlier.

One can now integrate on the Higgs triplet field and obtain an effective interaction at low energy given below

$$LLLL = \frac{1}{M}\epsilon_{abc}(Pf_1^uV)_{ij}(f_2^d)_{kl}(\tilde{u}_{Lbi}\tilde{d}_{Lcj}(\bar{e}^c_{Lk}(Vu_L)_{al} - \nu^c_k d_{Lal}) + ...) + H.c. \tag{14}$$

$$RRRR = -\frac{1}{M}\epsilon_{abc}(V^\dagger f^u)_{ij}(PVf^d)_{kl}(\bar{e}^c_{Ri}u_{Raj}\tilde{u}_{Rck}\tilde{d}_{Rbl} + ...) + H.c. \tag{15}$$

where V is the CKM matrix and f_i are the Yukawa couplings which are related to the quark masses by

$$m_i^u = f_i^u(sin2\theta_W/e)M_Z sin\beta, \quad m_i^d = f_i^d(sin2\theta_W/e)M_Z sin\beta \tag{16}$$

and P_i are generational phases

$$P_i = (e^{i\gamma_i}), \ \sum_i \gamma_i = 0; \ i = 1,2,3 \tag{17}$$

Both LLLL and RRRR interactions must be taken into account in a full analysis and their relative strength depends on the part of the parameter space where their effects are computed[17, 24].

The operators of Eqs.(14) and (15) are dimension five operators which must be dressed via the exchange of gluinos, charginos and neutralinos. The dressings give rise to dimension six operators. These dimension six operators are then used in the computation of proton decay. In the dressings one takes into account the L-R mixings, where for the up squark mass matrix one has

$$\begin{pmatrix} m^2_{Ru} & m_u(A_u m_0 - \mu ctn\beta) \\ m_u(A_u m_0 - \mu ctn\beta) & m^2_{Lu} \end{pmatrix} \tag{18}$$

The mass diagonal states are given by

$$\begin{aligned} \tilde{u}_R = cos\delta_u\tilde{u}_1 + sin\delta_u\tilde{u}_2, \ \tilde{u}_L &= -sin\delta_u\tilde{u}_1 + cos\delta_u\tilde{u}_2 \\ sin2\delta_u &= -2m_u(A_u m_0 - \mu ctn\beta)/(\tilde{m}^2_{u1} - \tilde{m}^2_{u2}) \end{aligned} \tag{19}$$

and similarly for the down squarks and the leptons. In supergravity grand unification the values of the trilinear couplings A_u, A_d, A_e are all related to the single parameter A_0. After dressing of the dimension 5 by the gluino, the chargino and the neutralino exchange diagrams one finds baryon and lepton number violating dimension six operators with chiral structures LLLL, LLRR, RRLL and RRRR which enter in the proton decay analysis.

4.1. *Effective Lagrangian Approach*

The B and L violating interaction one gets from the fundamental SU(5) Lagrangian is in terms of quarks and leptons, while the p decays involve physical mesons and baryons. There are various techniques for bridging the gap, i.e., in going from the fundamental to the phenomenological interactions, such as the Bag model, lattice QCD, and effective Lagrangians. Currently the most efficient of these approaches is that of effecive lagrangians[25]. Here the basic technique consists in finding the effective interactions in terms of mesons and baryons with the same chiral structures as LLLL, LLRR, etc. that appear in the fundamental Lagrangian. This effort is facilitated by utilizing the transformation properties of these terms under $SU(3)_L \times SU(3)_R$ as shown below:

Chiral structure	$SU(3)_L \times SU(3)_R$ *rep*
LLLL	$(8,1)$
$LLRR$	$(3^*,3)$
$RRLL$	$(3,3^*)$
$RRRR$	$(1,8)$

In the effective lagrangian approach one finds combinations of mesonic and baryonic fields with the same chiral transformation properties as the dimension six B& L violating quark fields. For this purpose one defines first a pseudo-goldstone mass matrix

$$M = \begin{pmatrix} \frac{\pi^0}{\sqrt{2}} + \frac{\eta}{\sqrt{6}} & \pi^+ & K^+ \\ \pi^- & -\frac{\pi^0+}{\sqrt{2}} + \frac{\eta}{\sqrt{6}} & K^0 \\ K^- & \bar{K}^0 & -\sqrt{\frac{2}{3}}\eta \end{pmatrix} \tag{20}$$

Similarly, one defines a baryon mass matrix so that

$$B = \begin{pmatrix} \frac{\Sigma^0}{\sqrt{2}} + \frac{\Lambda}{\sqrt{6}} & \Sigma^+ & p \\ \Sigma^- & -\frac{\Sigma^0+}{\sqrt{2}} + \frac{\Lambda}{\sqrt{6}} & n \\ \Xi^- & \Xi^0 & -\sqrt{\frac{2}{3}}\Lambda \end{pmatrix} \tag{21}$$

Defining

$$\xi = e^{i\frac{M}{f}} \tag{22}$$

one can find combinations of mesons and baryon mass matrices with the following transformations

meson – baryon structure	$SU(3)_L \times SU(3)_R$ *rep*
$\xi B \xi^\dagger$	$(8,1)$
$\xi^\dagger B \xi^\dagger$	$(3^*,3)$
$\xi B \xi$	$(3,3^*)$

$$\xi^\dagger B\xi \qquad (1,8)$$

Using the above technique one can write an effective Lagrangian with the same chiral transformation properties as the fundamental Lagriangian in terms of quarks. Currently the effective lagrangian approach is the most reliable approach to the computation of proton decay amplitudes.

In the minimal SU(5) model the dominant decay modes of the proton involve pseudo-scalar bosons and anti-leptons, i.e.,

$$\begin{gathered}\bar{\nu}_i K^+, \bar{\nu}_i \pi^+; i = e, \mu, \tau \\ e^+K^0, \mu^+K^0, e^+\pi^0, \mu^+\pi^0, e^+\eta, \mu^+\eta\end{gathered} \qquad (23)$$

The relative strength of these decay modes depends on various factors, such as quark masses, CKM factors, and third generation effects in the loop diagrams etc. denoted by y_1^{tk} below. The various decay modes and some of the factors that control these decays modes are summarized in Table 2.

Table 2: lepton + pseudoscalar decay modes of the proton

Mode	quark factors	CKM factors
$\bar{\nu}_e K$	$m_d m_c$	$V_{11}^\dagger V_{21} V_{22}$
$\bar{\nu}_\mu K$	$m_s m_c$	$V_{21}^\dagger V_{21} V_{22}$
$\bar{\nu}_\tau K$	$m_b m_c$	$V_{31}^\dagger V_{21} V_{22}$
$\bar{\nu}_e \pi, \bar{\nu}_e \eta$	$m_d m_c$	$V_{11}^\dagger V_{21}^2$
$\bar{\nu}_\mu \pi, \bar{\nu}_\mu \eta$	$m_s m_c$	$V_{21}^\dagger V_{21}^2$
$\bar{\nu}_\tau \pi, \bar{\nu}_\tau \eta$	$m_b m_c$	$V_{31}^\dagger V_{21}^2$
eK	$m_d m_u$	$V_{11}^\dagger V_{12}$
μK	$m_s m_u$	
$e\pi, e\eta$	$m_d m_u$	
$\mu\pi, \mu\eta$	$m_s m_u$	$V_{11}^\dagger V_{21}^\dagger$

The order of magnitude estimates can be gotten using

$$m_u V_{11} : m_c V_{21} : m_t V_{31} \approx 1 : 50 : 500 \qquad (24)$$

Typically the most dominant mode is $\bar{\nu}K$. It is governed by the interaction

$$\begin{gathered}L_6(N \to \bar{\nu}_i K) = ((\alpha_2)^2 (2MM_W^2 sin2\beta)^{-1} P_2 m_c m_i^d V_{i1}^\dagger V_{21} V_{22}) \\ (F(\tilde{c}, \tilde{d}_i, \tilde{W}) + F(\tilde{c}\tilde{d}_i, \tilde{W})[(1 + y_i^{tK} + (y_{\tilde{g}} + y_{\tilde{Z}})\delta_{i2} + \Delta_i^K)\alpha_i^L \\ +(1 + y_i^{tK} - (y_{\tilde{g}} y_{\tilde{Z}})\delta_{i2} + \Delta_i^K)\beta_i^L + (y_1(R)\alpha_3^R + y_2^{(R)}\beta_3^R)\delta_{i3}])\end{gathered} \qquad (25)$$

where

$$\alpha_i^L = \epsilon_{abc}(d_{aL}\gamma^0 u_{bL})(s_{cL}\gamma^0 \nu_{iL})$$
$$\alpha_i^R = \alpha_i^L(d_L, u_L \rightarrow d_R, u_R)$$
$$\beta_i^{L,R} = \alpha_i^{L,R}(d \longleftrightarrow s) \tag{26}$$

In the above y_i^{tk} is the third generation contribution

$$y_i^{tK} = \frac{P_2}{P_3}(\frac{m_t V_{31} V_{32}}{m_c V_{21} V_{22}})(\frac{F(\tilde{t}, \tilde{d}_i, \tilde{W}) + F(\tilde{t}, \tilde{e}_i, \tilde{W})}{F(\tilde{c}, \tilde{d}_i, \tilde{W}) + F(\tilde{c}, \tilde{e}_i, \tilde{W})}) \tag{27}$$

where F are dressing loop integrals and $y_{\tilde{g}}$ are corrections from the gluino exchange, and $y_{\tilde{Z}}$ are corrections from neutralino exchange. The second and the third generation squark loop contributions can interfere both constructively and destructively. The p lifetime is enhanced when there is destructive interference. Further, there are situations when the $\bar{\nu}\pi^+$ mode can be significantly enhanced so that it becomes comparable to the $\bar{\nu}K^+$ mode[17].

4.2. The $\bar{\nu}K$ Mode

We discuss now the details of the $\bar{\nu}K$ mode which as already pointed out is most often the most dominant mode in the nucleon decay in the minimal SU(5) model. The decay width of the $p \rightarrow \bar{\nu}_i K$ mode is given by

$$\Gamma(p \rightarrow \bar{\nu}_i K^+) = (\frac{\beta_p}{M_{H_3}})^2 |A|^2 |B_i| C \tag{28}$$

Here the factors A and B_i are given by

$$A = \frac{\alpha_2^2}{2M_W^2} m_s m_c V_{21}^\dagger V_{21} A_L A_S \tag{29}$$

$$B_i = \frac{1}{sin2\beta}\frac{m_i^d V_{i1}^\dagger}{m_s V_{21}^\dagger}[P_2 B_{2i} + \frac{m_t V_{31} V_{32}}{m_c V_{21} V_{22}} P_3 B_{3i}] \tag{30}$$

$$B_{ji} = F(\tilde{u}_i, \tilde{d}_j, \tilde{W}) + (\tilde{d}_j \rightarrow \tilde{e}_j) \tag{31}$$

$$F(\tilde{u}_i, \tilde{d}_j, \tilde{W}) = [E cos\gamma_- sin\gamma_+ \tilde{f}(\tilde{u}_i, \tilde{d}_j, \tilde{W}_1) + cos\gamma_+ sin\gamma_- \tilde{f}(\tilde{u}_i, \tilde{d}_j, \tilde{W}_1)]$$
$$-\frac{1}{2}\frac{\delta_{i3} m_i^u sin2\delta_{ui}}{\sqrt{2} M_W sin\beta}[E sin\gamma_- sin\gamma_+ \tilde{f}(\tilde{u}_{i1}, \tilde{d}_j, \tilde{W}_1)$$
$$-cos\gamma_- cos\gamma_+ \tilde{f}(\tilde{u}_{i1}, \tilde{d}_j, \tilde{W}_2) - (\tilde{u}_{i1} \rightarrow \tilde{u}_{i2})] \tag{32}$$

where $\tilde{f}$ is given by

$$\tilde{f}(\tilde{u}_i, \tilde{d}_j, \tilde{W}_k) = sin^2\delta_{ui}\tilde{f}(\tilde{u}_{i1}, \tilde{d}_j, \tilde{W}_k) + cos^2\delta_{ui}\tilde{f}(\tilde{u}_{i2}, \tilde{d}_j, \tilde{W}_k) \quad (33)$$

and where

$$f(a,b,c) = \frac{m_c}{m_b^2 - m_c^2}\left[\frac{m_b^2}{m_a^2 - m_b^2}ln(\frac{m_a^2}{m_b^2}) - (m_a \to m_c)\right] \quad (34)$$

In Eq.(30) $\gamma_\pm = \beta_+ \pm \beta_-$ and

$$\begin{aligned} sin2\beta_\pm &= \frac{(\mu \pm \tilde{m}_2)}{[4\nu_\pm^2 + (\mu \pm \tilde{m}_2)^2]^{1/2}} \\ \sqrt{2}\nu_\pm &= M_W(sin\beta \pm cos\beta) \\ sin2\delta_{u3} &= -\frac{-2(A_t + \mu ctn\beta)m_t}{m_{\tilde{t}_1}^2 - m_{\tilde{t}_2}^2} \\ E &= 1\ , sin2\beta > \mu\tilde{m}_2/M_W^2 \\ &= -1, sin2\beta < \mu\tilde{m}_2/M_W^2 \end{aligned} \quad (35)$$

C in Eq.(28) is a current algebra factor and is given by

$$C = \frac{m_N}{32\pi f_\pi^2}[(1 + \frac{m_N(D+F)}{m_B})(1 - \frac{m_K^2}{m_N^2})]^2 \quad (36)$$

where $f_\pi, D, F, ..$ etc are the chiral Lagrangian factors with the numerical values: $f_\pi = 139$ MeV, D=0.76, F=0.48, m_N=938 MeV, m_K=495 MeV, and m_B=1154. Finally, in Eq.(28) β_p is defined by

$$\beta_p U_L^\gamma = \epsilon_{abc}\epsilon_{\alpha\beta} < 0|d_{aL}^\alpha u_{bL}^\beta u_{cL}^\gamma|p> \quad (37)$$

where the lattice gauge analysis gives[26]

$$\beta_p = (5.6 \pm 0.5) \times 10^{-3}GeV^3 \quad (38)$$

4.3. Vector Meson Decays Modes of the Proton

In addition to the nucleon decay modes involving pseudo-scalar bosons and anti-leptons, one also has in general decay modes involving vector bosons and anti-leptons. The source of these modes are the same baryon number violating dimension six quark operators that give rise to the decay modes that give rise to pseudoscalar and anti–lepton modes. The vector decay modes of the proton are

$$\begin{aligned} &\bar{\nu}_i K^*, \bar{\nu}_i\rho, \bar{\nu}_i\omega; i = e, \mu, \tau \\ &eK^*, \mu K^*, e\rho, \mu\rho, e\omega, \mu\omega \end{aligned} \quad (39)$$

However, the vector meson decay modes have generally smaller branching ratios than the corresponding pseudo-scalar decay modes[27].

5. Supergravity Analysis

We shall work here in the framework of supergravity models where supersymmetry is broken in the hidden sector by a superhiggs phenomenon and the breaking communicated gravitationally to the physical sector[28, 29]. The simplest case corresponds to when the superhiggs coupling are generation blind. Here after breaking of supersymmetry and of the gauge group and after integrating out the superhiggs fields and the heavy fields of the theory, one finds that the supersymmetry breaking potential below the GUT scale is given by

$$V_{SB} = m_0^2 z_a z_a^\dagger + (A_0 W^{(3)} + B_0 W^{(2)} + h.c.) \tag{40}$$

and in addition one has a universal gaugino mass term

$$L^\lambda_{mass} = -m_{1/2}\bar{\lambda}^\alpha \lambda^\alpha \tag{41}$$

In the above, $W^{(3)}$ is the trilinear part of the superpotential and $W^{(2)}$ is the bilinear part which under the constaint of R parity invariance is given by $W^{(2)} = \mu_0 H_1 H_2$. Thus we find that the supersymmetry breaking sector contains just four parameters.

Supergravity unification possesses the remarkable feature that the electro-weak symmetry breaking can be manufactured by radiative effects. The radiative electro-weak symmetry breaking is governed by the following Higgs potential

$$\begin{aligned} V_H = m_1^2(t)|H_1|^2 + m_2^2(t)|H_2|^2 - m_3^2(t)(H_1 H_2 + h.c.) \\ + \frac{1}{8}(g^2 + g_y^2)(|H_1|^2 - |H_2|^2)^2 + \Delta V_1 \end{aligned} \tag{42}$$

where ΔV_1 is the one loop correction to the Higgs potential and the parameters $m_i^2(t)$ satisfy

$$m_i^2(0) = m_0^2 + \mu_0^2; i = 1, 2; \;\; m_3^2(0) = -B_0\mu_0 \tag{43}$$

while the gauge coupling constants satisfy the GUT relation $\alpha_3(0)= \alpha_2(0) = \alpha_G = (5/3)\alpha_Y(0)$. The breaking of the electroweak symmetry is accomplished by a satisfaction of the constraints ($\tan\beta =< H_2 > / < H_1 >$):

$$\begin{aligned} \frac{1}{2}M_Z^2 = (\mu_1^2 - \mu_2^2 tan^2\beta)/(tan^2\beta - 1) \\ sin2\beta = (2m_3^2)/(\mu_1^2 + \mu_2^2) \\ \mu_i^2 \equiv m_i^2 + \Sigma_i \end{aligned} \tag{44}$$

where Σ_i is the one loop correction from ΔV_1. On using the radiative breaking constraint the low energy SUSY parameters can be chosen to be

$$m_0, m_{1/2}, A_0, tan\beta, sign\mu \tag{45}$$

Supergravity unification exhibits the phenomenon of scaling over most of the parameter space of the model. This arises because over most of the parameter space of the model one has $\mu^2 >> M_Z^2$ which gives[30]

$$\begin{aligned} m_{\tilde{W}_1} \sim \frac{1}{3} m_{\tilde{g}} \ (\mu < 0); \ \ m_{\tilde{W}_1} \sim \frac{1}{4} m_{\tilde{g}} \ (\mu > 0) \\ 2m_{\tilde{Z}_1} \sim m_{\tilde{W}_1} \sim m_{\tilde{Z}_2}; \ \ m_{\tilde{Z}_3} \sim m_{\tilde{Z}_4} \sim m_{\tilde{W}_2} >> m_{\tilde{Z}_1} \end{aligned} \tag{46}$$

In addition one also has $m_{H^0} \sim m_A \sim m_{H^\pm} >> m_h$. Corrections to scaling are $O(M_Z/\mu)$. While these corrections are small over most of the parameter space of the model, they can become significant in the region of small μ.

5.1. Non-universalities of Soft SUSY Breaking Parameters

Universalities of the soft SUSY breaking parameters at the GUT scale are a consequence of the assumption of a generation independent Kahler potential in the supergravity analysis. However, more generally one can have generational dependence[31] and string based analyses indeed show such a dependence. From the phenomenological view point, these generational dependences cannot be arbitrary but are severly constrained by flavor changing neutral currents (FCNC). However, it is possible to introduce significant amounts of non-universalities in the Higgs sector and in the third generation sector without upsetting the FCNC constraints[32, 33, 34]. Thus, for example, Higgs sector non-universalities can be parametrized at the GUT scale so that

$$m_{H_1}^2 = m_0^2(1+\delta_1), m_{H_2}^2 = m_0^2(1+\delta_2) \tag{47}$$

where a reasonable range for δ_i is $|\delta_i| \leq 1$. Similarly one can parametrize the non-universalities in the third generation sector.

5.2. p Decay in mSUGRA

We give here a numerical analysis of the maximum p lifetime for the minimal supergravity model. The upper bounds on the proton lifetime for different values of m_0 as a function of the gluino mass are given in Fig.1. These upper limits are gotten by integrating over the allowed range of the parameter space, i.e., $\tan\beta$ and A_0 for fixed values of m_0 and $m_{\tilde{g}}$, consistent with radiative breaking of the electro-weak symmetry. The analysis is carried out under the constraint on the Higgs triplet mass such that

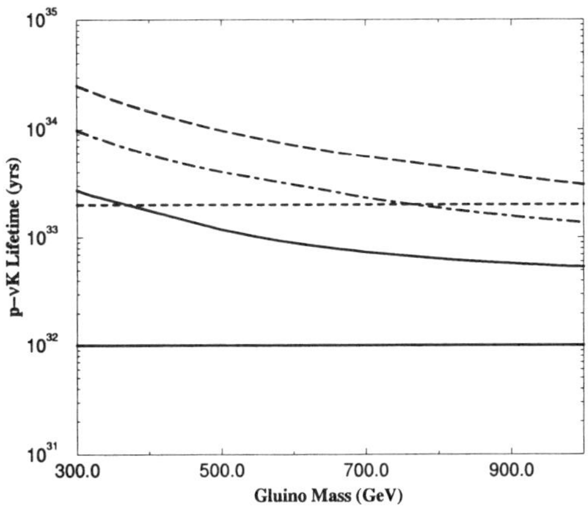

Figure 1. Plots of the maximum $\tau(p \to \bar{\nu}K)$ lifetime in the minimal SU(5) supergravity model with universal soft breaking as a function of the gluino mass. The analysis is for the naturalness limits on m_0 of 1 TeV (solid), 1.5 TeV (dashed-dot) and 2 TeV (dashed). The horizontal solid line is the lower limit from Kamionkande and the horizontal dashed line is the lower limit expected from Super K. (Taken from ref.[35]).

$M_{H_3} \leq 10M_G$. The theoretical upper bounds may be compared with the current experimental lower bound on the $p \to \bar{\nu}K^+$ mode[11] of

$$\tau(p \to \bar{\nu}K) > 5.5 \times 10^{32}\ yr \tag{48}$$

and the lower limits that the super Kamiokande (SuperK) hopes to achieve in the future for this mode which is[12]

$$\tau(p \to \bar{\nu}K) > 2 \times 10^{33}\ yr \tag{49}$$

The analysis shows that the current lower limit on the $\bar{\nu}K^+$ mode does not exhaust the parameter space for the naturalness limit of $m_0 \leq 1$ TeV. However, most of the parameter space for this naturalness limit will be exhausted when SuperK reaches its expected sensitivity of 2×10^{33} yr.

6. B and L Violation in Non-minimal Models

In this section we will discuss B and L violation in various non-minimal models such as the $(SU(3))^3$ models, unflipped and flipped no-scale models, and SO(10) models.

6.1. $(SU(3))^3$ Models

$(SU(3))^3$ models arise naturally in many Calabi-Yau string model constructions which below the compactification scale have the gauge structure $E_6 \times N = 1$ *supergravity*. After Wilson line breaking, the $E(6)$ can break to $(SU(3))^3$ and one can make contact with low energy physics. The most studied models of this type are the three generation models such as $CP^3 \times CP^3/Z_3$, and $CP^2 \times CP^3/Z_3 \times Z_3'$. In models of this type the massless sector of the theory falls into $27 + \bar{27}$ of E_6. The 27-plet decomposes under $SU(3)_C \times SU(3)_L \times SU(3)_R$ so that (see, e.g, refs.[36, 37, 38]),

$$27 = L(1,3,\bar{3}) + Q(3,\bar{3},1) + Q^c(\bar{3},1,3) \tag{50}$$

and one has a similar decomposition of $\bar{27}$ into $\bar{L}+ \bar{Q} + \bar{Q}^c$. The particle content of L, Q, Q^c is as follows

$$L = (l, H, H', e^c, \nu^c, N); \quad Q = (q, D); \quad Q^c = (q^c, D^c) \tag{51}$$

where $l, q, ..$ etc are the $SU(2)_L$ doublets and D and D^c are the $SU(2)_L$ singlet quarks. The coupling structure of the theory is given by

$$\begin{aligned}(27)^3 &= \lambda_1 duD + \lambda_2 u^c d^c D^c + \lambda_3[-HH'N - H\nu^c l + H' e^c l] \\ &+\lambda_4[DND^c - De^c u^c + D^a \nu^c d^c + qlD^c - qHu^c - qH' d^c]\end{aligned} \tag{52}$$

There are several sources of proton instabiltiy in the these. We list the two dominant sources among these: (a) From B and L violating interactions which contain a D or a D^c. They lead to nucleon decay from the B and L violating dimension 5 operators after D and D^c terms are eliminated; (b) $D-d$ mixing arising from N and ν^c VEV growth. This interaction provides the other dominant source of p decay in this theory. The superpotential with the D-d mixing after N and ν^c VEV growth has the form

$$W_3^{D-d} = [M_1 DD^c + M_2 Dd^c + M_3 dd^c] \tag{53}$$

Diagonalization of the mass matrix leads to interactions which lead to proton decay[39].

Superstring models of Calabi-Yau type with $(SU(3))^3$ gauge group can generate neutrinoless double beta decay $(\beta\beta)_{0\nu}$ after R parity violation. Thus as discussed ν^c VEV growth violates lepton number and R parity and one can obtain an effective Lagrangian of the type

$$L_{eff} = g\bar{\chi}_1^0 e_L \bar{d}_a u_L^a \tag{54}$$

Since $\bar{\chi}_1^0$ is Majorana, the interaction allows a neutrinoless double beta decay. The predictions give[40]

$$g^2/m_{\chi_1^0} < 10^{-15} GeV^{-5} \tag{55}$$

while the current experimental limits correspond to

$$g^2/m_{\chi_1^0} < 10^{-13} GeV^{-5} \tag{56}$$

Thus the predicted level of double beta decay in Calabi-Yau models lies below the current level of sensitivity of experiment by about two orders of magnitude.

6.2. P Decay in No Scale Models

An important sub-class of supergravity models are the so-called no scale models. In such models the F term contribution to the scalar potential[29] of the models, i.e.,

$$V = e^{-G}[G^{-1,B}_{,A} G_{,B} G^{,A} - 3] + \frac{1}{2} Re F^{-1}_{\alpha\beta} D^{\alpha} D^{\beta} \tag{57}$$

vanishes because one in on an Einstein manifold. In a class of no-scale models one has $m_0 = 0 = A_0$ and SUSY breaking is driven by the universal gaugino mass term $m_{1/2}$. In this class of models one typically finds $m_{\tilde{q}} < m_{\tilde{g}}$. Unfortunately unflipped no-scale models of the above type violate experimental p lifetime limits[41] in the interesting part of the parameter space since proton decay is governed roughly by the ratio $m_{\tilde{g}}/m^2_{\tilde{q}}$, and the constraint $m_{\tilde{q}} < m_{\tilde{g}}$ tends to destalilize the proton.

6.3. Proton Decay in the Flipped SU(5) Models

In the flipped SU(5) model[42] some particles assigned to the $\bar{5}$-plets ($\bar{5}_i; i = 1,2,3$) and the 10-plets ($T_i; i = 1,2,3$)) are flipped, i.e.,

$$u^c \longleftrightarrow d^c, \quad e^c \longleftrightarrow \nu^c \tag{58}$$

Thus one has three generations of matter of the type

$$\bar{F}_i = (u_i^c, L_i = (e_i, \nu_{e_i})); \quad T_i = (q_i = (u_i, d_i), d_i^c, \nu_i^c); \quad e_i^c; \quad i = 1 - 3 \tag{59}$$

The Higgs sector consists of the usual one pair of $5 + \bar{5}$ of Higgs $h, \bar{h}$, i.e.,

$$h = (H, D), \quad \bar{h} = (\bar{H}, \bar{D}) \tag{60}$$

and two pairs of $10, \bar{10}$ of Higgs $H_i, \bar{H}_i$(i=1,2) so that

$$H_i = (q_{H_i}, d^c_{H_i}, \nu^c_{H_i}), \quad \bar{H}_i = (q_{\bar{H}_i}, d^c_{\bar{H}_i}, \nu^c_{\bar{H}_i}); \quad i = 1,2 \tag{61}$$

The effective potential in this theory is[43]

$$\begin{aligned} W = \lambda_1 TTh + \lambda_2 T\bar{F}\bar{h} + \lambda_3 \bar{F} l^c h + \lambda_4 HHh \\ + \lambda_5 \bar{H}\bar{H}\bar{h} + \lambda_6 F\bar{H}\phi + \mu h\bar{h} \\ + \delta_1 HTh + \delta_2 H\bar{F}\bar{h} + M'H\bar{H} + M''\phi\phi \end{aligned} \tag{62}$$

where $\phi's$ are the singlet fields. Symmetry breaking occurs via VEV formation of $H, \bar{H}$ fields

$$< \nu^c_{H_i} >= M_i, < \nu^c_{\bar{H}_i} >= \bar{M}_i \tag{63}$$

The Higgs doublet mass matrix arises from

$$\mu h\bar{h} \rightarrow \mu H\bar{H}, \quad \delta_2 H\bar{F}\bar{h} \rightarrow \delta_2 ML\bar{H} \tag{64}$$

and thus has the matrix form (with column labelled by $\bar{H}$ and rows by H and L):

$$M_{doub} = \begin{pmatrix} \mu \\ \delta_2 M \end{pmatrix} \tag{65}$$

The usual scenarios consider the hierarchy: $M_i \sim \bar{M}_i \sim M >> M' >> \mu$. The light Higgs is gotten by setting $\delta_2 = 0$. Regarding the Higgs triplet sector, in addition to the D and $\bar{D}$ one also has color triplets from the 10 and $\bar{10}$ of Higgs. Thus there is mixings between the D and $\bar{D}$ terms and the color triplet terms from the 10 and $\bar{10}$ of SU(5). After symmetry breaking mass contributions arise as follows:

$$\begin{aligned} \mu h\bar{h} &\rightarrow \mu D\bar{D} \\ \delta_1 TFh \rightarrow \delta_1 Md^cD, \lambda_4 TTh &\rightarrow Md^c_H D \\ \lambda_5 \bar{H}\bar{H}\bar{h} \rightarrow \lambda_5 \bar{m} d^c_{\bar{H}}\bar{D}, M'H\bar{H} &\rightarrow M'd^c_H d^c_{\bar{H}} \end{aligned} \tag{66}$$

Often one sets $\delta_1 = 0$. With these assumptions the mass matrix in the Higgs triplet sector is (columns: labelled by $\bar{D}$, d^c_H; rows: labelled by $D, d^c_{\bar{H}}$):

$$M_{trip} = \begin{pmatrix} \mu & \lambda_4 M \\ \lambda_5 M & M' \end{pmatrix} \tag{67}$$

Diagonalization gives the Higgs triplets a superheavy mass O(M). The effective dimension five operator arising from $\mu D\bar{D}$ is of size

$$\mu/M^2 << 1/M \tag{68}$$

Thus in the flipped model the dimension five operators which mediate proton decay would be suppressed by a μ/M^2 rather than the usual $1/M$ factor as in the standard SU(5) model. However, it is possible to upset this suppression in some cases.

6.4. Proton Decay in SO(10) Models

SO(10) models have several interesting features. One of these is that SO(10) allows for the doublet-triplet splitting via VEV alignment. To achieve this one considers a Higgs sector of the type[22]

$$W_{d-t} = 10_1 45_1 10_2 + 10_2^2 \phi \tag{69}$$

where 10_1 is the Higgs that couples with matter and ϕ is a singlet. 45_1 gets a VEV $O(M_G)$ in the B-L direction giving the mass matrices

$$M_{doub} = \begin{pmatrix} 0 & 0 \\ 0 & \phi \end{pmatrix} \tag{70}$$

$$M_{trip} = \begin{pmatrix} 0 & a_1 \\ -a_1 & \phi \end{pmatrix} \tag{71}$$

The Higgs doublet is light but the higgs triplets develop an effective mass

$$M_{PD}^{-1} = M_{11}^{-1} = \phi / a_1^2 \tag{72}$$

In proton decay analyses in SO(10) models one has in addition to the matrix element β_p defined by Eq.(36) also the matrix element α_p defined by

$$\alpha_p U_L^{\gamma} = \epsilon_{abc}\epsilon_{\alpha\beta} < 0|\bar{d}_{aL}^{\alpha} \bar{u}_{bL}^{\beta} u_{cL}^{\gamma}|p > \tag{73}$$

where $|\alpha_p| = |\beta_p|$ but the phase of α_p/β_p remains to be specified. This phase can affect proton decay rates[44].

SO(10) models generally involve large $\tan\beta$. Because of that there are some features specific to these models. Thus, for example, one finds that the gluino contributions can be very significant and comparable to the contributions from the chargino exchange. Further, it is found that the contributions of the dimension six LLRR operators may be comparable to or even dominate the LLLL contribution in these models[44]. However, there is a potential problem in SO(10) regarding proton stability vs unification of gauge couplings using LEP data. To see the problem one notices that the mass scale necessary to suppress p decay to the current experimental value is[18]

$$(M^{-1})_{11} > tan\beta(0.57 \times 10^{16})\ GeV \tag{74}$$

which for $\tan\beta \sim 50$ requires a GUT mass of 2.5×10^{17} for suppression of p decay to the current experimental limit. However, a mass scale of this size upsets unification of gauge couplings and one needs large threshold corrections to get agreement with experiment[45, 46]. Fig.2 illustrates the problem of the high scale in SO(10). One finds from Fig.2 that for the experimental value of $\sin^2\theta_W(M_Z)_{\overline{MS}}$ which is 0.23122 ± 0.00002 the value of α_s corresponding to $M_{PD} = 2.7 \times 10^{17}$ is many standard deviations away from the current experimental limit of $\alpha_s(M_Z) = 0.118 \pm 0.003$. Thus as stated large threshold corrections are necessary to get consistency. There are a great variety of SO(10) models, and details of the model affect critically the nature of supersymmetric signals. Thus in a class of SO(10) models discussed recently[47] the charged lepton modes of the proton such as $l^+\pi^0$, l^+K^0 and $l^+\eta$ ($l = e, \mu$), can become prominent[47]. There are

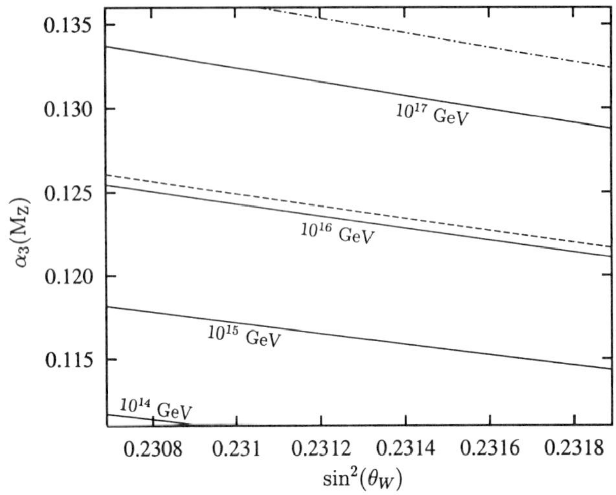

Figure 2. Plot in the $sin^2\theta_W - \alpha_3(M_Z)$ plane for various M_{PD} values. The dashed line is for the case of the lower bound on $M_{PD} = 1.2 \times 10^{16}$ GeV in the minimal SU(5) model. The dot-dashed line corresponds to the lower bound on $M_{PD} = 2.7 \times 10^{17}$ GeV in the minimal SO(10) model. (Taken from ref.[45]).

also other approaches in the non-minimal extensions where proton decay is correlated directly with the ansatz on the Yukawa couplings of the dimension five operators[48].

7. Effects of Textures on P Decay

Gut models give poor predictions for quark-lepton mass ratios. In SU(5) m_b/m_τ is in good agreement with experiment but m_s/m_μ and m_d/m_e are not. One needs textures, i.e.,

$$W_d = H_1 l A^E e^c + H_1 d^c A^D q + H_2 u^c A^U q \tag{75}$$

where $A^E, A^D, \& A^U$ are the texture matrices. The simplest choice for these are the the Georgi-Jarlskog choice[49]

$$A^E = \begin{pmatrix} 0 & F & 0 \\ F & -3E & 0 \\ 0 & 0 & D \end{pmatrix}, A^D = \begin{pmatrix} 0 & Fe^{i\phi} & 0 \\ Fe^{-i\phi} & E & 0 \\ 0 & 0 & D \end{pmatrix} \tag{76}$$

$$A^U = \begin{pmatrix} 0 & C & 0 \\ C & 0 & B \\ 0 & B & A \end{pmatrix} \tag{77}$$

where A-F have a hierarchy of the type

$$A \sim O(1), \quad B, D \sim O(\epsilon)$$
$$C, E \sim O(\epsilon^2), \quad F \sim O(\epsilon^3), \quad \epsilon << 1 \tag{78}$$

A possible origin of the parameter ϵ is from the ratio of mass scales, e.g.,

$$\epsilon = \frac{M_{GUT}}{M_{str}} \tag{79}$$

In the context of supergravity unified models this ratio can arise from higher dimensional operators. In the energy domain below the string scale after integration over the heavy modes of the string one has an effective theory of the type

$$W = W_3 + W_4 + W_5 + ... \tag{80}$$

where $W_n(n > 3)$ are suppressed by the string (Planck) scale and in general contain the adjoints which develop VEVs$\sim O(M_{GUT})$. After VEV formation of the heavy fields

$$W_n \sim O(\frac{M_{GUT}}{M_{string}})^{n-3} \times \ operators \ in \ W_3 \tag{81}$$

With the above one can generate mass heirarchy with $\lambda_{yuk} \sim O(1)$. One can also compute textures in the Higgs triplet sector

$$W_t = H_1 l B^E q + H_2 u^c B^U e^c + \epsilon_{abc}(H_1 d_b^c B^D u_c^c + H_2^a u_b^c C^U d_c) \tag{82}$$

However, the Planck scale expansion is not unique and consequently the Higgs triplet texture are not unque, and a dynamical principle is needed to achieve uniqueness. One such principle is the assumption of an exotic sector wherein fields in a minimal vector like representation couple to both the hidden sector fields and fields in the physical sector. If the exotic fields gain superheavy masses their elimination will lead to a spcific set of Planck scale interactions. Such an assumption indeed leads to unique Higgs triplet textures of the form[50]

$$B^E = \begin{pmatrix} 0 & aF & 0 \\ a^*F & \frac{16}{3}E & 0 \\ 0 & 0 & \frac{2}{3}D \end{pmatrix} \tag{83}$$

$$B^D = \begin{pmatrix} 0 & -\frac{8}{27}F & 0 \\ \frac{-8}{27}F & -\frac{4}{3}E & 0 \\ 0 & 0 & -\frac{2}{3}D \end{pmatrix} \tag{84}$$

$$B^U = \begin{pmatrix} 0 & \frac{4}{9}C & 0 \\ \frac{4}{9}C & 0 & -\frac{2}{3}B \\ 0 & -\frac{2}{3}B & A \end{pmatrix} \tag{85}$$

$$C^U = B^U, \quad a = (-\frac{19}{27} + e^{i\phi}) \tag{86}$$

Inclusion of textures gives a moderate modification of p decay branching ratios. The p lifetime for the $\bar{\nu}K^+$ mode is enhanced by a factor of $\sim (\frac{9}{8}\frac{m_s}{m_\mu})^2$. p decay modes hold important information on GUT physics. Further, the textures affect in a differential way the various decay modes which in turn can be used to test theories of textures[50, 51].

8. Effects of Dark Matter on Proton Stability

In this section we discuss the effects of dark matter constraints on proton stability. There is now a great deal of convincing evidence for the existence of dark matter in the universe and most of this dark matter must be non-baryonic. In supersymmetric theories with R parity invariance the least massive supersymmetric particle (LSP) is absolutely stable and hence a candidate for dark matter. A priori there are many possible LSP candidates for such non-baryonic dark matter in supersymmetry such as the gravitino, sneutrino, neutralino, etc. However, detailed analyses show that over most of the parameter space of the SUGRA models, it is the lightest neutralino (χ_1) which is the LSP and thus a candidate for cold dark matter. The LSP neutralino is an admixture of four neutral states, i.e.,

$$\tilde{\chi}_1 = n_1\tilde{W}_3 + n_2\tilde{B} + n_3\hat{H}_1 + n_4\tilde{H}_2 \tag{87}$$

where $\tilde{B}$ is the Bino, $\tilde{W}_3$ is the neutral Wino, and $\tilde{H}_1, \tilde{H}_2$ are the neutral Higgsinos. In the scaling region when $\mu^2 >> M_Z^2$ one finds

$$\begin{aligned} n_1 \simeq -\frac{M_Z^2}{2m_{\chi_1\mu}} sin2\theta_W sin\beta, n_2 \sim 1 - \frac{1}{2}\frac{M_Z^2}{\mu^2} sin^2\theta_W \\ n_3 \sim -\frac{M_Z}{\mu} sin\theta_W sin\beta, n_4 \sim \frac{M_Z}{\mu} sin\theta_W cos\beta \end{aligned} \tag{88}$$

The above analysis shows that the LSP is mostly a Bino in the scaling region which is most of the parameter space of the model. In the region when μ is small scaling breaks down, and one can have a large Higgsino component for the LSP. However, recent analyses show that this possibility may be close to being eliminated[52].

8.1. Cosmological Constraints

The amount of dark matter in the universe puts rather stringent constraints on supersymmetric models[53, 54]. These constraints arise from the allowed range of $\Omega_{\chi_1^0}h^2$ where $\Omega_{\chi_1^0} = \rho^0_{\chi_1}/\rho_c$. Here $\rho^0_{\chi_1}$ is the neutralino matter density, and ρ_c is the critical matter density

$$\rho_c = 3H_0^3/8\pi G_N = 1.88h_0^2 \times 10^{-29} gm/cm^3 \tag{89}$$

and h_0 is the Hubble parameter H_0 in units of 100 km/s.Mpc. The number density of χ_1's obeys the Boltzman equation

$$\frac{dn}{dt} = -3Hn - < \sigma v > (n^2 - n_0^2) \tag{90}$$

where $< \sigma v >$ is the thermal average of the neutralino annihilation cross-section and v is the relative neutralino velocity. At the "freeze-out" temperature T_f the χ_1 decouple from the background, and integration from T_f to the current temperature gives

$$\Omega_{\tilde{\chi}_1^0}h^2 \cong 2.48 \times 10^{-11} \left(\frac{T_{\tilde{\chi}_1^0}}{T_\gamma}\right)^3 \left(\frac{T_\gamma}{2.73}\right)^3 \frac{N_f^{1/2}}{J(x_f)} \tag{91}$$

where $J\ (x_f) = \int_0^{x_f} dx\ \langle\ \sigma v\ \rangle\ (x) GeV^{-2}$, N_f is the number of massless degrees of freedom at the freezeout, $x_f = kT_f/m_{\tilde{\chi}_1}$, T_γ is the current background temperature, and $(T_{\tilde{\chi}_1^0}/T_\gamma)^3$ is the reheating factor. In the anaysis below we shall assume the relic density constraint of

$$0.1 \leq \Omega_{\chi_1^0}h^2 \leq 0.4 \tag{92}$$

which encompasses a wide range of cosmological models. The imposition of the dark matter constraint reduces the proton lifetime by a very significant amount for values of the gluino mass greater than about 500 GeV[35]. The reduction factors as a function of the gluino mass are listed in Table 3 (taken from ref.[35]). One finds that the reduction factors lie in the range 10-30. Taking into account corrections due to Yukawa textures, and other uncertainties in the calculations, SuperK should be able to test the region $m_{\tilde{g}} > 500$ GeV when it achieves a sensitivity of 1×10^{33}. These results are exhibited in Fig.3 for the minimal supergravity model.

8.2. Effects of Proton Lifetime Constraint on Event Rates

The imposition of proton lifetime constraint has an important effect on the event rates in dark matter analysis[55]. One finds that in addition to the drastic reduction in the allowed range of the gluino mass, there is also a very significant reduction in the magnitude of the maximum event rate curves. In Fig.4 we exhibit the maximum event rate as a function of the neutralino mass. We find that the reduction of the maximum event rate curve can be a factor of 10 to 100. This reduction arises from the fact that the proton lifetime constraint eliminates certain parts of the parameter space which give rise to large event rates.

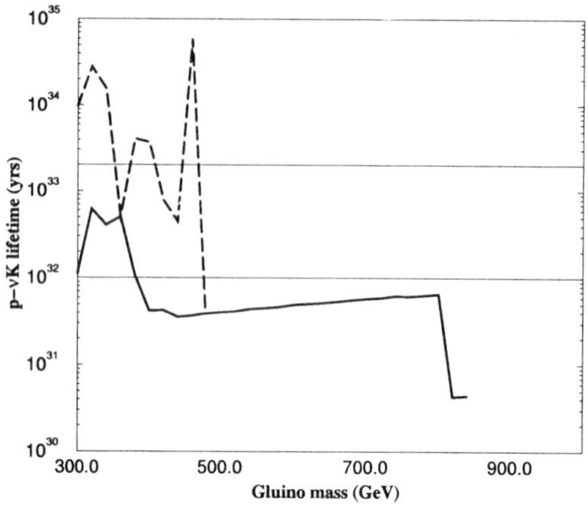

Figure 3. Plots of the maximum $\tau(p \to \bar{\nu}K)$ lifetime in the minimal SU(5) supergravity model with universal soft breaking as a function of the gluino mass with the inclusion of relic density constraint on neutralino dark matter of $0.1 < \Omega_{\chi_1^0}h^2 < 0.4$. The analysis is for the naturalness limits on m_0 of 1 TeV (solid), and 5 TeV (dashed). The lower horizontal solid line is the Kamionkande limit and the upper horizontal solid line is the lower limit expected from Super K. (Taken from ref.[35]).

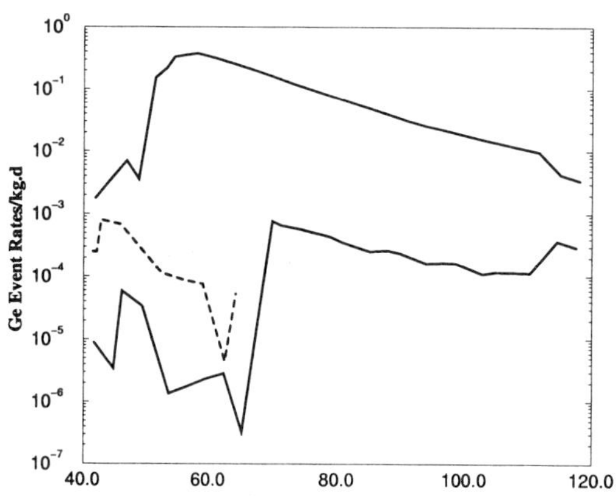

Figure 4. Plots of the the maximum and the minimum of event rates for the scattering of neutralinos off germanium target as a function of the neutralino mass. The relic density constraint $0.1 < \Omega_{\chi_1^0}h^2 < 0.4$ is imposed. The solid curves are for the case of $m_0 \leq 1$ TeV and without p decay constraint. The dashed curves are for the case with the same relic density constraint, but with $m_0 \leq 5$ TeV, and under the proton lifetime constraint $\tau(p \to \bar{\nu}K) > 1 \times 10^{32}$ yr. The minimum dashed curve coincides with the lower solid curve. [taken from Ref.[35]).

Table 3: Reduction factor from CDM constraint[35].

gluino mass (GeV)	reduction factor
500	29.5
550	23.2
600	18.6
650	15.3
700	12.9
750	11.3
800	9.8

9. Proton Stability in Gauge Mediated Breaking of Supersymmetry

In the simplest model of gauge mediated supersymmetry breaking [56](GMSB), one has that SUSY breaking arises from VEV formation of a chiral superfield $\hat{S}$ which couples to messenger chiral superfield $(\phi_i, \bar{\phi}_i)$ in vector like representations of the MSSM group with a superpotential of the form

$$W = \sum C_i S \phi_i \bar{\phi}_i \tag{93}$$

where in SU(5) one has $(\phi, \bar{\phi}) = (5, \bar{5})$, or $(10, \bar{10})$. The information on the breaking of supersymmetry is communicated by gauge interactions to the physical sector. Masses for the gauginos $(\tilde{M}_a)$ arise at the one loop level via gauge interactions, i.e.,

$$\tilde{M}_a = \frac{\alpha_a}{4\pi}\Lambda, \quad \Lambda = \frac{< F >}{< S >} \sim 10^2 TeV \tag{94}$$

where $\sqrt{F}$ is the SUSY breaking scale, Λ is the cutoff and typically $\sqrt{F} \sim \Lambda$. Masses for the scalars arise at the two loop level, i.e,

$$\tilde{m}_i^2 = \sum_a 2C_{iF}^a (\frac{\alpha_a}{4\pi})^2 \Lambda^2 \tag{95}$$

where C_{iF}^a are the Casimir co-efficients for the field i.

There are many extensions of this simplest GMSB version including models with the messenger fields in incomplete multiplets[57]. However, in these scenarios it is difficult to generate μ and $B\mu$ terms (where $B\mu$ is the value of $B_0\mu_0$ at the electro-weak scale), and one needs non-gauge interactions for their generation. Further, because of the low value of supersymmetry breaking scale $\sqrt{F}$ in these theories, i.e., $\sqrt{F} \sim 10^2$ TeV, the gravitino is the LSP with a mass $m_{3/2} < 1$ Kev[58]. Thus the gravitino cannot be a CDM in these models.

There are significant constraints that arise in these models in a GUT framework. First it is noted that the unification of the gauge couplings using the LEP data along with the $b-\tau$ unification puts severe constraints on the models[59]. This type of constraint, however, can be softened by inclusion of the Planck scale corrections as in Refs.[3, 4]. However, it is further found that the limits on $tan\beta$ from proton stability and the limits on it from radiative breaking of the electro-weak symmetry under the constraint that the bilinear and the trilinear soft couplings vanish at the messenger scale eliminate all GUT models considered in ref.[60] except for one or two isolated cases.

10. Planck Effects and Proton Decay

In the R-G analysis of the gauge coupling constants α_i there is an overlap of GUT threshold effects and of the Planck scale effects. For example, for SU(5) and for $Q \sim M_G$ one has[3, 4]

$$\alpha_i^{-1}(Q) = \alpha_G^{-1} + C_{ia} ln(\frac{M_a}{Q}) + \frac{cM}{2M_P}\alpha_G^{-1} n_i \tag{96}$$

where M_a are the superheavy GUT thresholds, and $n_i = (-3, -1, 2)$ for the SU(2), U(1) and $SU(3)_C$ sectors. It is easily seen that the Planck effects characterized by n_i can be absorbed in the GUT thresholds by a rescaling. Thus for the minimal SU(5) model rescaling gives

$$\alpha_i^{-1}(Q) = \alpha_G^{eff-1} + C_{ia} ln(\frac{M_a^{eff}}{Q}) \tag{97}$$

where

$$M_a^{eff} = M_a e^{k_a C_P}; k_a = (-\frac{3}{5}(\Sigma), \frac{3}{10}(V), 5(H_3)) \tag{98}$$

and where

$$\alpha_G^{eff-1} = \alpha_G^{-1} - \frac{15}{2\pi} C_P, \quad C_P = \frac{\pi c M}{\alpha_G M_P} \tag{99}$$

While the RG analysis involves the effective parameters M_V^{eff} and M_{H3}^{eff}, the proton decay lifeitme is determined by M_V and by M_{H_3}. Thus the RG analysis along with proton lifetime measurements can allow one to measure the size of the Planck correction, i.e., the value of c. For the $p \to \bar{\nu} K^+$ one finds[4]

$$p \to \bar{\nu} K^+, \quad c = \frac{\alpha_G}{10} \frac{M_P}{\pi M_V} ln \frac{M_{H_3}^{eff}}{M_{H_3}} \tag{100}$$

and for the $p \to e^+\pi^0$ mode one finds[3]

$$p \to e^+\pi^0, \quad c = \frac{100}{3}\sqrt{\frac{2}{3}} \alpha_G^{3/2} \frac{M_P}{M_V} ln \frac{M_V}{M_V^{eff}} \tag{101}$$

11. Exotic p Decay Modes

p decay modes discussed so far are all of the type where a proton decays into an anti-lepton and a meson, i.e.,

$$\begin{aligned} p &\to e^+\pi^0 \\ p &\to \bar{\nu}K^+(\pi^+) \\ p &\to \mu^+K^0 \end{aligned} \tag{102}$$

These decay modes arise in the minimal SU(5) and SO(10) models where the interactions obey the $B-L$ conservation. However, it is possible to include interactions where $B-L$ coservation is violated. Thus, for example, one may consider an interaction of the type[61],

$$5_M 5_M \bar{1}0_H \tag{103}$$

which can generate $\Delta(B-L)=2$ transitions. Such interactions will allow proton decay modes with a lepton and mesons such as

$$\begin{aligned} d+d+s &\to \mu^- \\ n &\to \mu^- K^+ \\ p &\to \mu^-\pi^+K^+, .. \end{aligned} \tag{104}$$

Thus p decay modes distinguish among the varieties of GUT interactions and can provide important insights into GUT physics.

12. Connection with String/M Theory

Proton stability is a very strong constraint on string model building. Most string models contain interactions which violate R parity and can generate rapid proton decay. However, even with R parity invariance models with string GUTs will show generically the same typical B and L violating interactions that one has in ordinary GUT models. Recently, there has been an exhaustive study of string theories which perturbatively allow grand unification. One finds that indeed it is possible to find string models with interesting unified gauge groups such as SU(5), SO(10), E(6) etc[62]. However, simultaneous satisfaction of other desirable properties such as $N=1$ space time supersymmetry, three chiral families, and massless ajoints needed to break the gauge symmetry is not so easy, and one needs to go to models with higher Kac-Moody levels for their satisfaction. Recently, there has been a great deal of work on models of this type and explicit models at Kac-moody level three with the above properties have been constructed[63].

Unfortunately, there are several problems of a phenomenological level still to be overcome before such models can become viable. String Guts is one of the many ways in which one can reconcile LEP data and the unification of gauge couplings within string theory. Without string GUTS one will have the Standard Model gauge group emerging directly at the string scale, and to reconcile the LEP data on the gauge coupling constants with this high scale, one needs some extra effects, such as string thresholds and extra vector like representations at an intermediate state below the string scale[64].

Horava and Witten[65, 66] have suggested a new possibility for the unification of the gauge couplings within the framework of string/M theory. It is within the framework of the conjecture that the strongly coupled limit of the $E_8 \times E_8$ heterotic string is M theory on $R_{10} \times S_1/Z_2$ with gravity propagating in the bulk and the gauge fields living on the hyperplanes[65, 67]. If M theory compactifies on an $M_4 \times CY \times S_1/Z_2$ the unification of gauge coupling constants can arise at the GUT scale with the MSSM spectrum with the appropriate identification of the Calabi-Yau compatification radius with the inverse of the GUT scale[66]. One has in addition unification of gauge coupling constants with gravity arising at this scale because of an extra running of the gravitational coupling due to the opening up of the fifth dimension at a scale which is an order of magnitude below the GUT scale. This picture is very close to the supergravity GUT picture as far as the particle sector of the theory is concerned.

13. Conclusions/Prospects

LEP data appears to support ideas of both grand unification and of supersymmetry. Thus SUSY/SUGRA GUT may be an important way station to the Planck scale where unification of all interactions occurs. In SUSY/SUGRA GUTS one needs R parity invariance to eliminate B and L violating dimension 4 operators which lead to rapid proton decay. B and L violating dimension 5 operators of GUT models allow one to probe via proton decay a majority of the parameter space of the minimal SUGRA model within the naturalness constraint of $m_0 \leq 1$ TeV and $m_{\tilde{g}} \leq 1$ TeV if the Super-K and ICARUS experiments can reach the expected sensitivity of 2×10^{33} y and 10^{34} y respectively for the $\bar{\nu}K^+$ mode. With the inclusion of dark matter constraints one finds that if the proton decay is not observed the gluino mass must lie below 500 GeV within any reasonable naturalness constraint. The simultaneous p stability and dark matter constraints will be tested in the near future in p decay experiments as well by experiments at the Tevatron, LEP2 and the LHC.

Acknowledgments

This work was supported in part by NSF grant number PHY-96020274 and PHY-9722090.

References

[1] For reviews of supersymmetric grand unification see,
G. Ross, "Grand Unified Theories", Frontiers in Physics, 1984;
R. Arnowitt and P. Nath, Proc. of VII J.A. Swieca Summer School ed. E. Eboli (World Scientific, Singapore, 1994);
R. Mohapatra, TASI Lectures 1997.

[2] P. Langacker, Proc. PASCOS 90-Symposium, Eds. P. Nath and S. Reucroft (World Scientific, Singapore 1990); J. Ellis, S. Kelley and D.V. Nanopoulos, Phys. Lett. **B249**, 441(1990); **B260**, 131(1991); C. Ciunti, C.W. Kim and U.W. Lee, Mod. Phys. Lett.**A6**, 1745(1991); U. Amaldi, W. de Boer and H. Furstenau, Phys. Lett. **B260**, 447(1991); P. Langacker and M. Luo, Phys. Rev. **D44**, 817(1991).

[3] T. Dasgupta, P. Mamales and P. Nath, Phys. Rev. **D52**, 5366(1995).

[4] D. Ring, S. Urano and R. Arnowitt, Phys. Rev. **D52**, 6623(1995).

[5] C.T. Hill, Phys. Lett. **B135**, 47(1984); Q. Shafi and C. Wetterich, Phys. Rev. Lett. **52**, 875(1984).

[6] A. Dedes and K. Tamvakis, IOA-05-97; J. Hisano, T. Moroi, K. Tobe and T. Yanagida, Phys. Lett. **B342**, 138(1995).

[7] For a review of the experimental situation of double beta decay see
H.V. Klapdor-Kleingrothaus and H. Paes, in Proceedings of the 6th International Symposium on Particles, Strings and Cosmology (PASCOS-98), Boston, MA, 22-27 Mar 1998, to be published by World Scientific; hep-ph/9808350

[8] For a review of theoretical analyses on $n\bar{n}$ oscillation see,
R. Mohapatra, in Proc. of the International Workshop on Future Prospects of Baryon Instability Search in p-Decay and $n\bar{n}$ Oscillation Experiments, Oak Ridge, Tenessee, U.S.A., March 28-30, 1996, ORNL-6910, Edited by S.J. Ball and Y.A. Kamyshkov, p.73.

[9] For a review of the experimental search for $n\bar{n}$ oscillation see, Y.A. Kamyshkov, in Proc. of the International Workshop on Future Prospects of Baryon Instability Search in p-Decay and $n\bar{n}$ Oscillation Experiments, Oak Ridge, Tenessee, U.S.A., March 28-30, 1996, ORNL-6910, Edited by S.J. Ball and Y.A. Kamyshkov, p.281.

[10] B. Marciano, in Proc. of SUSY97, Philadelphia, U.S.A., 27-31 May 1997.

[11] M. Takita, talk at the XXIX International Conference on High Energy Physics, July, 1998, Vancouver.

[12] Y.Totsuka, Proc. XXIV Conf. on High Energy Physics, Munich, 1988,Eds. R.Kotthaus and J.H. Kuhn (Springer Verlag, Berlin, Heidelberg,1989).

[13] G. Gilbert, N.P.**B328**, 59(1989).

[14] Krauss and F.Wilczek, Phys.Lett.**181B**,380(1986).

[15] L.E. Ibanez and G.G. Ross, Phys. Lett. **B260**, 291(1998).

[16] S. Weinberg, Phys. Rev. **D26**, 287(1982); N. Sakai and T. Yanagida, Nucl. Phys.**B197**, 533(1982); S. Dimopoulos, S. Raby and F. Wilczek, Phys.Lett. **112B**, 133(1982); J. Ellis, D.V. Nanopoulos and S. Rudaz, Nucl. Phys. **B202**, 43(1982); S. Chadha, G.D. Coughlan, M. Daniel and G.G. Ross, Phys. Lett.**149B**, 47(1984).

[17] P. Nath, R. Arnowitt and A.H. Chamseddine, Phys. Rev. **32D**, 2348(1985); J. Hisano, H. Murayama and T. Yanagida, Nucl. Phys. **B402**, 46(1993).

[18] R. Arnowitt and P. Nath, Phys. Rev. **D49**, 1479 (1994).

[19] M.E. Gomez, J. Rizos, and K. Tamvakis, hep-ph/9807289

[20] For a recent discussion of this mechanism see, S.M. Barr, Phys. Rev. **D57**, 198(1998).

[21] B. Grinstein, Nucl. Phys. **B206**, 387(1982); H. Georgi, Phys. Lett. **B115**, 380(1982).

[22] S.Dimopoulos and F.Wilczek, Report No.NSF-ITP-82-07(1981) (unpublished).

[23] K. Inoue, A. Kakuto and T. Tankano, Prog. Theor. Phys. **75**, 664(1986); A. Anselm and A. Johansen, Phys. Lett. **B200**, 331(1988); A. Anselm, Sov. Phys.JETP **67**,663(1988); R. Barbieri, G.Dvali and A. Strumia, Nucl. Phys. **B391**,487(1993); Z. Bereziani and G. Dvali, Sov. Phys. Lebedev Inst. Report 5, 55(1989); Z. Bereziani, C. Csaki and L.Randall, Nucl. Phys. **B44**, 61(1995).

[24] T. Goto and T. Nihei, hep-ph/9808255.

[25] S. Chadha and M. Daniel, Phys. Lett.**137B**, 374(1984); M. Claudson, W.B. Wise and L.J. Hall, Nucl.Phys. **B195**, 43(1982).

[26] M.B. Gavela et al, Nucl.Phys. **B312**,269(1989).

[27] T.C.Yuan, Phys.Rev. **D33**,1894(1986).

[28] A.H. Chamseddine, R. Arnowitt and P. Nath, Phys. Rev. Lett. **49**, 970 (1982); R. Barbieri, S. Ferrara and C.A. Savoy, Phys. Lett. **B119**, 343(1982); L. Hall, J. Lykken and S. Weinberg, Phys. Rev. **D27**, 2359(1983); P. Nath, R. Arnowitt and A.H. Chamseddine, Nucl. Phys. **B227**, 121(1983).

[29] For reviews see, P. Nath, R. Arnowitt and A.H. Chamseddine, "Applied N =1 Supergravity" (World Scientific, Singapore, 1984); H.P. Nilles, Phys. Rep. **110**, 1(1984); H. Haber and G.L. Kane, Phys. Rep. bf 117, 195(1985).

[30] R. Arnowitt and P. Nath, Phys. Rev. Lett. **69**, 725(1992); P. Nath and R. Arnowitt, Phys. Lett. **B289**, 368(1992).

[31] S. Soni and A. Weldon, Phys. Lett. **B126**, 215(1983); V.S. Kaplunovsky and J. Louis, Phys. Lett. **B306**, 268(1993).

[32] D. Matalliotakis and H. P. Nilles, Nucl. Phys. **B435**, 115(1995); M. Olechowski and S. Pokorski, Phys. Lett. **B344**, 201(1995); N. Polonski and A. Pomerol, Phys. Rev. **D51**, 6532(1995);

[33] P. Nath and R. Arnowitt, Phys. Rev. **D56**, 2820(1997);

[34] A. Datta, A. Datta, and M.K. Parida, Trieste Preprint IC/97/207

[35] R. Arnowitt and P. Nath, hep-ph/9801246, to appear in Phys. Lett. B.

[36] B.R. Greene, K.H. Kirklin, P.J. Miron, and G.G. Ross, Nucl. Phys. **B292**, 606(1987).

[37] D. Gepner, Nucl.Phys. **D296**, 757(1988); R. Schimmirigk, Phys. Lett. **B193**, 175(1987).

[38] P. Nath and R. Arnowitt, Phys.Rev. **D39**, 2006(1989); Phys. Rev. Lett. **62**, 1437(1989); R. Arnowitt and P. Nath, Phys. Rev, **D40**, 191(1989).

[39] For a discussion of p decay in this type of three generation superstring models see, R. Arnowitt and P. Nath, Phys. Rev. Lett. **62**, 2225(1989).

[40] R. Arnowitt and P. Nath, in Proc. of the Neutrino 88 conference, Boston (Medford), June 5-11, 1998, World Scientific, Edited by J. Schneps et. al.

[41] P.Nath and R. Arnowitt, Phys. Lett. **B287**, 3282(1992); K. Inoue, M. Kawasaki, M. Yamagouchi an T. Yanagida, Phys. Rev. **D45**, 328(1992).

[42] I. Antoniadis, J. Ellis, J.S. Hagelin and D.V. Nanopoulos, Phys. Lett. **B231**, 65(1987); ibid, **B205**, 459(1988).

[43] For a review of the flipped models see, J. Lopez, D.V. Nanopoulos, and A. Zichichi, in Proc. of the International School of Subnuclear Physics 30th Course, Erice, Italy, July 14-22, 1992.

[44] V. Lucas and S. Raby, Phys. Rev. **D54**, 2261 (1996); ibid. bf D55, 6988(1997).

[45] S. Urano and R. Arnowitt, hep-ph/9611389.

[46] For a recent discussion of the problem of α_s in SO(10) models see, S.M. Barr, hep-ph/9806217.

[47] K.S. Babu, J.C. Pati, and F. Wilczek, Phys.Lett. **B423**, 337(1998).

[48] K.S. Babu and M.J. Strassler, hep-ph/9808447.

[49] H. Georgi and C. Jarlskog, Phys. Lett. **B86**, 297(1979); J. Harvey, P. Ramond and D. Reiss, Phys. Lett. **B92**, 309 (1980).

[50] P. Nath, Phys. Rev. Lett. **76**, 2218(1996); Phys. Lett. **B381**, 147(1996).

[51] K.S. Babu, and S.M. Barr, Phys.Lett. **B381**,137(1996).

[52] J. Ellis, T.Falk, G. Ganis, K.A. Olive, and M. Schmitt, hep-ph/9801445.

[53] J.L. Lopez, D.V. Nanopoulos, and K. Yuan, Phys. Lett. **B267**, 219(1991); J. Ellis and L. Roszkowski, Phys. Lett. **B283**, 252(1992); M. Kawasaki and S. Mizuta, Phys. Rev. **D46**, 1634(1992); M. Drees and M.M. Nojiri, Phys. Rev. **D47**, 376(1993); G.L. Kane, C. Kolda,L. Roszkowski, and J.D. Wells, Phys. Rev. **D49**, 6173(1993).

[54] R. Arnowitt and P. Nath, Phys. Lett. **B299**, 58(1993) and Erratum ibid **B303**, 403(1993); P. Nath and R. Arnowitt, Phys. Rev. Lett. **70**, 3696(1993); M. Drees and A. Yamada, Phys. Rev. **D53**, 1586(1996); H. Baer and M. Brhlik, Phys. Rev.**D53**, 1586(1996); V. Barger and Kao, Phys. Rev. **D57**, 3131(1998).

[55] For recent analyses of event rates in dark matter detectors see, R. Arnowitt and P. Nath, Mod. Phys. Lett. **A** **10**,1257(1995); P. Nath and R. Arnowitt, Phys. Rev. Lett. **74**, 4592(1995); R. Arnowitt and P. Nath, Phys. Rev. **D54**, 2374(1996); A. Bottino et.al., Astropart. Phys. **1**, 61 (1992); **2**, 77 (1994); V.A. Bednyakov, H. V. Klapdor-Kleingrothaus and S.G. Kovalenko, Phys. Rev. **D50**, 7128(1994); L. Bergstrom and P. Gondolo, Astropart. Phys. 5:263-278 (1996); J.D. Vergados, J. Phys. **G22**, 253(1996).

[56] G.F. Giudice and R. Rattazzi, hep-ph/9801271.

[57] P. Martin, Phys. Rev. **D56**, 4207(1997).

[58] H. Pagels and J. Primack, Phys. Rev. Lett. **48**, 223(1982).

[59] C.D. Carone and H. Murayama, Phys. Rev. **D53**, 1658(1996); B. Blok, C.-D. Lu, and D.-X. Zhang, Phys. Lett. **B386**, 146(1996).

[60] H. Hamidian, K. Huitu, K. Puolamaki, and D.-X. Zhang, hep-ph/9803228.

[61] F. Wilczek, and A. Zee, Phys.Lett. **88B**, 311(1979).

[62] D.C. Lewellen, Nucl. Phys. **B337**, 61(1990); J. A. Schwarz, Phys. Rev. Phys. Rev. **D42**, 1777(1990); S. Chaudhuri, S.-W. Chung, G. Hockney, and J.D. Lykken, Nucl. Phys.**452**, 89(1995); G.B. Cleaver,Nucl. Phys. **B456**, 219(1995).

[63] Z. Kakushadze and S.H.H. Tye, Phys. Rev. **D55**, 7896(1997); ibid, **D56**, 7878(1997); Phys. Lett. **392**, 325(1997); Z. Kakushadze, G. Shiu, S.H.H. Tye, and Y. Vtorov-Karevsky, Nucl. Phys. **501**, 547(1997).

[64] See, e.g., K. Dienes, Phys. Rept. **287**, 447(1997).

[65] P. Horava and E. Witten, Nucl. Phys. **B460**, 506(1996).

[66] E. Witten, Nucl. Phys. **B471**, 135(1996).

[67] The issue of p stability in M theory model building is discussed in J. Ellis, A. E. Faraggi, and D.V. Nanopoulos, Phys.Lett. **B419**, 123(1998).

Agut Masses

C D Froggatt† and H B Nielsen‡

† Department of Physics and Astronomy, University of Glasgow, Glasgow G12 8QQ, UK

‡ Niels Bohr Institute, Blegdamsvej 17, Copenhagen ϕ, Denmark

Abstract. From considerations of the number of matrix elements of different orders of magnitude in the quark and charged lepton mass matrices, we suggest that the underlying gauge group responsible for the spectrum should have several—actually of the order of 7 to 10—cross product factors. This is taken to support our AGUT gauge group $SMG^3 \times U(1)_f$ which, under certain simple conditions, is the maximal group transforming the known 45 Weyl quark-lepton fields into each other. We describe the AGUT fit to the charged fermion mass spectrum, and briefly discuss baryogenesis and the neutrino mass problem.

1. Introduction

What is the origin of the well-known pattern of large ratios between the quark and lepton masses and of the small quark mixing angles? This is the problem of the hierarchy of Yukawa couplings in the Standard Model (SM). We suggest [1] that the natural resolution to this problem is the existence of some approximately conserved chiral charges beyond the SM. These charges, which we assume to be gauged, provide selection rules forbidding the transitions between the various left-handed and right-handed fermion states (except for the top quark).

For example, we suppose that there exists some charge (or charges) Q for which the quantum number difference between left- and right-handed Weyl states is larger for the electron than for muon:

$$|Q_{eL} - Q_{eR}| > |Q_{\mu L} - Q_{\mu R}| \tag{1}$$

It then follows that the SM Yukawa coupling for the electron g_e is suppressed more than that for the muon g_μ, when Q is taken to be approximately conserved. This is what is required if we want to explain the electron-muon mass ratio.

In section 2 we give arguments motivating our identification of the above chiral gauge charges with those of the anti-grand unification theory (AGUT) based on the non-simple gauge group $SMG^3 \times U(1)_f$, where $SMG \equiv SU(3) \times SU(2) \times U(1)$. We also give a crude statistical argument for the number of cross-product factors in the gauge group beyond the SM, suggested by the observed fermion spectrum. In section 3 we discuss the structure of the AGUT gauge group and how it can be rather simply characterized, as the maximal gauge group satisfying a few simple principles. The Higgs fields responsible for breaking the AGUT gauge group $SMG^3 \times U(1)_f$ to the diagonal SMG subgroup, identified as the SM gauge group, are considered in section 4. The structure of the resulting fermion mass matrices are presented in section 5, together with details of a fit to the charged fermion spectrum. In sections 6 and 7, we briefly discuss the problems of baryogenesis and neutrino oscillations respectively. Finally we present our conclusions in section 8.

2. Motivation for Anti-GUT

As pointed out in the introduction, the quark-lepton mass matrices—written in a basis of what we can call proto-flavours—have matrix elements typically suppressed relative to the electroweak scale ($< \phi_{WS} >= 246$ GeV) by rather large factors. We shall take the point of view that, in the fundamental theory beyond the SM, the Yukawa couplings allowed by gauge invariance are all of order unity and, similarly, all the mass terms allowed by gauge invariance are of order the fundamental mass scale of the theory—say the Planck scale. Then, apart from the matrix element responsible for the top quark mass, the quark-lepton mass matrix elements are only non-zero due to the presence of other Higgs fields having vacuum expectation values (VEVs) smaller (typically by one order of magnitude) than the fundamental scale. These Higgs fields will, of course, be responsible for breaking the fundamental gauge group G down to the SM group. In order to generate a particular effective SM Yukawa coupling matrix element, it is necessary to break the symmetry group G by a combination of Higgs fields with the appropriate quantum number combination $\Delta\vec{Q}$. When this "$\Delta\vec{Q}$" is different for two matrix elements, they will typically deviate by a large factor. If we want to explain the observed spectrum of quarks and leptons in this way, it is clear that we need charges which—possibly in a complicated way—separate the generations and, at least for $t-b$ and $c-s$, also quarks in the same generation. Just using the usual simple $SU(5)$

GUT charges does not help, because both (μ_R and e_R) and (μ_L and e_L) have the same $SU(5)$ quantum numbers. So we prefer to keep each SM irreducible representation in a separate irreducible representation of G and introduce extra gauge quantum numbers distinguishing the generations, by adding extra cross-product factors to the SM gauge group.

In order to be guided into a specific model and to convince ourselves that something like the model we propose should represent nature, we should like to estimate how complicated the gauge group G should be. In principle we could estimate over how many different orders of magnitude the various matrix elements in the mass matrices should distribute themeselves, because these matrix elements are to a large extent accessible to phenomenological—almost experimental—measurement. We could then adjust the degree of complication of the group - say the number of cross product factors in the gauge group G that should be used.

We consider the three charged particle mass matrices - since we have only rather uncertain information on the neutrinos anyway. With 3 generations they clearly contain together $3 \times 9 = 27$ different matrix elements. The number of matrix elements that are essentially measurable in practice is estimated as ca 12, corresponding to the meaurement of 3×3 masses and three mixing angles. How many different order of magnitude classes of matrix elements then needed will, of course, depend somewhat on the attitude as to when a couple of estimated matix elements deviate by more than of "order of magnitude unity", but we take roughly the number of classes to be around 7. We get this number 7 by saying that, of the "observable" 12 matrix elements in our model to be presented below, we have used 8 different orders of magnitude, but that there is one clear case in which there are used two different orders of magnitude in principle, although the experimental numbers give that the two matrix elements have the same order of magnitude: The matrix element giving the 2nd to 3rd generation mixing has, see eq.(33), the order T^3, while the one dominating the b-quark mass is of order WT^2, whereas they are experimentally numerically almost equal. If now we could guess that the non-observed matrix elements would distribute themselves into classes with the same order of magnitude in much the same way—i.e. with much the same number of matrix elements in the same class—then we would come to around $\frac{7}{12} \times 27 \approx 16$ classes for the elements in all three matrices. This number, in succession, must now depend on how many combination possibilities there are for the breaking of the symmetries, in various ways, by means of the Higgs fields. Obviously there should be more classes the more pieces the gauge group consists of. The connection between the number of cross product factors n, say in the gauge group $G = A_1 \times A_2 \times \cdots \times A_n$, and the number of order of magnitude classes of mass matrix elements may be seen crudely estimated in fig. 1.

This figure was constructed using three examples and a hand drawn

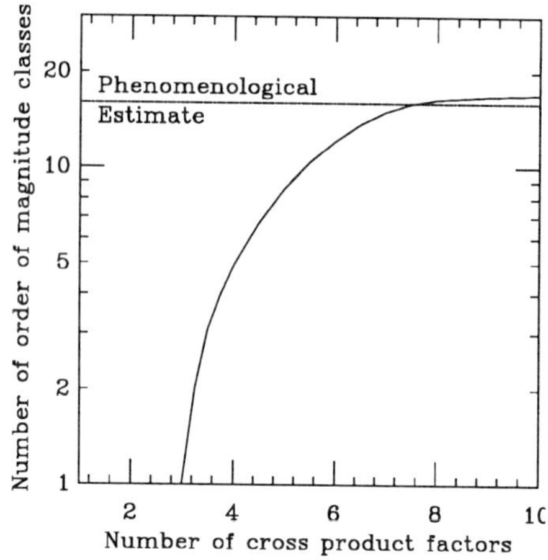

Figure 1. The connection between the number of cross product factors in the extended group G and the number of order of magnitude classes of quark-lepton mass matrix elements

curve through the three points, based on a vague expectation of how it should be curved. The examples we chose were the SM itself, the maximal AGUT group model and the reduced AGUT model (in which one of the Higgs fields S has a VEV of order unity in fundamental units) which we actually use to fit the experimental data. In the pure SM there are no suppression factors, and all 27 mass matrix elements are expected to belong to the same class and be of order the electroweak scale. Thus the SM corresponds to the point with the number of order of magnitude classes equal to one and number of cross product factors $n = 3$ on fig. 1. The other end of the curve corresponds to the maximal AGUT group $SMG^3 \times U(1)_f$, having $n = 10$ cross product factors. In this case the matrix elements are much more separated into different order of magnitude classes. As we shall see below the corresponding diagonal elements in each of the 3 Yukawa coupling matrices Y_U, Y_D and Y_E have the same order of magnitude. Furthermore we have $(Y_U)_{12} \simeq (Y_D)_{12}$ and $(Y_D)_{32} \simeq (Y_E)_{32}$, giving a total of 8 constraints and hence $27 - 6 - 2 = 19$ in principal different order of magnitude classes. If we take into account a couple of extra numerical coincidences in order of magnitude, the number of classes is reduced to 17. The third point on the curve corresponds to the AGUT model, in which the maximal AGUT group is replaced by the group to which it is broken by the non-suppressing Higgs field S, namely $SMG^2 \times U(1)$. The latter group has 7 factors in the cross product, and the reduced Yukawa matrices satisfy 2 extra constraints: $(Y_U)_{12} = (Y_U)_{21}$ and $(Y_D)_{12} = (Y_E)_{21}$. Thus the number of different order of magnitude classes in this case is $17 - 2 = 15$.

The idea now is to estimate the number of cross product factors in the "true" gauge group underlying the fermion mass spectrum, by looking up in figure 1 what number of factors would give the number of order of magnitude classes found phenemenologically. We get about 7 to 8 factors being suggested, not surprisingly since we used our model with seven cross product factors and the fact that it fits rather well to determine one point on the curve. In spite of this way of constructing the curve and what we consider the phenomenological number, we do not consider it an empty excercise to use our curve to estimate with what accuracy we must expect that a true model would have to have around the 7 cross product factors. Our point, of course, is to argue that only by a rather strange accident should it be possible for models, with a very different number of cross product factors, to fit the experimental data. In particular if the model had significantly less than 7 factors, there should be too many mass matrix elements not separated by the orders of magnitude found experimentally. It is anyway not possible to have more cross product factors than in the maximal AGUT model ($n = 10$) and still have an anomaly free group.

At the end of the day, of course, the real motivation for considering the rather specific AGUT gauge group close to the Planck scale, with order unity fundamental Yukawa couplings, is its success in fitting the masses of the quarks and leptons and the quark mixing angles. This fit is discussed in section 5.

3. The "maximal" AGUT gauge group

The AGUT model is based on extending the SM gauge group $SMG = S(U(2) \times U(3))$ in a similar way to grand unified $SU(5)$, but rather to the non-simple $SMG^3 \times U(1)_f$ group. Here we follow Michel and O'Raifeartaigh [2], in using not only the Lie algebra but even the Lie *group* for a Yang-Mills theory. The global properties of the SMG group imply that the SM particle representations of the $SU(3) \times SU(2) \times U(1)$ Lie algebra should obey the charge quantisation rule

$$\frac{y}{2} + \frac{1}{2}\text{"duality"} + \frac{1}{3}\text{"triality"} \equiv 0 \pmod 1 \qquad (2)$$

which expresses the somewhat complicated way in which the electric charge is quantised in the SM. Here y is the conventional weak hypercharge, and "duality" is 1 for the fundamental representation of SU(2) (the doublet) and 0 for the singlet. Similarly "triality" is 1 for the SU(3) triplet, -1 for the anti-triplet and 0 for the singlet.

The existence of the $SMG^3 \times U(1)_f$ group means that, near the Planck scale, there are three sets of SM-like gauge particles. Each set only couples to its own proto-generation [e.g. the proto- u, d, e and ν_e particles], but not to the other two proto-generations [e.g. the proto- c, s, μ, ν_μ, t, b, τ and ν_τ particles]. There is also an extra abelian $U(1)_f$ gauge boson, giving altogether $3 \times 8 = 24$ gluons, $3 \times 3 = 9$ W's and $3 \times 1 + 1 = 4$ abelian gauge bosons. The couplings of the $SMG_i = S(U(2) \times U(3))_i \approx SU(3)_i \times SU(2)_i \times U(1)_i$ group to the i'th proto-generation are identical to those of the SM group. Consequently we have a charge quantisation rule, analogous to eq. (2), for each of the three proto-generation weak hypercharge quantum numbers y_i.

To first approximation—namely in the approximation that the quark mixing angles V_{us}, V_{cb}, V_{ub} are small—we may ignore the prefix "*proto-*". However we really introduce in our model some "proto-fields" characterized by their couplings to the 37 gauge bosons of the $SMG^3 \times U(1)_f$ group. The physically observed u-quark, d-quark etc. are then superpositions of the proto-quarks (or proto-leptons), with the same named proto-particle dominating. Actually there is one deviation from this first approximation rule that proto-particles correspond to the same named physical particles. In the AGUT fit to the quark-lepton mass spectrum, discussed below, we find that to first approximation the right-handed components of the top and the charm quarks must be permuted:

$$c_R \text{ proto} \approx t_R \text{ physical} \qquad t_R \text{ proto} \approx c_R \text{ physical} \tag{3}$$

But for all the other components we have:

$$t_L \text{ proto} \approx t_L \text{ physical} \qquad b_R \text{ proto} \approx b_R \text{ physical} \tag{4}$$

and so on.

The AGUT group breaks down an order of magnitude or so below the Planck scale to the SM group, as the diagonal subgroup of its SMG^3 subgroup. The gauge coupling constants do not, of course, unify, but their values have been successfully calculated using the so-called multiple point principle [3].

At first sight, this $SMG^3 \times U(1)_f$ group with its 37 generators seems to be just one among many possible SM gauge group extensions. However, we shall now argue it is not such an arbitrary choice, as it can be uniquely specified by postulating 4 reasonable requirements on the gauge group G beyond the SM. As a zeroth postulate, of course, we require that the gauge group extension must contain the Standard Model group as a subgroup $G \supseteq SMG$. In addition it should obey the following 4 postulates:

1. G should transform the presently known (left-handed, say) Weyl particles into each other, so that $G \subseteq U(45)$. Here $U(45)$ is the group of

all unitary transformations of the 45 species of Weyl fields (3 generations with 15 in each) in the SM.

2. No anomalies, neither gauge nor mixed. We assume that only straightforward anomaly cancellation takes place and, as in the SM itself, do not allow for a Green-Schwarz type anomaly cancellation [4].

3. The fifteen irreducible representations of Weyl fields for the SM group remain irreducible under G. This is the most arbitrary of our assumptions about G. It is motivated by the observation that combining SM irreducible representations into larger unified representations introduces symmetry relations between Yukawa coupling constants, whereas the particle spectrum doess not exhibit any exact degeneracies (except possibly for the case $m_b = m_\tau$). In fact AGUT only gets the naive $SU(5)$ mass predictions as order of magnitude relations: $m_b \approx m_\tau$, $m_s \approx m_\mu$, $m_d \approx m_e$.

4. G is the maximal group satisfying the other 3 postulates. We already argued, in the previous section, that the large number of order of magnitude classes of fermion mass matrix elements indicates the need for a large number of cross product factors in G.

With these four postulates a somewhat cumbersome calculation shows that, modulo permutations of the various irreducible representations in the Standard Model fermion system, we are led to our gauge group $SMG^3 \times U(1)_f$. Furthermore it shows that the SM group is embedded as the diagonal subgroup of SMG^3, as in our AGUT model. The most difficult part of the calculation is to decide which identifications of the abelian groups have to be made in order to avoid anomalies, i.e. how big a subgroup of $U(1)^{15}$ can avoid having anomalies and be allowed in G. In searching for the generators of an allowed subgroup, one may expand them in terms of the generators for these 15 $U(1)$'s and they have to obey some first order (linear) relations for the coefficients, in order to avoid anomalies involving also the non-abelian or gravitational fields. Also there are third order relations that have to be satisfied, in order that there be no anomalies involving only the subspace of abelian generators. It turns out that, with the three proto-generations of fermions, there are too many constraints to be solved with an abelian subgroup of dimension higher than 4. It is found that three of the allowed abelian generators in G can be taken to be the 3 weak hypercharges, each defined to act on only one proto-generation. After that choice the scheme becomes so tight that, apart from various rewritings and permutations of the particle names, there is a unique fourth $U(1)$ allowed and that is what we call $U(1)_f$. Several of the anomalies involving this $U(1)_f$ are cancelled by assigning equal and opposite values of the $U(1)_f$ charge to the analogous particles belonging to second and third proto-generations, while the first proto-generation particles have just zero charge [5]. In fact the $U(1)_f$ group does not couple to the left-handed particles and the $U(1)_f$ quantum numbers can be chosen as follows for the proto-states:

$$Q_f(\tau_R) = Q_f(b_R) = Q_f(c_R) = 1 \tag{5}$$

$$Q_f(\mu_R) = Q_f(s_R) = Q_f(t_R) = -1 \tag{6}$$

Thus the quantum numbers of the quarks and leptons are uniquely determined in the AGUT model. However we do have the freedom of choosing the gauge quantum numbers of the Higgs fields responsible for the breaking the $SMG^3 \times U(1)_f$ group down to the SM gauge group. These quantum numbers are chosen with a view to fitting the fermion mass and mixing angle data [6], as discussed in the next section.

4. Symmetry breaking by Higgs fields

There are obviously many different ways to break down the large group G to the much smaller SMG. However, we can first greatly simplify the situation by assuming that, like the quark and lepton fields, the Higgs fields belong to singlet or fundamental representations of all non-abelian groups. The non-abelian representations are then determined from the $U(1)_i$ weak hypercharge quantum numbers, by imposing the charge quantisation rule eq. (2) for each of the SMG_i groups. So now the four abelian charges, which we express in the form of a charge vector

$$\vec{Q} = \left(\frac{y_1}{2}, \frac{y_2}{2}, \frac{y_3}{2}, Q_f\right)$$

can be used to specify the complete representation of G. The constraint, that we must eventually recover the SM group as the diagonal subgroup of the SMG_i groups, is equivalent to the constraint that all the Higgs fields (except for the Weinberg-Salam Higgs field which of course finally breaks the SMG) should have charges y_i satisfying:

$$y = y_1 + y_2 + y_3 = 0 \tag{7}$$

in order that their SM weak hypercharge y be zero.

We wish to choose the charges of the Weinberg-Salam (WS) Higgs field, so that it matches the difference in charges between the left-handed and right-handed physical top quarks. This will ensure that the top quark mass in the SM is not suppressed relative to the WS Higgs field VEV. However, as we remarked in the previous section, it is necessary to associate the physical right-handed top quark field not with the corresponding third proto-generation field t_R, but rather with the right-handed field c_R of the second proto-generation. Otherwise we cannot suppress the bottom quark and tau lepton masses. This is because, for the proto-fields, the charge differences between t_L and t_R are the same as between b_L and b_R and also between τ_L and τ_R. So now it is simple to calculate the quantum numbers of the WS Higgs field ϕ_{WS}:

$$\vec{Q}_{\phi_{WS}} = \vec{Q}_{c_R} - \vec{Q}_{t_L} = \left(0, \frac{2}{3}, 0, 1\right) - \left(0, 0, \frac{1}{6}, 0\right) = \left(0, \frac{2}{3}, -\frac{1}{6}, 1\right) \quad (8)$$

This means that the WS Higgs field will in fact be coloured under both $SU(3)_2$ and $SU(3)_3$. After breaking the symmetry down to the SM group, we will be left with the usual WS Higgs field of the SM and another scalar which will be an octet of $SU(3)$ and a doublet of $SU(2)$. This should not present any phenomenological problems, provided this scalar doesn't cause symmetry breaking and doesn't have a mass less than the few TeV scale. In particular an octet of $SU(3)$ cannot lead to baryon decay.

We can now choose the charges of the other Higgs fields in our model, by considering the charge differences between left-handed and right-handed fermions with the inclusion of the WS Higgs. Since we have the constraint of eq. (7), the charges of these Higgs fields must be chosen to span a 3 dimensional vector space of charges represented, for example, by $\frac{y_1}{2}$, $\frac{y_3}{2}$ and Q_f with $\frac{y_2}{2}$ being determined by eq. (7). This means that we will need at least 3 Higgs fields to break the gauge group down to the SMG. This gives us a lot of freedom, so we will choose the charges on these Higgs fields by considering phenomenological relations between fermion masses.

Since we are assuming that the fundamental Yukawa couplings are of order 1 but not exactly 1, we can only produce order of magnitude results. So we wish to choose, for example, 2 fermions with similar masses but not order of magnitude equal masses. We can then assume that the lighter fermion is suppressed relative to the heavier fermion by 1 Higgs with a VEV given approximately by the ratio of the 2 fermion masses. For example we would say that the bottom quark and tau lepton masses were of the same order of magnitude (remembering that we take all relations at the Planck scale). However we can take the following 2 ratios of effective Yukawa couplings to be significantly different from 1:

$$\frac{g_c}{g_b} \equiv \frac{< W >}{M_F} \approx \frac{1}{5} \quad (9)$$

$$\frac{g_\mu}{g_b} \equiv \frac{< T >}{M_F} \approx \frac{1}{13} \quad (10)$$

where we have defined 2 Higgs fields, W and T, to have appropriate VEVs, relative to the fundamental mass scale M_F of the theory, to cause m_c and m_μ to be suppressed relative to m_b.

First we define $\vec{b}$ to be the difference in charges between b_L and b_R proto-fields with the inclusion of the WS Higgs field. So we have:

$$\vec{b} = \vec{Q}_{b_L} - \vec{Q}_{b_R} - \vec{Q}_{WS} \tag{11}$$

Similarly we define $\vec{c}$ and $\vec{\mu}$ to be:

$$\vec{c} = \vec{Q}_{c_L} - \vec{Q}_{t_R} + \vec{Q}_{WS} \tag{12}$$
$$\vec{\mu} = \vec{Q}_{\mu_L} - \vec{Q}_{\mu_R} - \vec{Q}_{WS} \tag{13}$$

Note that $\vec{c}$ has been defined using the t_R proto-field, since we have essentially swapped the right-handed charm and top quarks. Also the charges of the WS Higgs field are added rather than subtracted for up-type quarks. We observe that:

$$\vec{b} + \vec{c} + \vec{\mu} = \vec{0} \tag{14}$$

Now we can express these charges in terms of those of the Higgs fields. We can define:

$$\vec{b} = \alpha\vec{Q}_W + \beta\vec{Q}_T + \vec{Q}_X \tag{15}$$

where we have chosen the overall sign of the charges on the Higgs fields W and T so that α and β are not negative. $\vec{Q}_X$ is the total charges of all other Higgs fields used to suppress m_b relative to m_t. We will assume that $\vec{Q}_X$ cannot be expressed as a linear combination of $\vec{Q}_W$ and $\vec{Q}_T$. Now eqs. (9) and (10) require that:

$$\vec{c} = \pm(\alpha + 1)\vec{Q}_W \pm \beta\vec{Q}_T \pm \vec{Q}_X \tag{16}$$
$$\vec{\mu} = \pm\alpha\vec{Q}_W \pm (\beta + 1)\vec{Q}_T \pm \vec{Q}_X \tag{17}$$

The presence of the $\pm$ signs is due to the fact that we can use the fields $W^\dagger$ and $T^\dagger$ as well as W and T.

So we can rewrite eq. (14) as:

$$\begin{pmatrix} 3\alpha+1 \\ \alpha+1 \\ \alpha-1 \\ -\alpha-1 \end{pmatrix} \vec{Q}_W + \begin{pmatrix} 3\beta+1 \\ \beta-1 \\ \beta+1 \\ -\beta-1 \end{pmatrix} \vec{Q}_T + \begin{pmatrix} 3 \\ 1 \\ 1 \\ -1 \end{pmatrix} \vec{Q}_X = \vec{0} \tag{18}$$

where the 4 coefficients for each term correspond to the 4 combinations of signs in front of the terms in eqs. (16) and (17), giving 64 cases altogether.

All possible choices of coefficient of $\vec{Q}_X$ are non-zero and, by assumption, $\vec{Q}_X$ is linearly independent of $\vec{Q}_W$ and $\vec{Q}_T$; so eq. (18) cannot hold. We must therefore conclude that there are no Higgs fields, other than W and T, used to suppress m_b relative to m_t. So we must set $\vec{Q}_X = \vec{0}$. We can now use the fact that α and β are not negative, along with the assumption that $\vec{Q}_T$ is not directly proportional to $\vec{Q}_W$, to conclude that eq. (18) requires that:

$$\alpha = \beta = 1 \tag{19}$$

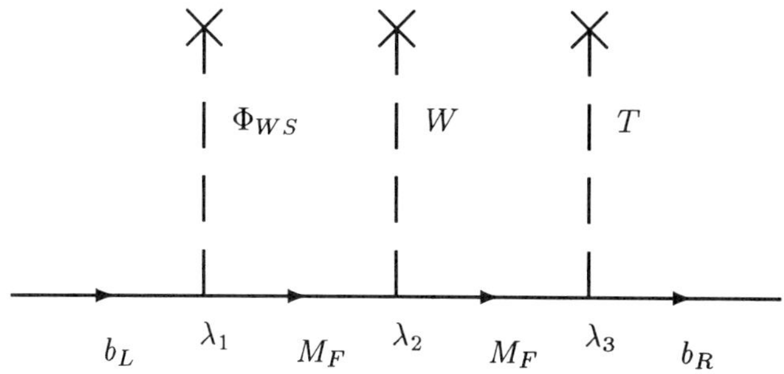

Figure 2. Feynman diagram for bottom quark mass in the AGUT model. The crosses indicate the couplings of the Higgs fields to the vacuum.

and that the combination of signs is chosen so that:

$$\vec{b} = \vec{Q}_W + \vec{Q}_T \tag{20}$$
$$\vec{c} = -2\vec{Q}_W + \vec{Q}_T \tag{21}$$
$$\vec{\mu} = \vec{Q}_W - 2\vec{Q}_T \tag{22}$$

We note that this immediately implies the reasonably good Planck scale relation:

$$g_b = \frac{<W>}{M_F}\frac{<T>}{M_F} \approx \frac{1}{65} \tag{23}$$

This expression for the effective SM bottom quark Yukawa coupling constant arises from the Feynman diagram in figure 2. Here we have assumed the existence of a rich spectrum of vector-like Dirac fermions, with unsuppressed masses of the order of the fundamental mass scale $M_F = M_{Planck}$, which provides the required intermediate states. Also the fundamental Yukawa couplings λ_i are taken of order unity.

It is now a simple matter to calculate the charges of the Higgs fields W and T. We have:

$$\vec{Q}_W = \frac{1}{3}(2\vec{b} + \vec{\mu}) = \left(0, -\frac{1}{2}, \frac{1}{2}, -\frac{4}{3}\right) \tag{24}$$

From this we can then calculate:

$$\vec{Q}_T = \vec{b} - \vec{Q}_W = \left(0, -\frac{1}{6}, \frac{1}{6}, -\frac{2}{3}\right) \tag{25}$$

We notice that the charges of W and T do not cover the 2 dimensional space of charges $\frac{y_3}{2}$ and Q_f, since only even Q_f charges can be constructed with integer numbers of these Higgs fields. Therefore, since both W and T have $\frac{y_1}{2} = 0$, we will need at least 2 more Higgs fields to fully cover the 3 dimensional charge space required to break G down to the SM group. We will now choose 2 more Higgs fields which, together with W and T, will fully cover this space.

Another parameter in the SM, which is within one order of magnitude from unity, is the mixing matrix element between the 1st and 2nd generations, which we associate with another Higgs field VEV:

$$V_{us} \equiv <\xi> \approx 0.2 \tag{26}$$

With the mass matrix texture in our model, V_{us} is approximately given by the ratio of the mass matrix transition element from d_L to s_R to the transition from s_L to s_R. This means that we must have:

$$\vec{Q}_\xi = \vec{Q}_{d_L} - \vec{Q}_{s_L} = \left(\frac{1}{6}, 0, 0, 0\right) - \left(0, \frac{1}{6}, 0, 0\right) = \left(\frac{1}{6}, -\frac{1}{6}, 0, 0\right) \tag{27}$$

From the well-known Fritzsch relation [7] $V_{us} \simeq \sqrt{\frac{m_d}{m_s}}$, it is suggested that the two off-diagonal mass matrix elements connecting the d-quark and the s-quark be equally big. We achieve this approximately in our model by introducing a special Higgs field S, with quantum numbers equal to the difference between the quantum number differences for these 2 matrix elements in the down quark matrix. Then we postulate that this Higgs field has a VEV of order unity in fundamental units, so that it does not cause any suppression but rather ensures that the two matrix elements get equally suppressed. Henceforth we will consider the VEVs of the new Higgs fields as measured in units of M_F, and so we have:

$$< S >= 1 \tag{28}$$

and the charges of S are given by:

$$\begin{aligned}
\vec{Q}_S &= [\vec{Q}_{s_L} - \vec{Q}_{d_R}] - [\vec{Q}_{d_L} - \vec{Q}_{s_R}] \\
&= \left[\left(0, \frac{1}{6}, 0, 0\right) - \left(-\frac{1}{3}, 0, 0, 0\right)\right] - \left[\left(\frac{1}{6}, 0, 0, 0\right) - \left(0, -\frac{1}{3}, 0, -1\right)\right] \\
&= \left(\frac{1}{6}, -\frac{1}{6}, 0, -1\right)
\end{aligned} \tag{29}$$

The existence of a non-suppressing field S means that we cannot control phenomenologically when this S-field is used. Thus the quantum numbers of the other Higgs fields W, T, ξ and ϕ_{WS} given above have only been determined modulo those of the field S.

5. Mass matrices, predictions

We define the mass matrices by considering the mass terms in the SM to be given by:

$$\mathcal{L} = Q_L M_u U_R + Q_L M_d D_R + L_L M_l E_R + \text{h.c.} \tag{30}$$

The mass matrices can be expressed in terms of the effective SM Yukawa matrices and the WS Higgs VEV by:

$$M_f = Y_f \frac{< \phi_{WS} >}{\sqrt{2}} \tag{31}$$

We can now calculate the suppression factors for all elements in the Yukawa matrices, by expressing the charge differences between the left-handed and right-handed fermions in terms of the charges of the Higgs fields. They are given by products of the small numbers denoting the VEVs, in fundamental units, of the fields W, T, ξ and the of order unity VEV of S. In the following matrices, we simply write W instead of $< W >$ etc. for the VEVs. With the quantum number choice given above, the resulting matrix elements are—but remember that "random" order unity factors are supposed to multiply all the matrix elements—for the uct-quarks:

$$Y_U \simeq \begin{pmatrix} S^\dagger W^\dagger T^2 (\xi^\dagger)^2 & W^\dagger T^2 \xi & (W^\dagger)^2 T \xi \\ S^\dagger W^\dagger T^2 (\xi^\dagger)^3 & W^\dagger T^2 & (W^\dagger)^2 T \\ S^\dagger (\xi^\dagger)^3 & 1 & W^\dagger T^\dagger \end{pmatrix} \tag{32}$$

the dsb-quarks:

$$Y_D \simeq \begin{pmatrix} SW(T^\dagger)^2 \xi^2 & W(T^\dagger)^2 \xi & T^3 \xi \\ SW(T^\dagger)^2 \xi & W(T^\dagger)^2 & T^3 \\ SW^2 (T^\dagger)^4 \xi & W^2 (T^\dagger)^4 & WT \end{pmatrix} \tag{33}$$

and the charged leptons:

$$Y_E \simeq \begin{pmatrix} SW(T^\dagger)^2 \xi^2 & W(T^\dagger)^2 (\xi^\dagger)^3 & (S^\dagger)^2 W T^4 \xi^\dagger \\ SW(T^\dagger)^2 \xi^5 & W(T^\dagger)^2 & (S^\dagger)^2 W T^4 \xi^2 \\ S^3 W (T^\dagger)^5 \xi^3 & (W^\dagger)^2 T^4 & WT \end{pmatrix} \tag{34}$$

We can now set $S = 1$ and fit the nine quark and lepton masses and three mixing angles, using 3 parameters: W, T and ξ. That really means we have effectively omitted the Higgs field S, and replaced the maximal AGUT gauge group $SMG^3 \times U(1)_f$ by the reduced AGUT group $SMG_{12} \times SMG_3 \times U(1)$, which survives the spontaneous breakdown due to S. In order to find the best possible fit, we must use some function which measures how good a fit is. Since we are expecting an order of magnitude fit, this function should depend only on the ratios of the fitted masses to the experimentally determined masses. The obvious choice for such a function is:

$$\chi^2 = \sum \left[\ln \left(\frac{m}{m_{\mathrm{exp}}} \right) \right]^2 \tag{35}$$

where m are the fitted masses and mixing angles and m_{exp} are the corresponding experimental values. The Yukawa matrices are calculated at the

Table 1. Best fit to conventional experimental data. All masses are running masses at 1 GeV except the top quark mass which is the pole mass.

	Fitted	Experimental
m_u	3.6 MeV	4 MeV
m_d	7.0 MeV	9 MeV
m_e	0.87 MeV	0.5 MeV
m_c	1.02 GeV	1.4 GeV
m_s	400 MeV	200 MeV
m_μ	88 MeV	105 MeV
M_t	192 GeV	180 GeV
m_b	8.3 GeV	6.3 GeV
m_τ	1.27 GeV	1.78 GeV
V_{us}	0.18	0.22
V_{cb}	0.018	0.041
V_{ub}	0.0039	0.0035

fundamental scale, which we take to be the Planck scale. We use the first order renormalisation group equations (RGEs) for the SM to calculate the matrices at lower scales.

We cannot simply use the 3 matrices given by eqs. (32)–(34) to calculate the masses and mixing angles, since only the order of magnitude of the elements is defined. Therefore we calculate statistically, by giving each element a random complex phase and then finding the masses and mixing angles. We repeat this several times and calculate the geometrical mean for each mass and mixing angle. In fact we also vary the magnitude of each element randomly, by multiplying by a factor chosen to be the exponential of a number picked from a Gaussian distribution with mean value 0 and standard deviation 1.

We then vary the 3 free parameters to find the best fit given by the χ^2 function. We get the lowest value of χ^2 for the VEVs:

$$\langle W \rangle = 0.179 \tag{36}$$

$$\langle T \rangle = 0.071 \tag{37}$$

$$\langle \xi \rangle = 0.099 \tag{38}$$

The fitted value of $\langle \xi \rangle$ is approximately a factor of two smaller than the estimate given in eq. (26). This is mainly because there are contributions to V_{us} of the same order of magnitude from both Y_U and Y_D. The result [6] of the fit is shown in table 1. This fit has a value of:

$$\chi^2 = 1.87 \tag{39}$$

Table 2. Best fit using alternative light quark masses extracted from lattice QCD. All masses are running masses at 1 GeV except the top quark mass which is the pole mass.

	Fitted	Experimental
m_u	1.9 MeV	1.3 MeV
m_d	3.7 MeV	4.2 MeV
m_e	0.45 MeV	0.5 MeV
m_c	0.53 GeV	1.4 GeV
m_s	327 MeV	85 MeV
m_μ	75 MeV	105 MeV
M_t	192 GeV	180 GeV
m_b	6.4 GeV	6.3 GeV
m_τ	0.98 GeV	1.78 GeV
V_{us}	0.15	0.22
V_{cb}	0.033	0.041
V_{ub}	0.0054	0.0035

This is equivalent to fitting 9 degrees of freedom (9 masses + 3 mixing angles - 3 Higgs VEVs) to within a factor of $\exp(\sqrt{1.87/9}) \simeq 1.58$ of the experimental value. It is better than what one might have expected from an order of magnitude fit.

We can also fit to different experimental values of the 3 light quark masses by using recent results from lattice QCD, which seem to be consistently lower than the conventional phenomenological values. The best fit in this case [6] is shown in table 2. The values of the Higgs VEVs are:

$$\langle W \rangle = 0.123 \tag{40}$$

$$\langle T \rangle = 0.079 \tag{41}$$

$$\langle \xi \rangle = 0.077 \tag{42}$$

and this fit has a larger value of:

$$\chi^2 = 3.81 \tag{43}$$

But even this is good for an order of magnitude fit.

6. Baryogenesis

A very important check of our model is whether or not it can be consistent with baryogenesis. In our model we have just the SM interactions up to about one or two orders of magnitude under the Planck scale. So we have no way, at the electroweak scale, to produce the baryon number in the universe. There is insufficient CP violation in the SM. Furthermore, even if created, the baryon number would immediately be washed out by sphaleron transitions after the electroweak phase transition. Our only chance to avoid the baryon number being washed out at the electroweak scale is to have a non-zero $B - L$ (i.e. baryon number minus lepton number) produced from the high, i.e. Planck, scale action of the theory. That could then in turn give rise to the baryon number at the electroweak scale. Now, in our model, the $B - L$ quantum number is broken by an anomaly involving the $U(1)_f$ gauge group. This part of the gauge group in turn is broken by the Higgs field ξ which, in Planck units, is fitted to have an expectation value around 1/10. The anomaly keeps washing out any net $B - L$ that might appear, due to CP-violating forces from the Planck scale physics, until the temperature T of the universe has fallen to $\xi = 1/10$. The $U(1)_f$ gauge particle then disappears from the thermal soup and thus the conservation of $B - L$ sets in. The amount of $B - L$ produced at that time should then be fixed and would essentially make itself felt, at the electroweak scale, by giving rise to an amount of baryon number of the same order of magnitude.

The question now is whether we should expect, in our model, to have a sufficient amount of time reversal symmetry breaking, at the epoch when $B-L$ settles down to be conserved, that the amount of $B-L$ relative to the entropy (essentially the amount of 3 degree Kelvin background radiation) becomes large enough to agree with the well-known phenomenological value of the order of 10^{-9} or 10^{-10}. We shall use purely dimensional arguments, assuming all couplings are generically of order unity, to estimate the effects of Planck scale physics. At the time of the order of the Planck scale, when the temperature was also of the order of the Planck temperature, we expect even the CP or time reversal violations were of order unity (in Planck units). So at that time there existed particles, say, with order of unity CP-violating decays. However, they had also, in our purely dimensional approximation, lifetimes of the order of the Planck scale too. Thus the $B - L$ biased decay products would mainly be dumped at time 1 in Planck units, rather than at the time of $B - L$ conservation setting in. In a radiation dominated universe, as we shall assume, the temperature is proportional to 1/a where a is the radius parameter—the size or scale parameter of the universe. Now the time goes as the square of this size parameter a. Thus the time in Planck units is given as the temperature to the negative second power:

$$t = \frac{0.3}{\sqrt{g_*} \times T^2} \tag{44}$$

where [8] g_* is the number of degrees of freedom—counted as 1 for bosons but as 7/8 per fermion degree of freedom—entering into the radiation density. In our model g_* gets a contribution of $\frac{7}{8} \times 45 \times 2$ from the fermions and

2×37 from the gauge bosons, and in addition there is some contribution from the Higgs particles. So we take g_* to be of order 100, in our crude estimate of the time t corresponding to the temperature $T = \xi = \frac{1}{10}$ in Planck units, when $B - L$ conservation sets in:

$$t \simeq \frac{0.3}{\sqrt{100}} \times \left(\frac{1}{10}\right)^{-2} = 3 \tag{45}$$

By that time we expect a fraction of the order of $\exp(-3)$ of the particles from the Planck era is still present and able to dump its CP-violating decay products. Of course here the uncertainty of, say, an order of magnitude in t would appear in the exponent, meaning a suppression factor anywhere between $\exp(0)$ and $\exp(-30)$, which could thus easily be in agreement with the value wanted for baryogenesis of order 5×10^{-10}. This result is encouraging, but clearly a more careful analysis is required.

7. Neutrino oscillations, a problem?

We expect the neutrinos to get a mass in the AGUT model, by the exchange of the WS Higgs field ϕ_{WS} twice—the see-saw mechanism [9, 10]—or the exchange of a weak isotriplet Higgs field Δ [11]. In general it will also be necessary to exchange other AGUT Higgs scalars, in order to balance the AGUT gauge quantum numbers beyond the SM. This mechanism naturally generates an effective three generation light neutrino mass matrix M_ν:

$$\mathcal{L}_m = (M_\nu)_{ij} \nu_{L_i} C \nu_{L_j} + \text{h.c.} \tag{46}$$

The neutrino mass matrix elements arise from Feynman diagrams similar to those of figure 2, but they involve two WS Higgs fields and the transitions are between ν_{iL} and ν^c_{jR}. It follows that

$$M_\nu = H_\nu \frac{\langle \phi_{WS} \rangle^2}{2M_F} \tag{47}$$

where the dimensionless coupling matrix H_ν is analogous to the quark-lepton Yukawa matrices Y_U, Y_D and Y_E, with elements expressed as products of the AGUT Higgs field VEVs like in eqs. (32)-(34).

Since the fundamental scale of the AGUT model is $M_F = M_{Planck}$, the overall neutrino mass scale in the model, as given by eq. (47), is:

$$\frac{\langle \phi_{WS} \rangle^2}{2M_{Planck}} \sim 3 \times 10^{-6} \text{ eV} \tag{48}$$

This overall mass scale essentially provides an upper limit to the neutrino masses. So there is a basic problem for the AGUT model, since this mass scale is too small to explain neutrino oscillation phenomenology, except possibly for "just-so" vacuum oscillations of solar neutrinos.

The neutrino mass matrix M_ν is, by its very definition eq. (46), symmetric. Also, in models like AGUT with approximately conserved chiral $U(1)$ charges, the matrix elements are generally of different orders of magnitude, due to the presence of various suppression factors. Thus the generic structure for M_ν is a matrix in which the various elements typically each have their own order of magnitude, except in as far as they are forced to be equal by the symmetry $M_\nu = M_\nu^T$. The largest neutrino mass eigenvalue is then given by the largest matrix element of M_ν. If it happens to be one of a pair of equal off-diagonal elements, we get two very closely degenerate states as the heaviest neutrinos with essentially maximal mixing; the third neutrino will be much lighter and, in first approximation, will not mix with the other two. In fact this is what happens in the AGUT model, if we assume that no new Higgs fields are introduced.

We can calculate the contributions to the neutrino mass matrix from the Higgs fields already introduced in our model and find:

$$H_\nu \sim \begin{pmatrix} (S^\dagger)^2(W^\dagger)^2T^4(\xi^\dagger)^4 & (S^\dagger)^2(W^\dagger)^2T^4\xi^\dagger & (W^\dagger)^2T(\xi^\dagger)^3 \\ (S^\dagger)^2(W^\dagger)^2T^4\xi^\dagger & W(T^\dagger)^5 & (W^\dagger)^2T \\ (W^\dagger)^2T(\xi^\dagger)^3 & (W^\dagger)^2T & S^2(W^\dagger)^2(T^\dagger)^2(\xi^\dagger)^2 \end{pmatrix} \tag{49}$$

where, as usual, we assume that all fundamental Yukawa couplings are of order 1. Clearly all the elements of H_ν are suppressed. The largest element is off-diagonal and of order $\langle W\rangle^2\langle T\rangle$. We find the following eigenvalues for H_ν:

$$h_{\nu_\mu} \simeq h_{\nu_\tau} \simeq \langle W\rangle^2\langle T\rangle \simeq 2.3 \times 10^{-3} \tag{50}$$

$$h_{\nu_e} \simeq \langle W\rangle^2\langle T\rangle^4\langle\xi\rangle^4 \simeq 7.8 \times 10^{-11} \tag{51}$$

There is almost maximal mixing ($\sin^2 2\theta = 1$) between ν_μ and ν_τ with very small mixing ($\sin^2 2\theta \simeq \langle\xi\rangle^6 \simeq 10^{-6}$) of ν_e. This is not suitable for observable solar neutrino oscillations and, although we do have the correct mixing structure for atmospheric neutrino oscillations, the masses

$$m_{\nu_\mu} \simeq m_{\nu_\tau} \simeq h_{\nu_\tau}\frac{\langle\phi_{WS}\rangle^2}{2M_F} \simeq 7 \times 10^{-9}\ \mathrm{eV} \tag{52}$$

are far too small.

So we predict no observable neutrino oscillations, unless we modify our AGUT model and introduce a new mass scale into the theory. Either some intermediate mass see-saw fermions or a weak isotriplet Higgs field Δ is required. The latter could acquire a vacuum expectation value of say $\langle\Delta^0\rangle \sim 1$ eV, via its interaction with two WS Higgs fields ϕ_{WS} and the other Higgs fields W, T, ξ and S; but only if, for some as yet unknown reason, it has a very small coefficient of Δ^2 in the Higgs potential compared to M^2_{Planck}. Furthermore, it is difficult to explain both the solar and atmospheric neutrino problems in the above scenario where a single pair of equal off-diagonal elements dominates M_ν. Although it can provide the large ν_μ - ν_τ mixing needed for the atmospheric neutrino problem, their quasi-degeneracy in mass implies that their mass differences with ν_e are too large to explain the solar neutrino problem [12]. In fact it appears necessary for M_ν to have at least two independent large elements being of the same order of magnitude. Such an "accidental" order of magnitude degeneracy is perhaps not so unlikely, in light of our discussion, in section 2, of the number of order of magnitude degeneracies among the AGUT charged fermion mass matrix elements.

8. Conclusions

We have tried to motivate the AGUT model from the characteristic features of the quark-lepton masses and mixing angles, which point rather strongly to proto-flavour mass matrices having elements typically of different orders of magnitude. We considered roughly how many different orders of magnitude would be represented by the proto-flavour mass matrices, using experimental data and interpolating by theoretical considerations. In this way, we estimated that the number of these order of magnitude classes should be around 16. In turn this estimate suggests around 7 to 8 cross product factors in the gauge group responsible for the generation splittings etc.

The largest anomaly-free gauge group acting on just the 45 SM Weyl fermions, without any unification of the SM irreducible representations, is the AGUT group $SMG^3 \times U(1)_f$, which has 10 cross-product factors giving 37 generators in all. This is broken down to the reduced AGUT group $SMG^2 \times U(1)$ with 7 cross product factors (which we actually use to fit the mass spectrum), by the Higgs field S with its VEV of order unity in Planck units.

Now, inspired by the experimental data, we introduced three more Higgs fields W, T and ξ, with appropriate quantum numbers made to break the AGUT group down to the SM group. We then presented a rather good fit, that in principle should only work to order of magnitude accuracy, to the charged quark-lepton masses and mixing angles, using three parameters corresponding to the VEVs of W, T and ξ. The most characteristic feature is that, apart from the t and c quarks, the masses of the particles in the same generation are predicted to be degenerate, but only order of magnitude-wise, at the Planck scale. The worst feature of the fit is the deviation by a factor of about 2 between the fitted and experimental values for m_s and V_{cb}. However this is what can be expected in an order of magnitude fit.

With this promising fit to the charged fermion masses, we considered the predictions of the model for the baryon number of the universe and neutrino masses. Using a crude dimensional argument, together with the expected number of degrees of freedom being of order 100, we obtain results for baryogenesis consistent with the observed baryon number, but with the uncertainty occurring inside an exponent. If anything we tend to get too high a prediction for the baryon number. However our predictions for the neutrino masses are set by a Planck mass see-saw mechanism, and are therefore too low to give observable neutrino oscillations. This suggests we may have to modify our model and introduce a new mass scale into the theory.

Apart from the neutrino puzzle, the AGUT model is successful in explaining the quark-lepton masses and mixing angles.

References

[1] Froggatt C D and Nielsen H B 1979 *Nucl. Phys.* B **147** 277

[2] Michel L 1964 *Group Theoretical Concepts and Methods in Elementary Particle Physics, Lectures of the Istanbul Summer School of Theoretical Physics 1962*, ed F Gursey (Gordon and Breach) p 135
O'Raifeartaigh L 1986 *Group Structure of Gauge Theories* (Cambridge University Press)

[3] Bennett D L, Froggatt C D and Nielsen H B 1995 *Proc. of the 27th Int. Conf. on High Energy Physics* (Glasgow, 1994) ed P Bussey and I Knowles (IOP Publishing Ltd) p 557; 1995 *Perspectives in Particle Physics '94* ed D Klabučar *et al* (World Scientific) p 255, hep-ph/9504294

[4] Green M B and Schwarz J 1984 *Phys. Lett.* B**149** 117

[5] Davidson A, Koca M and Wali K C 1979 *Phys. Rev. Lett.* **43** 92; 1979 *Phys. Rev.* D **20** 421

[6] Froggatt C D, Nielsen H B and Smith D J 1996 *Phys. Lett.* B **385** 150;
Froggatt C D, Gibson M, Nielsen H B and Smith D J *Int. J. Mod. Phys.* A (to be published), hep-ph/9706212.

[7] Fritzsch H 1977 *Phys. Lett.* B**70** 436

[8] Sarkar U *These Proceedings*

[9] Froggatt C D and Nielsen H B 1979 *Nucl. Phys.* B **164** 114

[10] Gell-Mann M, Ramond P and Slansky R 1980 *Supergravity* ed P van Nieuwenhuizen and D Z Freedman (New York: North Holland) p 315
Yanagida T 1979 *Proc. of the Workshop on Unified Theory and Baryon Number in the Universe* ed A Sawada and H Sugawara, (KEK, Tsukuba-Gu, Ibaraki-ken, Japan) p 95

[11] Gelmini G B and Roncadelli M 1981 *Phys. Lett.* B **99** 411

[12] Froggatt C D, Gibson M and Nielsen H B 1997 *Phys. Lett.* B **409** 305

Lepton number and lepton flavour violation in left-right symmetric models

Martti Raidal[1]

DESY, Theory Group, Notkestraße 85, 22603 Hamburg, Germany

Abstract. The sources of lepton flavour and lepton number violation in the left-right symmetric models are reviewed and the most sensitive processes to these violations are discussed. Both the non-SUSY and SUSY versions of the theories are considered. Present and future experimental constraints at low as well as at high energies are discussed. Some emphasis has been put on the doubly charged particle production at future colliders.

1. Introduction

The conservations of lepton number and lepton flavours are among the most stringently tested laws of physics [1]. In the standard model (SM) of the weak and electromagnetic interactions all three lepton flavours are exact global symmetries and conserved separately. This is a consequence of the vanishing neutrino masses in the SM. Therefore, if the SM is the ultimate theory of Nature then lepton flavour violation will never be observed experimentally.

The SM has been extremely successful in describing the experimental data. Nevertheless, there are both theoretical and experimental problems which cannot be explained in the SM. On the theoretical side we mention just a few of them which are relevant for our further considerations: unexplained maximal parity violation of the weak interaction, massless neutrinos, CP problems, hierarchy problem etc. On the experimental side the most serious result indicating for the physics beyond the SM is the

[1] E-mail: raidal@mail.desy.de

recent Super Kamiokande mesurement [2] of the flux of atmospheric neutrinos which clearly points in the direction of neutrino masses. Similarly the solar neutrino problem [3], the LSND measurements [4] and the COBE sattellite results [5] on the existence of the hot component of dark matter give indications of small neutrino masses.

Most of the above listed problems can be explained in the framework of the left-right symmetric models [6] based on the gauge group $SU_L(2) \times SU_R(2) \times U_{B-L}(1)$. In these models left- and right-handed components of the fields are treated on the same footing. The left-right symmetry was originally proposed in order to understand the origin of the parity violation as a consequence of the spontaneous symmetry breaking. It was then used to develop models of CP violation [7] as well as to discuss the strong CP problem [8]. If the Higgs sector of the models is choosen so that the right-handed symmetry is spontaneously broken by triplets then the model gives rise to small neutrino masses naturally, via the see-saw mechanism [9]. In this type of models there are two sources of lepton number violation, Majorana masses of the neutrinos and Yukawa interactions of the physical triplet Higgses.

Similar to the SM the non-supersymmetric left-right models suffer from the hierarchy problem. Supersymmetric left-right models [10] share all the important features of the non-susy left-right models, and in addition solve many problems of the minimal SUSY extension of the SM, MSSM, like automatic conservation of R-parity and SUSY and strong CP problems. These points have been extensively discussed in the reviews [11] by R. Mohapatra. In addition, the recent works on the vacuum structure of the SUSY left-right models have shown [12, 13] that certain Higgs bosons and their SUSY partners should be relatively light giving rise to interesting low energy as well as to collider phenomenology.

In the present talk we discuss the most stringently tested lepton number and lepton flavour violating processes in the context of left-right models. First we present some necessary details of the non-SUSY as well as the SUSY left-right models and point out the sources of the lepton number violation. Then we consider in some detail low energy experiments like $\mu \to e\gamma$ and $\mu - e$ conversion which provide the main probes of the muon number conservation. Finally we review the collider phenomenology of doubly charged Higgses and higgsinos.

2. Left-right symmetric models

2.1. Non-supersymmetric theories

In the minimal $SU(2)_L \times SU(2)_R \times U(1)_{B-L}$ model with a left-right discrete symmetry each generation of quarks and leptons is assigned to the multiplets

$$Q = \begin{pmatrix} u \\ d \end{pmatrix} \; ; \; \psi = \begin{pmatrix} \nu \\ l \end{pmatrix} \; ; \tag{1}$$

with the quantum numbers $(T_L; T_R; B-L)$

$$\begin{aligned} Q_L &: \left(\frac{1}{2}; 0; \frac{1}{3}\right) \; ; \quad \psi_L : \left(\frac{1}{2}; 0; -1\right); \\ Q_R &: \left(0; \frac{1}{2}; \frac{1}{3}\right) \; ; \quad \psi_R : \left(0; \frac{1}{2}; -1\right). \end{aligned} \tag{2}$$

Concerning the Higgs sector, in order to give masses to fermions, all the left-right models should contain a bidoublet

$$\phi = \begin{pmatrix} \phi_1^0 & \phi_1^+ \\ \phi_2^- & \phi_2^0 \end{pmatrix} = \left(\frac{1}{2}; \frac{1}{2}^*; 0\right). \tag{3}$$

Since the bidoublet does not break the right-handed symmetry the Higgs sector has to be enlarged. This procedure is not unique but interesting models are obtained by adding the scalar triplets,

$$\Delta_{L,R} = \begin{pmatrix} \frac{\Delta^+_{L,R}}{\sqrt{2}} & \Delta^{++}_{L,R} \\ \Delta^0_{L,R} & \frac{-\Delta^+_{L,R}}{\sqrt{2}} \end{pmatrix} \; ; \tag{4}$$

with the quantum numbers $\Delta_L : (1, 0, 2)$ and $\Delta_R : (0, 1, 2)$, respectively. In addition, we require the full Lagrangian of the model to be manifestly left-right symmetric i.e. invariant under the discrete symmetry

$$\psi_L \longleftrightarrow \psi_R \; ; \quad \Delta_L \longleftrightarrow \Delta_R \; ; \quad \phi \longleftrightarrow \phi^\dagger. \tag{5}$$

This symmetry plays a role in minimizing the Higgs potential and breaking of parity spontaneously. In general, the symmetry breaking would be triggered by the vevs

$$\langle\phi\rangle = \begin{pmatrix} \frac{k_1}{\sqrt{2}} & 0 \\ 0 & \frac{k_2}{\sqrt{2}} \end{pmatrix} \quad ; \quad \langle\Delta_{L,R}\rangle = \begin{pmatrix} 0 & 0 \\ \frac{v_{L,R}}{\sqrt{2}} & 0 \end{pmatrix}. \tag{6}$$

The vev v_R of the right triplet breaks the $SU(2)_R \times U(1)_{B-L}$ symmetry to $U(1)_Y$ and gives masses to new right-handed particles. Since the right-handed currents have not been observed, v_R should be sufficiently large [14, 15]. Further, the vevs k_1 and k_2 of the bidoublet break the SM symmetry and, therefore, are of the order of electroweak scale. The vev v_L of the left triplet, which contributes to the ρ parameter, is quite tightly bounded by experiments [15] and should be below a few GeV-s. Thus, the following hierarchy should be satisfied: $|v_R| \gg |k_1|, |k_2| \gg |v_L|$. In principle, due to the underlying symmetry two of the vevs can be chosen to be real but two of them can be complex leading to the spontaneous CP violation [7].

The most general Yukawa Lagrangian for leptons invariant under the gauge group is given by

$$\begin{aligned} -\mathcal{L}_Y &= f_{ij}\bar{\psi}^i_L\phi\psi^j_R + g_{ij}\bar{\psi}^i_L\tilde{\phi}\psi^j_R + \text{h.c.} \\ &\quad + i(h)_{ij}\left(\psi^{iT}_L C\tau_2\Delta_L\psi^j_L + \psi^{iT}_R C\tau_2\Delta_R\psi^j_R\right) + \text{h.c.}, \end{aligned} \tag{7}$$

where f, g and h are matrices of Yukawa couplings. The left-right symmetry (5) requires f and g to be Hermitian. The Majorana couplings h can be taken to be real and positive due to our ability to rotate ψ_L and ψ_R by a common phase without affecting f and g.

The Lagrangian (7) is the most relevant expression for our discussion because it gives rise to lepton number violation in the left-right models. The origin of it is twofold. Firstly, triplet Higgses which carry two units of lepton number can mediate lepton number or lepton flavour violating processes. Secondly, Lagrangian (7) gives to neutrinos Majorana masses as follows.

Neutrino masses derive both from the f and g terms, which lead to Dirac mass terms, and from the h term, which leads to large Majorana mass terms. Defining, as usual, $\psi^c \equiv C(\bar{\psi})^T$, the mass Lagrangian following from Eq.(7) can be written in the form

$$-\mathcal{L}_{mass} = \frac{1}{2}(\bar{\nu}^c_L\, M\, \nu_R + \bar{\nu}_R\, M^*\, \nu^c_L), \tag{8}$$

where $\nu^c_L = (\nu_L, \nu^c_R)^T$ and $\nu_R = (\nu^c_L, \nu_R)^T$ are six dimensional vectors of neutrino fields. The neutrino mass matrix M is complex-symmetric and can be written in the block form

$$M = \begin{pmatrix} M_L & M_D \\ M_D^T & M_R \end{pmatrix}, \tag{9}$$

where the entries are 3×3 matrices given by

$$M_L = \sqrt{2}hv_L, \quad M_D = h_D k_+, \quad M_R = \sqrt{2}hv_R. \tag{10}$$

Here we have defined $h_D = (fk_1 + gk_2)/(\sqrt{2}k_+)$, where $k_+^2 = k_1^2 + k_2^2$. The masses of charged leptons are given by $M_l = (gk_1 + fk_2)/\sqrt{2}$ and, therefore, without fine tuning of f and g one has $M_D \simeq M_l$. Moreover, on the basis of avoiding possible fine tunings it is natural to assume that all the Yukawa couplings h, h_D are of similar magnitude for a certain lepton family. In this case the mass matrix (9) has a strong hierarchy between different blocks which is set by the hierarchy of vevs. It is convenient to employ the self-conjugated spinors

$$\nu \equiv \frac{\nu_L + \nu^c_L}{\sqrt{2}}, \quad N \equiv \frac{\nu_R + \nu^c_R}{\sqrt{2}}. \tag{11}$$

In a single generation case these are also the approximate mass eigenstates with masses

$$
\begin{aligned}
m_\nu &\simeq \sqrt{2}\left(hv_L - \frac{h_D^2 k_+^2}{2hv_R}\right), \\
m_N &\simeq \sqrt{2}hv_R,
\end{aligned}
\qquad (12)
$$

respectively. The mixing between them depends on the ratio of the masses as $\sin^2\theta \sim m_\nu/m_N$.

Since the neutrino mass matrix is symmetric it can be diagonalized by the complex orthogonal transformation

$$U^T\, M\, U = M^d, \qquad (13)$$

where M^d is the diagonal neutrino mass matrix. If we denote

$$U = \begin{pmatrix} U_L^* \\ U_R \end{pmatrix}, \qquad (14)$$

then in the basis where the charged lepton mass matrix is diagonal (we can choose this basis without loss of generality) the physical neutrino mixing matrices which appear in the left- and right-handed charged currents are simply given by U_L and U_R, respectively. Indeed, U_L, U_R relate the left-handed and right-handed neutrino flavour eigenstates $\nu_{L,R}$ with the mass eigenstates ν_m according to

$$\nu_{L,R} = U_{L,R}\,\nu_m\,. \qquad (15)$$

2.2. Supersymmetric theories

SUSY left-right models have got a lot of attention recently. This is because they offer solutions [11] to the problems occuring in the MSSM. In SUSY versions of the left-right models all the qualitative discussion above is valid. In particular, the see-saw mechanism is active but the expression for the light neutrino mass gets its kanonical form

$$m_\nu \simeq -\frac{h_D^2 k_+^2}{\sqrt{2}hv_R}.$$

The important feature of a class of SUSY left-right models is that R-parity is automatically conserved as suggested by the stringent proton decay limits [16] while in the MSSM one has to tune R-parity violating couplings to be small.

The superfield content of the minimal SUSY left-right model is not identical to the particle content of the model considered in the last section. In order to cancel chiral anomalies the number of triplets should be doubled

Table 1. Field content of the SUSY left-right model

Fields	$SU(2)_L \times SU(2)_R \times U(1)_{B-L}$ representation
Q	$(2,1,+\frac{1}{3})$
Q^c	$(1,2,-\frac{1}{3})$
L	$(2,1,-1)$
L^c	$(1,2,+1)$
$\Phi_{1,2}$	$(2,2,0)$
Δ	$(3,1,+2)$
$\bar{\Delta}$	$(3,1,-2)$
Δ^c	$(1,3,+2)$
$\bar{\Delta}^c$	$(1,3,-2)$

by adding new multiplets with the opposite $B-L$ quantum numbers. Also, to get a nontrivial quark mixing matrix the number of bidoublets should be doubled (supersymmetry forbids $\tilde{\Phi}$ in the superpotential). The superfields present in the minimal SUSY left-right model together with their quantum numbers are listed in Table 1.

The superpotential for the theory containing these fields can be given by (we have suppressed the generation index):

$$\begin{aligned} W &= g_q^{(i)} Q^T \tau_2 \Phi_i \tau_2 Q^c + f_l^{(i)} L^T \tau_2 \Phi_i \tau_2 L^c \\ &+ i(h L^T \tau_2 \Delta L + h_c L^{cT} \tau_2 \Delta^c L^c) \\ &+ M_\Delta [\mathrm{Tr}(\Delta\bar{\Delta}) + \mathrm{Tr}(\Delta^c \bar{\Delta}^c)] + \mu_{ij} \mathrm{Tr}(\tau_2 \Phi_i^T \tau_2 \Phi_j). \end{aligned} \quad (16)$$

However, studies of the ground state of the SUSY left-right model have shown that keeping only the renormalizable terms in the superpotential, and with the given field content (authors of Ref.[12] actually consider a model with one additional singlet) the ground state of the model breaks the electric charge unless sneutrinos obtain nonzero vevs $< \tilde{\nu}^c >$. These vevs break R-parity spontaneously and give rise to lepton number violating processes. Such a situation is not as dangerous as the explicit R-parity breaking because baryon number is now conserved and proton is stable. Therefore the most stringent constraints on the R-parity violation are satisfied. The non-observation of lepton number violating processes constrain also such a scenario quite stringently. Nevertheless interesting new SUSY phenomenology at high energy collider experiments is allowed [17].

An important consequence of the scenario with sneutrino vevs is that the mass of the right-handed gauge boson W_R should be of the order of SUSY breaking scale $\mathcal{O}(1)$ TeV. The most stringent experimental constraint coming from the $K_L - K_S$ mass difference is $M_{W_R} \gtrsim 1.6$ TeV [14] while at LHC the right-handed gauge boson masses as high as 7 TeV will be probed. Therefore this scenario with non-zero sneutrino vevs will be probed in the near future. At the moment there is no experimental evidence in favour of it.

Another way to ensure the conservation of the electric charge by the ground state of the SUSY left-right model is to add to the superpotential non-renormalizable terms

$$W_{NR} = A[\mathrm{Tr}(\Delta^c \bar{\Delta}^c)]^2 + B\mathrm{Tr}(\Delta^c \Delta^c)\mathrm{Tr}(\bar{\Delta}^c \bar{\Delta}^c), \tag{17}$$

arising from higher scale physics such as grand unified theories or Planck scale effects (A and B are of the order $1/M_{Planck}$). It has been shown [13] that in this case the right-handed breaking scale and thus the mass of the right-handed gauge bosons must be very high, $v_R \gtrsim 10^{10}$ GeV. At such a scale it is natural that higher scale operators may play a role. In this scenario the right-handed currents have no importance for the low energy phenomenology. However, the following phenomenological consequences make this scenario very attractive.

1. R-parity is an exact symmetry. No lepton number violation from this sector.

2. The masses of the light neutrinos obtained by the see-saw mechanism by choosing the mass matrix Dirac entries to be of the order of the charged lepton masses and $M_N = 10^{10}$ GeV are exactly in the correct range to provide a solution to the solar and atmospheric neutrino problems [2, 3]. Neutrinoless double beta decay is possible.

3. There are light doubly charged Higgses and higgsionos in the model whose masses may be of the order of 10^2 GeV. Interesting new lepton number violating collider phenomenology possible.

The latter point deserves some explanation. Due to the vacuum structure of SUSY theories there may be light fields in the models despite of the very high right-handed breaking scale. If SUSY is unbroken there are flat directions in the vacuum and the particles corresponding to these flat directions are massless. When SUSY gets broken the flat directions are lifted and the previously massless fields obtain masses of the order v_R^2/M_{Planck}.

Motivated by these results we will concentrate in the following on the phenomenological aspects of the Majorana neutrinos and doubly charged Higgses and higgsinos.

3. Low and high energy phenomenology

3.1. Neutrinoless double beta decay and its inverse

Neutrinoless double beta decay violates the lepton number by two units [18]. Its observation would necessarily imply that neutrinos are Majorana particles [19]. In the left-right symmetric models the neutrinoless double beta decay receives contribution from three sources [20, 21]: from the t-channel exchange of the light and heavy Majorana neutrinos and from the s-channel exchange of the doubly charged Higgs bosons. The latter contribution is a necessary ingredient to preserve the unitarity of the cross section [22]. Due to the nuclear effects the heavy and light particle contributions factorize. Therefore the Heidelberg-Moscow experiment searching for the neutrinoless double beta decay in ^{76}Ge puts the upper limit on the effective mass of the light Majorana neutrino as

$$< m_\nu > \lesssim 0.46 \text{ eV}. \tag{18}$$

The heavy right-handed particle contribution to the process has been studied by several groups. The latest analyses [21] constrain the mass of the right-handed gauge boson to be

$$M_{W_R} \gtrsim 1.1 \text{ TeV}, \tag{19}$$

for the fixed effective right-handed neutrino mass $M_N = 1$ TeV and $M_\Delta = \infty$. In the case of relatively light, TeV scale, doubly charged Higgses the bound may be increased by a few TeV.

The inverse of the neutrinoless double beta decay [23, 24], the process

$$e^- e^- \to W^-_{L,R} W^-_{L,R}, \tag{20}$$

can be studied in the future $e^- e^-$ options of the linear colliders. The cross section for the right gauge boson production is proportional to the heavy neutrino mass M_N and therefore large [24]. However, most likely the collision energy of the future linear colliders will not allow to produce on-shell W_R-s. Therefore one has to look for the ordinary gauge boson W_L pair production which may proceed through the heavy-light neutrino mixing. This topic has been extensively studied in last years [23]. For a single neutrino generation case the non-observation of the neutrinoless double beta decay suppresses the cross section below the observable limit. In the case of three generations, and allowing for the opposite CP parities of the heavy neutrinos, there is still some part of the parameter space left for which the observable cross section at linear colliders is allowed. This conclusion is essentially independent of the further improvements of the sensitivity of the double beta decay experiments (GENIUS) [25] and should be tested at future colliders.

3.2. $\mu - e$ conversion, $\mu \to e\gamma$ and other low energy processes

The precision reached in the low experiments looking for the violation of the muon number exceeds considerably the experimental precision in the searches for the tau lepton number violation [26]. In this talk we consider the most stringent experimental bounds on the muon number violation only. Also, we do not consider the possibility of having SUSY particles in the loops (for these studies see [27]).

In the left-right models the muon flavour violating processes $\mu \to e\gamma$ and $\mu - e$ conversion in nuclei can occur at one loop level. There are two types of loop diagrams contributing to these processes: with neutrinos and charged gauge bosons in the loops and with singly and doubly charged triplet Higgses and leptons in the loops. Let us first discuss the first possibility.

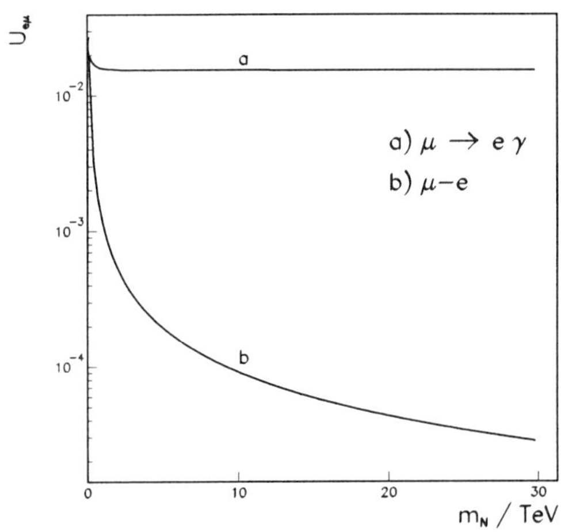

Figure 1. Constraints on the light-heavy neutrino mixings from $\mu \to e\gamma$ and $\mu - e$ conversion against the heavy neutrino mass.

In the case of massive neutrinos in the loops the muon flavour can be violated due to the mixings of neutrinos of different generations. Because of the very small masses of the light neutrinos they contribution to the processes is suppressed and the main contribution comes from the exchange of the heavy neutrinos and W_L. Therefore both $\mu \to e\gamma$ and $\mu - e$ conversion

Table 2. Upper bounds on the diagonal doubly charged Higgs couplings for $M_\Delta = 1$ TeV.

Process	Combination of couplings	Upper bound
Møller	h^{ee}	4
$(g-2)_\mu$	$h^{\mu\mu}$	10
$M-\bar{M}$	$h^{ee}h^{\mu\mu}$	0.2

are proportional to the quantity

$$|U_L|^{e\mu} = \left(\sum_i U_L^{ei} U_L^{i\mu\dagger}\right)^{\frac{1}{2}}. \tag{21}$$

Here the summation goes over the heavy neutrinos only which means that U_L bounds the light-heavy mixings.

The general $\bar{e}\mu\gamma$ vertex can be parametrized in terms of form factors as follows

$$j^\rho = \bar{e}\left[\gamma_\mu (f_{E0} + f_{M0}\gamma_5)\left(g^{\mu\rho} - \frac{q^\mu q^\rho}{q^2}\right) + (f_{M1} + f_{E1}\gamma_5) i\sigma^{\rho\nu}\frac{q_\nu}{m_\mu}\right]\mu, \tag{22}$$

where q is the momentum transferred by the photon. While $\mu \to e\gamma$ is induced only by the form factors proportional to $\sigma^{\rho\nu}$ term then the *photonic* $\mu{-}e$ conversion rate is proportional to $(|f_{E0}+f_{M1}|^2+|f_{M0}+f_{E1}|^2)$ [28]. On the other hand, $\mu - e$ conversion can also be induced by the *non-photonic* conversion mechanism, for example by the effective vertex $\bar{e}\mu Z_L$. This mode is somewhat suppressed by the mass of Z_L but, on the other, hand it is proportional to the heavy neutrino mass. Using the present experimental bounds on the branching ratios of the processes $B(\mu \to e\gamma) < 4.9 \cdot 10^{-11}$ and $B(\mu - e) < 4 \cdot 10^{-12}$ and the results of Ref. [29] we plot in Fig. 1 the constrains on $|U_L|^{e\mu}$ from $\mu \to e\gamma$ and $\mu - e$ conversion against the heavy neutrino mass. The $\mu - e$ conversion bound becomes more stringent for larger neutrino masses but it is orders of magnitude smaller that the constraints predicted by the see-saw mechanism (in the see-saw the heavy-light mixing is given by $U^2 \sim m_\nu/M_N$). Therefore these results are meaningful only in the models where neutrino masses are generated by other mechanisms than the see-saw [29, 30].

The situation is quite different for the couplings of the triplet Higgs bosons Δ. Due to large masses and unknown couplings to leptons one can presently only constrain their effective couplings of a generic form $G =$

Table 3. Upper bounds on the doubly charged Higgs couplings for $M_\Delta = 1$ TeV from various low energy leptonic processes.

Process	Combination of couplings	Upper bound
$\mu \to 3e$	$h^{\mu e}h^{ee}$	$4 \cdot 10^{-5}$
$\mu \to e\gamma$	$(hh)^{\mu e}$	$3 \cdot 10^{-3}$
$\mu - e$	$(hh)^{\mu e}$	$6 \cdot 10^{-4}$

$\sqrt{2}h^2/(8M_\Delta^2)$. Most stringent constraints on the diagonal couplings of the triplet Higgses [31] are summarized in Table 2. These are obtained from the Møller scattering, $(g-2)_\mu$ studies as well as from the searches for the muonium-antimuonium conversion. Here and in the following the masses are always taken in units of TeV. As can be seen, for the scale of new physics $\mathcal{O}(1)$ TeV only the muonium-antimuonium conversion experiment constrains diagonal couplings in a meaningful way. To date there is no constraints on $h^{\tau\tau}$ without involving the off-diagonal elements.

Most stringent constraints on the triplet couplings h from the muon number violating processes are summarized in Table 3. As expected, the strongest limit derives from $\mu \to 3e$ since it can occur at tree level. However, it constraints only a particular combination of the couplings. On the other hand, $\mu \to e\gamma$ and $\mu - e$ conversion limits apply on a sum of the coupling constant products and, therefore, provide complementary information. One should note here that the $\mu - e$ conversion is more sensitive to the Higgs interactions than $\mu \to e\gamma$. This is because the form factors f_{E0} and f_{M0} in Eq. (22) are enhanced by large $\ln(m_\ell^2/M_\Delta^2) \sim \mathcal{O}(10)$ while the form factors f_{E1} and f_{M1} are not [32]. Consequently $\mu - e$ conversion is enhanced while $\mu \to e\gamma$ is not. The enhancement arises from the diagrams in which the doubly charged Higgs Δ^{++} runs in the loop and the the photon is attached to the charged lepton line.

3.3. Collider phenomenology

The most distinctive signatures of the new physics at collider experiments would be produced by the decays of the doubly charged Higgses $\Delta^{++}_{L,R}$ and higgsinos $\tilde{\Delta}^{++}_{L,R}$. If these particles are light enough to be produced at future colliders then the most stringent constraints on their couplings will be obtained from these experiments. The production of Δ^{++} at linear and muon colliders as well as at the LHC is already extensively studied [33, 34, 35, 36, 37, 38]. The production of their supersymmetric partners, $\tilde{\Delta}^{++}$, has received somewhat less attention [39, 40]. We shall discuss the

$\tilde{\Delta}^{++}$, production later in this Subsection and start with the Higgs mediated processes.

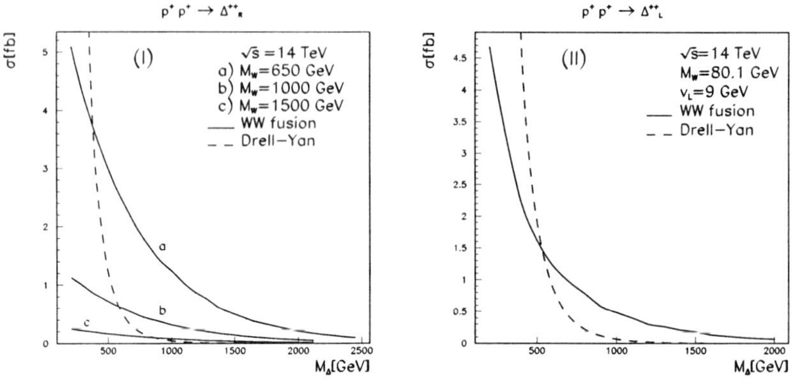

Figure 2. Production cross sections of $\Delta^{++}_{L,R}$ at LHC as functions of the doubly charged Higgs mass.

While $\Delta^{++}_{L,R}$ do not couple to quarks they still can be produced at LHC in pairs (Drell-Yan) or singly (WW fusion). The latter process is kinematically more favoured but it depends on the unknown model parameters: M_{W_R} for the Δ^{++}_{R} production and v_L for the Δ^{++}_{L} production. On the other hand, the Drell-Yan pair production is almost model independent due to the $\Delta^{++}_{L,R}$ couplings to the photon and Z_L. Using the results of Ref. [34] we plot in Fig. 2 the production cross sections of the Drell-Yan and WW fusion processes at LHC for the parameters indicated in the figure. In the pair production the doubly charged Higgses with masses below 600 GeV can be discovered at LHC. For the optimistic choice of the unknown model parameters the WW fusion may extend the discovery reach up to 1-2 TeV or so.

At the future linear and muon colliders the doubly charged Higgses with masses $M_\Delta \lesssim \sqrt{s}/2$ can be pair produced in e^+e^- collisions due to their couplings to photon and Z [35]. However, the most appropriate for studying $\Delta^{++}_{L,R}$-s are e^-e^- and $\mu^-\mu^-$ collision modes in which the resonant s-channel production of them via the process

$$e^-e^-(\mu^-\mu^-) \to \ell_i^- \ell_i^-, \tag{23}$$

$i = e, \mu, \tau$, is possible. It has been shown that for the realistic machine and beam parameters sensitivity of

$$h^{ij} \lesssim 5 \cdot 10^{-5} \tag{24}$$

can be acheved [36].

Despite of this extraordinary sensitivity it could happend that due to small h's or high Higgs masses no positive signal will be detected at future colliders. However, this may not be the case if neutrinos are massive enough. If the sum of light neutrino masses exceeds ~ 90 eV at least one of them has to be unstable. In order not to overclose the Universe the lifetime and mass of such an unstable neutrino ν_l must satisfy the requirement [41]

$$\tau_{\nu_l} \lesssim 8.2 \cdot 10^{31}\ \mathrm{MeV}^{-1} \left(\frac{100\ \mathrm{keV}}{m_{\nu_l}}\right)^2 . \tag{25}$$

The radiative decay modes $\nu_l \to \nu_f \gamma$, $\nu_f \gamma\gamma$ are highly suppressed [42] and cannot satisfy the constraint (25). The same is also true for Z' contribution to $\nu_l \to 3\nu_f$ decay [43]. The only possibility which lefts over is the decay $\nu_l \to 3\nu_f$ due to neutrino mixings induced by the neutral component of triplet Higgs Δ_L^0.

Clearly, since Δ_L^0 and Δ_L^{--} belong to the same multiplet the reaction (23) can be related to decay $\nu_l \to 3\nu_f$. Since the latter decay rate is bounded from below by the constraint (25) and limits on neutrino mixings also the cross section of (23) has a lower limit. Let us consider numerically the most conservative case, $\nu_\tau \to 3\nu_e$. From the constraints presented above one obtains the following bound on $G^{e\tau}$ which induces the process $e^-e^- \to \tau^-\tau^-$ [37]

$$G^{e\tau} \gtrsim 2 \cdot 10^{-3} \frac{h_{\tau\tau}}{h_{ee} + h_{\tau\tau}}\ \mathrm{TeV}^{-2} , \tag{26}$$

where $G^{e\tau} = \sqrt{2} h^{ee} h^{\tau\tau} / (8 M_\Delta^2)$. On the other hand, studies of the process (23) at linear colliders give that far off the resonance the minimal testable G^{ij} are $G^{ij}(\min) = 1.4 \cdot 10^{-4}/s\ \mathrm{TeV}^{-2}$, approximately the same for all relevant i, j. Therefore, the process $e^-e^- \to \tau^-\tau^-$ should be detected at the 1 TeV linear collider unless $h_{\tau\tau}/h_{ee} \lesssim 10^{-1}$. On the other hand, if this is the case then

$$G^{ee} \gtrsim 2 \cdot 10^{-3}\ \mathrm{TeV}^{-2} , \tag{27}$$

and the excess of the electron pairs due to the s-channel Higgs exchange will be detected. Note that the positive signal should be seen if $\sqrt{s} \gtrsim 0.3$ TeV which is below the planned initial energy of the linear collider.

Similar argumentation applies to all possible neutrino decays with even higher lower limits. Therefore, if one of the neutrino mass, indeed, exceeds 90 eV one should detect at least one of the processes (23) at future colliders.

If the e^-e^- option of the linear collider will not be available then single production of $\Delta_{L,R}^{++}$ in e^+e^- and $e^-\gamma$ collisions allowes one to probe masses up to $\sqrt{s}$ and test all the couplings h^{ij} more sensitively than any of the present low energy experiments [38]. In $e^-\gamma$ mode the production reaction

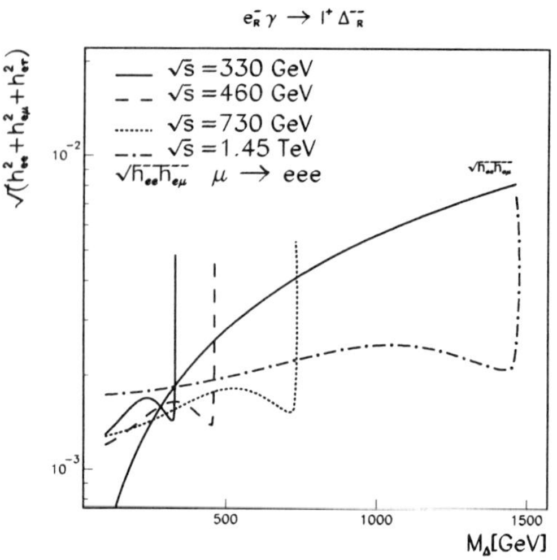

Figure 3. Achievable constraints of the triplet Yukawa couplings from the process (28) as functions of the Higgs mass. The present bound from $\mu \to 3e$ is also shown.

is

$$e^-\gamma \quad \to \quad l^+\Delta^{--} \tag{28}$$

and in e^+e^- mode

$$e^+e^- \quad \to \quad e^+l^+\Delta^{--}. \tag{29}$$

Let us first consider the reaction (28). The primary lepton created in the process will remain undetected as it is radiated almost parallel to the beam axis. One cannot tell whether this particle is a positron, antimuon or antitau. Therefore, the quantity which one can test in the reaction is actually the sum $h_{ee}^2 + h_{e\mu}^2 + h_{e\tau}^2$. The upper bound obtained for this sum is, of course, the upper bound of each individual term of the sum separately.

Assuming the integrated luminosities of $e^-\gamma$ collisions to be $L = 5, 10, 20, 40$ fb^{-1} and that for the discovery of Δ_R^{--} one needs ten events, we obtain the upper bounds plotted in Fig. 3. As one can see from the figure, the sensitivity of the linear collider on the quantity $(h_{ee}^2 + h_{e\mu}^2 + h_{e\tau}^2)^{1/2}$ is on the level of 10^{-3} almost up to the threshold value of the Δ_R^{--} mass. In other words,

$$h_{ee}, h_{e\mu}, h_{e\tau} \lesssim 10^{-3} \tag{30}$$

for $M_{\Delta^{--}} \lesssim \sqrt{s_{e\gamma}}$. Among the present constraints only the one from the $\mu \to 3e$ competes with these bounds and does so only at the low mass values. For the coupling $h_{e\tau}$ no bounds exist from the present experiments.

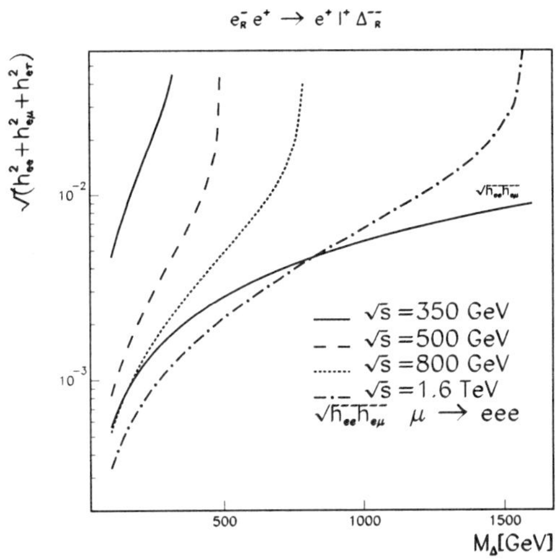

Figure 4. Achievable constraints of the triplet Yukawa couplings from the process (29) as functions of the Higgs mass. The present bound from $\mu \to 3e$ is also shown.

For the same Δ_R^{--} mass, the cross section of the process (29) is roughly two orders of magnitude smaller than the cross section of the process (28), implying that the constraints obtained for $(h_{ee}^2 + h_{e\mu}^2 + h_{e\tau}^2)^{1/2}$ are correspondingly weaker, although the higher luminosities $L = 20, 50, 100, 200$ fb^{-1} of e^+e^- slightly compensate the lack in cross section. The resulting bounds are presented in Fig. 4.

The production of SUSY partners of the doubly charged Higgses, $\tilde{\Delta}^{++}$, at linear colliders has been previously studied in Ref.[39]. Since in the model we consider the R-parity is conserved then the SUSY particles can be produced in pairs only. Therefore the simplest process to study is the $\tilde{\Delta}^{++}$, $\tilde{\Delta}^{--}$ pair production in e^+e^- collisions. This reaction is mediated by the s-channel photon and Z exchange and t-channel selectron exchange. Therefore the process is sensitive to the triplet coupling h. In Fig. 5 we plot the pair production cross section of $\tilde{\Delta}_L^{++}$ against the collision energy. The choice of the parameters is indicated in the figure. The cross section is large and depends strongly on the coupling h.

The produced higgsinos will decay fast. We assume here that their dom-

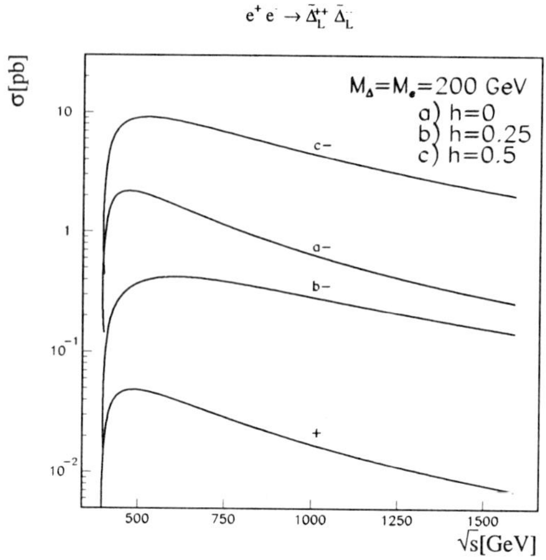

Figure 5. The polarized cross sections for the pair production of doubly charged higgsino $\tilde{\Delta}^{++}$ against the collision energy. The choice of parameters and beam polarizations are indicated in figure.

inant decay modes are $\tilde{\Delta}^{++} \to l^+l^+\tilde{\chi}^0$ which are mediated by the selectron. Since the $\tilde{\Delta}^{++}$ lifetime is very small, only the correlated production and decay can be observed experimentally:

$$
\begin{array}{l}
e^+e^- \to \tilde{\Delta}^{--}\tilde{\Delta}^{++} \\
\qquad\qquad\quad \hookrightarrow \tilde{\chi}^0 + (l^+l^+) \\
\qquad\qquad \hookrightarrow \tilde{\chi}^0 + (l^-l^-)
\end{array}
$$

The analysis is complicated as the two invisible neutralinos in the final state do not allow for a complete reconstruction of the events. In particular, it is not possible to measure the $\tilde{\Delta}^{++}$ production angle Θ; this angle can be determined only up to a two–fold ambiguity. Therefore one cannot measure the fundamental interactions of the theory directly from the primary production process.

The way out of this problem is to study the spin correlations of the final state higgsinos which will be reflected in the distributions of the final state leptons. Indeed, one has shown [40] that by measuring the final state lepton distributions one can actually measure the total cross section

σ and the following combinations of the helicity amplitudes of the primary process:

$$\mathcal{P} = \frac{1}{4}\int \mathrm{d}\cos\Theta \sum_{\pm} \left[|\langle\pm;++\rangle|^2 + |\langle\pm;+-\rangle|^2 - |\langle\pm;-+\rangle|^2 - |\langle\pm;--\rangle|^2 \right],$$
$$\mathcal{Q} = \frac{1}{4}\int \mathrm{d}\cos\Theta \sum_{\pm} \left[|\langle\pm;++\rangle|^2 - |\langle\pm;+-\rangle|^2 - |\langle\pm;-+\rangle|^2 + |\langle\pm;--\rangle|^2 \right],$$
$$\mathcal{Y} = \frac{1}{2}\int \mathrm{d}\cos\Theta \sum_{\pm} \mathrm{Re}\Big\{ \langle\pm;--\rangle\langle\pm;++\rangle^* \Big\}. \qquad (31)$$

In the helicity amplitudes $\langle\pm;\pm\pm\rangle$ the first $\pm$ denotes the helicity of the electron and the last ones the helicities of $\tilde{\Delta}^{++}$ and $\tilde{\Delta}^{--}$, respectively. The measurements of the cross section at an energy $\sqrt{s}$, and the ratios $\mathcal{P}^2/\mathcal{Q}$ or $\mathcal{Y}/\mathcal{Q}$ give enough information on the interactions of the doubly charged particles. Thus the quantum numbers of the doubly charged higgsino and its coupling h to lepton-slepton pair can be determined from the experiment.

To demonstrate that let us assume that at $\sqrt{s} = 500$ GeV the experimentally "measured" cross section σ, $\mathcal{P}^2/\mathcal{Q}$ and $\mathcal{Y}/\mathcal{Q}$ are

$$\sigma(e^+e^- \to \tilde{\Delta}^{++}\tilde{\Delta}^{--}) = 0.106 \text{ pb}, \qquad \frac{\mathcal{P}^2}{\mathcal{Q}} = -2.0, \qquad \frac{\mathcal{Y}}{\mathcal{Q}} = 0.26\,. \quad (32)$$

Let us also assume that all the particle masses are known (M_Δ can be measured from the threshold cross section). In Fig. 6 we plot the contour lines in (h, I_3) plane corresponding to the measured σ, $\mathcal{P}^2/\mathcal{Q}$ and $\mathcal{Y}/\mathcal{Q}$ keeping I_3, the isospin, as a continous parameter. Since I_3 can be only an integer or a half integer we see from Fig. 6 that the cross section measurement alone determines $I_3 = 1$. Because the negative interference between $s-$ and $t-$channel contributions does not allow to determine h unambigously from one measurement then f may still have two solutions, $h = 0.25$ or 0.31. However, the three contour lines derived from the three measurements meet at a single point ($I_3 = 1, h = 0.25$) which gives the physical values of I_3 and h. Since the $\mathcal{Y}/\mathcal{Q}$ curve crosses the σ curve at angle almost $\pi/2$ then the measurement can be carried our with high precision. In fact, the third $\mathcal{P}^2/\mathcal{Q}$ curve does not give any new information in this case and can be used just for the cross check.

4. Conclusions

In this talk we have discussed the sources of lepton number and lepton flavour violation in the non-SUSY and SUSY left-right symmetric models. After reviewing the basics of the models we have considered low and high

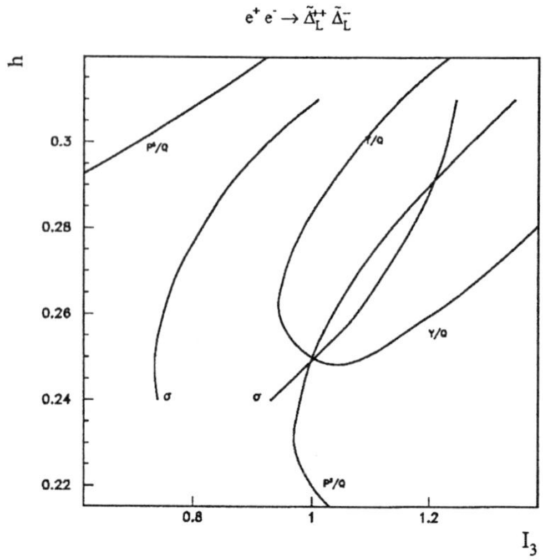

Figure 6. Contour lines from the measurements of σ, $\mathcal{P}^2/\mathcal{Q}$ and $\mathcal{Y}/\mathcal{Q}$ determining the coupling constant h and the weak isospin I_3 of the doubly charged higgsino.

energy phenomenology of the massive neutrinos and doubly charged Higgses and higgsinos. Recent works have shown that in SUSY models the doubly charged particles should be relatively light. In particular we have discussed neutrinoless double beta decay, $\mu \to e\gamma$, $\mu - e$ conversion in nuclei and future experiments at LHC and linear and muon colliders. The present constraints on the new particle masses and interactions allow for the lepton number violating signals in all these experiments. We emphasise the role of the future colliders for testing the left-right models.

Acknowledgments

I would like to thank organizers for creating pleasant atmosphere during the conference and U. Sarkar for several discussions. I am greatful for the A. von Humboldt Foundation for the grant.

References

[1] C. Caso et al., Particle Data Group, Eur. Phys. J. **C 3** (1998) 1.

[2] Y. Fukuda et al., Super-Kamiokande Collaboration, hep-ex/9807003.

[3] For the recent analyses see, J.N. Bahcall, P.I. Krastev and A.Yu. Smirnov, hep-ph/9807216.

[4] A. Athanassopoulos et al., LSND Collaboration, Phys. Rev. Lett. **77** (1996) 3082.

[5] G.F. Smoot et al., Ast. J. **396** (1992) L1.

[6] J.C. Pati and A. Salam, Phys. Rev. **D10** (1974) 275: R.N. Mohapatra and J.C. Pati, Phys. Rev. **D11** (1975) 566, 2558; G. Senjanović and R.N. Mohapatra, Phys. Rev. **D12** (1975) 1502; R.N. Mohapatra and G. Senjanović, Phys. Rev. **D 23**, (1981) 165.

[7] G. Senjanović, Nucl. Phys. **B153**, 334 (1979) 334; G. Senjanović and P. Senjanović, Phys. Rev. **D 21**, 3253 (1980); D. Chang, Nucl. Phys. **B214**, 435 (1983); H. Harari and M. Leurer, Nucl. Phys. **B233**, 221 (1984); G. Ecker and W. Grimus, Nucl. Phys. **B258**, 328 (1985); M. Leurer, Nucl. Phys. **B266**, 147 (1986); J.-M. Frère et al., Phys. Rev. **D 46**, 337 (1992); G. Barenboim, J. Bernabéu and M. Raidal, Nucl. Phys. **B478**, 527 (1996); B. Barenboim, J. Bernabéu and M. Raidal, Nucl. Phys. **B511**, 577 (1998); B. Barenboim, J. Bernabéu and M. Raidal, Phys. Rev. Lett. **80** (1998) 4625.

[8] R. Kuchimanchi, Phys. Rev. Lett. **76** (1996) 3486; R.N. Mohapatra and A. Rasin, Phys. Rev. Lett. **76** (1996) 3490; R.N. Mohapatra, A. Rasin and G. Senjanović, Phys. Rev. Lett. **79** (1997) 4744. For a review see, e.g., R.N. Mohapatra, hep-ph/9801235.

[9] M. Gell-Mann, P. Ramon and R. Slansky, in *Supergravity,* ed. P. van Niewenhuizen and D. Freedman (North-Holland, Amsterdam, 1979), p. 315; T. Yanagida, in *Proceedings of the Workshop on the Unified Theory and the Baryon Number in the Universe,* ed. O. Sawada and A. Sugamoto (KEK, Tsukuba, Japan, 1979); R.N. Mohapatra and G. Senjanović, Phys. Rev. Lett. **44** (1980) 912.

[10] For the early studies of SUSY left-right models see, M. Cvetic and J. Pati, Phys. Lett. **B135** (1984) 57; Y. Ahn, Phys. Lett. **B149** (1984) 337; R.M. Francis, M. Frank and C.S. Kalman, Phys. Rev. **D43** (1991) 2369; R.M. Francis, C.S. Kalman and H.N. Saif, Z. Phys. **C59** (1993) 655; K. Huitu, J. Maalampi and M. Raidal, Nucl. Phys. **B420** (1994) 449.

[11] R.N. Mohapatra, hep-ph/9707518; hep-ph/9806520.

[12] R. Kuchimanchi and R.N. Mohapatra, Phys. Rev. **D48** (1993) 4352; Phys. Rev. Lett. **75** (1995) 3989; K. Huitu and J. Maalampi, Phys. Lett. **B344** (1995) 217.

[13] C.S. Aulakh, K. Benakli and G. Senjanović, Phys. Rev. Lett. **79** (1997) 2188; C.S. Aulakh, A. Melfo and G. Senjanović, Phys. Rev. **D57** (1998) 4174; Z. Chacko and R.N. Mohapatra, hep-ph/9712359; C.S. Aulakh, A. Melfo, A. Raśin and G. Senjanović, hep-ph/9712551; Z. Chacko and R.N. Mohapatra, hep-ph/9802388.

[14] G. Beall, M. Bander and A. Soni, Phys. Rev. Lett. **48**, (1982) 848; G. Barenboim, J. Bernabéu, J. Prades and M. Raidal, Phys. Rev. **D 55**, (1997) 4213.

[15] P. Langacker and S. Uma Sankar, Phys. Rev. **D 40**, (1989) 1569.

[16] For reviews see, e.g., G. Bhattacharyya, Nucl. Phys. Proc. Suppl. **52A** (1997) 83; H. Dreiner, hep-ph/9707435.

[17] K. Huitu, J. Maalampi and K. Puolamäki, hep-ph/9705406, hep-ph/9708491.

[18] For the review of new physics in neutinoless double beta decay see, e.g., R.N. Mohapatra, hep-ph/9808284; and references therein.

[19] J. Schechter and J.W. Valle, Phys. Rev. **D 25** (1982) 2951.

[20] R.N. Mohapatra, Phys. Rev. **D 34** (1986) 909; M. Doi and T. Kotani, Progr. Theor. Phys. **89** (1993) 139.

[21] M. Hirsch, H.V. Klapdor-Kleingrothaus and O. Panella, Phys. Lett. **B374** (1996) 7.

[22] T.G. Rizzo, Phys. Lett. **B116** (1982) 23.

[23] C.A. Heusch and P. Minkowski, Nucl.Phys.**B416** (1994) 3; J. Gluza and M. Zrałek, Phys.Rev.**D52** (1995) 6238; J. Gluza and M. Zrałek, Phys. Lett.**B362** (1995) 148; G. Belanger, F. Boudjema, D. London and H. Nadeau, Phys.Rev. **D53** (1996) 6292; T. G. Rizzo, Int.J.Mod.Phys. **A11** (1996) 1613; J. Gluza and M. Zrałek, Phys. Lett.**B372** (1996) 259; C.A. Heusch and P. Minkowski,hep-ph/9611353; P. Duka, J. Gluza and M. Zrałek, hep-ph/9804372.

[24] P. Helde, K. Huitu, J. Maalampi and M. Raidal, Nucl. Phys. **B437** (1995) 305.

[25] H.V. Klapdor-Kleingrothaus, these proceedings and hep-ex/9802007.

[26] For the experimental prospects see, e.g., A. Czarnecki, hep-ph/9710425.

[27] M. Frank and H. Hamidian, Phys. Rev. **D54**(1996) 6790; G. Couture, M. Frank and H. Konig Phys. Rev. **D56** (1997) 4219; G. Couture, M. Frank, H. Konig and M. Pospelov, hep-ph/9701299.

[28] H.C. Chiang et al., Nucl. Phys. **A559** (1993) 526; A. Czarnecki, W. Marciano and K. Melnikov, hep-ph/9801218.

[29] G. Barenboim and M. Raidal, Nucl. Phys. **B484** (1997) 63.

[30] D. Wyler and L. Wolfenstein, Nucl. Phys. **B218** (1983) 205; R.N. Mohapatra and J.W.F. Valle, Phys. Rev. **D34** (1986) 1642; E. Witten, Nucl. Phys. **B268** (1986) 79; J. Bernabeu et al., Phys. Lett. **B187** (1987) 303; J.L. Hewett and T.G. Rizzo, Phys. Rep. **183** (1989) 193; P. Langacker and D. London, Phys. Rev. **D38** (1988) 886,907; W. Buchmüller and C. Greub, Nucl. Phys. **B363** (1991) 345; E. Nardi, Phys. Rev. **D48** (1993) 3277; D. Tommasini, G. Barenboim, J. Bernabeu and C. Jarlskog, Nucl. Phys. **B444** (1995) 451; J. Gluza, J. Maalampi, M. Raidal and M. Zrałek, Phys. Lett. **B407** (1997) 45.

[31] F. Cuypers, S. Davidson, Eur. Phys. J. **C2** (1998) 503.

[32] W.J. Marciano and A.I. Sanda, Phys. Rev. Lett. **38** (1977) 1512; M. Raidal and A. Santamaria, Phys. Lett. **B421** (1998) 250; and references therein.

[33] T. Rizzo, Phys. Rev. **D25** (1982) 1355, Phys. Rev. **D27** (1983) 657; M. Lusignoli and S. Petrarca, Phys. Lett. **B226** (1989) 397; M.D. Swartz, Phys. Rev. **D40** (1989) 1521; J.A. Grifols, A. Mendez and G.A. Schuler, Mod. Phys. Lett. **A4** (1989) 1485; J.F. Gunion, J.A. Grifols, A. Mendez, B. Kayser and F. Olness, Phys. Rev. **D40** (1989) 1546; J. Maalampi, A. Pietilä and M. Raidal, Phys. Rev. **D48** (1993) 4467; N. Lepore et al., Phys. Rev. **D50** (1994) 2031; E. Accomando and S. Petrarca, Phys. Lett. **B323** (1994) 212; S. Chakrabarti, D. Choudhuri, R.M. Godoble and B. Mukhopadhyaya, hep-ph/9804297.

[34] K. Huitu, J. Maalampi, A. Pietilä and M. Raidal, Nucl. Phys. **B487** (1997) 27.

[35] J.F. Gunion, C. Loomis and K.T. Pitts, hep-ph/9610327.

[36] J.F. Gunion, Int. Jour. Mod. Phys. **A11** (1996) 1551; F. Cuypers and M. Raidal, Nucl. Phys. **B501** (1997) 3.

[37] M. Raidal, Phys. Rev. **D57** (1998) 2013.

[38] G. Barenboim, K. Huitu, J. Maalampi and M. Raidal, Phys. Lett. **B394** (1997) 132.

[39] K. Huitu, J. Maalampi and M. Raidal, Nucl. Phys. **B420** (1994) 449, Phys. Lett. **B328** (1994) 60; B. Dutta and R.N. Mohapatra, hep-ph/9804277.

[40] M. Raidal and P. Zerwas, in preparation.

[41] E. Kolb and M. Turner, Phys. Rev. Lett. **67** (1991) 5; P. Herczeg and R.N. Mohapatra, Phys. Rev. Lett. **69** (1992) 2475.

[42] A. De Rujula and S.L. Glashow, Phys. Rev. Lett. **45** (1980) 942; P.B. Pal and L. Wolfenstein, Phys. Rev. **D 25** (1982) 766.

[43] Y. Hosotani, Nucl. Phys. **B 191** (1981) 411; J. Schechter and J.W. Valle, Phys. Rev. **D 25** (1982) 774.

On the Flavour Problem in SUSY

A. Masiero

SISSA, Via Beirut 2-4, 34013 Trieste, Italy and INFN, sez. di Trieste, Padriciano 99, 34012 Trieste, Italy

L. Silvestrini

Technische Universität München, Physik Department, D-85748 Garching, Germany

Abstract. We discuss the prospects for manifestation of low-energy supersymmetric extensions of the standard model in processes with flavour changing neutral currents and CP violation. In the first part of the talk we critically review the status of the minimal supersymmetric model to discuss the chances that direct and indirect supersymmetric searches may have before the LHC start. In the second part we point out what we consider to be the most promising grounds where departures from the standard model prediction may signal the presence of new physics, possibly of supersymmetric nature. We argue that the often invoked complementarity of direct and indirect searches of low energy supersymmetry is becoming even more true in the pre-LHC era.

1. Introduction

Baryon and Lepton number violations as well as the so-called flavour problem constitute potential threats for any extension of the Standard Model (SM) with an energy scale close to the electroweak scale, i.e. of O(1TeV). By flavour problem we mean the difficulty to keep under control the Flavour Changing Neutral Current (FCNC) and CP violating effects while allowing for a mixing of the fermions compatible with phenomenology. Extended technicolor theories are an example of how this potential danger can become very actual up to the point of ruling out the SM extension itself.

The flavour problem has been recognised as a severe constraint on low energy supersymmetric (SUSY) extensions of the SM long time ago, in the early 80's, soon after the phenomenological relevance of low energy SUSY started being appreciated.

In this talk we would like to review some of the main aspects of this issue. Let us state clearly that if, on one hand, it is true that the flavour problem represents a worry for generic low energy SUSY theories, at the same time it constitutes also a great hope in shedding more light on SUSY. Indeed, the possibility of getting some indirect signal of SUSY through departures from the SM expectations in some FCNC and/or CP violating processes represents a concrete tool to look for SUSY before the advent of LHC. In addition, the FCNC constraints constitute important indications on the kind of underlying physics from which low energy SUSY should come from. In this contribution we mainly focus on the former argument of the positive relevance of FCNC and CP violation in our search for SUSY.

It is not rare to hear the following gloomy forecast: if no supersymmetric signal is seen at LEP, then we have nothing else to do but wait for LHC. We do not agree with this statement. Apart from the fact that even for direct searches one should take into account the relevant potentialities of Tevatron in the LEP-LHC time interval, one should not neglect that indirect searches for new physics signals are going to be flourishing before 2005. We refer to processes exploring flavour physics (with or without CP violation) where new particles can play an active role being exchanged in the loop contributions and to several new astroparticle observations which may constitute privileged places to obtain information on physics beyond SM.

We wish to present here a brief overview (which is necessarily biased by our theoretical prejudices) of what we consider most promising in this effort of looking for indirect signals of low-energy SUSY before the LHC advent. First we will review the status and prospects for direct SUSY searches, then we will discuss the role that SUSY may play in FCNC and CP violating phenomena. Finally we will briefly comment on searches for the lightest SUSY particle in experiments looking for Dark Matter (DM).

2. Status of the MSSM

It is known that, even asking for the minimal content of superfields which are necessary to supersymmetrize the SM and imposing R parity, one is still left with more than 100 free parameters most of which are in the flavour sector. It is also true that very large portions on this enormous SUSY parameter space are already ruled out by the present phenomenology (in particular FCNC and CP constraints). If one wants to reduce the number of free parameters one has to make assumptions on what lies well beyond low-energy SUSY, in particular on the quite unknown issue of the origin of SUSY breaking. The two most popular drastic reductions of free SUSY parameters are provided by minimal supergravity (SUGRA) [1] (with the further assumption of unification of gauge couplings and gaugino masses at some grand unification scale) and by the models of Gauge-Mediated SUSY Breaking (GMSB) [2]-[4]. In minimal SUGRA and the minimal version of GMSB we have only 3 or 4 parameters in addition to those of the SM and so we can become much more predictive.

In the context of the minimal supergravity model (with electroweak radiative breaking), we ask the following relevant questions for direct SUSY searches: i) given the present experimental lower bounds on the masses of SUSY particles, how much room have we got in the SUSY parameter space to explore, or, in other words, when should we give up with SUSY if searches are fruitless? ii) is there any experimental signature of low-energy SUSY which is independent from the choice of the SUSY parameters in particular of the soft breaking sector? iii) are the electroweak precision tests telling us something relevant on low-energy SUSY?

i) SUSY must be a low-energy symmetry if it has to deal with the issue of the gauge hierarchy problem. This fact is usually translated into the statement that SUSY particle masses should not be significantly larger than $O(1\ \text{TeV})$ given that SUSY breaking should not exceed this energy scale to realize a suitable "protection" of the mass of the scalar Higgs responsible for the electroweak breaking. Actually one may try to be more quantitative [5]. First one relates the Z mass to the value of the 4 parameters of the minimal SUGRA run from the large scale, at which the soft breaking terms originate, down to the electroweak scale. Then one establishes a degree of naturalness corresponding to the amount of fine tuning of the initial SUSY parameters which is needed to reproduce the correct Z mass for increasing values of the low-energy SUSY masses. For instance it is clear that to have all SUSY particles with a mass of $O(1\ \text{TeV})$ would require a severe fine tuning of the boundary conditions.

As for all naturalness criteria, also in this case there is a large amount of subjectivity, but one message emerges quite clearly: already now, in particular with the lower bound on chargino masses exceeding 90 GeV, we are entering an area of parameter space where a certain degree of fine tuning is needed. Hence we are already at the stage where we may expect "naturally" to find SUSY particles. Moreover such naturalness analyses confirm that LHC represents kind of "definitive" machine for SUSY direct searches: if no SUSY particle is discovered at LHC, the degree of fine tuning becomes so severe that it is hard to still defend the idea of low-energy SUSY. Finally, an important comment on the degree at which different SUSY masses are constrained by such naturalness criteria: due to the large difference in the Yukawa couplings of the third (heaviest) generation with respect to the first two generations, it turns out that only the sbottoms and stops are required not to be very heavy, whilst squarks of the first two generations can be quite heavy (say tens of TeV) without severely affecting the correct electroweak breaking. This observations may play a relevant role in tackling the FCNC problem in SUSY (see below).

ii) If one allows the SUSY parameters to take larger and larger values, all the SUSY particles become heavier and heavier with only one remarkable exception. In the Higgs mass spectrum of SUSY models the lightest scalar always remains light. The mass of the light CP-even neutral Higgs in the MSSM is calculable at tree level in terms of two SUSY parameters of the Higgs potential. At this level it is smaller than the mass of the Z. When radiative corrections are included, the mass of the light Higgs becomes a function also of the other SUSY parameters and its upper bound increases significantly [6]. However, even varying the MSSM parameters as much as one wishes, it is not possible to exceed $130 - 135$ GeV for its mass. Indeed, taking $m_t = 175$ GeV and for a stop lighter than 1 TeV one obtains that the upper bound on the lightest Higgs is 125 GeV allowing for "maximal" mixing in the top squark sector (the bound decreases for smaller stop mixing).

It is not easy to significantly evade the above upper bound on the mass of the lightest Higgs even if one gives up the minimality of the SUSY model. For instance, if one adds a singlet to the two Higgs doublets (i.e., one goes to the so called Next-to-Minimal SUSY Standard Model, NMSSM), then a new parameter shows up in the scalar potential: the coupling of the singlet with the two doublets. If one imposes that all couplings remain perturbative up to the Planck scale, then the consequent upper bound on this new coupling implies that the lightest Higgs should not be heavier than 150 GeV or so [7].

Obviously having a possibly "exotic" Higgs below 150 GeV does not necessarily mean that it can be seen at LHC. While the lightest Higgs of the MSSM seems to be detectable at LHC, there may still be some significant loopholes for searches of the light Higgs in the NMSSM context.

iii) It is known that the MSSM is a decoupling theory. In the limit where we send the SUSY parameters to infinity all SUSY masses become infinite, with only the lightest Higgs remaining light and coinciding with the usual SM Higgs. In this limit we would recover the SM. It turns out that as far as electroweak precision tests are concerned, the decoupling of the MSSM is quite fast: already for SUSY masses above 200 – 300 GeV the effects due to the exchange of SUSY particles in radiative corrections to the electroweak observables become negligible. Notice that this is not true if, instead of electroweak precision tests, we consider FCNC and CP tests. In this latter case, the decoupling may be much slower with squarks and gluinos of 1 TeV still providing sizeable contributions in loop diagrams to some rare processes.

Obviously the SM fit of electroweak precision data is now so good that there is no point in trying to improve it by the addition of the several degrees of freedom represented by the SUSY particles. The situation was different a couple of years ago when the discrepancy between the SM prediction and the data in the decay of the Z into a b quark pair resulted in a SM fit which could be significantly improved. Now the goal of the game has changed: one looks for regions of the SUSY parameter space where (some) SUSY masses are sufficiently small so that virtual SUSY contributions to electroweak observables are sizeable [8]. Some of these regions may cause unbearably high departures from the SM predictions and hence they can be ruled out. In this way it is possible to exclude some (limited) portions of the MSSM parameter space which would be otherwise allowed by the limits on SUSY parameters coming from direct searches of SUSY particles.

Finally, we make a comment related to the prediction of one low-energy parameter (the electroweak angle or, as it is the case nowadays, the value of the strong coupling at the Z mass scale) when one asks for the unification of the gauge coupling constants in the MSSM. The value predicted for $\alpha_S(m_Z)$ in the MSSM is a couple of standard deviations higher than the experimental value. We do not consider this as a problem for the MSSM. Indeed, high energy thresholds generated from the masses of superheavy GUT particles may conceivably produce corrections able to account for such discrepancy. Taking into account the uncertainties in the dynamics at the GUT scale, we consider the argument of unification of couplings as a support to the existence of low-energy SUSY.

Before starting our discussion of indirect searches of SUSY, let us emphasise that direct production and detection of SUSY particles remain the only way to definitely prove the existence of low-energy SUSY. However it is true that if LEP II is not going to find a SUSY signal and unless some surprise possibly comes from Tevatron, we will have to wait almost ten years to obtain an answer from such direct searches. In view of this fact and of what we said in this section we think that indirect searches of SUSY in the pre-LHC era deserve a very special attention.

3. FCNC and SUSY

The generation of fermion masses and mixings ("flavour problem") gives rise to a first and important distinction among theories of new physics beyond the electroweak standard model.

One may conceive a kind of new physics which is completely "flavour blind", i.e. new interactions which have nothing to do with the flavour structure. To provide an example of such a situation, consider a scheme where flavour arises at a very large scale (for instance the Planck mass) while new physics is represented by a supersymmetric extension of the SM with supersymmetry broken at a much lower scale and with the SUSY breaking transmitted to the observable sector by flavour-blind gauge interactions [2]-[4]. In this case one may think that new physics does not cause any major change to the original flavour structure of the SM, namely that the pattern of fermion masses and mixings is compatible with the numerous and demanding tests of flavour changing neutral currents.

Alternatively, one can conceive a new physics which is entangled with the flavour problem. As an example consider a technicolor scheme where fermion masses and mixings arise through the exchange of new gauge bosons which mix together ordinary and technifermions. Here we expect (correctly enough) new physics to have potential problems in accommodating the usual fermion spectrum with the adequate suppression of FCNC. As another example of new physics which is not flavour blind, take a more conventional SUSY model which is derived from a spontaneously broken N=1 supergravity and where the SUSY breaking information is conveyed to the ordinary sector of the theory through gravitational interactions. In this case we may expect that the scale at which flavour arises and the scale of SUSY breaking are not so different and possibly the mechanism itself of SUSY breaking and transmission is flavour-dependent. Under these circumstances we may expect a potential flavour problem to arise, namely that SUSY contributions to FCNC processes are too large.

The potentiality of probing SUSY in FCNC phenomena was readily realized when the era of SUSY phenomenology started in the early 80's [9].

In particular, the major implication that the scalar partners of quarks of the same electric charge but belonging to different generations had to share a remarkably high mass degeneracy was emphasised.

Throughout the large amount of work in this last decade it became clearer and clearer that generically talking of the implications of low-energy SUSY on FCNC may be rather misleading. In minimal SUGRA FCNC contributions can be computed in terms of a very limited set of unknown new SUSY parameters. Remarkably enough, this minimal model succeeds to pass all the set of FCNC tests unscathed. To be sure, it is possible to severely constrain the SUSY parameter space, for instance using $b \to s\gamma$, in a way which is complementary to what is achieved by direct SUSY searches at colliders.

However, the MSSM is by no means equivalent to low-energy SUSY. A first sharp distinction concerns the mechanism of SUSY breaking and transmission to the observable sector which is chosen. As we mentioned above, in models with gauge-mediated SUSY breaking (GMSB models [2]-[4]) it may be possible to avoid the FCNC threat "ab initio" (notice that this is not an automatic feature of this class of models, but it depends on the specific choice of the sector which transmits the SUSY breaking information, the so-called messenger sector). The other more "canonical" class of SUSY theories that was mentioned above has gravitational messengers and a very large scale at which SUSY breaking occurs. In this talk we will focus only on this class of gravity-mediated SUSY breaking models. Even sticking to this more limited choice we have a variety of options with very different implications for the flavour problem.

First, there exists an interesting large class of SUSY realizations where the customary R-parity (which is invoked to suppress proton decay) is replaced by other discrete symmetries which allow either baryon or lepton violating terms in the superpotential. But, even sticking to the more orthodox view of imposing R-parity, we are still left with a large variety of extensions of the MSSM at low energy. The point is that low-energy SUSY "feels" the new physics at the superlarge scale at which supergravity (i.e., local supersymmetry) broke down. In this last couple of years we have witnessed an increasing interest in supergravity realizations without the so-called flavour universality of the terms which break SUSY explicitly. Another class of low-energy SUSY realizations which differ from the MSSM in the FCNC sector is obtained from SUSY-GUT's. The interactions involving superheavy particles in the energy range between the GUT and the Planck scale bear important implications for the amount and kind of FCNC that we expect at low energy.

Given a specific SUSY model it is in principle possible to make a full computation of all the FCNC phenomena in that context. However, given the variety of options for low-energy SUSY (even confining ourselves here to models with R matter parity), it is important to have a way to extract from the whole host of FCNC processes a set of upper limits on quantities which can be readily computed in any chosen SUSY frame.

The best model-independent parameterisation of FCNC effects is the so-called mass insertion approximation [10]. It concerns the most peculiar source of FCNC SUSY contributions that do not arise from the mere supersymmetrization of the FCNC in the SM. They originate from the FC couplings of gluinos and neutralinos to fermions and sfermions [11]. One chooses a basis for the fermion and sfermion states where all the couplings of these particles to neutral gauginos are flavour diagonal, while the FC is exhibited by the non-diagonality of the sfermion propagators. Denoting by Δ the off-diagonal terms in the sfermion mass matrices (i.e. the mass terms relating sfermion of the same electric charge, but different flavour), the sfermion propagators can be expanded as a series in terms of $\delta = \Delta/\tilde{m}^2$, where $\tilde{m}$ is the average sfermion mass. As long as Δ is significantly smaller than $\tilde{m}^2$, we can just take the first term of this expansion and, then, the experimental information concerning FCNC and CP violating phenomena translates into upper bounds on these δ's [12]-[14].

Obviously the above mass insertion method presents the major advantage that one does not need the full diagonalisation of the sfermion mass matrices to perform a test of the SUSY model under consideration in the FCNC sector. It is enough to compute ratios of the off-diagonal over the diagonal entries of the sfermion mass matrices and compare the results with the general bounds on the δ's that we provide here from all available experimental information.

There exist four different Δ mass insertions connecting flavours i and j along a sfermion propagator: $(\Delta_{ij})_{LL}$, $(\Delta_{ij})_{RR}$, $(\Delta_{ij})_{LR}$ and $(\Delta_{ij})_{RL}$. The indices L and R refer to the helicity of the fermion partners. Instead of the dimensional quantities Δ it is more useful to provide bounds making use of dimensionless quantities, δ, that are obtained dividing the mass insertions by an average sfermion mass.

Let us first consider CP-conserving $\Delta F = 2$ processes. The amplitudes for gluino-mediated contributions to $\Delta F = 2$ transitions in the mass-insertion approximation have been computed in refs. [13, 14]. Imposing that the contribution to $K-\bar{K}$, $D-\bar{D}$ and $B_d-\bar{B}_d$ mixing proportional to each single δ parameter does not exceed the experimental value, we obtain the constraints on the δ's reported in table 1, barring accidental cancellations [14] (for a QCD-improved computation of the constraints coming from $K-\bar{K}$ mixing, see ref. [15]).

We then consider the process $b \to s\gamma$. This decay requires a helicity flip. In the presence of a $\left(\delta^d_{23}\right)_{LR}$ mass insertion we can realize this flip in the gluino running in the loop. On the contrary, the $\left(\delta^d_{23}\right)_{LL}$ insertion requires the helicity flip to occur in the external b-quark line. Hence we expect a stronger bound on the $\left(\delta^d_{23}\right)_{LR}$ quantity. Indeed, this is what happens: $\left(\delta^d_{23}\right)_{LL}$ is essentially not bounded, while $\left(\delta^d_{23}\right)_{LR}$ is limited to be $< 10^{-3} - 10^{-2}$ according to the average squark and gluino masses [14].

Given the upper bound on $\left(\delta^d_{23}\right)_{LR}$ from $b \to s\gamma$, it turns out that the quantity x_s of the $B_s - \bar{B}_s$ mixing receives contributions from this kind of mass insertions which are very tiny. The only chance to obtain large values of x_s is if $\left(\delta^d_{23}\right)_{LL}$ is large, say of $O(1)$. In that case x_s can easily jump up to values of $O(10^2)$ or even larger.

Then, imposing the bounds in table 1, we can obtain the largest possible value for BR$(b \to d\gamma)$ through gluino exchange. As expected, the $\left(\delta^d_{13}\right)_{LL}$ insertion leads to very small values of this BR of $O(10^{-7})$ or so, whilst the $\left(\delta^d_{13}\right)_{LR}$ insertion allows for BR$(b \to d\gamma)$ ranging from few times 10^{-4} up to few times 10^{-3} for decreasing values of $x = m^2_{\tilde{g}}/m^2_{\tilde{q}}$. In the SM we expect BR$(b \to d\gamma)$ to be typically $10 - 20$ times smaller than BR$(b \to s\gamma)$, i.e. BR$(b \to d\gamma) = (1.7 \pm 0.85) \times 10^{-5}$. Hence a large enhancement in the SUSY case is conceivable if $\left(\delta^d_{13}\right)_{LR}$ is in the 10^{-2} range. Notice that in the MSSM we expect $\left(\delta^d_{13}\right)_{LR} < m^2_b/m^2_{\tilde{q}} \times V_{td} < 10^{-6}$, hence with no hope at all of a sizeable contribution to $b \to d\gamma$.

An analysis similar to the one of $b \to s\gamma$ decays can be performed in the leptonic sector where the masses $m_{\tilde{q}}$ and $m_{\tilde{g}}$ are replaced by the average slepton mass $m_{\tilde{l}}$ and the photino mass $m_{\tilde{\gamma}}$ respectively. The most stringent bound concerns the transition $\mu \to e\gamma$ with $\left(\delta^l_{12}\right)_{LR} < 10^{-6}$ for slepton and photino masses of O(100 GeV) [14].

4. CP and SUSY

The situation concerning CP violation in the MSSM case with $\Phi_A = \Phi_B = 0$ and exact universality in the soft-breaking sector can be summarised in the following way: the MSSM does not lead to any significant deviation from the SM expectation for CP-violating phenomena as d^e_N, ε, ε' and CP violation in B physics; the only exception to this statement concerns a small portion of the MSSM parameter space where a very light $\tilde{t}$ ($m_{\tilde{t}} < 100$ GeV) and χ^+ ($m_\chi \sim 90$ GeV) are present. In this latter particular situation sizeable SUSY contributions to ε_K are possible and, consequently, major restrictions in the $\rho - \eta$ plane can be inferred (see, for instance, ref. [16]). Obviously, CP violation in B physics becomes a crucial test for this MSSM case with very light $\tilde{t}$ and χ^+. Interestingly enough, such low values of SUSY masses are at the border of the detectability region at LEP II.

We now turn to CP violation in the model-independent approach that we are proposing here. For a detailed discussion we refer the reader to our general study [14]. Here we just summarise the situation in the following three points:

i) ϵ provides bounds on the imaginary parts of the quantities whose real part was limited by the K mass difference which are roughly one order of magnitude more severe than the corresponding ones derived from Δm_K.

ii) The nature of the SUSY contribution to CP violation is generally superweak, since the constraints from ε are always stronger (in the left-left sector) or at least equal (in the left-right sector) to the ones coming from ε'/ε.

iii) the experimental bound on the electric dipole moment of the neutron imposes very stringent limits on $\mathrm{Im}\left(\delta^d_{11}\right)_{LR}$ (of $O(10^{-6})$ for an average squark and gluino mass of 500 GeV.) In conclusion, although technically it is conceivable that some SUSY extension may provide a sizable contribution to ε'/ε, it is rather difficult to imagine how to reconcile a relatively large value of $\mathrm{Im}\left(\delta^d_{12}\right)_{LR}$ with the very strong constraint on the flavour-conserving $\mathrm{Im}\left(\delta^d_{11}\right)_{LR}$ from d^e_N.

We now move to the next frontier for testing the unitarity triangle in general and in particular CP violation in the SM and its SUSY extensions: B physics. We have seen above that the transitions between 1st and 2nd generation in the down sector put severe constraints on $\mathrm{Re}\,\delta^d_{12}$ and $\mathrm{Im}\,\delta^d_{12}$ quantities. To be sure, the bounds derived from ε and ε' are stronger than the corresponding bounds from ΔM_K. If the same pattern repeats itself in the transition between 3rd and 1st or 3rd and 2nd generation in the down sector we may expect that the constraints inferred from $B_d - \bar{B}_d$ oscillations or $b \to s\gamma$ do not prevent conspicuous new contributions also in CP violating processes in B physics. We are going to see below that this is indeed the case ad we will argue that measurements of CP asymmetries in several B-decay channels may allow to disentangle SM and SUSY contributions to the CP decay phase.

New physics can modify the SM predictions on CP asymmetries in B decays by changing the phase of the B_d–$\bar{B}_d$ mixing and the phase and absolute value of the decay amplitude. The general SUSY extension of the SM that we discuss here affects both these quantities.

The crucial question is then: where and how can one possibly distinguish SUSY contributions to CP violation in B decays [17]?

In terms of the decay amplitude A, the CP asymmetry reads

$$\mathcal{A}(t) = \frac{(1-|\lambda|^2)\cos(\Delta M_d t) - 2\mathrm{Im}\lambda\sin(\Delta M_d t)}{1+|\lambda|^2} \tag{1}$$

with $\lambda = e^{-2i\phi^M}\bar{A}/A$. In order to be able to discuss the results model-independently, we have labeled as ϕ^M the generic mixing phase. The ideal case occurs when one decay amplitude only appears in (or dominates) a decay process: the CP violating asymmetry is then determined by the total phase $\phi^T = \phi^M + \phi^D$, where ϕ^D is the weak phase of the decay. This ideal situation is spoiled by the presence of several interfering amplitudes.

We summarise the results in table 2 which is taken from the recent analysis of ref. [18]. We refer the interested reader to our work [18] for all the details of how our computation in the SM and in SUSY is carried out. Φ^D_{SM} denotes the decay phase in the SM; for each channel, when two amplitudes with different weak phases are present, we indicate the SM phase of the Penguin (P) and Tree-level (T) decay amplitudes. For $B \to K_S\pi^0$ the penguin contributions (with a vanishing phase) dominate over the tree-level amplitude because the latter is Cabibbo suppressed. For the channel $b \to s\bar{s}d$ only penguin operators or penguin contractions of current-current operators contribute. The phase γ is present in the penguin contractions of the $(\bar{b}u)(\bar{u}d)$ operator, denoted as u-P γ in table 2. $\bar{b}d \to \bar{q}q$ indicates processes occurring via annihilation diagrams which can be measured from the last two channels of table 2. In the case $B \to K^+K^-$ both current-current and penguin operators contribute. In $B \to D^0\bar{D}^0$ the contributions from the $(\bar{b}u)(\bar{u}d)$ and the $(\bar{b}c)(\bar{c}d)$ current-current operators (proportional to the phase γ) tend to cancel out.

SUSY contributes to the decay amplitudes with phases induced by δ_{13} and δ_{23} which we denote as ϕ_{13} and ϕ_{23}. The ratio of A_{SUSY}/A_{SM} for SUSY masses of 250 GeV are reported in the r_{250} column of table 2.

We now draw some conclusions from the results of table 2. In the SM, the first six decays measure directly the mixing phase β, up to corrections which, in most of the cases, are expected to be small. These corrections, due to the presence of two amplitudes contributing with different phases, produce uncertainties of $\sim$ 10% in $B \to K_S\pi^0$, and of $\sim$ 30% in $B \to D^+D^-$ and $B \to J/\psi\pi^0$. In spite of the uncertainties, however, there are cases where the SUSY contribution gives rise to significant changes. For example, for SUSY masses of O(250) GeV, SUSY corrections can shift the measured value of the sine of the phase in $B \to \phi K_S$ and in $B \to K_S\pi^0$ decays by an amount of about 70%. In $B \to J/\psi K_S$ and $B \to \phi\pi^0$ decays, SUSY effects are only about 10% but SM uncertainties are negligible. In $B \to K^0\bar{K}^0$ the larger effect, $\sim$ 20%, is partially covered by the indetermination of about 10% already existing in the SM. Moreover the rate for this channel is expected to be rather small. In $B \to D^+D^-$ and $B \to K^+K^-$, SUSY effects are completely obscured by the errors in the estimates of the SM amplitudes. In $B^0 \to D^0_{CP}\pi^0$ the asymmetry is sensitive to the mixing angle ϕ_M only because the decay amplitude is unaffected by SUSY. This result can be used in connection with $B^0 \to K_s\pi^0$, since a difference in the measure of the phase is a manifestation of SUSY effects.

Turning to $B \to \pi\pi$ decays, both the uncertainties in the SM and the

SUSY contributions are very large. Here we witness the presence of three independent amplitudes with different phases and of comparable size. The observation of SUSY effects in the $\pi^0\pi^0$ case is hopeless. The possibility of separating SM and SUSY contributions by using the isospin analysis remains an open possibility which deserves further investigation. For a thorough discussion of the SM uncertainties in $B \to \pi\pi$ see ref. [19].

In conclusion, our analysis shows that measurements of CP asymmetries in several channels may allow the extraction of the CP mixing phase and to disentangle SM and SUSY contributions to the CP decay phase. The golden-plated decays in this respect are $B \to \phi K_S$ and $B \to K_S\pi^0$ channels. The size of the SUSY effects is clearly controlled by the the non-diagonal SUSY mass insertions δ_{ij}, which for illustration we have assumed to have the maximal value compatible with the present experimental limits on B_d^0–$\bar{B}_d^0$ mixing.

5. DM and SUSY: a brief comment

We have strong indications that ordinary matter (baryons) is insufficient to provide the large amount of non-shining matter which has been experimentally proven to exist in galactic halos and at the level of clusters of galaxies [20]. In a sense, this might constitute the "largest" indication of new physics beyond the SM. This statement holds true even after the recent stunning developments in the search for non-shining baryonic objects. In September 1993 the discovery of massive dark objects ("machos") was announced. After five years of intensive analysis it is now clear that in any case machos cannot account for the whole dark matter of the galactic halos.

It was widely expected that some amount of non-shining baryonic matter could exist given that the contribution of luminous baryons to the energy density of the Universe $\Omega = \rho/\rho_{cr}$ ($\rho_{cr} = 3H_0^2/8\pi G$ where G is the gravitational constant and H_0 the Hubble constant) is less than 1%, while from nucleosynthesis we infer $\Omega_{baryon} = \rho_{baryon}/\rho_{cr} = (0.06 \pm 0.02)h_{50}^{-2}$, where $h_{50} = H_0/50$ Km/s Mpc. On the other hand, we have direct indications that Ω should be at least 20% which means that baryons can represent not more than half of the entire energy density of the Universe [20].

We could make these considerations on the insufficiency of the SM to obtain a large enough Ω more dramatic if we accept the theoretical input that the Universe underwent some inflationary era which produced Ω =1. In that case, at least 90% of the whole energy density of the Universe should be provided by some new physics beyond the SM.

Before discussing possible particle physics candidates, it should be kept in mind that DM is not only called for to provide a major contribution to Ω, but also it has to provide a suitable gravitational driving force for the primordial density fluctuations to evolve into the large-scale structures (galaxies, clusters and superclusters of galaxies) that we observe today [20]. Here we encounter the major difficulties when dealing with the two "traditional" sources of DM: Cold (CDM) and Hot (HDM) DM.

Light neutrinos in the eV range are the most typical example of HDM, being their decoupling temperature of O(1 MeV). On the other hand, the Lightest Supersymmetric Particle (LSP) in the tens of GeV range is a typical CDM candidate. Taking the LSP to be the lightest neutralino, one obtains that when it decouples it is already non-relativistic, being its decoupling temperature typically one order of magnitude below its mass.

Both HDM and CDM have some difficulty to correctly reproduce the experimental spectrum related to the distribution of structures at different scales. The conflict is more violent in the case of pure HDM. Neutrinos of few eV's tend to produce too many superlarge structures. The opposite problem arises with pure CDM: we obtain too much power in the spectrum at low mass scales (galactic scales).

A general feature is that some amount of CDM should be present in any case. A possibility which has been envisaged is that after all the whole Ω could be much smaller than one, say 20% or so and then entirely due to CDM. However, if one keeps on demanding the presence of an inflationary epoch, then it seems unnatural to have Ω so different from unity (although lately some variants of inflationary schemes leading to Ω smaller than one have been proposed). Another possibility is that CDM provides its 20% to Ω, while all the rest to reach the unity value is given by a non-vanishing cosmological constant.

Finally, the possibility which encounters quite some interest is the so-called Mixed Dark Matter (MDM) [21], where a wise cocktail of HDM and CDM is present. An obvious realization of a MDM scheme is a variant of the MSSM where neutrinos get a mass of few eV's. In that case the lightest neutralino (which is taken to be the LSP) plays the role of CDM and the light neutrino(s) that of HDM. With an appropriate choice of the parameters it is possible to obtain contributions to Ω from the CDM and HDM in the desired range.

In the MSSM with R parity the lightest SUSY particle (LSP) is absolutely stable. For several reasons the lightest neutralino is the favourite candidate to be the LSP fulfilling the role of CDM [22].

The neutralinos are the eigenvectors of the mass matrix of the four neutral fermions partners of the W_3, B, H_1^0 and H_2^0. There are four parameters entering this matrix: M_1, M_2, μ and $\tan\beta$. The first two parameters denote the coefficients of the SUSY breaking mass terms $\tilde{B}\tilde{B}$ and $\tilde{W}_3\tilde{W}_3$ respectively. μ is the coupling of the H_1H_2 term in the superpotential. Finally $\tan\beta$ denotes the ratio of the VEV's of the H_2 and H_1 scalar fields.

In general M_1 and M_2 are two independent parameters, but if one assumes that grand unification takes place, then at the grand unification scale $M_1 = M_2 = M_3$, where M_3 is the gluino mass at that scale. Then at M_W one obtains:

$$M_1 = \frac{5}{3}\tan^2\theta_w M_2 \simeq \frac{M_2}{2}, \qquad M_2 = \frac{g_2^2}{g_3^2} m_{\tilde{g}} \simeq \frac{m_{\tilde{g}}}{3}, \tag{2}$$

where g_2 and g_3 are the SU(2) and SU(3) gauge coupling constants, respectively.

The above relation between M_1 and M_2 reduces to three the number of independent parameters which determine the lightest neutralino composition and mass: $\tan\beta, \mu$ and M_2. Hence, for fixed values of $\tan\beta$ one can study the neutralino spectrum in the (μ, M_2) plane. The major experimental inputs to exclude regions in this plane are the request that the lightest chargino be heavier than M_Z and the limits on the invisible width of the Z hence limiting the possible decays $Z \to \chi\chi, \chi\chi'$.

Let us focus now on the role played by χ as a source of CDM. χ is kept in thermal equilibrium through its electroweak interactions not only for $T > m_\chi$, but even when T is below m_χ. However for $T < m_\chi$ the number of $\chi's$ rapidly decrease because of the appearance of the typical Boltzmann suppression factor exp $(-m_\chi/T)$. When T is roughly $m_\chi/20$ the number of χ diminished so much that they do not interact any longer, i.e. they decouple. Hence the contribution to Ω_{CDM} of χ is determined by two parameters: m_χ and the temperature at which χ decouples (T_D). T_D fixes the number of $\chi's$ which survive. As for the determination of T_D itself, one has to compute the χ annihilation rate and compare it with the cosmic expansion rate [23].

Several annihilation channels are possible with the exchange of different SUSY or ordinary particles, $\tilde{f}$, H, Z, etc. Obviously the relative importance of the channels depends on the composition of χ. For instance, if χ is a pure gaugino, then the $\tilde{f}$ exchange represents the dominant annihilation mode.

Quantitatively [24], it turns out that if χ results from a large mixing of the gaugino ($\tilde{W}_3$ and $\tilde{B}$) and Higgsino ($\tilde{H}_1^0$ and $\tilde{H}_2^0$) components, then the annihilation is too efficient to allow the surviving χ to provide Ω large enough. Typically in this case $\Omega < 10^{-2}$ and hence χ is not a good CDM candidate. On the contrary, if χ is either almost a pure Higgsino or a pure gaugino then it can give a conspicuous contribution to Ω

In the case χ is mainly a gaugino (say at least at the 90% level) what is decisive to establish the annihilation rate is the mass of $\tilde{f}$. If sfermions are light the χ annihilation rate is fast and the Ω_χ is negligible. On the other hand, if $\tilde{f}$ (and hence $\tilde{l}$, in particular) is heavier than 150 GeV, the annihilation rate of χ is sufficiently suppressed so that Ω_χ can be in the right ballpark for Ω_{CDM}. In fact if all the $\tilde{f}'s$ are heavy, say above 500 GeV and for $m_\chi \ll m_{\tilde{f}}$, then the suppression of the annihilation rate can become even too efficient yielding Ω_χ unacceptably large.

In the minimal SUSY standard model there are five new parameters in addition to those already present in the non–SUSY case. Imposing the electroweak radiative breaking further reduces this number to four. Finally, in simple supergravity realizations the soft parameters A and B are related. Hence we end up with only three new, independent parameters. One can use the constraint that the relic χ abundance provides a correct Ω_{CDM} to restrict the allowed area in this 3–dimensional space. Or, at least, one can eliminate points of this space which would lead to $\Omega_\chi > 1$, hence overclosing the Universe. For χ masses up to 150 GeV it is possible to find sizable regions in the SUSY parameter space where Ω_χ acquires interesting values for the DM problem. A detailed discussion on this point is beyond the scope of this talk. The interested reader can find a thorough analysis in the review of Ref. [22] and the original papers therein quoted.

There exist two ways to search for the existence of relic neutralinos. First we have direct detection: neutralinos interact with matter both through coherent and spin dependent effects. Only coherent effects are currently accessible to direct detection. The sensitivity of the direct detection experiment has reached now an area of the SUSY parameter space of the MSSM which is of great interest for neutralinos in the 50 GeV - 200 GeV range.

The indirect detection is based on the search for signals coming from pair annihilation of neutralinos. Such annihilation may occur inside celestial bodies (Earth, Sun, etc.) where neutralinos may be gravitationally captured. The signal is then a flux of muon neutrinos which can be detected as up-going muons in a neutrino telescope. Another possibility is that the neutralino annihilation occurs in the galactic halo. In this case the signal consists of photon, positron and antiproton fluxes. They can be observed by detectors placed on balloons or satellites. The computation of these fluxes is strongly affected by the composition of the lightest neutralino. In any case also these indirect searches for relic neutralinos are now probing interesting areas of the MSSM parameter space.

A very different prospect for DM occurs in the GMSB schemes. In this case the gravitino mass ($m_{3/2}$) loses its role of fixing the typical size of soft breaking terms and we expect it to be much smaller than what we have in models with a hidden sector. Indeed, given the well-known relation [1] between $m_{3/2}$ and the scale of SUSY breaking $\sqrt{F}$, i.e. $m_{3/2} = O(F/M)$, where M is the reduced Planck scale, we expect $m_{3/2}$ in the KeV range for a scale $\sqrt{F}$ of $O(10^6$ GeV) that has been proposed in models with low-energy SUSY breaking in a visible sector.

A gravitino of that mass behaves as a Warm Dark Matter (WDM) particle, that is, a particle whose free streaming scale involves a mass comparable to that of a galaxy, $\sim 10^{11-12} M_{\odot}$.

However, critical density models with pure WDM are known to suffer for serious troubles [25]. Indeed, a WDM scenario behaves much like CDM on scales above λ_{FS}. Therefore, we expect in the light gravitino scenario that the level of cosmological density fluctuations on the scale of galaxy clusters ($\sim 10\, h^{-1}$ Mpc) to be almost the same as in CDM. As a consequence, the resulting number density of galaxy clusters is predicted to be much larger than what observed.

We have recently considered different variants of a light gravitino DM dominated model. It seems that in all cases there exist difficulties to account correctly for cosmic structure formation. This provides severe cosmological constraints on the GMSB models [26].

In conclusion SUGRA models with R parity offer the best candidate for CDM. It is remarkable that as a by-product of the MSSM we obtain a lightest neutralino which can provide the correct amount of DM in a wide area of the SUSY parameter space. Even more interesting, we are now experimentally approaching the level of sensitivity which is needed to explore (directly or indirectly) large portions of this area of parameter space. The complementarity of this exploration to that performed by using FCNC and CP tests and direct collider SUSY searches looks promising.

Acknowledgements

We are grateful to our "FCNC collaborators" M. Ciuchini, E. Franco, F. Gabbiani, E. Gabrielli and G. Martinelli and our "DM collaborators" S. Borgani, E. Pierpaoli and M. Yamaguchi who contributed to most of the material presented in this talk. A.M. thanks the organisers for the stimulating settling in which the workshop took place. The work of A.M. was partly supported by the TMR project "Beyond the Standard Model" contract number ERBFMRX CT96 0090. L.S. acknowledges the support of the German Bundesministerium für Bildung and Forschung under contract 06 TM 874 and DFG Project Li 519/2-2.

References

[1] For a phenomenologically oriented review, see:
P. Fayet and S. Ferrara, *Phys. Rep.* **32C** (1977) 249;

H.P. Nilles, *Phys. Rep.* **110** (1984) 1.
H.E. Haber and G.L Kane, *Phys. Rep.* **117** (1987) 1;
For spontaneously broken N=1 supergravity, see:
E. Cremmer, S. Ferrara, L. Girardello and A. Van Proeyen, *Nucl. Phys.* **B 212** (1983) 413 and references therein.
P. Nath, R. Arnowitt and A.H. Chamseddine, *Applied N=1 Supergravity* (World Scientific, Singapore, 1984);
A.G. Lahanas and D.V. Nanopoulos, *Phys. Rep.* **145** (1987) 1.

[2] M. Dine, W. Fischler and M. Srednicki, *Nucl. Phys.* **B 189** (1981) 575;
S. Dimopoulos and S. Raby, *Nucl. Phys.* **B 192** (1981) 353;
M. Dine and W. Fischler, *Phys. Lett.* **B 110** (1982) 227;
M. Dine and M. Srednicki, *Nucl. Phys.* **B 202** (1982) 238;
M. Dine and W. Fischler, *Nucl. Phys.* **B 204** (1982) 346;
L. Alvarez-Gaumé, M. Claudson and M. Wise, *Nucl. Phys.* **B 207** (1982) 96;
C. Nappi and B. Ovrut, *Phys. Lett.* **B 113** (1982) 175;
S. Dimopoulos and S. Raby, *Nucl. Phys.* **B 219** (1983) 479.

[3] A. Nelson and M. Dine, *Phys. Rev.* **D 48** (1993) 1277;
M. Dine, A. E. Nelson and Y. Shirman, *Phys. Rev.* **D 51** (1995) 1362;
M. Dine, A. Nelson, Y. Nir and Y. Shirman, *Phys. Rev.* **D 53** (1996) 2658.

[4] E. Poppitz and S. Trivedi, *Phys. Rev.* **D 55** (1997) 5508;
N. Arkani-Hamed, J. March-Russel and H. Murayama, *Nucl. Phys.* **B 509** (1998) 3;
H. Murayama, *Phys. Rev. Lett.* **79** (1997) 18;
S. Dimopoulos, G. Dvali, G. Giudice and R. Rattazzi, *Nucl. Phys.* **B 510** (1998) 12 ;
S. Dimopoulos, G. Dvali and R. Rattazzi, *Phys. Lett.* **B 413** (1997) 336;
M. Luty, *Phys. Lett.* **B 413** (1997) 71;
T. Hotta, K.-I. Izawa and T. Yanagida, *Phys. Rev.* **D 55** (1997) 415;
N. Haba, N. Maru and T. Matsuoka, *Nucl. Phys.* **B 497** (1997) 31;
L. Randall, *Nucl. Phys.* **B 495** (1997) 37;
Y. Shadmi, *Phys. Lett.* **B 405** (1997) 99;
N. Haba, N. Maru and T. Matsuoka, *Phys. Rev.* **D 56** (1997) 4207;
C. Csaki, L. Randall and W. Skiba, *Phys. Rev.* **D 57** (1998) 383;
Y. Shirman, *Phys. Lett.* **B 417** (1998) 281;
For a complete review, see: G. Giudice and R. Rattazzi, hep-ph/9801271.

[5] R. Barbieri and G. Giudice, *Nucl. Phys.* **B 306** (1988) 63;
G. Anderson and D. Castano, *Phys. Rev.* **D 52** (1995) 1693;
P.H. Chankowski, J. Ellis and S. Pokorski, *Phys. Lett.* **B 423** (1998) 327;
R. Barbieri and A. Strumia, *Phys. Lett.* **B 433** (1998) 63.

[6] H. Haber and R. Hempfling, *Phys. Rev. Lett.* **66** (1991) 1815;
J. Ellis, G. Ridolfi and F. Zwirner, *Phys. Lett.* **B 257** (1991) 83;
Y. Okada, M. Yamaguchi and T. Yanagida, *Prog. Theor. Phys.* **85** (1991) 1.

[7] S. Dimopoulos and S. Thomas, *Nucl. Phys.* **B 465** (1996) 23.

[8] P.H. Chankowski and S. Pokorski, hep-ph/9707497;
D. Pierce and J. Erler, *Nucl. Phys. Proc. Suppl.* **62** (1998) 97.

[9] J. Ellis and D.V. Nanopoulos, *Phys. Lett.* **B 110** (1982) 44;
R. Barbieri and R. Gatto, *Phys. Lett.* **B 110** (1982) 211.

[10] L.J. Hall, V.A. Kostelecky and S. Raby, *Nucl. Phys.* **B 267** (1986) 415.

[11] M.J. Duncan, *Nucl. Phys.* **B 221** (1983) 285;
J.F. Donoghue, H.P. Nilles and D. Wyler, *Phys. Lett.* **B 128** (1983) 55;
A. Bouquet, J. Kaplan and C.A. Savoy, *Phys. Lett.* **B 148** (1984) 69.

[12] F. Gabbiani and A. Masiero, *Nucl. Phys.* **B 322** (1989) 235;
J.S. Hagelin, S. Kelley and T. Tanaka, *Nucl. Phys.* **B 415** (1994) 293.

[13] E. Gabrielli, A. Masiero and L. Silvestrini, *Phys. Lett.* **B 374** (1996) 80.

[14] F. Gabbiani, E. Gabrielli, A. Masiero and L. Silvestrini, *Nucl. Phys.* **B 477** (1996) 321.

[15] J.A. Bagger, K.T. Matchev and R.-J. Zhang, *Phys. Lett.* **B 412** (1997) 77.

[16] M. Misiak, S. Pokorski and J. Rosiek, hep-ph/9703442.

[17] For some recent discussions tackling this question, see:
N. Deshpande, B. Dutta and S. Oh, *Phys. Rev. Lett.* **77** (1996) 4499;
J. Silva and L. Wolfenstein, *Phys. Rev.* **D 53** (1997) 5331;
A. Cohen, D. Kaplan, F. Leipentre and A. Nelson, *Phys. Rev. Lett.* **78** (1997) 2300;
Y. Grossman and M. Worah, *Phys. Lett.* **B 395** (1997) 241;
Y. Grossman, Y. Nir and R. Rattazzi, hep-ph/9701231;
M. Ciuchini, E. Franco, G. Martinelli, A. Masiero and L. Silvestrini, *Phys. Rev. Lett.* **79** (1997) 978;
Y. Grossman, Y. Nir and M. Worah, *Phys. Lett.* **B 407** (1997) 307;
R. Barbieri and A. Strumia, *Nucl. Phys.* **B 508** (1997) 3.

[18] M. Ciuchini, E. Franco, G. Martinelli, A. Masiero and L. Silvestrini, in ref. [17].

[19] M. Ciuchini, E. Franco, G. Martinelli and L. Silvestrini, *Nucl. Phys.* **B 501** (1997) 271.

[20] For an introduction to the DM problem, see, for instance:
R. Kolb and S. Turner, *The Early universe* (addison-Wesley, New York, N.Y.) 1990;
Dark Matter, ed. by M. Srednicki (North-Holland, Amsterdam) 1989;
J. Primack, D. Seckel and B. Sadoulet, *Ann. Rev. Nucl. Part. Sci.* **38** (1988) 751.

[21] Q. Shafi and F.W. Stecker, *Phys. Lett. B* **53** (1984) 1292;
S.A. Bonometto and R. Valdarnini, *Astroph. J.* **299** (1985) L71;
S. Achilli, F. Occhionero and R. Scaramella, *Astroph. J.* **299** (1985) L77;
J.A. Holtzman, *Astroph. J. Suppl.* **71** (1981) 1;
A.N.Taylor and M. Rowan-Robinson, *Nature* **359** (1992) 396;
J.A. Holtzman and J. Primack, *Astroph. J.* **396** (1992) 113;

x	$\sqrt{\left\|\mathrm{Re}\,(\delta^d_{12})^2_{LL}\right\|}$	$\sqrt{\left\|\mathrm{Re}\,(\delta^d_{12})^2_{LR}\right\|}$	$\sqrt{\left\|\mathrm{Re}\,(\delta^d_{12})_{LL}\,(\delta^d_{12})_{RR}\right\|}$
0.3	1.9×10^{-2}	7.9×10^{-3}	2.5×10^{-3}
1.0	4.0×10^{-2}	4.4×10^{-3}	2.8×10^{-3}
4.0	9.3×10^{-2}	5.3×10^{-3}	4.0×10^{-3}
x	$\sqrt{\left\|\mathrm{Re}\,(\delta^d_{13})^2_{LL}\right\|}$	$\sqrt{\left\|\mathrm{Re}\,(\delta^d_{13})^2_{LR}\right\|}$	$\sqrt{\left\|\mathrm{Re}\,(\delta^d_{13})_{LL}\,(\delta^d_{13})_{RR}\right\|}$
0.3	4.6×10^{-2}	5.6×10^{-2}	1.6×10^{-2}
1.0	9.8×10^{-2}	3.3×10^{-2}	1.8×10^{-2}
4.0	2.3×10^{-1}	3.6×10^{-2}	2.5×10^{-2}
x	$\sqrt{\left\|\mathrm{Re}\,(\delta^u_{12})^2_{LL}\right\|}$	$\sqrt{\left\|\mathrm{Re}\,(\delta^u_{12})^2_{LR}\right\|}$	$\sqrt{\left\|\mathrm{Re}\,(\delta^u_{12})_{LL}\,(\delta^u_{12})_{RR}\right\|}$
0.3	4.7×10^{-2}	6.3×10^{-2}	1.6×10^{-2}
1.0	1.0×10^{-1}	3.1×10^{-2}	1.7×10^{-2}
4.0	2.4×10^{-1}	3.5×10^{-2}	2.5×10^{-2}

Table 1. Limits on $\mathrm{Re}\,(\delta_{ij})_{AB}\,(\delta_{ij})_{CD}$, with $A, B, C, D = (L, R)$, for an average squark mass $m_{\tilde{q}} = 500\mathrm{GeV}$ and for different values of $x = m^2_{\tilde{g}}/m^2_{\tilde{q}}$. For different values of $m_{\tilde{q}}$, the limits can be obtained multiplying the ones in the table by $m_{\tilde{q}}(\mathrm{GeV})/500$.

D. Pogosyan and A. Starobinski, *Astroph. J.* **447** (1995) 465; A. Klypin, J. Holtzman, J. Primack and E. Regos, *Astroph. J.* **415** (1993) 1.

[22] G. Jungman, M. Kamionkowski and K. Griest, *Phys. Rep.* **267** (1996) 195, and references therein.

[23] J. Ellis, J.S. Hagelin, D.V. Nanopoulos, K. Olive and M. Srednicki, *Nucl. Phys.* **B 238** (1984) 453.

[24] A. Bottino, F. Donato, N. Fornengo and S. Scopel, hep-ph/9710295.

[25] S. Colombi, S. Dodelson and L.M. Widrow, astro-ph/9505029.

[26] E. Pierpaoli, S. Borgani, A. Masiero and M. Yamaguchi, *Phys. Rev.* **D 57** (1998) 2089.

Incl.	Excl.	$\phi^D_{\rm SM}$	$r_{\rm SM}$	$\phi^D_{\rm SUSY}$	r_{250}
$b \to c\bar{c}s$	$B \to J/\psi K_S$	0	–	ϕ_{23}	$0.03 - 0.1$
$b \to s\bar{s}s$	$B \to \phi K_S$	0	–	ϕ_{23}	$0.4 - 0.7$
$b \to u\bar{u}s$		P 0			
	$B \to \pi^0 K_S$		$0.01 - 0.08$	ϕ_{23}	$0.4 - 0.7$
$b \to d\bar{d}s$		T γ			
$b \to c\bar{u}d$		0			
	$B \to D^0_{CP}\pi^0$		0.02	–	–
$b \to u\bar{c}d$		γ			
	$B \to D^+D^-$	T 0	$0.03 - 0.3$		$0.007 - 0.02$
$b \to c\bar{c}d$				ϕ_{13}	
	$B \to J/\psi\pi^0$	P β	$0.04 - 0.3$		$0.007 - 0.03$
	$B \to \phi\pi^0$	P β	–		$0.06 - 0.1$
$b \to s\bar{s}d$				ϕ_{13}	
	$B \to K^0\bar{K}^0$	u-P γ	$0 - 0.07$		$0.08 - 0.2$
$b \to u\bar{u}d$	$B \to \pi^+\pi^-$	T γ	$0.09 - 0.9$	ϕ_{13}	$0.02 - 0.8$
$b \to d\bar{d}d$	$B \to \pi^0\pi^0$	P β	$0.6 - 6$	ϕ_{13}	$0.06 - 0.4$
	$B \to K^+K^-$	T γ	$0.2 - 0.4$		$0.04 - 0.1$
$b\bar{d} \to q\bar{q}$				ϕ_{13}	
	$B \to D^0\bar{D}^0$	P β	only β		$0.01 - 0.03$

Table 2. CP phases for B decays. ϕ^D_{SM} denotes the decay phase in the SM; T and P denote Tree and Penguin, respectively; for each channel, when two amplitudes with different weak phases are present, one is given in the first row, the other in the last one and the ratio of the two in the r_{SM} column. ϕ^D_{SUSY} denotes the phase of the SUSY amplitude, and the ratio of the SUSY to SM contributions is given in the r_{250} column for SUSY masses of 250 GeV.

Neutron–antineutron oscillations: theory

W.M. Alberico

Dipartimento di Fisica Teorica dell'Università di Torino
and INFN, Sezione di Torino
via P.Giuria 1, 10125 Torino, Italy

Abstract. We review the problematics connected with the occurrence of $n\bar{n}$ oscillations both in free space and inside nuclei, a process entailing baryon number violation and leading to a $\Delta B = 2$ nuclear instability mode. A few experimental lower bounds have been till now established, both in attempts to measure directly the $n\bar{n}$ oscillation time in a beam of free neutrons, and in setting indirect limits for specific channels of nuclear instability. The analysis of indirect measurements relies on theoretical assumptions: providing the same mechanism is active as for the free neutron conversion, the influence of the nuclear medium turns out to be crucial in determining the global nuclear lifetime. The newly formed antineutron state feels a considerably different mean field, with an imaginary absorptive part which accounts for the annihilation. As a result, the $n\bar{n}$ conversion is substantially suppressed, in the nuclear medium, with respect to the free process. Various theoretical approaches are revisited here and the corresponding estimates for the suppression factor are confronted, finding rather stable results. Thus both direct and indirect measurements can be employed in the search for $n\bar{n}$ oscillations.

1. Introduction

Among the topics related to baryon number violation, the $n\bar{n}$ oscillation has received renewed attention in recent years, both from the theoretical point of view, with the development of Grand Unified Schemes allowing for this process within reasonable mass scale parameters[1], and from experimental proposals which could realistically set up new limits on free $n\bar{n}$ oscillations (ORNL[2]) or intranuclear processes (Superkamiokande[3], Icarus[4]).

Accurate theoretical estimates of the nuclear medium effects on the $\bar{n}$ propagation following an $n\bar{n}$ oscillation in nuclei are of great relevance, since they allow to provide a bridge between the measurements of nuclear instability rates and the free oscillation time for an $\bar{n}$ conversion occurring in a beam of free neutrons. It is thus appropriate to review here the existing theoretical attempts to set precise relationships between the $n\bar{n}$ oscillation time and the nuclear lifetime.

After recalling the main features of the considered process, we shall shortly review the experimental limits in section 2 which are presently known both from direct and indirect searches. Section 3 illustrates the evaluation of the nuclear medium effects to be taken into account, should the $n\bar{n}$ conversion occur to a neutron in a nucleus. Various nuclear models are discussed: they can be mainly grouped into two different many–body frameworks, namely the optical model approach (subsection 3.1) and the S–matrix approach (subsection 3.2), which rely on different approximations of the nuclear dynamics involved in the process. The results obtained in both cases are confronted and consistently compared with the present experimental limits for free $n\bar{n}$ oscillations. In section 4 we finally resume some debate and criticisms which has occurred, during the last years, about the reliability of the expected nuclear medium effects on intranuclear transitions and we draw our conclusions on the intrinsic interest of both direct and indirect search for $n\bar{n}$ oscillations.

2. Direct and indirect limits on the $n\bar{n}$ oscillation time

The fundamental interaction which produces the process we are interested in can be described by the phenomenological $\Delta B = 2$ Hamiltonian

$$H_{n\bar{n}} = \epsilon_{n\bar{n}} \int d^3x \, \bar{\Psi}(x) \sigma_1 \Psi(x) \tag{1}$$

$\epsilon_{n\bar{n}}$ being the transition mass (sometimes referred to as δm), and σ_1 the usual Pauli matrix acting on the two–dimensional $(n, \bar{n})$ space spanned by the vectors Ψ:

$$\Psi(x) = \begin{pmatrix} \psi_n(x) \\ \psi_{\bar{n}}(x) \end{pmatrix}, \tag{2}$$

the antineutron field being the customary charge conjugated of the neutron one: $\psi_{\bar{n}}(x) = \psi_n^C(x)$.

The $n\bar{n}$ oscillation time for a process occurring in free space is then defined as the inverse transition mass ($\hbar = c = 1$):

$$\tau_{n\bar{n}} = \frac{1}{\epsilon_{n\bar{n}}} \tag{3}$$

Two different approaches have been conceived in order to experimentally determine $\tau_{n\bar{n}}$:

1. the *direct* measurement of an $\bar{n}$ component in a freely propagating beam of neutrons (or spallation sources)

2. indirect limits stemming from $n\bar{n}$ oscillations occurring in nuclei and detectable by general purposes experiments, set up for nuclear instability search.

In the direct search, the probability of finding at time t an antineutron component in an initially pure neutron state turns out to be

$$P(\bar{n}, t) = \frac{\epsilon_{n\bar{n}}^2}{\epsilon_{n\bar{n}}^2 + \Delta E^2} \sin^2 \left\{ t\sqrt{\epsilon_{n\bar{n}}^2 + \Delta E^2} \right\} \tag{4}$$

$2\Delta E$ being the energy gap between the n and $\bar{n}$ states, eventually due to external perturbative fields (as, for example, the Earth magnetic field). Should ΔE be made negligibly small with respect to $\epsilon_{n\bar{n}}$, then the above probability more simply reduces to $P(\bar{n}, t) \approx \epsilon_{n\bar{n}}^2 t^2$.

The present experimental limit on $\tau_{n\bar{n}}$ has been set by the measurements at ILL (Grenoble)[5, 6] to be

$$\tau_{n\bar{n}} > 0.86 \times 10^8 sec \qquad (90\% CL) \tag{5}$$

A significant improvement by two or possibly three orders of magnitude is expected to be possible with the proposed experiment at ORNL[7, 8], which should reach a sensitivity up to $10^9 \div 10^{10}$ sec.

In addition to specific and direct measurements of the free $n\bar{n}$ oscillation time, several multipurpose experiments (some of them already operating since several years, like Kamiokande[16] and Frejus[24]), have attempted to reveal unusual nuclear decays like the ones induced either by proton decay or by $n\bar{n}$ oscillations. Apart from the intrinsic technical difficulties in disentangling the occurrence of such processes from the overwhelming background, the extraction of the $n\bar{n}$ oscillation time is, in this case, somewhat complicated by its occurrence inside the nuclear medium, which strongly interacts with both the initial neutron and the antineutron resulting from the oscillation process.

Indeed the $n\bar{n}$ conversion, induced by the $\Delta B = 2$ Hamiltonian (1), followed by the $\bar{n}$ annihilation with the surrounding nucleons into a nucleus with mass number decreased by 2 units and a certain number of pions (which carry away a total energy of about 2 GeV), provides a source for nuclear instability which could be revealed by the above mentioned experiments.

This argument was suggested and debated since the 80's [9, 10, 11, 12, 13]. Expressed in simple terms, it relies on the following view: in the nuclear medium one can consider the following $n\bar{n}$ mass matrix:

$$M = \begin{pmatrix} M_n + U_n & \epsilon_{n\bar{n}} \\ \epsilon_{n\bar{n}} & M_{\bar{n}} + U_{\bar{n}} + iW_{\bar{n}} \end{pmatrix} \quad (6)$$

which contains the interaction potentials of the neutron (U_n) and of the antineutron ($U_{\bar{n}} + iW_{\bar{n}}$) with the residual nucleus, the latter being a complex quantity with an imaginary (absorptive) part related to the $\bar{n}$ annihilation. The neutron (M_n) and antineutron ($M_{\bar{n}}$) free masses are obviously identical, their difference in binding energy being embodied in the specific optical potentials.

Diagonalization of (6) leads to two mass eigenstates $N_\pm$ with eigenvalues

$$\begin{aligned} m_+ &= M_n + < U_n > + \frac{\epsilon^2_{n\bar{n}} (< U_n - U_{\bar{n}} >)}{< U_n - U_{\bar{n}} >^2 + < W_{\bar{n}} >^2} \\ & \quad - i\epsilon^2_{n\bar{n}} \frac{< W_{\bar{n}} >}{< U_n - U_{\bar{n}} >^2 + < W_{\bar{n}} >^2} \quad (7) \\ m_- &= M_{\bar{n}} + < U_{\bar{n}} > - i < W_{\bar{n}} > \quad (8) \end{aligned}$$

where appropriate nuclear mean values of the interaction potentials have been introduced. The $|N_+ >$ state, belonging to the eigenvalue (7), contains the conversion of a neutron into an antineutron and the subsequent nuclear decay, with lifetime

$$\tau_+^{-1} \equiv T_A^{-1} = \epsilon^2_{n\bar{n}} \frac{< W_{\bar{n}} >}{< U_n - U_{\bar{n}} >^2 + < W_{\bar{n}} >^2} \equiv \epsilon^2_{n\bar{n}} T_R^{-1} \quad (9)$$

which allows us to define the socalled "nuclear factor" T_R. Eq.(9) can be equivalently rewritten in the form

$$T_A = \tau^2_{n\bar{n}} T_R. \quad (10)$$

This relationship clearly shows the different time scales involved by the $n\bar{n}$ oscillation time ($\tau_{n\bar{n}}$) and the nuclear decay lifetimes (T_A). Indeed, as it will be shown in the following Sections, the nuclear quenching factor T_R turns out to be rather large, involving several orders of magnitude (it is typically of the order of 10^{23} sec^{-1}). According to (10), experimental limits on nuclear stability (say $T_A > T_0^{exp}$), together with the theoretical evaluation of T_R, allow one to establish lower limits on $\tau_{n\bar{n}}$ as well.

Along these lines, the Kamiokande experiment[16] has provided the following lower limit for the nuclear lifetime of ^{16}O in this channel:

$$T_0 = 4.3 \times 10^{31} y \ (90\% CL) \quad (11)$$

while the Frejus experiment[24] gives, for ^{56} Fe, a slightly higher limit, namely:

$$T_0 = 6.5 \times 10^{31} y \ (90\% CL). \tag{12}$$

These number must be employed together with the specific evaluations of the nuclear factor T_R for the two nuclei and, as we will see, the limit attainable on $\tau_{n\bar{n}}$ is practically the same in both cases.

This process continues to be investigated in experiments like Superkamiokande[3] and is embodied in the program of the forthcoming Icarus experiment[4], in which the ^{40} Ar nucleus is involved.

3. Theoretical approaches

The existing estimates of T_R can be roughly divided into two categories, the first one being based on a potential model approach, the second one directly stemming from the S–matrix formulation along with two–body NN and $N\bar{N}$ interactions. One should, however, keep in mind that the two formulations are essentially equivalent, notwithstanding that they are focussed on different aspects of the many–body treatment of the problem.

3.1. The optical model approach

After the first rough estimates of the nuclear factor T_R, Dover *et al.*[14, 15] afforded an exact calculation based on the available N–nucleus and $\bar{N}$–nucleus optical potentials. The starting point is the (single–particle) nuclear Hamiltonian

$$H = \frac{p^2}{2M} + U_n P_+ + (U_{\bar{n}} + iW_{\bar{n}})P_- + H_{n\bar{n}} \equiv H_0 + H_{n\bar{n}} \tag{13}$$

acting onto the two–dimensional $n\bar{n}$ space. In eq.(13) $P_\pm = \frac{1}{2}(1 \pm \sigma_3)$ are the projection operators onto n ($\bar{n}$) states and $H_{n\bar{n}}$ is the $n\bar{n}$ transition Hamiltonian given in eq.(1). The eigenvalue equation for H,

$$(H_0 + H_{n\bar{n}}) \begin{pmatrix} \psi_n \\ \psi_{\bar{n}} \end{pmatrix} = E \begin{pmatrix} \psi_n \\ \psi_{\bar{n}} \end{pmatrix} \tag{14}$$

provides two coupled differential equations for the neutron and antineutron states:

$$\left(\frac{p^2}{2M} + U_n(\vec{x}) - E_\alpha\right) \psi_{n,\alpha}(\vec{x}) \ \simeq \ 0 \tag{15}$$

$$\left(\frac{p^2}{2M} + U_{\bar{n}}(\vec{x}) + iW_{\bar{n}}(\vec{x}) - E_\alpha\right) \psi_{\bar{n},\alpha}(\vec{x}) \ = \ -\epsilon_{n\bar{n}}\psi_{n,\alpha}(\vec{x}) \tag{16}$$

where the small coupling term $(-\epsilon_{n\bar{n}}\psi_{\bar{n}})$ to the $\bar{n}$–state is usually neglected in solving the wave equation for the neutron, while the inhomogeneous term is essential in equation (16), in order to justify the creation of an $\bar{n}$ inside the nuclear medium.

The n–nucleus optical potential $U_n(\vec{x})$ included in the Hamiltonian is choosen to reproduce the neutron binding energies, E_α, in the considered nucleus, while the complex $\bar{n}$–nucleus optical potential, $U_{\bar{n}}(\vec{x}) + iW_{\bar{n}}(\vec{x})$ is adapted to fit energy shifts and widths of the $\bar{p}$–atom levels as well as low–energy $\bar{p}$–nucleus scattering data measured at LEAR. Its imaginary part accounts for the annihilation width of the antiparticle inside the nuclear medium. For an $\bar{n}$ in the state $\psi_{\bar{n},\alpha}$ the latter turns out to be:

$$\Gamma_\alpha = -2 \int d^3x \, |\psi_{\bar{n},\alpha}(\vec{x})|^2 \, W_{\bar{n}}(\vec{x}). \tag{17}$$

The total nuclear annihilation width, obtained by averaging (17) over all possible occupied states,

$$< \Gamma_A >= \frac{< \Gamma_\alpha >}{N} = \frac{1}{N} \sum_\alpha n_\alpha \Gamma_\alpha, \tag{18}$$

N being the number of neutrons in the nucleus and n_α the occupation number of the state α, gives the inverse nuclear lifetime ($T_A^{-1} =< \Gamma_A >$).

From eq.(16) it appears that $\psi_{\bar{n},\alpha} \propto \epsilon_{n\bar{n}}$, so that $\Gamma_\alpha \propto \epsilon_{n\bar{n}}^2 = \tau_{n\bar{n}}^{-2}$. It is thus appropriate to explicitly factorize this factor in $< \Gamma_A >$ or, according to (10), to set:

$$T_R^{-1} = \frac{\tau_{n\bar{n}}^2}{N} \sum_\alpha n_\alpha \Gamma_\alpha. \tag{19}$$

This quantity is the above mentioned nuclear quenching factor and depends *only* on nuclear physics aspects, but not on the strength of the $n\bar{n}$ conversion.

Typical $\bar{n}$–nucleus optical potentials utilized in ref.[14, 15] are of Woods-Saxon form:

$$U_{\bar{n}}(r) = -\frac{V_0}{1 + e^{(r-R_V)/a_V}} \tag{20}$$

$$W_{\bar{n}}(r) = -\frac{W_0}{1 + e^{(r-R_W)/a_W}} \tag{21}$$

($r = |\vec{x}|$). The best fits to the available $\bar{p} - N$ data are obtained by employing potentials with a shallow real part and a deep (strong) imaginary part; it is worth mentioning, however, that equivalently good fits can be obtained with set of parameters (in particular V_0, W_0) sometimes differing by one order of magnitude.

Dover *et al.*[15] estimated T_R [eq.(19)] for the ^{16}O nucleus, for which the Kamiokande experiment[16] provided in '86 the following lower limit of the nuclear lifetime:

$$T_A > T_0^{exp} = 4.3 \times 10^{31} y. \tag{22}$$

Eq.(10) translates then (22) into a lower limit for the $n\bar{n}$ oscillation time:

$$\tau_{n\bar{n}} \geq \sqrt{\frac{T_0^{exp}}{T_R}}. \tag{23}$$

By employing various models for the $\bar{n}$–nucleus potential, the authors of ref.[15] found the following range of T_R values (which expresses the "uncertainty" on the theoretical estimate within this approach):

$$0.46 \leq T_R \leq 1.93 \quad (\times 10^{-23} sec^{-1}). \tag{24}$$

Then, by employing (22), one gets:[1]

$$0.84 \leq \tau_{n\bar{n}} \leq 1.72 \quad (\times 10^{8} sec). \tag{25}$$

We remark that the uncertainty in the T_R value, stemming from the different models for the nuclear optical potential employed in the calculation, is appreciably attenuated in $\tau_{n\bar{n}}$, due to the power 1/2 which relates the two quantities.

A simpler approach, still based on the potential model, starts from eq.(9) and evaluates T_R in terms of averages over the nuclear density[18]; for example, the average of the real part of the $\bar{n}$–nucleus potential is given by:

$$< U_{\bar{n}} >= \frac{\int_0^\infty drr^2 \rho(r) U_{\bar{n}}(r)}{\int_0^\infty drr^2 \rho(r)} \tag{26}$$

When the same $\bar{n}$–nucleus potential is employed, this approximation yields T_R values which are approximately twice the "exact" result[15]; thus the resulting limits on $\tau_{n\bar{n}}$ turn out to be somewhat overestimated.

3.2. S–matrix approach

This method can be developed not only starting from an optical model (single particle) hamiltonian, but also from the full many-body nuclear Hamiltonian

$$H = H_{SM} + V_{NN} + V_{N\bar{N}} + H_{n\bar{n}} \tag{27}$$

[1] The numbers in [15] differ somewhat from (25) since in '85 the authors were using a previous (lower) limit for the nuclear lifetime, provided by the IMB collaboration[17].

where H_{SM} is a suitably choosen shell model Hamiltonian, $V_{NN(\bar{N})}$ are short-cut notations for the two-body interaction terms, referring to the nucleon–nucleon (nucleon–antinucleon) potentials, respectively (e.g. $V_{NN} = \sum_{i>j} V_{NN}(i,j)$) and $H_{n\bar{n}}$ is the transition Hamiltonian (1).

The time evolution of a nuclear system described by (27) embodies a finite probability for the process:

$$(A, Z) \quad \Longrightarrow \quad (A-1, Z, \bar{n}) \quad \Longrightarrow \quad (A-2, Z) + n\pi \qquad (28)$$

where a certain number n of pions (most probably 4 or 5) are produced in the final state, by annihilation of the $\bar{n}$ in the nucleus. The S–matrix for the $n - \bar{n}$ conversion and the subsequent evolution of the system up to a certain time t reads:

$$S(t,0) = U(t,0) T \left\{ e^{-i \int_0^t dt' H_{n\bar{n}}(t')} \right\} \qquad (29)$$

where the evolution operator is

$$U(t,0) = e^{-i \int_0^t dt' [V_{NN} + V_{N\bar{N}}]} \qquad (30)$$

and

$$H_{n\bar{n}}(t) = U^{-1}(t,0) H_{n\bar{n}} U(t,0). \qquad (31)$$

The decay probability of the initial nuclear state $|\Psi_A>$ is then determined by:

$$|< \Psi_A | S(t,0) | \Psi_A >|^2 = 1 - \Gamma_A t + \ldots \qquad (32)$$

where the nuclear width, neglecting terms of order $O(\epsilon^4_{n\bar{n}})$, reads:

$$\Gamma_A \equiv \epsilon^2_{n\bar{n}} T_R^{-1} = \epsilon^2_{n\bar{n}} \sum_i |a_i|^2 \frac{\Gamma_i}{(\Delta M_i)^2 + \Gamma_i^2/4} \,. \qquad (33)$$

In the above the coefficients $|a_i|^2$ represent the probability that the wavefunction resulting from the $n\bar{n}$ conversion contains the quasi–nuclear state $|(A-1, Z, \bar{n})_i>$, with energy $E_{i(\bar{n})}$ and annihilation width Γ_i into the nuclear state $|A-2, Z>$. Using two–body NN and (complex) $N\bar{N}$ interaction potentials:

$$\Delta M_i \equiv E_{i(n)} - E_{i(\bar{n})} = < i|V_{NN}|i> - < i|Re\, V_{N\bar{N}}|i> \qquad (34)$$

$$\Gamma_i = -2 < i|Im\, V_{N\bar{N}}|i> . \qquad (35)$$

Closure approximation

Since $\sum_i |a_i|^2 = 1$, if Γ_i and ΔM_i have only a weak dependence upon the state $|i>$, then eq.(33) reduces to

$$\Gamma_A \simeq \epsilon^2_{n\bar{n}} \frac{<\Gamma>}{<\Delta M>^2 + <\Gamma>^2/4}, \qquad (36)$$

where quantities averaged over the available quasi–nuclear states have been introduced.

We have adopted the S–matrix approach [19, 20, 21] within the framework of two different nuclear models:

(a) A nuclear matter description, which starts from the Hartree–Fock basis implemented by the local density approximation;

(b) A standard nuclear shell model, together with the above mentioned closure approximation.

In case (a) the states $|i>$ are expressed through single particle wave functions labelled by the wave number $\vec{k}$ and the coefficients $a_i \equiv a(\vec{k})$ are constants. With appropriate normalization and *without* resorting to the closure approximation, equation (33) can then be rewritten as:

$$T_R^{-1} = \frac{4}{A}\frac{\Omega}{(2\pi)^3}\int d^3k\theta(k_F - k)\frac{\Gamma_a(k)}{[\Delta M(k)]^2 + \Gamma_a^2(k)/4} \tag{37}$$

where (s, t are spin, isospin quantum numbers)

$$\Delta M(k) = V_n(k,s,t) - V_{\bar{n}}(k,s,t) \tag{38}$$

$V_{n(\bar{n})}(k,s,t)$ being the Hartree–Fock (Hartree only for the antineutron) fields felt by the neutron (antineutron) as a result of the two–body interaction potentials contained in H [eq.(27)], while $\Gamma_a(k)$ is the annihilation width of an antineutron in the state $(\vec{k}, s, t)$, stemming from $ImV_{N\bar{N}}$ (for detailed definitions see ref.[20]). The local density approximation is then applied by averaging (37) over the nuclear volume with a locally defined Fermi momentum

$$k_F(r) = \left\{\frac{3\pi^2}{2}\rho(r)\right\}^{1/3}, \tag{39}$$

$\rho(r)$ being the phenomenological nuclear density distribution (e.g. the usual Fermi parameterization).

In this approach there are two crucial ingredients: the nuclear model description and the two–body NN and $N\bar{N}$ interactions. Estimates of T_R made in the past[19, 20] utilized a schematic OBEP (One Boson Exchange Potential) for V_{NN} and $Re\,V_{N\bar{N}}$, together with the short range complex $N\bar{N}$ potential of Richard and Sainio[22]. The T_R values (in units of 10^{23} sec^{-1}) obtained for ^{16}O and ^{56}Fe are, within the allowed range of the model parameters:

$$^{16}\mathrm{O}: \quad 1.80 \leq T_R \leq 2.12 \tag{40}$$

$$^{56}\mathrm{Fe}: \quad 2.49 \leq T_R \leq 4.76, \tag{41}$$

the first one being comparable with the estimate of Dover *et al.*[15], previously reported in subsection 3.1.

This method has mainly two advantages: it avoids the closure approximation and, at the same time, it allows a sensible estimate of T_R for any mass number A, by simply changing the nuclear density parameters. Obviously more specific shell effects on the dynamics of the neutron and antineutron involved in the process cannot be accounted for. One could notice that T_R turns out to be a rather smooth function of A, growing from about 1 to 2.5 in the mass range up to 60, and then asymptotically reaching a value of about 3 for the largest mass numbers (all values are in units of 10^{23} sec^{-1}).

Concerning point (b), we have evaluated[21] the r.h.s. of (33) within the nuclear shell model description, implemented by the closure approximation; the average quantities entering into (36) are thus defined as:

$$< V_n > \;=\; \frac{1}{2A}\sum_{\{\alpha\}} < \alpha_1\alpha_2|V_{NN}|\alpha_1\alpha_2 >_{\mathcal{A}} \qquad (42)$$

$$< V_{\bar{n}} > \;=\; \frac{1}{2A}\sum_{\{\alpha\}} < \alpha_1\alpha_2|Re\,V_{N\bar{N}}|\alpha_1\alpha_2 >_{\mathcal{D}} \qquad (43)$$

$$< \Gamma_a > \;=\; -\frac{1}{A}\sum_{\{\alpha\}} < \alpha_1\alpha_2|Im\,V_{N\bar{N}}|\alpha_1\alpha_2 >_{\mathcal{D}} \qquad (44)$$

where the α_i's label all quantum numbers (n, l, j, τ) which identify a single particle state in the Harmonic Oscillator shell model and the summations are extended to the set $\{\alpha\}$ of states which are occupied in the (initial) ground state of the considered nucleus.

Concerning the nucleon–nucleon (antinucleon) interaction, in ref.[21] the most recent parameterization of the NN and $N\bar{N}$ Paris potentials has been employed (with the customary spin, spin–orbit, tensor and quadratic spin–orbit contributions): the latter was fitted to the whole available set of $p\bar{p}$ data, including polarization observables.

Calculations have been performed again in ^{16}O and ^{56}Fe (in order to update previous estimates and compare with the above mentioned nuclear matter approach) and also in the ^{40}Ar nucleus, in view of the future measurements of nuclear instability expected from ICARUS[23]. The results are the following (the range for the estimated T_R values stemming from different sets of parameters of the $N\bar{N}$ Paris potential, which fit the $p\bar{p}$ data with comparable accuracy):

$$^{16}\mathrm{O} \qquad T_R = (1.7 \div 2.6) \times 10^{23}\mathrm{sec}^{-1} \qquad (45)$$

$$\text{which, for } T_A > 4.3 \times 10^{31}\mathrm{yr} \quad (90\% CL, ref.[16])$$

$$\text{ensues}: \quad \tau_{n\bar{n}} > (0.7 \div 0.9) \times 10^{8}\mathrm{sec}; \qquad (46)$$

$$
\begin{array}{llcll}
{}^{56}\mathrm{Fe} & T_R & = & (2.2 \div 3.4) \times 10^{23}\mathrm{sec}^{-1} & (47) \\
\text{which, for} & T_A & > & 6.5 \times 10^{31}\mathrm{yr} \quad (90\% CL, ref.[24]) & \\
\text{ensues :} & \tau_{n\bar{n}} & > & (0.8 \div 1.0) \times 10^{8}\mathrm{sec}; & (48)
\end{array}
$$

and finally

$$
{}^{40}\mathrm{Ar} \qquad T_R = (2.1 \div 3.2) \times 10^{23}\mathrm{sec}^{-1} \tag{49}
$$

which will provide limits on $\tau_{n\bar{n}}$ once the first ICARUS data will be available for this channel.

In concluding this section one important remark concerns the relative stability of the estimated nuclear suppression factor: different models and approximations for both the nuclear model description and the neutron/antineutron interactions with the nuclear medium provide T_R values which *do not differ* by more than $10 \div 15\%$. Even by allowing for somewhat larger deviations, one can see that the expected lower limit for the $n\bar{n}$ oscillation time is now of the order of 10^8 sec, which makes it quite interesting the possibility of exploring free neutron oscillation times up to one or two orders of magnitude larger than in previous experiments.

4. Criticism and conclusions

In recent years there have been some unjustified rumors about the nuclear suppression factor being of order unity, thus leading to limits for the $n\bar{n}$ oscillation time of the same order as the nuclear instability limits[25], which would make hopless any direct search for free neutron–antineutron oscillations.

This author claims that the S–matrix approach is inconsistent with the potential model approach, thus pointing to an incorrect use of the latter for the evaluation of $n\bar{n}$ oscillations in the nucleus; this "inconsistency" has been clearly disproved by Krivoruchenko[26], with a careful analysis of the nuclear decay probability as determined by the time dependent S–matrix evolution of the nuclear states.

However an even more unfortunate mistake occurs in ref.[25] within the attempt of establishing a comparison with the potential model results: he obtains the following probability of $n\bar{n}$ conversion with subsequent $\bar{n}$–nucleus decay:

$$
W(t) = \epsilon_{n\bar{n}}^2 t^2 \left(\frac{1}{2} + \frac{1}{\Gamma t} + \frac{1}{\Gamma^2 t^2} \left[e^{-\Gamma t} - 1 \right] \right) \tag{50}
$$

where the nuclear annihilation width is (more or less correctly) assumed to be $\Gamma \simeq 100$ MeV. Then the two extreme situations are considered:

- $\Gamma t << 1$, identified with some low density limit (and thus free space oscillations), in which case (50) reduces to $W(t) \simeq \epsilon_{n\bar{n}}^2 t^2$;

- $\Gamma t >> 1$, which should refer to the decay in the nuclear medium and gives, from (50), $W(t) \simeq \frac{1}{2}\epsilon_{n\bar{n}}^2 t^2$.

According to this calculation the nuclear suppression factor would turn out to be of order 1/2 and $\tau_{n\bar{n}}$ of the same order as the nuclear decay lifetime (10^{31} yr)!

But, as clearly pointed out in ref.[27], formula (50) is **wrong**: within the same assumptions of [25] the correct expression should read:

$$W(t) = \frac{4\epsilon_{n\bar{n}}}{\Gamma} t \left[1 - \frac{2}{\Gamma t}\left(1 - e^{-\Gamma t/2}\right)\right], \tag{51}$$

from which the limit $\Gamma t >> 1$ (oscillations in the medium) turns out to be:

$$W(t) \simeq \frac{4\epsilon_{n\bar{n}}^2}{\Gamma} t \tag{52}$$

and the corresponding nuclear lifetime:

$$T_A = \frac{\Gamma}{4}\tau_{n\bar{n}}^2 \equiv \frac{\Gamma}{4\epsilon_{n\bar{n}}}\tau_{n\bar{n}} >> \tau_{n\bar{n}}, \tag{53}$$

in qualitative (and semiquantitative) agreement with all previous evaluations in the existing literature.

A more recent calculation[28], based on the non–relativistic diagram technique, has challenged the existing estimates of the intranuclear transition rates, suggesting quite smaller nuclear quenching factors T_R and thus a lower limit for the $n\bar{n}$ oscillation time which is about two orders of magnitude larger than, e.g., (25). Should this calculation be correct, this lower limit would be comparable with the planned sensitivity of the proposed ORNL experiment[2], thus seriously affecting the interest in pursuing it. The calculation is carried out in deuterium and then extended to heavier nuclei. It was suggested[29] that the approach of ref.[28] neglects the important effect of the surface peacking of antineutron wavefunctions in nuclei. Moreover the nuclear effects on the antineutron are only taken into account through an average binding energy of about 8 MeV, which largely underestimate the $\bar{n}$ energy eigenvalue obtained with realistic potentials. Indeed Hüfner and Kopeliovic[30], using the same approach as in ref.[28], find for the annihilation rate of antineutrons in a nucleus the following expression:

$$\frac{\Gamma_A}{N} = \frac{\epsilon_{n\bar{n}}^2}{T_R} = \frac{M\epsilon_{n\bar{n}}^2}{A\pi Im a_{\bar{n}N}} \int d^2x \frac{\rho(\vec{x})}{\tilde{\rho}(\vec{x})} \frac{1}{1+K^2(\vec{x})}, \tag{54}$$

where

$$K(\vec{x}) = \frac{\Delta E + \omega}{(2\pi/M)\tilde{\rho}(\vec{x})|Im a_{\bar{n}N}|} \,. \tag{55}$$

In the above the dominant dependence upon the imaginary part of the $\bar{n}-N$ scattering length ($a_{\bar{n}N}$) turns out to be the explicit inverse proportionality rather than one embodied in K and averaged over the nuclear density. Hence the results of ref.[30] are in close agreement with the ones of [15] or [21] and re-establish a lower limit for $\tau_{n\bar{n}}$ in the range $(1 \div 2) \times 10^8$ sec.

A more serious criticism can be developed about the extraction of the "free" $n\bar{n}$ oscillation time from measurements of nuclear lifetime: the oscillation process indeed could be intrinsically modified in the presence of the nuclear medium. This was qualitatively pointed out long ago by Kabir[31] and semi–quantitatively considered by Dover *et al.*[15].

i) Based on the analogy between $n\bar{n}$ and β decay processes (both being induced by essentially zero range operators at very low momentum transfers) one can argue that the possible medium modifications are of long range character, like the ones, e.g., induced by two–body pionic correlations. The estimated correction to the transition mass should not exceed $15 \div 30\%$[15], which would not be a dramatic effect.

ii) Another source of uncertainty for the evaluation of $n\bar{n}$ oscillations in the medium can be ascribed to the occurrence of alternative direct $\Delta B = 2$ processes, which are suppressed in free space because of energy conservation[32]; typically one can envisage generic neutron–antibaryon oscillations like:

$$n \longrightarrow \bar{N}\pi \tag{56}$$

$$n \longrightarrow \bar{\Lambda} \tag{57}$$

which could lead to the same decay products but, in order to occur, require energy balance from the interactions with the nuclear medium. No quantitative estimates of the decay widths associated to these processes exist, but they should incoherently add to the one considered till now (both within the optical model and the S–matrix approach); as a result the limits on $\tau_{n\bar{n}}$ derived from nuclear decay observations would be less restrictive than presently believed, but still the estimated limit can be regarded as a conservative one.

In conclusion one can consider the existing estimate of the nuclear suppression factor as fairly reliable, although further improvements can still be pursued; but it is clear by now that these will not modify the order of magnitude of the effect. On the other hand the above mentioned ambiguities in the definition of $n\bar{n}$ transitions inside the nuclear medium can be clarified and serious theoretical estimates of these nuclear modifications should be provided. We strongly need further experiments to go on, both on the issue of *free* $n\bar{n}$ oscillations and of nuclear instability search: hopefully theory can provide a reliable bridge in both directions.

References

[1] Mohapatra R N 1996 *Proc. Int. Workshop on Future Prospects of Baryon Instability Search* (Oak Ridge, TN, 28–30 Mar 1996) p 73–88; Mohapatra R N 1998, hep–ph/9802388

[2] Kamyshkov Yu A 1996 *Proc. Int. Workshop on Future Prospects of Baryon Instability Search* (Oak Ridge, TN, 28–30 Mar 1996)

[3] Totsuka Y 1996 *Nucl. Phys. Proc. Suppl.* **48** p 547–56

[4] Cline David B 1996 *Proc. Int. Workshop on Future Prospects of Baryon Instability Search* (Oak Ridge, TN, 28–30 Mar 1996) p 199–210

[5] Baldo–Ceolin M *et al* 1990 *Phys. Lett.* **B263** p 95

[6] Baldo–Ceolin M *et al* 1994 *Z. Phys.* **C 63** p 409

[7] Kamyshkov Yu A *et al* 1996 *Nucl. Phys. Proc. Suppl.* **48** p 460–62

[8] Kamyshkov *et al* 1995 PSI-Proc.95-02, P-10

[9] Kuzmin V A 1970 *JEPT Lett.* **12** p 228

[10] Mohapatra R N *et al* 1980 *Phys. Rev. Lett.* **44** p 1316; ——1980 *Phys. Lett.* **B94** p 183

[11] Chetyrkin K G *et al* 1981 *Phys. Lett.* **99B** p 358

[12] Riazuddin 1981 VPI-HEP–81/4

[13] Sandars P G 1980 *J. Phys. G: Nucl. Phys.* **6** p L161

[14] Dover CB, Gal A and Richard JM 1983 *Phys. Rev.* **D27** p 1090

[15] Dover CB, Gal A and Richard JM 1985 *Phys. Rev.* **C31** p 1423

[16] Takita H *et al* (Kamiokande) 1986 *Phys. Rev.* **D34** p 902

[17] Jones TW *et al* 1984 *Phys. Rev. Lett.* **52** p 720

[18] Wong CY, Kerman AK, Satchler GR and MacKellar AD 1984 *Phys. Rev.* **C29** p 574

[19] Alberico W M, Bottino A and Molinari A 1982 *Phys. Lett.* **114B** p 266

[20] Alberico W M, Bernabeu J, Bottino A and Molinari A 1984 *Nucl. Phys.* **A429** p 445

[21] Alberico W M, De Pace A and Pignone M 1991 *Nucl. Phys.* **A523** p 488

[22] Richard J M and Sainio M E 1982 *Phys. Lett.* **110B** p 349

[23] see the contribution of Mauri F (ICARUS collaboration) in these Proceedings

[24] Berger Ch *et al* (Fréjus coll.) 1989 *Phys. Lett.* **B240** p 237

[25] Nazaruk V I 1995 *Phys. Lett.* **337B** p 328

[26] Krivoruchenko M I Preprint ITEP–95–37/March 1995

[27] Dover C B, Gal A and Richard J M 1995 *Phys. Lett.* **B344** p 433

[28] Kondratyuk L A 1996 *JETP Lett.* **64** p 495

[29] Richard J M 1996 *Proc. Int. Workshop on Future Prospects of Baryon Instability Search* (Oak Ridge, TN, 28–30 Mar 1996) p 241–43

[30] Hüfner J and Kopeliovich Boris Z 1998 *Proc. of the Workshop on Future Directions in Quark Nuclear Physics*, Adelaide, March 9–20, 1998

[31] Kabir P K 1983 *Phys. Rev. Lett.* **51** p 231

[32] Basecq J and L. Wolfenstein L 1983 *Nucl. Phys.* **B224** p 21

PART 2

B and L Violation, Neutrino Mass and Oscillation, Proton Decay

The Status of Neutrino Physics

José W. F. Valle†[1]

†Instituto de Física Corpuscular - C.S.I.C.
Departament de Física Teòrica, Universitat de València
46100 Burjassot, València, Spain
http://neutrinos.uv.es

Abstract.
I review the status of neutrino physics post-Neutrino 98, including the implications of solar and atmospheric neutrino data, which strongly indicate nonzero neutrino masses. LSND and the possible role of neutrinos as hot dark matter (HDM) are also mentioned. The simplest schemes proposed to reconcile these requirements invoke a light sterile neutrino in addition to the three active ones, two of them at the MSW scale and the other two maximally-mixed neutrinos at the HDM/LSND scale. The latter scale arises at one-loop, while the solar and atmospheric parameters $\Delta m^2{}_\odot$ & $\Delta m^2{}_{atm}$ appear at the two-loop level. The lightness of the sterile neutrino, the nearly maximal atmospheric neutrino mixing, and the generation of $\Delta m^2{}_\odot$ & $\Delta m^2{}_{atm}$ follow naturally from the assumed lepton-number symmetry and its breaking. There are two basic schemes which can be distinguished at future solar & atmospheric neutrino experiments and have different cosmological implications.

1. Introduction

Neutrinos are the only fermions which the Standard Model (SM) predicts to be massless. This ansatz was justified due to the apparently masslessness of neutrinos in most experiments. However, the situation has changed due to

[1] E-mail: valle@flamenco.ific.uv.es

the important impact of underground experiments, since the pioneer geochemical experiments of Davis and collaborators, to the more recent Gallex, Sage, Kamiokande and SuperKamiokande experiments [1, 2, 3, 4, 5]. Altogether they provide solid evidence for the solar and the atmospheric problems, two milestones in the search for physics beyond the SM. Of particular importance has been the recent confirmation by the SuperKamiokande collaboration [3] of the atmospheric neutrino zenith-angle dependent deficit, which has marked a turning point in our understanding of neutrinos, providing a strong evidence for neutrino mass. In addition to the neutrino data from underground experiments there is also some possible indication for neutrino oscillations from the LSND experiment [6]. To this we may add the possible role of neutrinos in the resolution of the dark matter problem hinted at by structure formation considerations [7, 8, 9]. If one boldly insists in including also the last two requirements, together with the data on solar and atmospheric neutrinos, then we have *three mass scales* involved in neutrino oscillations. The simplest way to reconcile these requirements invokes the existence of a light sterile neutrino [10, 11, 12]. The prototype models proposed in [10, 11] enlarge the $SU(2) \otimes U(1)$ Higgs sector in such a way that neutrinos acquire mass radiatively, without unification nor seesaw. Out of the four neutrinos, two of them lie at the MSW scale and the other two maximal mixed neutrinos are at the HDM/LSND scale. The latter scale arises at one-loop, while the solar and atmospheric scales come in at the two-loop level. The lightness of the sterile neutrino, the nearly maximal atmospheric neutrino mixing, and the generation of the solar and atmospheric neutrino scales all result naturally from the assumed lepton-number symmetry and its breaking. Either ν_e - ν_τ conversions explain the solar data with ν_μ - ν_s oscillations accounting for the atmospheric deficit [10], or else the roles of ν_τ and ν_s are reversed [11]. These two basic schemes have distinct implications at future solar & atmospheric neutrino experiments, as well as cosmology.

2. Theories of Neutrino Mass

One of the most unpleasant features of the SM is that the masslessness of neutrinos is not dictated by an underlying *principle*, such as that of gauge invariance in the case of the photon: the SM simply postulates that neutrinos are massless by choosing a restricted multiplet content. *Why are neutrinos so special when compared with the other fundamental fermions?* If massive, neutrinos would present another puzzle: *Why are their masses so small compared to those of the charged fermions?* The fact that neutrinos are the only electrically neutral elementary fermions may hold the key to the answer, namely neutrinos could be Majorana fermions, the most fundamental ones. In this case the suppression of their mass could be associated to the breaking of lepton number symmetry at a very large energy scale within a *unification approach*, which can be implemented in many extensions of the SM. Alternatively, neutrino masses could arise from garden-variety *weak-scale physics* characterized by a scale $\langle\sigma\rangle = \mathcal{O}\ (m_Z)$ where $\langle\sigma\rangle$ denotes a $SU(2) \otimes U(1)$ singlet vacuum expectation value which owes its smallness to the symmetry enhancement which would result if $\langle\sigma\rangle$ and $m_\nu \to 0$.

If neutrinos are Majorana, then there should be neutrino-less double beta decay and vice-versa, according to a general theorem [13], discussed here in other talks. The present status of experimental searches as well as the prospects towards a large scale double beta decay experiment, named GENIUS, and its physics potential have been discussed in ref. [14] and presented here by Klapdor.

One should realize however that, although the physics of neutrinos can be rather different in various gauge theories of neutrino mass, there is in general no predictive power on masses and mixings, this is one of the aspects of the so-called flavour problem which is probably the toughest open problem in physics.

2.1. Unification Approach

An attractive possibility is to ascribe the origin of parity violation in the weak interaction to the spontaneous breaking of B-L symmetry in the context of left-right symmetric extensions such as the $SU(2)_L \otimes SU(2)_R \otimes U(1)$[15], $SU(4) \otimes SU(2) \otimes SU(2)$ [16] or $SO(10)$ gauge groups [17]. In this case the masses of the light neutrinos are obtained by diagonalizing the following mass matrix in the basis ν, ν^c

$$\begin{bmatrix} M_L & D \\ D^T & M_R \end{bmatrix} \tag{1}$$

where D is the standard $SU(2) \otimes U(1)$ breaking Dirac mass term and $M_R = M_R^T$ is the isosinglet Majorana mass that may arise from a 126 vacuum expectation value (vev) in $SO(10)$. The magnitude of the $M_L \nu\nu$ term [18] is also suppressed by the left-right breaking scale, $M_L \propto 1/M_R$ [15].

In the seesaw approximation, one finds

$$M_{\nu\, eff} = M_L - D M_R^{-1} D^T \ . \tag{2}$$

As a result one is able to explain naturally the relative smallness of neutrino masses since $m_\nu \propto 1/M_R$. Although M_R is expected to be large, its magnitude heavily depends on the model and it may have different possible structures in flavour space (so-called textures) [19]. As a result it is hard to make firm predictions for the corresponding light neutrino masses and mixings that are generated through the seesaw mechanism. In fact this freedom has been exploited in model building in order to account for an almost degenerate seesaw-induced neutrino mass spectrum [20].

One virtue of the unification approach is that it may allow one to gain a deeper insight into the flavour problem. There have been interesting attempts at formulating supersymmetric unified schemes with flavour symmetries and texture zeros in the Yukawa couplings. In this context a challenge is to obtain the large lepton mixing now indicated by the atmospheric neutrino data.

2.2. Weak-Scale Approach

Although very attractive, the unification approach is by no means the only way to generate neutrino masses. There are many schemes which do not require any large mass scale. The extra particles employed to generate the neutrino masses have masses $\mathcal{O}\ (m_Z)$ accessible to present experiments. There is a variety of such mechanisms, in which neutrinos acquire mass either at the tree level or radiatively. Let us look at some.

2.2.1. Tree Level For example, it is possible to extend the lepton sector of the $SU(2) \otimes U(1)$ theory by adding a set of *two* 2-component isosinglet neutral fermions, denoted ${\nu^c}_i$ and S_i, $i = e,\ \mu$ or τ in each generation. In this case one can consider the mass matrix (in the basis ν, ν^c, S) [21]

$$\begin{bmatrix} 0 & D & 0 \\ D^T & 0 & M \\ 0 & M^T & \mu \end{bmatrix} \tag{3}$$

The Majorana masses for the neutrinos are determined from

$$M_L = DM^{-1}\mu M^{T^{-1}} D^T \tag{4}$$

In the limit $\mu \to 0$ the exact lepton number symmetry is recovered and will keep neutrinos strictly massless to all orders in perturbation theory, as in the SM. The corresponding texture of the mass matrix has been suggested in various theoretical models [22], such as superstring inspired models [23]. In the latter the zeros arise due to the lack of Higgs fields to provide the usual Majorana mass terms. The smallness of neutrino mass then follows from the smallness of μ. The scale characterizing M, unlike M_R in the seesaw scheme, can be low. As a result, in contrast to the heavy neutral leptons of the seesaw scheme, those of the present model can be light enough as to be produced at high energy colliders such as LEP [24] or at a future Linear Collider. The smallness of μ is in turn natural, in t'Hooft's sense, as the symmetry increases when $\mu \to 0$, i.e. total lepton number is restored. This scheme is a good alternative to the smallness of of neutrino mass, as it bypasses the need for a large mass scale, present in the seesaw unification approach. One can show that, since the matrices D and M are not simultaneously diagonal, the leptonic charged current exhibits a non-trivial structure that cannot be rotated away, even if we set $\mu \equiv 0$. The phenomenological implication of this, otherwise innocuous twist on the SM, is that there is neutrino mixing despite the fact that light neutrinos are strictly massless. It follows that flavour and CP are violated in the leptonic currents, despite the masslessness of neutrinos. The loop-induced lepton flavour and CP non-conservation effects, such as $\mu \to e + \gamma$ [25, 26], or CP asymmetries in lepton-flavour-violating processes such as $Z \to e\bar{\tau}$ or $Z \to \tau\bar{e}$ [27] are precisely calculable. The resulting rates may be of experimental interest [28, 29, 30], since they are not constrained by the bounds on neutrino mass, only by those on universality, which are relatively poor. In short, this is a conceptually simple and phenomenologically rich scheme.

Another remarkable implication of this model is a new type of resonant neutrino conversion mechanism [31], which was the first resonant mechanism to be proposed after the MSW effect [32], in an unsuccesful attempt to bypass the need for neutrino mass in the resolution of the solar neutrino problem. According to the mechanism, massless neutrinos and anti-neutrinos may undergo resonant flavour conversion, under certain conditions. Though these do not occur in the Sun for the ν_e to ν_μ or ν_τ channels, they can be realized in the chemical environment of supernovae [33] or, in media with normal chemical composition like the Earthen for the ν_μ to ν_τ channel. Recently it has been pointed out how the resonant ν_e to ν_τ conversions may provide an elegant approach for explaining the observed velocity of pulsars [34], while ν_μ to ν_τ conversions in the Earth provide an elegant explanation of the atmospheric zenith-angle-dependent neutrino deficit [35]. For details on the simplest explanation of the recent SK data in terms of massless neutrinos see ref. [35].

2.2.2. Radiative Level There is also a large variety of *radiative* models, where the $SU(2) \otimes U(1)$ multiplet content is extended in order to generate neutrino masses. The prototype one-loop scheme is the one proposed by Zee [36]. Supersymmetry with explicitly broken R-parity also provides an alternative one-loop mechanism to generate neutrino mass. These arise, for example, from scalar quark or scalar lepton contributions, as shown in Fig. (1) [37].

A two-loop scheme to induce neutrino mass was suggested by Babu [38]. The relevant diagram is shown in Fig. (2). [2].

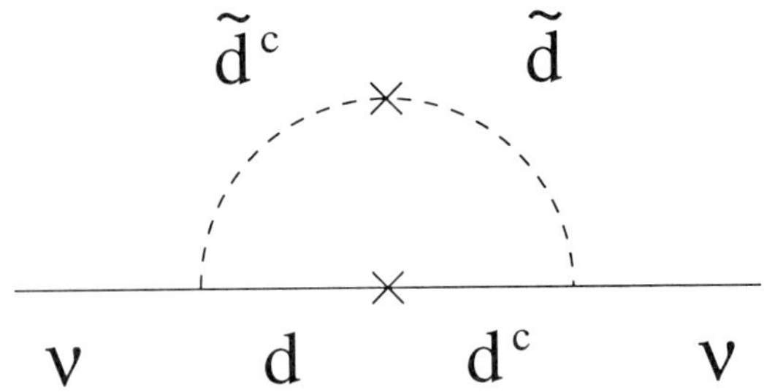

Figure 1. One-loop-induced Neutrino Mass.

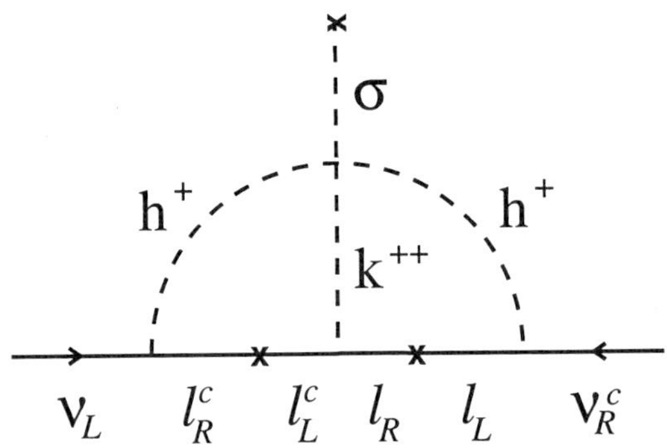

Figure 2. Two-loop-induced Neutrino Mass.

In the above examples active neutrinos acquire radiative mass. One can also employ the radiative approach to construct models including sterile neutrinos, such as those in ref. [10, 11]. In this case some new Feynman diagram topologies are encountered.

2.3. A Hybrid Approach

I now describe an interesting mechanism of neutrino mass generation that combines seesaw and radiative mechanisms. It invokes supersymmetry with broken R-parity, as the origin of neutrino mass and mixings [40]. The simplest model is a unified minimal supergravity model with universal soft breaking parameters (MSUGRA) and bilinear breaking of R–parity [40, 41]. Contrary to a popular misconception, the bilinear violation of R–parity implied by the ϵ_3 term in the superpotential is physical, and can not be

[2] Note here that I have used the slight variant of the Babu model suggested in ref. [39], which incorporates the idea of spontaneous, rather than explicit lepton number violation

rotated away [42]. It leads also by a minimization condition, to a non-zero sneutrino vev, v_3. It is well-known [43] that in such models of broken R–parity the tau neutrino ν_τ acquires a mass, due to the mixing between neutrinos and neutralinos. It comes from the matrix

$$\begin{bmatrix} M_1 & 0 & -\frac{1}{2}g'v_d & \frac{1}{2}g'v_u & -\frac{1}{2}g'v_3 \\ 0 & M_2 & \frac{1}{2}gv_d & -\frac{1}{2}gv_u & \frac{1}{2}gv_3 \\ -\frac{1}{2}g'v_d & \frac{1}{2}gv_d & 0 & -\mu & 0 \\ \frac{1}{2}g'v_u & -\frac{1}{2}gv_u & -\mu & 0 & \epsilon_3 \\ -\frac{1}{2}g'v_3 & \frac{1}{2}gv_3 & 0 & \epsilon_3 & 0 \end{bmatrix} \tag{5}$$

where the first two rows are gauginos, the next two Higgsinos, and the last one denotes the tau neutrino. The v_u and v_d are the standard MSUGRA vevs, $g's$ are gauge couplings and $M_{1,2}$ are the gaugino mass parameters. Since the ϵ_3 and the v_3 are related, the simplest (one-generation) version of this model contains only one extra free parameter in addition to those of the MSUGRA model. The universal soft supersymmetry-breaking parameters at the unification scale m_X are evolved via renormalization group equations down to the weak scale $\mathcal{O}\,(m_Z)$. This induces an effective non-universality of the soft terms *at the weak scale* which in turn implies a non-zero sneutrino vev v_3' given as

$$v_3' \approx \frac{\epsilon_3 \mu}{m_Z{}^4}\left(v_d' \Delta M^2 + \mu' v_u \Delta B\right) \tag{6}$$

where the primed quantities refer to a basis in which we eliminate the ϵ_3 term from the superpotential (but reintroduce it, of course, in other sectors of the theory).

The scalar soft masses and bilinear mass parameters obey $\Delta M^2 = 0$ and $\Delta B = 0$ at m_X. However at the weak scale they are calculable from radiative corrections as

$$\Delta M^2 \quad \approx \quad \frac{3h_b^2}{8\pi^2} m_Z^2 \ln \frac{M_{GUT}}{m_Z} \tag{7}$$

Note that eq. (6) implies that the R–parity-violating effects induced by v_3' are *calculable* in terms of the primordial R–parity-violating parameter ϵ_3. It is clear that the universality of the softs plays a crucial role in the calculability of the v_3' and hence of the resulting neutrino mass [40]. Thus eq. (5) represents a new kind of see-saw scheme in which the M_R of eq. (1) is the neutralino mass, while the rôle of the Dirac entry D is played by the v_3', which is induced radiatively as the parameters evolve from m_X to the weak scale. Thus we have a *hybrid* see-saw mechanism, with naturally suppressed Majorana ν_τ mass induced by the mixing between the weak eigenstate tau neutrino and the *zino*.

Let me now turn to estimate of the expected ν_τ mass. For this purpose let me first determine the tau neutrino mass in the most general supersymmetric model with bilinear breaking of R-parity, *without imposing soft*

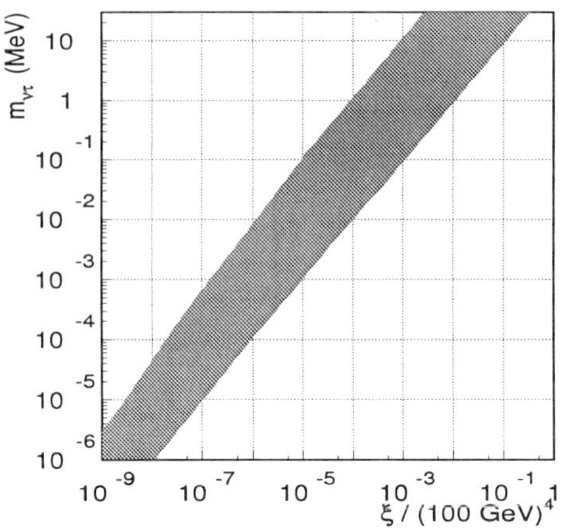

Figure 3. Tau neutrino mass versus $\xi \equiv (\epsilon_3 v_d + \mu v_3)^2$, from ref. [40]

universality. The ν_τ mass depends quadratically on an effective parameter ξ defined as $\xi \equiv (\epsilon_3 v_d + \mu v_3)^2 \propto {v'_3}^2$ characterizing the violation of R–parity. The expected m_{ν_τ} values are illustrated in Fig. (3). The band shown in the figure is obtained through a scan over the parameter space requiring that the supersymmetric particles are not too light. Let us now compare this with the cosmologically allowed values of the tau neutrino mass. The cosmological critical density bound $m_{\nu_\tau} \lesssim 92 \Omega h^2$ eV only holds if neutrinos are stable. In the present model (with 3-generations) the ν_τ can decay into 3 neutrinos, via the neutral current [18, 44], or by slepton exchanges. This decay will reduce the relic ν_τ abundance to the required level, as long as ν_τ is heavier than about 100 keV or so. On the other hand primordial Big-Bang nucleosynthesis implies that ν_τ is lighter than about an Mev or so [45].

However, *if one adopts a SUGRA scheme where universality of the soft supersymmetry breaking terms at m_X is assumed*, then the ν_τ mass is theoretically *predicted* in terms of h_b and can be small in this case due to a natural cancellation between the two terms in the parameter ξ, which follows from the assumed universality of the softs at m_X. One can verify that m_{ν_τ} may easily lie in the electron-volt range, in which case ν_τ could be a component of the hot dark matter of the Universe.

Notice that ν_e and ν_μ remain massless in this approximation. They get masses either from scalar loop contributions in Fig. (1) [37] or by mixing with singlets in models with spontaneous breaking of R-parity [46]. It is important to notice that even when m_{ν_τ} is small, many of the corresponding R-parity violating effects can be sizeable. An obvious example is the fact that the lightest neutralino decay will typically decay inside the detector, unlike standard R-parity-conserving supersymmetry. This leads to a vastly unexplored plethora of phenomenological possibilities in supersymmetric physics [47].

In conclusion I can say that, other than the seesaw scheme, none of the above models requires a large mass scale. As a result they lead to a potentially rich phenomenology, since the extra particles required have masses at scales that could be accessible to present experiments. In the simplest versions of these models the neutrino mass arises from the explicit violation of lepton number. Their phenomenological potential gets richer if one generalizes the models so as to implement a spontaneous violation scheme. This brings me to the next section.

2.4. Weak-scale majoron

If lepton number (or B-L) is an ungauged symmetry and if it is arranged to break spontaneously, the generation of neutrino masses will be accompanied by a the existence of a physical Goldstone boson that we generically call majoron. Except for the left-right symmetric unification approach, in which B-L is a gauge symmetry, in all of the above schemes one can implement the spontaneous violation of lepton number. One can also introduce it in an $SU(2) \otimes U(1)$ seesaw framework [48], as originally proposed, but I do not consider this case here, see ref. [49] for a review. Here I will mainly concentrate on weak-scale physics. In all models I consider the lepton-number breaks at a scale given by a vacuum expectation value $\langle\sigma\rangle \sim m_{weak}$. Such scale arises as the most natural one since in all of these models, as already mentioned, we have that the neutrino masses vanish as the lepton-breaking scale $\langle\sigma\rangle \to 0$ [50].

It is also clear that in any acceptable model one must arrange for the majoron to be mainly an $SU(2) \otimes U(1)$ singlet, ensuring that it does not affect the invisible Z decay width, well-measured at LEP. In models where the majoron has L=2 the neutrino mass is proportional to an insertion of $\langle\sigma\rangle$, as indicated in Fig. (2). In the supersymmetric model with broken R-parity the majoron is mainly a singlet sneutrino, which has lepton number L=1, so that $m_\nu \propto \langle\sigma\rangle^2$, where $\langle\sigma\rangle \equiv \langle\widetilde{\nu^c}\rangle$, with $\widetilde{\nu^c}$ denoting the singlet sneutrino. The presence of the square, just as in the parameter ξ in Fig. (3), reflects the fact that the neutrino gets a Majorana mass which has lepton number L=2. The sneutrino gets a vev at the effective supersymmetry breaking scale $m_{susy} = m_{weak}$.

The weak-scale majorons may have remarkable phenomenological implications, such as the possibility invisibly decaying Higgs bosons [50]. Unfortunately I have no time to discuss it here (see, for instance [47]).

If the majoron acquires a KeV mass (natural in weak-scale models) from gravitational effects at the Planck scale [51] it may play a role in cosmology as dark matter [52]. In what follows I will just focus on two examples of how the underlying physics of weak-scale majoron models can affect neutrino cosmology in an important way.

2.4.1. Heavy neutrinos and the Universe Mass Neutrinos of mass less than $\mathcal{O}$ (100 KeV) or so, are cosmologically stable if they have only SM interactions. Their contribution to the present density of the universe implies [53]

$$\sum m_{\nu_i} \lesssim 92\,\Omega_\nu h^2\; eV\;, \tag{8}$$

where the sum is over all isodoublet neutrino species with mass less than $\mathcal{O}$ (1 MeV). The parameter $\Omega_\nu h^2 \leq 1$, where h^2 measures the uncertainty in the present value of the Hubble parameter, $0.4 \lesssim h \lesssim 1$, while $\Omega_\nu = \rho_\nu/\rho_c$, measures the fraction of the critical density ρ_c in neutrinos. For the ν_μ and ν_τ this bound is much more stringent than the laboratory limits.

In weak-scale majoron models the generation of neutrino mass is accompanied by the existence of a physical majoron, with potentially fast majoron-emitting decay channels such as [49, 54]

$$\nu' \to \nu + J\;. \tag{9}$$

as well as new annihilations to majorons,

$$\nu' + \nu' \to J + J\;. \tag{10}$$

These could eliminate relic neutrinos and therefore allow neutrinos of higher mass, as long as the rates are large enough to allow for an adequate red-shift of the heavy neutrino decay and/or annihilation products. While the annihilation involves a diagonal majoron-neutrino coupling g, the decays proceed only via the non-diagonal part of the coupling, in the physical mass basis. A careful diagonalization of both mass matrix and coupling matrix is essential in oder to avoid wild over-estimates of the heavy neutrino decay rates, such as that in ref. [48]. The point is that, once the neutrino mass matrix is diagonalized, there is a danger of simultaneously diagonalizing the majoron couplings to neutrinos. That would be analogous to the GIM mechanism present in the SM for the couplings of the Higgs to fermions. Models that avoid this GIM mechanism in the majoron-neutrino couplings have been proposed, e.g. in ref. [54]. Many of them are weak-scale majoron models [21, 50, 46]. A general method to determine the majoron couplings to neutrinos and hence the neutrino decay rates in any majoron model was first given in ref. [44]. For an estimate in the model with spontaneously broken R-parity [55] see ref. [46].

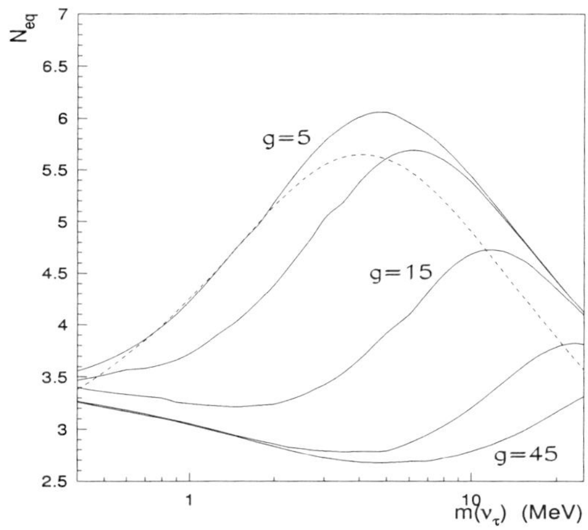

Figure 4. The dashed line shows the effective number of massless SM neutrinos equivalent to the heavy ν_τ ($g = 0$). Depending on the value of g (in units of 10^{-5}) one can lower N_ν below the canonical SM value $N_\nu = 3$ due to the effect of ν_τ annihilations. From ref. [59]

In short one may say that neutrino lifetimes can be shorter than required by the cosmological mass bound, for all values of the masses which are presently allowed by laboratory experiments.

2.4.2. Heavy neutrinos and Cosmological Nucleosynthesis Similarly, the number of light neutrino species is also restricted by cosmological Big Bang Nucleosynthesis (BBN). Due to its large mass, an MeV stable (lifetime longer than $\sim$ 100 sec) tau neutrino would be equivalent to several SM massless neutrino species and would therefore substantially increase the abundance of primordially produced elements, such as ^4He and deuterium [56, 57, 58]. This can be converted into restrictions on the ν_τ mass. If the bound on the effective number of massless neutrino species is taken as $N_\nu < 3.4 - 3.6$, one can rule out ν_τ masses above 0.5 MeV [45]. If we take $N_\nu < 4.5$ [58] the m_{ν_τ} limit loosens accordingly, as seen from Fig. (4), and allows a ν_τ of about an MeV or so.

In the presence of ν_τ annihilations the BBN m_{ν_τ} bound is substantially weakened or eliminated [59]. In Fig. (4) we also give the expected N_ν value for different values of the coupling g between ν_τ's and J's, expressed in units of 10^{-5}. Comparing with the SM $g = 0$ case one sees that for a fixed N_ν^{max}, a wide range of tau neutrino masses is allowed for large enough values of g. No ν_τ masses below the LEP limit can be ruled out, as long as g exceeds a few times 10^{-4}. One can also see from the figure

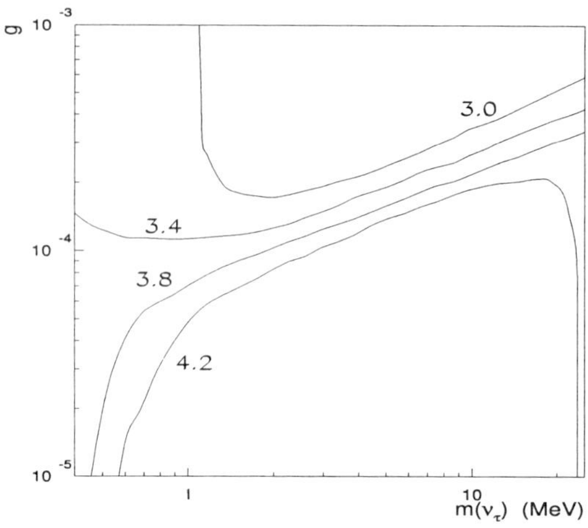

Figure 5. The region above each curve is allowed for the corresponding N_ν^{max}. From ref. [59]

that N_ν *can also be lowered below the canonical SM value* $N_\nu = 3$ due to the effect of the heavy ν_τ annihilations to majorons. These results may be re-expressed in the $m_{\nu_\tau} - g$ plane, as shown in figure 5. We note that the required values of $g(m_{\nu_\tau})$ fit well with the theoretical expectations of many weak-scale majoron models.

The above discussion has been on the effect of ν_τ annihilations to majorons in BBN. In some weak-scale majoron models decays in eq. (9) may lead to short enough ν_τ lifetimes that they may also play an important rôle in BBN [60].

Before concluding the discussion on majorons, let me comment that the majoron may be realized even in the context of models where B-L is a gauge symmetry, such as left-right-symmetric models, by suitably implementing a spontaneously broken global U(1) symmetry similar to lepton number. It plays an interesting role in such models as it allows the left-right scale to be relatively low [61].

3. Indications for Neutrino Mass

The most solid indications in favour of nonzero neutrino masses come from underground experiments on solar and atmospheric neutrinos. I will provide a theorist's sketch of the present experimental situation.

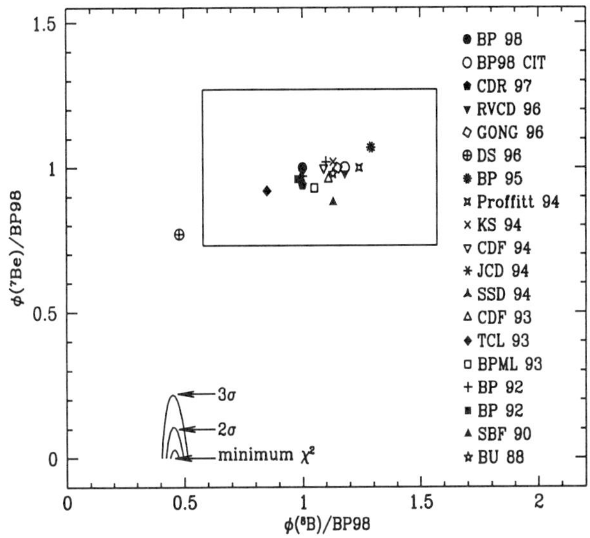

Figure 6. Recent SSM predictions, from ref. [62]

3.1. Solar Neutrinos

The puzzle posed by the data collected by the Homestake, Kamiokande, and the geochemical Gallex and Sage experiments still defy an explanation in terms of the Standard Model. The most recent data on rates are summarized as: 2.56 ± 0.23 SNU (chlorine), 72.2 ± 5.6 SNU (Gallex and Sage gallium experiments sensitive to the *pp* neutrinos), and $(2.44 \pm 0.10) \times 10^6 \mathrm{cm}^{-2}\mathrm{s}^{-1}$ (^{8}B flux from SuperKamiokande) [1]. This has been re-confirmed by the 504 days data sample now collected by the SuperKamiokande (SK) collaboration and reported at Neutrino 98 [5]. In Fig. (6) one can see the predictions of various standard solar models in the plane defined by the ^{7}Be and ^{8}B neutrino fluxes, normalized to the predictions of the BP98 solar model [63]. Abbreviations such as BP95, identify different solar models, as given in ref. [64]. The rectangular error box gives the 3σ error range of the BP98 fluxes. The values of these fluxes indicated by present data on neutrino event rates are also shown by the contours in the figure. The best-fit ^{7}Be neutrino flux is negative! Possible non-standard astrophysical solutions are strongly constrained by helioseismology studies [62] [65]. Within the standard solar model approach, the theoretical predictions clearly lie far from the best-fit solution, and even far from the 3σ contour, leading us to conclude that new particle physics is the only way to account for the data [66].

The most likely possibility is to assume the existence of neutrino conver-

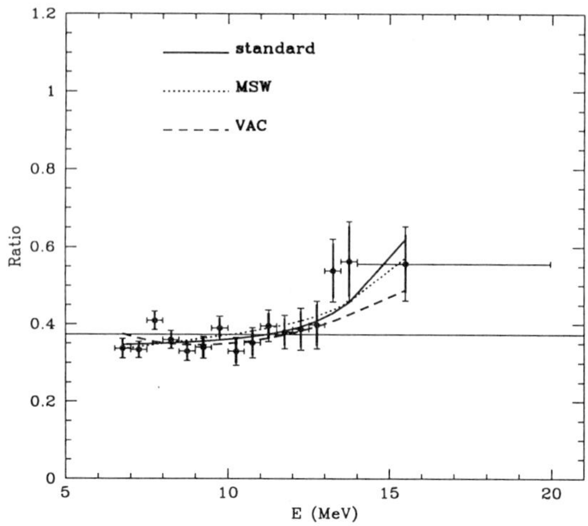

Figure 7. Combined ^{8}B plus *hep* energy spectrum from ref. [70]. The total flux of *hep* neutrinos was varied so as to obtain the best-fit for each scenario.

sions involving very small neutrino masses. The most attractive theoretical schemes are the MSW effect [32], vacuum neutrino oscillations or *just-so solution* and, possibly, the Spin-Flavour Precession mechanism proposed in ref. [67], aided by the Resonant enhancement due to matter effects in the Sun found in ref. [68]. The resulting RSFP mechanism still provides a viable solution to the solar neutrino problem [69].

The recent SK data updates the 300 days situation we had before Neutrino 98 [4] without major surprises, except that the SK collaboration has now given the first detailed report of the recoil energy spectrum produced by solar neutrino interactions [5]. The measured spectrum they reported at Neutrino 98 shows more events at the highest bins than would have been expected from the most popular neutrino oscillation parameters discussed previously. At first sight this might seem bad news for the oscillation scenarios. However, Bahcall and Krastev have noted that if the low energy cross section for ^{3}He $+\ p\ \to\ ^4$He $+\ e^+\ +\ \nu_e$, the so-called *hep* reaction, is $\gtrsim 20$ times larger than the best (but uncertain) theoretical estimates, then this reaction could significantly influence the electron energy spectrum produced by solar neutrino interactions in the high recoil region. This would hardly have any effect at lower energies. They compare the predicted energy spectra for different assumed *hep* fluxes and different neutrino oscillation scenarios with the one measured at SuperKamiokande. Fig. 7 shows the ratio of the measured [5] to the calculated number of

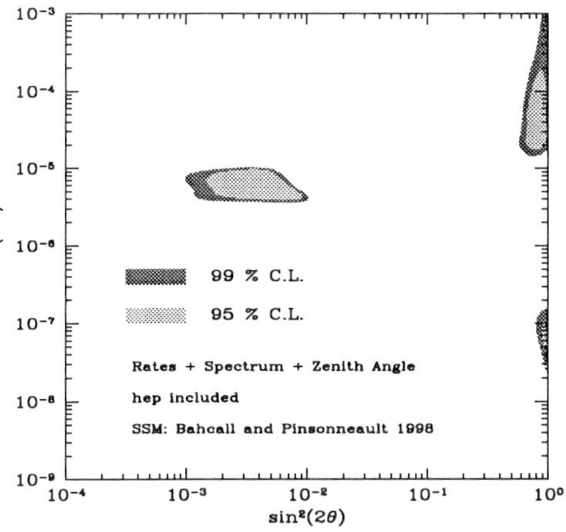

Figure 8. Presently allowed MSW solar neutrino parameters for 2-flavour active neutrino conversions with an enhanced *hep* flux, from ref. [70]

events with electron recoil energy E. The crosses are the recent SK measurements [5], while the calculated curves are global fits to all of the data. The horizontal line at Ratio = 0.37 represents the ratio of the total event rate measured by SuperKamiokande to the predicted event rate [63] with no oscillations and only ^{8}B neutrinos. One sees how the spectra with enhanced *hep* contributions provide better fits to the SK data, suggesting that these neutrinos may be playing a rôle.

One can determine the required solar neutrino parameters Δm^2 and $\sin^2 2\theta$ through a χ^2 fit of the experimental data. In Fig. (8) we show the allowed two-flavour regions obtained in an updated MSW global fit analysis of the solar neutrino data for the case of active neutrino conversions. The data include the chlorine, Gallex, Sage [1] and SK total event rates [5], the SK energy spectrum [5], as well as the SK day-night asymmetry [5], which would be expected in the MSW scheme due to regeneration effects at the Earth. The data also includes the recent SK 504 days sample. The analysis uses the BP98 model but with an arbitrary *hep* neutrino flux [71]. One notices from the analysis that rate-independent observables, such as the electron recoil energy spectrum and the day-night asymmetry (zenith angle distribution), play an important role in ruling out large regions of MSW parameters.

A theoretical issue which has raised some interest recently is the study of the possible effect of random fluctuations in the solar matter density [72, 73, 74]. The possible existence of noise fluctuations at a few percent

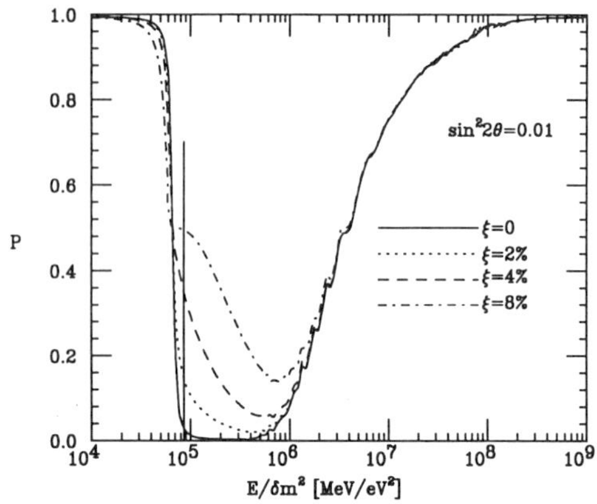

Figure 9. Solar neutrino survival probability in the presence of random density fluctuations, ref. [73]

level is not excluded by present helioseismology studies. In Fig. (9) we show averaged solar neutrino survival probability as a function of $E/\Delta m^2$, for $\sin^2 2\theta = 0.01$. This figure was obtained via a numerical integration of the MSW evolution equation in the presence of noise, using the density profile in the Sun from BP95 in ref. [64], and assuming that the correlation length L_0 (which corresponds to the scale of the fluctuation) is $L_0 = 0.1\lambda_m$, where λ_m is the neutrino oscillation length in matter. An important assumption in the analysis is that $l_{free} \ll L_0 \ll \lambda_m$, where $l_{free} \sim 10$ cm is the mean free path of the electrons in the solar medium. The fluctuations may strongly affect the ^{7}Be neutrino component of the solar neutrino spectrum so that the Borexino experiment should provide an ideal test, if sufficiently small errors can be achieved. The potential of Borexino in probing the level of solar matter density fluctuations provides an additional motivation for the experiment [75]. This is discussed in more detail in ref. [73].

The most popular alternative solution to the solar neutrino problem is the *vacuum oscillation solution* which clearly requires large neutrino mixing and *just-so* adjustment of the oscillation length so as to coincide roughly with the Earth-Sun distance. This solution fits with simplistic see-saw insipred-numerology and has attractive features, as recently advocated in ref. [76]. Fig. 10 shows the regions of just-so oscillation parameters obtained in a recent global fit of the data including the 504 days SK data sample, both the rates and the recoil energy spectrum. Seasonal effects are expected in this scenario and could potentially be used to further constrain the parameters, as described in ref. [77], and aslo to help discriminating it

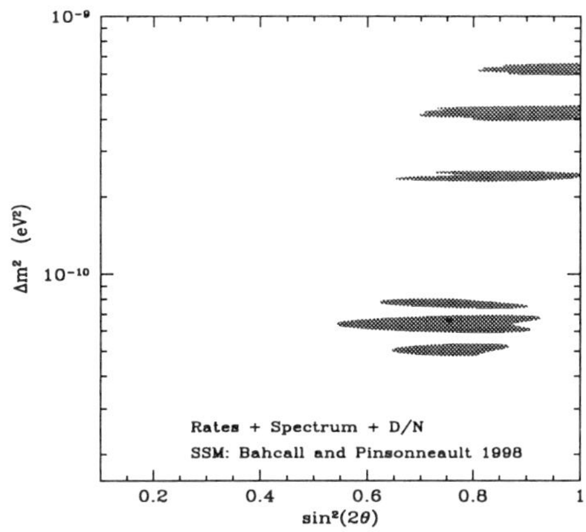

Figure 10. Presently allowed vacuum oscillation parameters, from ref. [71]

from the MSW scenario.

3.2. Atmospheric Neutrinos

Showers initiated when primary cosmic rays hit the Earth's atmosphere originate secondary mesons, mostly pions and kaons, which decay producing ν_e 's, ν_μ 's as well as $\bar{\nu}_e$'s and $\bar{\nu}_\mu$'s [78]. There has been a long-standing discrepancy between the predicted and measured ν_μ /ν_e ratio of the atmospheric neutrino fluxes [2]. The anomaly was found both in water Cerenkov experiments, such as Kamiokande, SuperKamiokande and IMB [4], as well as in the iron calorimeter Soudan2 experiment. Negative experiments, such as Frejus and Nusex have much larger errors.

Although individual ν_μ or ν_e fluxes are only known to within 30% accuracy, the ν_μ /ν_e ratio is known to 5%. The most important feature of the atmospheric neutrino 535-day data sample reported by the SK collaboration at Neutrino 98 [3] is that it exhibits a *zenith-angle-dependent* deficit of muon neutrinos which is inconsistent with expectations based on calculations of the atmospheric neutrino fluxes. For recent analyses see ref. [79, 80, 81]. Experimental biases and uncertainties in the prediction of neutrino fluxes and cross sections are unable to explain the data.

In Fig. (11) I show the measured zenith angle distribution of electron-like and muon-like sub-GeV and multi-GeV events, as well as the one predicted in the absence of oscillation. I also give the expected distribution in various neutrino oscillation schemes. The thick-solid histogram is the theoretically expected distribution in the absence of oscillation, while the

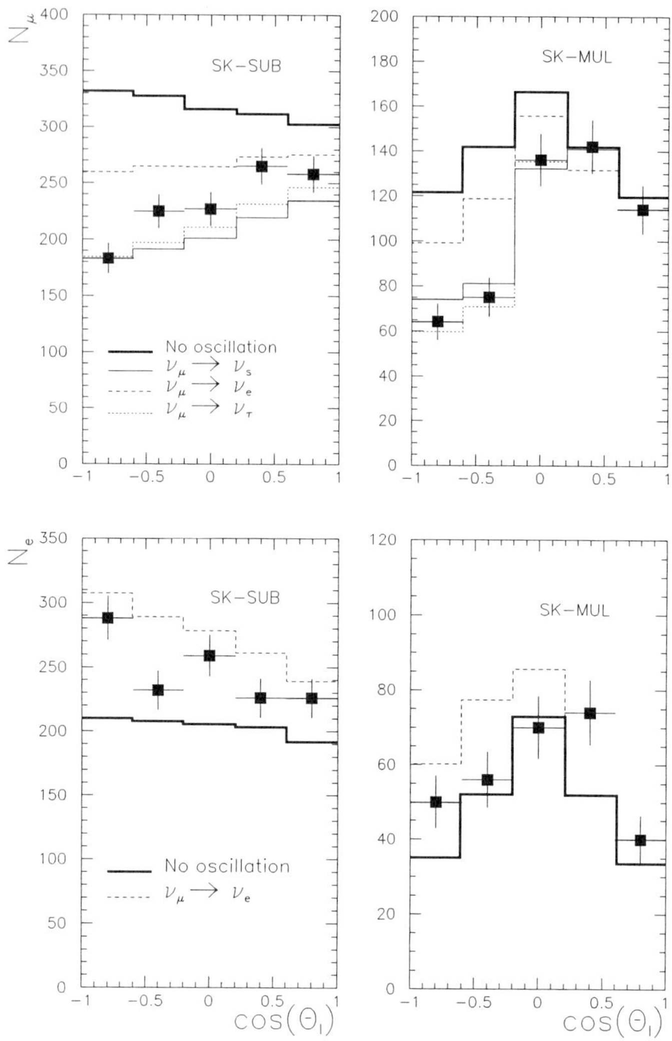

Figure 11. Theoretically expected zenith angle distributions for SK electron and muon-like sub-GeV and multi-GeV events in the SM (no-oscillation) and for the best-fit points of the various oscillation channels, from ref. [79, 80]. The crosses correspond to the SK observations reported at Neutrino 98.

predictions for the best-fit points of the various oscillation channels is indicated as follows: for $\nu_\mu \to \nu_s$ (solid line), $\nu_\mu \to \nu_e$ (dashed line) and $\nu_\mu \to \nu_\tau$ (dotted line). The error displayed in the experimental points is only statistical.

In the theoretical analysis we have used the latest improved calculations of the atmospheric neutrino fluxes as a function of zenith angle, including the muon polarization effect and took into account a variable neutrino production point [82]. Clearly the data are not reproduced by the no-oscillation hypothesis, adding substantially to our confidence that the atmospheric neutrino anomaly is real.

In Fig. (12) I show the allowed neutrino oscillation parameters obtained in a recent global fit of the sub-GeV and multi-GeV (vertex-contained) atmospheric neutrino data [79, 80] including the recent data reported at Neutrino 98, as well as all other experiments combined at 90 (thick solid line) and 99 % CL (thin solid line) for each oscillation channel considered. The two lower panels Fig. (12) differ in the sign of the Δm^2 which was assumed in the analysis of the matter effects in the Earth for the $\nu_\mu \to \nu_s$ oscillations. We found that $\nu_\mu \to \nu_\tau$ oscillations give a slightly better fit than $\nu_\mu \to \nu_s$ oscillations. At present the atmospheric neutrino data cannot distinguish between the ν_μ to ν_τ and ν_μ to ν_s channels. It is well-know that the neutral-to-charged current ratios are important observables in neutrino oscillation phenomenology, which are especially sensitive to the existence of singlet neutrinos, light or heavy [18]. The atmospheric neutrinos produce isolated neutral pions (π^0-events) mainly in neutral current interactions. One may therefore study the ratios of π^0-events and the events induced mainly by the charged currents, as recently advocated in ref. [83]. This has the virtue of minimizing uncertainties related to the original atmospheric neutrino fluxes. In fact the SK collaboration has already tried to do this by estimating the double ratio of π^0 over e-like events in their sample [3] and found $R = 0.93 \pm 0.07 \pm 0.19$. This is consistent both with ν_μ to ν_τ or ν_μ to ν_s channels, with a slight preference for the former. The situation should improve in the future.

We also display in Fig. (12) the sensitivity of present accelerator and reactor experiments, as well as that expected at future long-baseline (LBL) experiments in each channel. The first point to note is that the Chooz reactor [84] data already excludes the region indicated for the $\nu_\mu \to \nu_e$ channel when all experiments are combined at 90% CL.

From the upper-left panel in Fig. (12) one sees that the regions of $\nu_\mu \to \nu_\tau$ oscillation parameters obtained from the atmospheric neutrino data analysis cannot be fully tested by the LBL experiments, as presently designed. One might expect that, due to the upward shift of the Δm^2 indicated by the fit for the sterile case (due to the effects of matter in the Earth) it would be possible to completely cover the corresponding region of

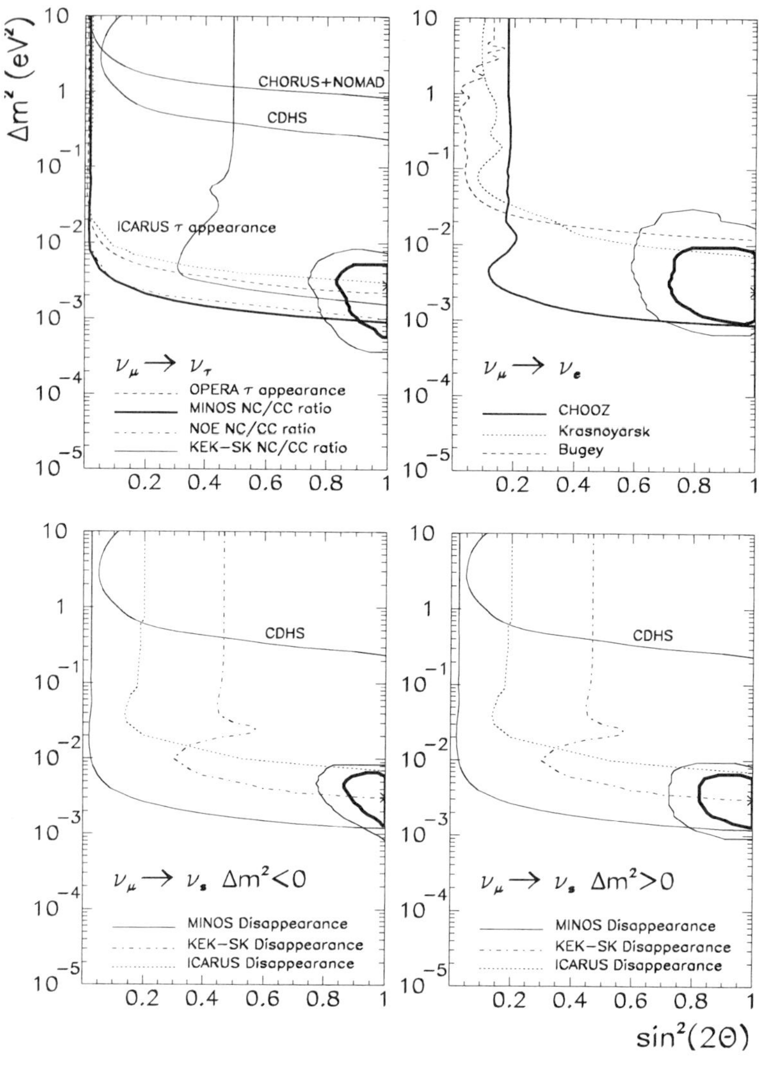

Figure 12. Allowed atmospheric oscillation parameters for all experiments including the SK data reported at Neutrino 98, combined at 90 (thick solid line) and 99 % CL (thin solid line) for all possible oscillation channels, from ref. [79, 80]. In each case the best-fit point is denoted by a star and always corresponds to maximal mixing, a feature which is well-reproduced by the theoretical predictions of the models proposed in ref. [10, 11]. The sensitivity of the present accelerator and reactor experiments as well as the expectations of upcoming long-baseline experiments is also displayed.

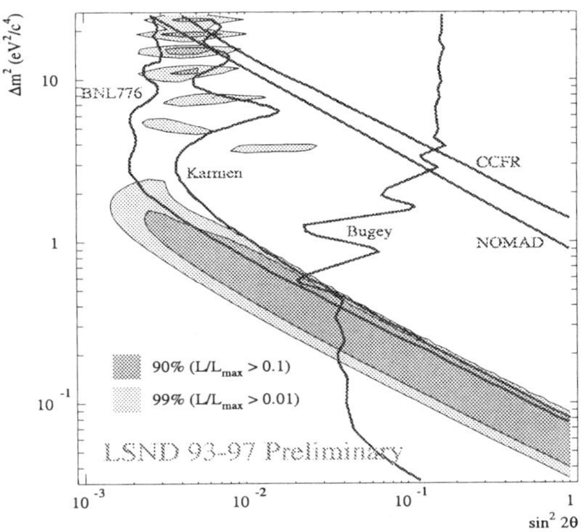

Figure 13. Allowed LSND oscillation parameters versus competing experiments [85]

oscillation parameters. Although this is the case for the MINOS disappearance test, in general most of the LBL experiments can not completely probe the region of oscillation parameters allowed by the $\nu_\mu \to \nu_s$ atmospheric neutrino analysis. This is so irrespective of the sign of Δm^2 assumed. For a discussion of the various potential tests that can be performed at the future LBL experiments in order to unravel the presence of oscillations into sterile channels see ref. [80].

3.3. LSND, Dark Matter & Pulsars

LSND

A search for $\bar{\nu}_\mu \to \bar{\nu}_e$ oscillations has been conducted at the Los Alamos Meson Physics Facility by using $\bar{\nu}_\mu$ from μ^+ decay at rest [6]. The $\bar{\nu}_e$'s are detected via the reaction $\bar{\nu}_e\, p \to e^+\, n$, correlated with a γ from $np \to d\gamma$ (2.2 MeV). The use of tight cuts to identify e^+ events with correlated γ rays yields 22 events with e^+ energy between 36 and 60 MeV and only 4.6 ± 0.6 background events. A fit to the e^+ events between 20 and 60 MeV yields a total excess of $51.8^{+18.7}_{-16.9} \pm 8.0$ events. If attributed to $\bar{\nu}_\mu \to \bar{\nu}_e$ oscillations, this corresponds to an oscillation probability of $(0.31^{+0.11}_{-0.10} \pm 0.05)\%$ and leads to the oscillation parameters shown in Fig. (13). The shaded regions are the favoured likelihood regions given in ref. [6]. The curves show the 90 % and 99 % likelihood allowed ranges from LSND, and compares them to limits from BNL776, KARMEN1, Bugey, CCFR, and NOMAD. A search

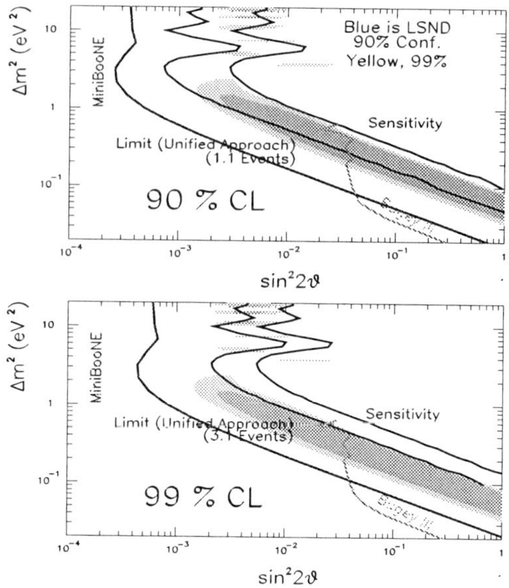

Figure 14. Expected sensitivity of the proposed MiniBooNE experiment [85]

for $\nu_\mu \to \nu_e$ oscillations has also been conducted by the LSND collaboration. Using ν_μ from π^+ decay in flight, the ν_e appearance is detected via the charged-current reaction $C(\nu_e\ ,e^-)X$. Two independent analyses are consistent with the above signature, after taking into account the events expected from the ν_e contamination in the beam and the beam-off background. If interpreted as an oscillation signal, the observed oscillation probability of $2.6 \pm 1.0 \pm 0.5 \times 10^{-3}$ is consistent with the $\bar{\nu}_\mu \to \bar{\nu}_e$ oscillation evidence describe above. Fig. 14 compares the LSND region with the expected sensitivity from MiniBooNE, which was recently approved to run at Fermilab [85]. A possible confirmation of the LSND anomaly would be a discovery of far-reaching implications.

Dark Matter

The research on the nature of the cosmological dark matter and the origin of galaxies and large scale structure in the Universe within the standard theoretical framework of gravitational collapse of fluctuations as the origin of structure in the expanding universe has undergone tremendous progress recently. Indeed the observations of cosmic background temperature anisotropies on large scales performed by the COBE satellite [7] combined with cluster-cluster correlation data e.g. from IRAS [9] can not be reconciled with the simplest cold dark matter (CDM) model. Barring a non-zero cosmological constant and high value of the Hubble parameter ($h \gtrsim 0.7$) the simplest model that have a chance to work is Cold + Hot Dark Matter (MDM, for mixed dark matter), if the Hubble parameter and age parameter allow for an $\Omega = 1$ cosmology [8], suggested by inflation. Electron-volt mass neutrinos are the most well-motivated HDM candidate. This mass scale is similar to that indicated by the hints reported by the LSND experiment [6].

However it is too early to be confident on the MDM scenario, and one should ffor the moment keep an open mind. For example, I note that an MeV range (unstable) tau neutrino is an interesting possibility to consider from the point of view of dark matter. If such neutrino decays before the matter dominance epoch, its decay products would add energy to the radiation, thereby delaying the time at which the matter and radiation contributions to the energy density of the universe become equal. Such delay would allow one to reduce the density fluctuations on the smaller scales purely within the standard cold dark matter scenario [86].

Future sky maps of the cosmic microwave background radiation (CMBR) with high precision at the upcoming MAP and PLANCK missions should bring more light into the nature of the dark matter and the possible rôle of neutrinos [87].

Pulsars

One of the most challenging problems in modern astrophysics is to find a consistent explanation for the high velocity of pulsars. Observations [88] show that these velocities range from zero up to 900 km/s with a mean value of 450 ± 50 km/s. An attractive possibility is that pulsar motion arises from an asymmetric neutrino emission during the supernova explosion. In fact, neutrinos carry more than 99% of the new-born proto-neutron star's gravitational binding energy so that even a 1% asymmetry in the neutrino emission could generate the observed pulsar velocities. One possible explanation to this puzzle may reside in the interplay between the parity non-conservation present in weak interactions and the strong magnetic fields which are expected during a SN explosion. Possible realizations of this idea in the framework of the Standard Model (SM) have been proposed [89, 90] However, it has recently been noted [91] that no asymmetry in neutrino emission can be generated in thermal equilibrium, even in the presence of parity violation. This suggests an that alternative mechanism is at work. Several neutrino conversion mechanisms in matter have been invoked as a possible engine for powering pulsar motion. They all share in common the feature that neutrino propagation properties are affected by the *polarization* [92] of the SN medium which is provided by the strong magnetic fields 10^{15} Gauss present during a SN explosion. This would give rise to some angular dependence of the matter-induced neutrino potentials leading to a deformation of the "neutrino-sphere" for, say, tau neutrinos and hence to an anisotropic neutrino emission. As a consequence, in the presence of non-vanishing ν_τ mass and mixing the resonance sphere for the $\nu_e - \nu_\tau$ conversions is distorted. If the resonance surface lies between the ν_τ and ν_e neutrino spheres, such a distortion would induce a temperature anisotropy in the flux of the escaping tau-neutrinos produced by the conversions, hence a recoil kick of the proto-neutron star. This mechanism was realized in ref. [93] invoking MSW conversions [32] with $m_{\nu_\tau} \gtrsim 100$ eV or so, assuming a negligible ν_e mass. This is necessary in order for the resonance surface to be located between the two neutrino-spheres. It should be noted, however, that such requirement is at odds with cosmological bounds on neutrinos masses unless the τ-neutrino is unstable. On the other hand in ref. [94] a realization was proposed in the resonant spin-flavour precession scheme (RSFP) [68]. Here the magnetic field not only affects the medium properties, but also induces the spin-flavour precession through its coupling to the neutrino transition magnetic moment [67]. Perhaps the simplest suggestion was proposed in ref. [34] where the required pulsar velocities would arise from anisotropic neutrino emission induced by resonant conversions of massless neutrinos (hence no magnetic moment) [31]. This mechanism arises in the model described in eq. (3) and has been shown to be of potential relevance for SN physics [33].

Very recently, however, Raffelt and Janka [95] have claimed that the asymmetric neutrino emission effect was vastly overestimated, because the variation of the temperature over the deformed neutrino-sphere is not an adequate measure for the anisotropy of the neutrino emission. This would invalidate the oscillation mechanisms, leaving the pulsar velocity problem without any known viable solution. The only potential way out of their criticism would invoke conversions into sterile neutrinos, since the conversions would take place deeper in the star. However, it is too early to tell whether or not it works [96].

4. Reconciling the neutrino puzzles

It is easy to accommodate the solar and atmospheric neutrino data by themselves in a general gauge theory of neutrino mass, since it lacks predictivity. One could even have a situation where three-neutrino mixing could be *bi-maximal*, i.e. maximal in both the atmospheric as well as solar neutrino transitions, if the solution chosen by nature is *just-so* [76]. The challenge to reconcile these two requirement arises mainly if one wishes to do that in a predictive quark-lepton *unification* scheme that relates lepton and quark mixing angles. This especially so since the latter are small, in contrast to the lepton mixing indicated by the SK atmospheric data. The story gets more complicated if one wishes to account also for the LSND anomaly and for the hot dark matter. There has been a lot of effort to solve the bigger puzzle posed by the inclusion of any of these additional hints [10, 11, 12]. As we have seen the atmospheric neutrino data requires Δm^2_{atm} which is much larger than the scale $\Delta m^2{}_{\odot}$ which is indicated by the solar neutrino data, either in the context of the MSW mechanism or the just-so solution. These two experiments fix two different scales for neutrino mass differences, so that with just the three known neutrinos and without discarding any experimental data, there is no room to include the LSND scale indicated in Fig. (13), nor the HDM scale which is roughly similar [3].

Reconciling the neutrino puzzles may be attempted within the *unification approach* or the *weak-scale approach* to the theory of neutrino mass. I will concentrate mostly on the latter, because it is an interesting and simpler alternative to the former.

4.1. Almost Degenerate Neutrinos

The only possibility to fit solar, atmospheric and HDM scales in a world with just the three known neutrinos is if all of them have nearly the same mass [12], of about ~ 1.5 eV or so in order to provide the right amount of HDM [8] (all three active neutrinos contribute to HDM). There is no room in this case to accommodate the LSND anomaly. This can be arranged in the unification approach discussed in sec. 2 using the M_L term present in general in seesaw models. With this in mind one can construct, e.g. unified $SO(10)$ seesaw models where all neutrinos lie at the above HDM mass scale (~ 1.5 eV), due to a suitable horizontal symmetry, while the parameters $\Delta m^2{}_{\odot}$ & $\Delta m^2{}_{atm}$ appear as symmetry breaking effects. An interesting fact is that the ratio $\Delta m^2{}_{\odot}$ / $\Delta m^2{}_{atm}$ appears as $m_c{}^2/m_t{}^2$ [20].

4.2. Four-Neutrino Models

The simplest way to open the possibility of incorporating the LSND scale is to invoke a sterile neutrino, i.e. one whose interaction with standard model particles (such as the W and the Z) is much weaker than the SM weak interaction. It must come in as an $SU(2) \otimes U(1)$ singlet ensuring that it does not affect the invisible Z decay width, well-measured at LEP. The sterile neutrino ν_s must also be light enough in order to participate in the oscillations involving the three active neutrinos. The theoretical challenges we have are:

[3] I will ignore the pulsar velocity problem since there is no clear working-model at the moment.

- to understand why the sterile neutrino is so light (it is clear that if a sterile neutrino is introduced into the SM, the $SU(2) \otimes U(1)$ gauge symmetry allows it to have a bare mass, which could be large)
- to account for the maximal neutrino mixing indicated by the atmospheric data
- to account for the three scales Δm^2_{atm}, $\Delta m^2{}_{\odot}$ and $\Delta m^2_{LSND/HDM}$ from first principles

With this in mind we have formulated the simplest and first schemes [10, 11] which provide an answer to the above points. I will denote them, $(e\tau)(\mu\ s)$ [10] and $(es)(\mu\tau)$ [11], respectively. One should realize that a given phenomenological scheme (mainly determined by the structure of the leptonic charged current) may be realized in more than one theoretical model. For example, an alternative to the model in [11] was suggested in ref. [12]. There have been many attempts to reproduce the above phenomenological scenarios from different theoretical assumptions, as has been discussed here [97, 98].

These two basic schemes are characterized by a very symmetric mass spectrum in which there are two ultra-light neutrinos at the solar neutrino scale and two maximally mixed almost degenerate eV-mass neutrinos (LSND/HDM scale), split by the atmospheric neutrino scale [10, 11]. The HDM problem requires the heaviest neutrinos at about 2 eV mass [99]. These scales are generated radiatively due to the additional Higgs bosons which are postulated, as follows: $\Delta m^2_{LSND/HDM}$ arises at one-loop, while Δm^2_{atm} and $\Delta m^2{}_{\odot}$ are two-loop effects. Since this proposal pre-dated the LSND results, it naturally focused on accounting for the HDM problem, rather than LSND. However, it has been realized that the LSND oscillation effects may be accounted for in its framework. These are the simplest theories based only on weak-scale physics, in which one *explains* the lightness of the sterile neutrino, the large lepton mixing required by the atmospheric neutrino data, as well as the generation of the mass splittings responsible for solar and atmospheric neutrino conversions. These follow naturally from the underlying lepton-number-like symmetry and its breaking [10, 11].

These models are minimal in the sense that they add a single $SU(2) \otimes U(1)$ singlet lepton to the SM. Before breaking the symmetry the heaviest neutrinos are exactly degenerate, while the other two which will be responsible for the explanation of the solar neutrino problem are still massless [100]. After the global U(1) lepton symmetry breaks the massive ones split and the light ones get mass. The models differ according to whether the ν_s lies at the dark matter scale or at the solar neutrino scale. In the $(e\tau)(\mu\ s)$ scheme the ν_s lies at the LSND/HDM scale, as illustrated in Fig. (15) while in the alternative $(es)(\mu\tau)$ model, ν_s is at the solar neutrino scale

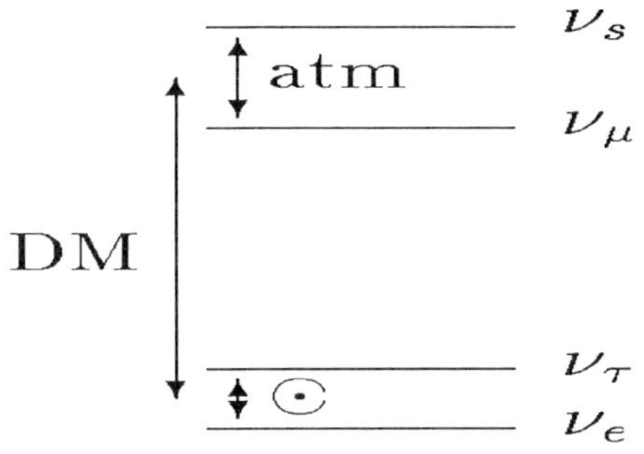

Figure 15. $(e\tau)(\mu\ s)$ scheme: ν_e - ν_τ conversions explain the solar neutrino data and ν_μ - ν_s oscillations account for the atmospheric deficit, ref. [10].

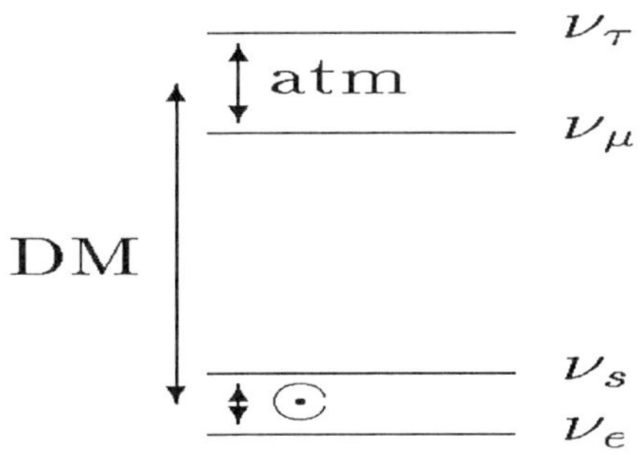

Figure 16. $(es)(\mu\tau)$ scheme: ν_e - ν_s conversions explain the solar neutrino data and ν_μ - ν_τ oscillations account for the atmospheric deficit, ref. [11].

as shown in Fig. (16) [11] [97]. In the $(e\tau)(\mu\ s)$ case the atmospheric neutrino puzzle is explained by ν_μ to ν_s oscillations, while in $(es)(\mu\tau)$ it is explained by ν_μ to ν_τ oscillations. Correspondingly, the deficit of solar neutrinos is explained in the first case by ν_e to ν_τ conversions, while in the second the relevant channel is ν_e to ν_s. The two models are therefore clearly inequivalent. In both cases it is possible to fit all present observations together.

I now turn to the consistency of the models with BBN. The presence of additional weakly interacting light particles, such as our light sterile neutrino ν_s , is constrained by BBN since the ν_s would enter into equilibrium with the active neutrinos in the early Universe (and therefore would contribute to N_ν^{max}) via neutrino oscillations [101], unless $\Delta m^2 sin^4 2\theta \lesssim 3\times10^{-6}\ eV^2$ where Δm^2 denotes the mass-square difference of the active and sterile species and θ is the vacuum mixing angle. However, systematical uncertainties in the derivation of BBN bounds still caution us not to take them too literally. For example, it has been argued in [58] that present observations of primordial Helium and deuterium abundances can allow up to $N_\nu = 4.5$ neutrino species if the baryon to photon ratio is small. Adopting this as a limit, clearly both models described above are consistent. Should the BBN constraints get tighter, e.g. $N_\nu^{max} < 3.5$ they could rule out the $(e\tau)(\mu\ s)$ model, and leave out only the competing scheme as a viable alternative. For recent work on this see ref. [102].

The two models would be distinguishable both from the analysis of future solar as well as atmospheric neutrino data. For example they may be tested in the SNO experiment [103] once they measure the solar neutrino flux (Φ_ν^{NC}) in their neutral current data and compare it with the corresponding charged current value (Φ_ν^{CC}). If the solar neutrinos convert to active neutrinos, as in the $(e\tau)(\mu\ s)$ model, then one expects $\Phi_\nu^{CC}/\Phi_\nu^{NC} \simeq .5$, whereas in the $(es)(\mu\tau)$ scheme (ν_e conversion to ν_s), the above ratio would be nearly $\simeq 1$. Looking at pion production via the neutral current reaction $\nu_\tau + N \to \nu_\tau + \pi^0 + N$ in atmospheric data might also help in distinguishing between these two possibilities [83], since this reaction is absent in the case of sterile neutrinos, but would exist in the $(es)(\mu\tau)$ scheme.

If light sterile neutrinos indeed exist, as suggested by the current solar and atmospheric neutrino data, together with the LSND experiment, one can show that in some four-neutrino scenarios, neutrinos would contribute to a cosmic hot dark matter component and to an increased radiation content at the epoch of matter-radiation equality. These effects leave their imprint in sky maps of the cosmic microwave background radiation (CMBR) and may thus be detectable with the precision measurements of the upcoming MAP and PLANCK missions as noted recently in ref. [87].

4.3. Mev Tau Neutrino

In ref. [104] a model was presented where an unstable MeV Majorana tau neutrino naturally reconciles the cosmological observations of large and small-scale density fluctuations with the cold dark matter picture (CDM). The model assumes the spontaneous violation of a global lepton number symmetry at the weak scale. The breaking of this symmetry generates the cosmologically required decay of the ν_τ with lifetime $\tau_{\nu_\tau} \sim 10^2 - 10^4$ sec, as well as the masses and oscillations of the three light neutrinos ν_e , ν_μ and ν_s. One can also verify that the BBN constraints can be satisfied. The cosmological attractiveness of this scheme should encourage one to check whether one can indeed account for the present solar and atmospheric data through oscillations among the three light neutrinos, after taking into account the recent SKdata.

5. In conclusion

A major news has been the re-confirmation of an angle-dependent atmospheric neutrino deficit by the SK collaboration, providing a strong evidence for neutrino masses, similar to that offered by the solar neutrino data. Unfortunately future LBL experiments do not all probe the full region indicated by the atmospheric data. If the LSND result stands the test of time, this would be a puzzling indication for the existence of a light sterile neutrino. *Who ordered it?* The two most attractive schemes to reconcile these observations invoke either ν_e - ν_τ conversions to explain the solar data, with ν_μ - ν_s oscillations accounting for the atmospheric deficit, or the other way around. These two basic schemes have distinct implications at future solar & atmospheric neutrino experiments. SNO and SuperKamiokande have the potential to distinguish them due to their neutral current sensitivity.

How about heavy neutrinos? Although cosmological bounds are a fundamental tool to restrict neutrino masses, in many theories heavy neutrinos will either decay or annihilate very fast, thereby loosening or evading the cosmological bounds. From this point of view, *neutrinos can have any mass presently allowed by laboratory experiments*, and it is therefore important to search for manifestations of heavy neutrinos at the laboratory.

Last but not least, though most of the recent excitement comes from underground experiments, one should note that models of neutrino mass may lead to a plethora of new signatures which may be accessible also at accelerators, thus illustrating the complementarity between the two approaches in unraveling the properties of neutrinos and probing for signals beyond the SM.

Acknowledgements

I thank John Bahcall, Plamen Krastev and Bill Louis, for making their postcript figures available to me, and to Thomas Janka and Eligio Lisi for correspondence. I thank all my collaborators, especially Hiroshi Nunokawa for going over the first draft of this manuscript critically. This work was supported by DGICYT grant PB95-1077 and by the EEC under the TMR contract ERBFMRX-CT96-0090.

References

[1] Invited alks by Lande, Gavrin, Kirsten, and Suzuki at Neutrino 98.

[2] NUSEX Collaboration, M. Aglietta *et al.*, Europhys. Lett. **8**, 611 (1989); Fréjus Collaboration, Ch. Berger *et al.*, Phys. Lett. **B227**, 489 (1989); IMB Collaboration, D. Casper *et al.*, Phys. Rev. Lett. **66**, 2561 (1991);

R. Becker-Szendy *et al.*, Phys. Rev. **D46**, 3720 (1992); Kamiokande Collaboration, H. S. Hirata *et al.*, Phys. Lett. **B205**, 416 (1988) and Phys. Lett. **B280**, 146 (1992); Kamiokande Collaboration, Y. Fukuda *et al.*, Phys. Lett. **B335**, 237 (1994); Soudan Collaboration, W. W. M Allison *et al.*, Phys. Lett. **B391**, 491 (1997).

[3] T. Kajita, Invited talk at Neutrino 98, Takayama, Japan; SuperKamiokande Collaboration, Y. Fukuda et al, hep-ex/9803006, hep-ex/9805006 and hep-ex/9807003; See also talk by Chang Kee Jung in these Proceedings.

[4] SuperKamiokande Collaboration, Y. Fukuda et al. Phys. Rev. Lett. (accepted for publication, hep-ex/9805021); C. Yanagisawa, talk at the *International Workshop on Physics Beyond The Standard Model: from Theory to Experiment*, Valencia, October 13-17; to appear in the Proceedings (World Scientific, 1998), http://flamenco.uv.es//val97.html

[5] SuperKamiokande Collaboration, Y. Suzuki, Proceedings Neutrino 98, of the XVIII International Conference on Neutrino Physics and Astrophysics, Takayama, Japan, 4–9 June 1998, Y. Suzuki, Y. Totsuka (Eds.). To be published in Nucl. Phys. B (Proc. Suppl.)

[6] C. Athanassopoulos, *Phys. Rev. Lett.* **75** (1995) 2650; *Phys. Rev. Lett.* **77** (1996) 3082 ; C. Athanassopoulos et al. nucl-ex/9706006.

[7] G. F. Smoot et al., *Astrophys. J.* **396** (1992) L1-L5; E.L. Wright et al., *Astrophys. J.* **396** (1992) L13

[8] J.R. Primack, astro-ph/9707285 & astro-ph/9610078, E. Gawiser and J. Silk, astro-ph/9806197; Science, **280**, 1405 (1998), and references therein.

[9] R. Rowan-Robinson, in *Cosmological Dark Matter*, (World Scientific, 1994), ed. A. Perez, and J. W. F. Valle, p. 7-18, ISBN 981-02-1879-6

[10] J. T. Peltoniemi, D. Tommasini, and J W F Valle, *Phys. Lett.* **B298** (1993) 383

[11] J. T. Peltoniemi, and J W F Valle, *Nucl. Phys.* **B406** (1993) 409

[12] D.O. Caldwell and R.N. Mohapatra, *Phys. Rev.* **D48** (1993) 3259

[13] J. Schechter, J.W.F. Valle, *Phys. Rev.* **D25** (1982) 2951

[14] H.V. Klapdor-Kleingrothaus, M. Hirsch, *Zeit. fur Physik* **A359** (1997) 361

[15] See, e.g. R. N. Mohapatra and G. Senjanovic, *Phys. Rev.* **D23** (1981) 165.

[16] J.C. Pati, A. Salam. *Phys. Rev.* **D8** (1973) 1240

[17] M Gell-Mann, P Ramond, R. Slansky, in *Supergravity*, eds. P. van Niewenhuizen and D. Freedman (North Holland, 1979); T. Yanagida, in *KEK lectures*, ed. O. Sawada and A. Sugamoto (KEK, 1979)

[18] J. Schechter and J. W. F. Valle, *Phys. Rev.* **D22** (1980) 2227

[19] A. Yu. Smirnov, *Phys. Rev.* **D48** (1994) 3264

[20] A. Ioannissyan, J. W. F. Valle, *Phys. Lett.* **B332** (1994) 93-99; B. Bamert, C.P. Burgess, *Phys. Lett.* **B329** (1994) 289; D. Caldwell and R. N. Mohapatra, *Phys. Rev.* **D50** (1994) 3477; D. G. Lee and R. N. Mohapatra,

Phys. Lett. **B329** (1994) 463; A. S. Joshipura, *Zeit. fur Physik* **C64** (1994) 31

[21] M. C. Gonzalez-Garcia, J. W. F. Valle, *Phys. Lett.* **B216** (1989) 360.

[22] D. Wyler, L. Wolfenstein, *Nucl. Phys.* **B218** (1983) 205

[23] R. Mohapatra, J. W. F. Valle, *Phys. Rev.* **D34** (1986) 1642; J. W. F. Valle, *Nucl. Phys. B (Proc. Suppl.)* **B11** (1989) 118-177

[24] M. Dittmar, M. C. Gonzalez-Garcia, A. Santamaria, J. W. F. Valle, *Nucl. Phys.* **B332** (1990) 1; M. C. Gonzalez-Garcia, A. Santamaria, J. W. F. Valle, *ibid.* **B342** (1990) 108; J . Gluza, J. Maalampi, M. Raidal, M. Zralek, *Phys. Lett.* **B407** (1997) 45; J. Gluza, M. Zralek; *Phys. Rev.* **D55** (1997) 7030

[25] J. Bernabeu, A. Santamaria, J. Vidal, A. Mendez, J. W. F. Valle, *Phys. Lett.* **B187** (1987) 303; J. G. Korner, A. Pilaftsis, K. Schilcher, *Phys. Lett.* **B300** (1993) 381

[26] M. C. Gonzalez-Garcia, J. W. F. Valle, *Mod. Phys. Lett.* **A7** (1992) 477; erratum *Mod. Phys. Lett.* **A9** (1994) 2569; A. Ilakovac, A. Pilaftsis, *Nucl. Phys.* **B437** (1995) 491; A. Pilaftsis, *Mod. Phys. Lett.* **A9** (1994) 3595

[27] G. C. Branco, M. N. Rebelo, J. W. F. Valle, *Phys. Lett.* **B225** (1989) 385; N. Rius, J. W. F. Valle, *Phys. Lett.* **B246** (1990) 249

[28] M. Dittmar, J. W. F. Valle, contribution to the High Luminosity at LEP working group, yellow report CERN-91/02, p. 98-103

[29] R. Alemany et. al. hep-ph/9307252, published in ECFA/93-151, ed. R. Aleksan, A. Ali, p. 191-211

[30] Opal collaboration, *Phys. Lett.* **B254** (1991) 293 and *Zeit. fur Physik* **C67** (1995) 365; L3 collaboration, *Phys. Rep.* **236** (1993) 1-146; *Phys. Lett.* **B316** (1993) 427, Delphi collaboration, *Phys. Lett.* **B359** (1995) 411.

[31] J. W. F. Valle, *Phys. Lett.* **B199** (1987) 432

[32] M. Mikheyev, A. Smirnov, *Sov. J. Nucl. Phys.* **42** (1986) 913; L. Wolfenstein, *Phys. Rev.* **D20** (1979) 2634.

[33] H. Nunokawa, Y.Z. Qian, A. Rossi, J. W. F. Valle, *Phys. Rev.* **D54** (1996) 4356-4363, hep-ph/9605301

[34] D. Grasso, H. Nunokawa, J. W. F. Valle, *Phys. Rev. Lett.* **81** (1998) 2412, [astro-ph/9803002]

[35] M. C. Gonzalez-Garcia, et al, [hep-ph/9809531]

[36] A. Zee, *Phys. Lett.* **B93** (1980) 389

[37] R. Hempfling, *Nucl. Phys.* **B478**, 3 (1996)

[38] K. S. Babu, *Phys. Lett.* **B203** (1988) 132

[39] J. T. Peltoniemi, and J. W. F. Valle, *Phys. Lett.* **B304** (1993) 147

[40] M.A. Díaz, J.C. Romão, and J. W. F. Valle, *Nucl. Phys.* **B524** 23-40 (1998), hep-ph/9706315.

[41] F. Vissani and A. Yu. Smirnov, Nucl. Phys. **B460**, 37-56 (1996); R. Hempfling, *Nucl. Phys.* **B478**, 3 (1996), and hep-ph/9702412; H.P. Nilles and N. Polonsky, *Nucl. Phys.* **B484**, 33 (1997); B. de Carlos, P.L. White, Phys.Rev. **D55** 4222-4239 (1997); E. Nardi, Phys. Rev. **D55** (1997) 5772; S. Roy and B. Mukhopadhyaya, Phys. Rev. **D55**, 7020 (1997); A.S. Joshipura and M. Nowakowski, *Phys. Rev. D* **51**, 2421 (1995); T. Banks, Y. Grossman, E. Nardi, and Y. Nir, *Phys. Rev. D* **52**, 5319 (1995); F.M. Borzumati, Y. Grossman, E. Nardi, Y. Nir, *Phys. Lett. B* **384**, 123 (1996); A. Faessler, S. Kovalenko, F. Simkovic, hep-ph/9712535; M. Carena, S. Pokorski, and C.E.M. Wagner, hep-ph/9801251; M.E. Gómez and K. Tamvakis, hep-ph/9801348.

[42] M.A. Díaz, hep-ph/9711435, hep-ph/9712213; J.C. Romão, hep-ph/9712362 and J. W. F. Valle, hep-ph/9808292.

[43] G.G. Ross, J. W. F. Valle. *Phys. Lett.* **151B** 375 (1985); John Ellis, G. Gelmini, C. Jarlskog, G.G. Ross, J. W. F. Valle, *Phys. Lett.* **150B** 142 (1985); A. Santamaria, J. W. F. Valle, *Phys. Lett.* **195B** 423 (1987).

[44] J. Schechter, J. W. F. Valle, *Phys. Rev.* **D25** (1982) 774

[45] See, e.g. A.D. Dolgov, S. Pastor, and J. W. F. Valle, *Phys. Lett.* **B383** (1996) 193-198; [hep-ph/9602233] Steen Hannestad, Jes Madsen, *Phys. Rev. Lett.* **76** (1996) 2848-2851, Erratum-ibid. **77** (1996) 5148; J.B. Rehm, G. G. Raffelt, A. Weiss *Astron. & Astrophys.* **327** (1997) 443-452, A. D. Dolgov, S. H. Hansen, D. V. Semikoz, Nucl. Phys. **B524** (1998) 621-638

[46] J. C. Romão, J. W. F. Valle, *Nucl. Phys.* **B381** (1992) 87-108

[47] For reviews see J. W. F. Valle, hep-ph/9712277 and hep-ph/9603307; for detailed analyses of the related signatures see M. A. Diaz, M. A. Garcia-Jareno, D. A. Restrepo, J. W. F. Valle, Nucl. Phys. **B527** (1998) 44-60 and F. de Campos, O. J. P. Eboli, J. Rosiek, J. W. F. Valle, Phys. Rev.**D55** (1997) 1316-1325, hep-ph 9601269.

[48] Y. Chikashige, R. Mohapatra, R. Peccei, *Phys. Rev. Lett.* **45** (1980) 1926

[49] For reviews see J. W. F. Valle, *Gauge Theories and the Physics of Neutrino Mass*, *Prog. Part. Nucl. Phys.* **26** (1991) 91-171

[50] A. Joshipura and J. W. F. Valle, *Nucl. Phys.* **B397** (1993) 105; J. C. Romao, F. de Campos, and J. W. F. Valle, *Phys. Lett.* **B292** (1992) 329

[51] R. Barbieri, J. Ellis & M. K. Gaillard, Phys. Lett. **90 B**, 249 (1980); E. Akhmedov, Z. Berezhiani & G. Senjanović, Phys. Rev. Lett. **69**, 3013 (1992).

[52] V. Berezinskii, J. W. F. Valle Phys.Lett **B318** 360-366,1993, [hep-ph/9309214]

[53] E. Kolb, M. Turner, *The Early Universe*, Addison-Wesley, 1990, and references therein

[54] J. W. F. Valle, *Phys. Lett.* **B131** (1983) 87; G. Gelmini, J. W. F. Valle, *Phys. Lett.* **B142** (1984) 181; J. W. F. Valle, *Phys. Lett.* **B159** (1985) 49; A. Joshipura, S. Rindani, *Phys. Rev.* **D46** (1992) 3000

[55] A Masiero, J. W. F. Valle, *Phys. Lett.* **B251** (1990) 273; J. C. Romao, C. A. Santos, and J. W. F. Valle, *Phys. Lett.* **B288** (1992) 311

[56] R.F. Carswell, M. Rauch, R.J. Weynman et al., *MNRAS* **268** (1994) L1; A. Songalia, L.L. Cowie, C. Hogan and M. Rugers, *Nature* **368** (1994) 599; D. Tytler and X.M. Fan, *Bull. Am. Astr. Soc.* **26** (1994) 1424; D. Tytler, talk at the Texas Symposium, December 1996.

[57] N. Hata et al., *Phys. Rev. Lett.* **75** (1995) 3977; C.J. Copi, D.N. Schramm and M.S. Turner, *Science* **267** (1995) 192 and *Phys. Rev. Lett.* **75** (1995) 3981; K. A. Olive and G. Steigman, *Phys. Lett.* **B354** (1995) 357-362; S. Sarkar, *Rep. on Prog. in Phys.* **59** (1997) 1493

[58] S. Sarkar, Rep. Prog. Phys. **59**, 1493 (1996); P. J. Kernan and S. Sarkar, *Phys. Rev.* **D 54** (1996) R3681

[59] A.D. Dolgov, S. Pastor, J.C. Romão and J. W. F. Valle, *Nucl. Phys.* **B496** (1997) 24-40, hep-ph/9610507.

[60] S. Hannestad, *Phys. Rev.* **D57** (1998) 2213-2218; M. Kawasaki, K. Kohri, K. Sato, *Phys. Lett.* **B430** (1998) 132-139; for a review see G. Steigman; in *Cosmological Dark Matter*, p. 55, World Scientific, 1994, ISBN 981-02-1879-6

[61] E. Kh. Akhmedov, A. S. Joshipura, S. Ranfone & J. W.F. Valle, Nucl. Phys. **B441** (1995) 61-75. [hep-ph 9501248]

[62] J. N. Bahcall, astro-ph/9808162

[63] J. N. Bahcall, S. Basu and M. H. Pinsonneault, Phys. Lett. B 433 (1998) 1.

[64] (GONG) J. Christensen-Dalsgaard et al., GONG Collaboration, Science 272 (1996) 1286; (BP95) J. N. Bahcall and M. H. Pinsonneault, Rev. Mod. Phys. 67 (1995) 781; (KS94) A. Kovetz and G. Shaviv, Astrophys. J. 426 (1994) 787; (CDF94) V. Castellani, S. Degl'Innocenti, G. Fiorentini, L.M. Lissia and B. Ricci, Phys. Lett. B 324 (1994) 425; (JCD94) J. Christensen-Dalsgaard, Europhys. News 25 (1994) 71; (SSD94) X. Shi, D.N. Schramm and D.S.P. Dearborn, Phys. Rev. D 50 (1994) 2414; (DS96) A. Dar and G. Shaviv, Astrophys. J. 468 (1996) 933; (CDF93) V. Castellani, S. Degl'Innocenti and G. Fiorentini, Astron. Astrophys. 271 (1993) 601; (TCL93) S. Turck-Chièze and I. Lopes, Astrophys. J. 408 (1993) 347; (BPML93) G. Berthomieu, J. Provost, P. Morel and Y. Lebreton, Astron. Astrophys. 268 (1993) 775; (BP92) J.N. Bahcall and M.H. Pinsonneault, Rev. Mod. Phys. 64 (1992) 885; (SBF90) I.-J. Sackman, A.I. Boothroyd and W.A. Fowler, Astrophys. J. 360 (1990) 727; (BU88)

J.N. Bahcall and R.K. Ulrich, Rev. Mod. Phys. 60 (1988) 297; (RVCD96) O. Richard, S. Vauclair, C. Charbonnel and W.A. Dziembowski, Astron. Astrophys. 312 (1996) 1000; (CDR97) F. Ciacio, S. Degl'Innocenti and B. Ricci, Astron. Astrophys. Suppl. Ser. 123 (1997) 449.

[65] J.N. Bahcall, M.H. Pinsonneault, S. Basu and J. Christensen-Dalsgaard, Phys. Rev. Lett. 78 (1997) 171.

[66] J. N. Bahcall, *Phys. Lett.* **B338** (1994) 276; V. Castellani, *et al Phys. Lett.* **B324** (1994) 245; N. Hata, S. Bludman, and P. Langacker, *Phys. Rev.* **D49** (1994) 3622; V. Berezinsky, Comm. on Nucl. and Part. Phys. **21** (1994) 249

[67] J. Schechter, J. W. F. Valle, Phys. Rev. **D24** 1883, (1981), Err. ibid.**D25** 283, (1982).

[68] E.Kh. Akhmedov, *Phys. Lett.* **B213** (1964-68) 88; C. S. Lim and W. Marciano, *Phys. Rev.* **D37** (191368) 88

[69] E.Kh. Akhmedov, *The neutrino magnetic moment and time variations of the solar neutrino flux*, hep-ph/9705451, talk given at the *4th International Solar Neutrino Conference*, Heidelberg, Germany, April 1997

[70] J. N. Bahcall, P. I. Krastev, hep-ph 9807525

[71] J. N. Bahcall, P. Krastev, A.Yu. Smirnov, hep-ph/9807216.

[72] A.B. Balantekin, J.M. Fetter and F.N. Loreti, *Phys. Rev.* **D54** (1996) 3941-3951; F. N. Loreti and A. B. Balantekin, Phys. Rev. **D50** (1994) 4762; F. N. Loreti *et al.*, Phys. Rev. **D52** (1996) 6664.

[73] H. Nunokawa, A. Rossi, V. Semikoz, J. W. F. Valle, *Nucl. Phys.* **B472** (1996) 495-517 [see also talk at Neutrino 96, hep-ph/9610526]

[74] C.P. Burgess and D. Michaud, Ann. Phys. (NY) **256** (1997) 1; P. Bamert, C.P. Burgess and D. Michaud, Nucl.Phys. B513 (1998) 319-342, hep-ph/9707542; P. Bamert, hep-ph/9708224; C.P. Burgess, hep-ph/9711425; C.P. Burgess, D. Michaud hep-ph/9611368; E. Torrente-Lujan, hep-ph/9602398

[75] C. Arpesella at al., Proposal of the Borexino experiment (1991).

[76] V. Barger, S. Pakvasa, T. Weiler, K. Whisnant, hep-ph/9806387; S. L. Glashow, P. J. Kernan, L. M. Krauss, hep-ph/9808470; S. Davison, S. F. King, hep-ph/9808296

[77] B. Faid, G. L. Fogli, E. Lisi, D. Montanino, hep-ph/9805293

[78] For a review, see for e.g. T.K. Gaisser, F. Halzen and T. Stanev, Phys. Rep. 258, 174 (1995) and references therein.

[79] M. C. Gonzalez-Garcia, H. Nunokawa, O. L. G. Peres, T. Stanev, J. W. F. Valle, *Phys. Rev.* **D58** (1998) 033004 [hep-ph 9801368]; for details on the 535 days data sample update (Neutrino 98 adat), and the comparison of active and sterile channels see M. C. Gonzalez-Garcia et al, hep-ph 9807305, to be published in *Nucl. Phys.* **B** (1998) xx.

[80] M.C. Gonzalez-Garcia, hep-ph/9808203, and references therein.

[81] R. Foot, R. R. Volkas, O. Yasuda, TMUP-HEL-9801. For recent 3-generation analyses see O. Yasuda, hep-ph/9804400 and G. L. Fogli, E. Lisi, A. Marrone, G. Scioscia, hep-ph/9808205

[82] V. Agrawal *et al.*, Phys. Rev. **D53**, 1314 (1996); L. V. Volkova, Sov. J. Nucl. Phys. **31**, 784 (1980); T. K. Gaisser and T. Stanev, Phys. Rev. **D57** 1977 (1998).

[83] A. Smirnov and F. Vissani, Phys.Lett. **B432** (1998) 376; J. G. Learned, S. Pakvasa and J. Stone, hep-ph/9805343; H. Murayama and L. Hall, hep-ph/9806218

[84] M. Appollonio et al. Phys. Lett. **B420** 397(1998), hep-ex/9711002

[85] B. Louis, private communication

[86] J. Bardeen, J. Bond and G. Efstathiou, *Astrophys. J.* **321** (1987) 28; J. Bond and G. Efstathiou, *Phys. Lett.* **B265** (1991) 245; M. Davis et al., *Nature* **356** (1992) 489; S. Dodelson, G. Gyuk and M. Turner, *Phys. Rev. Lett.* **72** (1994) 3754; H. Kikuchi and E. Ma, *Phys. Rev.* **D51** (1995) 296; H. B. Kim and J. E. Kim, *Nucl. Phys.* **B433** (1995) 421; M. White, G. Gelmini and J. Silk, *Phys. Rev.* **D51** (1995) 2669 A. S. Joshipura and J. W. F. Valle, *Nucl. Phys.* **B440** (1995) 647.

[87] G. Raffelt, these Proceedings, S. Hannestad, G. Raffelt astro-ph/9805223.

[88] A.G. Lyne and D.R. Lorimer, *Nature* **369** (1994) 127.

[89] N.N. Chugai, *Pis'ma Astron. Zh.***10**, 87 (1984).

[90] A. Vilenkin, *Astrophys. J.* **451** (1995) 700; Dong Lai, Y.-Z. Qian, astro-ph/9712043

[91] A. Kusenko, G. Segrè, A. Vilenkin, astro-ph/9806205

[92] H. Nunokawa, V.B. Semikoz, A.Yu. Smirnov and J. W. F. Valle, *Nucl. Phys.* **B501** (1997) 17

[93] A. Kusenko, G. Segrè, *Phys. Rev. Lett.* **77** (1996) 4872 & *Phys. Rev. Lett.* **79** (1997) 2751; Y.Z. Qian, *Phys. Rev. Lett.* **79** (1997) 2750

[94] E.Kh. Akhmedov, A. Lanza and D.W. Sciama, *Phys. Rev.* **D56** (1997) 6117

[95] H.-T. Janka, G. G. Raffelt, astro-ph/9808099

[96] D. Grasso, H. Nunokawa, A . Rossi, A. Yu. Smirnov, J. W. F. Valle, in preparation

[97] Q. Y. Liu, A. Yu. Smirnov, Nucl. Phys. **B524** (1998) 505-523; V. Barger, K. Whisnant and T. Weiler, Phys.Lett. **B427** (1998) 97-104; S. Gibbons, R. N. Mohapatra, S. Nandi and A. Raichoudhuri, Phys. Lett. **B430** (1998) 296-302; Nucl.Phys. **B524** (1998) 505-523; S. Bilenky, C. Giunti and W. Grimus, Eur. Phys. J. **C 1**, 247 (1998); S. Goswami, Phys. Rev. **D 55**, 2931 (1997); N. Okada and O. Yasuda, Int. J. Mod. Phys. **A12** (1997) 3669-3694

[98] P. Langacker, hep-ph/9805281; M. Bando and K. Yoshioka, hep-ph/9806400; E. J. Chun, A. Joshipura and A. Smirnov, in *Elementary Particle Physics: Present and Future* (World Scientific, 1996), ISBN 981-02-2554-7

[99] J. R. Primack, et al. Phys. Rev. Lett. **74** (1995) 2160

[100] J. Schechter and J. W. F. Valle, *Phys. Rev.* **D21** (1980) 309

[101] R. Barbieri and A. Dolgov, Phys. Lett. **B 237**, 440 (1990); K. Enquist, K. Kainulainen and J. Maalampi, Phys. Lett. **B 249**, 531 (1992); D. P. Kirilova and M. Chizov, hep-ph/9707282.

[102] R. Foot, R.R. Volkas, *Phys. Rev.* **D55** (1997) 5147-5176

[103] SNO collaboration, E. Norman et al. *Proc. of The Fermilab Conference: DPF 92* ed. C. Albright, P. H. Kasper, R. Raja and J. Yoh (World Scientific), p. 1450.

[104] A. Joshipura, J. W. F. Valle, *Nucl. Phys.* **B440** (1995) 647.

Exotic mechanisms for neutrino masses

Zurab Berezhiani [1]

Università di Ferrara, Via Paradiso 12, I-44100 Ferrara, Italy,
Institute of Physics, Georgian Academy of Sciences, Tbilisi, Georgia

Abstract. The problem of neutrino masses and mixings is discussed in the context of the standard model and its extensions. The recent hints of neutrino oscillations are reviewed as are the solar and atmospheric neutrino deficits and the LSND anomaly. If all these will be confirmed by future statistics, this would require existence of extra sterile neutrino. The origin of the latter can be naturally explained if it is identified to the neutrino of the hypothetical mirror world which is a complete duplicate to the visible one, except that its weak scale v' could be somewhat different from the ordinary weak scale. The mirror neutrinos can mix the ordinary ones through the Planck scale induced higher order operators, which gives rise to the oscillation of ν_e into its mirror partner ν_e' with parameters naturally in the Just-so range if $v' \sim v$, or in the MSW range if $v' \sim 30v$. We also discuss the cosmological and astrophysical implications of of mirror neutrinos and mirror baryons.

1. Neutrino masses in the standard model and beyond

1.1. In the standard model

The Standard Model (SM), being an internally consistent renormalizable gauge theory, has been extremely successfull in overcoming all experimental precission tests. It accomodates the observed quarks and leptons in a consistent way. Three families sharing the same quantum numbers under

[1] E-mail: berezhiani@fe.infn.it

the $SU(3) \times SU(2) \times U(1)$ gauge symmetry are introduced as an anomaly free set of chiral left-handed (LH) fermions:

$$q_i = \begin{pmatrix} u_i \\ d_i \end{pmatrix} \sim (3, 2, +1/3), \quad u_i^c \sim (\bar{3}, 1, -4/3), \quad d_i^c \sim (\bar{3}, 1, 2/3),$$

$$l_i = \begin{pmatrix} \nu_i \\ e_i \end{pmatrix} \sim (1, 2, -1), \qquad e_i^c \sim (1, 1, 2) \tag{1}$$

where $i = 1, 2, 3$ is a family index. (The weak isosinglet states u^c, d^c, e^c are the C-conjugates to the right-handed (RH) components u_R, d_R, e_R). A remarkable feature of the SM is that the fermion and the gauge boson $W^\pm, Z$ masses have a common origin, namely the Higgs mechanism. In fact, fermions would remain massless as far as gauge symmetry is unbroken. They get masses through the Yukawa couplings to the Higgs doublet ϕ:

$$\mathcal{L}_{\text{Yuk}} = \lambda_{ij}^u q_i C u_j^c \tilde{\phi} + \lambda_{ij}^d q_i C d_j^c \phi + \lambda_{ij}^e l_i C e_j^c \phi \qquad (\tilde{\phi} = i\tau_2 \phi^*) \tag{2}$$

So, the fermion masses are related to the weak scale $\langle\phi\rangle = v = 174$ GeV. The coupling constant matrices $\hat{\lambda}^{u,d,e}$ and correspondingly the fermion mass matrices $\hat{m}^f = \hat{\lambda}^f v$ $(f = u, d, e)$ can be brought to the physical basis of the mass eigenstates by the unitary transformations:

$$V_f \hat{m}^f V_f' = \hat{m}_{diag}^f \,. \tag{3}$$

where

$$\begin{aligned} \hat{m}_{diag}^u &= \text{diag}(m_u, m_c, m_t) = v \cdot \text{diag}(\lambda_u, \lambda_c, \lambda_t) \\ \hat{m}_{diag}^d &= \text{diag}(m_d, m_s, m_b) = v \cdot \text{diag}(\lambda_d, \lambda_s, \lambda_b) \\ \hat{m}_{diag}^e &= \text{diag}(m_e, m_\mu, m_\tau) = v \cdot \text{diag}(\lambda_e, \lambda_\mu, \lambda_\tau) \end{aligned} \tag{4}$$

The quark mass eigenstates are mixed in the charged current interactions:

$$\frac{g}{\sqrt{2}} \overline{(u_1, u_2, u_3)_L} \gamma^\mu W_\mu^+ \begin{pmatrix} d_1 \\ d_2 \\ d_3 \end{pmatrix}_L = \sqrt{2} g\, \overline{(u, c, t)} \gamma^\mu (1 + \gamma^5) W_\mu^+ V \begin{pmatrix} d \\ s \\ b \end{pmatrix} \tag{5}$$

where $V = V_u^+ V_d$ is the Cabibbo-Kobayashi-Maskawa (CKM) matrix. In a convenient parametrization it has a form

$$V = \begin{pmatrix} c_{12}c_{13} & s_{12}c_{13} & s_{13}e^{-i\delta} \\ -s_{12}c_{23} - c_{12}s_{23}s_{13}e^{i\delta} & c_{12}c_{23} - s_{12}s_{23}s_{13}e^{i\delta} & s_{23}c_{13} \\ s_{12}s_{23} - c_{12}c_{23}s_{13}e^{i\delta} & -c_{12}s_{23} - s_{12}c_{23}s_{13}e^{i\delta} & c_{23}c_{13} \end{pmatrix} \tag{6}$$

where c_{ij} and s_{ij} respectively stand for the 'cos' and 'sin' of the mixing angles θ_{12}, θ_{23} and θ_{13}, and δ is a CP-violating phase.

However, in the SM the Yukawa constants remain arbitrary: $\hat{\lambda}^{u,d,e}$ are general complex 3×3 matrices. This constitutes the one of the key puzzles in modern particle physics, dubbed the problem of fermion flavours (or families). As a whole, this problem has different aspects, questioning origin of the family replication (why three families?), quark and lepton mass spectrum and mixing pattern, CP violation in weak interactions, etc.

One is of the main aspects of the flavour problem is related to pattern of neutrino masses and mixing. It is rather natural to expect that in the SM neutrinos have the masses of the Majorana nature, as a consequence of the absence of the weak isosinglet RH neutrino states $N_i^c(1,1,0)$ among the fermions (1). By this reason, the SM contains no renormalizable couplings that could generate the neutrino masses. In other words, as far as renormalizable interactions are concerned, the lepton and baryon number conservations arise as accidental symmetries of the theory. However, giving up the old fashion requirement that only the renormalizable couplings can present in the low energy effective theory, then the neutrino masses could emerge from the higher order operators cutoff by some physical scale $M \gg v$ at which the lepton number conservation is violated explicitly or spontaneously [1]. Namely, the lowest order coupling relevant for the neutrino masses is the $d = 5$ one (see Fig. 1):

$$\mathcal{L}_\nu = \frac{\alpha_{ij}}{M}(l_i\tilde{\phi})C(l_j\tilde{\phi}), \qquad \alpha^\nu_{ij} = \alpha^\nu_{ji} \tag{7}$$

The neutrino mass matrix $\hat{m}^\nu = \hat{\alpha}(v^2/M)$ can be brought to the diagonal basis of mass eigenstates by unitary transformation:

$$V_\nu \hat{m}^\nu V_\nu^T = \hat{m}^\nu_{diag}, \tag{8}$$

$$\hat{m}^\nu_{diag} = \mathrm{diag}(m_1, m_2, m_3) = \frac{v^2}{M} \cdot \mathrm{diag}(\alpha_1, \alpha_2, \alpha_3) \tag{9}$$

and the neutrino mass eigenstates $\nu_{1,2,3}$ in general should be different from the so called neutrino flavour eigenstates $\nu_{e,\mu,\tau}$ defined in the basis where the charged leptons are diagonal. Thus, in the case of massive neutrinos a nontrivial mixing matrix $V_{\mathrm{Lept}} = V_\nu^+ V_e$ similar to the CKM matrix emerges also in the lepton sector. In the following we parametrize the lepton mixing matrix in a manner similar to (6) with the mixing angles defined in the flavour basis ($\hat{m}^e = \hat{m}^e_{diag}$) respectively as $\theta_{e\mu}$, $\theta_{\mu\tau}$ and $\theta_{e\tau}$.

If the SM would be a literally true theory all the way up to planckian energies (i.e. if besides quarks and leptons (1), the Higgs doublet ϕ and the $SU(3) \times SU(2) \times U(1)$ gauge bosons the theory contains no more particles with masses less than M_P), then only the Planck scale could be envisaged as a natural cutoff scale for the non-renormalizable operator (7). Such operator with $M \sim M_P$ could effectively emerge due to the non-perturbative quantum gravitational effects [2, 3]. In this view, the value

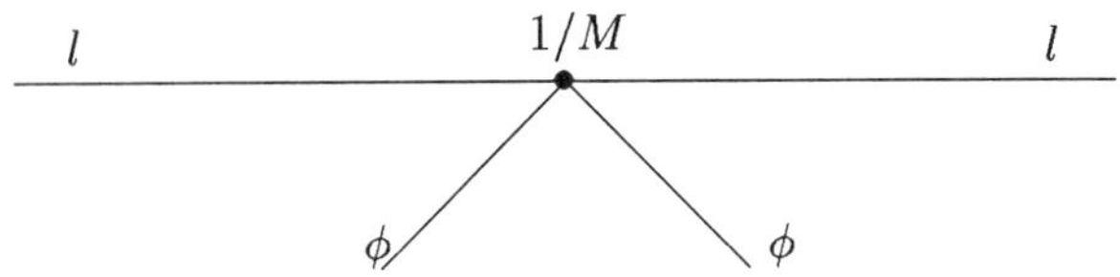

Figure 1. The effective operator for neutrino mass

$\overset{o}{m} = v^2/M_P \sim 10^{-5}$ eV can be regarded as a natural unit of the neutrino masses in the SM.[2]

However, if at higher energies the theory contains some relevant heavy states (e.g. the RH neutrino states N^c with the Majorana mass $M \leq M_P$), then in the low energy effective theory the operator (7) could emerge from the renormalizable Lagrangian, as a result of integrating out the heavy degrees of freedom. In particular, in the seesaw picture [4] this operator is induced by exchange of the heavy RH neutrinos (see Fig. 2) with the cutoff scale M defined by the mass scale of the latter which in principle could be less than M_P.

Indeed, one can introduce the RH neutrino states N_i^c, and consider the most general renormalizable Lagrangian

$$\mathcal{L}_\nu = \lambda_{ij}^\nu l_i N_j^c \tilde{\phi} + M_{ij} N_i^c N_j^c \tag{10}$$

where λ_{ij}^ν are the typical Yukawa constants similar to the ones (2) for the charged fermions, and $M_{ij} = g_{ij} M$ is the Majorana mass matrix parametrized by some dimensionless (Yukawa) constants g_{ij} and the typical mass scale $M \gg v$ of the lepton number violation. (In particular, this scale could emerge as a VEV of some scalar spontaneously violating the lepton number conservation, in which case the particle spectrum would contain the Goldstone boson – Majoron [47]). Clearly, at energies $E \ll M$ the diagram of Fig. 2 reduces, after decoupling the heavy states N_i^c, to the effective higher order operator (7) with the coupling constant matrix

$$\hat{\alpha} = \hat{\lambda}_\nu \hat{g}^{-1} \hat{\lambda}_\nu^T. \tag{11}$$

Any other known mechanism for the neutrino masses, e.g. the models of Zee loop [6] or the heavy isotriplet [7], effectively ends in producing such operators with $M < M_P$ at lower energies.

[2] Without knowing exactly to what is the basic scale of the theory, under the M_P we imply the reduced Planck mass $2 \cdot 10^{18}$ GeV.

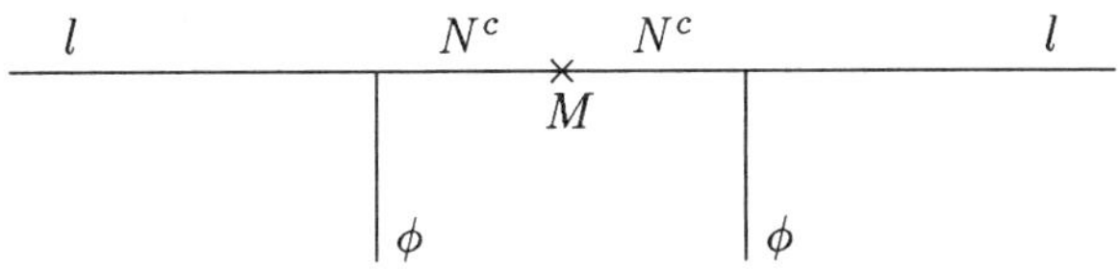

Figure 2. The seesaw diagram giving rise to the effective higher order operator for neutrino mass

1.2. Beyond the Standard Model

As noted above, the SM provides no explanation to the observed hierarchy of the fermion masses and mixing. In fact, the fermion mass and mixing problem can be phrased as a problem of the Yukawa coupling matrices $\hat{\lambda}^f$ which remain arbitrary in the SM. Therefore, there is no understanding to what is the origin of a strong hierarchy between the Yukawa eigenvalues, why $\hat{\lambda}^u$ and $\hat{\lambda}^d$ are alligned so that the CKM mixing angles are small, etc.

It is attractive to think that at some intermediate scale between the electroweak and Planck scales there exists a more fundamental theory which determines the fermion mass and mixing structures, or at least somehow constrains them. (For a review of the fermion mass and mixing problem in the SM and beyond see e.g. [8].) It would be plausible to think that the scale of the lepton number violation – i.e. the cutoff scale of the effective operator (7) is also related to the physical scale of such a theory.

Nowdays the most promising ideas beyond the SM are related to the concepts of supersymmetry (SUSY) and grand unification theories (GUT). Below we briefly outline the features of the neutrino mass generation mechanisms within these concepts.

• *SU*(3) × *SU*(2) × *U*(1) *MSSM* [9]. In the minimal supersymmetric standard model (MSSM) the superpotential terms responsible for the charged fermion masses are just the direct extensions of the SM couplings (2):

$$\mathcal{W}_{\rm Yuk} = \lambda^u_{ij} q_i u^c_j \phi_2 \, + \, \lambda^d_{ij} q_i d^c_j \phi_1 \, + \, \lambda^e_{ij} l_i e^c_j \phi_1, \tag{12}$$

where now $(q, l, u^c, d^c, e^c)_i$, $i = 1, 2, 3$ are the quark and lepton superfields. In the MSSM the Yukawa matrices $\lambda^{u,d,e}$ still remain arbitrary, while presence of two Higgses ϕ_1 and ϕ_2 with VEVs $v_1 = v\cos\beta$ and $v_2 = v\sin\beta$ ($v = 174$ GeV) involves also an additional parameter $\tan\beta = v_2/v_1$.

However, there is a dramatic difference from the SM – in the MSSM *per se* there is no more accidental conservation of the baryon and lepton numbers, and the latter can be violated by the renormalizable terms

$$W_{\text{R-viol}} = \lambda_{ijk} l_i l_j e^c_k + \lambda'_{ijk} l_i d^c_j q_k + \lambda''_{ijk} d^c_i d^c_j u^c_k \qquad (13)$$

These, so called R-violating terms would induce the neutrino masses by loops involving the fermions and their superpartners (the first two terms violating the lepton number are relevant). However, the presence of the third term would violate the baryon number and bring to catastrophycally short proton lifetime unless the constants λ''_{ijk} are strongly suppressed. The implications of the R-violating terms for the neutrino masses were discussed in a number of papers. The situation can be briefly discussed as follows: once the soft SUSY breaking scale is $\sim TeV$, i.e. essentially the same as the electroweak scale v, the cutoff scale of the effective operators (7) induced by sfermion loops will be essentially the same, and therefore the smallnes of the neutrino masses can be understood only at the price of the very small coupling constants α_{ij} – i.e. constants R-violating constants λ_{ijk} or λ'_{ijk} (the loop supression factor is by no means enough). The smallness of these constants might be related to the same flavour physics which explains the hierarchy of the Yukawa constants in (2), however no explicit mechanism is known. In the following we do not address the R-violating terms and accept the standard paradigm to suppress these terms by imposing R-parity, or so called matter parity under which the fermion superfields change the sign while the Higgs superfields $\phi_{1,2}$ are invariant.

In this case the lowest order operator violating the lepton number in the superpotential is

$$\mathcal{W}_\nu = \frac{\alpha^\nu_{ij}}{M}(l_i \phi_2)(l_j \phi_2) \qquad (14)$$

which in fact is the strightforward extension of the SM coupling (7) generating the neutrino masses. Therefore, with $M \sim M_P$ one cannot expect the neutrino masses being larger than the natural unit $\overset{o}{m}$.

• *SU*(5) *unification* [10]. In this model the quarks and leptons of each family fit into the supermultiplets

$$\bar{5}_i = (d^c, l)_i, \qquad 10_i = (u^c, q, e^c)_i; \qquad i = 1, 2, 3 \qquad (15)$$

The Higgs sector consists of chiral superfields in adjoint (24 dimensional) representation Σ, which at the scale $M_X \sim 10^{16}$ GeV breaks the $SU(5)$ symmetry down to $SU(3) \times SU(2) \times U(1)$ due to the VEV $\langle\Sigma\rangle = V_\Sigma \cdot \text{diag}(2, 2, 2, -3, -3)$, with $V_\Sigma \sim M_X$, and fundamental $(5 + \bar{5})$ representations

$$H = (T + \phi_2), \qquad \bar{H} = (\bar{T} + \phi_1) \qquad (16)$$

with $\phi_{1,2}$ being the MSSM Higgs doublets and $\bar{T}, T$ their colour triplet partners. The latter should be superheavy, with mass $\sim M_X$, in order to avoide catastrophically fast proton decay due to $d = 5$ operators mediated by the colour higgsino exchange [11], while the doublets $\phi_{1,2}$ should have mass $\sim v$. This is the content of the so called doublet-triplet splitting which in the context of the minimal $SU(5)$ demands the fine tuning of the parameters in the Higgs superpotential with the accuracy of about 10^{-14}.

The superpotential terms relevant for fermion masses become

$$\mathcal{W}_{\rm Yuk} = \lambda^u_{ij} 10_i H 10_j + \lambda^d_{ij} 10_i \bar{H} \bar{5}_j \, . \qquad (17)$$

At the GUT scale M_X these terms reduce to the MSSM couplings (12) with $\hat{\lambda}^e_{ij} = \hat{\lambda}^d_{ji}$, and hence $\lambda_{d,s,b} = \lambda_{e,\mu,\tau}$.

As far as the neutrino masses are concerned, the $SU(5)$ model, alike the SM, does not contain the RH neutrino states N^c. Therefore, if $SU(5)$ would be a fundamental theory up to the Planck scale M_P, the the neutrino masses could emerge only from the $d = 5$ operator

$$\mathcal{W}_\nu = \frac{\alpha^\nu_{ij}}{M_P} (\bar{5}_i H)(\bar{5}_j H) \qquad (18)$$

and the natural value of the neutrino masses would be rather $\overset{o}{m} = v^2/M_P \sim 10^{-5}$ eV.

The following remark is in order. The $\lambda_b = \lambda_\tau$ unification is definitely a great success of the SUSY $SU(5)$ GUT. After accounting for the RG running of the Yukawa constants from M_X to lower scales, it provides the nice explanation of the mass value of the bottom quark and its intimate relation to the top mass (When I get to the bottom I go back to the top [12]). However, the other predictions $\lambda_s = \lambda_\mu$ and $\lambda_d = \lambda_e$ are wrong: they imply $m_s/m_d = m_\mu/m_e \simeq 200$. In addition, there is no explanation neither for the fermion mass hierarchy nor for the CKM mixing pattern: the Yukawa matrices λ^u and λ^d remain arbitrary and there is no reason for their allignment.

However, one can consider the higher order terms with successively increasing dimensions involving the 24-plet Σ as being responsable for the observed fermion mass hierarchy. Indeed, there is no physical reason for suppressing such terms scaled by inverse power of M_P, which can exist on the same rights as a term (18) for the neutrino mass generation. In particular, one can consider the situation when the renormalizable terms (17) fix only the third family masses thus providing their masses $\sim v$ and maintaining the $\lambda_b = \lambda_\tau$ unification, while e.g. the operators like

$$\frac{1}{M_P} 10\,\Sigma H 10 + \frac{1}{M_P} 10\,\Sigma \bar{H} \bar{5}, \qquad \frac{1}{M_P^2} 10\,\Sigma^2 H 10 + \frac{1}{M_P^2} 10\,\Sigma^2 \bar{H} \bar{5}, \; \ldots \qquad (19)$$

are relevant for the fermion masses of the second and the first generations. Below the scale $\langle\Sigma\rangle \sim M_X$ they contribute to the Yukawa couplings (12) with the magnitudes $\sim \varepsilon_X; \varepsilon_X^2$ etc.; In this case one could understand the fermion mass hierarchy (25) in terms of the small ratio $\varepsilon_X = M_X/M_P$. At the same time this would avoide the *wrong* predictions $\lambda_{d,s} = \lambda_{e,\mu}$ of the minimal $SU(5)$, since the tensor product 24×5 contains also the effective 45-plet. In other words, one could say that the coefficients $\lambda^u ij$ and λ^d_{ij} in (17) are the Σ field dependent functions understood as expansion series $\lambda_{ij}(\frac{\Sigma}{M}) = \lambda_{ij}^{(0)} + \lambda_{ij}^{(1)}\frac{\Sigma}{M} + \dots$.

The higher order operators can be induced through the renormalizable interactions, as a result of integrating out the hypothetical superheavy particles. Namely, the quark and lepton masses can emerge through mixings with the superheavy fermions in representations $5 + \bar{5}$, $10 + \overline{10}$, etc., in which view the celebrated seesaw scenario for neutrinos represents only a particular case of such *universal seesaw* mechanism [14]. Such fermions with mass $\sim M_P$ can exist in string theories. Typically they emerge also in the context GUT theories larger than $SU(5)$: E.g., 27-plet of the E_6 theory, in addition to $\bar{5} + 10$ contains also extra $5 + \bar{5}$ states. In principle, heavy fermions can have masses less than Planck mass and their exchanges can induce effective operators like (19) cutoff by a scale $M < M_P$. On the other hand, they can be more instructive for the fermion mass model building than the *ad hoc* introduced non-renormalizable operators, since they can be produced in a rather selective way, depending on the heavy fermion representations, and thus can fix the Clebsch factors between the Yukawa constants.

• *SO(10) Unification* [15]. $SO(10)$ is a smallest group in which all the fermions in one family fit into one irreducible representation 16. In addition to the quark and lepton states of the SM, it includes also RH neutrino N^c (singlet of $SU(5)$):

$$SU(5): \quad 16_i = \bar{5}(d^c, l)_i + 10(u^c, q, e^c)_i + 1(N^c)_i, \tag{20}$$

As for the Higgs doublets $\phi_{1,2}$, they fit into the 10-plet H of $SO(10)$:

$$SU(5): \quad 10 = \bar{5}(\bar{T}, \phi_1) + 5(T, \phi_2) \tag{21}$$

The $SO(10)$ symmetry can break down to $SU(5)$ by VEVs of Higgs superfields $\psi \sim 16$, $\bar{\psi} \sim \overline{16}$ towards their $SU(5)$ singlet components. The symmetry can be further broken down to the standard $SU(3)\times SU(2)\times U(1)$ by a set of Higgses in representations $S \sim 54(= 24 + 15 + \overline{15})$ and $A \sim 45(= 1 + 24 + 10 + \overline{10})$, which can acquire the VEVs towards the $SU(5)$ adjoint as well as singlet components. It is worthwile to remark, that the DT problem in the $SO(10)$ model can be solved through the missing VEV mechanism [16]. In this way the Higgs doublets $\phi_{1,2}$ can be kept light while their colour triplet partners $\bar{T}, T$ acquire the $O(M_X)$ mass.

The fermion masses are generated via the Yukawa terms

$$W_{\rm Yuk} = \lambda_{ij} 16_i 16_j H \tag{22}$$

which define the pattern of the charged fermion Yukawa matrices $\hat{\lambda}^{u,d,e}$ (2) as well as of the Yukawa matrix $\hat{\lambda}^{\nu}$ which generates the Dirac mass term for neutrinos. in order to avoide the unphysical mass spectrum for quarks and leptons, one has to assume that in general the coefficients λ_{ij} are functions of the 45-plets, $\lambda_{ij} = \lambda_{ij}(\frac{A}{M})$, or in other words, that the expression (22) includes the higher order operators involving 45-plets and scaled by the inverse powers of M. Needless to say that such operators can be effectively induced by the universal seesaw mechanism [14] after integrating out the heavy fermions like $16 + \overline{16}$ etc., which is they key tool for the predictive $SO(10)$ models [17].

Thus, $SO(10)$ can in principle relate all Yukawa matrices $\hat{\lambda}^{u,d,e,\nu}$ by the $SO(10)$ Clebsch factors and thus reduce the number of fundamental parameters in the fermion sector. In particular, the predictive $SO(10)$ frameworks [17, 18] allow to calculate the neutrino Dirac masses, and their spectrum typically has the same shape as that of the charged fermions. However, this cannot suffice for deducing the mass pattern for the light Majorana neutrinos, since the $SO(10)$ theory contains the RH neutrinos N^c and the LH neutrino masses are generated by means of the seesaw mechanism [4]. Therefore, now the question mark is contained in the RH neutrinos Majorana mass matrix $M_{ij} = g_{ij}M$ (see eq. (10)). The latter in itself can emerge e.g. through the $d = 5$ operators involving the the Higgs 16-plets $\psi, \bar{\psi}$:

$$\frac{g_{ij}}{M_P}(16_i\bar{\psi})(\bar{\psi}16_j) \tag{23}$$

Thus we obtain $M \sim M_{10}^2/M_P$, where M_{10} is the $SO(10) \to SU(5)$ breaking scale and it can range from $M_X \simeq 10^{16}$ GeV to $M_P = 10^{18}$ GeV. For example, for $M_{10} \sim M_X$ we have $M \sim 10^{13-14}$ GeV, which corresponds to the mass of the heaviest neutrino (ν_τ) of about $0.1 - 1$ eV. More precise predictions are obscured by the unknown coupling constants g_{ij}, which however in the context of complete flavour models could be fixed by imposing some symmetry constraints (see e.g. in [18]).

2. Signals for neutrino mass and mixing

The mass spectrum of the quarks and charged leptons is spread over five orders of magnitude, from MeVs to 100 GeVs. It is divided into the following weigh groups (in units of the weak scale $v = 174$ GeV):

$$m_t \sim v, \qquad m_{c,b,\tau} \sim 10^{-2}v, \qquad m_{s,\mu} \sim 10^{-3}v, \qquad m_{u,d,e} \sim 10^{-5}v \tag{24}$$

We also observe that the *horizontal* hierarchy of quark masses exhibites the approximate scaling low (with $\varepsilon \sim 1/20$):

$$m_t : m_c : m_u \sim 1 : \varepsilon^2 : \varepsilon^4, \qquad m_b : m_s : m_d \sim 1 : \varepsilon : \varepsilon^2, \tag{25}$$

while for the charged leptons we have $m_\tau : m_\mu : m_e \sim 1 : \varepsilon : \varepsilon^3$.

The weak transitions dominantly occur inside the families, and are suppressed between different families. The CKM angles exhibit the following hierarchy:

$$\theta_{12} \sim \varepsilon^{1/2}, \quad \theta_{23} \sim \varepsilon, \quad \theta_{13} \sim \varepsilon^2, \tag{26}$$

which points to the correlations between the quark mass spectrum and mixing pattern. Moreover, there are intriguing relations between masses and mixing angles, such as the well-known formula for the Cabibbo angle $s_{12} = \sqrt{m_d/m_s}$.

As for the neutrinos, direct measurements show no evidence for any of them to be massive, providing only the upper bounds [19]:

$$m_{\nu_\tau} < 18 \text{ MeV}, \quad m_{\nu_\mu} < 170 \text{ keV}, \quad m_{\nu_e} < 15 \text{ eV} \tag{27}$$

Experimental limits on the neutrinoless double beta decay provide the upper bound for the (1,1) element of the neutrino Majorana mass matrix in the flavour basis [20]:

$$2\beta_{0\nu}: \quad m^\nu_{ee} < 0.5 \text{ eV} \tag{28}$$

In addition, one should mention the cosmological upper limit on the (stable) neutrino masses:

$$\sum m_\nu < 94h^2 \text{ eV}, \quad \nu = \nu_{e,\mu,\tau} \tag{29}$$

where h is the Hubble constant in units 100 km s^{-1} Mpc-1.

On the other hand, there have been indirect "positive" signals for neutrino masses and mixing accumulated during the past years. If all these hints will indeed be confirmed in future experiments, then three standard neutrinos ν_e, ν_μ and ν_τ would not suffice for their explanation. Since existence of the fourth active neutrino is excluded by the LEP measurements of the invisible decay width of Z-boson, one has to introduce an extra light sterile neutrino ν_s. These signals are:

A. *Solar neutrino problem* (SNP). The solar ν_e deficit indicated by the solar neutrino experiments [21] cannot be explained by nuclear or astrophysical reasons (see [22] and references therein). The most popular and natural solution is provided by the oscillation of ν_e into another neutrino ν_x ($\nu_x = \nu_\mu, \nu_\tau$ or ν_s). It can explain the experimental data in two following regimes (for the most recent analysis see ref. [23]):

Just-so [25]: long wavelength oscillation from Sun to Earth, with parameters in the range

$$\delta m^2_{ex} = (0.4 - 6) \cdot 10^{-10} \text{ eV}^2, \quad \sin^2 2\theta_{ex} = 0.6 - 1 \tag{30}$$

MSW [24]: resonant oscillation inside the solar medium, with the parameter range

$$\delta m^2_{ex} = (0.3 - 1) \cdot 10^{-5}\ \text{eV}^2, \qquad \sin^2 2\theta_{ex} = (0.2 - 2) \cdot 10^{-2} \tag{31}$$

B. *Atmospheric neutrino problem* (ANP). There has been a long-standing discrepance by almost a factor of 2 between the predicted and observed ν_μ/ν_e ratio of the atmospheric neutrino fluxes. Recently the Super-Kamiokande Collaboration has published the results of a high statistics experiment [26] which indicate the zenith angle/energy dependence of the depletion of the atmospheric ν_μ flux consistent with the oscillation $\nu_\mu - \nu_x$ in the following parameter range:

$$\delta m^2_{\mu x} = (0.5 - 6) \cdot 10^{-3}\ \text{eV}^2, \qquad \sin^2 2\theta_{\mu x} = 0.8 - 1. \tag{32}$$

where ν_x can be ν_τ or ν_s (the possibility of $\nu_x = \nu_e$, or admixture of ν_e with angle larger than 13° is disfavoured by the data of the CHOOZ experiment [27]).

C. *LSND result:* The results of the LSND experiment at Los Alamos indicate the $\bar\nu_\mu - \bar\nu_e$ oscillation with ν_μ coming from the π^+ decay at rest, and also the $\nu_\mu - \nu_e$ oscillation using the pion decays in flight [28], with the following parameter range:

$$\delta m^2_{e\mu} = (0.2 - 1)\ \text{eV}^2, \qquad \sin^2 2\theta_{e\mu} = (0.3 - 4) \cdot 10^{-2}. \tag{33}$$

The values of $\delta m^2_{e\mu}$ larger than 1 eV^2 are excluded by the results of the KARMEN experiment [30].

One can mention also the *Dark matter problem* (DMP). The "after COBE" status of the cosmological Large scale structures may suggest, among the other possible scenarios, the existence of some fraction of the hot dark matter (HDM), along with the cold dark matter (CDM) being a dominant component [31]. The role of HDM can be naturally played by neutrinos with mass of few eV's. As for the CDM, there are several candidates, e.g. the lightest supersymmetric particle (LSP) or the axion condensate. However, it is of certain interest to think of it as also consisting of some heavier (keV range) neutrinos with correspondingly small concentration, so called warm dark matter (WDM) [32].

Therefore, the present neutrino anomalies indicate that the relevant spectrum of δm^2 is divided into the following weight groups (in terms of the natural mass unit $\overset{o}{m} = v^2/M_P \sim 10^{-5}$ eV):

$$\begin{aligned} &[\delta m^2_{e\mu}]^{1/2} \sim 10^5 \overset{o}{m} && \text{(LSND)}, \\ &[\delta m^2_{\mu x}]^{1/2} \sim 5 \cdot 10^3 \overset{o}{m} && \text{(ANP)}, \\ &[\delta m^2_{ey}]^{1/2} \sim \overset{o}{m}\ \ \text{(Just-so)}\ \ \text{or}\ \ \sim 3 \cdot 10^2 \overset{o}{m}\ \ \text{(MSW)} \end{aligned} \tag{34}$$

whereas for the mixing angles we have (in terms of the hierarchy parameter $\varepsilon \sim 1/20$ as in (25)):

$$\begin{aligned} &\theta_{e\mu} \sim \varepsilon; && \text{(LSND)} \\ &\theta_{\mu x} \simeq \frac{\pi}{4}; && \text{(ANP)} \\ &\theta_{ey} \sim \frac{\pi}{4} \quad \text{(Just-so)} \quad \text{or} \quad \sim \varepsilon && \text{(MSW)} \end{aligned} \tag{35}$$

These indicate that the neutrino mass and mixing structures have no close similarity to the patterns (24), (25) and (26) for the charged fermions.

First of all, the ranges of δm^2 required by the SNP and ANP cannot be reconciled with the LSND data within three known neutrinos $\nu_{e,\mu,\tau}$. Indeed, if $\delta m^2_{e\mu} \sim 1$ eV^2 is fixed from LSND, and the oscillation $\nu_\mu - \nu_\tau$ is responsible for the ANP with $\delta m^2_{\mu\tau} \sim 10^{-3}$ eV^2, then the relation $\delta m^2_{e\mu} - \delta m^2_{\mu\tau} = \delta m^2_{e\tau}$ excludes the possibility for $\nu_e - \nu_\tau$ oscillation to be relevant for the SNP in either of Just-so or MSW regimes.

If one neglects the LSND data, then the SNP and ANP could be reconciled within $\nu_{e,\mu,\tau}$ states. E.g., one can assume that the $\nu_\mu - \nu_\tau$ mixing is maximal and the value $\delta m^2_{\mu\tau} \sim 3 \cdot 10^{-3}$ eV^2 demanded by the ANP is due to mass of the eigenstate $\nu_3 \simeq \frac{1}{\sqrt{2}}(\nu_\mu + \nu_\tau)$ order $m_3 \sim 5 \cdot 10^{-2}$ eV, while the mass of the orthogonal state $\nu_2 \simeq \frac{1}{\sqrt{2}}(\nu_\mu - \nu_\tau)$ is much smaller. In the context of the operators (7) this implies that the leading contribution comes form the one with the cutoff scale $M \sim 10^{14}$ GeV and the coupling constant matrix α_{ij} $(i, j = \mu, \tau)$ being completely democratic with the elements ~ 1. Then the SNP can be due to $\nu_e - \nu_2$ oscillation. The Just-so solution demands $\nu_{e,2}$ neutrino masses $\sim \overset{o}{m}$ and large mixing angle θ_{e2}. These could be induced from the subdominant operators (7) with $M \sim M_P$ and all $\alpha_{ij} \sim 1$ that could naturally emerge from the Planck scale physics [2, 3]. The MSW solution implies the mass of ν_2 order $2 \cdot 10^{-3}$ eV, which favours smaller scale $M \sim 10^{15-16}$ GeV, and the "hierarchial" form of the matrix α_{ij}, similar to the charged fermion Yukawa matrices $\lambda^{u,d,e}_{ij}$.

However, the pattern of the neutrino masses and mixing required for the reconciling of SNP and ANP seems to be rather different form the ones for the charged fermions.[3] This points that the flavour physics responsible for the fermion masses is very specific as far as neutrinos are concerned. Although the mass hierarchy $m_3 : m_2 \sim 1 : \varepsilon$ required by the ANP+MSW case is similar to $m_\tau : m_\mu$ or $m_b : m_s$ (c.f. (25), the ANP data point to the maximal mixing angle $\theta_{\mu\tau} \simeq \pi/4$ (45°), much different from the corrsponding angle in quark mixing $\theta_{23} \sim \varepsilon$ ($\theta_{23} = 2.3°$). On the opposite, the mixing angle $\theta_{e\mu}$ required by MSW is $\sim \varepsilon$ (1°–4°), whereas the Cabibbo angle is much larger: $\theta_{12} = 12.7°$.

[3] Several ansatzes for the neutrino masses have been considered in the literature for reconciling the SNP and ANP – see e.g. [33] and references therein.

In any case, the evidence for the neutrino masses larger than the natural unit $\overset{o}{m}$ points that the regulator scale M of the operator (7) should be much smaller than the Planck scale. It is by no means attractive to consider the truly non-renormalizable operators cutoff by the scale less than M_P, therefore one can consider (7) as an effective operators arising in the low energy limit after decoupling of some heavy states with the mass $\sim M \ll M_P$, the most plausible possibility given by the seesaw mechanism [4]. In particular, for having the neutrino mass in the ANP range – of the order of 0.1 eV, one has to take $M \sim 10^{14}$ GeV, which as we noted can be implemented in the $SO(10)$ model for the mass of ν_τ. Then, if the neutrino mass hierarchy is similar to the one of the charged fermion masses, the ν_μ mass can be $10^{-2.5}$ eV and thus the SNP can be due to the MSW mechanism via the oscillation $\nu_e \to \nu_\mu$.

Leaving aside for the moment the SNP, also the ANP and LSND can be easily reconciled within the $\nu_{e,\mu\tau}$ system. For example, one can assume that the neutrino mass matrix in flavour basis $\nu_{e,\mu,\tau}$ has a texture obeying the approximate $\bar{L} = L_e + L_\mu - L_\tau$ conservation [34, 35]:

$$\hat{m}_\nu = \begin{pmatrix} 0 & 0 & m\sin\theta_{e\mu} \\ 0 & 0 & m\cos\theta_{e\mu} \\ m\sin\theta_{e\mu} & m\cos\theta_{e\mu} & \varepsilon m \end{pmatrix} \tag{36}$$

with say $m \sim 1$ eV and $\varepsilon \simeq 10^{-3}$. Then it has one massless eigenstate $\nu_1 \simeq \nu_e$ and two almost degenerate eigenstates $\nu_{2,3} \simeq \frac{1}{\sqrt{2}}(\nu_\mu \pm \nu_\tau)$, with masses $m_{2,3} \approx m(1 \pm \frac{\varepsilon}{2})$. Then the ANP can be explained by the oscillation $\nu_\mu - \nu_\tau$, since $\sin 2\theta_{\mu\tau} \approx 1$ emerges naturally and $\delta m^2_{\mu\tau} \approx 2\varepsilon m^2 \sim 2 \cdot 10^{-3}$ eV^2 is fixed by choice of the ε. In addition, the two almost degerate states ν_μ and ν_τ will play a role of the HDM [36]. At the same time, the oscillation $\nu_\mu - \nu_e$ with $\delta m^2_{e\mu} = m^2 \sim 1\,\mathrm{eV}^2$ can explain the LSND result, if $\sin\theta_{e\mu} \simeq 0.1$. Thus, only the SNP remains unresolved.

One can built a simple seesaw model that could naturally implement the above structure [35]. Let us introduce beyond the left-handed lepton doublets $l_{e,\mu,\tau}$ and the right-handed charged leptons e, μ, τ of the standard model, also two right-handed neutrinos N and S (subscripts L and R are omitted). We prescribe the global lepton number $\bar{L} = 1$ to states $l_{e,\mu}$, e, μ and N, and $\bar{L} = -1$ to states l_τ, τ and S. Let us also introduce a gauge singlet scalar Φ with $\bar{L} = -2$. Then the relevant Lagrangian has a form

$$\begin{aligned}(g_e \bar{l}_e \phi e + g_\mu \bar{l}_\mu \phi \mu + g_\tau \bar{l}_\tau \phi \tau) + (g_1 \bar{l}_e \tilde{\phi} N + g_2 \bar{l}_\mu \tilde{\phi} N + g_3 \bar{l}_\tau \tilde{\phi} S) \\ + MNCS + g_N \Phi NCN + g_S \Phi^* SCS \,+\, \text{h.c.}\end{aligned} \tag{37}$$

where ϕ is a standard Higgs doublet with a VEV $v = 174$ GeV ($\tilde{\phi} = i\tau_2\phi^*$), and C is the charge conjugation matrix. We assume that the scalar Φ developes a nonzero VEV $\langle\Phi\rangle = V \ll M$, which spontaneously breaks the $\bar{L}$-invariance and gives rise to majoron in the particle spectrum. We also assume that the Yukawa constants $g_{N,S} \sim 1$ while $g_{1,2,3}$ have the same pattern as $g_{e,\mu,\tau}$, so that the Dirac mass terms of neutrinos emerge with approximately the same magnitudes as masses of the charged leptons: $m^D_{1,2,3} \sim m_{e,\mu,\tau}$. Clearly, this model provides a texture resembling (36). One neutrino eigenstate ν_1 (an admixture of ν_e and ν_μ with angle $\theta_{e\mu} = m^D_1/m^D_2 \sim 10^{-2}$) is left massless while two other eigenstates $\nu_{2,3}$ get masses $m \simeq m^D_2 m^D_3/M$ through the seesaw mixing to the heavy states N, S, with small splitting $\varepsilon \sim (m^D_3/m^D_2)(V/M)$.

Then m in the eV range corresponds to $M \sim 10^8$ GeV, while $\varepsilon \sim 10^{-3}$ demands that $V \sim 10^4$ GeV. Interestingly, the latter value satisfies the upper bound on the spontaneous lepton number breaking scale which arises if the the Planck scale effects on the majoron are taken into account [37].

Thus, the texture (36) can explain ANP and LSND and leaves open only the SNP.

3. Sterile neutrinos as the neutrinos of mirror world

3.1. Where do they all come from, and where do they all belong?

If all the SNP, ANP and LSND anomalies will indeed be confirmed in future experiments, then three standard neutrinos ν_e, ν_μ and ν_τ would not suffice for their explanation – three different ranges for the δm^2 (34) dictated by these data cannot be accomodated within this system. Since existence of the fourth active neutrino is excluded by the LEP measurements of the invisible decay width of Z-boson, one has to introduce an extra light sterile neutrino ν_s. It could be substantially mixed with ν_e (if the oscillation $\nu_e - \nu_s$ is responsible for the SNP, in which case the ANP could be due to the oscillation $\nu_\mu - \nu_\tau$), or maximally mixed with ν_μ if the ANP is due to the oscillation $\nu_\mu - \nu_s$.

It was shown [38] that only one possible texture is compatible with all the above mentioned data, which requires that $m_{\nu_{e,s}} \ll m_{\nu_{\mu,\tau}}$. In this case the SNP can be explained by the $\nu_e - \nu_s$ oscillation and the ANP by the $\nu_\mu - \nu_\tau$ oscillation. In addition, the ν_μ and ν_τ with mass ~ 1 eV could provide the cosmological HDM and can also explain the LSND result. The detailed mass matrix for the neutrinos in this case is essentially a block matrix with each block being 2×2 and with a small mixing between the ν_e and ν_μ to account for the LSND result.

Indeed, the $\nu_e - \nu_s$ oscillation in the small mixing angle MSW regime or large wavelength Just-so regime is the only natural place where the extra sterile neutrino could show up. Although the $\nu_\mu - \nu_s$ oscillation could be applied also to for explaining the ANP, in this case it would be hard to satisfy the primordial nucleosynthesis bound on the effective number of neutrino species $N_\nu^{eff} < 3.5$ [39], since for the parameter range required by the APN the sterile neutrino would come into the equilibrium [40] and lead to $N_\nu^{eff} = 4$.

One can immediately question, from where the sterile neutrino comes from and how it can be so light when it is allowed to have a large mass by the gauge symmetry of the standard model. In the SM context the sterile neutrinos are nothing but the RH neutrinos, and it is difficult to understand why they do not combine with LH neutrinos to form their Dirac masses as large as the ones of the charged fermions, and why they do not have large Majorana masses in the spirit of the seesaw scenario.

It was suggested in [41, 34] that the sterile neutrino is in fact a neutrino of a mirror world[4] which is the exact duplicate of our visible world, and the two worlds interact only through the gravity or possibly through some singlet fields of the type of moduli. The existence of such world can be motivated by fundamental superstring type theories based on the $E_8 \times E_8'$ symmetry, where the factor E_8 after the symmetry breaking at the compactification leads to our ordinary world while E_8' could lead to the mirror world. Thus, the mirror neutrinos $\nu'_{e,\mu,\tau}$ should be light by the same reason as the usual ones $\nu_{e,\mu,\tau}$: their mass terms are suppressed by the accidental B-L symmetry resulting from the gauge symmetry and field content of the theory. In addition, in the ref. [34] it was assumed that the mirror parity between two worlds can be broken and thus the 'electroweak' scale v' in the mirror sector does not necessarily coincide with the electroweak scale $v = 174$ GeV of the ordinary world. In particular, one could have the parameter $\zeta = v'/v$ in the range $1 - 30$.

This framework can provide a plausible solution to all present neutrino puzzles. One can suppose that the dominant entries ($\sim$ eV) in the neutrino mass matrix have origin in some intermediate scale physics which respects an approximate global ZKM-type lepton number $\bar{L} = L_e + L_\mu - L_\tau$ conservation in both sectors. This fixes the neutrino mass texture considered in the previous section, which can reconcile the ANP and the LSND oscillation, and also provide the cosmological HDM. In this texture the ν_e and ν'_e states are left massless. The masses and mixing of the latter and thereby the oscillation $\nu_e - \nu'_e$ then can be induced by the Planck scale effects [3], which for $\zeta \sim 1$ properly reproduce the parameter range needed for the Just-so solution to the SNP, and for $\zeta \sim 30$ the parameter range for the MSW solution. The question arises, what is a possible role played by two other mirror neutrinos $\nu'_{\mu,\tau}$. In the framework presented below they have masses in the keV range and constitute the WDM of the universe.

The concept of the hidden mirror world has been considered in several earlier papers [43]. A key difference of our approach [34, 44, 35] from the

[4] For some other proposals for light ν_s invoking extra global symmetries, see [42].

earlier ones is that we consider a case of the spontaneously broken mirror parity between two worlds, so that the weak scales v and v' are different. This can allow also to reconcile the Big Bang nucleosynthesis (BBN) constraints on the effective number of the light neutrinos [39]. The mirror photons and neutrinos could apriori give a large contribution considerably exceeding $N_\nu = 3$. Therefore, their abundances at the BBN epoch must be appropriately reduced. To achieve this goal, we assume an asymmetric inflationary reheating between the two universes, which can naturally occur once the mirror parity is spontaneously broken. In particular, if the mirror universe reheats to a lower temperature than our universe, then BBN constraint can be satisfied. We also address some cosmological implications of the mirror particles.

3.2. Mirror world and neutrino oscillations

Having in mind the $E_8 \times E_8'$ string theory, one can imagine that in the field-theoretical limit it reduces to a gauge theory given by the product of two identical groups $G \times G'$, where $G = SU(3) \times SU(2) \times U(1)$ stands for the standard model describing particles of the visible world: quarks and leptons q_i, u_i^c, d_i^c; l_i, e_i^c and the Higgs doublet ϕ, and $G' = [SU(3) \times SU(2) \times U(1)]'$ is its mirror counterpart with analogous particle content: q_i', $u_i'^c$, $d_i'^c$; l_i', $e_i'^c$ and ϕ' ($i = 1, 2, 3$ is a family index). Let us suppose also that there exists a discrete mirror parity $P(G \leftrightarrow G')$ interchanging all particles in corresponding representations of G and G'. It implies that all coupling constants (gauge, Yukawa, Higgs) have the same pattern in both sectors. Let us also assume that there is some mechanism that spontaneously breaks P parity at lower energies and thus allows the weak interaction scales $\langle\phi\rangle = v$ and $\langle\phi'\rangle = v'$ to be different. Thus the fermion mass and mixing pattern in the mirror world is completely analogous to that of the visible world, but all fermion masses are scaled up by the factor $\zeta = v'/v$. The W', Z' boson and mirror Higgs masses are also scaled up by factor ζ, while photons and gluons remain massless in both systems.

We suppose that two worlds communicate only through gravity and possibly also via some superheavy matter which is singlet with respect to the SM.[5] It is also essential that at higher energies the $SU(3) \times SU(2) \times U(1)$ factors in both sectors should be embedded into some simple gauge groups: otherwise, kinetic terms of the two $U(1)$ gauge fields can mix which would induce arbitrary electric charges to the particles [43]. E.g. one can consider (SUSY) GUT like $SU(5) \times SU(5)'$ without mixed representations, which then breaks down to $G \times G'$ at higher energies. We also assume that

[5] E.g. two sectors could also communicate through the common gauge horizontal symmetry group $SU(3)_H$ acting on fermion families [45], or through the common Peccei-Quinn symmetry [46].

the mirror parity P is not broken at the GUT scales and its breaking is essentially related to the electroweak symmetry breaking scales v and v'.

Concerning the strong interactions, it is clear that a big difference between the weak scales v' and v will not cause as big difference between the confinement scales in two worlds. As far as P parity is valid at higher scales, the strong coupling constant will evolve down in energies with same value in both sectors until it reaches threshold of the mirror-top (t') mass. Below it α_s' will have a different slope than α_s. It is then very easy to calculate the value of the scale Λ' at which α_s' becomes large. This value will of course depend on ζ. By taking $\Lambda = 200$ MeV for the ordinary QCD, then for $\zeta \sim 30$ we find $\Lambda' \simeq 300$ MeV. On the other hand, for $\zeta \sim 30$ the masses of 'light' mirror quarks $m'_{u,d} = \zeta m_{u,d}$ do not exceed the value of Λ' and thus they should develope condensates $\langle \bar{q}' q' \rangle$ as their visible partners do. So, the mirror π-mesons should have masses rather close to that of our K-mesons. Also the mirror nucleons are not much heavier than the usual proton and neutron, $m'_{p,n} \sim 1.5 m_{p,n}$. However, the mirror light quarks have ~ 200 MeV masses and thus we expect the mass difference between the mirror neutron and proton of 100 MeV or so, while the mirror electron mass is $m'_e = \zeta m_e \sim 15$ MeV. Then unlike in our world, in the mirror world all bound neutrons will be unstable against β decay and the mirror hydrogen will be the only stable nucleus.

As for neutrinos, they can acquire nonzero masses only via operators bilinear in the Higgs fields and cutoff by some large scale M. As far as the P parity breaking is a lower energy phenomenon, it should be respected by these operators. For example, one can directly extend the model of previous section, by introducing along with the heavy neutral states N, S also their mirror partners N', S', with the same mass $M \sim 10^8$ GeV. Then the neutrino mass operators in two sectors are:

$$\frac{h_{ij}}{M}(l_i \phi) C (l_j \phi) + \frac{h_{ij}}{M}(l'_i \phi') C (l'_j \phi') + \text{h.c.} \tag{38}$$

with constants h_{ij} obeying the approximate $L_e + L_\mu - L_\tau$ symmetry. In doing so, the $\nu_{2,3}$ states get almost degenerate masses $m \sim$ few eV and thus can constitute HDM, and the ANP and LSND problems are also solved. Then for $\zeta \sim 30$, mass of their mirror partners $\nu'_{2,3}$, $m' = \zeta^2 m$, is in the keV range and thus the latter could consitute the WDM of the universe.

The ν_e and ν'_e states are left massless. However, they can get masses from the Planck scale effects which explicitly violate the global lepton number, and can also induce the neutrino mixing between two sectors [3]. The relevant operators are:

$$\frac{\alpha_{ij}}{M_{Pl}}(l_i \phi) C (l_j \phi) + \frac{\alpha_{ij}}{M_{Pl}}(l'_i \phi') C (l'_j \phi') + \frac{\beta_{ij}}{M_{Pl}}(l_i \phi) C (l'_j \phi') + \text{h.c.} \tag{39}$$

with constants $\alpha, \beta \sim 1$. Then ν_e and ν'_e acquire masses respectively $\sim v^2/M_P = \overset{o}{m}$ and $\sim v'^2/M_P = \zeta^2 \overset{o}{m}$, and their mixing term is $\sim vv'/M_P = \zeta \overset{o}{m}$, where $\overset{o}{m} = v^2/M_{Pl} = 3 \cdot 10^{-6}$ eV. In other words, the Majorana mass matrix in the ν_e, ν'_e basis has a form:

$$\begin{pmatrix} \overset{o}{m} & \zeta \overset{o}{m} \\ \zeta \overset{o}{m} & \zeta^2 \overset{o}{m} \end{pmatrix} \tag{40}$$

(for simplicity we $\alpha = beta = 1$ is assumed). Diagonalizing this matrix, we obtain for the mixing angle θ of the mixing $\nu_e - \nu'_e$ and corresponding δm^2:

$$\sin^2 2\theta_{ee'} = \frac{4\zeta^2}{(\zeta^2+1)^2}; \qquad \delta m^2_{ee'} = (\zeta^2+1)^2 \cdot 10^{-11}\ \text{eV}^2 \tag{41}$$

Hence, for $\zeta \gg 1$ parameters of the oscillation $\nu_e - \nu'_e$ are in the range:

$$\delta m^2 \sim \left(\frac{\zeta}{30}\right)^4 \times 6 \cdot 10^{-6}\,\text{eV}^2; \qquad \sin^2 2\theta \sim \left(\frac{30}{\zeta}\right)^2 \times 5 \cdot 10^{-3} \tag{42}$$

which for $\zeta \sim 30$ perfectly fits the small mixing angle MSW solution to the SNP [24]. More generally, by taking into account the solar model uncertainties [22, 24], as well as the possible order of magnitude spread in constants α, β, the relevant range for ζ can be extended to $\zeta \sim 20 - 40$ [34]. Alternatively, for $\zeta \sim 1$ we get $\delta m^2 \sim 10^{-10}\,\text{eV}^2$ and $\sin^2 2\theta \sim 1$, which corresponds to the 'just-so' solution [25].

In either MSW or Just-so regimes, the critical prediction is that the SNP is due to oscillation of ν_e into sterile neutrino, which can be tested at neutrino detectors like SNO or BOREXINO.

3.3. Spontaneous parity breaking and asymmetric inflation

The simplest possibility to spontaneously break the mirror parity $P(G \leftrightarrow G')$ is to introduce a P-odd real scalar η with VEV $\langle \eta \rangle = \mu$ at some intermediate scale [47]. Then the Higgs potential has a form:

$$\begin{aligned} &\mathcal{V}(\eta) + m^2(|\phi|^2 + |\phi'|^2) + \lambda(|\phi|^4 + |\phi'|^4) \\ &\quad + g\mu\eta(|\phi|^2 - |\phi'|^2) + h\eta^2(|\phi|^2 + |\phi'|^2) \end{aligned} \tag{43}$$

so that after the non-zero VEV of η emerges, the effective mass terms of ϕ and ϕ' become different and their VEVs v and v' will be different as well. As for the gauge and Yukawa coupling constants, they will not be affected and thus will maintain the same pattern in both worlds. Hence, the particle spectrum in the mirror world will have the same shape as that of the visible one but scaled up by the factor $\zeta = v'/v$.

On the other hand, the P-parity breaking can be related to the possibility of asymmetric postinflationary "reheating" between two worlds.[6] In fact, it is natural to assume that the field η itself plays a role of the inflaton [34, 44], provided that potential $\mathcal{V}(\eta)$ is sufficiently flat and satisfies all necessary 'inflationary' conditions [49]. As far as P-parity is broken, then it should be violated also in the inflaton couplings to visible and mirror Higgses: e.g. constants of trilinear couplings of the inflaton to ϕ and ϕ' become respectively $(h \pm g)\mu$. Then the inflaton will decay into the visible and mirror particles with different rates, so that the two thermal bathes can be established having different temperatures T_R and T'_R. One has also to assume that already at the reheating stage two worlds are decoupled from each other. In particular, this means that couplings like $a|\phi|^2|\phi'|^2$, in principle allowed by symmetry, should be strongly suppressed: $a < 10^{-7}$. In this way, the initial cosmological abundance of the mirror particles can be smaller than that of their visible partners. For more detailed discussion of asymmetric reheating and possible realistic supersymmetric models see ref. [44].

Let us discuss now constraints on the difference of the reheating temperatures T_R and T'_R. The most serious bound emerges from the BBN constraint on the effective number of the light particle species [39]. As far as the two worlds are decoupled already at the inflationary reheating epoch, during the universe expansion they evolve with separately conserved entropy[7]. Then the T_R, T'_R are related to the temperatures T, T' respectively of the usual and mirror photons at the BBN era as

$$\frac{T'_R}{T_R} = \left(\frac{2 + 5.25x^3}{10.75}\right)^{\frac{1}{3}} \frac{T'}{T}; \qquad x = \frac{T'_\nu}{T'} \tag{44}$$

where T'_ν is a temperature of the mirror neutrinos.

In standard cosmology effective number of the light degrees of freedom at the BBN era is $g_* = 10.75$ as it is contributed by photons, electrons and three neutrino species $\nu_{e,\mu,\tau}$, in a remarkable agreement with the observed abundances of light elements [39]. In our case, the mirror photons γ' and neutrinos $\nu'_{e,\mu,\tau}$ would also contribute the effective number of the light particles, as

$$\Delta g_* = 1.75\Delta N_\nu = (2 + 5.25x^4)\left(\frac{T'}{T}\right)^4 \tag{45}$$

The value of x is determined by the temperature T'_D at which ν' decouple from the mirror plasma. It approximately scales as $T'_D \sim \zeta^{4/3} T_D$, where

[6] The idea of using inflation to provide a temperature difference between ordinary matter and mirror or other forms of hidden matter was first discussed in ref. [48].

[7] In fact, this applies if the expansion goes adiabatically in both sectors and only the second order phase transitions occur. At the presence of the first order phase transitions in both sectors this relation would change due to additional entropy production [44].

$T_D = 2 - 3$ MeV is the decoupling temperature of the usual neutrinos. For smaller values of ζ, when $T'_D < \Lambda'$, ν' decouple after the mirror QCD phase transition, so that the mirror electrons e' contribute the heating of γ' after the ν' freezing out and we arrive to the standard result $x = (4/11)^{1/3} = 0.71$. For sufficiently large ζ, T'_D can be larger than both the QCD scale Λ' and the light mirror quark masses $m'_{u,d} = \zeta m_{u,d}$. Then u', d' and the mirror gluons will also contribute and we obtain $x = (4/53)^{1/3} = 0.36$. Thus, for x in the interval $0.71 - 0.36$ from eqs. (45) and (44) we obtain:

$$\frac{T'}{T} < (0.85 - 0.96)(\Delta N_\nu)^{1/4} \quad \Longrightarrow \quad \frac{T'_R}{T_R} < (0.60 - 0.57)(\Delta N_\nu)^{1/4} \quad (46)$$

Therefore, by taking the conservative bound $\Delta N_\nu < 1$ or very strong limit $\Delta N_\nu < 0.1$, we obtain that reheating temperature in the mirror sector has to be about 2 or 3 times smaller than that of the visible world.

A somewhat stronger bound can be derived from the overclosure constraint of the universe. Since in our model almost degenerate ν_μ and ν_τ have masses $m \sim 1$ eV, they form the HDM of the universe with $\Omega_\nu = 2m/(94h^2 \text{ eV})$. Then their mirror partners $\nu'_{\mu,\tau}$ being ζ^2 times heavier would contribute the cosmological energy density as $\Omega'_\nu = r\zeta^2\Omega_\nu$, where $r = n'_\nu/n_\nu$ stands for the present abundance of mirror neutrinos relative to ordinary ones: $r = (xT'/T)^3$, and h is the Hubble constant in units 100 Km s^{-1} Mpc^{-1}. Then by taking rather conservatively that $\Omega_\nu + \Omega'_\nu < 1$ (bearing in mind that other particles like LSP or mirror baryons could also contribute the present energy density), we obtain

$$\frac{T'}{T} < \frac{1}{x\zeta^{2/3}}(\Omega_\nu^{-1} - 1)^{1/3} = 0.29\left(\frac{0.36}{x}\right)\left(\frac{30}{\zeta}\right)^{2/3}(\Omega_\nu^{-1} - 1)^{1/3} \quad (47)$$

Therefore, if $\Omega_\nu \simeq 0.2$ [31], then for $\zeta \approx 30$ we get that $T'/T < 0.46$ and $T'_R/T_R < 0.27$, which limits are comparable to that of eqs. (46) for $\Delta N_\nu = 0.05$!

3.4. Cosmological implications of mirror particles

Thus, the concept of the mirror universe suggests a natural possibility for solving the DMP with dark matter components entirely consisting of neutrinos. Indeed, in the most interesting case the electroweak scale v' in the mirror sector should be by factor $\zeta \sim 30$ larger than the standard electroweak scale $v = 174$ GeV, while the reheating temperature of the mirror universe should be 3-4 times smaller than that of the visible one. Hence, if the usual neutrinos with $m \sim$ few eV form the HDM component, then bulk of the dark matter can be the WDM component consisting of their mirror partners with $m' = \zeta^2 m \sim$ few keV. Clearly, the latter could form the halo dark matter even in dwarf spheroidal galaxies where the Tremaine-Gunn limit is most stringent ($m'_\nu > 0.3 - 0.5$ keV).

Implications of the WDM for the shape of the large scale structure are rather similar [32] to that of the currently popular CDM made upon the heavy ($m \sim 100$ GeV) particles or axion condensate. However, more detailed observational data on the distribution of matter in the universe may make it possible to discriminate between warm and cold dark matter. Moreover, dark matter consisting of sterile neutrinos invalidates direct searches of the CDM candidates via superconducting detectors or axion haloscopes. High energy neutrino fluxes from the Sun and from the Galactic center which are expected from the annihilation of LSP's if they dominate in the universe, will also be absent. In supersymmetric versions of our scheme, however, CDM as well could exist in the form of the LSP.

An interesting question is what is the amount and form of the mirror baryonic dark matter in the universe. Most likely, baryogenesis in the mirror universe proceeds through the same mechanism as in the visible one and we may expect that the baryon asymmetries in both worlds should be nearly the same. Since mirror nucleons are not much heavier than the usual ones, their fraction in the present energy density, $\Omega_{B'}$, would be about the same as Ω_B, that is around a few percent.

Let us discuss now cosmological evolution of mirror baryonic matter. Since the binding energy of the mirror hydrogen atom is thirty times larger than that of the ordinary hydrogen, its recombination occurs much before the usual recombination era. Hence, the evolution of density fluctuations in the mirror matter would be more efficient than in the visible one. (From the viewpoint of the visible observer mirror baryons behave as a dissipative dark matter.) As a result, one can expect that the distribution of mirror baryons in galactic discs should be more clumped towards the center. It is noteworthy that mirror dark matter may show antibiasing behaviour ($b < 1$) which is considered unphysical for normal dark matter. Recent data on the dark matter distribution in dwarf spiral galaxies obtained at smaller distances from the center and with a better resolution, do not agree with the assumption of purely collisionless dark matter and may indicate the existence of dissipative dark matter [50].

On the other hand, since mirror hydrogen is the only stable nucleus in the mirror world, nuclear burning could not be ignited and luminous (in terms of γ') mirror stars cannot be formed. Therefore, nothing can prevent the sufficiently big protostars to collapse and in dense galactic cores a noticeable fraction of mirror baryons should form the black holes. Recent observational data indeed suggest a presence of giant black holes with masses $\sim 10^{6-7} M_\odot$ in galactic centers. In addition, easier formation of mirror black holes may explain the early origin of quasars.

The remaining fraction of the mirror baryons could fragment into smaller objects like white dwarves (or possibly neutron stars) which can maintain stability due to the pressure of degenerate fermions. For the mirror stars consisting entirely of hydrogen, the Chandrasekhar limit is $M'_{\rm Ch} = 5.75(m_p/m'_p)^2 M_\odot \simeq 3M_\odot$. For smaller mirror objects the evaporation limit should be $2-3$ orders of magnitude smaller than for the visible ones because the Bohr radius of the mirror hydrogen is 30 times smaller than that of the usual one.

These mirror objects, being dark for the normal observer, could be observed as Machos in the gravitational microlensing experiments (for a review, see e.g. ref. [51]). In principle they can be distinguished from the Machos of the visible world. The latter presumably consist of the dim compact objects (brown dwarves) too light to burn hydrogen, with masses ranging from the evaporation limit ($\sim 10^{-7} M_\odot$) to the ignition limit ($\sim 10^{-1} M_\odot$) [52]. As for the mirror Machos, for large $\zeta \sim 30$ their mass spectrum can extend from the smaller evaporation limit $\sim 10^{-9} M_\odot$ up to the Chandrasekhar limit $\sim 3M_\odot$.

For small $\zeta \sim 1$, the the stellar astrophysics as well as the cosmological hystory of the mirror matter should be about the same as in ordinary one, and therefore any star with typical mass formed by the mirror baryons should be seen as dark Macho-like object for observations.

The present data on the microlensing events are still too poor to allow any firm conclusion on the presence of such heavy (or light) objects. An unambiguous determination of the Macho mass for each event is impossible, and only the most probable mass can be obtained, depending of the spatial and velocity distribution of Machos. The optical depth or the fraction of the sky covered by the Einstein disks of Machos, is nearly independent of their mass: the Einstein disk surface is proportional to M, while the number of deflectors for a given total mass decreases as M^{-1}. Therefore, the information on the mass spectrum of Machos can be deduced only from the typical time duration of the microlensing events. However, very interestingly, the event statistics accumulated over past years by the MACHO and EROS collaborations monitoring the target stars in LMC indicates the average mass of Machos $\bar{M} = (0.5 \pm 0.2) M_\odot$, which is larger than the ignition limit $0.08 M_\odot$ for the compact objects built upon the ordinary matter. Maybe this confusing result, if resists in future, will be the first indication for the existence of the mirror stars?

On the other hand, one can expect that mirror stars in our Galaxy would significantly contribute to the microlensing events towards the galactic bulge. Interestingly, the event rates in the galactic bulge observed by OGLE and MACHO experiments are about twice larger than the expected value deduced from the low mass star population in the Galactic disk [54]. Barring accidental conspiracies like a presence of bar (elongated dense stellar distribution along the line of sight), this can be explained by the contribution of mirror stars, which could naturally increase the optical depth towards the galaxy bulge by factor 2 or so.

Acknowledgments

I wish to thank H.V. Klapdor-Kleingrothaus and other organizers for the most pleasant hospitality during the B&L Meeting in Trento.

References

[1] S. Weinberg, *Phys. Rev. Lett.* **43** (1979) 1566.

[2] R. Barbieri, J. Ellis and M.K. Gaillard, *Phys. Lett.* **B90** (1980) 249.

[3] E. Akhmedov, Z. Berezhiani, G. Senjanović, *Phys. Rev. Lett.* **69** (1992) 3013

[4] M. Gell-Mann, P. Ramond and R. Slansky, in *Supergravity*, eds. D. Freedman et al., North-Holland, Amsterdam, 1979;
T. Yanagida, *Prog. Th. Phys.* **B135** (1979) 66;
R.N. Mohapatra and G. Senjanović, *Phys. Rev. Lett.* **44** (1980) 912.

[5] Y. Chikashige, R.N. Mohapatra and R.D. Peccei, *Phys. Lett.* **B98** (1981) 265.

[6] A. Zee, *Phys. Lett.* **B93** (1980) 389 and **B161** (1985) 141.

[7] J. Schrechter and J. Valle, *Phys. Rev.* **D25** (1982) 774.

[8] Z. Berezhiani, ICTP Series in Theoretical Physics – Vol. 12, Lectures of the 1995 Summer School on *High Energy Physics and Astrophysics*, eds. E. Gava et al., World Scientific, Singapore, 1996, p. 618 [hep-ph/9602325].

[9] For review, see e.g. H.E. Haber and G.L. Kane, Phys. Rep. **117** (1985) 75.

[10] H. Georgi and S.L. Glashow, *Phys. Rev. Lett.* **32** (1974) 438
H. Georgi, H.R. Quinn and S. Weinberg, *ibid.* **33** (1974) 451

[11] S. Weinberg, *Phys. Rev.* **D26** (1982) 287;
N. Sakai and T. Yanagida, *Nucl. Phys.* **B197** (1982) 533.

[12] J. Lennon and P. McCartney, "Helter Skelter" in *White Album*, vol. 2, ed. G. Martin, EMI, London, 1968.

[13] J. Ellis and M.K. Gaillard, *Phys. Lett.* **B84** (1979) 113.

[14] C.D. Froggatt and H.B. Nielsen, *Nucl. Phys.* **B147** (1979) 277;
Z.G. Berezhiani, *Phys. Lett.* **B129** (1983) 99 and *Phys. Lett.* **B150** (1985) 177;
S. Dimopoulos, *Phys. Lett.* **B129** (1983) 417.

[15] H. Georgi, in *Particles and Fields*, APS Proceedings, ed. C. Carlson, 1974;
H. Fritzsch and P. Minkowski, *Ann. Phys.* (N.Y.) **93** (1975) 193.

[16] S. Dimopoulos and F. Wilczek, Preprint NSF-ITP-82-07 (1982).

[17] G.W. Anderson et al., *Phys. Rev.* **D49** (1994) 3660;
K.S. Babu and S.M. Barr, *Phys. Rev.* **D48** (1994) 5354;
Z.G. Berezhiani, *Phys. Lett.* **B355** (1995) 178; hep-ph/9407264.

[18] Z. Berezhiani, Z. Tavartkiladze, *Phys. Lett.* **B409** (1997) 220.

[19] Particle Data Group, *European Physical Journal* **C 3** (1998) 1.

[20] Heidelberg-Moscow Collab., *Phys. Rev.* **D55** (1997) 54 and *Phys. Lett.* **B407** (1997) 219;
H.V. Klapdor-Kleingrothaus, J. Helming and M. Hirsh, *J. Phys. G: Nucl. Phys.* **24** (1998) 483.

[21] Homestake Collaboration, Nucl. Phys. B (Proc. Suppl.) **38** (1995) 47;
GALLEX Collaboration, *Phys. Lett.* **B342** (1995) 440 and **B388** (1996) 364;
SAGE Collaboration, V. Gavrin et al., Proc. XVIII Int. Conf. on *Neutrino Physics and Astrophysics*,, eds. Y. Suzuki and I. Totsuka;
Kamiokande Collaboration, *Phys. Rev. Lett.* **77** (1996) 1683;
SuperKamiokande Collaboration, *Phys. Rev. Lett.* **81** (1998) 1158.

[22] J.N. Bahcall, *Neutrino Astrophysics*, Cambridge University Press, 1989;
V. Castellani, S. Degl'Innocenti and G. Fiorentini, *Astron. & Astrophys.* **271** (1993) 601;
V.S. Berezinsky, *Comments Nucl. Part. Phys.* **21** (1994) 249.

[23] J. Bahcall, P. Krastev and A. Smirnov, hep-ph/9807216.

[24] S.P. Mikheyev and A.Yu. Smirnov, it Yad. Fiz. **42** (1995) 1441;
L. Wolfenstein, *Phys. Rev.* **D17** (1978) 2369;
for the later refs., see P. Krastev and A. Smirnov, *Phys. Lett.* **B338** (1994) 282;
V. Berezinsky, G. Fiorentini and M. Lissia, *Phys. Lett.* **B341** (1994) 38;
N. Hata and P. Langacker, *Phys. Rev.* **D50** (1994) 632.

[25] V. Gribov and B. Pontecorvo, *Phys. Lett.* **28** (1967) 493;
V. Barger, R. Phillips and S. Whisnant, *Phys. Rev.* **D24** (1981) 538;
S.L. Glashow and L. Krauss, *Phys. Lett.* **B190** (1987) 199; A. Acker, S. Pakvasa, J. Pantaleone, *Phys. Rev.* **D43** (1991) 1754;
for the later refs. see F. Calabresu et al., hep-ph/9507352;
P. Krastev and S. Petcov, *Nucl. Phys.* **B449** (1995) 605;
Z.G. Berezhiani and A. Rossi, *Phys. Rev.* **D51** (1995) 5229 and *Phys. Lett.* **B367** (1996) 219;
J.N. Bahcall and P. Krastev, *Phys. Rev.* **D53** (1996) 4211.

[26] SuperKamiokande Collaboration, Y. Fukuda et al., *Phys. Rev. Lett.* **81** (1998) 1562.

[27] CHOOZ Collaboration, M. Apolonio et al., hep-ex/9711002.

[28] LSND collaboration, C. Athanassopoulos et al., *Phys. Rev. Lett.* **75** (1995) 2650; *Phys. Rev.* **C54** (1996) 2685; nucl-ex/9709006.

[29] C. Athanassopoulos et al., (LSND collab.) *Phys. Rev. Lett.* **75** (1997) 2650 and *Phys. Rev.* **C54** (1997) 2685.

[30] K. Eitel for the KARMEN collab., hep-ex/9706023.

[31] R. Shaefer and Q. Shafi, *Nature* **359** (1992) 199;
E.L. Wright et al, *Ap. J.* **396** (1992) L13;
A. Klypin et al., *ibid.* **416** (1993) 1.

[32] S. Dodelson and L. Widrow, *Phys. Rev. Lett.* **72** (1994) 17;
R. Malaney, G. Starkman and L. Widrow, preprint CITA-95-9 (1995).

[33] R. Barbieri, L. Hall and A. Strumia, hep-ph/9808333.

[34] Z.G. Berezhiani and R.N. Mohapatra, *Phys. Rev.* **D52** (1995) 6607.

[35] Z. Berezhiani, *Acta Phys. Polon.* **B 27** (1996) 1503.

[36] J.R. Primack et al., PRL **74** (1995) 2160.

[37] E. Akhmedov, Z. Berezhiani, R.N. Mohapatra and G. Senjanović, *Phys. Lett.* **B299** (1993) 90;
I. Rothstein, K.S. Babu and D. Seckel, *Nucl. Phys.* **B403** (1993) 725;
J. Cline, K. Kainulainen and K.A. Olive, *Astroparticle Phys.* **1** (1993) 387.

[38] D. Caldwell and R.N. Mohapatra, *Phys. Rev.* **D48** (1993) 3259
J. Peltoniemi and J.W.F. Valle, *Nucl. Phys.* **B406** (1993) 409

[39] V.F. Schwartzmann, *JETP Lett.* **9** (1969) 184;
T. Walker et al, *Ap. J.* **376** (1991) 51;
K.A. Olive and G. Steigman, *Ap. J. Suppl.* **97** (1995) 49;
C.J. Copi, D.W. Schramm and M.S. Turner, *Ap. J.* **455** (1995) L95, and references therein.

[40] R. Barbieri and A. Dolgov, *Phys. Lett.* **B237** (1990) 440; *Nucl. Phys.* **B349** (1991) 743;
K. Kainulainen, Phys. Lett. **B244** (1990) 191;
K. Enqvist, K. Kainulainen and M. Thomson, *Nucl. Phys.* **B373** (1992) 498.

[41] R. Foot and R. Volkas, *Phys. Rev.* **D52** (1995) 6595.

[42] A.Yu. Smirnov and J.W.F. Valle, *Nucl. Phys.* **B375** (1992) 649;
J. Peltoniemi, D. Tommasini and J.W.F. Valle, *Phys. Lett.* **B298**, (1993) 383;
E. Akhmedov, Z. Berezhiani, G. Senjanović, Z. Tao, *Phys. Rev.* **D47** (1993) 3245;
E.J. Chun, A. Joshipura and A. Smirnov, *Phys. Lett.* **B357** (1995) 608 and hep-ph/9507371.
E. Ma, hep-ph/9507348.

[43] T.D. Li and C.N. Yang, *Phys. Rev.* **104** (1956) 254
A. Salam, *N. Cimento* **5** (1957) 299
Y. Kobzarev, L. Okun and I. Pomeranchuk, *Yad. Fiz.* **3** (1966) 1154
L. Okun, *ZhETF* **79** (1980) 694
S. Blinnikov and M. Khlopov, *Astron. Zh.* **60** (1983) 632
B. Holdom, *Phys. Lett.* **B166** (1985) 196
S.L. Glashow, *ibid.* **B167** (1986) 35
R. Foot, H. Lew, R. Volkas, *ibid.* **272** (1991) 67 and *Mod. Phys. Lett.* **A9** (1994) 169;

[44] Z.G. Berezhiani, A.D. Dolgov and R.N. Mohapatra, *Phys. Lett.* **B 375** (1996) 26.

[45] Z. Berezhiani, *Phys. Lett.* **B417** (1998) 287.

[46] V. Rubakov, it JETP Lett. **65** (1997) 621;
Z. Berezhiani, D. Comelli, A. Dolgov and A. Rossi, *in preparation.*

[47] D. Chang, R.N. Mohapatra, M. Parida, Phys. Rev. D **30** (1984) 1052.

[48] E.W. Kolb, D. Seckel and M.S. Turner, *Nature* **514** (1985) 415;
H.M. Hodges, *Phys. Rev.* **D47** (1993) 456.

[49] A. Linde, *Particle Physics and Inflationary Cosmology*, Harwood, Switzerland, 1990.

[50] B. Moore, *Nature* **370** (1994) 629;
R.A. Flores and J.R. Primack, *Ap. J.* **427** (1994);
A. Burkert, astro-ph/9504041.

[51] R. Ansari, *Nucl. Phys.* B *(Proc. Suppl.)* **43** (1995) 108.

[52] A. De Rujula et al., *Astron. and Astrophys.* **254** (1992) 99.

[53] MACHO Collaboration, *Phys. Rev. Lett.* **74** (1995) 2867; *Ap. J* **471** (1996) 774; *ibid.* **486** (1997) 697;
EROS Collaboration, *Astron. Astrophys.* **301** (1995) 1; **314** (1996) 94; **324** (1997) 69.

[54] A. Udalski et al., *Ap. J.* **426** (1994) L69;
Ch. Alcock et al., astro-ph/9506016.

Large Neutrino Flavor Mixings and Lepton Mass Matrices

Morimitsu TANIMOTO

Ehime University, Matsuyama 790-8577, JAPAN

Abstract.
Recent atmospheric neutrino data at Super-Kamiokande suggest the near-maximal flavor mixing. Models for the lepton mass matrix, which give the near-maximal flavor mixing, are discussed in this report. Mass matrix models are classified according to the mechanism providing the large flavor mixing, and those are reviewed briefly. "Naturalness" of the mass matrix is also discussed in order to select the neutrino mass matrix. Details of the mass matrix with the S_3 flavor symmetry are presented.

1. Introduction

The standard model (SM) has still many unexplained features. It is a remarkable property of the quark and the lepton mass spectra that the masses of successive particles increase by large factors. Mixings of the quark sector (CKM matrix) [1] seem also to have an hierarchical structure. Those features may provide an important basis for a new physics beyond the SM.

On the other hand, the flavor mixing of the lepton sector is very ambiguous. However, neutrino flavor oscillations provide information of the fundamental property of neutrinos such as masses, flavor mixings and CP violating phase. In these years, there is growing experimental evidences of the neutrino oscillations. The exciting one is the atmospheric neutrino deficit [2]~[4] and the solar neutrino deficit [5]. Super-Kamiokande [6] also presented the large neutrino flavor oscillation in atmospheric neutrinos. Furthermore, a new stage is represented by the long baseline(LBL) neutrino oscillation experiments. The first LBL reactor experiment CHOOZ has already reported a bound of the neutrino oscillation [8], which gives a strong constraint of the flavor mixing pattern. The LBL accelerator experiment K2K [9] is planned to begin taking data in 1999, whereas the MINOS [10] and ICARUS [11] experiments will start in the first year of the next century. Those LBL experiments will clarify masses, flavor mixings and CP violation of neutrinos.

The short baseline experiments may be helpful to understand neutrino masses and flavor mixings. The tentative indication has been already given by the LSND experiment [12], which is an accelerator experiment for $\nu_\mu \to \nu_e(\overline{\nu}_\mu \to \overline{\nu}_e)$. The CHORUS and NOMAD experiments [13] have reported the new bound for $\nu_\mu \to \nu_\tau$ oscillation, which has already improved the E531 result [14]. The KARMEN experiment [15] is also searching for the $\nu_\mu \to \nu_e(\overline{\nu}_\mu \to \overline{\nu}_e)$ oscillation as well as LSND. The Bugey [16] and Krasnoyarsk [17] reactor experiments and CDHS [18] and CCFR [19] accelerator experiments have already given bounds for the neutrino mixing parameters as well as E776 [20].

In this report, our starting point as to neutrino mixings is the large $\nu_\mu \to \nu_\tau$ oscillation of the atmospheric neutrino oscillation with

$$\Delta m^2_{\rm atm} = 10^{-3} \sim 10^{-2}{\rm eV}^2 , \qquad \sin^2 2\theta_{\rm atm} \geq 0.8 , \tag{1}$$

which are derived from the recent data of the atmospheric neutrino deficit at Super-Kamiokande [6]. Then, questions are raised: Why is there large flavour mixing in the lepton sector in contrast with the quark sector? Are there possible mechanisms providing a large mixing angle from the lepton mass matrices, which are consistent with the quark sector? Answer is "Yes". There are many mass matrix models to predict the near-maximal mixing.

2. Origin of Near-Maximal Mixing

In this section, we review lepton mass matrix models, which predict the near-maximal mixing between flavors. We classify mass matrix models according to the mechanism giving the large mixing.

A: See-saw enhancement

The see-saw mechanism of neutrino mass generation gives a very natural and elegant understanding for the smallness of neutrino masses [21]. This mechanism may play another important role, which is to reproduce the large flavor mixing. In the standpoint of the quark-lepton unification, the Dirac mass matrix of neutrinos is similar to the quark mass matrices. Therefore, the neutrino mixings turn out to be typically of the same order of magnitude as the quark mixings. However, the large flavor mixings of neutrinos could be obtained in the see-saw mechanism as a consequence of a certain structure of the right-handed Majorana mass matrix [22][23]. That is the so called see-saw enhancement of the neutrino mixing due to the cooperation between the Dirac and Majorana mass matrices.

Mass matrix of light Majorana neutrinos m_ν has the following form

$$m_\nu \simeq -m_D M_R^{-1} m_D^T \ , \tag{2}$$

where m_D is the neutrino Dirac mass matrix and M_R is the Majorana mass matrix of the right-handed neutrino components. Then, the lepton mixing matrix is [22]

$$V_\ell = S_\ell^\dagger \cdot S_\nu \cdot V_s \ , \tag{3}$$

where S_ℓ, S_ν are transformations which diagonalize the Dirac mass matrices of charged leptons and neutrinos, respectively. The V_s specifies the effect of the see-saw mechanism, i.e. the effects of the right-handed Majorana mass matrix. It is determined by

$$V_s^T m_{ss} V_s = diag(m_1, m_2, m_3) \ , \tag{4}$$

where

$$m_{ss} = -m_D^{diag} M_R^{-1} m_D^{diag} \ . \tag{5}$$

Here $m_i (i = 1, 2, 3)$ are the masses of light neutrinos and

$$m_D^{diag} \equiv diag(m_{1D}, m_{2D}, m_{3D}) \ , \tag{6}$$

is the diagonalized Dirac mass matrix of neutrinos. In the case of two generations, the mixing matrix V_s is easily investigated in terms of one angle θ_s as follows:

$$\tan 2\theta_s = \frac{\sin 2\theta_M \epsilon_D (1-\epsilon)}{\epsilon - \epsilon_D^2 + \sin^2\theta_M (1-\epsilon)(1+\epsilon_D^2)} \ , \tag{7}$$

$$\sin^2\theta_M = \frac{1}{(1-\epsilon)(1-\epsilon_D^2)} \left[\pm \left(1 + \frac{m_2}{m_3}\right) \sqrt{\epsilon_0 \epsilon} - \epsilon - \epsilon_D^2 \right] \ , \tag{8}$$

where

$$\epsilon \equiv \frac{M_2}{M_3} \ , \quad \epsilon_D \equiv \frac{m_{2D}}{m_{3D}} \ , \quad \epsilon_0 \equiv \frac{m_{2D}^2 m_3}{m_{3D}^2 m_2} \ . \tag{9}$$

Thus, the mixing angle is given in terms of the diagonal components in mass matrices. In the range $4\epsilon_D^4/\epsilon_0 \ll \epsilon \ll \epsilon_0$ with $m_2 \ll m_3$, the mixing can be approximately by [22]

$$\sin^2\theta_s \simeq \frac{\epsilon_D^2}{\sqrt{\epsilon_0 \epsilon}} \ . \tag{10}$$

The mixing becomes maximal value $\sin^2\theta_s = 1$ at $\epsilon = 4\epsilon_D^4/\epsilon_0$. That is the enhancement due to the see-saw mechanism. The rich structure of right-handed Majorana mass matrix can lead to the maximal flavor mixing of neutrinos.

In this estimate, mass matrices are assumed to be real. However, the Majorana mass matrix has a non-trivial phase even in the two generation model. The see-saw enhancement condition should be modified including CP violating phases [23]. It was found that the see-saw enhancement could be obtained due to the phase even if the Majorana mass matrix is proportional to the unit matrix.

Models which satisfy the see-saw enhancement were proposed on the early stage by Harvey, Reiss and Ramond [24], and Babu and Shafi [25] in the framework of $SO(10)$. Recent works based on the flavor $U(1)$ symmetry [26] are attractive examples for the see-saw enhancement. The $U(2)$ symmetry [27] was also studied focusing on the see-saw enhancement [28]. A successful mass matrix is also presented in the phenomenological point of view [29].

B: Type II see-saw model

The conventional see-saw mechanism for neutrino masses is implemented in gauge model such as $SO(10)$ or the left-right symmetric models. The general form of the see-saw mass matrix is

$$\begin{pmatrix} f v_L & m_D \\ m_D^T & f M_R \end{pmatrix} , \tag{11}$$

which gives

$$m_\nu = f v_L - m_D M_R^{-1} m_D^T \quad \text{with} \quad v_L \simeq \lambda v_{EW}^2 / v_R \ . \tag{12}$$

This is called the type II see-saw formula [30]. Recall that the conventional see-saw formula omits the first term. If due to some symmetry reasons, $f_{ab} = f_0 \delta_{ab}$, then a degenarate neutrino spectrum emerges. In this model, the flavor symmetry S_4 or the horizontal $SU(2)_H$ guarantee the degenerate f_{ab}. For instance, in the range of $v_R = 10^{13} - 10^{16}$GeV, the desired value $v_L \simeq 0.01 \sim 1$eV is quite reasonable. In the minimal SUSY $SO(10)$ model, the realistic neutrino mixings $\sin^2 2\theta_\odot \simeq 2.8 \times 10^{-2}$ and $\sin^2 2\theta_{\rm atm} \simeq 0.84$ have been obtained by putting experimental data of masses and CKM matrix elements of the quark sector.

C: Exotic fields

The new particle may be essential for the large flavor mixing. The Zee model is a typical model, in which charged gauge singlet Higgs boson plays important role to give the maximal mixing [31]. Anti-GUT model also need additional some Higgs bosons [32], which lead to the large flavor mixing.

The mixing between the ordinary fermions and the exotic ones may be an origin of the large flavor mixing [33][34]. The exogenous mixing based on $SO(10)$ has made possible to investigate the flavor mixing quantitatively [33].

D: Large evolution of mixings by RGE's

There may be another enhancement mechanism of the neutrino flavor mixing. The renormalization group equation(RGE) of the see-saw neutrino mass operators with dimension-5 has been investigated by some authors [35]~[37]. Babu, Leung and Pantaleone pointed out that the neutrino flavor mixing is enhanced by the RGE in the MSSM under the special conditions of the mass matrices. The numerical analyses have been given in ref.[37] focusing on recent experimental data of the atmospheric neutrino deficit.

E: Large mixing derived from the charged lepton mass matrix

In the standpoint of the quark-lepton unification, the charged lepton mass matrix is considered to be similar to the down quark one. Then, the mixing following from the charged lepton mass matrix may be considered to be small in the hierarchical base of the quark mass matrix. However, this expectation is not true if the mass matrix is non-Hermitian. In the $SU(5)$ model, the left(right)-handed down quark mixings are related to the right(left)-handed charged lepton mixings because these fermions belong to same representation 5^* such as

$$5^* : \psi_L = (d^{c1},\ d^{c2},\ d^{c3},\ e^-,\ \nu)_L\ , \tag{13}$$

and the Yukawa couplings are given by $5^*_i 10_j 5^*_H$(i,j=1,2,3). This feature was nicely taken into consideration in models of refs.[38] and [39].

It should be noticed that observed quark mass spectra and CKM matrix only constrains the down quark mass matrix as follows [40]:

$$m_{\text{down}} \sim \begin{pmatrix} \lambda^4 & \lambda^3 & \lambda^3 \\ & \lambda^2 & \lambda^2 \\ & & 1 \end{pmatrix} \quad \text{with} \quad \lambda = 0.22\ . \tag{14}$$

The three unknown entries are related to the left-handed lepton mixing in the $SU(5)$ model. Thefore, there is a room in the charged lepton mass matrix to provide a source of the large flavor mixing of the lepton sector.

F: Effective neutrino mass matrix

The democratic mass matrix [41] needs the large rotation in order to move to the diagonal base. In the quark sector, this large rotation is canceled each other between down quarks and up quarks. However, the situation of the lepton sector is very different from the quark sector if the effective neutrino mass matrix m^{ν}_{LL} is far from the democratic one. Details of the model is discussed in the section 4.

Since the result of the atmospheric neutrino at Super-Kamiokande excites the model building of the lepton mass matrix, new ideas and models will be presented in the near future. Our classification of models with the large flavor mixing will be not enough.

3. Naturalness and Near-Maximal Mixing Angle

Since there are many models, which predict the large flavor mixing, the "naturalness" of the mass matrix is helpful to select models. The idea of the natural mass matrix was proposed in order to restrict severely the arbitrariness in the construction of the quark mass matrices [42].

Let us consider the 2×2 Hermitian quark mass matrix $M_i(i = u,\ d)$. Assume for simplicity that it can be diagonalized by some orthogonal matrices $O_i(i = u,\ d)$ as follows:

$$O_i^T M_i O_i = M_i^{diag} \equiv \begin{pmatrix} m_{i1} & 0 \\ 0 & m_{i2} \end{pmatrix} , \qquad (i = u,\ d) , \tag{15}$$

with

$$O_i = \begin{pmatrix} \cos\theta_i & \sin\theta_i \\ -\sin\theta_i & \cos\theta_i \end{pmatrix} , \qquad V_q = O_u^T O_d = \begin{pmatrix} \cos\theta_c & \sin\theta_c \\ -\sin\theta_c & \cos\theta_c \end{pmatrix} , \tag{16}$$

where $\theta_c = \theta_d - \theta_u$. Then the mass matrices can be written as

$$M_i = \begin{pmatrix} c_i^2 m_{i1} + s_i^2 m_{i2} & c_i s_i (m_{i2} - m_{i1}) \\ c_i s_i (m_{i2} - m_{i1}) & s_i^2 m_{i1} + c_i^2 m_{i2} \end{pmatrix} , \qquad (i = u,\ d) , \tag{17}$$

where $c_i \equiv \cos\theta_i$ and $s_i \equiv \sin\theta_i$. The mixing matrix V_q is invariant under the changes

$$O_d \to O O_d , \qquad O_u \to O O_u , \tag{18}$$

where O is some arbitrary orthogonal matrix. Thus, using the fact that $\sin\theta_c \ll 1$, we can assume both $\theta_d \ll 1$ and $\theta_u \ll 1$ without loss of generality. Taking into account the quark mass hierarchy, we set

$$m_u = a\lambda^4 m_c , \qquad m_d = b\lambda^2 m_s , \tag{19}$$

where a and b are $\mathcal{O}(1)$ coefficients, and $\lambda \equiv \sin\theta_c \simeq 0.22$. Thus the mass matrices are expressed in terms of θ_u and θ_d as following:

$$M_u \simeq \begin{pmatrix} a\lambda^4 + \sin^2\theta_u & \sin\theta_u \\ \sin\theta_u & 1 \end{pmatrix} m_c, \quad M_d \simeq \begin{pmatrix} b\lambda^2 + \sin^2\theta_d & \sin\theta_d \\ \sin\theta_d & 1 \end{pmatrix} m_s . \tag{20}$$

There are three different options now for the angles θ_u and θ_d:

$$
\begin{aligned}
&1. \qquad \sin\theta_d \sim \lambda\ , \quad \sin\theta_u \sim \lambda\ , \\
&2. \qquad \sin\theta_d \sim \lambda\ , \quad \sin\theta_u \leq \lambda^2\ , \\
&3. \qquad \sin\theta_d \leq \lambda^2\ , \quad \sin\theta_u \sim \lambda\ .
\end{aligned}
$$

The options 1 and 3 are unnatural. Indeed, one would require a severe fine-tuning of the matrix element $[M_u]_{11}$ forcing $[M_u]_{11} \simeq ([M_u]_{12})^2$ to arrive at the large $m_u/m_c \sim \lambda^4$ hierarchy. On the other hand, option 2 gives a natural quark mass matrix without fine-tuning. Here the fine-tuning means a tuning of $\mathcal{O}(\lambda^2)$, which comes from the following inspection. There is a well known phenomenological relation between the CKM matrix element and the quark mass ratio as

$$|V_{us}| \simeq \sqrt{\frac{m_d}{m_s}} \simeq \lambda\ . \tag{21}$$

The down quark mass ratio dominates V_{us}, while the contribution of the up quark mass ratio is at most $\sqrt{m_u/m_c} \simeq \mathcal{O}(\lambda^2)$. This relation is consistent with the option 2. Thus, the criterion of $\mathcal{O}(\lambda^2)$ tuning is very useful in order to select the option 2 by excluding options 1 and 3. The extension of the natural mass matrices to the three family case in quark sector is also given in Ref.[42].

Naturalness is also expanded to the lepton sector [43]. If neutrino sector has the hierarchical mass structure, the naturalness argument is exactly the same as in the example just described for the quark sector. However, in the neutrino sector the inverse mass hierarchy for the flavor is still allowed by the constraints obtained in the disappearance experiments of the neutrino oscillations in Bugey [16], Krasnoyarsk [17], CDHS [18] and CCFR [19]. Since the neutrino mixing is chosen to be $\sin\theta_\nu \sim 1$ in this case, one cannot always guarantee to have both $\theta_\nu \ll 1$ and $\theta_E \ll 1$(charged lepton mixing) by the change in eq.(18). Therefore, we should reconsider the naturalness of the quark mass matrices in the lepton sector.

In the case of $s \simeq 1 (c \ll 1)$ with $m_1 \leq m_2$ in eq.(17), the mass matrix is expressed approximately:

$$M_\nu \simeq \begin{pmatrix} m_2 & cm_2 \\ cm_2 & m_1 + c^2 m_2 \end{pmatrix}\ . \tag{22}$$

The natural mass matrix without the fine-tuning requires

$$m_1 \geq c^2 m_2\ . \tag{23}$$

For example, if we take $c \simeq a'\lambda$, $m_1 \simeq a\lambda m_2$, which satisfies the condition(23), then matrix

$$M_\nu \simeq \begin{pmatrix} 1 & a'\lambda \\ a'\lambda & a\lambda \end{pmatrix} m_2\ , \tag{24}$$

gives us the masses m_2 and $a\lambda m_2$, correspondingly. Thus, the naturalness condition follows from the (2,2) entry in the case of the inverse mass hierarchy of the flavor, while in the quark sector this condition follows from the (1,1) entry.

It is necessary to clarify the concepts of naturalness for the large mixing angle. Furthermore, the recent experiments [2]~[6] also indicate the large flavor mixing in the neutrinos. Let us consider the 2×2 symmetric matrix M_ν. Again, assume for simplicity that it can be diagonalized by some orthogonal matrix O_ν:

$$O_\nu^T M_\nu O_\nu = M_\nu^{diag} \equiv \begin{pmatrix} m_1 & 0 \\ 0 & m_2 \end{pmatrix}. \qquad (25)$$

Here the orthogonal matrix O_ν is

$$O_\nu = \begin{pmatrix} c & s \\ -s & c \end{pmatrix}, \qquad c \equiv \cos\theta_\nu \simeq \frac{1}{\sqrt{2}}, \qquad s \equiv \sin\theta_\nu \simeq \frac{1}{\sqrt{2}}. \qquad (26)$$

Then M_ν is expressed in terms of the mass eigenvalues and mixing angle as follows:

$$M_\nu = O_\nu M_\nu^{diag} O_\nu^T = \begin{pmatrix} m_1 c^2 + m_2 s^2 & cs(m_2 - m_1) \\ cs(m_2 - m_1) & m_1 s^2 + m_2 c^2 \end{pmatrix}. \qquad (27)$$

If the neutrino masses have the hierarchy such as $m_1/m_2 = \epsilon \ll 1$ with the large mixing angle, the mass matrix M_ν is

$$M_\nu \simeq \begin{pmatrix} \epsilon c^2 + s^2 & cs \\ cs & c^2 + \epsilon s^2 \end{pmatrix} m_2 \, . \qquad (28)$$

In the (1,1) entry, ϵc^2 should be fine tuned against s^2 in order to arrive at $m_1/m_2 \sim \epsilon$. For example, taking $c \simeq s \simeq 1/\sqrt{2}$, the matrix

$$M_\nu \simeq \begin{pmatrix} \frac{1}{2} & \frac{1}{2} \\ \frac{1}{2} & \frac{1}{2} \end{pmatrix} m_2 \, , \qquad (29)$$

gives $m_1 \simeq 0$ and m_2. Further if we assume that $\epsilon \simeq a\lambda^2$, the matrix M_ν in (29) should be replaced by the following matrix

$$M_\nu \simeq \begin{pmatrix} \frac{1+a\lambda^2}{2} & \frac{1}{2} \\ \frac{1}{2} & \frac{1+a\lambda^2}{2} \end{pmatrix} m_2 \, . \qquad (30)$$

The latter gives the mass $m_1 = O(\lambda^2) m_2$ provided that the order of $O(\lambda^2)$ is fine-tuned. Thus, the hierarchical neutrino masses are unnatural in the case of the large mixing angle.

In the case when neutrino masses are approximately degenerate such as $m_1 \simeq m_2$ and $(m_2 - m_1)/m_2 = \epsilon \ll 1$, the mass matrix M_ν is

$$M_\nu \simeq \begin{pmatrix} 1 & cs\epsilon \\ cs\epsilon & 1 \end{pmatrix} m_2 \, . \qquad (31)$$

Hence, no fine tuning is required for any entry. We can explore the same argument when the mass eigenvalues are $-m_1$ and m_2, in which case

$$M_\nu \simeq \begin{pmatrix} c^2\epsilon & 2cs \\ 2cs & s^2\epsilon \end{pmatrix} m_2 \ . \tag{32}$$

Therefore in the case of the large mixing angle we call eqs.(31,32) the natural mass matrix.

In the sequel the naturalness of the mass matrix for the lepton sector is investigated for the three family model. The conclusion in the case of three families as follows: the neutrino mass matrices with quasi degenerate masses and maximal mixing are the natural ones. Quasi degenerate masses means $m_3 \simeq m_2 \gg m_1$ or $m_3 \simeq m_2 \simeq m_1$.

In the next section, we present a texture with the S_3 symmetry, which is a typical natural mass matrix.

4. Neutrino Mass Matrix with S₃ Symmetry

One of the most attractive description of the quark sector in the phenomenological mass matrix approach starts with an $S_3(R) \times S_3(L)$ symmetric mass term (often called "democratic" mass matrix) [41]:

$$M_q = \frac{c_q}{3} \begin{pmatrix} 1 & 1 & 1 \\ 1 & 1 & 1 \\ 1 & 1 & 1 \end{pmatrix} , \tag{33}$$

where $q =$ up and down, and quarks belong to $\mathbf{3}=\mathbf{2}\oplus\mathbf{1}$ of $S_3(L)$ or $S_3(R)$. The same form is still available for the charged lepton sector. However, the neutrino mass matrix is different if they are Majorana particles. The $S_3(L)$ symmetric mass term is given as follows:

$$M_\nu = c_\nu \begin{pmatrix} 1 & 0 & 0 \\ 0 & 1 & 0 \\ 0 & 0 & 1 \end{pmatrix} + c_\nu r \begin{pmatrix} 0 & 1 & 1 \\ 1 & 0 & 1 \\ 1 & 1 & 0 \end{pmatrix} , \tag{34}$$

where r is an arbitrary parameter. The eigenvalues of this matrix are given as $c_\nu(1+2r,\ 1-r,\ 1-r)$, which means that there are at least two degenerate masses in the $S_3(L)$ symmetric Majorana mass matrix [44][45].

If three degenerate light neutrinos are required, the parameter r should be taken as $r = 0$ or $r = -2$. The first case was discussed in ref.[44] and the second case was discussed in ref.[45]. The difference of r leads to the difference in the CP property of neutrinos. If $r = -1/2$, one finds two massive neutrinos and one massless neutrino. So the $S_3(L)$ symmmetry could be reconciled with the LSND data [12] by including the symmetry breaking terms.

Alternative representation of the $S_3(L)$ symmetric mass matrix is given as

$$M_\nu = c_\nu \begin{pmatrix} e^{i\alpha} & 1 & 1 \\ 1 & e^{i\alpha} & 1 \\ 1 & 1 & e^{i\alpha} \end{pmatrix} , \tag{35}$$

which is based on the universal strength for Yukawa couplings (USY) hypothesis [46]. If $\alpha = 2\pi/3$, three neutrino masses are degenerate.

In order to reproduce the atmospheric neutrino deficit by the large neutrino oscillation, the symmetry breaking terms are required. Since results are almost same, we show the numerical analyses in ref.[44], where the LSND data is disregarded.

Let us start with discussing the following charged lepton mass matrix:

$$M_\ell = \frac{c_\ell}{3} \begin{pmatrix} 1 & 1 & 1 \\ 1 & 1 & 1 \\ 1 & 1 & 1 \end{pmatrix} + \begin{pmatrix} \delta_1^\ell & 0 & 0 \\ 0 & \delta_2^\ell & 0 \\ 0 & 0 & \delta_3^\ell \end{pmatrix} . \tag{36}$$

The first term is a unique representation of the $S_3(R) \times S_3(L)$ symmetric matrix and the second one is a symmetry braking matrix given by Koide [47]. This matrix is diagonalised as

$$U_\ell^\dagger M_\ell U_\ell = \mathrm{diag}(m_1^\ell, m_2^\ell, m_3^\ell) , \tag{37}$$

where

$$\begin{aligned} m_1^\ell &= (\delta_1^\ell + \delta_2^\ell + \delta_3^\ell)/3 - \xi^\ell/6 , \\ m_2^\ell &= (\delta_1^\ell + \delta_2^\ell + \delta_3^\ell)/3 + \xi^\ell/6 , \\ m_3^\ell &= c_\ell + (\delta_1^\ell + \delta_2^\ell + \delta_3^\ell)/3 , \end{aligned} \tag{38}$$

with

$$\xi^\ell = [(2\delta_3^\ell - \delta_2^\ell - \delta_1^\ell)^2 + 3(\delta_2^\ell - \delta_1^\ell)^2]^{1/2} . \tag{39}$$

The matrix that diagonalises $U_\ell = AB_\ell$ reads

$$A = \begin{pmatrix} 1/\sqrt{2} & 1/\sqrt{6} & 1/\sqrt{3} \\ -1/\sqrt{2} & 1/\sqrt{6} & 1/\sqrt{3} \\ 0 & -2/\sqrt{6} & 1/\sqrt{3} \end{pmatrix} , \tag{40}$$

$$B_\ell \simeq \begin{pmatrix} \cos\theta^\ell & -\sin\theta^\ell & \lambda^\ell \sin 2\theta^\ell \\ \sin\theta^\ell & \cos\theta^\ell & \lambda^\ell \cos 2\theta^\ell \\ -\lambda^\ell \sin 3\theta^\ell & \lambda^\ell \cos 3\theta^\ell & 1 \end{pmatrix} , \tag{41}$$

with

$$\tan 2\theta^\ell \simeq -\sqrt{3}\frac{\delta_2^\ell - \delta_1^\ell}{2\delta_3^\ell - \delta_2^\ell - \delta_1^\ell} , \quad \lambda_\ell = \frac{1}{\sqrt{2}}\frac{1}{3c_\ell}\xi^\ell . \tag{42}$$

It has been shown that all quark masses and mixing angles are successfully given by taking $\delta_1 = -\epsilon$, $\delta_2 = \epsilon$ and $\delta_3 = \delta$. Analogous to the quark sector, $\delta_1^\ell = -\epsilon_\ell$, $\delta_2^\ell = \epsilon_\ell$ and $\delta_3^\ell = \delta_\ell$ are taken. The three mass eigenvalues are then

$$m_1^\ell \simeq -\epsilon_\ell^2/2\delta_\ell, \quad m_2^\ell \simeq 2\delta_\ell/3 + \epsilon_\ell^2/2\delta_\ell, \quad m_3^\ell \simeq c_\ell + \delta_\ell/3 \,, \tag{43}$$

and the angle θ^ℓ is

$$\sin\theta^\ell \simeq -\sqrt{\left|\frac{m_1^\ell}{m_2^\ell}\right|} \,. \tag{44}$$

Let us turn to the neutrino mass matrix:

$$M_\nu = c_\nu \begin{pmatrix} 1 & 0 & 0 \\ 0 & 1 & 0 \\ 0 & 0 & 1 \end{pmatrix} + \begin{pmatrix} 0 & \epsilon_\nu & 0 \\ \epsilon_\nu & 0 & 0 \\ 0 & 0 & \delta_\nu \end{pmatrix} \,, \tag{45}$$

where the symmetry braking is given by a small term with two adjustable parameters. An alternative natural choice to lift the mass degeneracy may be diag$(-\epsilon_\nu, \epsilon_\nu, \delta_\nu)$, which we shall also discuss later. The mass eigenvalues are $c_\nu \pm \epsilon_\nu$, and $c_\nu + \delta_\nu$, and the matrix that diagonalises M_ν ($U^T M_\nu U$ =diagonal) is

$$U_\nu = \begin{pmatrix} 1/\sqrt{2} & 1/\sqrt{2} & 0 \\ -1/\sqrt{2} & 1/\sqrt{2} & 0 \\ 0 & 0 & 1 \end{pmatrix} \,. \tag{46}$$

That is, our M_ν represents three degenerate neutrinos, with the degeneracy lifted by a small parameters.

The lepton mixing angle as defined by $V_\ell = (U_\ell)^\dagger U_\nu = (AB_\ell)^\dagger U_\nu$ is thus given by

$$V_\ell \simeq \begin{pmatrix} 1 & -(1/\sqrt{3})\sqrt{(m_e/m_\mu)} & (2/\sqrt{6})\sqrt{(m_e/m_\mu)} \\ \sqrt{(m_e/m_\mu)} & 1/\sqrt{3} & -2/\sqrt{6} \\ 0 & 2/\sqrt{6} & 1/\sqrt{3} \end{pmatrix} \,, \tag{47}$$

where the neutrino mass parameters do not appear. The parameters $\epsilon_\nu c_\nu$ and $\delta_\nu c_\nu$ are fixed by the neutrino mass difference explored by the oscillation effect. The normalisation c_ν is not fixed unless one of the neutrino masses is known, but it is not important for this argument, since the lepton mixing matrix is almost independent of the details of these parameters except for the m_e/m_μ ratio. If we retain all small terms, the lepton mixing angle is predicted to be

$$V_\ell = \begin{pmatrix} 0.998 & -0.045 & 0.05 \\ 0.066 & 0.613 & -0.787 \\ 0.005 & 0.789 & 0.614 \end{pmatrix} , \tag{48}$$

which leads the large $\nu_\mu - \nu_\tau$ oscillation $\sin^2 2\theta_{\rm atm} \simeq 8/9$. For the $\nu_e - \nu_\mu$ oscillation $\sin^2 2\theta_\odot \simeq 8 \times 10^{-3}$, which also agrees with the neutrino mixing corresponding to the small angle solution of the MSW scenario [48] for the solar neutrino problem [49]. It is remarked that predicted $V_{\ell e3} \simeq 0.05$ is stable against the symmetry breaking parameters. In the future, this prediction will be tested in the following long baseline experiments $\nu_\nu \to \nu_e$ and $\nu_e \to \nu_\tau$:

$$\begin{aligned} P(\nu_\mu \to \nu_e) &\simeq 4V_{\mu3}^2 V_{e3}^2 \sin^2 \frac{\Delta m_{31}^2 L}{4E} , \\ P(\nu_e \to \nu_\tau) &\simeq 4V_{e3}^2 V_{\tau3}^2 \sin^2 \frac{\Delta m_{31}^2 L}{4E} . \end{aligned} \tag{49}$$

A very important constraint comes from neutrinoless double beta decay experiments. The latest result on the lifetime of ^{76}Ge$\to^{76}$Se, $\tau_{1/2} > 1.1 \times 10^{25}$ yr [50] yields an upper limit on the Majorana neutrino mass 0.4 eV [51] to 1.1 eV [52] depending on which nuclear model is adopted for nuclear matrix elements. We are then left with quite a narrow window for the neutrino mass $0.1\text{eV} \le m_{\nu_e} \simeq m_{\nu_\mu} \simeq m_{\nu_\tau} \le 1\text{eV}$ for the present scenario to be viable. It will be most interesting to push down the lower limit of neutrinoless double beta decay lifetime. If the limit on neutrino mass is lowered by one order of magnitude the our degenerate neutrino mass scenario will be ruled out.

The argument we have made above is of course by no means unique, and a different assumption on the matrix leads to a different mass-mixing relation. Let us briefly discuss the consequence of the other matrices we have encountered in the line of our argument above. If we adopt the symmetry breaking term alternative to eq.(45),

$$\begin{pmatrix} -\epsilon_\nu & 0 & 0 \\ 0 & \epsilon_\nu & 0 \\ 0 & 0 & \delta_\nu \end{pmatrix} , \tag{50}$$

in parallel to the charged lepton and quark sectors, we obtain the lepton mixing matrix to be

$$V_\ell \simeq \begin{pmatrix} 1/\sqrt{2} & -1/\sqrt{2} & 0 \\ 1/\sqrt{6} & 1/\sqrt{6} & -2/\sqrt{6} \\ 1/\sqrt{3} & 1/\sqrt{3} & 1/\sqrt{3} \end{pmatrix} . \tag{51}$$

This is identical to the matrix presented by Fritzsch and Xing [53]. For this case one gets

$$\sin^2 2\theta_\odot \simeq 1, \qquad \sin^2 2\theta_{\rm atm} \simeq 8/9 . \tag{52}$$

This case can accommodate the "just-so" scenario for the solar neutrino problem due to neutrino oscillation in vacuum [54]. This matrix has been investigated in detail [55] focusing on recent data at Super-Kamiokande [56]~[59]. The "just so solution" leads to the bi-maximal flavor mixing, which may be interesting for the theoretical origin [60][61].

5. Summary

Atmospheric neutrino deficit at Super-Kamiokande suggests the near-maximal flavor mixing, which have excited the model building of the lepton mass matrices. Mass matrix models are classified according to the mechanism providing the large(maximal) flavor mixing. However, more quantitative studies are needed in order to understand the origin of the large flavor mixings deeply. The studies of the lepton mass matrices will give clues of new symmetry such as the flavor symmetry and will indicate the particular directions for the unification of matter. Furthermore, the structute of the neutrino mass matrix will give strong impact on other fields such as the leptogenesis [62]. The CP violating phase structure in the neutrino mass matrix is also a very attractive subject as well as the CP violation of the quark sector.

Acknowledgements

I thank M. Fukugita and T. Yanagida for collaboration on the lepton mass matrix model with the S_3 symmetry, and M. Matsuda for the collaboration on the naturalness of the lepton mass matrices. This research is supported by the Grant-in-Aid for Science Research, Ministry of Education, Science and Culture, Japan(No.10140218, No.10640274).

References

[1] M. Kobayashi and T. Maskawa, Prog. Theor. Phys. **49** (1973) 652.

[2] K. S. Hirata et al., Phys. Lett. **205B** (1988) 416; **280B** (1992) 146;
D. Casper et al., Phys. Rev. Lett. **66** (1991) 2561;
R. Becker-Szendy et al., Phys. Rev. **D46** (1992) 3720.

[3] NUSEX Collaboration, M. Aglietta et al., Europhys. Lett. **15** (1991) 559;
SOUDAN2 Collaboration, W. W. M. Allison et al., Phys. Lett. **391B** (1997) 491;
Fréjus Collaboration, K. Daum, Z. Phys. **C66** (1995) 417;
MACRO Collaboration, S. Ahlen et al., Phys. Lett. **357B** (1995) 481.

[4] Y. Fukuda et al., Phys. Lett. **335B** (1994) 237.

[5] GALLEX Collaboration, Phys. Lett. **327B** (1994) 377; **388B** (1996) 384;
SAGE Collaboration, J. N. Abdurashitov et al., Phys. Lett. **328B** (1994) 234;
Homestake Collaboration, Nucl. Phys. **B38** (Proc. Suppl.) (1995) 47;
Kamiokande Collaboration, Nucl. Phys. **B38** (Proc. Suppl.) (1995) 55;
Super-Kamiokande Collaboration, Y. Fukuda et al, hep-ex/9805021.

[6] Super-Kamiokande Collaboration, Y. Fukuda et al, hep-ex/9805006;
T. Kajita, invited talk at *Neutrino 98*, 4-9 June, 1998, Takayama, Japan.

[7] Y. Suzuki, invited talk at *Neutrino 98*, 4 - 9 June, 1998, Takayama, Japan.

[8] The CHOOZ Collaboration, M. Apollonio et al., Phys. Lett. **420B** (1998) 397.

[9] Y. Suzuki, Proc. of *Neutrino 96*, Helsinki, June 1996, edited by K. Enqvist et al., p.237 (World Scientific, Singapore, 1997).

[10] S. G. Wojcicki, Proc. of *Neutrino 96*, Helsinki, June 1996, edited by K. Enqvist et al., p.231 (World Scientific, Singapore, 1997).

[11] ICARUS Collaboration, P. Cennini et al., LNGS-94/99-I, May 1994;
F. Pietropaola, invited talk at *Neutrino 98*, 4 - 9 June, 1998, Takayama, Japan.

[12] LSND Collaboration, C. Athanassopoulos et al., Phys. Rev. Lett. **75** (1995) 2650;
77 (1996) 3082; Phys. Rev. **C54** (1996) 2685; nucl-ex/9706006(1997);
J. E. Hill, Phys. Rev. Lett. **75** (1995) 2654.

[13] NOMAD Collaboration, J. Altegöer et al., CERN-EP-98-057 (1998);
CHORUS Collaboration, E. Eskut et al., Phys. Lett. **424B** (1998) 202;
O. Sato, invited talk at *Neutrino 98*, 4 - 9 June, 1998, Takayama, Japan;
J.J. Gomez-Cadenas, invited talk at *Neutrino 98*, 4 - 9 June, 1998, Takayama, Japan.

[14] E531 Collaboration, N. Ushida et al., Phys. Rev. Lett. **57** (1986) 2897.

[15] KARMEN Collaboration, Nucl. Phys. **B38** (Proc. Suppl.) (1995) 235;
B. Zeitnitz, invited talk at *Neutrino 98*, 4 - 9 June, 1998, Takayama, Japan.

[16] B. Achkar et al., Nucl. Phys. **B434** (1995) 503.

[17] G.S. Vidyakin et al., Pis'ma Zh.Eksp. Thor. Fiz. **59** (1994) 364;
JETP Lett. **59** (1994) 390.

[18] CDHS Collaboration, F. Dydak et al., Phys. Lett. **134B** (1984) 281.

[19] CCFR Collaboration, I. E. Stockdale et al., Phys. Rev. Lett. **52** (1984) 1384;
Z. Phys. **C27** (1985) 53;
CCFR Collaboration, K. S. McFarland et al., Phys. Rev. Lett. **75** (1995) 3993.

[20] E776 Collaboration, L. Borodovsky et al., Phys. Rev. Lett. **68** (1992) 274.

[21] M. Gell-Mann, P. Ramond and R. Slansky, in *Supergravity*, Proceedings of the Workshop, Stony Brook, New York, 1979, edited by P. van Nieuwenhuizen and D. Freedmann, North-Holland, Amsterdam, 1979, p.315;

T. Yanagida, in *Proceedings of the Workshop on the Unified Theories and Baryon Number in the Universe*, Tsukuba, Japan, 1979, edited by O. Sawada and A. Sugamoto, KEK Report No. 79-18, Tsukuba, 1979, p.95.

[22] A.Yu. Smirnov, Phys. Rev. **D48** (1993) 3264; Nucl. Phys. **B466** (1996) 25.

[23] M. Tanimoto, Phys. Lett. **345B** (1995) 477.

[24] J.A. Harvey, D.B. Reiss and P. Ramond, Nucl. Phys. **B199** (1982) 223.

[25] K.S. Babu and Q. Shafi, Phys. Lett. **294B** (1992) 235.

[26] P. Binétruy, S. Lavignac and P. Ramond, Phys. Lett. **350B** (1995) 49; Nucl. Phys. **B477** (1996) 353;
P. Binétruy, S. Lavignac, S.T. Petcov and P. Ramond, Nucl. Phys. **B496** (1997) 3;
N. Irges, S.Lavignac and P. Ramond, hep-ph/9802334v2.

[27] R. Barbieri, G. Dvali and L. J. Hall, Phys. Lett. **377B** (1996) 76.

[28] M. Tanimoto, Phys. Rev. **D57** (1998) 1983;
G. Eyal, hep-ph/9807308.

[29] M. Bando, T. Kugo and K. Yoshioka, Phys. Rev. Lett. **80** (1998) 3004.

[30] D.G. Lee and R.N. Mohapatra, Phys. Lett. **329B** (1994) 463;
A. Ioannissyan and J.W.F. Valle, Phys. Lett. **332B** (1994) 93;
B. Brahmachari and R.N. Mohapatra, Phys. Rev. **D58** (1998) 15003.

[31] A. Zee, Phys. Lett. **93B** (1980) 389; ibidem **161B** (1985) 141;
S. T. Petcov, Phys. Lett. **115B** (1982) 401;
S. M. Bilenky and S. T. Petcov, Rev. Mod. Phys. **59** (1987) 671;
L. Wolfenstein, Nucl. Phys. **B175** (1980) 93;
A. Y. Smirnov and Z. Tao, Nucl. Phys. **B426** (1994) 415;
L. Wolfenstein, CMU-HEP95-05, hep-ph/9506352 (1995);
A. Y. Smirnov and M. Tanimoto, Phys. Rev. **D55** (1997) 1665;
E. Ma, hep-ph/9807386 v2.

[32] C. D. Froggatt G. Lowe and H. B. Nielsen, Nucl. Phys. **B414** (1994) 579;
C. D. Froggatt, H. B. Nielsen and D.J. Smith, Phys. Lett. **385B** (1996) 150;
C. D. Froggatt, M. Gibson and H. B. Nielsen, Phys. Lett. **409B** (1997) 305.

[33] K.S. Babu and S.M. Barr, Phys. Lett. **381B** (1996) 202.

[34] N. Haba and T. Matsuoka, Prog. Theor. Phys. **99** (1998) 831.

[35] P.H. Chankowski and Z. Pluciennik, Phys. Lett. **316B** (1993) 312.

[36] K.S. Babu, C.N. Leung and J. Pantaleone, Phys. Lett. **319B** (1993) 191.

[37] M. Tanimoto, Phys. Lett. **360B** (1995) 41.

[38] J. Sato and T. Yanagida, hep-ph/9710516.

[39] C.H. Albright, K.S. Babu and S.M. Barr, hep-ph/9802314, hep-ph/9805266.

[40] J.K. Elwood, N. Irges and P. Ramond, hep-ph/9807228.

[41] H. Harari, H. Haut and J. Weyers, Phys. Lett. **78B** (1978) 459.

[42] R.D. Peccei and K. Wang, Phys. Rev. **D53** (1996) 2712.

[43] M. Matsuda and M. Tanimoto, hep-ph/9711294, Phys. Rev. **D**, in press.

[44] M. Fukugita, M. Tanimoto and T. Yanagida, Phys. Rev. **D57** (1998) 4429.

[45] K. Kang and S. Kang, hep-ph/9802328.

[46] G.C. Branco, M.N. Rebelo and J.I. Silva-Marcos, Phys. Lett. **428B** (1998) 136.

[47] Y. Koide, Phys. Rev. D **28** (1983) 252; **39** (1989) 1391.

[48] S.P. Mikheyev and A.Yu. Smirnov, Yad. Fiz. **42** (1985) 1441;
L. Wolfenstein, Phys. Rev. **D17** (1987) 2369;
E.W. Kolb, M.S. Turner and T.P. Walker, Phys. Lett. **175B** (1986) 478;
S.P. Rosen and J.M. Gelb, Phys. Rev. **D34** (1986) 969;
J.N. Bahcall and H.A. Bethe, Phys. Rev. Lett. **65** (1990) 2233;
N. Hata and P. Langacker, Phys. Rev. **D50** (1994) 632;
P.I. Krastev and A. Yu. Smirnov, Phys. Lett. **338B** (1994) 282.

[49] J.N. Bahcall, P.I. Krastev and A.Yu. Smirnov, hep-ph/9807216.

[50] H.V. Klapdor-Kleingrothaus, in Proceedings of the 17th International Conference on Neutrino Physics and Astrophysics, Helsinki, Finland, ed. K. Enqvist et al., World Scientific, 317 (1996).

[51] T. Tomoda, A. Faessler, K. W. Schmid and F. Grümmer, Nucl. Phys. A **452** (1986) 591.

[52] J. Engel, P. Vogel, X. Ji and S. Pittel, Phys. Lett. **225B** (1989) 5.

[53] H. Fritzsch and Z. Xing, Phys. Lett. **372B** (1996) 265. Z. Xing, DPNU-98-12, hep-ph/9804433.

[54] E.g., V. Barger, R. J. N. Phillips and K. Whisnant, Phys. Rev. Lett. **69**, (1992) 3135.

[55] M. Tanimoto, hep-ph/9807283.

[56] S.P. Mikheyev and A.Yu. Smirnov, IC/97/113, hep-ph/9708403.

[57] B. Faïd, G.L. Fogli, E. Lisi and M. Montanino, BARI-TH/299-98, hep-ph/9805293.

[58] P. Osland and G. Vigdel, CERN-TH/98-200, hep-ph/9806339.

[59] A.J. Baltz, A.S. Goldhaber and M. Goldhaber, hep-ph/9806540.

[60] V. Barger, S. Pakvasa, T.J. Weiler and K. Whisnant, MADPH-98-1063, hep-ph/9806387 v3.

[61] Y. Nomura and T. Yanagida, hep-ph/9807325.

[62] M. Fukugita and T. Yanagida, Phys. Lett. **174B** (1986) 45.

Solar Neutrinos, Atmospheric Neutrinos and Proton Decays in Super-Kamiokande, and KamLAND Project

Atsuto Suzuki[1]

Research Center for Neutrino Science,
Tohoku University,
Aoba, Sendai, 980-8578, JAPAN

Abstract. Results on solar neutrinos, atmospheric neutrinos and proton decays from about one year operation of Super-Kamiokande are summarized. The solar neutrino deficit and atmospheric neutrino anomaly are reconfirmed with high statistics data. In particular a strong suggestion of neutrino oscillations has been obtained in the zenith-angle distributions of atmospheric neutrinos. No candidate events of proton decays are observed in the present data sample.
Shown also here the project of a 1000 ton liquid scintillator experiment, KamLAND.

1. Introduction

Experimental verification of GUT's is one of the most revolutionary subjects in the present-day particle physics and cosmology. In Kamiokande prospective indication concerning the finite neutrino masses were obtained as solutions to the solar neutrino deficit and atmospheric neutrino anomaly. As for proton decays the simple SU(5) GUT was already rejected, and the stringent constrainrs was also imposed on the minimal SUSY SU(5) GUT.

Under such an experimental situation, 50,000 ton water Cherenkov experiment, Super-Kamiokande started on schedule in April 1, 1996. Much

[1] E-mail: suzukia@awa.tohoku.ac.jp

expectation to detect the evidence of GUT predictions is focused on the Super-Kamiokande high statistic data. About one year data sample which is used in the present analysis, gives high quality results exceeding the Kamiokande final results for the same subjects shown here.

Evidence on the finite neutrino mass would give rise to further studies of what neutrinos are: the masses of three neutrinos, Dirac-type or Majorana-type, the neutrino charge, the neutrino magnetic moment and so on. A 1,000 ton liquid scintillator experiment was proposed to follow these neutrino studies. Construction of the detector has already begun. The target-day of data taking is scheduled to be January 1, 2001.

2. Solar Neutrino Observation

Results on the solar neutrino analysis in Super-Kamiokande have been obtained by 374.2 days data sample taken in May 31, 1996 through October 20, 1997. The 22.5 kton fiducial volume and the analysis energy range of 6.5 to 20.0 MeV are applied to reduce background events.

The angular distribution of recoiling electrons in $\cos\theta_{\mathrm{SUN}}$ is shown in Fig. 1, where $\cos\theta_{\mathrm{SUN}} = 1$ corresponds to the expected direction to the Earth from the Sun. The histogram in this figure shows the best fit of expected solar neutrino signals plus flat background distribution to the data.

Figure 1. Angular distribution in $\cos\theta_{\mathrm{SUN}}$ of the 374.2 live-day events with $E_e \geq 6.5$ MeV.

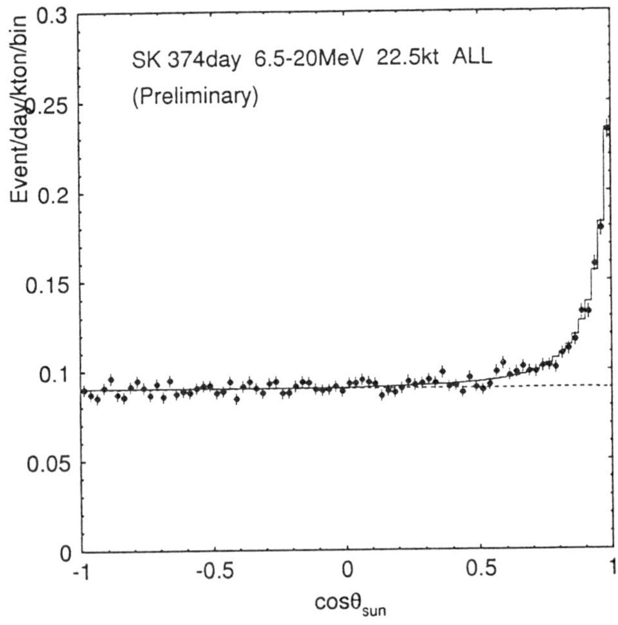

A total of 4952^{+118}_{-111}(stat.)$^{+198}_{-144}$(syst.) events are observed for the ^{8}B solar neutrinos, which corresponds to the flux of $[2.37^{+0.06}_{-0.05}(\text{stat.})^{+0.09}_{-0.07}(\text{syst.})] \times 10^6$ cm^{-2}s^{-1}. Ratio of the observation to the prediction based on the Standard Solar Model of BP95 [1], DATA/SSM_{BP95} is equal to $0.358^{+0.009}_{-0.008}$(syst.)$^{+0.014}_{-0.010}$(syst.). Consequently the solar neutrino deficit is reconfirmed with 38 σ significance.

The most plausible solution to the solar neutrino deficit is brought by the neutrino oscillations in matter (the MSW oscillation) or in vacuum (the "Just So" oscillation). The MSW solutions were given by Hata and Langacker [2], combining the flux observations of Kamiokande, Super-Kamiokande, Homestake, SAGE and GALLEX, and assuming SSM_{BP95}. Fig.2 shows the 95 % C.L. allowed region of the MSW oscillations. So-called "small angle solution" and "large angle solution" are still alive.

Figure 2. The MSW allowed parameter space at 95 % C.L., assuming SSM_{BP95}. (From [2])

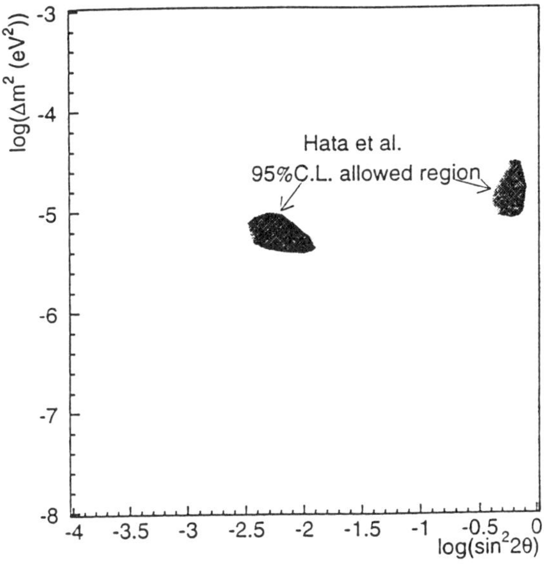

A flux difference in observations during the daytime and nighttime provides a sensitive probe on the MSW oscillations inside the Sun and/or the Earth. In Fig. 3 the observed total flux, daytime flux and nighttime flux are plotted, together with the nighttime fluxes obtained in different nighttime intervals (N1~N5) defined in the same figure. No significant change in the daytime (D) and nighttime (N) fluxes is found. The difference defined by D-N/D+N is equal to be -0.03 ± 0.024(stat.) ± 0.014(syst). This result gives the excluded space of the MSW parameters as shown in Fig. 4. A lower part of the MSW large angle solution is rejected at 95 % C.L.

Figure 3. The ^{8}B solar neutrino fluxes during the daytime and nighttime. The nighttime data sample is divided into five subsamples in terms of the direction of the Sun .

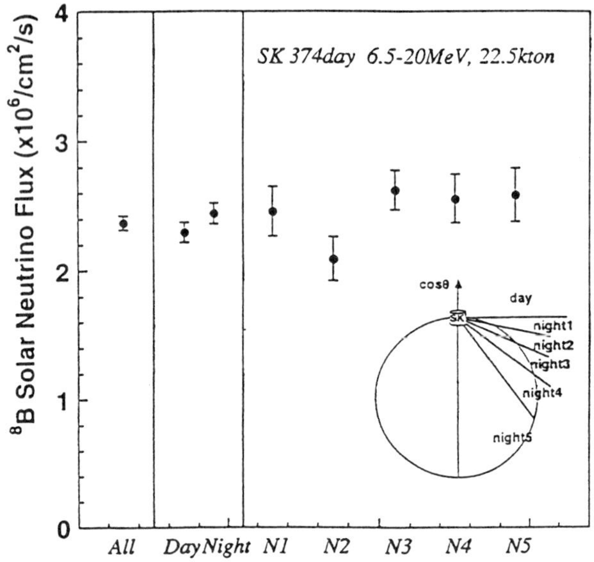

Figure 4. Region excluded at 95 % C.L. in the MSW $\Delta m^2 - \sin^2 2\theta$ space by the null day-night result (crosshatched region). The MSW small and large angle solutions are also shown here [2].

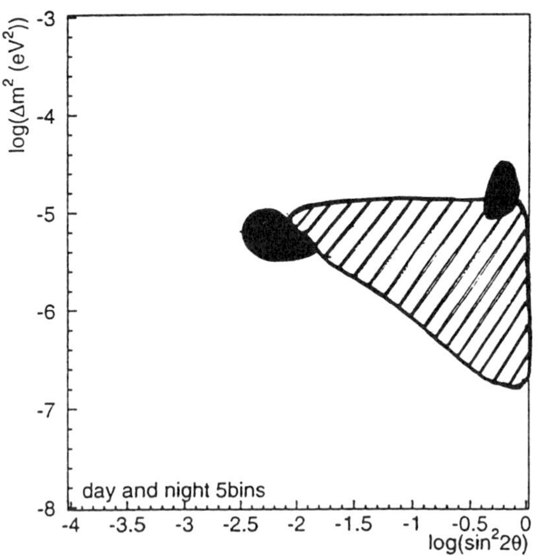

The recoiling electron energy spectrum is also sensitive to the oscillation parameters. To analyze being free from uncertainties of the SSM flux calculation, the ratio of the observed energy spectrum to the prediction from SSM_{BP95} in each energy bin is plotted in Fig. 5. Deviation from a flat distribution tells us an indication of some new physics in the solar neutrinos. In Fig. 5 also shown are the predictions from the MSW two solutions of the small angle ($\Delta m^2 = 6.3 \times 10^{-6} eV^2, \sin^2 2\theta = 0.91 \times 10^{-2}$) and large angle ($\Delta m^2 = 2.8 \times 10^{-5} eV^2, \sin^2 2\theta = 0.66$) by a solid line and a dashed line, respectively. No significant deviation from a flat distribution can be seen.

Figure 5. Points are the ratio of the recoiling electron energy spectrum to the expected spectrum. The lines are the distorted electron spectra due to neutrino oscillations with two representative parameters (see text).

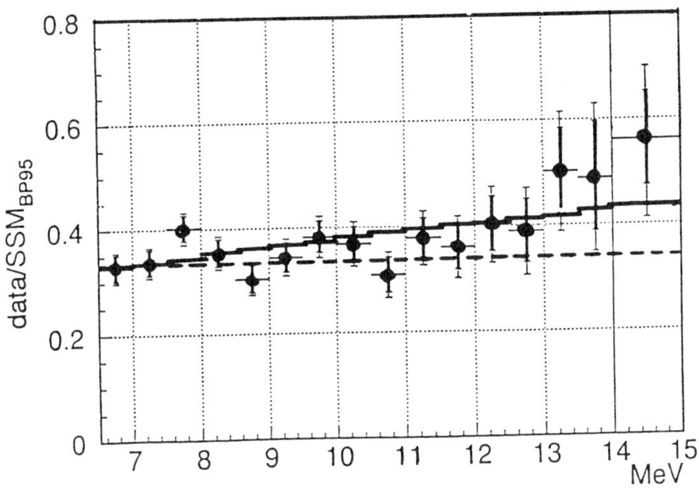

The "Just So" oscillation is tested by using Fig. 5 again. The 95 % C.L. allowed space in the "Just So" oscillation parameters is obtained in Fig. 6. Comparing with the Hata and Langacker result [2] based on the observed fluxes of so far all solar neutrino experiments gives a strongly restricted allowed space.

It is essential to go down to lower energies of < 5 MeV in order to find out a large discrepancy between the energy spectra expected from the small angle MSW oscillation and the "Just So" oscillation as shown in Fig. 7. In Super- Kamiokande much effort has been paid in lowering the analysis threshold energy. Very preliminary results on the $\cos\theta_{SUN}$ distributions for the energy intervals of 5.5 - 6.5 MeV has been obtained as shown in Fig. 8, where bumps around $\cos\theta_{SUN} = 1$ are clearly seen in both distributions. Fluxes in these energy intervals will be presented soon.

Figure 6. Region allowed at 95 % C.L. in the "Just So" oscillation. Hatched region is the result of Hata and Langacker [2].

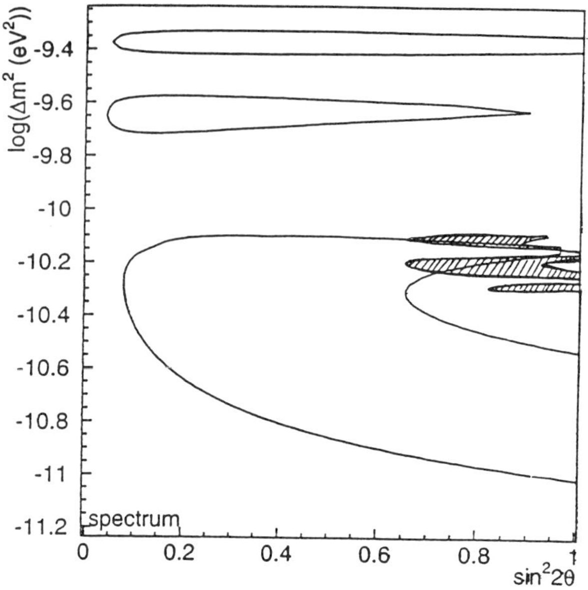

Figure 7. Survival probability of solar neutrinos as a function of the neutrino energy for the vacuum oscillation (VAC), the MSW small angle oscillation (SMA) and the MSW large angle oscillation (LMA).

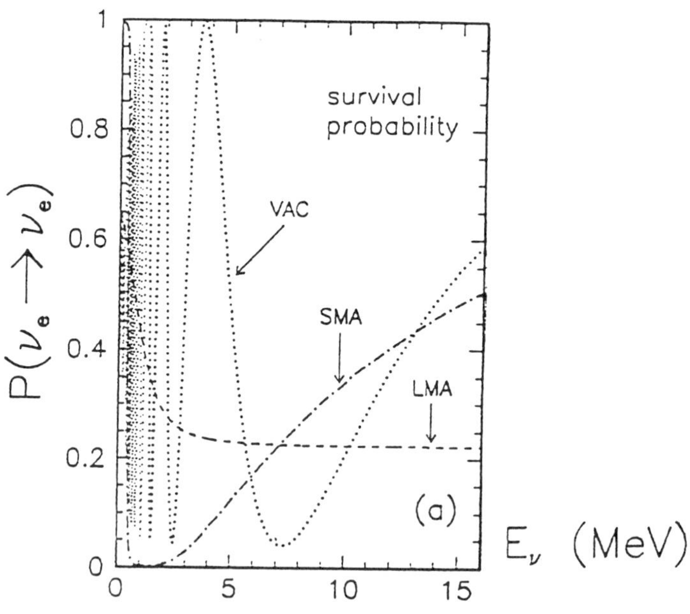

Figure 8. Very preliminary results on the electron angular distributions in $\cos\theta_{SUN}$ of 200 days events with $E_e = (6 - 6.5)$ MeV and $E_e = (5.5 - 6)$ MeV.

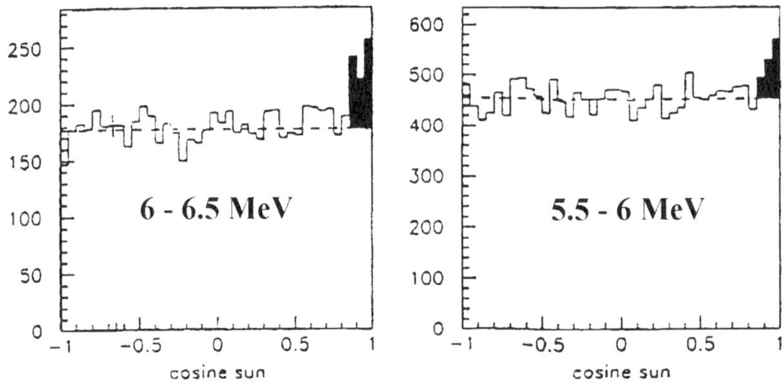

3. Atmospheric Neutrino Observation

Atmospheric neutrino studies have been carried out by detecting neutrino interactions with single Cherenkov ring events occurred in a detector. The reason why only single Cherenkov ring events are used, is that more than 94% of single-ring events are arisen from charged leptons ($l^{\pm}$) produced in the reactions, $\nu_l(\bar{\nu}_l) + H_2O \rightarrow l^{\mp} + X$. Hence if leptons are identified as $\mu^{\mp}$ or $e^{\mp}$, the flavor of incident neutrinos can be determined as $\nu_\mu(\bar{\nu}_\mu)$ or $\nu_e(\bar{\nu}_e)$. An essential task in the analysis is to identify the particle type (μ or e), using the characteristic Cherenkov-ring patterns of a showering track (e) and a non-showering track (μ). Less than 5 % misidentification probability is obtained for $p_\mu > 200$ MeV/c and $p_e > 100$ MeV/c by the Monte Carlo estimation.

Single ring events are classified into fully- and partially-contained events, depending on whether all produced particles stop inside a detector or not. Since 98 % of partially-contained events were estimated to be ν_μ charged-current interactions, all partially-contained events are treated as μ-like events. Events are also grouped into "sub-GeV" and "multi-GeV", depending on whether the energy is less or greater than 1.33 GeV. In the "sub-GeV" group fully-contained events are dominant, and fully- and partially-contained events are equally mixed in the "multi-GeV" events. The present analysis is based on the data sample of 25.52 kton·year. The 22.5 kton fiducial volume is set, requiring to be > 2 m from the PMT surface.

Observed event numbers of "sub-GeV" and "multi-GeV" are listed in Table 1, together with the Monte Carlo prediction based on the Honda *et al.* atmospheric neutrino fluxes [3]. Comparing the data with the MC

Table 1. Summary of the "sub-GeV" and "multi-GeV" atmospheric neutrino data.

Type of Event	Type of Group			
	"sub - GeV"		"multi - GeV"	
Single Ring	Data	MC	Data	MC
Fully-contained				
total	1883	1974.3	394	411.6
e-like	983	788.9	214	182.7
μ-like	900	1185.4	176	229.0
Partially- contained				
total			230	287.7
R(μ/e)	$0.61 \pm 0.03 \pm 0.05$		$0.66 \pm 0.06 \pm 0.08$	

prediction, enhancement of e-like events and reduction of μ-like events can be seen both in the "sub-GeV" and "multi-GeV" groups. However, since uncertainty of the absolute atmospheric neutrino flux calculations is about 30 %, the ratio R(μ/e) defined by $(\mu/e)_{\text{data}}$ (observed number of μ-like events / observed number of e-like events) to $(\mu/e)_{\text{MC}}$ (Monte Carlo expected number of μ-like events / Monte Carlo expected number of e-like events) is used. R(μ/e) is necessarily equal to 1, if no new physics is embedded in the atmospheric neutrino phenomena. From this table R(μ/e)'s of 0.61 ± 0.03(stat.) ± 0.05(syst.) and 0.66 ± 0.06(stat.) ± 0.08(syst.) are obtained for the "sub-GeV" and "multi-GeV" groups, respectively. The deviation from the expected value of 1 which is called the atmospheric neutrino anomaly was reconfirmed with the present high statistic data sample.

Momentum distributions of the e-like and μ-like events, and R(μ/e) momentum dependence are shown in Fig.9. Comparing the data with the prediction (histograms in this figure), it is found that enhancement for e-like events and reduction for μ-like events exist in the whole momentum range except for the highest momentum data in "multi-GeV" with low statistics. This tells R(μ/e) is almost constant in momentum, thus the atmospheric neutrino anomaly seems to be independent on neutrino energies.

The zenith angle distributions of e-like and μ-like events are shown in Fig. 10, separately for the "sub-GeV" and "multi-GeV" groups. The observed shape of e-like events seems to be consistent with the prediction (hatched histograms). On the other hand that of μ-like events deviates from the prediction particularly in $\cos\theta_{\text{zenith}} < 0$. The dependence of the

Figure 9. Momentum distributions of e-like (a) and μ-like (b) events, and the momentum dependence of R(μ/e) (R) both in the "sub-GeV" and "multi-GeV" groups. Histograms are the Monte Carlo prediction.

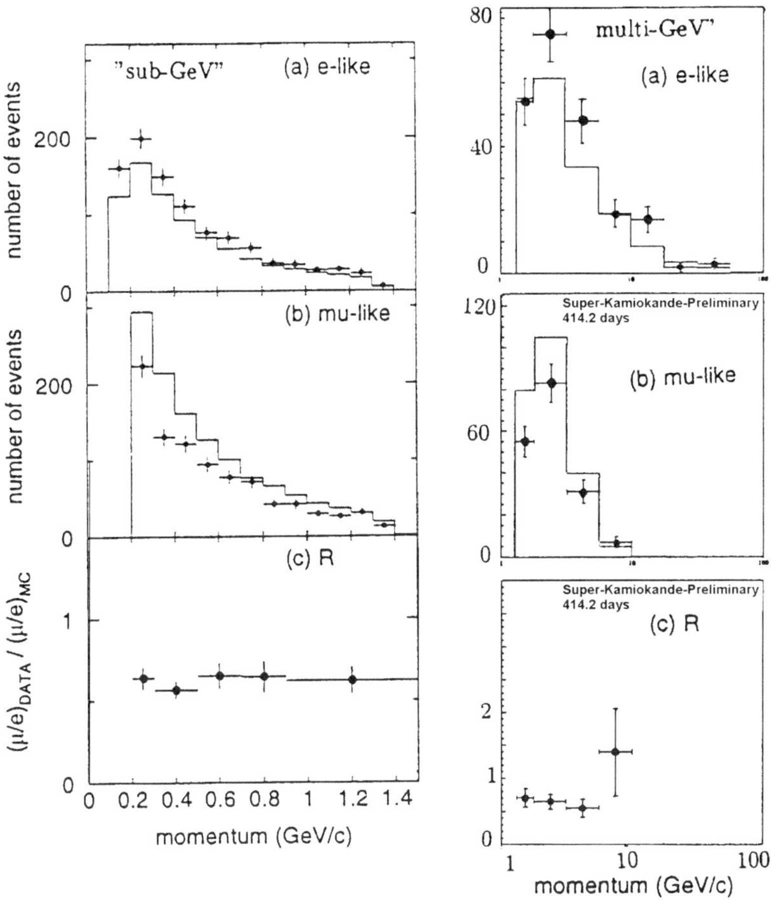

ratio R (μ/ e) on $\cos\Theta_{\text{zenith}}$ shown also in Fig. 10, shows this feature more clearly, especially for the "multi-GeV" group. This is because better angular correlation (r.m.s. = $15^o \sim 20^o$) between the incident neutrinos and the produced charged-leptons is expected in the multi-GeV events, comparing with that (r.m.s. = $5^o \sim 60^o$) of the sub-GeV events. A clear zenith angle dependence of R(μ/e) is found, where it decreases as the zenith angle increases. So the atmospheric neutrino anomaly seems to originate from the difference of neutrino flight pass up to reaching to a detector from their generated atmospher.

Neutrino oscillations are a natural solution to understand the atmospheric neutrino anomaly, considering the apparent zenith angle dependence of R(μ/e). The $\nu_\mu \to \nu_\tau$ oscillation predicts the zenith angle distributions of e-like and μ-like events shown in Fig.11 by solid lines, together with no oscillation case by hatched histograms. The oscillation parameters of $(\Delta m^2, \sin^2 2\theta) = (5 \times 10^{-3}, 1)$ gives the best fitting to all the distributions. $\Delta m^2 \sim (3 \times 10^{-4} - 10^{-2})\mathrm{eV}^2$ and $\sin^2 2\theta > 0.7$ are obtained as the allowed oscillation parameter space at 99% C.L.

Figure 10. Zenith angle distributions of e-like (a) and μ-like (b) events, and the zenith angle dependence of R(μ/e) (c) both in the "sub-GeV" and "multi-GeV" groups. The Monte Carlo prediction is shown by hatched histograms.

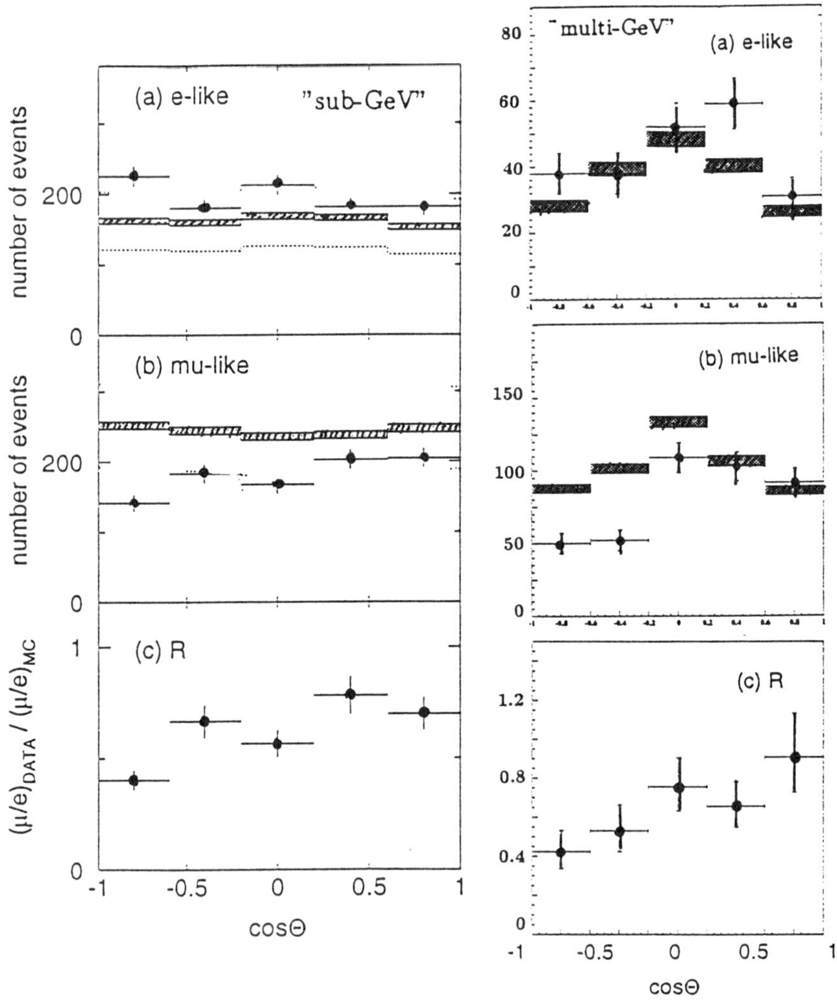

Figure 11. The best fit zenith angle distributions (solid lines) of e-like and μ-like events for the "sub-GeV" and "multi-GeV" groups obtained by the $\nu_\mu \rightarrow \nu_\tau$ oscillation. Hatched histogram shows the distribution without oscillations.

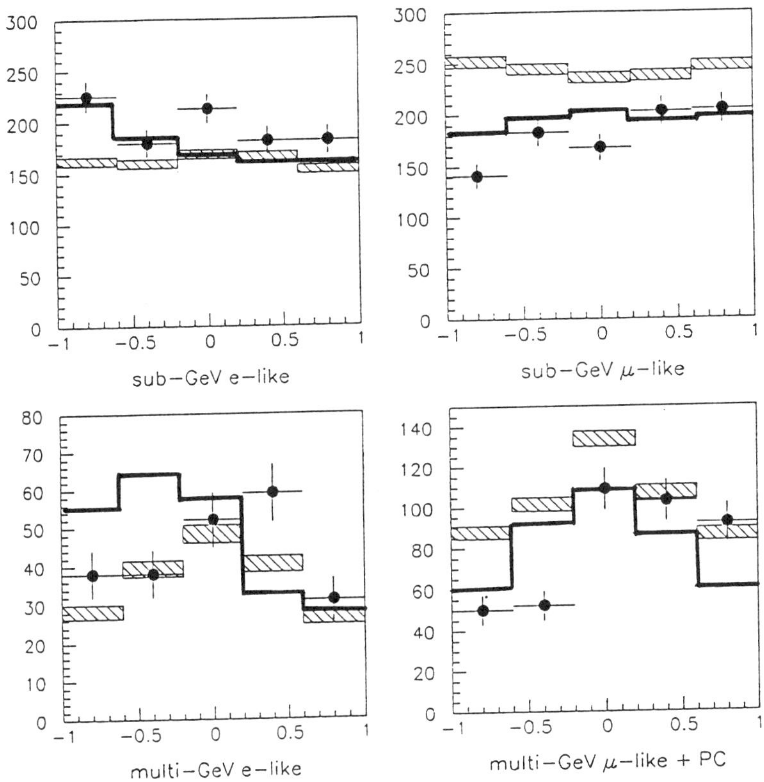

4. Proton Decays

Searching for proton decays has been carried out for the $p \rightarrow e^+\pi^0$ and the $p \rightarrow \bar{\nu}K^+$ modes with 25.5 kton· year data sample, and for the $p \rightarrow \mu^+\pi^0$ mode with 20 kton·year data sample. Candidate events of the $p \rightarrow e^+\pi^0$ mode are selected by the following event reduction processes: (1) choosing events with visible energy (Q_{tot} of (6800 - 9500) photoelectrons (p.e), (2) having 2 or 3 shower-type rings, (3) π^0 invariant mass M_{π^0} of (85 - 185) MeV for 3-ring events, (4) having no decaying electrons, (5) the momentum balance (P_{tot}) of less than 250 MeV/c; and (6) the proton invariant mass, M_p of (800 - 1050) MeV. Fig. 12 is a plot in the proton invariant mass and the momentum balance plane for events survived through other selection criteria. No events are remained inside the box drawn by the above-mentioned criteria. O candidate event against 0.1 estimated background events gives the lower limit of a partial lifetime of the $p \to e^+\pi^0$ mode 1.6×10^{33} yrs at 90 % C.L.

Figure 12. Plot of the invariant mass and the total momentum for events survived up to the final event selection of the $p \to e^+\pi^0$ mode. The 25.5 kton (414 days live-time) data sample is used.

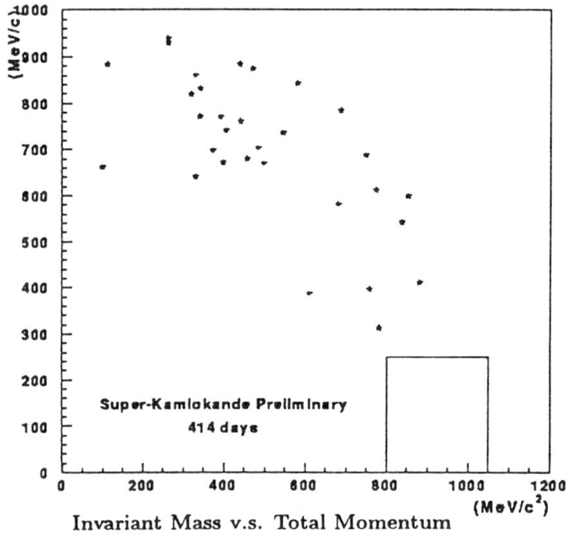

For the $p \to \mu^+\pi^0$ mode the following event selections are applied: (1) 5000 p.e $< Q_{tot} <$ 7800 p.e, (2) 2 or 3 rings with 1 μ-type and 1 or 2 shower-type rings, (3) 1 decaying electron, (4) 85 MeV $< M_{\pi^0} <$ 185 MeV for 3-ring events, (5) 800 MeV $< M_p <$ 1050 MeV; and (6) $P_{tot} <$ 250 MeV/c. Fig. 13 shows the same plot as Fig. 12 for the $p \to \mu^+\pi^0$ mode. 0 candidate event is found inside the box while the number of background events is estimated to be less than 0.06. The lower limit for this decay mode is 1.1×10^{33} yrs at 90 % C.L.

The $p \to \bar{\nu}K^+$ mode was searched for in three decay processes of $p \to \bar{\nu}K^+(\to \nu\mu^+(\to e^+\nu\bar{\nu}))$, $p \to \bar{\nu}K^+(\to \pi^0(\to 2\gamma)\pi^+(\to \nu\mu^+(\to e^+\nu\bar{\nu})))$, and $^{16}O \to ^{15}N^*(\to \gamma ^{15}O)\bar{\nu}K^+(\to \nu\mu^+(\to e^+\nu\bar{\nu}))$ with a γ-tagging. Fig. 14 is the momentum distribution of single μ - ring events. The observed distribution is consistent with the prediction from atmospheric neutrino interactions multiplying 0.76 to the event rate due to the atmospheric neutrino anomaly. The result combined 3 decay processes gives 3.3×10^{32} yrs at 90 % C.L.

Figure 13. The same plot as Fig. 12 for the p $\rightarrow \mu^+\pi^0$ mode. The 20 kton (326 days live-time) data sample is used.

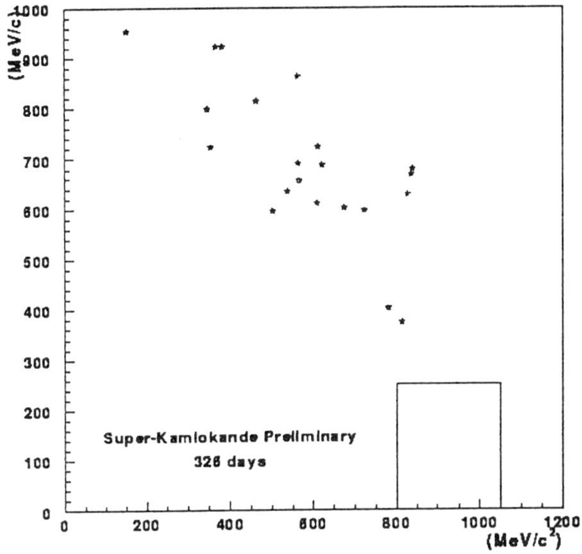

Invariant Mass v.s. Total Momentum

Figure 14. Momentum distribution of single-ring μ-like events. The solid histogram shows a 90 % C.L. limit of the p $\rightarrow \bar{\nu}K^+$ mode candate events. The dashed histogram is the prediction from atmospheric neutrino interactions (see text). The 25.5 kton (414 days live-time) data sample is used.

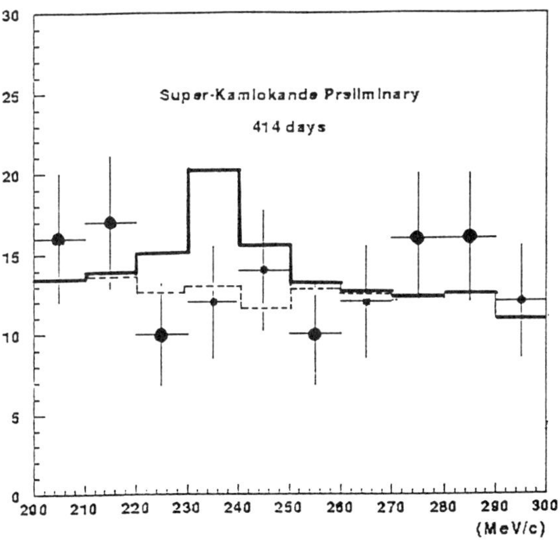

5. KamLAND Project

To aim at being a frontier of physics and astrophysics of neutrinos by means of detecting lower energy terrestrial and extra-terrestrial neutrinos and anti-neutrinos, the project of 1000 ton liquid scintillator detector experiment, KamLAND was started in 1997. This type of experiment was originally proposed by Raghavan and Pakvasa [4], although physics aims are somewhat different. Primary targets of physics in KamLAND are to search for neutrino oscillations by measuring anti-neutrinos produced in nearby power plant reactors, to measure the ^{8}B solar neutrinos (1st phase experiment) and ^{7}Be solar neutrinos (2nd phase experiment), to detect geoneutrinos, and to detect supernova burst neutrinos.

5.1. Detector Concept

The KamLAND detector is constructed at the old Kamiokande site, that is 1000 m (2700 m.w.e.) underground at the Kamioka mine,dismantling the Kamiokande detector components. The detector consists of 3-layers as illustrated in Fig. 15. The outer layer is a water Cherenkov anti-counter with a cylindrical shape of 20 m diameter and 20 m height. The counter inside is segmented into small rooms by light reflection sheets so as to identify an entrance and exit points of cosmic-ray muons quickly. The old Kamiokande 20" photomultiplier tubes (PMT's) are used here.

The middle layer is a stainless steel spherical buffer tank with 18 m diameter in which mineral oil is filled. 1280 pieces of newly developed 17" PMT's are installed on to the inner surface of this tank. Photosensitive coverage is 21 % of the entire inner surface. 17" PMT's have good timing and pulse-height performance : for single photoelectron illumination, Transit Time Spread $\sim$ 1.2 ns for 1 σ; and Peak to Valley ratio $\sim$ (3-5).

The innermost layer is 1200 m^3 ($\sim$ 1000 ton) liquid scintillator plastic balloon. To make a balloon tough mechanically and chemically, a multi-layer film of nylon + EVOH (poly vinyl alcohol) film + nylon is under development. Liquid scintillator materials are determined , considering high light output ($>$ 50 % of Anthracene), long light attenuation length ($>$ 14 m), good particle identification of n/γ, a high quenching effect for α-particles and α/γ by a pulse shape discrimination, low U, Th and K contamination and so on. A candidate scintillator in KamLAND is made of 80% of isoparaffin or normal paraffin, 20 % of pseudocumene and fluor, PPO (2g/l). In KamLAND we expect $\sim$ 110 photoelectron/MeV in utilizing the above-mentioned detector components.

5.2. Target physics

5.2.1. Searching for reactor anti-neutrino oscillations

Reactor anti-neutrinos are measured by detecting the inverse β-decay pro-

Figure 15. Schematic view of the KamLAND detector.

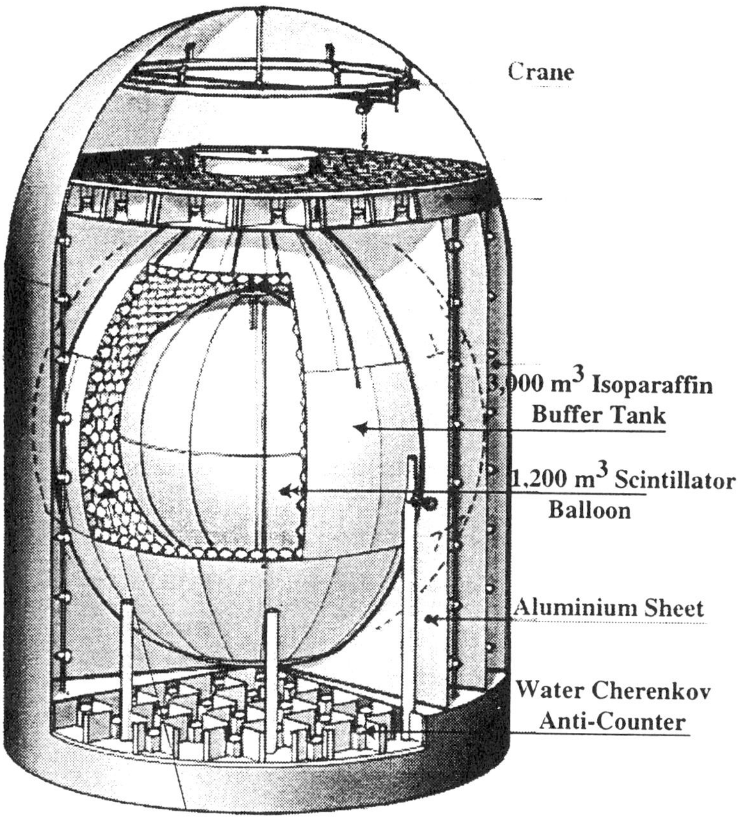

cess (E_{th} = 1.8 MeV), $\bar{\nu}_e + p \to e^+ + n$ with the aid of a delayed timing coincidence between a prompt e^+ signal and a delayed 2.2 MeV - γ produced by a thermal neutron capture on a proton. Contamination of background events is considerably reduced by this method. There are many power-plant reactors around Kamioka. More than 95 % of events are contributed by nearby reactors with the distance of < 200 km. About 450 anti-neutrino events are expected for one year running and the 600 ton fiducial mass. On the other hand the background event rate is estimated to be 8 in a year with assumptions of 10^{14} g/g for U and Th, 10^{12} g/g for K and 0.5 mBq/m^3 for Rn concentrations in liquid scintillator. The expected positron energy spectrum and that of background events are shown in Fig. 16. One can see that no significant effect from background events is brought in the $\bar{\nu}_e \to \bar{\nu}_x$ oscillation analysis. An accessible oscillation parameter space with 90 % C.L. significance is depicted in Fig. 20 for 3 years data taking. Region of the MSW large angle solution ($\Delta m^2 \sim 10^{-5}$ eV^2) can be examined by the KamLAND reactor experiment.

Figure 16. Expected energy spectra of positrons (reactor anti-neutrino events) and background events.

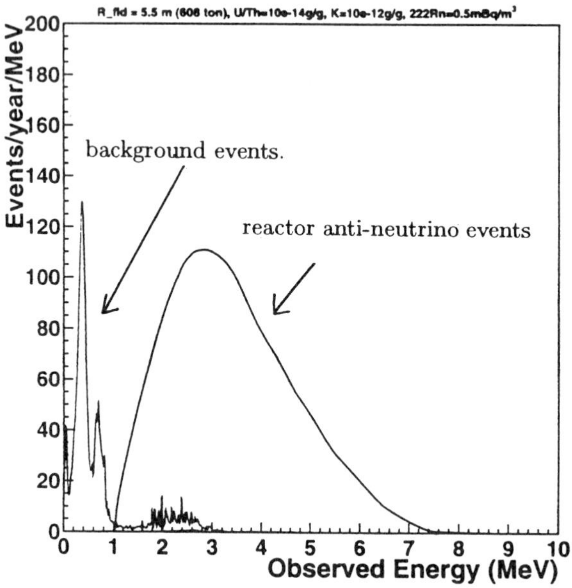

5.2.2. Measuring solar neutrinos

The energy measurement sensitivity of $\sim$ 110 p.e./MeV in KamLAND makes it possible to set a detection threshold energy down to $\sim$ 300 keV. Hence the ^{8}B and ^{7}Be solar neutrinos can be principally measured. Of course tremendous background phenomena are waiting for us in the "sub-MeV" region, because solar neutrinos produce only single electron signal through the $\nu + e \rightarrow \nu' + e'$ reaction. Consequently at the first stage in KamLAND we aim to measure the ^{8}B solar neutrinos with higher energies. Fig. 19 shows the expected electron energy spectrum of ^{8}B neutrino events. 940 recoiling electrons with energies > 4 MeV are detected for the 560 ton fiducial mass in one year operation. Background spectra coming from U, Th and Rn contamination in the detector materials and surrounding rocks are also shown in this figure., where concentrations of 10^{-16} g/g for U and Th and 10 μBq/m^3 for Rn in liquid scintillator are assumed. As mentioned before, measuring the low energy electron spectrum in the energy range of < 5 MeV is essential to find out whether the solar neutrino deficit is caused by the MSW small angle or "Just So" oscillation. Fig. 19 tells us the energy spectrum down to 3.5 MeV is detectable in KamLAND.

Figure 17. 90 % C.L. parameter constraints on $\nu_e \to \nu_x$ oscillations.

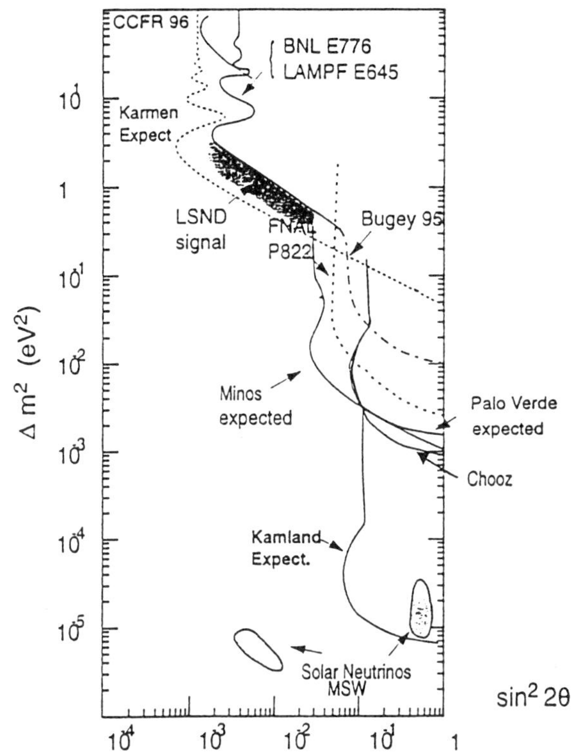

Figure 18. Expected energy spectra of recoiling electrons (^{8}B solar neutrino events) and background events.

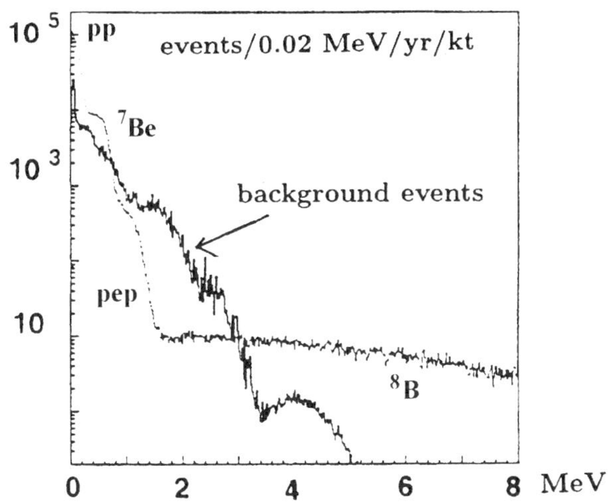

5.2.3. Measuring geoneutrinos

A first chance for terrestrial $\bar{\nu}_e$ (geoneutrino) search can be expected in KamLAND. A basic factor in the interior dynamics and the evolution of the present Earth is the radiogenic heat mainly from U and Th decays inside the Earth. Measuring the U and Th concentrations in the Earth interior shed new light on geophysics. Calculations of geoneutrino fluxes strongly depends on models concerning the abundances of U and Th in the continental crust, oceanic crust, upper-mantle and lower mantle. Fig. 19 is an example of the flux calculation by Krauss *et al.* [5]. Geoneutrinos are detected by the same method as reactor anti-neutrinos with a delayed coincidence technique. From Fig. 19 U and Th components can be measured separately in the energy range above the inverse β-decay threshold energy, 1.8 MeV. Positron energy spectra expected by several models on the U and Th abundance are shown in Fig. 20 in addition to the reactor signals [6]. One can see clear and sharp peaks of the U and U+Th components standing on the continuous reactor signals. More than several tens of geoneutrinos can be expected for one year operation in KamLAND.

Figure 19. Geoneutrino flux calculation [5].

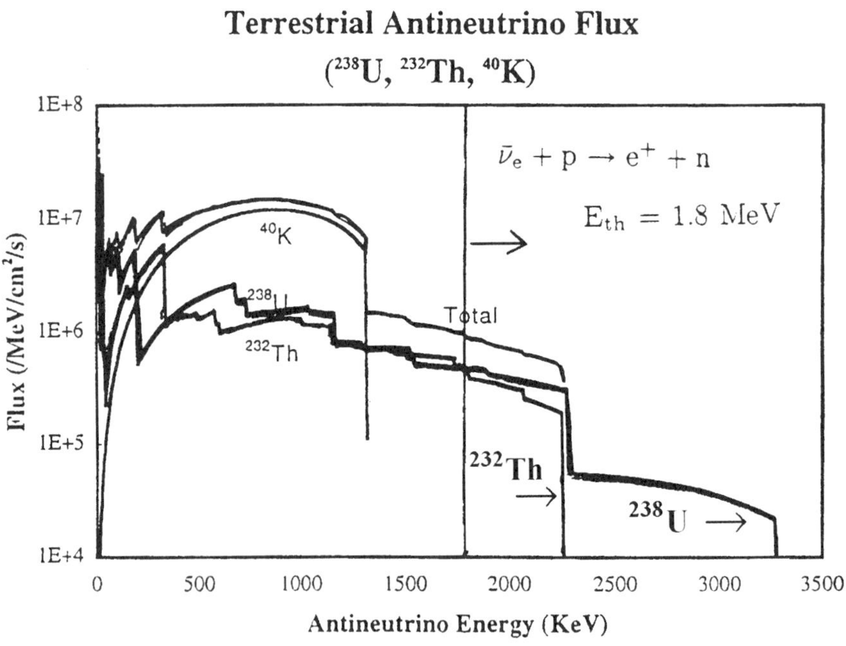

Figure 20. Positron energy spectra induced by geoneutrinos and reactor antineutrinos [6].

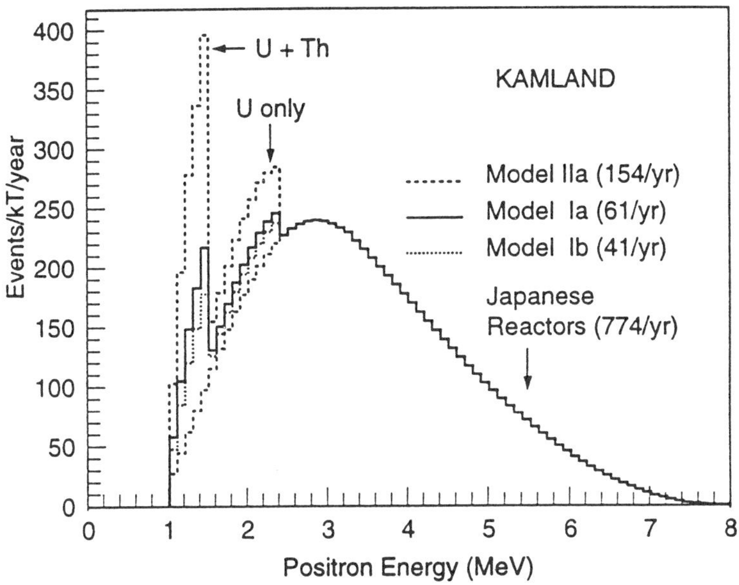

5.2.4. Detecting a supernova neutrino burst

Characteristic measurements of supernova burst neutrinos are expected in KamLAND. Event rates for a galactic supernova burst was calculated by the following supernova parameters: the distance of 10 kpc, the released energy of 3×10^{53} erg, the temperatures of 3.5 MeV ($< E_\nu > = 11$ MeV) for ν_e, 5 MeV ($< E_\nu > = 16$ MeV) for $\bar{\nu}_e$; and 8 MeV ($< E_\nu > = 25$ MeV) for $\nu_\mu, \bar{\nu}_\mu, \nu_\tau, \bar{\nu}_\tau$ [7]. Results are 8 events for the $\nu_x(\bar{\nu}_x)+e^- \rightarrow \nu_x(\bar{\nu}_x)+e^-$ channel and 330 events for the $\bar{\nu}_e + p \rightarrow e^+ + n$ channel. The first observation of supernova $\bar{\nu}_e$ energy spectrum can be expected (see Fig. 21) by the delayed coincidence method. Nuclear excitations of ^{12}C by all types of neutrinos provide a powerful tool to the oscillation analysis. Although the event rates for charged current excitations of the $\nu_e +^{12} C \rightarrow e^- +^{12} N$ channel and the $\bar{\nu}_e +^{12} C \rightarrow e^+ +^{12} B$ channel are small, 2 and 7 events, respectively, the MSW oscillation predicts 27 events and 7.4 events, and the vacuum oscillation 14.5 events and 13.8 events. Such enhancement gives a chance to detect neutrino oscillations. Moreover these two reactions are identified with a time and space coincidence between the prompt e^- (e^+) and the delayed decaying $e^+(e^-)$ from the unstable nucleus, ^{12}N (^{12}B). Thus they are essentially background free. 58 events are expected in neutrino current excitations on ^{12}C which accompany 15.11 MeV monochromatic γ-rays shown in Fig. 21. This gives a sensitive monitor on the supernova neutrino temperature.

Figure 21. Energy spectrum from "standard" supernova explosion [7].

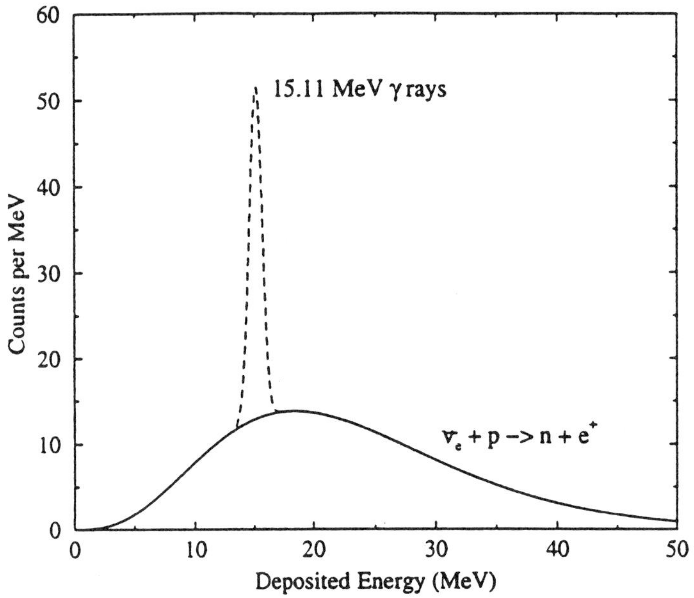

6. Conclusions

KamLAND was approved by a 5 year project since 1997. Hence we have to show some physics results until 2002. The construction schedule is so tight in the following:

1. In 1998
   (a) to remove the Kamiokande detector and expand the tunnel size and detector doom depth,
   (b) to install new power station and illumination, and
   (c) to construct the outer-layer (water Cherenkov anti-counter).

2. In 1999

   (a) to construct the middle-layer (stainless steel spherical buffer tank) and install the innermost-layer (1200 m^3plastic balloon),
   (b) to construct the water and liquid scintillator purification systems, and
   (c) to install 1280 17" PMT's into the buffer tank.

3. In 2000

   (a) to install data taking electronics and computers, and
   (b) to fill 1000 ton liquid scintillator inside the balloon.

4. In 2001

   (a) to start data taking in January 1.

KamLAND has been performing as the Japan-US collaboration since April 1998.

7. Acknowledgments

We wish to thank Prof. H. V. Klapdor-Kleingrothaus, the chairman of this workshop for giving the opportunity to visit the small beautiful town. We also would like to thank the Super-Kamiokande collaborators and the KamLAND collaborators for discussions.

References

[1] J. N. Bahcall and M. H. Pinsonneault, *Rev. Mod. Phys.* **67**, 781 (1995).

[2] N. Hata and P. Langacker, *Phys. Rev.* **D56**, 6107 (1997).

[3] M. Honda, T. Kajita, K. Kasahara and S. Midorikawa, *Phys. Rev.* **D52**, 4985 (1995).

[4] R. S. Raghavan and S. Pakvasa, *Phys. Rev.* **D37**, 849 (1988); R. S. Raghavan, AT&T Bell Labs Report 11131-880114-02TM (1988).

[5] L. M. Krauss, S. L. Glashow and D. N. Schrammd, *Nature*, **310**, 191 (1984).

[6] R. S. Raghavan, S. Schoenert, S. Enomoto, J. Shirai, F. Suekane and A.Suzuki, *Phys. Rev. Lett.* **80**, 635 (1998).

[7] The U.S KamLAND proposal (1998)

BOREXINO: a Real Time Detector for Low Energy Solar Neutrinos

Marco G. Giammarchi[1]

Istituto Nazionale di Fisica Nucleare - Milano

Via Celoria 16 - 20133 Milano, Italy

On Behalf of the BOREXINO Collaboration [1]

Abstract. After a short introduction to solar neutrino physics, this paper describes the conceptual design and expected performance of the BOREXINO underground detector at Gran Sasso. Low background results from the Counting Test Facility, the first phase of the BOREXINO project, are summarized. Indicative sensitivities to non-standard neutrino physics scenarios are presented.

1. Introduction

One of the most important and challenging goals of neutrino physics is the identification of a possible non-zero rest mass. Some evidence in this direction has been accumulated during the last decades by experiments using solar, atmospheric and laboratory (reactors, accelerators) neutrinos. As the most recent result, the SuperKamiokande collaboration has presented evidence for oscillation of atmospheric neutrinos coming from more than 500 days of exposure of their water Čerenkov detector [2].

The identification of a non-zero neutrino rest mass would imply the existence of physics not included in the Standard Model of Electroweak Interactions and, at the same time, provide an important component of the (hot) Dark Matter pervading the Universe.

[1] E-mail: marco.giammarchi@mi.infn.it.

If neutrinos have masses they can naturally mix [3], so that flavor (ν_e, ν_μ, ν_τ) and mass (ν_1, ν_2, ν_3) eigenstates are related by a non-diagonal mixing matrix:

$$\nu_l = \sum_{i=1}^{3} U_{il}\nu_i \qquad (1)$$

Since the mass eigenstates will propagate differently, one will observe different mixtures of weak eigenstates at a given distance from the neutrino generating source. In a two-flavor oscillation model, the transition probability from a flavor eigenstate (say ν_e) to another one (say ν_μ) is given by:

$$p(\nu_e \rightarrow \nu_\mu) = sin^2(2\theta)sin^2(\pi L/L_0) \qquad (2)$$

where $L_0 = 4\pi E/\Delta m^2$. Experiments having different E/L (ν energy/baseline) have different sensitivities to the mixing parameters in the bidimensional θ (mixing angle) / Δm^2 (square mass difference) plane. In the case of solar neutrinos - given the distance and the ν energy range - the oscillation sensitivity is in the $\Delta m^2 \sim 10^{-11}\ eV^2$ range, when the oscillation wavelength is comparable with the Earth-Sun yearly distance variation (vacuum oscillations, $\lambda \simeq 5 \times 10^9\ m$). The experimental signature of vacuum oscillations are an energy dependent depletion and a seasonal variation of the ν_e solar flux.

The sensitivity of Solar Neutrino experiments is vastly extended above the vacuum oscillation value due to the possibility of matter enhanced oscillations [4]. This process (hereafter called MSW from the initials of Mikheyev, Smirnov and Wolfenstein) is a resonant amplification of the vacuum oscillation, generated by the fact that a beam of electron neutrinos and a beam of muon neutrinos propagate differently in matter. Because of this, even with a small mixing angle, a total conversion can take place (in the Sun):

$$tg(2\theta_m) = \frac{tg(2\theta)}{1 \pm (L_0/L_e)sec(2\theta)} \qquad (3)$$

where $L_e = \sqrt{2}\pi/n_e G_F$. n_e is the local electron density and G_F the Fermi constant, so that L_e is the neutrino-electron interaction length in the solar matter. This equation relates the vacuum oscillation angle θ to the matter oscillation angle θ_m. If the critical density is met during the propagation in the solar core, then a total conversion can take place even if θ is small. Because of this mechanism, sensitivity is gained to much higher values of the mass (up to about $10^{-4}\ eV^2$). The experimental signatures associated to the MSW oscillations are an energy dependent reduction of the solar neutrino ν_e flux and (for some choice of the oscillation parameters), a day/night effect due to the inverse process: regeneration of the neutrino beam in the Earth's core and mantle.

In this paper only solar neutrinos, the detection goal of BOREXINO, will be considered. In §2 we will summarize the predictions and measurements of the solar neutrino fluxes and shortly discuss the so called *Solar Neutrino Problem*. §3 describes the BOREXINO detector concept and expected performance while §4 summarizes the radiopurity results of the BOREXINO prototype. Finally §5 discusses the sensitivity of BOREXINO to non standard neutrino physics.

2. The solar neutrino flux measurement

The sun is a powerful source of neutrinos, with a predicted total flux on Earth of $\simeq 6\times10^{10}$ particles $cm^{-2}sec^{-1}$. Since the typical cross sections for neutrinos are only $\sim 10^{-46}$ cm^2, the characteristic reaction rate per unit target will be $\sim 10^{-36}$ per second, a special unit often called SNU (solar neutrino unit). These numbers imply that solar neutrinos physics deals with low event rates in massive (and massively shielded) detectors. For instance BOREXINO is begin built in the Hall C of the at the Laboratori Nazionali del Gran Sasso (LNGS) where the rock overburden (of about 3500 mwe) guarantees a 10^6 reduction factor for cosmic muons relative to the sea level.

Solar neutrino fluxes have been calculated by several authors (see for instance [5]), and are generally considered to be reliable, especially after the confirmations coming from the helioseismology measurements [6]. The reputation of solar models is nowadays so high as to deserve, in its most complete forms, the title of Standard Solar Model (or SSM, usually considered the first model in ref. [5] and now being updated [7]).

The shape and magnitude of the predicted solar neutrino spectrum is shown in fig. 1. About 90% of the flux comes from the $pp \rightarrow de^+\nu_e$ reaction, abbreviated to pp.

On the experimental side, solar neutrino experiments have been operating since 1968, when the Chlorine underground detector at the Homestake mine started taking data. They can be either radiochemical (Homestake, GALLEX, SAGE) or real-time (Kamiokande, SuperKamiokande); having different thresholds, they are sensitive to different parts of the neutrino spectrum.

The results of the flux measurements are summarized in table 1.

The experimentally measured fluxes are considerably lower than the theoretical (SSM) predictions (also shown in table 1), a discrepancy which

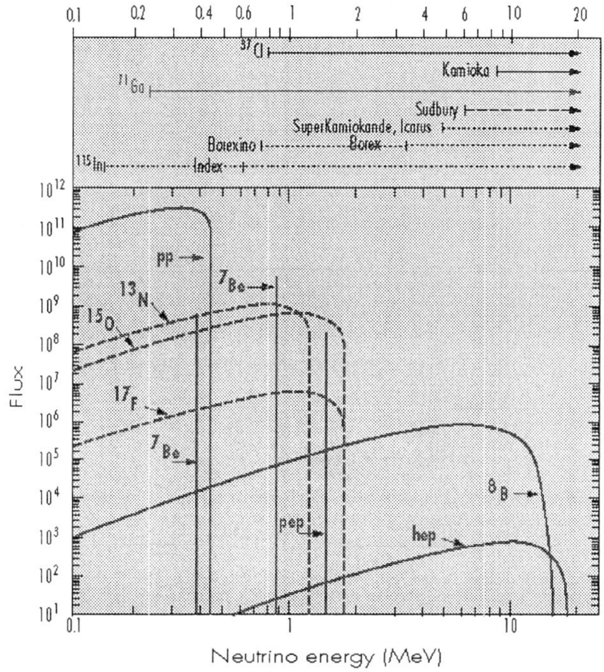

Figure 1. *Solar neutrino spectra predicted by the Standard Solar Model. On the top of the figure are experimental detection thresholds of existing or planned solar neutrino experiments.*

consitutes the so-called *Solar neutrino problem*, one of the strongest indicators of physics beyond the Standard Model of Electroweak Interactions.

This energy-dependent discrepancy could in principle be due to a bad understanding of the Sun physics (or a bad choice of the physical input parameters used in modelling the Sun). This is sometimes called the *astrophysical* solution of the *Solar Neutrino Problem*; it implies that the neutrino flux predictions of the SSM are incorrect.

The other possibility involves neutrinos oscillations and masses. This is the *particle physics* solutione of the *Solar Neutrino Problem*; it implies that the Standard Model of Electroweak Interactions is incomplete.

The analysis of the available data performed by several authors indicates rather strongly that:

Table 1. Summary of solar neutrino flux measurements (July 1998). The units are SNU's for Homestake, SAGE and GALLEX and $10^6\ cm^{-2}s^{-1}$ for Kamiokande and SuperKamiokande.

Experiment	Threshold (MeV)	Data	Theory (SSM, [7])
Homestake	0.814	$2.56 \pm 0.16 \pm 0.16$	$7.7^{+1.2}_{-1.0}$
Kamiokande	6.5	$2.80 \pm 0.19 \pm 0.33$	$5.15^{+1.0}_{-0.7}$
SAGE	0.235	$66.6^{+7.8}_{-8.1}$	129^{+8}_{-6}
GALLEX	0.235	$77.5 \pm 6.2^{+4.3}_{-4.7}$	129^{+8}_{-6}
SuperKamiokande	7.5	$2.44 \pm 0.05^{+0.09}_{-0.07}$	$5.15^{+1.0}_{-0.7}$

- Changes in Solar Models (or non Standard Solar Models) cannot accomodate the experimental results [8]. This difficulty can be recast in the statement that experimental observations indicate a severe suppression of the ^{7}Be solar ν flux.

- Non Standard Solar Models have severe difficulties in predicting other well known Sun properties. The most recent and important indication in this direction comes from the helioseismological measurements.

To summarize, the *astrophysical* solution of the *Solar Neutrino Problem* is disfavored. Reference [9] is a complete and recent review of the situation.

It is to be noted that the integral flux measurement is not the only tool in hand for solar neutrino experiments. Spectral distorsion of the ^{8}B signal (which extends up to about 15 MeV, see fig. 1), zenith-angle dependence of rates and day/night effects are also used as sensitive indicators of new physics. For a more systematic presentation of the investigation tools used by the new solar neutrino experiments see [10] and references cited therein.

On the other hand, good fits to all the experimental data can be obtained if one assumes non-standard neutrino physics (neutrino oscillations) within allowed ranges of the MSW or vacuum oscillation scenarios. In the two-flavor oscillation model, this implies determining the oscillation parameters θ and Δm^2 that best fit the available data (rates, spectral shape, zenith-angle and day/night dependence of the signal). The most recent of these analyses can be found in reference [11]; the main result are summarized in table 2 (see also fig. 2). There are four main allowed regions in the parameter space: MSW Small Mixing Angle (SMA), MSW Large Mixing Angle (LMA), MSW Low Δm^2 (LOW) and vacuum oscillations.

A more general analysis has been developed which takes into account the (probably more realistic) possibility of a full three-flavor neutrino oscillation [12]. In this formalism the parameter space features two angles and a

Table 2. Best fit points for the two-flavor neutrino oscillation [11].

Scenario	Δm^2 (eV^2)	$sin^2(2\theta)$	
MSW Small Angle (SMA)	5×10^{-6}	5.5×10^{-3}	acceptable at 7% C.L.
MSW Large Angle (LMA)	1.8×10^{-5}	0.76	disfavored
MSW Low (LOW)	7.9×10^{-8}	0.96	disfavored
Vacuum Oscillations	6.5×10^{-11}	0.75	acceptable at 6% C.L.

δm^2 (between the two light mass eigenstates ν_1 and ν_2); roughly speaking, a solar neutrino experiment tests the mass composition (ν_1, ν_2, ν_3) of the ν_e. Nevertheless in the following we will be using the two flavor oscillation model, for reasons of simplicity and availability of material. Most of the semi-quantitative conclusions drawn in the frame of the more intuitive two-flavor model remain essentially valid when extrapolated to the three flavor case (see [12]).

From the analysis of several authors (see for instance [8]) it clearly emerges the importance of a direct measurements of the intensity of the ^{7}Be neutrino line at 0.862 MeV (see fig. 1). The ^{7}Be line is predicted by all solar models to be the second most important neutrino production reaction in the Sun (after the basic *pp* reaction). The flux of ^{7}Be neutrinos is predicted much more accurately (uncertainly less than 10%) and is about 1000 times larger than the ^{8}B neutrino flux that is being measured by the SuperKamiokande detector. Also, since ^{7}Be decay produces only neutrino lines, the theoretical predictions of neutrino oscillations are more unique for ^{7}Be than for the ^{8}B neutrinos, which have a broad (0-15 MeV) energy spectrum.

The detection of this signal is the main goal of the BOREXINO experiment, under construction at the Laboratori Nazionali del Gran Sasso (LNGS), in central Italy.

3. The BOREXINO detector and the expected background

The ^{7}Be 0.862 MeV neutrino line originates from the electron capture $^7Be+e^- \rightarrow^7 Li+\gamma+\nu_e$ reaction in the Sun core. These neutrinos can be detected by means of the $\nu e^- \rightarrow \nu e^-$ elastic scattering reaction, which originates recoil electrons in the BOREXINO detector with a maximum energy of 664 keV (see fig. 4).

The ^{7}Be line is predicted by the SSM to give rise to 46 events/day (recoiling electrons) in the 100 tonnes BOREXINO Fiducial Volume, with energy in the 250/700 keV observational window.

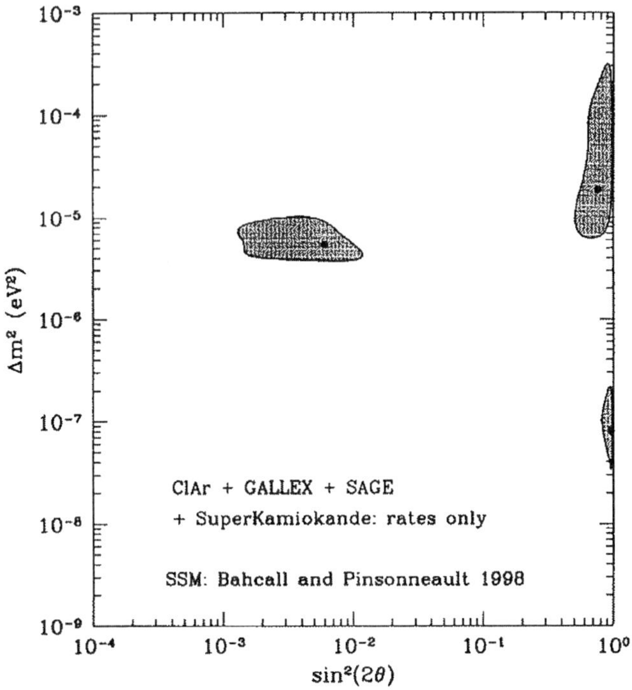

Figure 2. *MSW solutions when only the rate information from the solar neutrino experiments are considered. The three shaded areas represents the SMA, LMA and LOW parameter regions, allowed at 99% C.L. (from reference [11]).*

BOREXINO is an unsegmented liquid scintillator detector featuring 300 tons of well shielded ultrapure scintillator viewed by 2200 photomultipliers (fig. 3). The detector core is a transparent spherical vessel (Nylon Sphere), 8.5 m of diameter, filled with 300 tonnes of liquid scintillator and surrounded by 1000 tons of a high-purity buffer liquid. The scintillator mixture is composed by pseudocumene (PC, 1-2-4 trimethylbenzene) as the solvent with PPO (1.5 g/l) as a fluor, while the buffer liquid will be PC alone (with the possible addition of a light quencher, DMP). The photomultipliers are supported by a steel sphere which also separates the inner part of the detector from the external shielding, provided by 2400 tons of pure water (water buffer). An additional containment vessel (Nylon film Radon barrier) is interposed between the Scintillator Nylon Sphere and the photomultipliers, to the goal of reducing Radon diffusion towards the internal part of the detector.

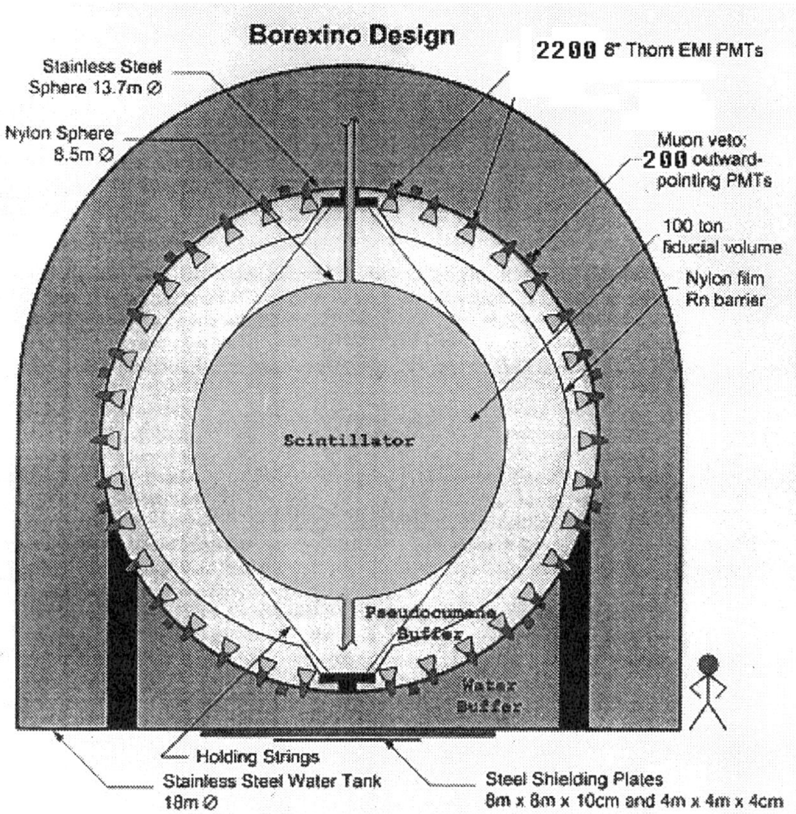

Figure 3. *Schematics of the BOREXINO detector. See text.*

The outer water shield is instrumented with 200 photomultipliers serving as a veto for penetrating muons, the only significant remaining cosmic ray background at the Gran Sasso depth. In addition the 2200 photomultipliers viewing the internal part of the detector are divided in two sets: 1800 photomultipliers are equipped with light cones so that they see light only from the Inner Vessel region, while the remaining 400 PM's are sensitive to the total volume of the stainless steel sphere. This design greatly increases the capability of the system to identify muons crossing the PC buffer and not the scintillator.

The detection of the ^{7}Be signal in the 100 tonnes of the BOREXINO Fiducial Volume requires a background of less than about 20 events/day in the same energy window. This event rate can be achieved if the radiopurity of the scintillator is below 5×10^{-16} g/g of ^{238}U, ^{232}Th equivalent [13] and no other significant backgrounds are present.

The BOREXINO design is based on the concept of a *graded shield* of progressively lower intrinsic radioactivity as one approaches the sensitive volume of the detector; this culminates in the use of 200 tons of the ultralow background scintillator to shield the 100 tonnes innermost Fiducial Volume.

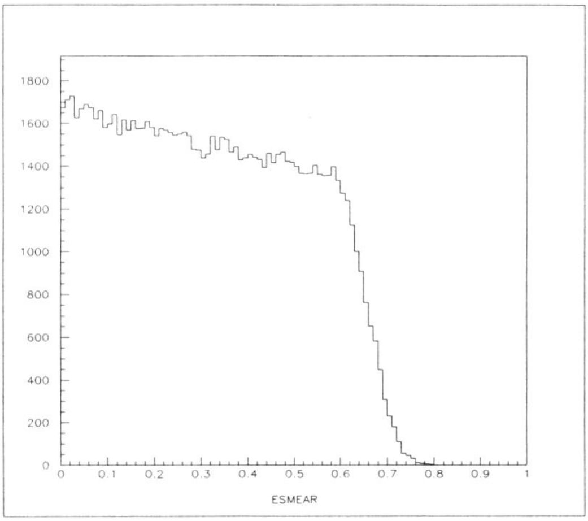

Figure 4. *Recoil electron spectrum produced by 862 keV neutrins. Due to energy resolution effects (included in the figure) the end-points extends somewhat more than the kinematical limit. Energies are in MeV. Vertical scale is in arbitrary units.*

The background source to consider in BOREXINO are the cosmic muons and muon-induced activity, the material radioactivity and the intrinsic radiopurity of the scintillator. The general strategy is to minimize all possible *external* background contributions so that only the intrinsic purity of the scintillator will be a significant background in the Fiducial Volume.

Cosmic muons arrive in hall C at a rate of about 1 m^{-2} hr^{-1}, inducing backgrounds that can fall in either of these three categories:

- Direct prompt signals, induced by muons at a rate of about 4500 events/day. A combined rejection [14] based on the total apparent energy (muons crossing the scintillator will release very big pulses), the spatial reconstruction (the track will not be reconstructed as a pointlike event in the Nylon Sphere), the pulse height (even for muons not crossing the Nylon Sphere, the Čerenkov light can be collected), PMT's timing pattern (typical of a particle coming from outside the stainless steel sphere) and the external veto cut (Čerenkov light in the water shield seen by the 200 external PMT's), will reduce this component to less than 1 event/day. Some of these cuts have been tested in the Counting Test Facility detector described below.
- Gamma rays at 2.2 MeV are released as a result of cosmogenically produced neutrons. This rate (about 100 event/day) can be easily reduced to below one event per day by using an energy cut and exploiting the muon/neutron time coincidence ($\sim$ 200 μs) [14].
- Cosmogenic radionuclides can be produced starting from the carbon nuclei in the C_9H_{12} pseudocumene molecule; the raw rate will be of about 30 events/day. Most of them (^{8}Li, ^{9}Li, ^{11}Be, ^{12}Be, ^{8}B, ^{9}C, ^{11}C) have a short lifetime (less than 1 sec) and very high beta endpoint, so they can be easily rejected by a 3 sec exclusion time cut correlated with a preceding muon, introducing a negligible dead time [14] [2] .

In conclusion, it is estimated that the whole muon related background will contribute at most 2 events/day in the 250/700 keV energy range.

Coming now to the detector construction materials, as a general statement, the choice of the dimensions, materials and radiopurity levels of any spherical shell of the detector follows the basic concept that the background produced by the shell cannot exceed the background induced in its position by more external sources. This choice could be enforced everywhere, except for the case of the photomultipliers, for which a careful material selection procedure was implemented. Table 3 summarizes the construction materials radiopurities obtained so far.

The use of these criteria for all the detector equipment and the predicted efficiency of the muon veto, leads to the conclusion that the ultimate limiting background will be the *instrinsic* background of the scintillator itself in the Fiducial Volume, all the other backgrounds (muon-related and construction-materials related) being less than 2 counts/day in the 250/700

[2] This is not true for ^{11}C and ^{7}Be. However, the conclusions drawn above are still valid; see detailed discussion in [14].

Table 3. Summary of a few indicative radiopurity results (in g/g of U-238 equivalent) of BOREXINO construction materials.

	U-238	Rn-222	Th-232	Knat
Nylon	10^{-11}		10^{-11}	
Water	10^{-14}	10^{-12}	10^{-14}	10^{-14}
Steel	10^{-9}		10^{-9}	
PMT's	10^{-8}		10^{-8}	

keV window (see discussions in [13]Nimpaper and data from Counting Test Facility in [16]).

The intrinsic purity of the scintillator, a critical part in the BOREXINO design, has been the object of an intense development during the last decade. A part of this R&D development is described in [13]. However, the most important indication that intrinsic purities below 10^{-15} g/g U,Th can be reached in a large scale detector comes from the Counting Test Facility to be described below.

4. The Counting Test facility

The basic concepts of the BOREXINO detector have been thoroughly studied in a smaller scale BOREXINO-like detector, called Counting Test Facility (CTF for short). The CTF has served both as a test bench for the technical choices to be adopted in BOREXINO and as a high sensitivity device to measure radioactive scintillator contaminations in the 10^{-16} g/g U equivalent range.

In the CTF detector [15] 4 tons of liquid scintillator are shieldied by 1000 tons of high purity water and viewed by 100 photomultipliers located on a support steel structure (see fig. 5). The apparatus is coupled to a water purification system, to produce the low radioactivity water shield (10^{-14} g/g U equivalent), and to a scintillator purification system. In addition a 1.8 m^2 muon detector has been positioned on the top of the tank so as to test the muon identification capability of the PMT system.
The results of the CTF in terms of U,Th and ^{14}C contamination (internal background), discussed in [16], indicate that the radiopurity goals of the experiment have been reached.

Uranium chain contamination (more appropriately ^{226}Ra) has been measured through the delayed coincidence method, by exploiting the 214-mass Bi-Po 236 μsec coincidence. The results of this analysis, which will not be discussed here, yield an upper limit of $3.5 \pm 1.3 \times 10^{-16}$ g/g.

Figure 5. *The Counting Test Facility experimental setup viewed from inside the water tank. The scintillator is contained in the nylon balloon at the center.*

Thorium contamination was also studied with the delayed time tag given by the 212-mass Bi-Po 432 *nsec* coincidence. The results of the analysis indicate an ultimate purity of $4.4^{+1.5}_{-1.2} \times 10^{-16}$ g/g.

C-14 poses a special case, since Carbon is a constituent of the scintillator itself, which contains ^{12}C. C-14 gives a beta decay spectrum with end-point of 156 keV, which can enter the observational window because of finite energy resolution effects. This contamination limits the possibility of BOREXINO to explore the very low energy region of the spectrum. The CTF measured value [17] of $1.94 \pm 0.09 \times 10^{-18}$ of $^{14}C/^{12}C$ is fully compatible with the BOREXINO requirements for a low energy threshold

Table 4. ^{7}Be neutrino line in BOREXINO. Summary of predicted suppression factors (with respect to SSM) and time variability information.

	SMA	LMA	LOW	Vacuum
Rate reduction	$0.23^{+0.24}_{-0.01}$	$0.59^{+0.15}_{-0.18}$	$0.59^{+0.06}_{-0.08}$	$0.45^{+0.33}_{-0.11}$
Seasonal tag	No	No	No	up to 30%
Day/Night	No	No	Yes	No

of 250 keV.

To further confirm this view, the untagged count rate of the device (featureless counts in the 250/800 keV energy window) has been demonstrated to be consistent with zero (-9 ± 47 count/day, only statistical error) once quality cuts are applied, muon events are accounted for and background of external origin is subtracted [16]. The error on this measurement comes mainly from the presence of a sizeable background of external origin; events originating outside the scintillator can release energy in the scintillator and therefore simulate an internal background. This effect is important in CTF since the scintillator vessel has only one meter of radius.

The results of the Counting Test Facility strongly indicate that an extreme high purity can be reached and that BOREXINO can be used as a neutrino telescope with sensitivity down to 250 keV. The Counting Test Facility itself has set noteworthy records in the world of radioactive background control: even without any event quality cuts, the specific rate was ~ 0.05 counts $kg^{-1}keV^{-1}yr^{-1}$ in the 0.25/2.5 MeV energy range.

5. The BOREXINO physics program

BOREXINO will be the only solar neutrino experiment in the next decade that will measure the solar neutrino flux below 5 MeV for a specific solar neutrino producing reaction. ^{7}Be neutrinos can be studied both using the rate information and the time variability of the detected signal.

The pure ^{7}Be rate is the first information to be considered. The different scenarios have in fact different suppression factors for the intensity of the recoil electron signal, as summarized in the first line of table 4 (the numbers are from [11]).

The discriminating power of the pure ^{7}Be rate information can be appreciated when the rate-only information is combined with the results from the other new solar neutrino experiments SuperKamiokande and SNO. The analysis performed in [18] assumes that the solar neutrino signals are measured with a 5% 1σ error by all experiments. Under this assumption it is shown that the combination of the rate measurements by SuperKamiokande, SNO and BOREXINO can - for instance - fully eliminate the vacuum and (almost all the) LMA scenario if the SMA is the correct solution.

Moreover, in the supposed case in which vacuum oscillations is the correct scenario, the rate-only information can rule out most of the MSW possibilities (in particular the SMA).

In addition to the rate information, the time variability of the signal has to be considered. Seasonal variation is a tool of primary importance in this respect, since the detection of such a signal (besides the $1/R^2$ geometrical factor due to the Earth's orbit eccentricity) is a clear indication of the vacuum oscillation scenario. The predicted size of this effect is at the level of about ±20% depending on the specific value of the oscillation parameters. The analysis, conducted in [19], shows that BOREXINO should either detect seasonal variations or essentially rule out the vacuum oscillation mechanism.

The second kind of time variability in BOREXINO is the day/night effect, specifically associated with the LOW solution. In the case of the ^{7}Be signal, this effect is at the 50-100% level due to the Earth regeneration effect [20] and will be easily detectable in the experiment [3] .

In summary, BOREXINO has different tools to understand the ^{7}Be signal and to discriminate the different oscillation scenario regions.

While the experimental program is mainly focused on the detection of ^{7}Be neutrinos, BOREXINO is a large scale low background neutrino detector which can be used for a variety of additional goals, which are under consideration by the Collaboration. We will now briefly comment on these additional studies.

5.1. More solar neutrinos: ^8B spectrum and CNO studies.

^{8}B neutrinos have a spectrum extending up to 15 MeV; they will produce a ~1 event/day signal in BOREXINO above 4 MeV, in a region where the background is expected to be of the same magnitude. This signal can be useful as a cross check of the detector performance and also as a potential window for new physics, since the distorsion of the ^{8}B spectrum predicted by different scenarios are much more pronounced at the ~ 4 MeV energy than above 6.5 MeV [13]. An excess production of CNO cycle neutrinos can also be directly ruled out in BOREXINO.

[3] Similar conclusions hold for the zenith-angle dependence.

5.2. Neutrino magnetic moment

The presence of an anomalous (electromagnetic) magnetic moment can be seen in ν-e scattering by studying the event rate and shape at low energies [13]. The electromagnetic part of the differential cross section is of the form:

$$\mu_\nu \frac{\pi \alpha^2}{mc^2}(1/T - 1/q) \tag{4}$$

where μ_ν is the (Dirac) magnetic moment, q is the neutrino energy and T is the kinetic energy of the recoiling electron. The presence of this term will therefore raise the event rate at low T making possible the identification of a non-zero magnetic moment. This study can be performed either by using a ^{51}Cr radioactive source [21] [22] or a Sr-Y 90 source [22] positioned close to the BOREXINO detector. An indicative sensitivity of 10^{-11} μ_B is predicted for a typical 100 days exposure of a 60×10^{15} Bq source placed at 9 m from the detector centre.

5.3. Antineutrino studies: Sun and Earth

Antineutrino searches can be performed through the $\overline{\nu}p \rightarrow ne^+$ reaction, with a 1.8 MeV energy threshold. This reaction features a powerful delayed coincidence signature (about 200 μs) between the prompt positron capture and the 2.2 MeV γ signal from neutron absorption on hydrogen. This can be used to search for $\nu_e \rightarrow \overline{\nu}_e$ transitions in the case of solar neutrinos, for which the predicted sensitivity in terms of transition proability is about 10^{-3} (see [23]).

This method is also interesting for studying the Earthly emitted antineutrinos to the goal of a better understanding of the Earth composition [23]. This study can be done by exploting the fact that Earth radioactivity component will give signals at lower energy (and with a different shape) than the main antineutrino background (dominated by European reactors).

5.4. Double beta decay

Double beta-decay searches can be made by using candidate nuclei that can be loaded in the BOREXINO scintillator [24]. This is the case, for instance, of ^{136}Xe, an important double-beta decay candidate that has high solubility in organic solvents. The experimental technique consists in looking for the double beta peak (summed energy of the two electrons) at 2480 keV [13] [24]. The predicted sensitivity depends on several factors, included the actual solubility of Xenon in the pseudocumene, the possible enrichment and the final background in the relevant energy window. The indicative sensitivity value is in the $T_{1/2} \simeq 10^{24}$ yr range.

6. Conclusion

The solar neutrino experiment BOREXINO will be the first real-time detector to explore the sub-MeV energy region. This fact is made possible by the ultralow background of the target material, by the detector shielding and by the cosmic background rejection capability.

The main goal of the experiment is the detection of ^{7}Be neutrinos from the Sun. In addition, a variety of other physics goals can be investigated with this detector, including other aspects of solar neutrino physics, neutrino magnetic moment, antineutrino searches, double beta decay.

References

[1] The BOREXINO Collaboration: G. Alimonti, C. Arpesella, M. Balata, A. de Bari, G. Bellini, J. Benziger, S. Bonetti, B. Caccianiga, L. Cadonati, F. P. Calaprice, G. Cecchet, M. Chen, M. Deutsch, R. Dossi, F. Elisei, F. von Feilitzsch, R. Fernholz, C. Galbiati, F. Gatti, M.G. Giammarchi, D. Giugni, A. Golubchikov, A. Goretti, C. Hagner, T. Hagner, F.X. Hartmann, R. von Hentig, G. Heusser, A. Ianni, M. Johnson, T. Kirsten, G. Korga, G. Korschinek, M. Laubenstein, P. Lombardi, S. Magni, S. Malvezzi, J. Maneira, I. Manno, G. Manuzio, F. Masetti, U. Mazzucato, E. Meroni, P. Musico, M. Neff, A. Nostro, L. Oberauer, L. Perasso, A. Perotti, G. Pieri, R.S. Raghavan, G. Ranucci, W. Rau, E. Resconi, C. Salvo, R. Scardaoni, S. Schönert, K.H. Schuhbeck, O. Smirnov, A. Sotnikov, R. Tartaglia, G. Testera, R.B. Vogelaar, S. Vitale, M. Wojcik, O. Zaimidoroga, Y. Zakharov.

[2] Y. Fukuda et al., Evidence for oscillation of atmospheric neutrinos, July 1988, hep-ex/9807003, submitted to *Phys. Rev. Lett.*

[3] B. Pontecorvo, *Chalk River Report* PD-205 (1946). See F. Boehm and P. Vogel, *Physics of massive neutrinos*, Cambridge University Press (1987) for a complete discussion on the subject.

[4] S.P. Mikheyev and A. Smirnov, *Sov. J. Nucl. Phys.* 42 (1985) 913. L. Wolfenstein *Phys. Rev. D* 17 (1978) 2369.

[5] J.N. Bahcall and M.H. Pinsonneault, *Rev. Mod. Phys.* 67 (1995) 781. S. Turck-Chièze et al., *Phys. Rep.* 230 (1993) 57.

[6] J.N. Bahcall et al., *Phys. Rev. Lett.* 78 (1997) 171. B. Ricci et al., *Phys. Lett. B* 407 (1997) 155. J. Christensen-Dalsgaard et al., *Science* 272 (1996) 1286.

[7] J.N. Bahcall, S. Basu and M.H. Pinsonneault, to be published in *Phys. Lett. B*, July 1998, astro-ph/9805135.

[8] See for instance N. Hata and P. Langacker, *Phys. Rev. D* 56 (1997) 6107.

[9] V. Castellani et al., *Physics Reports*, 281 (1997) 310.

[10] M.G. Giammarchi, *Nuclear Physics B. (Proc. Suppl.)*, 65 (1998) 165.

[11] J.N. Bahcall, P.I. Krastev and A.Yu. Smirnov, August 1998, hep-ph/9807216v2.

[12] G.L. Fogli, E. Lisi and D. Montanino, *Phys. Rev. D*, 54 (1996) 2048.

[13] C. Arpesella et al., *BOREXINO: proposal for a real time detector for low energy solar neutrinos*. Ed. by G. Bellini, M. Campanella, D. Giugni, R.S. Raghavan. Milano 1991. G. Ranucci et al., *Nuclear Instruments and Methods A*, 315 (1992) 229. S. Bonetti et al., *Nuclear Phys. B (Proc. Suppl.)*, 28A (1992) 486. G. Ranucci et al., *Nuclear Phys. B (Proc. Suppl.)*, 32 (1993) 149.

[14] M. Deutsch et al., *Proposal for a cosmic ray detection system*. MIT 1996.

[15] G. Alimonti et al., *Nuclear Instr. and Methods A*, 406 (1998) 411.

[16] G. Alimonti et al., *Astroparticle Physics*, 8 (1998) 141.

[17] G. Alimonti et al., *Phys. Lett. B*, 422 (1998) 349.

[18] J.N. Bahcall and P.I. Krastev, *Phys. Rev. D*, 53 (1996) 4211.

[19] E. Calabresu et al., *Astroparticle Physics*, 4 (1995) 159.

[20] A.J. Baltz and J. Weneser, *Phys. Rev. D*, 50 (1994) 5971.

[21] N. Ferrari, G. Fiorentini, B. Ricci, *Phys.Lett. B*, 387 (1996) 427.

[22] A. Ianni and D. Montanino, Borex Internal Memo, 1998. INFN/AE-98/12 1 July 1998.

[23] R.S. Raghavan et al., *Phys. Rev. Lett.*, 80 (1998) 635. R. S. Raghavan and S. Schönert, *Antineutrinos in BOREXINO from the Earth and the Sun*, BOREX Internal Memo, 1997.

[24] R.S. Raghavan, *Phys. Rev. Lett.*, 72 (1994) 1411. B. Caccianiga and M.G. Giammarchi, *Double Beta Decay with Xe-136 in BOREXINO and CTF2*, BOREX Internal Memo, 1998.

Searches for neutrino oscillations: LSND

G T Garvey,[1] representing the LSND Collaboration

Physics Division, Los Alamos National Laboratory, Los Alamos, NM 87545

Abstract. The LSND experiment at Los Alamos has conducted searches for $\bar{\nu}_\mu \to \bar{\nu}_e$ oscillations using $\bar{\nu}_\mu$ from μ^+ decay at rest and for $\nu_\mu \to \nu_e$ oscillations using ν_μ from π^+ decay in flight. For the $\bar{\nu}_\mu \to \bar{\nu}_e$ search, a total excess of $82.8 \pm 23.7 \pm 12.0$ events is observed with e^+ energy between 20 and 60 MeV, while for the $\nu_\mu \to \nu_e$ search, a total excess of $18.1 \pm 6.6 \pm 4.0$ events is observed with e^- energy between 60 and 200 MeV. If attributed to neutrino oscillations, these excesses correspond to oscillation probabilities (averaged over the experimental energies and spatial acceptances) of $(0.31 \pm 0.09 \pm 0.05)\%$ and $(0.26 \pm 0.10 \pm 0.05)\%$, respectively. For the future, the BooNE experiment at Fermilab could conclusively prove that neutrino oscillations occur and make precision measurements of Δm^2 and $\sin^2 2\theta$.

1. Introduction

One of the few ways to probe small neutrino masses is to search for neutrino oscillations, where a neutrino of one type (e.g., $\bar{\nu}_\mu$) spontaneously transforms into a neutrino of another type (e.g., $\bar{\nu}_e$). For this phenomenon to occur, neutrinos must be massive and the apparent conservation law of lepton family number must be violated. In 1995 the LSND experiment [1] published data showing an excess of events consistent with $\bar{\nu}_\mu \to \bar{\nu}_e$ oscillations [2]. Additional results have been published in 1996 and 1998 for both $\bar{\nu}_\mu \to \bar{\nu}_e$ oscillations [3] and $\nu_\mu \to \nu_e$ oscillations [4]. The two oscillation searches have completely different backgrounds and systematics, and together provide strong evidence for neutrino oscillations.

[1] E-mail: garvey@lanl.gov.

2. Detector

The Liquid Scintillator Neutrino Detector (LSND) experiment at Los Alamos [5] was designed to search with high sensitivity for $\bar{\nu}_\mu \to \bar{\nu}_e$ oscillations from μ^+ decay at rest. The LANSCE accelerator is a most intense source of low energy neutrinos due to its 1 mA proton intensity and 800 MeV energy. For the 1993–1995 running period the beam interacted with a 30-cm long water target (20-cm in 1993) one meter in front of a water-cooled Cu beam dump, while for the 1996–1997 running period the water target was replaced by a close-packed high-z target for testing tritium production. The muon decay-at-rest neutrino flux with this new configuration is 2/3 of that from the old beam stop; however, the pion decay-in-flight neutrino flux is reduced to 1/2 the original flux, so that the 1996–1997 data serve as a systematic check on the earlier results. The neutrino source is well understood because almost all neutrinos arise from π^+ or μ^+ decay; π^- and μ^- are readily captured in the Fe of the shielding and Cu of the beam stop [6]. The production of kaons and heavier mesons is negligible at these energies. The $\bar{\nu}_e$ rate is calculated to be only 4×10^{-4} relative to $\bar{\nu}_\mu$ in the $36 < E_\nu < 52.8$ MeV energy range, so that the observation of a significant $\bar{\nu}_e$ rate would be evidence for $\bar{\nu}_\mu \to \bar{\nu}_e$ oscillations.

The LSND detector consists of an approximately cylindrical tank 8.3 m long by 5.7 m in diameter. The center of the detector is 30 m from the neutrino source. On the inside surface of the tank 1220 8-inch Hamamatsu phototubes provide 25% photocathode coverage. The tank is filled with 167 metric tons of liquid scintillator consisting of mineral oil and 0.031 g/l of b-PBD. This low scintillator concentration allows the detection of both Čerenkov light and scintillation light and produces a relatively long attenuation length of more than 20 m for wavelengths greater than 400 nm [7]. A typical 45 MeV electron created in the detector produces a total of ~ 1500 photoelectrons, of which ~ 280 photoelectrons are in the Čerenkov cone. The phototube time and pulse height signals are used to reconstruct a track with an average r.m.s. position resolution of ~ 30 cm, an angular resolution of ~ 12 degrees, and an energy resolution of $\sim 7\%$. The Čerenkov cone for relativistic particles and the time distribution of the light arriving at the phototubes, which is significantly broader for scintillation light, give excellent particle identification. Surrounding the detector is a veto shield [8] which tags cosmic ray muons going through the detector.

3. Cross Sections Measured In LSND

An advantage of accelerator based neutrino experiment relative to those experiments measuring neutrinos created in the upper atmosphere or via

Table 1.

Process	LSND Measurement Cross Section (10^{-42} cm^2)	Predicted Cross Section (10^{-42} cm^2)
$\nu_e + {}^{12}\text{C} \to {}^{12}\text{N(g.s.)} + e$	$9.1 \pm 0.4 \pm 0.9$	9.3
$\nu_\mu + {}^{12}\text{C} \to {}^{12}\text{N(g.s.)} + \mu$	$66 \pm 10 \pm 10$	63.0
$\nu + e \to \nu + e$	$4.2 \pm 0.6 \pm 0.4$	3.97

nuclear processes in the sun is that the neutrino flux created at the accelerator is well characterized. As a result there are several known cross sections that can be measured to test the detector performance and many of overall systematics associated with accelerator based neutrino experiments.

For example the cross section for neutrinos for the process $\nu_e + {}^{12}\text{C} \to {}^{12}\text{N}$(ground state) $+ e$ is known to better than 5% for ν_e originating from μ^+ decay at rest because all the relevant form factors have been measured in other electroweak processes. The process has a unique signature as the ^{12}N ground state is the only particle stable level in ^{12}N. It decays via an energetic positron decay (endpoint; 16.32 MeV) with a life time of 15.87 ms so is readily identified in a detector such as LSND. The same holds true for the process $\nu_\mu + {}^{12}\text{C} \to {}^{12}\text{N}$(ground state) $+ \mu$, which we have also measured. The various processes, $\nu + e \to \nu + e$ are also known as they are predicted via the standard model and are readily identified in the detector by their extreme peaking along the neutrino direction. Table 1 lists the LSND measured cross sections and the corresponding predicted cross sections. There is good agreement between all the reliably predicted cross sections and the LSND measurements. This gives a measure of confidence that the neutrino fluxes, and the LSND detector efficiencies are established to the order of 10%.

4. $\bar{\nu}_\mu \to \bar{\nu}_e$ Oscillation Data

The signature for a $\bar{\nu}_e$ interaction in the detector is the reaction $\bar{\nu}_e p \to e^+ n$ followed by $np \to d\gamma$ (2.2 MeV). A likelihood ratio, R, is employed to determine whether the event subsequent to the detected e^+ is a 2.2 MeV photon correlated with a positron or is from an accidental coincidence. R is the likelihood that the γ is correlated, divided by the likelihood that it is accidental (uncorrelated). As shown in Fig. 1, R depends on the number of hit phototubes for the γ, the reconstructed distance between the positron and the γ, and the relative time between the γ and positron. Figure 2 shows

Table 2. Numbers of events with $R > 30$ from the entire 1993-1997 data sample.

e^+ Energy	Events Beam On	Events Beam Off	ν Background	Total Excess
$20 < E_e < 60$ MeV	61	15.6 ± 1.0	11.5 ± 1.5	33.9 ± 8.0
$36 < E_e < 60$ MeV	29	5.2 ± 0.6	3.0 ± 0.6	20.8 ± 5.4

Table 3. Numbers of excess events and the corresponding oscillation probabilities from fits to the R distributions for the running periods 1993–1995, 1996–1997, and 1993–1997.

Data Sample	Fitted Excess	ν Background	Total Excess	Oscillation Probability
1993–1995	63.5 ± 20.0	12.5 ± 2.9	51.0 ± 20.2	$(0.31 \pm 0.12 \pm 0.05)\%$
1996–1997	35.1 ± 14.7	4.8 ± 1.1	30.3 ± 14.8	$(0.32 \pm 0.15 \pm 0.05)\%$
1993–1997	100.1 ± 23.4	17.3 ± 4.0	82.8 ± 23.7	$(0.31 \pm 0.09 \pm 0.05)\%$

the R distribution, beam on minus beam off, for events with positrons in the $20 < E < 60$ MeV energy range. Shown in the figure is the best fit to the data together with the expectations for uncorrelated and correlated γs. There is a clear excess of events at high R that corresponds to events with a correlated 2.2 MeV photon from the capture of the neutron. Table 2 lists the number of events with $R > 30$, while Table 3 gives the number of excess events and corresponding oscillation probabilities from fits to the R distribution.

Figure 3 shows the electron energy distribution, beam on minus beam off excess, for events with an associated γ with $R > 30$. For this latter requirement, the total 2.2 MeV γ detection efficiency is 23% and the probability that an event has an accidental γ in coincidence is 0.6%. Shown in the figure are the beam-excess data, estimated neutrino background, expected distribution for neutrino oscillations at large Δm^2 plus estimated neutrino background, and expected distribution for neutrino oscillations at small Δm^2 plus estimated neutrino background. There are 61 events beam on in the $20 < E < 60$ MeV energy range and a total estimated background (beam off plus neutrino-induced background) of 27.1 ± 1.8 events, which leads to a total excess of 33.9 ± 8.0 events. Table 4 gives the background estimate for events in the $20 < E_e < 60$ MeV energy range with $R \geq 0$ and $R > 30$. The observed average value of $\cos\theta_b$, the angle between the

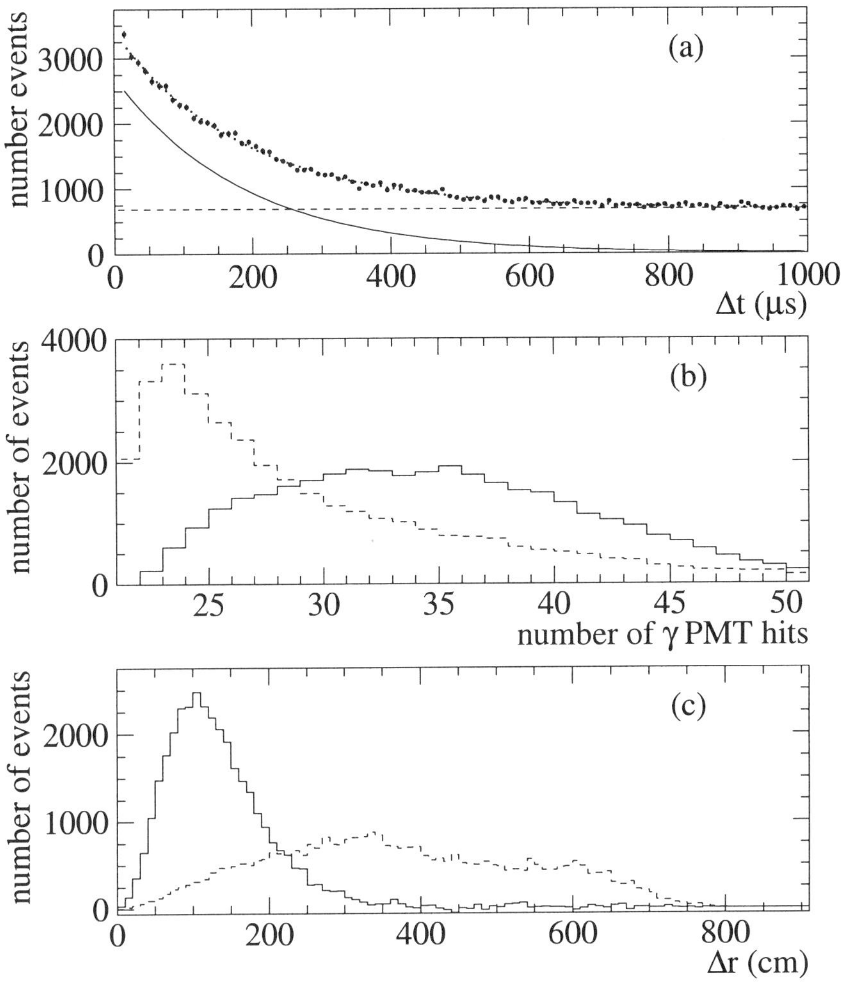

Figure 1. Associated γ distributions for (a) time, (b) PMT hits, and (c) distance distributions for associated (solid) and accidental (dashed) γs.

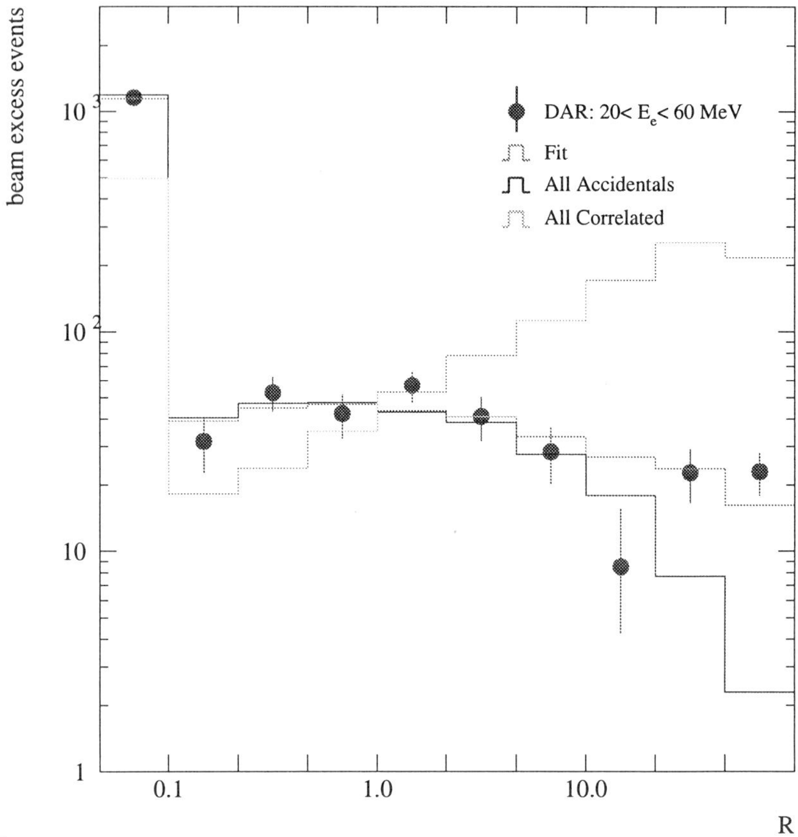

Figure 2. The R distribution, beam on minus beam off excess, for events that have energies in the range $20 < E_e < 60$ MeV. Shown in the figure are the best fit to the data together with the expectations for uncorrelated and correlated γs.

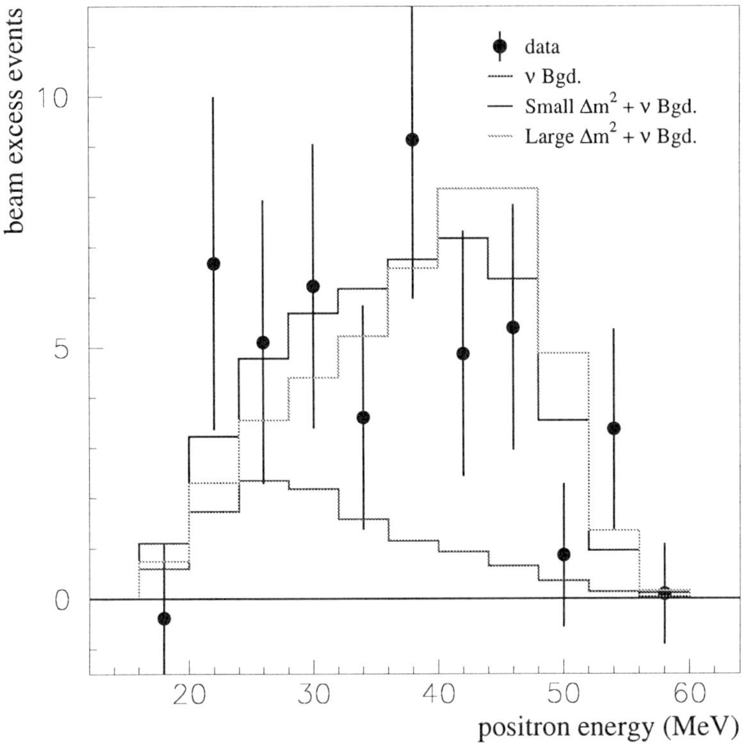

Figure 3. The energy distribution for events with $R > 30$. Shown in the figure are the beam-excess data, estimated neutrino background, expected distribution for neutrino oscillations at large Δm^2 plus estimated neutrino background, and expected distribution for neutrino oscillations at small Δm^2 plus estimated neutrino background.

neutrino direction and the reconstructed positron direction, is 0.21 ± 0.12, in agreement with the expected value of 0.16 for $\bar{\nu}_e p$ interactions. Figures 4 and 5 show the $\cos\theta_b$ and spatial distributions for the beam on-off excess events with $R > 30$ and $36 < E < 60$ MeV. Table 4 shows for various distributions the χ^2 per degree of freedom for events that have $R > 30$ and $20 < E < 60$ MeV. Within statistics, all of the distributions are consistent with what is expected from $\bar{\nu}_\mu \to \bar{\nu}_e$ oscillations followed by $\bar{\nu}_e p \to e^+ n$ scattering.

Assuming the observed excess to be due to neutrino oscillations, Fig.

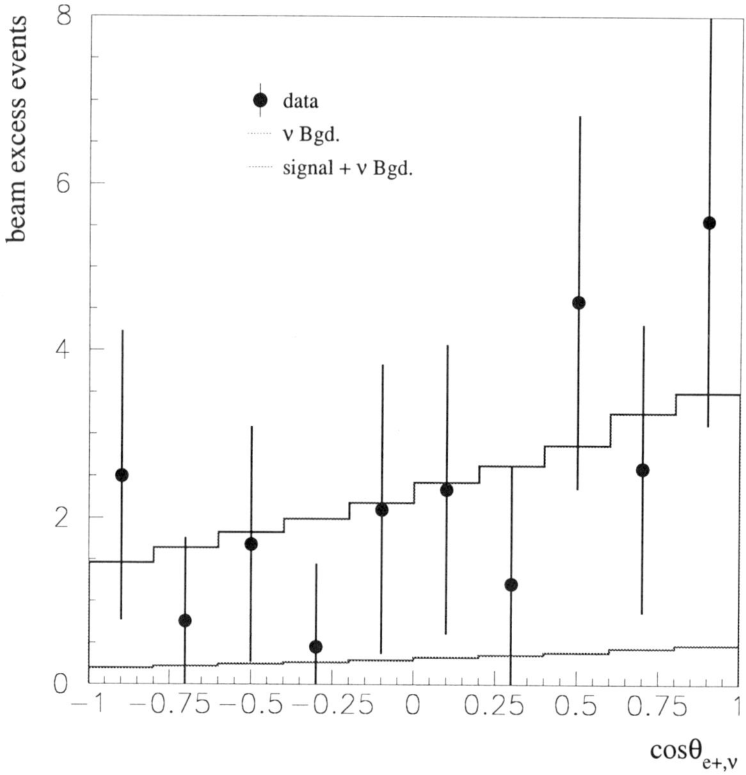

Figure 4. The $\cos\theta_b$ distribution for events with $R > 30$ and $36 < E < 60$ MeV, where $\cos\theta_b$ is the angle between the neutrino direction and the reconstructed positron direction.

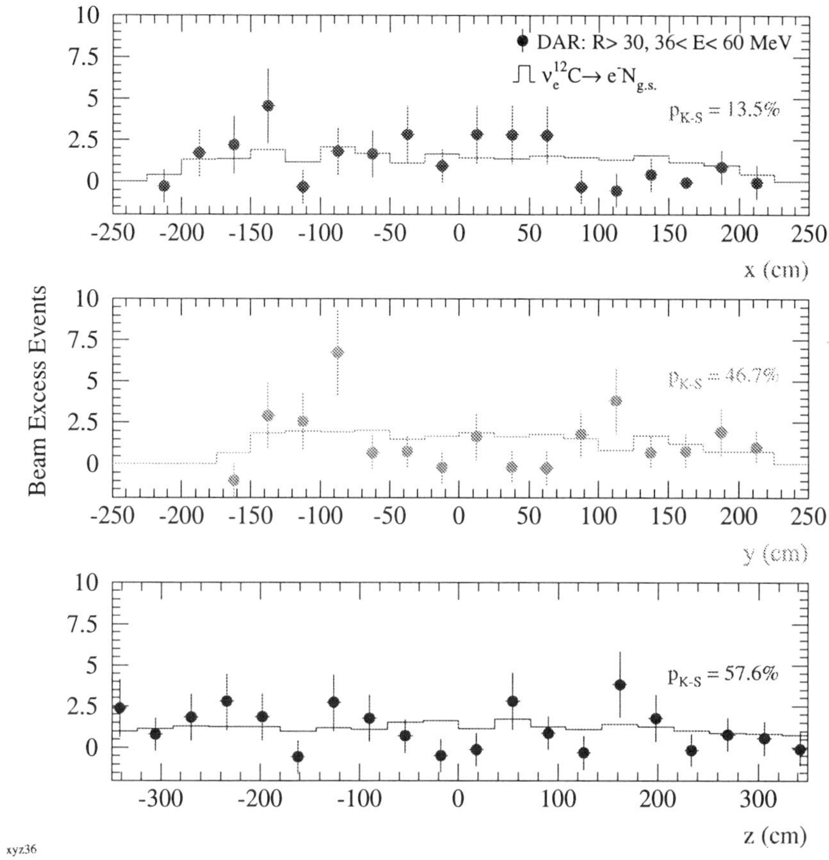

Figure 5. The spatial distributions for events with $R > 30$ and $36 < E < 60$ MeV.

Table 4. A list of all backgrounds with the expected number of background events in the $20 < E_e < 60$ MeV energy range for $R \geq 0$ and $R > 30$. The neutrinos are from either π and μ decay at rest (DAR) or decay in flight (DIF). Also shown are the number of events expected for 100% $\bar{\nu}_\mu \to \bar{\nu}_e$ transmutation.

Background	Source	$R \geq 0$	$R > 30$
Beam off		1306.1 ± 8.8	15.6 ± 1.0
Beam neutrons		< 26.1	< 0.3
DAR			
$\bar{\nu}_e p \to e^+ n$	$\mu^- \to e^- \nu_\mu \bar{\nu}_e$	12.9 ± 2.6	3.0 ± 0.6
DIF			
$\bar{\nu}_\mu p \to \mu^+ n$	$\pi^- \to \mu^- \bar{\nu}_\mu$	5.7 ± 2.9	1.4 ± 0.6
$\bar{\nu}_e p \to e^+ n$	$\pi \to e\nu$ and $\mu \to e\nu\bar{\nu}$	0.1 ± 0.1	0
Total		18.7 ± 4.4	4.4 ± 0.9
DAR			
$\nu_e\ ^{12}C \to e^-\ ^{12}N$	$\mu^+ \to e^+ \bar{\nu}_\mu \nu_e$	1000.1 ± 200.0	6.0 ± 1.2
$\nu_e\ ^{13}C \to e^-\ ^{13}N$	$\mu^+ \to e^+ \bar{\nu}_\mu \nu_e$	68.4 ± 13.7	0.5 ± 0.1
$\nu e \to \nu e$	$\mu^+ \to e^+ \bar{\nu}_\mu \nu_e$	85.1 ± 8.6	0.5 ± 0.1
$\nu_e \mathrm{C} \to e^- X$	$\pi \to e \nu_e$	39.0 ± 1.5	0
DIF			
$\nu_\mu \mathrm{C} \to \mu^- X$	$\pi^+ \to \mu^+ \nu_\mu$	17.0 ± 8.4	0.1 ± 0.1
$\nu e \to \nu e$	$\pi \to \mu \nu_\mu$	12.6 ± 2.6	0.1 ± 0.1
$\nu_\mu \mathrm{C} \to \pi X$	$\pi \to \mu \nu_\mu$	0.5 ± 0.1	0
$\nu_e \mathrm{C} \to e^- X$	$\pi \to e\nu$ and $\mu \to e\nu\bar{\nu}$	1.4 ± 0.3	0
Total		1192.5 ± 200.9	7.2 ± 1.2
Grand Total		2517.3 ± 201.1	27.2 ± 1.8
100% Transmutation	$\mu^+ \to e^+ \bar{\nu}_\mu \nu_e$	25005 ± 2500	5745 ± 690

Table 5. The χ^2 per degree of freedom for various distributions for events that have $R > 30$ and $20 < E < 60$ MeV. E is the event energy, θ_b is the angle between the neutrino direction and the outgoing positron, *dist* is the distance from the reconstructed vertex to the nearest phototube, *r.dr* is the dot product of the event radius vector and the event unit vector direction, R is the γ likelihood ratio, and *veto* is the number of veto hits. For the spatial distribution, the detector fiducial volume was divided up into 8 regions corresponding to $\pm x$, $\pm y$, and $\pm z$.

Distribution	χ^2 per Degree of Freedom
E	8.3/9
$\cos\theta_b$	5.6/9
dist	11.3/19
r.dr	9.0/10
R	12.4/8
veto	1.5/3
Spatial	1.7/7
Total	49.8/65

6 shows the allowed region (90% and 99% likelihood regions) of $\sin^2 2\theta$ vs. Δm^2 from a maximum likelihood fit to the L/E distribution of the entire data sample. Some of the allowed region is excluded by the KARMEN1 experiment at ISIS [9], the E776 experiment at BNL [10], the Bugey reactor experiment [11], the CCFR experiment at FNAL [12], and the NOMAD experiment at CERN [13]. The KARMEN2 experiment is now taking data with an improved veto shield, and its sensitivity is shown in Fig. 7, KARMEN2.

5. $\nu_\mu \to \nu_e$ Oscillation Data

The signature for $\nu_\mu \to \nu_e$ oscillations is an electron from the reaction $\nu_e C \to e^- X$ in the energy range $60 < E_e < 200$ MeV. Using two independent analyses [4], a total of 40 beam-related events and 175 beam-unrelated events are observed, corresponding to a beam on-off excess of 27.7 ± 6.4 events. The neutrino-induced backgrounds are dominated by $\mu^+ \to e^+ \bar{\nu}_\mu \nu_e$ and $\pi^+ \to e^+ \nu_e$ decays-in-flight in the beam-stop and are estimated to be 9.6 ± 1.9 events. Therefore, a total excess of $18.1 \pm 6.6 \pm 3.5$ events is observed above background. Fig. 8 shows the energy distribution of the beam on-off excess, along with the expectation for the backgrounds, for an oscillation signal at large Δm^2 and for the sum of the two. Fig. 9 shows the angular distribution for the beam on-off excess and the expectation from $\nu_e C \to e^- N$ scattering.

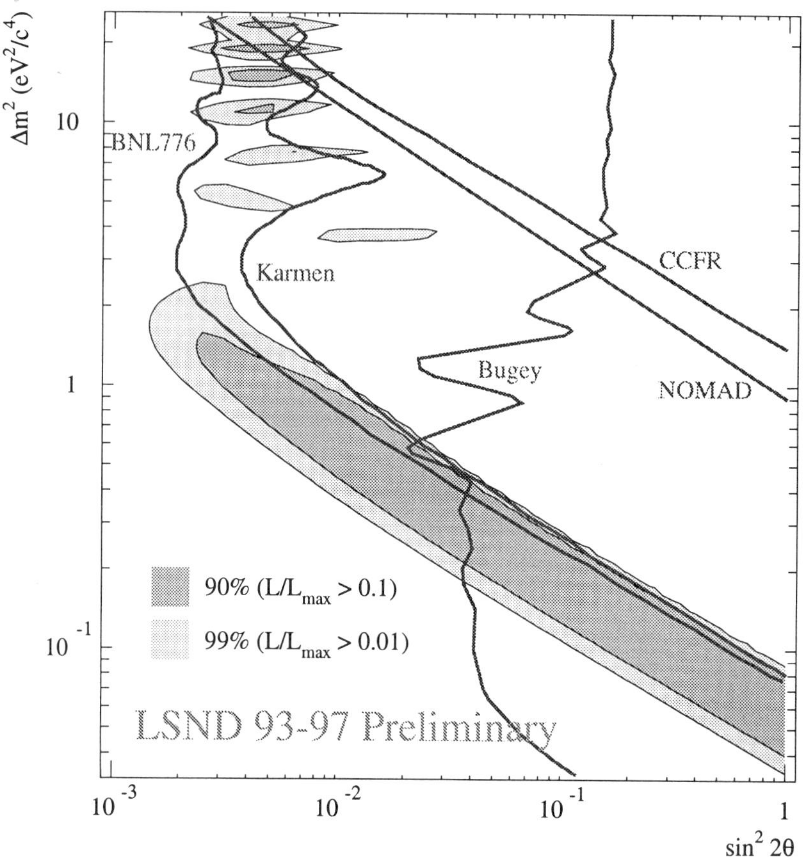

Figure 6. Plot of the LSND Δm^2 *vs* $\sin^2 2\theta$ favored regions. They correspond to 90% and 99% likelihood regions after the inclusion of the effects of systematic errors. Also shown are 90% C.L. limits from KARMEN1 at ISIS, E776 at BNL, the Bugey reactor experiment, the CCFR experiment at FNAL, and the NOMAD experiment at CERN.

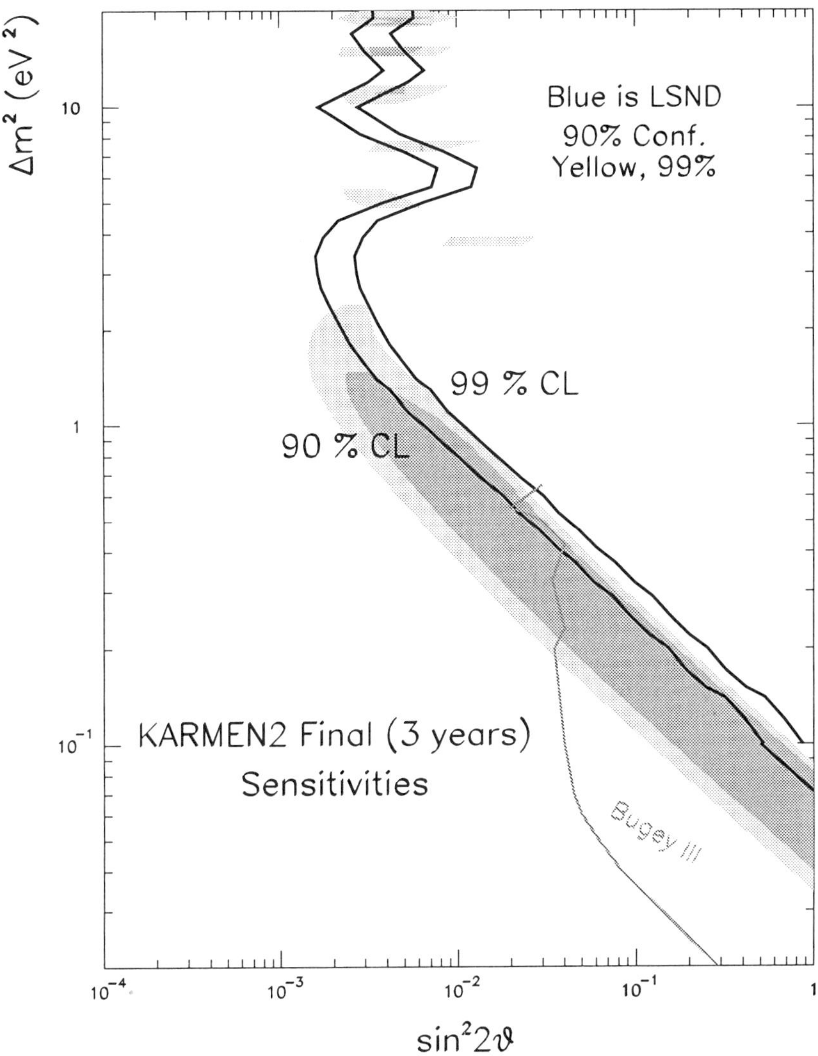

Figure 7. Plot of the LSND Δm^2 *vs* $\sin^2 2\theta$ favored regions, corresponding to 90% and 99% likelihood regions after the inclusion of the effects of systematic errors. Also shown are the expected 90% and 99% C.L. sensitivities from KARMEN2 at ISIS.

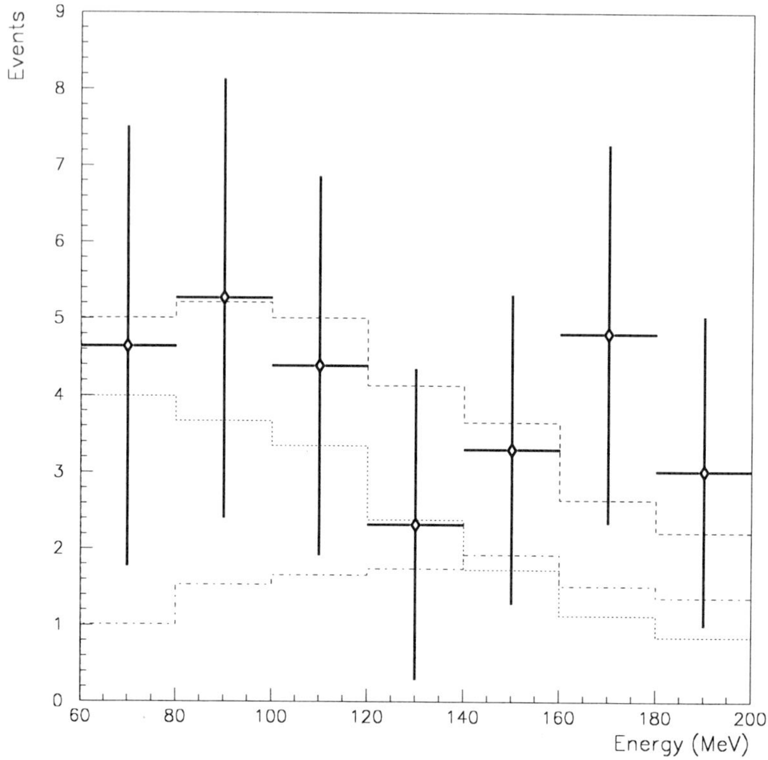

Figure 8. The energy distribution for the final beam-excess decay-in-flight events. The expectation for backgrounds (dot-dashed histogram), the oscillation signal for large values of Δm^2 (dotted histogram), and the sum of the two (dashed histogram) are shown.

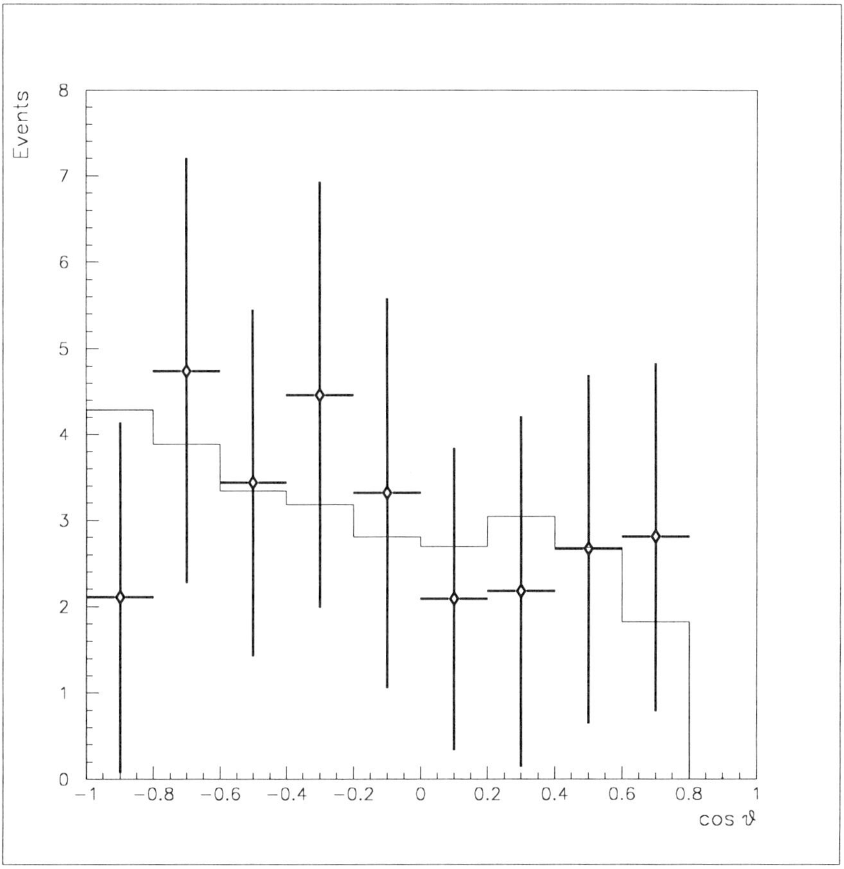

Figure 9. The angular distribution for the final beam-excess decay-in-flight events. The expectation for neutrino background and an oscillation signal at large values of Δm^2 (solid histogram) is shown.

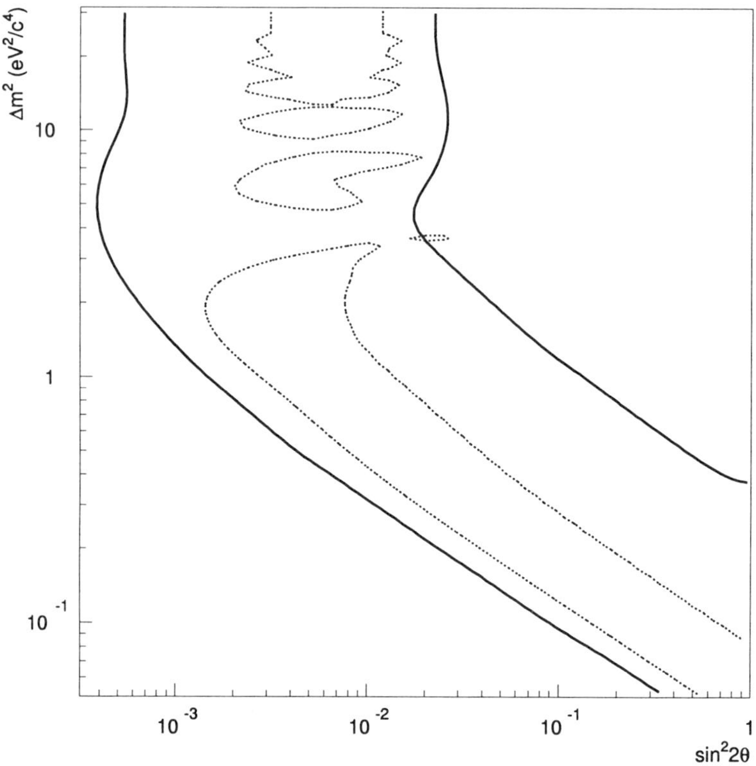

Figure 10. The 95% confidence region for $\nu_\mu \rightarrow \nu_e$ oscillations (solid curve) along with the favored regions for $\bar{\nu}_\mu \rightarrow \bar{\nu}_e$ oscillations (dotted curve).

The excess events are consistent with $\nu_\mu \rightarrow \nu_e$ oscillations with an oscillation probability of $(0.26 \pm 0.10 \pm 0.05)\%$. A fit to the event distributions yields the allowed region in the $(\sin^2 2\theta, \Delta m^2)$ parameter space shown in Fig. 10, which is consistent with the allowed region from the $\bar{\nu}_\mu \rightarrow \bar{\nu}_e$ search. These two searches have completely different backgrounds and systematic errors, and together they provide strong evidence that the observed event excesses are indeed due to neutrino oscillations.

6. The BooNE Experiment at Fermilab

Motivated by the LSND results, a proposal [14] has been submitted to and approved by Fermilab to construct a Booster Neutrino Experiment (BooNE) to search for $\nu_\mu \to \nu_e$ oscillations and ν_μ disappearance over the region of the LSND excess. Figs. 11 and 12 show the expected sensitivities for $\nu_\mu \to \nu_e$ appearance and ν_μ disappearance after one calendar year of operation. Assuming that the LSND signal is verified, BooNE will be able to measure with precision the Δm^2 and $\sin^2 2\theta$ oscillation parameters and to search for CP violation in the lepton sector by comparing the $\nu_\mu \to \nu_e$ and $\bar{\nu}_\mu \to \bar{\nu}_e$ oscillation rates. Fig. 13 shows the number of events expected from BooNE in 1 calendar year of running for the low Δm^2 favored region of LSND (shaded), while Fig. 14 shows how well BooNE could measure the oscillation parameters if, for example, $\Delta m^2 = 2.0$ eV^2.

The BooNE experiment (phase 1) will begin taking data towards the end of 2001. By using phototubes and electronics from the LSND experiment, the BooNE Detector is relatively inexpensive, $\sim \$3M$, and can be constructed on a short time scale. The detector, as shown in Fig. 15, will consist of a 12-meter-diameter spherical tank covered on the inside by the 1220 8-inch phototubes from LSND (10% coverage) and filled with 750 t of mineral oil, resulting in a 450 t fiducial volume. The volume outside of the phototubes will serve as a veto shield for identifying particles both entering and leaving the detector. The detector will be located 500 m from the neutrino source. The second phase of the experiment can add an additional detector, that is identical to the first, at a distance of $\sim$ 1000 m from the neutrino source. The ratio of $\nu_e C \to e^- N$ events in the two detectors as a function of energy will allow the unambiguous determination of the oscillation parameters with much reduced systematic uncertainties.

The neutrino beam will be produced by the 8 GeV proton Booster at FNAL. The neutrino beam line is to consist of a production target followed by a focusing system and a $\sim$50 m long pion decay volume. The low energy, high intensity, and 1 μs time-structure of the Booster neutrino beam will be ideal for this experiment. The sensitivities shown in Figs. 11 and 12 assume the experiment receives 5 Hz of proton bunches over one calendar year (2×10^7 s).

This Booster experiment is compatible with the operation of the Fermilab collider or the fixed-target MI programs. The FNAL Booster is capable of running at 15 Hz (5×10^{12} protons per pulse), or 30 Booster batches per 2 s Main Injector Cycle. The antiproton stacking requires only 6 Booster batches at the start of the Main Injector cycle. In principle, this means the BooNE beam line could receive 12 Hz from the Booster, well above the expectation on which the sensitivities are based.

The BooNE experiment represents an opportunity to resolve two out-

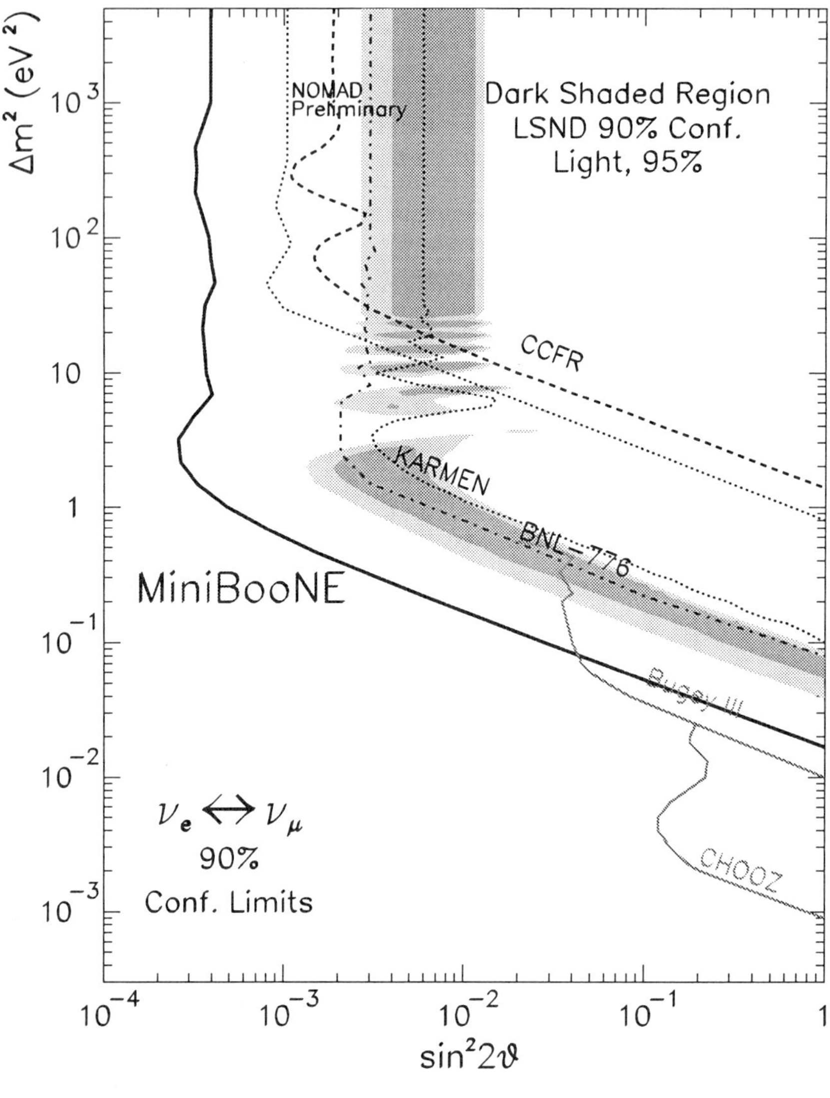

Figure 11. The 90% C.L. limit expected from BooNE for $\nu_\mu \to \nu_e$ appearance after one calendar year of running, including systematic errors, if LSND signal is not observed (solid line). Summary of results from past experiments and expectations for the future MINOS experiment are also shown.

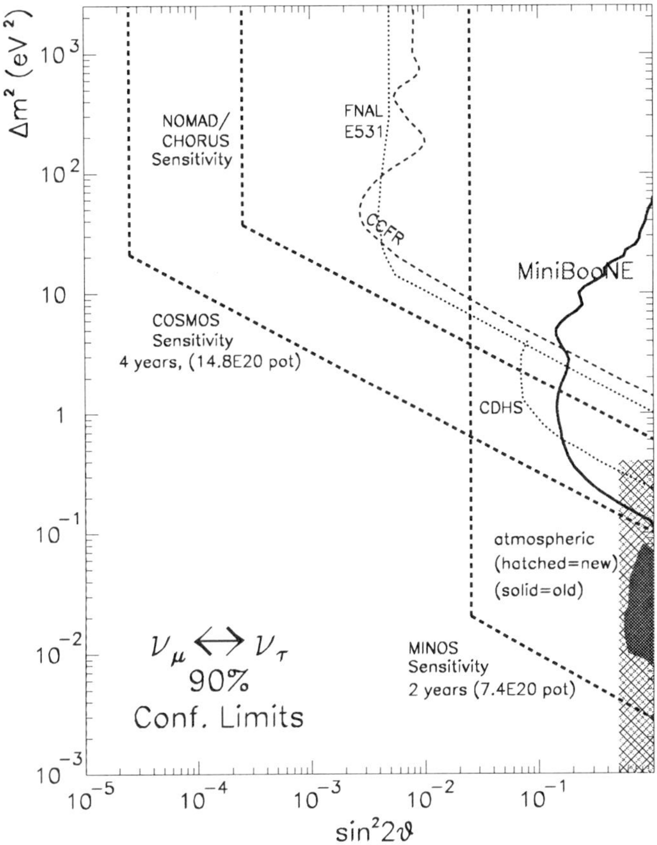

Figure 12. Summary of results from past experiments (narrow, dashed and dotted), future approved experiments (wide, dashed) and 90% C.L. limit expected for BooNE (solid) for ν_μ disappearance after one calendar year of running at 0.5 km. Solid region indicates the favored region for the atmospheric neutrino deficit from the Kamioka experiment. A result from Kamiokanda indicating no zenith angle dependence extends the favored region to higher Δm^2 as indicated by the hatched region.

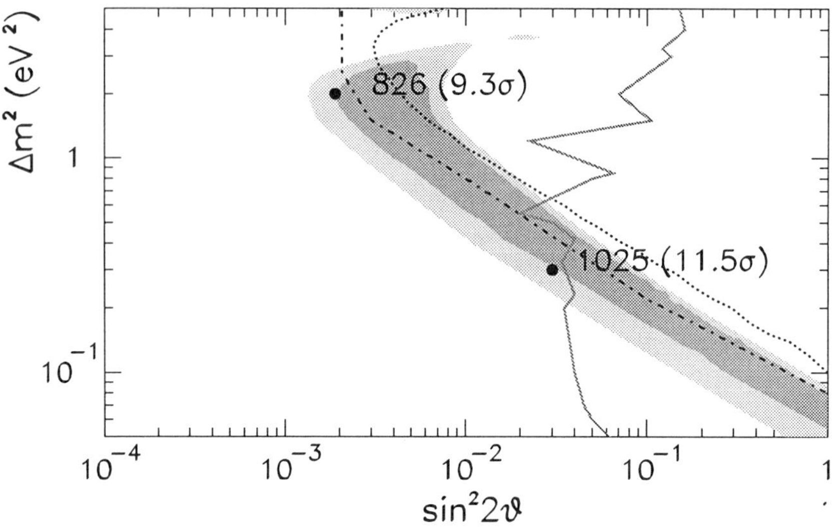

Figure 13. If the LSND signal is observed, this plot shows the number of events expected in 1 calendar year of running with BooNE for the low Δm^2 favored region of LSND (shaded). Lines indicate regions excluded by past experiments.

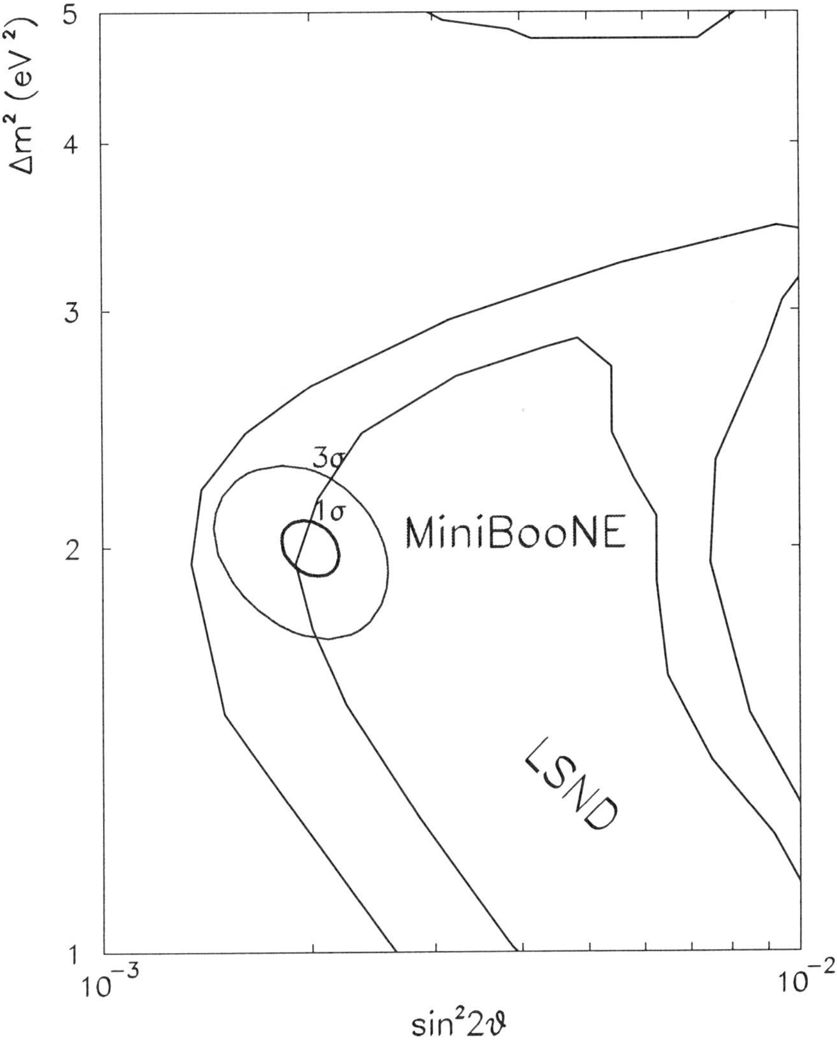

Figure 14. If the LSND signal is observed, this plot shows how well BooNE could measure the oscillation parameters if, for example, $\Delta m^2 = 2.0$ eV^2.

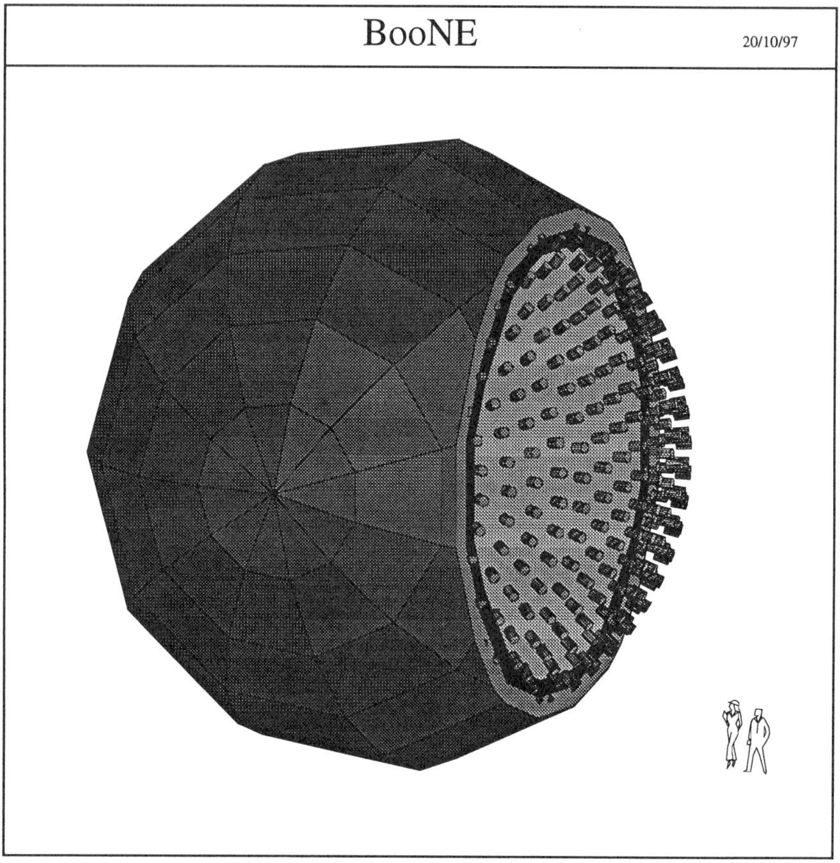

Figure 15. Schematic of the proposed BooNE detector.

standing neutrino oscillation questions on a short-time scale. Within the upcoming five years, no existing or approved experiments will be able to address conclusively the LSND signal region. Thus BooNE represents an important and unique addition to the worldwide search for neutrino oscillations.

7. Conclusion

In summary, the LSND experiment observes excesses of events for both the $\bar{\nu}_\mu \to \bar{\nu}_e$ and $\nu_\mu \to \nu_e$ oscillation searches, corresponding to oscillation probabilities of $(0.31 \pm 0.09 \pm 0.05)\%$ and $(0.26 \pm 0.10 \pm 0.05)\%$, respectively. These two searches have completely different backgrounds and systematics and together provide strong evidence for neutrino oscillations in the range $0.2 < \Delta m^2 < 2.0$ eV^2. This implies that at least one neutrino has a mass greater than 0.4 eV. For the future, the BooNE experiment at Fermilab is capable of making precision measurements of the Δm^2 and $\sin^2 2\theta$ oscillation parameters. If neutrino oscillations have in fact been observed, then the minimal standard model will need to be modified and neutrinos may have sufficient mass to influence cosmology and the evolution of the universe.

References

[1] The LSND Collaboration presently consists of the following people and institutions: Church E D, Stancu I, Strossman W, VanDalen G J (Univ. of California, Riverside); Vernon W (Univ. of California, San Diego); Caldwell D O, Yellin S (Univ. of California, Santa Barbara); Smith D (Embry-Riddle Aeronautical Univ.); Burman R L, Donahue J B, Garvey G T, Louis W C (spokesman), Mills G B, Sandberg V, Tayloe R, White D H (Los Alamos National Laboratory); Imlay R, Kim H J, Malik A, Metcalf W, Sung M K, Wadia N (Louisiana State Univ.): Johnston K (Louisiana Tech Univ.); Fazely A, Gunasingha R M (Southern Univ.); Auerbach L B, Majkic R (Temple Univ.).

[2] Athanassopoulos C *et al.* 1995 *Phys. Rev. Lett.* **75** 2650

[3] Athanassopoulos C *et al.* 1996 *Phys. Rev. C* **54** 2685; Athanassopoulos C *et al.* 1996 *Phys. Rev. Lett.* **77** 3082

[4] Athanassopoulos C *et al.* 1998 *Phys. Rev. Lett.* **81** 1774; Athanassopoulos C *et al.* 1998 submitted to *Phys. Rev. C* Los Alamos National Laboratory document LA-UR-97-1998

[5] Athanassopoulos C *et al.* 1997 *Nucl. Instrum. Methods A* **388** 149

[6] Burman R L, Potter M E and Smith E S 1990 *Nucl. Instrum. Methods A* **291** 621; Burman R L, Dodd A C and Plischke P 1996 *Nucl. Instrum. Methods in Phys. Res. A* **368** 416

[7] Reeder R A *et al.* 1993 *Nucl. Instrum. Methods A* **334** 353

[8] Napolitano J J *et al.* 1989 *Nucl. Instrum. Methods A* **274** 152

[9] Bodmann B *et al.* 1991 *Phys. Lett. B* **267** 321; Bodmann B *et al.* 1992 *Phys. Lett. B* **280** 198; Zeitnitz B *et al.* 1994 *Prog. Part. Nucl. Phys.* **32** 351

[10] Borodovsky L *et al.* 1992 *Phys. Rev. Lett.* **68** 274

[11] Achkar B *et al.* 1995 *Nucl. Phys.* **B434** 503

[12] Romosan A *et al.* 1997 *Phys. Rev. Lett.* **78** 2912

[13] Autiero D, talk presented at the *1998 International Conference on High Energy Physics*, Vancouver, Canada

[14] Church E *et al.*, "A proposal for an experiment to measure $\nu_\mu \to \nu_e$ oscillations and ν_μ disappearance at the Fermilab Booster: BooNE," Los Alamos National Laboratory document LA-UR-98-352, Fermilab experiment 898

Double Beta and Dark Matter Search - Window to New Physics beyond the Standard Model of Particle Physics

H.V. Klapdor–Kleingrothaus

Max–Planck–Institut für Kernphysik
P.O.Box 10 39 80, D–69029 Heidelberg, Germany

Abstract. Nuclear double beta decay provides an extraordinarily broad potential to search for beyond Standard Model physics, probing already now the TeV scale, on which new physics should manifest itself. These possibilities are reviewed here.
First, the results of present generation experiments are presented. The most sensitive one of them – the Heidelberg–Moscow experiment in the Gran Sasso – probes the electron mass now in the sub eV region and will reach a limit of ~ 0.1 eV in a few years. Basing to a large extent on the theoretical work of the Heidelberg Double Beta Group in the last two years, results are obtained also for SUSY models (R–parity breaking, sneutrino mass), leptoquarks (leptoquark-Higgs coupling), compositeness, right–handed W boson mass, test of special relativity and equivalence principle in the neutrino sector and others. These results are comfortably competitive to corresponding results from high–energy accelerators like TEVATRON, HERA, etc. One of the enriched ^{76}Ge detectors also yields the most stringent limits for cold dark matter (WIMPs) to date by using raw data.
Second, future perspectives of $\beta\beta$ research are discussed. A new Heidelberg experimental proposal (GENIUS) is described which would allow to increase the sensitivity for Majorana neutrino masses from the present level of at best 0.1 eV down to 0.01 or even 0.001 eV. Its physical potential would be a breakthrough into the multi-TeV range for many beyond standard models. Its sensitivity for neutrino oscillation parameters would be larger than of all present terrestrial neutrino oscillation experiments

and of those planned for the future. It could probe directly the atmospheric neutrino problem and the large angle, and for almost degenerate neutrino mass scenarios even the small angle solution of the solar neutrino problem. It would further, already in a first step using only 100 kg of natural Ge detectors, cover almost the full MSSM parameter space for prediction of neutralinos as cold dark matter, making the experiment competitive to LHC in the search for supersymmetry.

1. Introduction – Motivation for the search for double beta decay – and a future perspective: GENIUS

Double beta decay yields – besides proton decay – the most promising possibilities to probe beyond standard model physics beyond accelerator energy scales. Propagator physics has to replace direct observations. That this method is very effective, is obvious from important earlier research work and has been stressed, e.g. by [Rub96], etc.. Examples are the properties of W and Z bosons derived from neutral weak currents and β–decay, and the top mass deduced from LEP electroweak radiative corrections.

The potential of double beta decay includes information on the neutrino and sneutrino mass, SUSY models, compositeness, leptoquarks, right–handed W bosons, Lorentz invariance and the equivalence principle in the neutrino sector, and others (see Table 1). The recent results of the Heidelberg–Moscow experiment, which will be reported here (see also [Kla98a]), have demonstrated that $0\nu\beta\beta$ decay probes already now the TeV scale on which new physics should manifest itself according to present theoretical expectations. To give just one example, inverse double beta decay $e^-e^- \rightarrow W^-W^-$ requires an energy of at least 4 TeV for observability, according to present constraints from double beta decay [Bel96, Bel98]. Similar energies are required to study, e.g. leptoquarks [Buc91, H195, Hir96a, Bav95, Leu94, Cho94, Blü94].

To increase by a major step the present sensitivity for double beta decay and dark matter search, we describe here a new project proposed recently [Kla98a, Kla98b] which would operate one ton of 'naked' enriched **GE**rmanium detectors in liquid **NI**trogen as shielding in an **U**nderground **S**etup (GENIUS). It would improve the sensitivity from the present potential of at best ~ 0.1 eV to neutrino masses down to 0.01 eV, a ten ton version even to 0.001 eV. The first version would allow to test a $\nu_e \rightarrow \nu_\mu$ explanation of the atmospheric neutrino problem, the second directly the large angle solution of the solar neutrino problem, and, for degenerate neutrinos even the small angle solution. The sensitivity for neutrino oscillation parameters would be larger than for all present accelerator neutrino oscillation experiments, or those planned for the future. GENIUS would further allow one to test the recent hypothesis of a sterile neutrino and the underlying idea of a shadow world (see section 2). Both versions of GENIUS would definitely be a breakthrough into the multi-TeV range for many beyond standard models currently discussed in the literature, and the sensitivity would be comparable or even superior to LHC for various quantities such as right–handed W–bosons, R–parity violation, leptoquark or compositeness searches.

Another issue of GENIUS is the search for Dark Matter in the universe. The full MSSM parameter space for predictions of neutralinos as cold dark matter could be covered already in a first step of the full experiment using only 100 kg of ^{76}Ge or even natural Ge, making the experiment competitive to LHC in the search for supersymmetry.

2. Double beta decay and particle physics

We present a brief introductory outline of the potential of $\beta\beta$ decay for some representative examples, including some comments on the status of the required nuclear matrix elements. The potential of double beta decay for probing neutrino oscillation parameters will be addressed in section 4.2.

Double beta decay can occur in several decay modes (Figs. 1–3)

$$^A_ZX \rightarrow ^A_{Z+2}X + 2e^- + 2\overline{\nu}_e \tag{1}$$

$$^A_ZX \rightarrow ^A_{Z+2}X + 2e^- \tag{2}$$

$$^A_ZX \rightarrow ^A_{Z+2}X + 2e^- + \phi \tag{3}$$

$$^A_ZX \rightarrow ^A_{Z+2}X + 2e^- + 2\phi \tag{4}$$

the last three of them violating lepton number conservation by $\Delta L = 2$. Fig. 3 shows the corresponding spectra, for the neutrinoless mode (2) a sharp line at $E = Q_{\beta\beta}$, for the two–neutrino mode and the various Majoron–accompanied modes classified by their spectral index, continuous spectra. Important for particle physics are the decay modes (2)–(4).

The neutrinoless mode (2) needs not be necessarily connected with the exchange of a virtual neutrino or sneutrino. *Any* process violating lepton number can in principle lead to a process with the same signature as usual $0\nu\beta\beta$ decay. It may be triggered by exchange of neutralinos, gluinos, squarks, sleptons, leptoquarks,... (see below and [Päs97]). This gives rise to the broad potential of double beta decay for testing or yielding restrictions on quantities of beyond standard model physics (see Table 1), realized

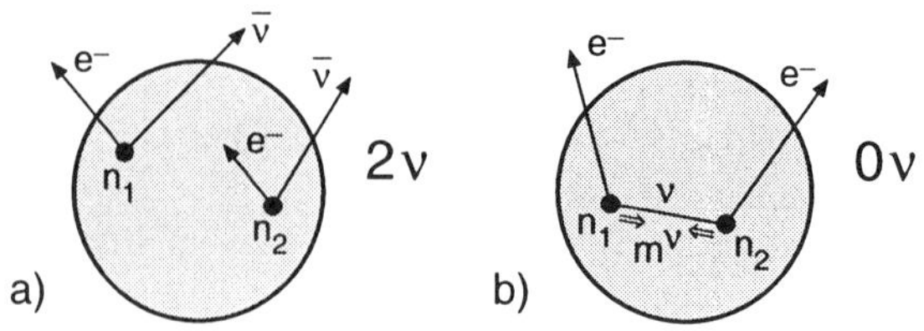

Fig. 1 *Schematic representation of* 2ν *and* 0ν *double beta decay.*

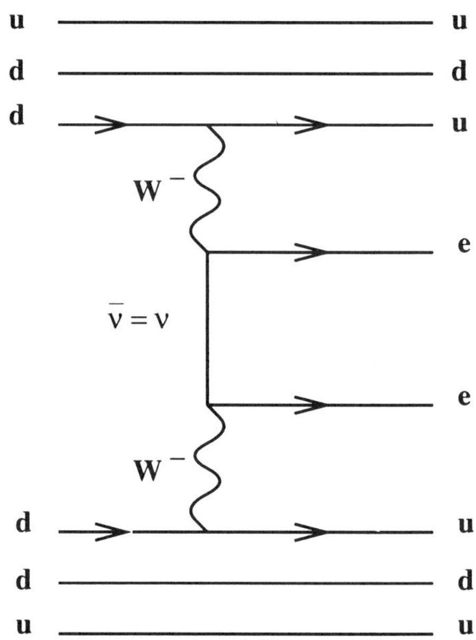

Fig. 2 *Feynman graph for neutrinoless double beta decay triggered by exchange of a left–handed light or heavy neutrino*

Observ.	Restrictions	Topics investigated
0ν:	via ν exchange:	Beyond the standard model and SU(5)
	Neutrino mass	model; early universe, matter–antimatter
	Light Neutrino	asymmetry, Dark matter
	Heavy Neutrino	L–R -symmetric models (e.g. SO(10)), compositeness
	Test of Lorentz invariance and equivalence principle	
	Right handed weak currents	$V + A$ interaction, $W_R^{\pm}$ masses
	via photino, gluino, zino (gaugino) or sneutrino exchange:	SUSY models: Bounds for parameter space beyond the range of accelerators
	R-parity breaking, sneutrino mass	
	via leptoquark exchange	leptoquark masses and models
	leptoquark-Higgs interaction	
$0\nu\chi$:	existence of the Majoron	Mechanism of (B-L) breaking -explicit -spontaneous breaking of the local/global B-L symmetry new Majoron models

Table 1 *$\beta\beta$ decay and particle physics*

and investigated to a large extent by the Heidelberg Double Beta Group in the last two years. There is, however, a generic relation between the amplitude of $0\nu\beta\beta$ decay and the $(B - L)$ violating Majorana mass of the neutrino. It has been recognized about 15 years ago [Sch81] that if any of these two quantities vanishes, the other one vanishes, too, and vice versa, if one of them is non–zero, the other one also differs from zero. This Schechter-Valle-theorem is valid for any gauge model with spontaneously broken symmetry at the weak scale, independent of the mechanism of $0\nu\beta\beta$ decay. A generalisation of this theorem to supersymmetry has been given recently [Hir97, Hir98a]. This Hirsch–Klapdor-Kleingrothaus–Kovalenko–theorem claims for the neutrino Majorana mass, the $B - L$ violating mass of the sneutrino and neutrinoless double beta decay amplitude: If one of them is non–zero, also the others are non–zero and vice versa, independent of the mechanisms of $0\nu\beta\beta$ decay and (s-)neutrino mass generation. This theorem connects double beta research with new processes potentially observable at future colliders like NLC (next linear collider) [Hir97, Hir98].

2.1. Mass of the (electron) neutrino

The neutrino is one of the best examples for the merging of the different disciplines of micro– and macrophysics. The neutrino plays, by its nature (Majorana or Dirac particle), and its mass, a key role for the structure of modern particle physics theories (GUTs, SUSYs, SUGRAs,...) [Kla95, Kla97, Lan88, Gro90, Moh91, Kla97a, Kla97c, Kla98b]. At the

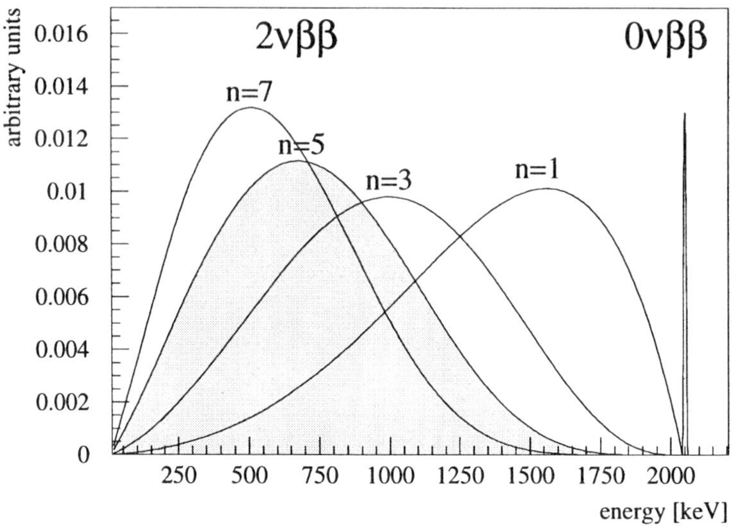

Fig. 3 *Spectral shapes of the different modes of double beta decay, n denotes the spectral index , n=5 for $2\nu\beta\beta$ decay (see text)*

same time it is candidate for non–baryonic hot dark matter in the universe, and the neutrino mass is connected – by the sphaleron effect – to the matter–antimatter asymmetry of the early universe [Kuz90]. Neutrino physics has entered an era of new actuality in connection with several possible indications of physics beyond the standard model (SM) of particle physics: A lack of solar (^7Be) neutrinos, an atmospheric ν_μ deficit and mixed dark matter models could all be explained simulaneously by non–vanishing neutrino masses. Recent GUT models, for example an extended SO(10) scenario with S_4 horizontal symmetry could explain these observations by requiring degenerate neutrino masses of the order of 1 eV [Lee94, Moh94, Pet94, Ioa94, Fri95, Moh95, Pet96, Val96]. For an overview see [Smi96a]. Such degenerate scenarios are the more general solution of the well-known see-saw mechanism, of which the often discussed strongly hierarchical neutrino mass pattern is just a special solution (see [Moh97]). If the atmospheric neutrino data are excluded but LSND [Ath95, Ath96], HDM and solar neutrino constraints are kept, they could be explained by an inverted mass texture [Raf96, Cal95], where $m_{\nu_e} \simeq m_{\nu_\tau} \simeq 2.4eV >> m_{\nu_\mu}$.

This brings double beta decay experiments into some key position, since with some second generation $\beta\beta$ experiments like the HEIDELBERG–MOSCOW experiment using large amounts of enriched $\beta\beta$–emitter material the predictions of or assumptions in such scenarios can now be tested. If the first of the above scenarios of neutrino mass textures is ruled out by tightening the double beta limit on m_{ν_e}, then the only way to understand *all* neutrino results may require an additional sterile neutrino [Cal93, Pel93], coupling only extremely weakly to the Z–boson. Then the solar neutrino puzzle would be explained by the $\nu_e - -\nu_S$ oscillation, and atmospheric neutrino data by $\nu_\mu - \nu_\tau$ oscillations, and the $\nu_{\mu,\tau}$ would constitute the hot dark matter (HDM) of the universe. The request for a light sterile neutrino would naturally lead to the concept of a shadow world [Ber95]. This assumes exact duplication of the Standard Model in both the gauge and the fermion content (the shadow sector), yielding three extra sterile neutrinos $\nu^{'}$, the only interaction connecting known and shadow sector being gravitation. Mixing of the ν and $\nu^{'}$ will occur by Planck scale effects. Such a scenario could explain all *four* present indications for non-vanishing neutrino mass [Moh97]. The expectation for the effective neutrino mass (see below) to be seen in double beta decay would be $\langle m_{\nu_e} \rangle \simeq 0.002eV$ [Moh97a]. Thus it could be checked by the new Genius project (see section 4.2.2). Interestingly in such a scenario the $\nu^{'}_\mu$, $\nu^{'}_\tau$ having masses of $\sim$ 2 keV could act as warm or cold dark matter in the universe [Moh97].

At present neutrinoless double decay is the most sensitive of the various existing methods to determine the mass of the electron neutrino. It further provides a unique possibility of deciding between a Dirac and a Majorana nature of the neutrino. Neutrinoless double beta decay can be triggered by exchange of a light or heavy left-handed Majorana neutrino (Figs. 1,2). For exchange of a heavy *right*–handed neutrino see section 2.3. The propagators in the first and second case show a different m_ν dependence: Fermion propagator $\sim \frac{m}{q^2 - m^2} \Rightarrow$

$$a) \quad m \ll q \rightarrow\sim m \quad \textit{'light' neutrino} \tag{5}$$

$$b) \quad m \gg q \rightarrow\sim \frac{1}{m} \quad \textit{'heavy' neutrino} \tag{6}$$

The half–life for $0\nu\beta\beta$ decay induced by exchange of a light neutrino is given by [Mut88]

$$[T^{0\nu}_{1/2}(0^+_i \rightarrow 0^+_f)]^{-1} = C_{mm}\frac{\langle m_\nu \rangle^2}{m_e^2} + C_{\eta\eta}\langle\eta\rangle^2 + C_{\lambda\lambda}\langle\lambda\rangle^2 + C_{m\eta}\frac{m_\nu}{m_e}$$
$$+ C_{m\lambda}\langle\lambda\rangle\frac{\langle m_\nu\rangle}{m_e} + C_{\eta\lambda}\langle\eta\rangle\langle\lambda\rangle \tag{7}$$

or, when neglecting the effect of right–handed weak currents, by

$$[T^{0\nu}_{1/2}(0^+_i \rightarrow 0^+_f)]^{-1} = C_{mm}\frac{\langle m_\nu \rangle^2}{m_e^2} = (M^{0\nu}_{GT} - M^{0\nu}_F)^2 G_1 \frac{\langle m_\nu \rangle^2}{m_e^2} \tag{8}$$

where G_1 denotes the phase space integral, $\langle m_\nu \rangle$ denotes an effective neutrino mass

$$\langle m_\nu \rangle = \sum_i m_i U^2_{ei}, \tag{9}$$

respecting the possibility of the electron neutrino to be a mixed state (mass matrix not diagonal in the flavor space)

$$|\nu_e\rangle = \sum_i U_{ei}|\nu_i\rangle \qquad (10)$$

The effective mass $\langle m_\nu \rangle$ could be smaller than m_i for all i for appropriate CP phases of the mixing coefficiants U_{ei} [Wol81]. In general not too pathological GUT models yield $m_{\nu_e} = \langle m_{\nu_e} \rangle$ (see [Lan88]).

η,λ describe an admixture of right–handed weak currents, and $M^{0\nu} \equiv M_{GT}^{0\nu} - M_F^{0\nu}$ denote nuclear matrix elements.

Nuclear matrix elements:

A detailed discussion of $\beta\beta$ matrix elements for neutrino induced transitions including the substantial (well–understood) differences in the precision with which 2ν and $0\nu\beta\beta$ rates can be calculated, can be found in [Gro90, Mut88, Mut89, Sta90]. After the major step of recognizing the importance of ground state correlations for the calculation of $\beta\beta$ matrix elements [Kla84, Gro86], in recent years the main groups used the QRPA model for calculation of $M^{0\nu}$. The different groups obtained very similar results for $M^{0\nu}$ when using a realistic nucleon–nucleon interaction [Mut89, Sta90, Tom87], consistent with shell model approaches [Mut91, Hax84], where the latter are possible. Some deviation is found only when a non–realistic nucleon–nucleon interaction is used (e.g. δ force, see [Vog86] and also [Vog96]). Also use of a by far too small configuration space like in recent shell model Monte Carlo (SMMC) calculations [Rad95] can hardly lead to reliable results. Recent so–called large scale shell model calculations [Cau96] also fail to fulfill the Ikeda sum rule by about 40-60%, thus predicting too small matrix elements. On the other hand refinements of the QRPA approach by going to higher order QRPA (see [Sto96, Suh96])lead only to minor changes for the $0\nu\beta\beta$ ground state transitions. The most recent QRPA calculations including renormalization [Sim97] do not fulfill the Ikeda sum rule by 30 %. The calculated matrix elements are (correspondingly ?) about 40 % smaller than earlier calculations fulfilling the sum rule properly [Mut89, Sta90]. The consequences of high-lying GT strength (in the GTGR and in the Δ resonance) have already been studied earlier [Gro86].

Since the usual QRPA approach does ignore deformation, some larger uncertainty in these approaches may occur in deformed nuclei. This shows up for example in different results obtained for ^{150}Nd by QRPA and by a pseudo SU(3) model as used by [Hir95d]. Calculation of matrix elements of all double beta emitters have been published by [Gro85, Sta90]. Typical uncertainties of calculated $0\nu\beta\beta$ rates originating from the limited knowledge of the particle–particle force, which is the main source of the uncertainty in those nuclei where this QRPA approach is applicable, are shown in [Sta90]. They are of the order of a factor of 2.

2.2. *Supersymmetry*

Supersymmetry (SUSY) is considered as prime candidate for a theory beyond the standard model, which could overcome some of the most puzzling questions of today's particle physics (see, e.g. [Hab93, Moh92, Kan97]). Accelerator experiments have hunted for signs of supersymmetric particles so far without success. Lower limits on masses of SUSY particles are at present in the range of 20–100 GeV [PDG96], mainly from experiments at LEP and TEVATRON.

Conservation of R–parity has been imposed ad hoc to the minimal supersymmetric extension of the standard model (MSSM) to ensure baryon number and lepton number conservation. SUSY particles differ then from usual particles not only in their masses but also in R–parity, assigned to be $R_P = 1$ for usual particles and $R_P = -1$ for SUSY particles. This assumption, however, is not guaranteed by supersymmetry or gauge invariance. Generally one can add the following R–parity violating terms to the usual superpotential [Hal84].

$$W_{\not R_P} = \lambda_{ijk} L_i L_j \overline{E}_k + \lambda'_{ijk} L_i Q_j \overline{D}_k + \lambda'' \overline{U}_i \overline{D}_j \overline{D}_k, \qquad (11)$$

where indices i, j, k denote generations. L,Q denote lepton and quark doublet superfields and $\overline{E}, \overline{U}, \overline{D}$ lepton and up, down quark singlet superfields. Terms proportional to λ, λ' violate lepton number, those proportional to λ'' violate baryon number. From proton decay limits it is clear that both types of terms cannot be present at the same time in the superpotential. On the other hand, once the λ'' terms being assumed to be zero, λ and λ' terms are not limited. $0\nu\beta\beta$ decay can occur within the $\not R_p$ -MSSM through Feynman graphs such as those of Fig. 4. In lowest order there are alltogether six different graphs of this kind. [Hir95, Hir95c, Hir96c]. Attention has, therefore , been focused also on SUSY theories with R–parity violation, in which $0\nu\beta\beta$ decay can proceed by exchange of supersymmetric particles like gluinos, photinos,... Thus $0\nu\beta\beta$ decay can be used to restrict R–parity violating SUSY models [Hir95, Hir96c, Moh91, Hir95c, Moh86]. From these graphs one derives [Hir95] under some assumptions

$$[T^{0\nu}_{1/2}(0^+ \to 0^+)]^{-1} \sim G_{01}\left(\frac{\lambda'^2_{111}}{m^4_{\tilde{q},\tilde{e}} m_{\tilde{g}\chi}} M\right)^2 \qquad (12)$$

where G_{01} is a phase space factor, $m_{\tilde{q}\tilde{e}\tilde{g}\chi}$ are the masses of supersymmetric particles involved: squarks, selectrons, gluinos, or neutralinos. λ'_{111} is the

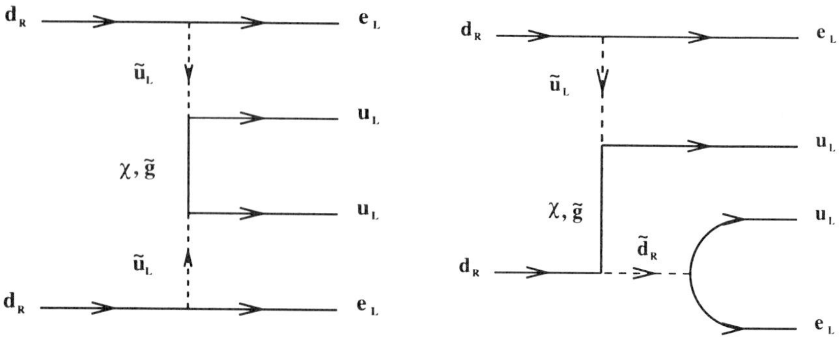

Fig. 4 *Examples of Feynman graphs for $0\nu\beta\beta$ decay within R–parity violating supersymmetric models (from [Hir95]).*

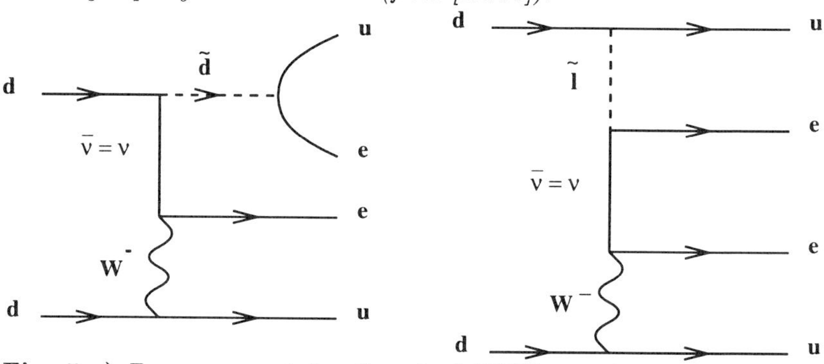

Fig. 5 a) *Feynman graph for the mixed SUSY–neutrino exchange mechanism of $0\nu\beta\beta$ decay. R–parity violation occurs through scalar quark exchange.* **b)** *As figure 1, but for scalar lepton exchange (from [Hir96]).*

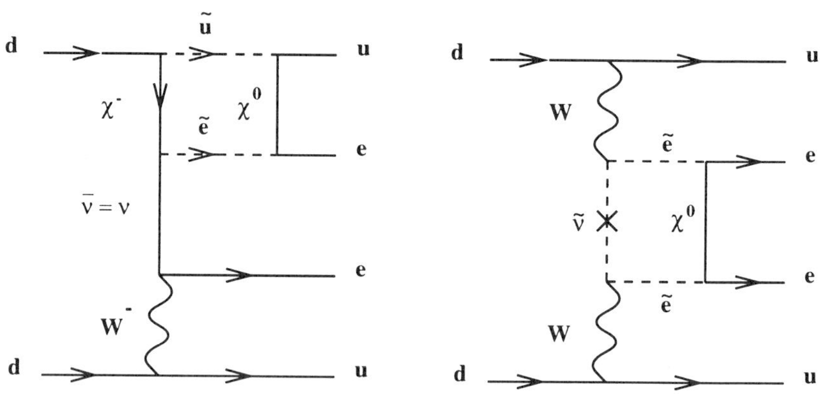

Fig. 6 *Examples of R_P conserving SUSY contributions to $0\nu\beta\beta$ decay (from [Hir97a]).*

strength of an R–parity breaking interaction (eq. 11), and M is a nuclear matrix element. For the matrix elements and their calculation see [Hir96c].

It is also worthwile to notice that $0\nu\beta\beta$ decay is not only sensitive to λ'_{111}. Taking into account the fact that the SUSY partners of the left and right–handed quark states can mix with each other, one can derive limits on different combinations of λ' [Hir96, Moh96a, Bab95]. Graphs allowing such information are as those shown in Fig. 5. The dominant diagram of this type is the one where the exchanged scalar particles are the $\tilde{b} - \tilde{b}^C$ pair. Under some assumptions (e.g. the MSSM mass parameters to be approximately equal to the "effective" SUSY breaking scale Λ_{SUSY}), one obtains [Hir96]

$$\lambda'_{11i} \cdot \lambda'_{1i1} \leq \epsilon'_i \left(\frac{\Lambda_{SUSY}}{100GeV}\right)^3 \qquad (13)$$

and

$$\Delta_n \lambda'_{311} \lambda_{n13} \leq \epsilon \left(\frac{\Lambda_{SUSY}}{100GeV}\right)^3 \qquad (14)$$

With the known values of the SM quark and lepton masses follows $\epsilon'_{1,2,3} = 6.4 \cdot 10^{-5}$, $3.2 \cdot 10^{-6}$ and $1.1 \cdot 10^{-7}$, and $\epsilon = 6.5 \cdot 10^{-8}$. For an overview on our knowledge on λ'_{ijk} from other sources we refer to [Kol97a] and [Bha97].

Also R–parity *conserving* softly broken supersymmetry can give contributions to $0\nu\beta\beta$ decay, via the $B - L$–violating sneutrino mass term, the latter being a generic ingredient of any weak–scale SUSY model with a Majorana neutrino mass [Hir97, Hir98]. These contributions are realized at the level of box diagrams [Hir98] (fig. 6). The $0\nu\beta\beta$ half-life for contributions from sneutrino exchange is found to be [Hir98]

$$[T_{1/2}^{0\nu\beta\beta}]^{-1} = G_{01} \frac{4m_p^2}{G_F^4} \left| \frac{\eta^{SUSY}}{m_{SUSY}^5} M^{SUSY} \right|, \qquad (15)$$

where the phase factor G_{01} is tabulated in [Doi85], η^{SUSY} is the effective lepton number violating parameter, which contains the $(B - L)$ violating sneutrino mass $\tilde{m}_M$ and M^{SUSY} is the nuclear matrix element [Hir96d].

2.3. Left–Right symmetric theories – Heavy neutrinos and right–handed W Boson

Heavy *right–handed* neutrinos appear quite naturally in left–right symmetric GUT models. They offer in some natural way via the see–saw mechanism explanation for the small neutrino masses compared to other fermions and can explain also naturally parity violation. However the symmetry breaking scale for the right–handed sector is not fixed by the theory and thus the mass of the right–handed W_R boson and the mixing angle between the mass eigenstates W_1, W_2 are free parameters. $0\nu\beta\beta$ decay taking into account contributions from both, left– and right–handed neutrinos have

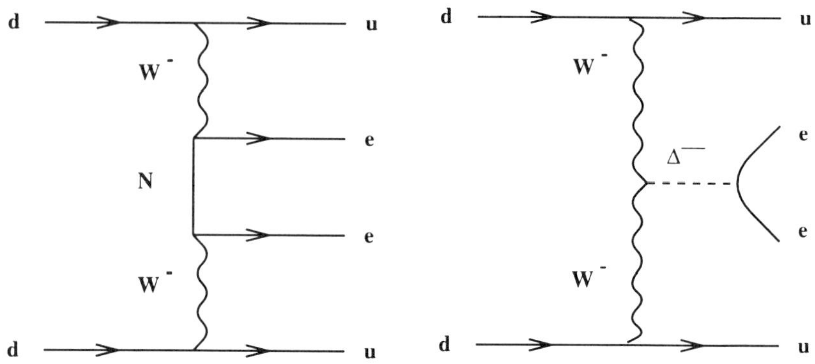

Fig. 7 a) *Heavy neutrino exchange contribution to neutrinoless double beta decay in left right symmetric models, and* **b)** *Feynman graph for the virtual exchange of a doubly–charged Higgs boson, see text (from [Hir96d]).*

been studied theoretically by [Hir96d, Doi93]. The former gives a more general expression for the decay rate than introduced earlier by [Moh86b].

$0\nu\beta\beta$ decay proceeds through the diagram shown in Fig. 7a, where N denotes the heavy right–handed partner of the ordinary neutrino. In order to preserve the unitarity of the cross section in *inverse* $0\nu\beta\beta$ decay, LR models must according to [Riz82] include an additional Higgs triplet. This then gives rise to a second contribution to $0\nu\beta\beta$ decay, shown in Fig. 7b. From the Feynman graphs of Fig. 7 it is obvious that the amplitude will be proportional to [Hir96d]

$$\left(\frac{m_{W_L}}{m_{W_R}}^4\right)\left(\frac{1}{m_N}+\frac{m_N}{m^2_{\Delta_R^{--}}}\right) \tag{16}$$

Eq. (16) and the experimental lower limit of $0\nu\beta\beta$ decay leads to a constraint limit within the 3–dimensional parameter space ($m_{W_R} - m_N - m_{\Delta_R^{--}}$). The most conservative (weakest) limit on m_{W_R} is obtained in the limit, where the mass of the Δ^{--} goes to infinity (see section 3 below). If adding information on the vacuum stability, an absolute lower limit on the mass of the right–handed W–boson can be obtained.

2.4. Compositeness

Although so far there are no experimental signals of a substructure of quarks and leptons, there are speculations that at some higher energy ranges beyond 1 TeV or so there might exist an energy scale Λ_C at which a substructure of quarks and leptons (preons) might become visible [Pan96, Moh92, Sou92](Fig. 8).

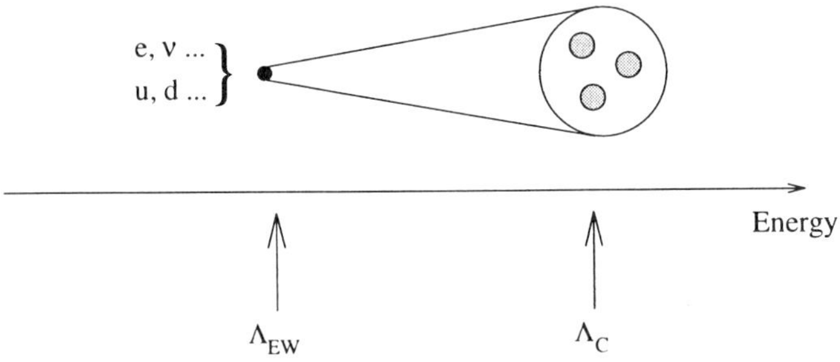

Fig. 8 *The idea of compositeness. At a (still unknown) energy scale Λ_C quarks and lepton might show an internal structure*

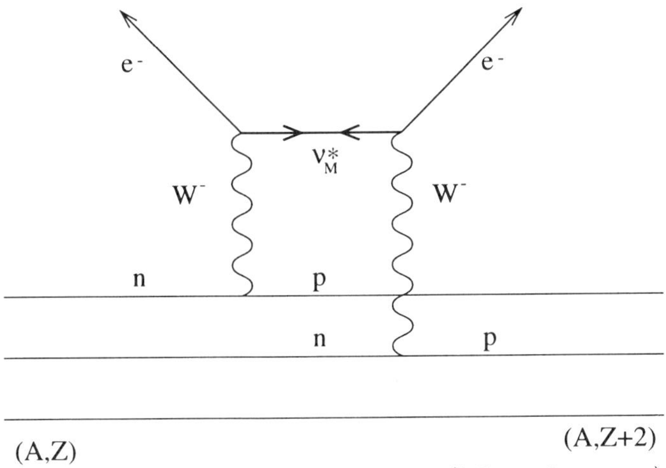

Fig. 9 *Neutrinoless double beta decay ($\Delta L = +2$ process) mediated by a composite heavy Majorana neutrino.*

The main consequences of compositeness of quarks and leptons are (1) modifications to the gauge boson propagators and the interaction vertices with fermions, and additional effective four–fermion interactions through constituent exchange (2) highly massive excited states which couple to the ordinary fermions through gauge interactions. This is discussed in detail in [Pan96]. Lower bounds on the compositeness scale have been deduced from accelerator experiments at LEP [ALE93], Fermilab [CDF91], HERA [H195] and from a theoretical analysis of the effect of contact interactions in the leptonic τ decay [Dia93]. They are all in the range of $\Lambda_C \geq 1.6$ TeV.

The *masses* of the excited leptons (l^*) and quarks (q^*) should not be lower than the compositeness scale Λ_C. Already in 1982 it was shown [Ren82] that precise measurements of the anomalous magnetic moment of the electron give bounds on the masses of the excited states and thus the compositeness scale.

Limits on the masses of excited leptons from accelerators are $m_{e^*} > 127$ GeV [Adr92], $m_{e^*,\nu^*} > 91$ GeV [ALE92], $m_{\nu^*} > 180$ GeV [Der95, Rau94] $m_{q^*} > 540$ GeV [Abe94].

A possible low energy manifestation of compositeness could be neutrinoless double beta decay, mediated by a composite heavy Majorana neutrino (Fig. 9), which then should be a Majorana particle.

Recent theoretical work shows (see [Pan96, Tak96, Pan97, Tak97]) that the mass bounds for such an excited neutrino which can be derived from double beta decay are at least of the same order of magnitude as those coming from the direct search of excited states in high energy accelerators (see also section 3).

2.5. Majorons

The existence of new bosons, so–called Majorons, can play a significant role in new physics beyond the standard model, in the history of the early universe, in the evolution of stellar objects, in supernovae astrophysics and the solar neutrino problem [Geo81, Fri88, Kla92]. In many theories of physics beyond the standard model neutrinoless double beta decay can occur with the emission of Majorons

$$2n \rightarrow 2p + 2e^- + \phi \tag{17}$$

$$2n \rightarrow 2p + 2e^- + 2\phi. \tag{18}$$

In the classical Majoron model invented by Gelmini and Roncadelli in '81 [Gel81], the Majoron is the Nambu–Goldstone boson associated with the spontaneous breaking of the $B - L$–symmetry and so generates Majorana masses of neutrinos. This was expected [Geo81] to give a sizeable contribution to double beta decay. It was, however, ruled out, as also the doublet Majoron [Aul82] by LEP [Ste91] since it should contribute the equivalent of two neutrino species to the width of the Z^0. On the other hand, Majoron models in which the Majoron is an electroweak isospin singlet [Chi81, Ber92] are still viable. The drawback of the singlet Majoron is that it requires a severe finetuning in order to preserve existing bounds on neutrino masses and at the same time get an observable rate for Majoron accompanied $0\nu\beta\beta$ decay.

To avoid such an unnatural fine–tuning in recent years several new Majoron models were proposed [Bur93, Bam95, Car93], where the term Majoron denotes in a more general sense light or massless bosons with couplings to neutrinos.

The main novel features of these "New Majorons" are that they can carry leptonic charge, that they need not be Goldstone bosons and that emission of two Majorons can occur. The latter can be scalar–mediated or fermion–mediated. Table 2 shows some features of the different Majoron models according to [Bam95, Car93]. L denotes the leptonic charge, n the spectral index defining the phase space of the emitting particles, M the nuclear matrix elements. For details we refer to [Päs96, Bur96].

The half–lifes are according to [Moh88, Doi85] in some approximation given by

$$[T_{1/2}]^{-1} = | < g_\alpha > |^2 \cdot |M_\alpha|^2 \cdot G_{BB_\alpha} \tag{19}$$

for $\beta\beta\phi$-decays, or

$$[T_{1/2}]^{-1} = | < g_\alpha > |^4 \cdot |M_\alpha|^2 \cdot G_{BB_\alpha} \tag{20}$$

for $\beta\beta\phi\phi$–decays. The index α indicates that effective neutrino–Majoron coupling constants g, matrix elements M and phase spaces G differ for different models.

Nuclear matrix elements:

According to Table 2 there are five different nuclear matrix elements. Of these M_F and M_{GT} are the same which occur in $0\nu\beta\beta$ decay. The other ones and the corresponding phase spaces have been calculated for the first time by [Päs96, Hir96b]. The calculation of the matrix elements show that the new models predict, as consequence of the small matrix elements very large half–lives and that unlikely large coupling constants would be needed to produce observable decay rates (see Table 3).

2.6. Sterile neutrinos

Introduction of sterile neutrinos has been claimed to solve simultaneously the conflict between dark matter neutrinos, LSND and supernova nucleosynthesis [Pel95] and light sterile neutrinos are part of popular neutrino mass textures for understanding the various hints for neutrino oscillations (see section 2.1) and [Moh96, Moh97, Moh97a]. Neutrinoless double beta decay can also investigate several effects of *heavy* sterile neutrinos [Bam95a].

If we assume having a light neutrino with a mass $\ll$ 1 eV, mixing with a much heavier (m $\geq$ 1 GeV) sterile neutrino can yield under certain conditions a detectable signal in current $\beta\beta$ experiments.

In models with two (or more) sterile neutrinos, the sterile neutrinos can mix appreciably even in the limit $m_{\nu_e} \to 0$ and so can be potentially visible in many processes [Pil93]. Neutrinoless double beta decay proceeds in these models through the virtual exchange of the heavier (i.e. GeV scale or higher) neutrinos. Fig. 10 shows the mass ranges leading to a $0\nu\beta\beta$ signal close to observability (shaded areas).

case	modus	Goldstone boson	L	n	Matrix element
IB	$\beta\beta\phi$	no	0	1	$M_F - M_{GT}$
IC	$\beta\beta\phi$	yes	0	1	$M_F - M_{GT}$
ID	$\beta\beta\phi\phi$	no	0	3	$M_{F\omega^2} - M_{GT\omega^2}$
IE	$\beta\beta\phi\phi$	yes	0	3	$M_{F\omega^2} - M_{GT\omega^2}$
IIB	$\beta\beta\phi$	no	-2	1	$M_F - M_{GT}$
IIC	$\beta\beta\phi$	yes	-2	3	M_{CR}
IID	$\beta\beta\phi\phi$	no	-1	3	$M_{F\omega^2} - M_{GT\omega^2}$
IIE	$\beta\beta\phi\phi$	yes	-1	7	$M_{F\omega^2} - M_{GT\omega^2}$
IIF	$\beta\beta\phi$	Gauge boson	-2	3	M_{CR}

Table 2 *Different Majoron models according to [Bam95]. The case IIF corresponds to the model of [Car93].*

model	$T_{1/2}(< g >= 10^{-4})$	$T_{1/2}(< g >= 1)$	$T_{1/2exp}$
IB,IC,IIB	$4 \cdot 10^{22}$	$4 \cdot 10^{14}$	$1.67 \cdot 10^{22}$
ID,IE,IID	10^{38-42}	10^{22-26}	$1.67 \cdot 10^{22}$
IIC,IIF	$2 \cdot 10^{28}$	$2 \cdot 10^{20}$	$1.67 \cdot 10^{22}$
IIE	10^{38-42}	10^{22-26}	$3.37 \cdot 10^{22}$

Table 3 *Comparison of half-lives calculated for different $< g >$–values for the new Majoron models with experimental best fit values, see section 3.1 (from [Hir96b])*

2.7. Leptoquarks

Interest on leptoquarks (LQ) has been renewed during the last few years since ongoing collider experiments have good prospects for searching these particles [Buc87]. LQs are vector or scalar particles carrying both lepton and baryon numbers and, therefore, have a well distinguished experimental signature. Direct searches of LQs in deep inelastic ep-scattering at HERA [H196] placed lower limits on their mass $M_{LQ} \geq 225 - 275$ GeV, depending on the LQ type and couplings.

In addition to the direct searches on LQs, there are many constraints which can be derived from the study of low-energy processes [Dav94]. Effective 4-fermion interactions, induced by virtual LQ exchange at energies much smaller than their masses, can contribute to atomic parity violation, flavour-changing neutral current processes, meson decays, meson-antimeson mixing and some rare processes.

To consider LQ phenomenology in a model-independent fashion one usually follows some general principles in constructing the Lagrangian of the LQ interactions with the standard model fields. In order to obey the stringent constraints from (c1) helicity-suppressed $\pi \to e\nu$ decay, from (c2) FCNC processes and from (c3) proton stability, the following assumptions

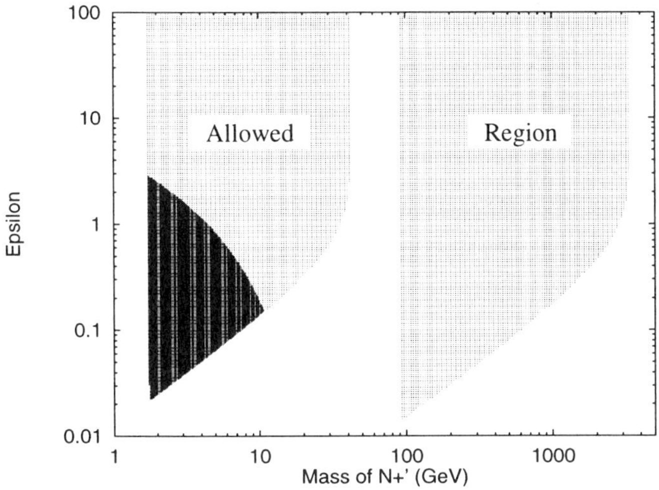

Fig. 10 *Regions of the parameter space ($\epsilon - M_{N'_+}$ plane yielding an observable signal (shaded areas) (from [Bam95a]). Darker area: 'natural' region, lighter shaded: Finetuning needed, to keep m_{ν_e} below 1 eV. $M_{N'_+}$: mass eigenstate, ϵ: strength of lepton number violation in mass matrix*

are commonly adopted: (a1) LQ couplings are chiral, (a2) LQ couplings are generation diagonal, and (a3) there are no diquark couplings.

Recently, however, it has been pointed out [Hir96a] that possible LQ-

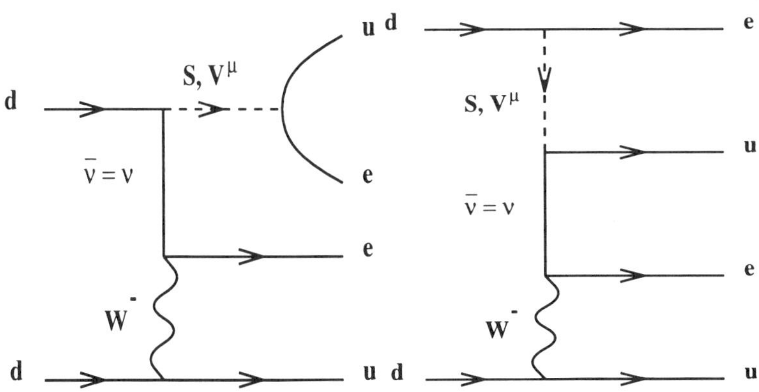

Fig. 11 *Examples of Feynman graphs for $0\nu\beta\beta$ decay within LQ models. S and V^μ stand symbolically for scalar and vector LQs, respectively (from [Hir96a]).*

Higgs interactions spoil assumption (a1): Even if one assumes LQs to be chiral at some high energy scale, LQ-Higgs interactions introduce after electro-weak symmetry breaking mixing between LQ states with different chirality. Since there is no fundamental reason to forbid such LQ-Higgs interactions, it seems difficult to get rid of the unwanted non-chiral interactions in LQ models.

In such LQ models there appear contributions to $0\nu\beta\beta$ decay via the Feynman graphs of Fig. 11. Here, S and V^{μ} stand symbolically for scalar and vector LQs, respectively. The half-life for $0\nu\beta\beta$ decay arising from leptoquark exchange is given by [Hir96a]

$$T_{1/2}^{0\nu} = |M_{GT}|^2 \frac{2}{G_F^2} [\tilde{C}_1 a^2 + C_4 b_R^2 + 2C_5 b_L^2]. \tag{21}$$

with $a = \frac{\epsilon_S}{M_S^2} + \frac{\epsilon_V}{M_V^2}$, $b_{L,R} = \frac{\alpha_S^{(L,R)}}{M_S^2} + \frac{\alpha_V^{(L,R)}}{M_V^2}$, $\tilde{C}_1 = C_1 \left(\frac{\mathcal{M}_1^{(\nu)}/(m_e R)}{M_{GT} - \alpha_2 M_F} \right)^2$.

For the definition of the C_n see [Doi85] and for the calculation of the matrix element $\mathcal{M}_1^{(\nu)}$ see [Hir96a]. This allows to deduce information on leptoquark masses and leptoquark–Higgs couplings (see section 3.2).

2.8. Special Relativity and Equivalence Principle

Special relativity and the equivalence principle can be considered as the most basic foundations of the theory of gravity. Many experiments already have tested these principles to a very high level of accuracy [Hug60] for ordinary matter - generally for quarks and leptons of the first generation. These precision tests of local Lorentz invariance – violation of the equivalence principle should produce a similar effect [Wil92] – probe for any dependence of the (non–gravitational) laws of physics on a laboratory's position, orientation or velocity relative to some preferred frame of reference, such as the frame in which the cosmic microwave background is isotropic.

A typical feature of the violation of local Lorentz invariance (VLI) is that different species of matter have a characteristical maximum attainable speed. This can be tested in various sectors of the standard model through vacuum Cerenkov radiation [Gas89a], photon decay [Col97], neutrino oscillations [Gla97, Gas89, Hal91, Hal96, But93] and K−physics [Ham98, Goo61]. These arguments can be extended to derive new constraints from neutrinoless double beta decay [Kla98f].

The equivalence principle implies that spacetime is described by unique operational geometry and hence universality of the gravitational coupling for all species of matter. In the recent years there have been attempts to constrain a possible amount of violation of the equivalence principle (VEP) in the neutrino sector from neutrino oscillation experiments

[Gas89, Hal91, Hal96, But93]. However, these bounds do not apply when the gravitational and the weak eigenstates have small mixing. In a recent paper [Kla98f] a generalized formalism of the neutrino sector has been given to test the VEP and it has been shown that neutrinoless double beta decay also constrains the VEP. VEP implies different neutrino species to suffer from different gravitational potentials while propagating through the nucleus and hence the effect of different eigenvalues doesn't cancel for the same effective momentum. The main result is that neutrinoless double beta decay can constrain the amount of VEP even when the mixing angle is zero, *i.e.*, when only the weak equivalence principle is violated, for which there does not exist any bound at present.

3. Double Beta Decay Experiments: Present Status and Results

3.1. Present Experimental Status

Fig. 12 shows an overview over measured $0\nu\beta\beta$ half–life limits and deduced mass limits. The largest sensitivity for $0\nu\beta\beta$ decay is obtained at present by active source experiments (source=detector), in particular ^{76}Ge [HM95, HM97, Kla94, Kla97] and ^{136}Xe [Ger96]. The main reason is that large source strengths can be used (simultaneously with high energy resolution), in particular when enriched $\beta\beta$ emitter materials are used. Geochemical experiments, though having contributed important information to double beta decay, have no more future in the sense that their inherent background from $2\nu\beta\beta$ decay cannot be eliminated.

Only a few of the present most sensitive experiments may probe the neutrino mass in the next years into the sub–eV region, the Heidelberg–Moscow experiment being the by far most advanced and most sensitive one, see Fig. 12b. No one of them will pass, however, below $\sim 0.1 - 0.2$ eV (see section 4.1) A detailed discussion of the various experimental possibilities can be found in [Kla95, Kla96, Kla96a]. A useful listing of existing data from the various $\beta\beta$ emitters is given in [Tre95].

3.2. Present limits on beyond standard model parameters

The sharpest limits from $0\nu\beta\beta$ decay are presently coming from the Heidelberg–Moscow experiment [Kla87, HM95, Kla94, HM97, Kla97, Kla98a, Kla98b]. They will be given in the following. With five enriched (86% of ^{76}Ge) detectors of a total mass of 11.5 kg taking data in the Gran Sasso underground laboratory, and with a background of at present 0.06 counts/kg year keV, the experiment has reached its final setup and is now exploring the sub–eV range for the mass of the electron neutrino. Fig. 13 shows the spectrum taken in a measuring time of 42 kg y.

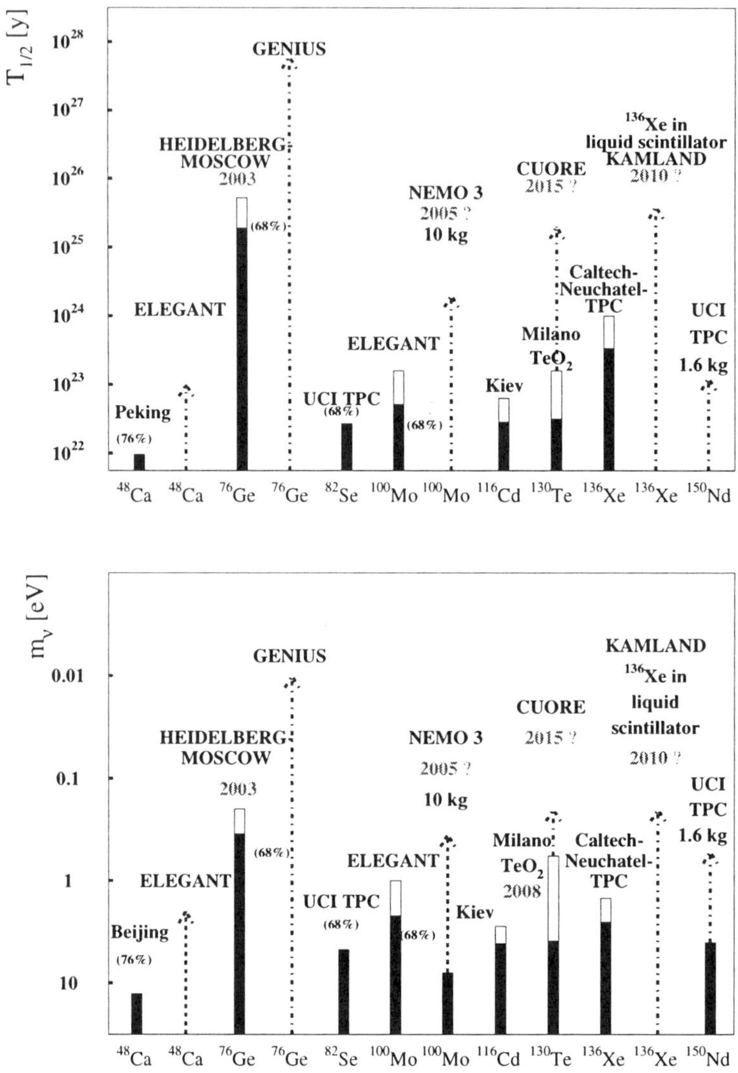

Fig. 12 *Present situation, 1998, and expectation for the near future and beyond, of the most promising $\beta\beta$-experiments concerning accessible half life (a) and neutrino mass limits (b). The filled bars correspond to the present status, open bars to expectations for running experiments, dashed lines to experiments under construction and dash-dotted lines to proposed experiments.*

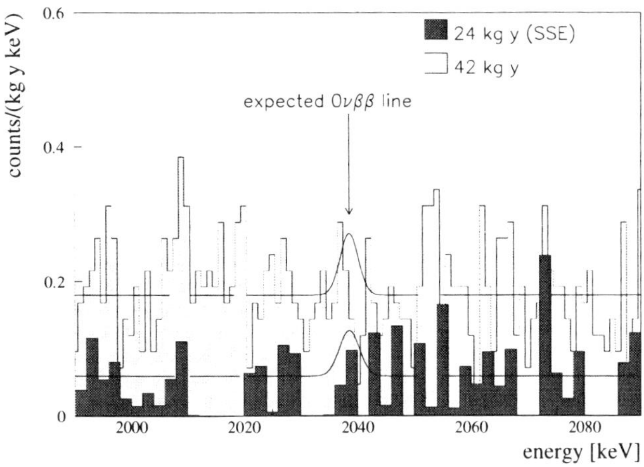

Fig. 13 *Integral spectrum in the region of interest after subtraction of the first 200 days of measurement of each detector, leaving 42 kg y of measuring time. The darkened histogram corresponds to data accumulated meanwhile using a new pulse shape analysis method (SSE) [Hel96] in a measuring time of 24 kg y. The two solid curves correspond to the signal excluded with 90%C.L. (by the 42 kg y measurement and the 24 kg y SSE measurement). They correspond to $T^{0\nu}_{1/2} > 1.3 \cdot 10^{25}$ y (upper curve) and $T^{0\nu}_{1/2} > 1.6 \cdot 10^{25}$ y, respectively.*

Half–life of neutrinoless double beta decay

The deduced half–life limit for $0\nu\beta\beta$ decay is, for the full data,

$$T^{0\nu}_{1/2} > 1.3 \cdot 10^{25} y \quad (90\% C.L.) \tag{22}$$

$$> 2.2 \cdot 10^{25} y \quad (68\% C.L.) \tag{23}$$

and for the 24 kg y of measurement with pulse shape analysis

$$T^{0\nu}_{1/2} > 1.6 \cdot 10^{25} y \quad (90\% C.L.) \tag{24}$$

$$> 2.8 \cdot 10^{25} y \quad (68\% C.L.) \tag{25}$$

Neutrino mass

Light neutrinos: The deduced upper limit of an (effective) electron neutrino Majorana mass is, with the matrix element from [Sta90]

$$\langle m_\nu \rangle < 0.42 eV \quad (90\% C.L.) \tag{26}$$

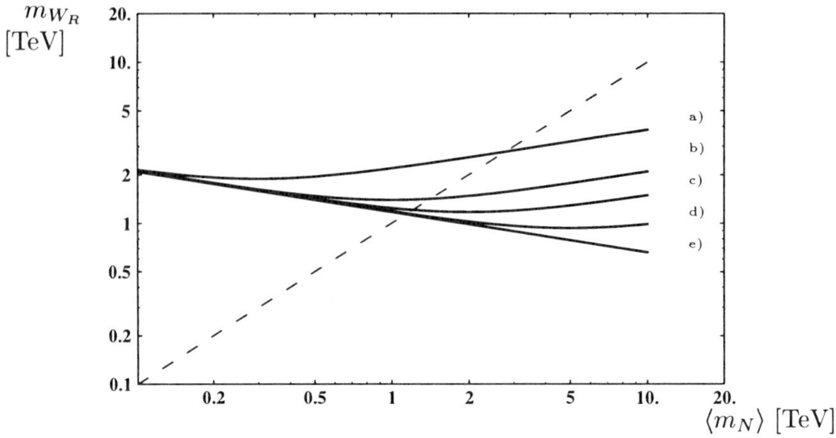

Fig. 14 *Limits on the mass of the right-handed W-boson from neutrinoless double beta decay (full lines) and vacuum stability (dashed line). The five full lines correspond to the following masses of the doubly charged higgs,* $m_{\Delta^{--}}$*: a) 0.3, b) 1.0, c) 2.0, d) 5.0 and e)* ∞ *[TeV] (from [Hir96d]).*

$$< 0.33 eV \quad (68\% C.L.) \tag{27}$$

and from the 24 kg y with pulse shape analysis (SSE)

$$\langle m_\nu \rangle < 0.38 eV \quad (90\% C.L.) \tag{28}$$

$$< 0.29 eV \quad (68\% C.L.) \tag{29}$$

This is the sharpest limit for a Majorana mass of the electron neutrino so far.

Superheavy neutrinos: For a superheavy *left*–handed neutrino we deduce [HM95] exploiting the mass dependence of the matrix element (for the latter see [Mut89, Bel96, Bel98]) a lower limit

$$\langle m_H \rangle \geq 100 TeV. \tag{30}$$

For a heavy *right*–handed neutrino the relation obtained to the mass of the right–handed W boson is shown in Fig. 14 (see [Hir96d]).

Right–handed W boson

For the right–handed W boson a lower limit of (Fig. 14)

$$m_{W_R} \geq 1.2TeV \tag{31}$$

is obtained [Hir96d].

SUSY parameters – R–parity breaking and sneutrino mass

The constraints on the parameters of the minimal supersymmetric standard model with explicit R–parity violation deduced [Hir95, Hir96c, Hir96] from the $0\nu\beta\beta$ half–life limit are more stringent than those from other low–energy processes and from the largest high energy accelerators (Fig. 15). The limits are

$$\lambda'_{111} \leq 3.9 \cdot 10^{-4}\Big(\frac{m_{\tilde{q}}}{100GeV}\Big)^2\Big(\frac{m_{\tilde{g}}}{100GeV}\Big)^{\frac{1}{2}} \tag{32}$$

with $m_{\tilde{q}}$ and $m_{\tilde{g}}$ denoting squark and gluino masses, respectively, and with the assumption $m_{\tilde{d}_R} \simeq m_{\tilde{u}_L}$. This result is important for the discussion of new physics in the connection with the high–Q^2 events seen at HERA. It excludes the possibility of squarks of first generation (of R–parity violating SUSY) being produced in the high–Q^2 events [Cho97, Alt97, Hir97b].

We find further [Hir96]

$$\lambda'_{113}\lambda'_{131} \leq 1.1 \cdot 10^{-7} \tag{33}$$

$$\lambda'_{112}\lambda'_{121} \leq 3.2 \cdot 10^{-6}. \tag{34}$$

For the $(B - L)$ violating sneutrino mass $\tilde{m}_M$ the following limits are obtained [Hir98a]

$$\tilde{m}_M \leq 2\Big(\frac{m_{SUSY}}{100GeV}\Big)^{\frac{3}{2}} GeV, \quad \chi \simeq \tilde{B} \tag{35}$$

$$\tilde{m}_M \leq 11\Big(\frac{m_{SUSY}}{100GeV}\Big)^{\frac{7}{2}} GeV, \quad \chi \simeq \tilde{H} \tag{36}$$

for the limiting cases that the lightest neutralino is a pure Bino $\tilde{B}$, as suggested by the SUSY solution of the dark matter problem [Jun96], or a pure Higgsino. Actual values for $\tilde{m}_M$ for other choices of the neutralino composition should lie in between these two values.

Another way to deduce a limit on the 'Majorana' sneutrino mass $\tilde{m}_M$ is to start from the experimental neutrino mass limit, since the sneutrino contributes to the Majorana neutrino mass m^{ν}_M at the 1–loop level proportional to $\tilde{m}^2_M$. This yields under some assumptions [Hir98a]

$$\tilde{m}_{M_{(i)}} \leq (60 - 125)\Big(\frac{m^{exp}_{\nu(i)}}{1eV}\Big)^{1/2} MeV \tag{37}$$

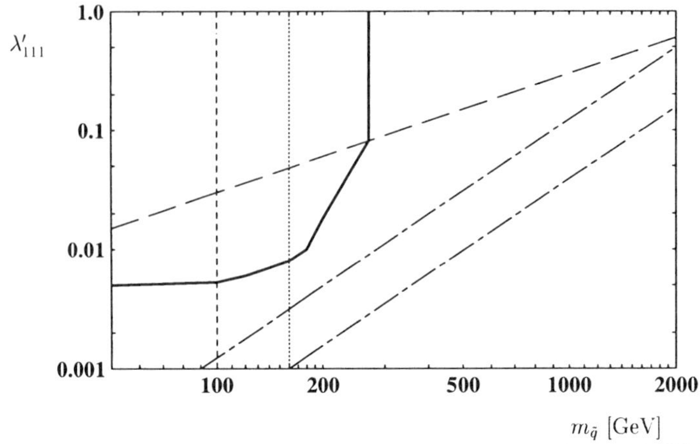

Fig. 15 *Comparison of limits on the R–parity violating MSSM parameters from different experiments in the λ'_{111}–$m_{\tilde{q}}$ plane. The dashed line is the limit from charged current universality according to [Bar89]. The vertical line is the limit from the data of Tevatron [Roy92]. The thick full line is the region which might be explored by HERA [But93a]. The two dash–dotted lines to the right are the limits obtained from the half–life limit for $0\nu\beta\beta$ decay of ^{76}Ge, for gluino masses of (from left to right) $m_{\tilde{g}}$ =1TeV and 100 GeV, respectively. The regions to the upper left of the lines are forbidden. (from [Hir95])*

Starting from the mass limit determined for the electron neutrino by $0\nu\beta\beta$ decay this leads to

$$\tilde{m}_{M_{(e)}} \leq 22MeV \tag{38}$$

This result is somewhat dependent on neutralino masses and mixings. A non–vanishing 'Majorana' sneutrino mass would result in new processes at future colliders, like sneutrino–antisneutrino oscillations. Reactions at the Next Linear Collider (NLC) like the SUSY analog to inverse neutrinoless double beta decay $e^-e^- \to \chi^-\chi^-$ (where χ^- denote charginos) or single sneutrino production, e.g. by $e^-\gamma \to \tilde{\nu}_e\chi^-$ could give information on the Majorana sneutrino mass, also. This is discussed by [Hir97, Hir98a, Hir98]. A conclusion is that future accelerators can give information on second and third generation sneutrino Majorana masses, but for first generation sneutrinos cannot compete with $0\nu\beta\beta$–decay.

Compositeness

Evaluation of the $0\nu\beta\beta$ half–life limit assuming exchange of excited Majorana neutrinos ν^* yields for the mass of the excited neutrino a lower bound of [Pan97, Tak97].

$$m_N \geq 3.4 m_W \tag{39}$$

for a coupling of order O(1) and $\Lambda_c \simeq m_N$. Here, m_W is the W–boson mass.

Leptoquarks

Assuming that either scalar or vector leptoquarks contribute to $0\nu\beta\beta$ decay, the following constraints on the effective LQ parameters (see section 2.7) can be derived [Hir96a]:

$$\epsilon_I \leq 2.8 \times 10^{-9} \left(\frac{M_I}{100\text{GeV}}\right)^2, \tag{40}$$

$$\alpha_I^{(L)} \leq 3.5 \times 10^{-10} \left(\frac{M_I}{100\text{GeV}}\right)^2, \tag{41}$$

$$\alpha_I^{(R)} \leq 7.9 \times 10^{-8} \left(\frac{M_I}{100\text{GeV}}\right)^2. \tag{42}$$

Here, different effective LQ couplings have been introduced. They are defined as:

$$\epsilon_I = 2^{-\eta_I} \left[\lambda_{I_1}^{(L)} \lambda_{\tilde{I}_{1/2}}^{(R)} \left(\theta_{43}^I(Q_I^{(1)}) + \eta_I \sqrt{2} \theta_{41}^I(Q_I^{(2)})\right)\right.$$
$$\left. - \lambda_{I_0}^{(L)} \lambda_{\tilde{I}_{1/2}}^{(R)} \theta_{13}^I(Q_I^{(1)})\right] \tag{43}$$

$$\alpha_I^{(L)} = \frac{2}{3+\eta_I} \lambda_{I_{1/2}}^{(L)} \lambda_{I_1}^{(L)} \theta_{24}^I(Q_I^{(2)}) \tag{44}$$

$$\alpha_I^{(R)} = \frac{2}{3+\eta_I} \lambda_{I_0}^{(R)} \lambda_{\tilde{I}_{1/2}}^{(R)} \theta_{23}^I(Q_I^{(1)}). \tag{45}$$

$\eta_{S,V} = 1, -1$ for scalar and vector LQs. $\theta_{kn}^I(Q)$ is a mixing parameter defined by

$$\theta_{kn}^I(Q) = \sum_l \mathcal{N}_{kl}^{(I)}(Q) \mathcal{N}_{nl}^{(I)}(Q) \left(\frac{M_I}{M_{I_l}(Q)}\right)^2, \tag{46}$$

where $\mathcal{N}^{(I)}(Q)$ are mixing matrix elements which diagonalize the LQ mass matrices for the scalar $I = S$ and vector $I = V$ LQ fields with electric charges $Q = -1/3, -2/3$, for complete definitions see [Hir96a]. Common mass scales M_S of scalar and M_V of vector LQs are introduced for convenience.

Since the LQ mass matrices appearing in $0\nu\beta\beta$ decay are (4×4) matrices [Hir96a], it is difficult to solve their diagonalization in full generality algebraically. However, if one assumes that only one LQ-Higgs coupling is present at a time, the (mathematical) problem is simplified greatly and one can deduce from, for example, eq. (40) that either the LQ-Higgs coupling must be smaller than $\sim 10^{-(4-5)}$ or there can not be any LQ with e.g. couplings of electromagnetic strength with masses below $\sim 250GeV$. These bounds from $\beta\beta$ decay are of interest in connection with recently discussed evidence for new physics from HERA [Hew97, Bab97, Kal97, Cho97]. Assuming that actually leptoquarks have been produced at HERA, double beta decay (the Heidelberg–Moscow experiment) would allow to fix the leptoquark–Higgs coupling to a few 10^{-6} [Hir97b]. It may be noted, that after the first consideration of leptoquark–Higgs coupling in [Hir96a] recently Babu et al. [Bab97b] noted that taking into account leptoquark–Higgs coupling reduces the leptoquark mass lower bound deduced by TEVATRON – making it more consistent with the value of 200 GeV required by HERA.

Special Relativity and Equivalence Principle

Violation of Lorentz invariance (VLI): The bound obtained from the Heidelberg–Moscow experiment is

$$\delta v < 4 \times 10^{-16} \quad \text{for} \quad \theta_v = \theta_m = 0 \tag{47}$$

where $\delta v = v_1 - v_2$ is the measure of VLI in the neutrino sector. θ_v and θ_m denote the velocity mixing angle and the weak mixing angle, respectively. In Fig. 16 (from [Kla98f]) the bound implied by double beta decay is presented for the entire range of $sin^2(2\theta_v)$, and compared with bounds obtained from neutrino oscillation experiments (see [Hal96]).

Violation of equivalence principle (VEP): Assuming only violation of the weak equivalence principle, there does not exist any bound on the amount of VEP. It is this region of the parameter space which is most restrictively bounded by neutrinoless double beta decay. In a linearized theory the gravitational part of the Lagrangian to first order in a weak gravitational field $g_{\mu\nu} = \eta_{\mu\nu} + h_{\mu\nu}$ ($h_{\mu\nu} = 2\frac{\phi}{c^2}diag(1,1,1,1)$) can be written as $\mathcal{L} = -\frac{1}{2}(1+g_i)h_{\mu\nu}T^{\mu\nu}$, where $T^{\mu\nu}$ is the stress-energy in the gravitational eigenbasis. In the presence of VEP the g_i may differ. We obtain [Kla98f] the following bound from the Heidelberg–Moscow experiment, for $\theta_v = \theta_m = 0$:

$$\begin{aligned} \phi\delta g &< 4 \times 10^{-16} \text{ (for } \bar{m} < 13\text{eV)} \\ \phi\delta g &< 2 \times 10^{-18} \text{ (for } \bar{m} < 0.08\text{eV).} \end{aligned} \tag{48}$$

Here $\bar{g} = \frac{g_1+g_2}{2}$ can be considered as the standard gravitational coupling, for which the equivalence principle applies. $\delta g = g_1 - g_2$. The bound on the VEP thus, unlike the one for VLI, will depend on the choice for the Newtonian potential ϕ.

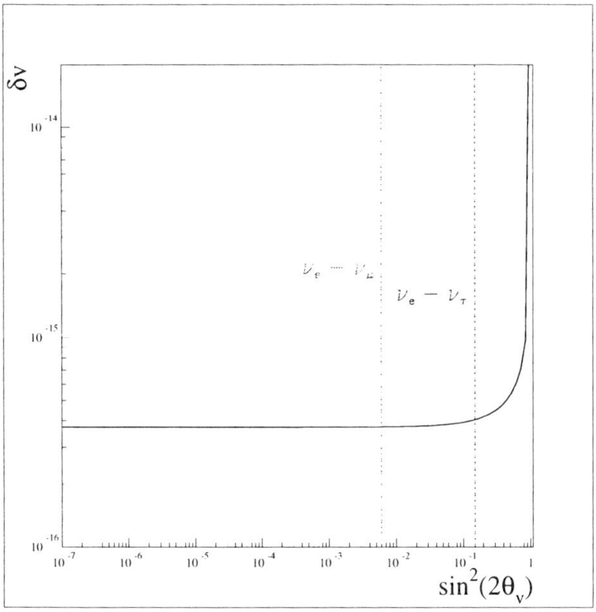

Fig. 16 *Double beta decay bound (solid line) on violation of Lorentz invariance in the neutrino sector, excluding the region to the upper left. Shown is a double logarithmic plot in the δv–$\sin^2(2\theta)$ parameter space. The bound becomes most stringent for the small mixing region, which has not been constrained from any other experiments. For comparison the bounds obtained from neutrino oscillation experiments (from [Hal96]) in the $\nu_e - \nu_\tau$ (dashed lines) and in the $\nu_e - \nu_\mu$ (dashed-dotted lines) channel, excluding the region to the right, are shown (from [Kla98f]).*

Half–life of $2\nu\beta\beta$ decay

The Heidelberg–Moscow experiment produced for the first time a high statistics $2\nu\beta\beta$ spectrum ($\sim$ 20000 counts, to be compared with the 40 counts on which the first detector observation of $2\nu\beta\beta$ decay by [Ell87] (for the decay of ^{82}Se) had to rely. The deduced half–life is [HM97]

$$T_{1/2}^{2\nu} = (1.77^{+0.01}_{-0.01}(stat.)^{+0.13}_{-0.11}(syst.)) \cdot 10^{21} y \tag{49}$$

This result brings $\beta\beta$ research for the first time into the region of 'normal' nuclear spectroscopy and allows for the first time statistically reliable investigation of Majoron–accompanied decay modes.

Majoron–accompanied decay

From simultaneous fits of the 2ν spectrum and one selected Majoron mode, experimental limits for the half–lives of the decay modes of the newly introduced Majoron models [Bur96] are given for the first time [Päs96, HM96].

The small matrix elements and phase spaces for these modes [Päs96, Hir96b] already determined that these modes by far cannot be seen in experiments of the present sensivity if we assume typical values for the neutrino–Majoron coupling constants around $\langle g \rangle = 10^{-4}$ (see table 3).

4. Double Beta Experiments: Future Perspectives – the GENIUS Project

4.1. The known experiments and proposals

Figs. 12a,b show in addition to the present status the future perspectives of the main existing $\beta\beta$ decay experiments and includes some ideas for the future which have been published. The HEIDELBERG–MOSCOW experiment will probe the neutrino mass within 5 years down to the order of 0.1 eV. The best presently existing limits besides the HEIDELBERG-MOSCOW experiment (filled bars in Fig. 12), have been obtained with the isotopes: ^{48}Ca [You95], ^{82}Se [Ell92], ^{100}Mo [Als89], ^{116}Cd [Dan95], ^{130}Te [Ale94], ^{136}Xe [Vui93] and ^{150}Nd [Moe94]. These and other double beta decay setups presently under construction or partly in operation such as NEMO [NEM94, Bar97], the Gotthard ^{136}Xe TPC experiment [Jör94], the ^{130}Te cryogenic experiment [Ale94], a new ELEGANT ^{48}Ca experiment using 30 g of ^{48}Ca [Kum96], a hypothetical experiment with an improved UCI TPC [Moe94] assumed to use 1.6 kg of ^{136}Xe, etc., will not reach or exceed the ^{76}Ge limits. The goal 0.3 eV aimed at for the year 2004 by the NEMO experiment (see [Piq96, Bar97] and Fig. 12) may even be very optimistic if claims about the effect of proton-neutron pairing on the $0\nu\beta\beta$ nuclear matrix elements by [Pan96b] will turn out to be true, and also if the energy resolution will not be improved considerably (see Fig. 1 in [Tre95]). Therefore, the conclusion given by [Bed97c] concerning the future SUSY potential of NEMO has no serious basis. As pointed out by Raghavan [Rag94], even use of an amount of about 200 kg of enriched ^{136}Xe or 2 tons of natural Xe added to the scintillator of the KAMIOKANDE detector or similar amounts added to BOREXINO (both primarily devoted to solar neutrino investigation) would hardly lead to a sensitivity larger than the present ^{76}Ge experiment. This idea is going to be realized at present by the KAMLAND experiment [Suz97]. An interesting future candidate was for some time a ^{150}Nd bolometer exploiting the relatively large phase space of this nucleus (see [Moe94]). The way outlined by [Moe91] proposing a

TPC filled with 1 ton of liquid enriched ^{136}Xe and identification of the daughter by laser fluorescence seems not be feasible in a straight-forward way. However, another way of using liquid ^{136}Xe may be more promising [Cli97].

It is obvious that, from the experiments and proposals, the HEIDELBERG-MOSCOW experiment will give the sharpest limit for the electron neutrino mass for the next decade. It is also obvious from Fig. 12 that *none* of the present experimental approaches, or plans or even vague ideas has a chance to surpass the border of 0.1 eV for the neutrino mass to lower values (see also [Nor97]). At present there is only one way visible to reach the domain of lower neutrino masses, suggested by [Kla98a] and meanwhile investigated in some detail concerning its experimental realization and and physics potential in [Kla97d, Hel97, Kla98b, Kla98c].

4.2. Genius – A Future Large Scale Double Beta and Dark Matter Experiment

The idea of GENIUS is to use a large amount of 'naked' enriched **GE**rmanium detectors in liquid **NI**trogen as shielding in an **U**nderground **S**etup. Use of 1 (in an extended version 10) tons of enriched ^{76}Ge will increase the source strength largely, removing all material from the vicinity of the detectors and shielding by liquid nitrogen will lead to a drastic background reduction compared to the present level. Using Ge detectors in liquid nitrogen has been discussed already earlier [Heu95]. That Ge detectors can be operated in liquid nitrogen has been demonstrated recently in the Heidelberg low level laboratory [Hel97]. The natural site for GENIUS would be the Gran Sasso underground laboratory. The cost of the project would be a minor fraction of detectors prepared for LHC physics as CMS or ATLAS. We give in the next two subsections some results of Monte Carlo simulations of the setup [Hel97] and some estimates of the physics potential [Kla97d] (see also [Kla98b, Kla98c]).

4.2.1. Realization and Sensitivity of GENIUS A simplified model of GENIUS is shown in Fig. 17 consisting of about 300 enriched ^{76}Ge detectors with a total of one ton mass in the center of a 9 m high liquid nitrogen tank with 9 m diameter. Figs. 18, 19 show the results of Monte Carlo simulations, using the CERN GEANT code, of the background [Hel97], starting from purity levels of the nitrogen being in general an order of magnitude less stringent than those already achieved in the CTF for the BOREXINO experiment. The influence of muons penetrating the Gran Sasso rock on the background can be reduced comfortably through coincidences between the Germanium detectors from the muon induced showers. The count rate in the region of interest for neutrinoless double beta decay is 0.04 counts/ keV · y · ton (Fig. 18). Below 100 keV the background count rate is about

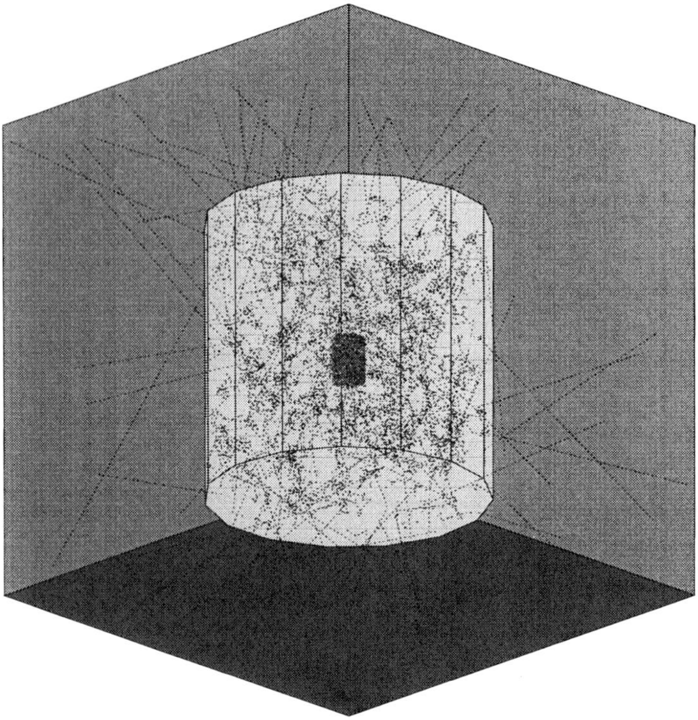

Fig. 17 *Simplified model of the GENIUS experiment: 288 enriched ^{76}Ge detectors with a total of one ton mass in the center of a 9 m high liquid nitrogen tank with 9 m diameter; GEANT Monte Carlo simulation of 1000 2.6 MeV photons randomly distributed in the nitrogen is also shown.*

10 counts/keV · y · ton. Two neutrino double beta decay would dominate the spectrum with $4 \cdot 10^6$ events per year.

Starting from these numbers, a lower half–life limit of

$$T_{1/2}^{0\nu} \geq 5.8 \cdot 10^{27} \qquad (68\% C.L.) \tag{50}$$

can be reached within one year of measurement (following the highly conservative procedure for analysis recommended by [PDG94], which has been used also in the derivation of the results given in section 3.2, but is not used in the analysis of several other $\beta\beta$ experiments). This corresponds – with the matrix elements of [Sta90] – to an upper limit on the neutrino mass of

$$\langle m_\nu \rangle \leq 0.02 eV \qquad (68\% C.L.) \tag{51}$$

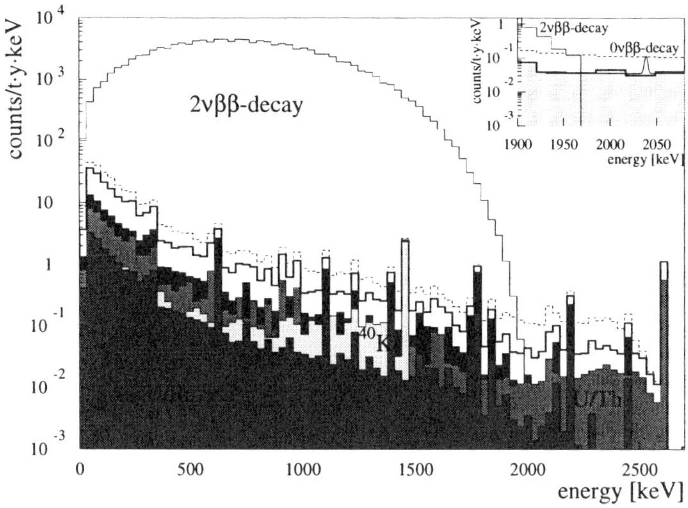

Fig. 18 *Monte Carlo simulation of the background of GENIUS. Simulations of U/Ra, U/Th and* ^{40}K *(shaded),* ^{222}Rn *(black histogram) activities in the liquid nitrogen; the sum of the activities is shown with anticoincidence between the 288 detectors (thick line) and without (dashed line); the* $2\nu\beta\beta$*-decay dominates the spectrum with 4 million events per year (from [Kla98c]).*

Figure 20 shows the obtainable limits on the neutrino mass in the case of zero background. This assumption might be justified since our assumed impurity concentrations are still more conservative than proved already now for example by Borexino. The final sensitivity of the experiment can be defined by the limit, which would be obtained after 10 years of measurement. For the one ton experiment this would be:

$$T^{0\nu}_{1/2} \quad \geq \quad 6.4 \cdot 10^{28}\, y \quad \text{(with 68\% C.L.)} \tag{52}$$

and

$$\langle m_\nu \rangle \quad \leq 0.006 eV \quad \text{(with 68\% C.L.)} \tag{53}$$

The ultimate experiment could test the $0\nu\beta\beta$ half life of ^{76}Ge up to a limit of $5.7{\cdot}10^{29}$y and the neutrino mass down to $2{\cdot}10^{-3}$eV using 10 tons of enriched Germanium.

4.2.2. The Physics Potential of GENIUS

Neutrino mass textures and neutrino oscillations: GENIUS will allow a

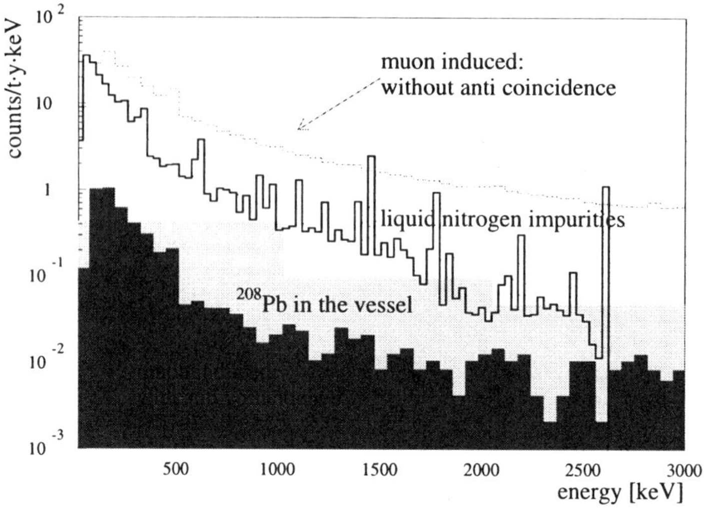

Fig. 19 *Background from outside the nitrogen: 200 GeV muons induced events (dashed line) and single hit events (filled histogram); decay of* ^{208}Tl *in the steel vessel (light shaded histogram) and the background originating from the nitrogen impurities for comparison (thick line) (from [Kla98c])*

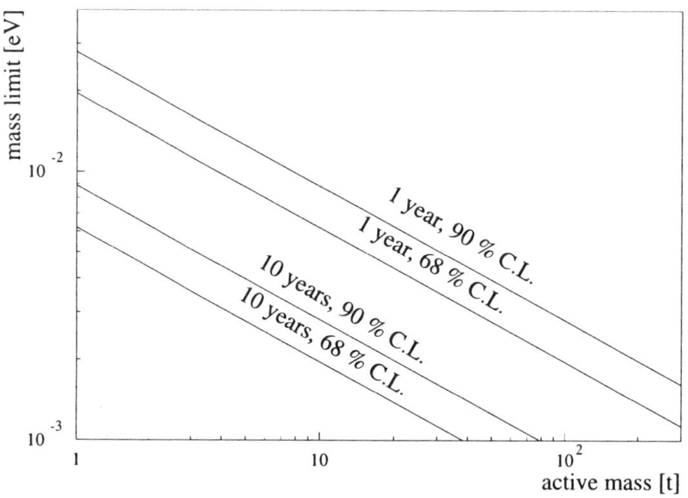

Fig. 20 *Mass limits on Majorana neutrino mass after one and ten years measuring time as function of the active detector mass; zero background is assumed (from [Hel97])*

large step in sensitivity for probing the neutrino mass. It will allow to probe the neutrino mass down to $10^{-(2-3)}$ eV, and thus surpass the existing neutrino mass experiments by a factor of 50-500. GENIUS will test the structure of the neutrino mass matrix and thereby also neutrino oscillation parameters [1] superior in sensitivity to the best proposed dedicated terrestrial neutrino oscillation experiments. Even in the first stage GENIUS will confirm or rule out degenerate or inverted neutrino mass scenarios, discussed in the literature as possible solutions of current hints to finite neutrino masses , and also test the $\nu_e \leftrightarrow \nu_\mu$ hypothesis of the atmospheric neutrino problem. If the 10^{-3} eV level is reached, GENIUS will even allow to test the large angle MSW solution of the solar neutrino problem. It will also allow to test the hypothesis of a shadow world underlying introduction of a sterile neutrino mentioned in section 2.1. The figures 21–25 show some examples of this potential. Fig. 21 compares the potential of GENIUS with the sensitivity of CHORUS/NOMAD and with the proposed future experiments NAUSIKAA–CERN and NAUSIKAA–FNAL, looking for $\nu_e \leftrightarrow \nu_\tau$ oscillations, for different assumptions on m_1/m_2.

Already in the worst case for double beta decay of $m_1/m_2 = 0$ GENIUS 1 ton is more sensitive than the running CERN experiments. For quasi–degenerate models, for example $R = 0.01$ already, GENIUS 1 ton would be more sensitive than all currently planned future accelerator neutrino experiments.

The situation of $\nu_e \leftrightarrow \nu_\mu$ oscillations (assuming $sin^2\theta_{13} = 0$) is shown in Fig. 22. The original figure is taken from [Gel95]. While the GENIUS 1 ton sensitivity is sufficient (even in the worst case of $m_{\nu_e} << m_{\nu_\mu}$) to extend to smaller values of Δm^2 at large mixing angles, GENIUS 10 ton would have a sensitivity better than all existing or planned oscillation experiments, at least at large $sin^2 2\theta$. In the quasi–degenerate models GENIUS would be much more sensitive – similar to the cases shown in Fig. 21. Fig. 23 (background from [Gel95]) compares the double beta worst

[1] The double beta observable, the effective neutrino mass (eq. 10), can be expressed in terms of the usual neutrino oscillation parameters, once an assumption on the ratio of m_1/m_2 is made. E.g., in the simplest two–generation case

$$\langle m_\nu \rangle = |c_{12}^2 m_1 + s_{12}^2 m_2 e^{2i\beta}|, \tag{54}$$

assuming CP conservation, i.e. $e^{2i\beta} = \eta = \pm 1$, and $c_{12}^2 m_1 << \eta s_{12}^2 m_2$,

$$\Delta^2_{m_{12}} \simeq m_2^2 = \frac{4\langle m_\nu \rangle^2}{1 - \sqrt{1 - sin^2 2\theta}} \tag{55}$$

A little bit more general, keeping corrections of the order (m_1/m_2) one obtains

$$m_2 = \frac{\langle m_\nu \rangle}{|(\frac{m_1}{m_2}) + \frac{1}{2}(1 - \sqrt{1 - sin^2 2\theta})(\pm 1 - (\frac{m_1}{m_2}))|}. \tag{56}$$

For the general case see [Kla97d].

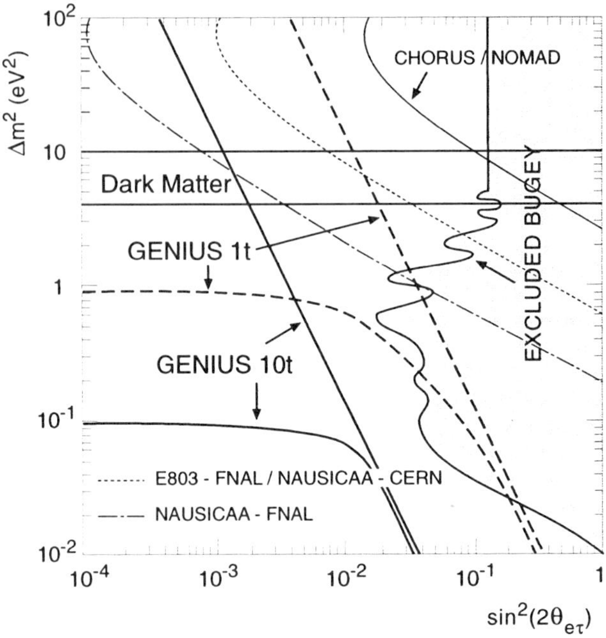

Fig. 21 *Current limits and future experimental sensitivity on $\nu_e - \nu_\tau$ oscillations. The shaded area is currently excluded from reactor experiments. The thin line is the estimated sensitivity of the CHORUS/NOMAD experiments. The dotted and dash-dotted thin lines are sensitivity limits of proposed accelerator experiments, NAUSICAA and E803-FNAL [Gon95]. The thick lines show the sensitivity of GENIUS (broken line: 1 t, full line: 10 t), for two examples of mass ratios. The straight lines are for the strongly hierarchical case (R=0), while the lines bending to the left assume R=0.01. (from [Kla97d])*

case of strong hierarchy ($m_1/m_2 = 0$), to the KAMIOKANDE allowed range for atmospheric neutrino oscillations. GENIUS 1 ton would already be able to test the $\nu_e \leftrightarrow \nu_\mu$ oscillation hypothesis. Fig. 24 shows the potential of GENIUS for checking the LSND indication for neutrino oscillations (original figure from [Ath96]). Under the assumption $m_1/m_2 \geq 0.02$ and $\eta = 1$, GENIUS 1 ton will be sufficient to find $0\nu\beta\beta$ decay if the LSND result is to be explained in terms of $\nu_e \leftrightarrow \nu_\mu$ oscillations. This might be of particular interest also since the upgraded KARMEN will not completely cover [Dre97] the full allowed LSND range. Fig. 25 shows a summary of currently known constraints on neutrino oscillation parameters (original taken from [Hat94]), but including the $0\nu\beta\beta$ decay sensitivities of GENIUS

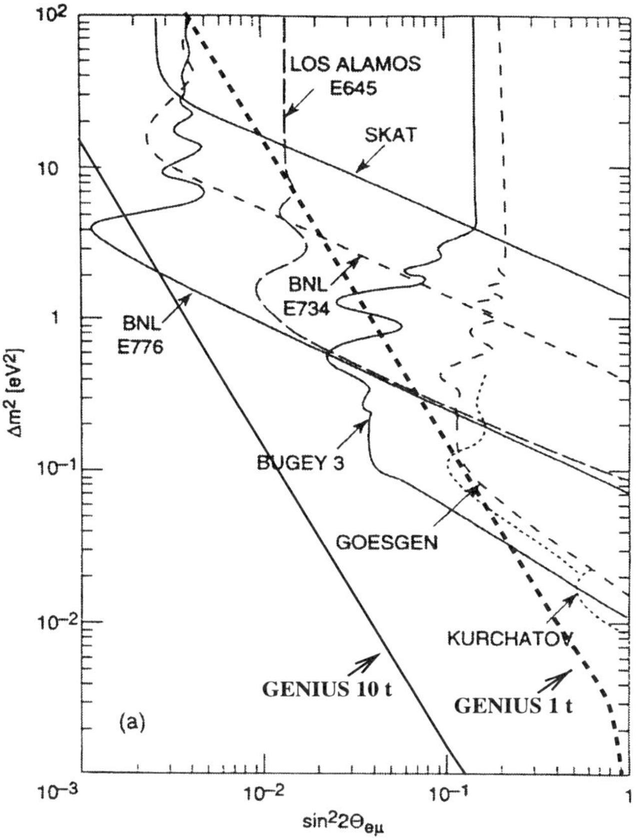

Fig. 22 *Current limits on $\nu_e - \nu_\mu$ oscillations. Various existing experimental limits from reactor and accelerator experiments are indicated, as summarized in ref. [Gel95]. In addition, the figure shows the expected sensitivities for GENIUS with 1 ton (thick broken line) and GENIUS with 10 tons (thick, full line) (from [Kla97d])*

1 ton and GENIUS 10 tons, for different assumptions on m_1/m_2 (and for $\eta^{CP} = +1$). It is seen that already GENIUS 1 ton tests all degenerate or quasi–degenerate ($m_1/m_2 \geq\sim 0.01$) neutrino mass models in any range where neutrinos are interesting for cosmology, and also the atmospheric neutrino problem, if it is due to $\nu_e \leftrightarrow \nu_\mu$ oscillations. GENIUS in its 10 ton version would directly test the large angle solution of the solar neutrino problem.

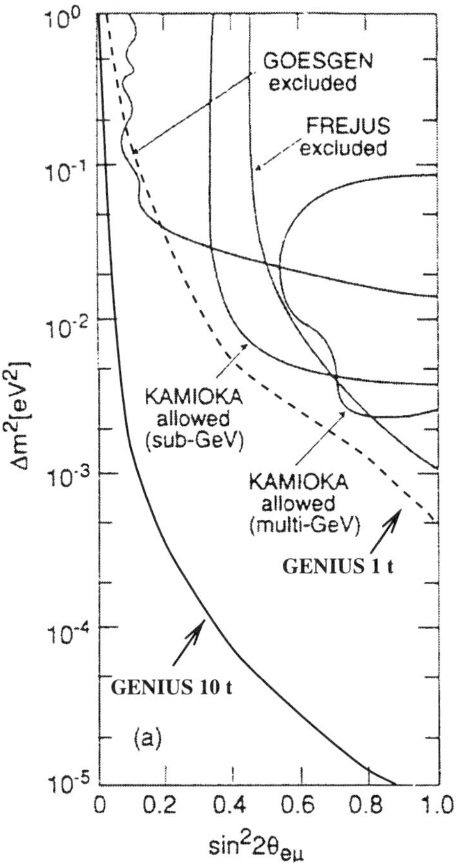

Fig. 23 *Oscillation parameters which solve the atmospheric neutrino problem for $\nu_e \leftrightarrow \nu_\mu$ oscillations. In addition the best currently existing reactor constraints are shown. GENIUS would be able to test the atmospheric neutrino problem already with 1 ton, already in the shown, worst strong hierarchy scenario ($m_1/m_2 = 0$) (from [Kla97d]).*

GENIUS and left–right symmetry: If GENIUS is able to reach down to $\langle m_\nu \rangle \leq 0.01$ eV, it would at the same time be sensitive to right-handed W-boson masses up to $m_{W_R} \geq 8$ TeV (for a heavy right-handed neutrino mass of 1 TeV) or $m_{W_R} \geq 5.3$ TeV (at $\langle m_N \rangle = m_{W_R}$). Such a limit would be comparable to the one expected for LHC, see for example [Riz96], which quotes a final sensitivity of something like 5 – 6 TeV. Note, however that in order to obtain such a limit the experiments at LHC need to accumulate

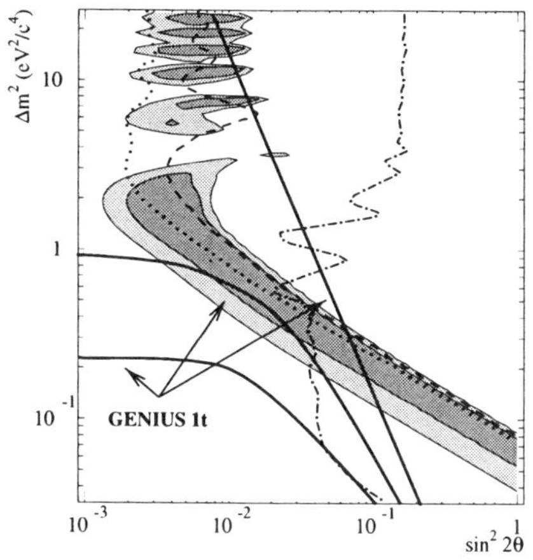

Fig. 24 *LSND compared to the sensitivity of GENIUS 1t for* $\eta^{CP} = +1$ *and three ratios* R_{12}*, from top to bottom* $R_{12} = 0, 0.01, 0.02$ *(from [Kla97d]*

about $100 fb^{-1}$ of statistics. A 10 ton version of GENIUS could even reach a sensitivity of $m_{W_R} \geq 18$ TeV (for a heavy right-handed neutrino mass of 1 TeV) or $m_{W_R} \geq 10.1$ TeV (at $\langle m_N \rangle = m_{W_R}$).

This means that already GENIUS 1 ton could be sufficient to definitely test recent supersymmetric left–right symmetric models having the nice features of solving the strong CP problem without the need for an axion and having automatic R–parity conservation [Kuc95, Moh96].

GENIUS and R_p–violating SUSY: The improvement on the R–parity breaking Yukawa coupling λ'_{111} (see section 2.2) is shown in Fig. 26, which updates Fig. 15. The full line to the right is the expected sensitivity of the LHC – in the limit of large statistics. The three dashed–dotted lines denote (from top to bottom) the current constraint from the Heidelberg–Moscow experiment and the sensitivity of GENIUS 1 ton and GENIUS 10 tons, all for the conservative case of a gluino mass of 1 TeV. If squarks would be heavier than 1 TeV, LHC could not compete with GENIUS. However, for typical squark masses below 1 TeV, LHC could probe smaller couplings. However, one should keep in mind, that LHC can probe squark masses up to 1 TeV only with several years of data taking.

GENIUS and R_p–conserving SUSY: Since the limits on a 'Majorana–like' sneutrino mass $\tilde{m}_M$ scale with $(T_{1/2})^{1/4}$, GENIUS 1 ton (or 10 tons) would test 'Majorana' sneutrino masses lower by factors of about 7(20), compared with present constraints [Hir97, Hir98a, Hir97b].

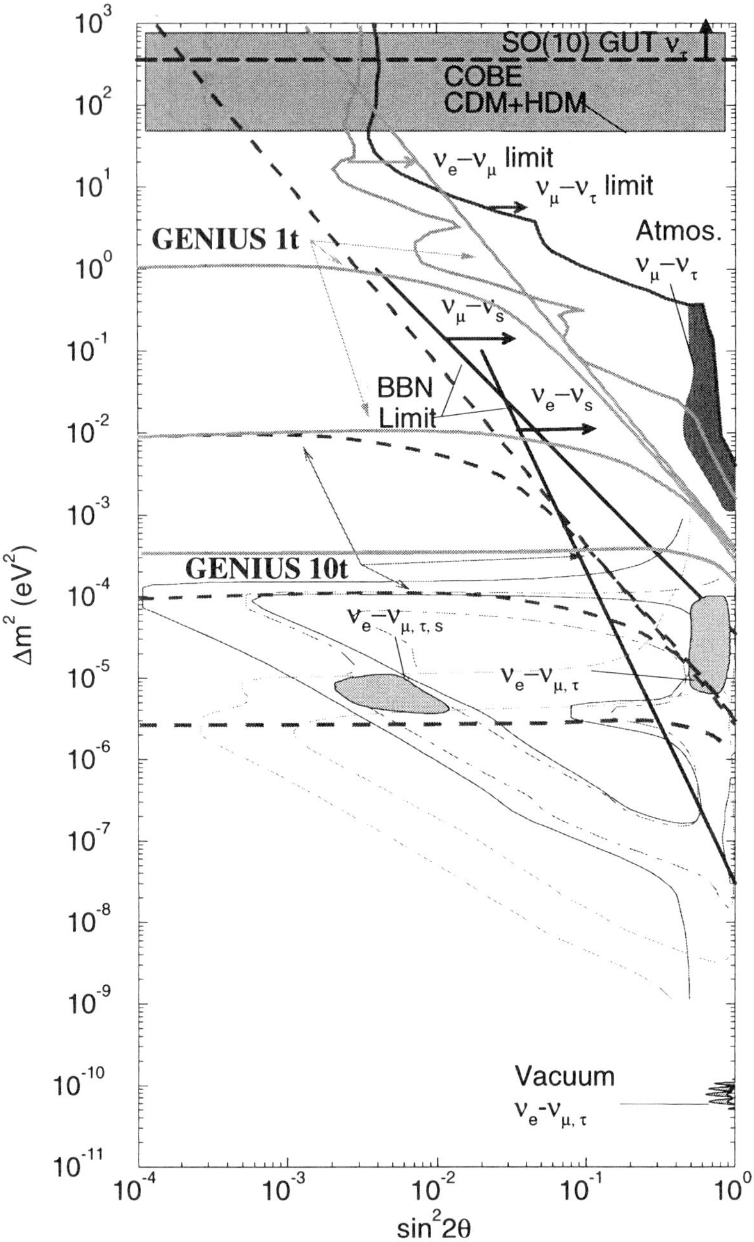
SO(10) GUT ν_τ
COBE
CDM+HDM
$\nu_e-\nu_\mu$ limit
$\nu_\mu-\nu_\tau$ limit
GENIUS 1t
Atmos.
$\nu_\mu-\nu_\tau$
$\nu_\mu-\nu_s$
BBN
Limit
$\nu_e-\nu_s$
GENIUS 10t
$\nu_e-\nu_{\mu,\tau,s}$
$\nu_e-\nu_{\mu,\tau}$
Vacuum
ν_e-$\nu_{\mu,\tau}$
Δm^2 (eV^2)
$\sin^2 2\theta$
10^3
10^2
10^1
10^0
10^{-1}
10^{-2}
10^{-3}
10^{-4}
10^{-5}
10^{-6}
10^{-7}
10^{-8}
10^{-9}
10^{-10}
10^{-11}
10^{-4}
10^{-3}
10^{-2}
10^{-1}
10^0

Fig. 25 *Summary of currently known constraints on neutrino oscillation parameters. The (background) figure without the $0\nu\beta\beta$ decay constraints can be obtained from http://dept.physics.upenn.edu/ www/neutrino/solar.html. Shown are the vacuum and MSW solutions (for two generations of neutrinos) for the solar neutrino problem, the parameter range which would solve the atmospheric neutrino problem and various reactor and accelerator limits on neutrino oscillations. In addition, the mass range in which neutrinos are good hot dark matter candidates is indicated, as well as limits on neutrino oscillations into sterile states from considerations of big bang nucleosynthesis. Finally the thick lines indicate the sensitivity of GENIUS (full lines 1 ton, broken lines 10 ton) to neutrino oscillation parameters for three values of neutrino mass ratios $R = 0, 0.01$ and 0.1 (from top to bottom). For GENIUS 10 ton also the contour line for $R = 0.5$ is shown. The region beyond the lines would be excluded. While already the 1 ton GENIUS would be sufficient to constrain degenerate and quasi-degenerate neutrino mass models, and also would solve the atmospheric neutrino problem if it is due to $\nu_e \leftrightarrow \nu_\mu$ oscillations, the 10 ton version of GENIUS could cover a significant new part of the parameter space, including the large angle MSW solution to the solar neutrino problem, even in the worst case of $R = 0$. For $R \geq 0.5$ it would even probe the small angle MSW solution (see [Kla98g, Kla98e]).*

GENIUS and Leptoquarks: Limits on the lepton–number violating parameters defined in sections 2.7, 3.2 improve as $\sqrt{T_{1/2}}$. This means that for leptoquarks in the range of 200 GeV LQ–Higgs couplings down to (a few) 10^{-8} could be explored. In other words, if leptoquarks interact with the standard model Higgs boson with a coupling of the order $\mathcal{O}(1)$, either $0\nu\beta\beta$ must be found, or LQs must be heavier than (several) 10 TeV.

GENIUS and composite neutrinos GENIUS in the 1(10) ton version would improve the limit on the excited Majorana neutrino mass deduced from the Heidelberg–Moscow experiment (eq. 32) to

$$m_N \geq\sim 1.1(2.3) \quad TeV \tag{57}$$

4.2.3. GENIUS, special relativity and equivalence principle in the neutrino sector

The already now strongest limits given by the Heidelberg–Moscow experiment discussed in section 3.2 would be improved by 1–2 orders of magnitude. It should be stressed again, that while neutrino oscillation bounds constrain the region of large mixing of the weak and gravitational eigenstates, these bounds from double beta decay apply even in the case of no mixing and thus probe a totally unconstrained region in the parameter space.

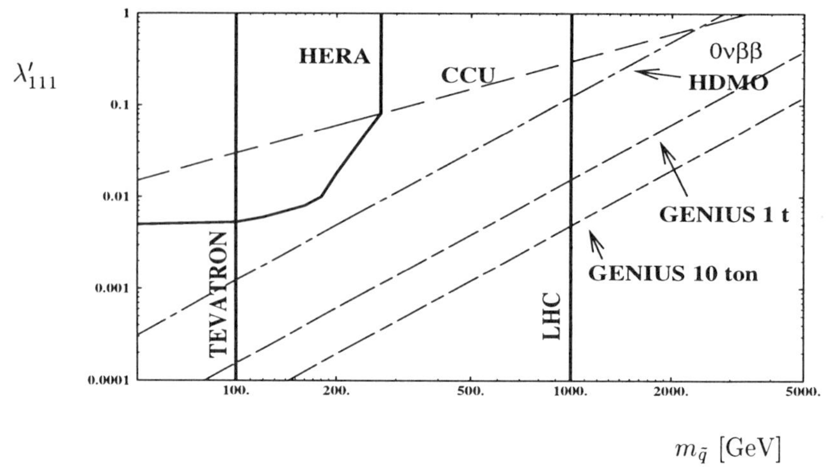

Fig. 26 *Comparison of sensitivities of existing and future experiments on* $R\!\!\!/_p$*SUSY models in the plane* $\lambda'_{111} - m_{\tilde{q}}$*. Note the double logarithmic scale! Shown are the areas currently excluded by the experiments at the TEVATRON, the limit from charged-current universality, denoted by CCU, and the limit from absence of* $0\nu\beta\beta$ *decay from the Heidelberg-Moscow collaboration (*$0\nu\beta\beta$ *HDMO). In addition, the estimated sensitivity of HERA and the LHC is compared to the one expected for GENIUS in the 1 ton and the 10 ton version. The figure is essentially an update of Fig. 15.*

4.2.4. GENIUS and dark matter

Neutrinos as hot dark matter

If neutrinos have masses in the range of a few eV, they would be good candidates for the hot dark matter in the universe. Of course, from the dark matter argument itself it does not follow which neutrino has to be in this mass range. Clearly, if a neutrino with sizeable mixing angle to the electron neutrino in this mass range exists, one expects GENIUS to find $0\nu\beta\beta$ decay.

However, if the ν_τ is in the eV range, the ν_e and ν_μ being lighter by at least factors of hundreds and the the $\nu_\tau - \nu_e$ mixing angle small at the same time GENIUS with 1 ton would not find double beta decay. In the case of quasidegenerate models or degenerate models, on the other hand, $0\nu\beta\beta$ decay should be found by GENIUS, unless the CP–phases between the different mass eigenstates take on some special combinations and have a relative minus sign, see the discussion in [Kla98c].

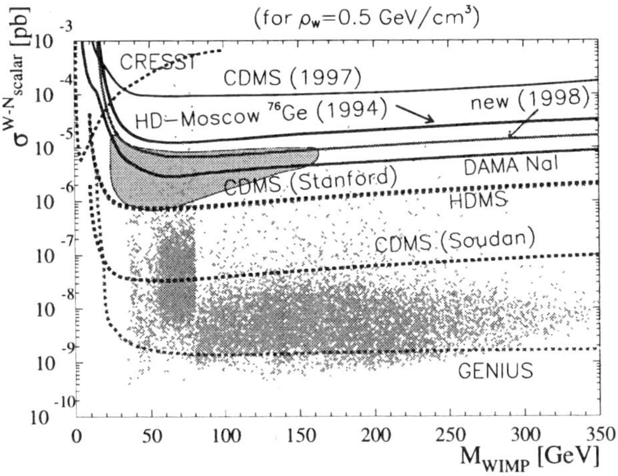

Fig. 27 *WIMP–nucleon cross section limits in pb for scalar interactions as function of the WIMP–mass in GeV. Regions beyond solid lines are excluded by experiment [HM94, HM98, Ber97, Ake97]. Further shown are expected sensitivities of experiments under construction (dashed lines for HDMS [Bau97, Kla97e], CDMS [Ake97], CRESST and for GENIUS). These limits are compared to theoretical expectations (scatter plot) for WIMP–neutralino cross sections calculated in the MSSM framework with non–universal scalar mass unification [Bed97b]. The 90 % allowed region claimed by [Ber97a] (light filled area), which is further restricted by indirect dark matter searches [Bot97] (dark filled area), could already be easily tested with a 100 kg version of the GENIUS experiment.*

Cold Dark Matter

Weakly interacting massive particles (WIMPs) are candidates for the cold dark matter in the universe. The favorite WIMP candidate is the lightest supersymmetric particle, presumably the neutralino. The expected detection rates for neutralinos of typically less than one event per day and kg of detector mass [Bed94, Bed97a, Bed97b, Jun96], however, make direct searches for WIMP scattering experimentally a formidable task.

Fig. 27 shows a comparison of existing constraints and future sensitivities of cold dark matter experiments, together with the theoretical expectations for neutralino scattering rates [Bed97b]. Obviously, GENIUS could easily cover the range of positive evidence for dark matter recently claimed by DAMA [Ber97a, Bot97]. It would also be by far more sensitive than all other dark matter experiments at present under construction or proposed, like the cryogenic experiment CDMS. Furthermore, obviously GENIUS will be the only experiment, which could seriously test the MSSM predictions over the whole SUSY parameter space. In this way, GENIUS could compete even with LHC in the search for SUSY, see for example the discussion in [Bae97]. It is important to note, that GENIUS could reach the sensitivity shown in Fig. 26 with only 100 kg of *natural* Ge detectors in a measuring time of three years [Kla98d].

It is interesting to note, that if WIMP scattering is found by GENIUS it could be used to constrain the amount of R-parity violation within supersymmetric models. The arguments are very simple [Hir97c]. Due to the fact that neutralinos are abound in the galaxy even today, neutralino decays via R-parity violating operators would have to be highly suppressed.

The details depend, of course, on the neutralino mass and composition. However, finding the neutralino with GENIUS would imply typical limits on R-parity violating couplings of the order of $10^{-(16-20)}$ for any of the λ_{ijk}, λ'_{ijk} or λ''_{ijk} in the superpotential (eq. 11). A positive result of the CDM search at hand, one could thus finally safely conclude that R-parity is conserved.

5. Conclusion

Double beta decay has a broad potential for providing important information on modern particle physics beyond present and future high energy accelerator energies which will be competitive for the next decade and more. This includes SUSY models, compositeness, left–right symmetric models, leptoquarks, the neutrino and sneutrino mass and tests of Lorentz invariance and equivalence principle in the neutrino sector. Based to a large extent on the theoretical work of the Heidelberg Double Beta group, results have been deduced from the HEIDELBERG–MOSCOW experiment for these topics and have been presented here. For the neutrino mass double beta decay now is particularly pushed into a key position by the recent possible indications of beyond standard model physics from the side of solar and atmospheric neutrinos, dark matter COBE results and others. New classes of GUTs basing on degenerate neutrino mass scenarios which could explain these observations, can be checked by double beta decay in near future. The HEIDELBERG–MOSCOW experiment has reached a leading position among present $\beta\beta$ experiments and as the first of them now yields results in the sub–eV range. We have described a new idea and proposal of a future double beta experiment (GENIUS) with highly increased sensitivity based on use of 1 ton or more of enriched 'naked' ^{76}Ge detectors in liquid nitrogen. This new experiment would be a breakthrough into the multi-TeV range for many beyond standard models. The sensitivity for the neutrino mass would reach down to 0.01 or even 0.001 eV. The experiment would be competitive to LHC with respect to the mass of a right–handed W boson, in search for R–parity violation and others, and would improve the leptoquark and compositeness searches by considerable factors. It would probe the Majorana electron sneutrino mass more sensitive than NLC (Next Linear Collider). It would yield constraints on neutrino oscillation parameters far beyond all present terestrial $\nu_e - \nu_x$ neutrino oscillation experiments and could test directly the atmospheric neutrino problem and the large and, for degenerate models, even the small angle solution of the solar neutrino problem. GENIUS would cover the full SUSY parameter space for prediction of neutralinos as cold dark matter and compete in this way with LHC in the search for supersymmetry. Even if SUSY would be first observed by LHC, it would still be fascinating to verify the existence and properties of neutralino dark matter, which could be achieved by GENIUS. Concluding GENIUS has the ability to provide a major tool for future particle– and astrophysics.

Finally it may be stressed that the technology of producing and using enriched high purity germanium detectors, which have been produced for the first time for the Heidelberg–Moscow experiment, has found meanwhile applications also in pre-GENIUS dark matter search [HM94, Fal94, Kla97e, Bau97] and in high–resolution γ-ray astrophysics, using balloons and satellites [Kla91, Kla94, Bar93, Bar94, Boc94, Kla97b].

References

[Abe94] F. Abe *et al.* (CDF Collab.), Phys. Rev. Lett. 72 (1994) 3004

[Adr92] O. Adriani *et al.*, Phys. Lett. **B 288** (1992) 404

[Ake97] D.S. Akerib et al. (CDMS collab.), preprint astro-ph/9712343

[ALE92] ALEPH Collab., Phys. Rep. 216 (1992) 343

[ALE93] ALEPH Collab., Z.Phys. **C 59** (1993) 215

[Ale94] A. Alessandrello et al., Phys. Lett **B 335** (1994) 519

[Als89] M. Alston-Garnjost et al., Phys. Rev Lett. **71** (1993) 831

[Alt97] G. Altarelli, J. Ellis, G.F. Guidice, S. Lola, M.L. Mangano, preprint hep-ph/9703276

[Ath95] C. Athanassopoulos *et al.* (LSND–Collaboration), Phys. Rev. Lett. **75** (1995) 2650 and D. Smith, in [Kla96]

[Ath96] C. Athanassopoulos *et al.*, LSND collab., Phys. Rev. **C 54** (1996) 2685, Phys. Rev. Lett. **77** (1996) 3082

[Aul82] C.S. Aulakh and R.N. Mohapatra, Phys. Lett. **B 119** (1982) 136

[Bab95] K.S. Babu, R.N. Mohapatra, Phys. Rev. Lett. **75** (1995) 2276

[Bab97] K.S. Babu et al., preprint hep-ph/9703299 (March 1997)

[Bab97b] K.S. Babu *et al.*, preprint hep-ph/9705414v2 (1997)

[Bae97] H. Baer. M. Bhrlik, hep-ph/9706509

[Bam95a] P. Bamert, C.P. Burgess, R.N. Mohapatra, Nucl. Phys. B 438 (1995) 3

[Bam95] P. Bamert, C.P. Burgess, R.N. Mohapatra, Nucl. Phys. **B 449** (1995) 25

[Bar89] V. Barger, G.F. Guidice, T. Han. Phys.Rev. **D 40** (1989) 2987

[Bar93] S.D. Barthelmy *et al.*, AIP Conf. Proc. 280 (1993) 1166

[Bar94] S. D. Barthelmy et al., Astrophys. J. **427** (1994) 519

[Bar96] P.B. Barnes *et al.*, Nucl. Instr. Meth. **A 370** (1996) 233

[Bar97] A.S. Barabash, Proc. NEUTRINO 96, Helsinki, June 1996, World Scientific, Singapore (1997), p. 374

[Bau97] L. Baudis, J. Hellmig, H.V. Klapdor–Kleingrothaus, A. Müller, F. Petry, Y. Ramachers, H. Strecker, Nucl. Inst. Meth. **A 385** (1997) 265

[Bav95] E. Baver, M. Leurer, Phys. Rev. **D 51** (1995) 260

[Bed94] V. Bednyakov, H.V. Klapdor–Kleingrothaus, S.G. Kovalenko, Phys. Lett. **B 329** (1994) 5 and Phys. Rev. **D 50** (1994) 7128

[Bed97a] V. Bednyakov, H.V. Klapdor–Kleingrothaus, S.G. Kovalenko, Phys. Rev. **D 55** (1997) 503

[Bed97b] V. Bednyakov, H.V. Klapdor–Kleingrothaus, S.G. Kovalenko, Y. Ramachers, Z. Phys. **A 357** (1997) 339

[Bed97c] V. Bednyakov, V.B. Brudanin, S.G. Kovalenko, Ts. D. Vylov, Mod. Phys. Lett **A 12** (1997) 233

[Bel96] G. Belanger, F. Boudjema, D.London, H. Nadeau, Phys. Rev. **D 53** (1996) 6292

[Bel98] G. Belanger, these Proceedings

[Ber92] Z.G. Berezhiani, A. Yu. Smirnov, J.W.F. Valle, Phys.Lett. **B 291** (1992) 99

[Ber95] Z. Berezhiani, R.N. Mohapatra, Phys. Rev. **D 52** (1995) 6607

[Ber97] R. Bernabei *et al.*, Phys. Lett. **B 389** (1997) 757

[Ber97a] R. Bernabei *et al.*, ROM2F/97/33; P. Belli at TAUP, Gran Sasso, Sept. 7 (1997)

[Bha97] G. Bhattacharyya, in [Kla98]

[Blü94] J. Blümlein, E. Boos, A. Pukhov, preprint DESY 94-072 (April 1994)

[Boc94] J. Bockholt, H.V. Klapdor–Kleingrothaus, Nucl. Phys. **B** (Proc. Suppl.) 35 (1994) 403

[Bot97] A. Bottino *et al.*, hep-ph/9709292

[Buc87] W. Buchmüller, R. Rückl and D. Wyler, Phys.Lett. B191 (1987) 442.

[Buc91] W. Buchmüller and G. Ingelman, Proc. Workshop Physics at HERA, Hamburg, Oct. 29–30 (1991)

[Bur93] C.P. Burgess, J.M. Cline, Phys. Lett. **B 298** (1993) 141; Phys. Rev. **D 49** (1994) 5925

[Bur96] C.P.Burgess, in [Kla96]

[But93] M. N. Butler *et al.*, Phys. Rev. **D47**, 2615 (1993); A. Halprin and C. N. Leung, Phys. Rev. Lett. **67**, 1833 (1991); J. Pantaleone, A. Halprin, and C. N. Leung, Phys. Rev. **D47**, R4199 (1993); K. Iida, H. Minakata and O. Yasuda, Mod. Phys. Lett. **A8** (1993) 1037.

[But93a] J. Butterworth, H. Dreiner, Nucl. Phys. **B397** (1993) 3 and H. Dreiner, P. Morawitz, Nucl. Phys. **B428** (1994) 31

[Cal93] D.O. Caldwell, R.N. Mohapatra, Phys. Rev. **D 48** (1993) 3259

[Cal95] D.O. Caldwell, R.N. Mohapatra, Phys. Lett. **354** (1995) 371

[Car93] C.D. Carone, Phys. Lett. **B 308** (1993) 85

[Cau96] E. Caurier et al., Phys. Rev. Lett. **77** (1996) 1954

[CDF91] CDF Collab., Phys. Rev. Lett. **67** (1991) 2148

[Chi81] Y. Chikashige, R.N. Mohapatra and R.D. Peccei, Phys. Lett. **B 98** (1981) 265; Phys. Rev. Lett. **45** (1980) 265

[Cho94] D. Choudhury, hep-ph/9408250 (1994)

[Cho97] D. Choudhury, S. Raychaudhuri, preprint hep-ph/9702392

[Cli97] D. Cline, in: Proceedings of the *International Workshop Dark Matter in Astro- and Particle Physics* (DARK96), Eds. H.V. Klapdor-Kleingrothaus, Y. Ramachers, World Scientific 1997, p. 479

[Col97] S. Coleman and S.L. Glashow, Phys. Lett. **B 405**, 249 (1997).

[Dan95] F. A. Danevich et al., Phys. Lett. **B 344** (1995) 72

[Dav94] S. Davidson, D. Bailey and A. Campbell, Z.Phys. C61 (1994) 613; M. Leurer, Phys.Rev.Lett. 71 (1993) 1324; Phys.Rev. D50 (1994) 536.

[Der95] M. Derrik, *et al.* (ZEUS Collab.), Z. Phys. **C 65** (1995) 627

[Dia93] J.L. Diaz Cruz, O.A. Sampayo, Phys. Lett. **B 306** (1993) 395; Phys. Rev. **D 49** (1994) R2149

[Doi85] M. Doi, T. Kotani, E.Takasugi, Progr. Theor. Phys. Suppl. **83** (1985) 1

[Doi93] M. Doi, T. Kotani, Progr. Theor. Phys. 89 (1993) 139

[Dre97] G. Drexlin, Proc. Internat. School on Neutrino Physics, Erice, Italy, Sept. 1997, to be publ. in Plenum Press

[Ell87] S.R. Elliot, A.A. Hahn, M.K. Moe, Phys. Rev. Lett. **59** (1987) 1649

[Ell92] S. R. Elliott et al., Phys. Rev. **C 46** (1992) 1535

[Fal94] T. Falk, A. Olive, M. Srednicki, Phys. Lett. **B339** (1994) 248

[Fri88] J.Friemann, H.Haber, K.Freese, Phys. Lett. **B 200** (1988) 115; J. Bahcall, S. Petcov, S. Toshev and J.W.F. Valle, Phys.Lett. **B 181** (1986) 369; Z. Berezhiani and M. Vysotsky, Phys. Lett. **B 199** (1988) 281.

[Fri95] H. Fritzsch and Zhi-zhong Xing, preprint hep-ph/9509389, Phys. Lett. **B 372** (1996) 265

[Gas89] M. Gasperini, Phys. Rev. **D38**, 2635 (1988); *ibid.* **D39**, 3606 (1989).

[Gas89a] M. Gasperini, Phys. Rev. Lett. **62**, 1945 (1989).

[Gel81] G.B. Gelmini and M. Roncadelli, Phys. Lett. **B 99** (1981) 411

[Gel95] G. Gelmini, E. Roulet, Rep. Progr. Phys. **58**(1995) 1207

[Geo81] H.M. Georgi, S.L.Glashow and S. Nussinov, Nuc. Phys. **B 193** (1981) 297

[Ger96] G. Gervasio, in [Kla96]

[Gla97] S.L. Glashow, A. Halprin, P.I. Krastev, C.N. Leung, and J. Panteleone, Phys. Rev. **D 56**, 2433 (1997).

[Gon95] M. Gonzalez–Garcia, hep-ph/9510419

[Goo61] M.L. Good, Phys. Rev. **121**, 311 (1961); O. Nachtmann, Acta Physica Austriaca, Supp. VI *Particle Physics* ed. P. Urban, (1969) p. 485; S.H. Aronson, G.J. Bock, H-Y Cheng and E. Fishbach, **48**, 1306 (1982); Phys. Rev. **D28**, 495 (1983); I.R. Kenyon, Phys. Lett. **B237**, 274 (1990); R.J. Hughes, Phys. Rev. **D46**, R2283 (1992);

[Gro85] K. Grotz, H. V. Klapdor, Phys. Lett. **B 157** (1985) 242

[Gro86] K. Grotz, H.V. Klapdor, Nucl. Phys **A 460** (1986) 395

[Gro90] K. Grotz, H.V. Klapdor, 'The Weak Interaction in Nuclear, Particle and Astrophysics' (Adam Hilger: Bristol, Philadelphia) 1990

[H195] H1 Collab., Phys. Lett. **B 353** (1995) 578, Z. Phys. **C 64** (1994) 545

[H196] H1 Collab., S. Aid et al., Phys. Lett. B 369 (1996) 173

[Hab93] H. E. Haber, in Proc. on Recent Advances in the Superworld, Houston, April 14-16 (1993), hep-ph/9308209

[Hal84] L. Hall, M. Suzuki, Nuclear Physics **B 231** (1984) 419

[Hal91] A. Halprin and C.N. Leung, Phys. Rev. Lett. **67**, 1833 (1991); Nucl. Phys. **B28A** (Proc. Supp.), 139 (1992); J. N. Bahcall, P. I. Krastev, and C. N. Leung, Phys. Rev. **D 52**, 1770 (1995); R.B. Mann and U. Sarkar, Phys. Rev. Lett **76** (1996) 865;

[Hal96] A. Halprin, C.N. Leung, J. Pantalone, Phys.Rev. **D 53** (1996) 5365

[Ham98] T. Hambye, R.B. Mann and U. Sarkar, Phys. Lett. **B 421**, 105 (1998); Phys. Rev. **D 58**, 025003 (1998).

[Hat94] N. Hata, P. Langacker, Phys. Rev. **D 50** (1994) 632

[Hax84] W.C. Haxton, G.J. Stephenson, Progr. Part. Nucl. Phys. 12 (1984) 409

[Hel96] J. Hellmig, PhD Thesis, Univ. of Heidelberg, 1996

[Hel97] J. Hellmig, H.V. Klapdor–Kleingrothaus, Z. Phys. **A 359** (1997) 351

[Hew97] J.L. Hewett, T.G. Rizzo, preprint hep-ph/9703337v3 (May1997)

[Heu95] G. Heusser, Ann. Rev. Nucl. Part. Sci. **45** (1995) 543

[Hir95] M. Hirsch, H.V. Klapdor–Kleingrothaus, S.G. Kovalenko, Phys. Rev. Lett. 75 (1995) 17

[Hir95b] M. Hirsch, H.V. Klapdor–Kleingrothaus, S.G. Kovalenko, in [Kla96]

[Hir95c] M. Hirsch, H.V. Klapdor–Kleingrothaus, S. Kovalenko, Phys. Lett. **B 352** (1995) 1

[Hir95d] J. Hirsch, O. Castanos, P.O. Hess, Nucl. Phys. **A 582** (1995) 124

[Hir96] M. Hirsch, H.V. Klapdor–Kleingrothaus, S.G. Kovalenko, Phys. Lett. **B 372** (1996) 181, Erratum: Phys. Lett. **B381** (1996) 488

[Hir96a] M. Hirsch, H.V. Klapdor–Kleingrothaus, S.G. Kovalenko, Phys. Lett. **B 378** (1996) 17 and Phys. Rev. **D 54** (1996) R4207

[Hir96b] M. Hirsch, H. V. Klapdor–Kleingrothaus, S. G. Kovalenko, H. Päs, Phys. Lett. **B 372** (1996) 8

[Hir96c] M. Hirsch, H.V. Klapdor–Kleingrothaus, S. Kovalenko, Phys. Rev. **D 53** (1996) 1329

[Hir96d] M. Hirsch, H.V. Klapdor–Kleingrothaus, in [Kla96]; M. Hirsch, H.V. Klapdor–Kleingrothaus, O. Panella, Phys. Lett. **B 374** (1996) 7

[Hir97] M. Hirsch, H.V. Klapdor–Kleingrothaus, S.G. Kovalenko, Phys. Lett. **B 398** (1997) 311 and **403** (1997) 291

[Hir97b] M. Hirsch, H.V. Klapdor–Kleingrothaus, S. Kovalenko, in [Kla98]

[Hir97c] M. Hirsch, H.V. Klapdor–Kleingrothaus, Proc. Int. Workshop on Dark Matter in Astro– and Particle Physics (DARK96), Heidelberg, Sept. 1996, Eds. H.V. Klapdor–Kleingrothaus and Y. Ramachers (World Scientific, Singapore) 1997, p. 640

[Hir98] M. Hirsch, H.V. Klapdor–Kleingrothaus, St. Kolb, S.G. Kovalenko, Phys. Rev. **D 57** (1998) 2020

[Hir98a] M. Hirsch, H.V. Klapdor–Kleingrothaus, S.G. Kovalenko, Phys. Rev. **D 57** (1998) 1947

[HM94] HEIDELBERG–MOSCOW collab., Phys. Lett. **B 336** (1994) 141

[HM95] HEIDELBERG–MOSCOW collab., Phys. Lett. **B 356** (1995) 450

[HM96] HEIDELBERG–MOSCOW collab., Phys. Rev. **D 54** (1996) 3641,

[HM97] HEIDELBERG–MOSCOW collab., Phys. Rev. **D 55** (1997) 54 and Phys. Lett. **B 407** (1997) 219

[HM98] HEIDELBERG–MOSCOW collab., Phys. Rev. **D** Dec 1998

[Hug60] V.W. Hughes, H.G. Robinson, and V. Beltran-Lopez, Phys. Rev. Lett. **4**, 342 (1960); R.W.P. Drever, Philos. Mag. **6**, 683 (1961); D. Newman, G.W. Ford, A. Rich and E. Sweetman, Phys. Rev. Lett. **40**, 1355 (1978); A. Brillet and J.L. Hall, Phys. Rev. Lett. **42**, 549 (1979); J.D. Prestage, J.J. Bollinger, W.M. Itano, and D.J. Wineland, Phys. Rev. Lett. **54**, 2387 (1985); S.K. Lamoureaux, J.P. Jacobs, B.R. Heckel, R.J. Raab, and E.N. Fortson, Phys. Rev. Lett. **57**, 3125 (1986).

[Ioa94] A. Ioanissyan, J.W.F. Valle, Phys. Lett **B 322** (1994) 93

[Jör94] V. Jörgens et al., Nucl. Phys. (Proc. Suppl.) **B 35** (1994) 378

[Jun96] G. Jungmann, M. Kamionkowski, K. Griest, Phys. Rep. 267 (1996) 195

[Kal97] J. Kalinowski et al., preprint hep-ph/9703288v2 (March 1997)

[Kan97] G. Kane, in [Kla98]

[Kla84] H.V. Klapdor, K. Grotz, Phys.Lett. **B142** (1984) 323

[Kla87] H.V. Klapdor–Kleingrothaus, MPI–H 1987, proposal

[Kla91] H.V. Klapdor–Kleingrothaus, Proc. Int. Symposium on γ-Ray Astrophysics, Paris 1990, AIP Conf. Proc. 232 (1991) 464

[Kla92] H.V. Klapdor–Kleingrothaus, K. Zuber, Phys. Bl. **48** (1992) 1017

[Kla94] H.V. Klapdor–Kleingrothaus, Progr. Part. Nucl. Phys. **32** (1994)261

[Kla95] H. V. Klapdor–Kleingrothaus, A. Staudt, Non–Accelerator Particle Physics, IOP Publ., Bristol, Philadelphia, 1995; and Teilchenphysik ohne Beschleuniger, Teubner Verlag, Stuttgart, 1995

[Kla96] H.V. Klapdor–Kleingrothaus and S. Stoica (Eds.), Proc. *Int. Workshop on Double Beta Decay and Related Topics*, Trento, 24.4.–5.5.95, World Scientific Singapore

[Kla96a] H.V. Klapdor–Kleingrothaus, in Proc. *Int. Workshop on Double Beta Decay and Related Topics*, Trento, 24.4.–5.5.95, World Scientific Singapore 1996, Ed.: H.V. Klapdor–Kleingrothaus and S. Stoica

[Kla97] H.V. Klapdor–Kleingrothaus, Invited talk at NEUTRINO 96, Helsinki, June 1996, World Scientific Singapore 1997, p. 317

[Kla97a] H.V. Klapdor–Kleingrothaus, in Proc. Int. Workshop on Non-Accelerator New Physics (NANP-97), Dubna, Moscow region, Russia, July 7-11, 1997 and in Proc. Int. School on Neutrinos, Erice, Italy, Sept. 1997, Progr. Part. Nucl. Phys. **40** (1998) 265-282

[Kla97b] H.V. Klapdor–Kleingrothaus, M.I. Kudravtsev, V.G. Stolpovski, S.I. Svertilov, V.F. Melnikov, I. Krivosheina, J. Moscow. Phys. Soc. 7 (1997) 41

[Kla97c] H.V. Klapdor–Kleingrothaus and K. Zuber, Particle Astrophysics, IOP Publ., Bristol, Philadelphia 1997; and Teilchenastrophysik, Teubner Verlag, Stuttgart, 1997

[Kla97d] H.V. Klapdor–Kleingrothaus, M. Hirsch, Z. Phys. **A 359** (1997) 361

[Kla97e] H.V. Klapdor–Kleingrothaus, Y. Ramachers, in: Proc. Int. Workshop on Dark Matter in Astro– and Particle Physics (DARK96) Sept. 1996, Eds. H.V. Klapdor–Kleingrothaus and Y. Ramachers, Heidelberg, (World Scientific Singapore) 1997, p. 459

[Kla98] H.V. Klapdor–Kleingrothaus and H. Päs (Eds.), Beyond the Desert – Accelerator- and Non–Accelerator Approaches, IOP, Bristol, 1998

[Kla98a] H.V. Klapdor–Kleingrothaus, in [Kla98]

[Kla98b] H.V. Klapdor–Kleingrothaus, Int. J. Mod. Phys. **A 13** (1998) 3953-3992

[Kla98c] H.V. Klapdor–Kleingrothaus, J. Hellmig, M. Hirsch, J. Phys. **G 24** (1998) 483-516

[Kla98d] H.V. Klapdor–Kleingrothaus, Y. Ramachers, Eur. Phys. J. **A 3** (1998) 85

[Kla98e] H.V. Klapdor–Kleingrothaus, H. Päs; in: Proc. of the 6th Symp. on Particles, Strings and Cosmology (PASCOS'98), Boston/USA, 1998

[Kla98f] H.V. Klapdor–Kleingrothaus, H. Päs, U. Sarkar; submitted to Phys. Rev. Lett. 1998

[Kla98g] H.V. Klapdor–Kleingrothaus, Proc. NEUTRINO'98 (Takayama,Japan,June 1998) World Scientific, Singapore

[Kol97] S. Kolb, M. Hirsch, H.V. Klapdor–Kleingrothaus, S. Kovalenko, in [Kla98]

[Kol97a] S. Kolb, M. Hirsch, H.V. Klapdor–Kleingrothaus, Phys. Rev. **D 56** (1997) 4161

[Kuc95] R. Kuchimanchi, R.N. Mohapatra, Phys. Rev. Lett. **75** (1995) 3939

[Kum96] K. Kume (ELEGANT collaboration), in [Kla96]

[Kuz90] V. Kuzmin, V. Rubakov, M. Shaposhnikov, Phys. Lett. **B 185** (1985) 36; M. Fukugita, T. Yanagida, Phys. Rev. **D 42** (1990) 1285; G. Gelmini, T. Yanagida, Phys. Lett. **B 294** (1992) 53; B. Campbell *et al.*, Phys. Lett. **B 256** (91) 457

[Lan88] P.Langacker, in 'Neutrinos' (Springer: Heidelberg, New York), 1988, ed. H.V. Klapdor, p. 71

[Lee94] D.G. Lee, R.N. Mohapatra, Phys. Lett. **B 329** (1994) 463

[Leu94] M. Leurer, Phys. Rev. **D 49** (1994) 333

[Moe91] M. K. Moe, Phys. Rev. **C 44** (1991) R931

[Moe94] M. K. Moe, Prog. Part. Nucl. Phys. **32** (1994) 247; Nucl. Phys. (Proc. Suppl.) **B 38** (1995) 36

[Moh86] R.N. Mohapatra, Phys. Rev. **D 34** (1986) 3457

[Moh86b] R.N. Mohapatra, Phys. Rev. **D 34** (1986) 909

[Moh88] R.N. Mohapatra and E. Takasugi, Phys. Lett. **B 211** (1988) 192

[Moh91] R.N. Mohapatra, P.B. Pal, 'Massive Neutrinos in Physics and Astrophysics' (World Scientific, Singapore) 1991

[Moh92] R.N. Mohapatra, Unification and Supersymmetry (Springer: Heidelberg, New York) 1986 and 1992

[Moh94] R.N. Mohapatra, Progr. Part. Nucl. Phys. **32** (1994) 187

[Moh95] R.N. Mohapatra, S. Nussinov, Phys. Lett. **B 346** (1995) 75

[Moh96] R.N. Mohapatra, A. Rasin, Phys. Rev. Lett. **76** (1996) 3490 and Phys. Rev. **D54** (1996) 5835

[Moh96a] R.N. Mohapatra, in [Kla96]

[Moh97] R. N. Mohapatra, Proc. *Neutrino 96*, Helsinki, 1996, World Scientific Singapore, 1997, p. 290

[Moh97a] R.N. Mohapatra, Proc. Int. School on Neutrinos, Erice, Italy, Sept. 1997, Progr. Part. Nuc. Phys. **40** (1998)

[Mut88] K. Muto, H.V. Klapdor, in 'Neutrinos (Springer: Heidelberg, New York) 1988, ed. H.V. Klapdor, p. 183

[Mut89] K. Muto, E. Bender, H.V. Klapdor, Z. Phys. **A 334** (1989) 177,187;

[Mut91] K. Muto, E.Bender, H.V. Klapdor–Kleingrothaus, Z. Phys **A** 339 (1991) 435

[NEM94] NEMO Collaboration, Nucl. Phys. (Proc. Suppl.) **B 35** (1994) 369

[Nor97] D. Normile, Science 276 (1997) 1795

[PDG94] Particle Data Group, Phys. Rev. **D 50** (1994)

[PDG96] Particle Data Group, Phys. Rev. **D 54** (1996) 1

[Päs96] H. Päs *et al.*, in [Kla96];

[Päs97] H. Päs, M. Hirsch, H.V. Klapdor–Kleingrothaus, S.G. Kovalenko, in [Kla98]

[Pan96] O.Panella, in [Kla96]

[Pan97] O. Panella, in [Kla98]

[Pan96b] G. Pantis, F. Simkovic, J.D. Vergados, A. Faessler, Phys. Rev. **C 53** (1996) 695

[Pel93] J. Peltoniemi, J. Valle, Nucl. Phys. **B 406** (1993) 409

[Pel95] J.T. Peltoniemi, preprint hep-ph/9506228

[Pet94] S.T. Petcov, A.Yu. Smirnov, Phys. Lett. **B 322** (1994) 109

[Pet96] S. Petcov, in [Kla96]

[Pil93] A. Pilaftsis, Phys. Rev. **D 49** (1993) 2398

[Piq96] F. Piquemal *et al.*, in [Kla96]

[Pri95] J.R. Primack, J. Holtzman, A. Klypin, D.O. Caldwell, Phys. Rev. Lett. 74 (1995) 2160

[Rad95] P.B. Radha *et. al*, preprint nucl-th/9510052 (Oct. 1995)

[Raf96] G. Raffelt, J. Silk, Phys. Lett. **B 366** (1996) 429

[Rag94] R. S. Raghavan, Phys. Rev. Lett. **72** (1994) 1411

[Rau94] F. Raupach (H1 Collab.), Proc. Int. Europhys. Conf. on High En. Phys., Marseille, 1993, Edition Frontieres, Gif–Sur–Yvette, France 1994. Eds J. Carr, M. Perrothet

[Ren82] F.M. Renard, Phys. Lett. **B 116** (1982) 264

[Riz82] T.G. Rizzo, Phys. Lett. **B 116** (1982) 23

[Riz96] T.G. Rizzo, hep/ph/9612440

[Roy92] D.P. Roy, Phys.Lett. **B 283** (1992) 270

[Rub96] C. Rubbia, Proc. TAUP 95, Toledo, Sept. 17–21 (1995), Nucl. Phys. **B** (Proc. Supl.) 48 (1996)172

[Sch81] J. Schechter, J.W.F. Valle, Phys. Rev. **D 25** (1982) 2951

[Sim96] F. Simkovic et al., Proc. TAUP 95, Toledo, 17–21 Sept. 1995; Nucl. Phys. **B** (Proc. Supplement) 48 (1996) 257

[Sim97] F. Simkovic *et al.*, Phys. Lett. **B 393** (1997) 267

[Smi96] P.F. Smith *et al.*, Phys. Lett. **B 379** (1996) 299

[Smi96a] A. Yu. Smirnov, Proc. Int. Conf. on High Energy Physics, Warsaw 1996, hep-ph/9611465v2 (Dec 1996)

[Sou92] I.A. D'Souza. C.S. Kalman, Preons, Models of Leptons , Quarks and Gauge bosons as Composite Objects (World Scientific, Singapore) 1992

[Sta90] A. Staudt, K.Muto, H.V. Klapdor–Kleingrothaus, Europhys. Lett. **13** (1990) 31

[Ste91] J. Steinberger, Phys. Rep. **203** (1991) 345

[Sto96] S. Stoica, in [Kla96]

[Suh96] J. Suhonen, in [Kla96]

[Suz97] A. Suzuki, priv. comm. 1997, and KAMLAND proposal, (in Japanese)

[Tak96] E.Takasugi, in [Kla96]

[Tak97] E. Takasugi, in [Kla98]

[Tom87] T.Tomoda, A. Faessler, Phys. Lett. **B199** (1987) 475

[Tre95] V.I. Tretyak, Yu. Zdesenko, At. Data Nucl. Data Tables **61** (1995) 43

[Val96] J.W.F. Valle, in [Kla96]

[Vog86] P.Vogel, M.R. Zirnbauer, Phys. Rev. Lett. **57** (1986)3148

[Vog96] P.Vogel, in [Kla96]

[Vui93] J.-C. Vuilleumier et al., Phys. Rev. **D 48** (1993) 1009

[Wil92] C. M. Will, *Theory and Experiment in Gravitational Physics*, 2nd edition (Cambridge University Press, Cambridge, 1992).

[Wol81] L. Wolfenstein, Phys. Lett **107 B** (1981) 77

[You95] Ke You et al., Phys. Lett. **B 265** (1995) 53

The CUORE project: a large observatory for Neutrinoless Double Beta Decay and other rare events

Andrea Giuliani [1], *on behalf of the CUORE* [2] *collaboration*

INFN, Sez. di Milano - Via Celoria 16 - I-20133 - Milano, Italy

Abstract. We propose constructing a 750 kg array of 1000 TeO_2 bolometers to search for neutrinoless Double Beta Decay, Cold Dark Matter and Solar Axions. The physical motivations and the technical feasibility of the project are briefly discussed.

1. Introduction

We propose a large, high granularity, close-packed array of bolometric detectors to be operated underground (Gran Sasso INFN Laboratories, Italy), named "CUORE" (which stands for Cryogenic Underground Observatory for Rare Events), aiming first of all at Double Beta Decay (DBD) [1] study of ^{130}Te: other interesting applications are searches for WIMPs [2] and solar axions undergoing coherent Bragg conversion to photons by the Primakoff effect [3]. Due to the high proposed mass (750 kg) and to the innovative technique (bolometry), this experiment represents a revolutionary approach to rare event investigation.

2. ^{130}Te: a promising candidate for Double Beta Decay study

Double Beta Decay [1] is a rare nuclear process consisting of the spontaneous transmutation of an even-even nuclide (A,Z) into a nuclide (A,Z+2),

[1] E-mail: Andrea.Giuliani@mi.infn.it
[2] WEB site: http://hpbbgs.lngs.infn.it/halla/index_CUORE.html

accompanied by the emission of two electrons and, in the standard electroweak frame, of two electron antineutrinos. An occurence of the decay without neutrino emission could be naturally interpreted in terms of new neutrino physics, implying a finite mass and a Majorana nature for the neutrino. Other mechanisms beyond the Standard Model could induce the neutrinoless transition [4].

Since no neutrinoless process has been observed up to now, results are expressed in terms of a limit on the half-life for a given nuclide: from this limit, through nuclear physics calculations, it is possible to extract a value for the relevant weak-interaction-physics parameter, the so-called *effective neutrino mass*, a mixture of neutrino-eigenstate masses usually indicated as $< m_{\nu_e} >$. The neutrinoless DBD rate scales as $< m_{\nu_e} >^2$. A systematic uncertainty of a factor typically 2-3 on $< m_{\nu_e} >$ comes from the different nuclear matrix evaluations.

The present technology, thanks to the experiments which study DBD of ^{76}Ge [5, 6], can push the limit on $< m_{\nu_e} >$ down to $\sim$ 0.1 eV. To improve substantially this goal, a new concept experiment (as in the case of the GENIUS project [4], based on ^{76}Ge) must be designed, by choosing carefully the candidate nucleus and the experimental technique. As we shall show in the following, we think that the study of ^{130}Te using the bolometric technique is one of the most promising approach for new generation DBD experiments.

2.1. ^{130}Te Double Beta Decay probability

The DBD rate is given by the product of three factors: $< m_{\nu_e} >^2$, a phase space term and a nuclear matrix element term. The phase space term, which is exactly evaluable, grows fast with the transition energy E_0 (roughly as E_0^5): therefore, high transition energies are advisable. As a consequence, all the candidates whose transition energy is below 2 MeV are not interesting from an experimental point of view. ^{130}Te has a transition energy of 2.528 MeV, leading to a phase space 6.9 times larger than that of ^{76}Ge.

When taking into account also the nuclear matrix elements, one must consider that some disagreement is present among calculations from different authors [7](see fig. 1). On the average ^{130}Te DBD remains anyway about four/five times faster than ^{76}Ge. One more important point is that for ^{130}Te there is no serious spread among the rates proposed by the various authors (unlike for example ^{100}Mo).

2.2. ^{130}Te availability

In a new generation DBD experiment, at least 10^{27}–10^{28} nuclei must be contained in the detector. At this level, the availability of the nuclide,

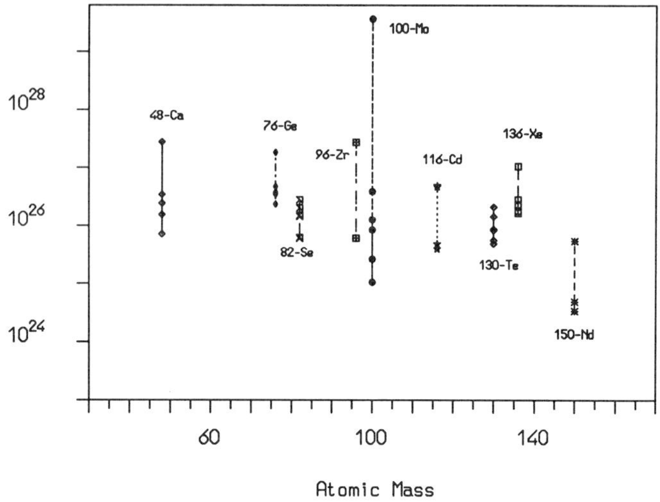

Figure 1. DBD half-life [y] evaluated assuming $< m_{\nu_e} >= 0.1$ eV for various nuclei according to different nuclear matrix calculations [7].

both from the practical and economical point of view, is a critical point. In this respect, ^{130}Te is in a unique position: its natural isotopic abundance is indeed 34%, much higher than for the other interesting nuclei. This implies that a significant experiment can be performed even *without isotopic enrichment*, which is often economically prohibitive in large quantities. Taking into account both isotopic abundance and decay rate, 750 kg of TeO_2 (which contain 202 kg of ^{130}Te) are equivalent to the effectiveness of between 700 and 1000 kg (depending on the nuclear matrix elements) of Ge detectors isotopically enriched to 86 % ^{76}Ge.

2.3. Te compatibility with the bolometric technique

A powerful method to investigate neutrinoless DBD is the so called "source = detector" technique [8], applied to several high sensitivity DBD experiments. In this approach, the isotopes under study are contained in the detector itself: neutrinoless DBD signature would be therefore a peak in the detector energy spectrum in correspondence of the DBD transition energy. Basic requirements for this technique are a high energy resolution and a very low background.

We propose the bolometric technique [9] for ^{130}Te study. A bolometer [10] is a device that measures the energy released to it by a single particle as a temperature rise of the whole detector, measured by means of an extremely sensitive thermometer thermally coupled to the detector absorbing part. In order to have high temperature signals in macroscopic amounts of material for nuclear energy depositions, it is necessary to operate the detector at very low temperatures ($\ll 1$ K), exploiting the strong reduction of specific heats with temperature in ordinary materials. Best materials for this application are dielectric diamagnetic single crystals, whose specific heat scales as T^3 at low temperatures according to the Debye law. We propose 750 g mass TeO_2 single crystals for the CUORE project. A crucial aspect of the device is the thermometric element, which should be in our case a semiconductor doped thermistor, suitable for a multiplication of the channels, mainly because it requires a standard, room temperature, reasonable cost electronics. Reliable and reproducible results are obtained with the Neutron Transmutation Doped (NTD) Ge thermistors developed by E. Haller in Berkeley.

3. Present performances of TeO_2 bolometers

The operation of bolometric detectors with large sensitive volumes requires high sensitivity thermistors, very low temperatures (~ 10 mK), long term stability and single detector reproducibility. The pluriennial work of the Milano group has shown that the bolometric technique is mature in all these aspects and that a large channel multiplication is now viable.

3.1. Single detector mass and resolution

Many tests were performed on 340 g TeO_2 detectors, operated with NTD Ge thermometric elements: presently, the state-of-art of detector construction and signal processing allows to reach in a reproducible way 1.5-2 keV FWHM intrinsic resolution. Only one test was up to now performed on a 750 g mass crystal, as foreseen for the CUORE modules. An energy resolution of 9 keV at 2615 keV was obtained, as can be appreciated in the calibration spectrum reported in fig. 2. The measurement was however performed in unsatisfactory microphonic noise condition, due to naive temporary cryogenic problems. We are therefore confident that this result can be improved in future tests.

3.2. Detector response stability

Cryogenic set-ups exhibit instabilities which can affect bolometer performances: an active mechanism to stabilize detector response is therefore

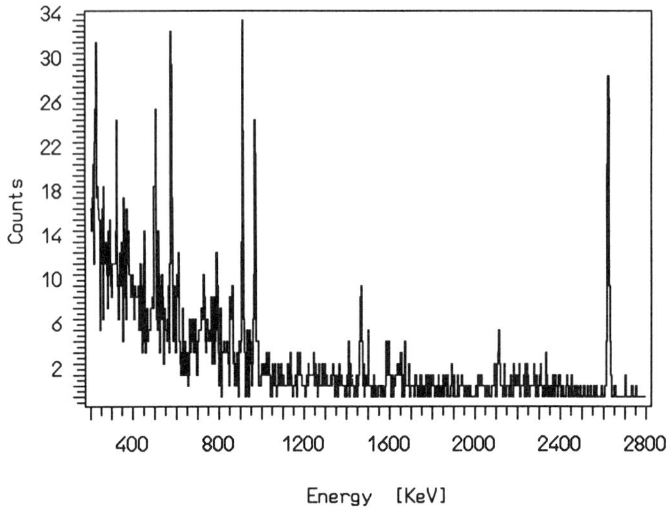

Figure 2. ^{232}Th calibration spectrum obtained with a 750 g TeO_2 detector.

necessary. We use a doped silicon resistance, with a metallic behaviour, as a heater which delivers calibrated energy amounts to the crystal through joule power. The heater pulses are identical to particle pulses: through a correlation between pulse height and detector operation point it is possible to correct the amplitude of all physical pulses in a very effective way, as explained in [11].

3.3. Reproducibility of the array elements

The Milano group is now operating a 20 detector array [12] whose single TeO_2 module has a mass of 340 g. The array performances are very satisfactory and show that the reproducibility problem is almost completely solved, thanks to the construction procedures described in [13]: the base temperature is 8 mK for all the elements within 1 mk spread; to optimize detector resolution, the mixing chamber is slightly heated up to about 10 mK; at the optimum bias point the operating temperatures of the various detectors range from 11 to 14 mK. In these conditions the typical voltage-to-energy conversion is around 250 μV/MeV with a spread of about a factor 2. The detector intrinsic resolution (which corresponds to the real resolution at low energies) is between 1.5 and 2.5 keV FWHM. At high energies (around

2 MeV) , the effective resolution is a factor 2-3 worse due to not yet completely understood systematic effects. This array, in a few week operation (corresponding to 0.35 kg·y), has allowed to set a limit on neutrinoless DBD half-life of 7.7×10^{22} y, corresponding to effective neutrino mass limits ranging from 2.5 to 5.2 eV [12, 7].

4. The future: CUORICINO and CUORE

CUORE constitutes the natural expansion of the present 20 detector array experiment. It should consist of a close-packed array of 1000 TeO_2 bolometers, with a mass of 750 g each, for a total of 0.75 tons. CUORE represents a severe technical challenge, most of all from the cryogenic and bolometric point of views: a high power dilution refrigerator must be developed on the purpose, similar to the ones required for cooling down multi-ton gravitational antennas. CUORE is proposed by a still open collaboration, including at the moment: Milano, Gran Sasso and Firenze, Italy; Berkeley and South Carolina, USA; Leiden, the Netherlands; Neuchatel, Switzerland; Zaragoza, Spain. The total cost of CUORE should be around 8 M$, most of which for the crystals (6 M$), while the rest is mainly due to the refrigerator and the electronics. The CUORE approach is flexible: other materials could be easily studied in addition to or in place of the TeO_2 crystals.

4.1. Preliminary structure of CUORICINO and CUORE

In a very preliminary approach, CUORE should consist of seventeen towers. Each tower consists of a stack of 15 modules. Each module, which is the smallest independent unit, contains 4 crystals. The whole arrangement is such that the inert material is minimized and the crystals are separated each other only by a few millimiters. The detector is included in a 110 cm high, 75 cm diameter cylindrical volume, which will corresponds to the evacuated experimental space of the dilution refirgerator. The operation temperature should be around 10 mK.

A single CUORE tower could be cooled down in the refrigerator presently housing the 20 element array, representing at the same time a test of the CUORE principle and a new very powerful DBD experiment, containing 5.7×10^{25} ^{130}Te nuclei. The CUORE collaboration is proposing this project with the name "CUORICINO", which means "small CUORE".

4.2. Potential of CUORE in Double Beta Decay search

For quoting CUORE sensitivity in DBD search, a reliable background level should be estimated. A Monte Carlo evaluation of the background on the basis of our preliminary structure and of the typical contamination levels of the employed materials (mainly copper and PTFE) is in progress. For the moment, we prefer to discuss the sensitivity *as a function* of the background level. If we had in CUORE the same background as in the 20 detector array (0.5 counts/keV kg day in DBD transition energy region), the effective neutrino mass 5 y sensitivity would be around 0.2 eV. This is however a very pessimistic assumption, since there is still a lot of work to do about radiopurity of cryostat components; furthermore, the CUORE structure, due to a much better volume-to-surface ratio, provides a natural self-shielding. One has to consider also that the detector granularity could help very much in background suppression, by operating the modules in anticoincidence.If we could improve the present background by a factor 100 (it would remain anyway 50 times worse than the background claimed by GENIUS [4]), a 5 y sensitivity down to 6×10^{-2} eV could be reached for effective neutrino mass.

As far as WIMPs and axion searches are concerned, the CUORE potential [14] will depend mainly on the threshold and the low energy background achievable. In the present 20 detector array, a threshold of 5 keV looks possible. Analyses are in progress on these topics.

References

[1] Moe M K and Vogel P 1994 *Ann. Rev. Nucl. Part. Sci.* **44** 247

[2] Mosca L 1996 *Nucl. Phys.* (Proc. Suppl.) **48** 34

[3] Creswick R J *Phys. Lett.* (in press)

[4] Hellmig J and Klapdor- Kleingrothaus H V 1997 *Z. Phys.* **A359** 361-372

[5] Baudis L et al. 1997 *Phys. Lett.* **B407** 219

[6] Aalseth C E et al. 1996 *Proc. XVII Int. Conf. on Neutrino Phys. and Astrophys., Helsinki, Finland* (World Scientific) p 361

[7] Suhonen J and Civitarese O *Phys. Rep.* (in press)

[8] Dell'Antonio G F and Fiorini E 1960 *Suppl. Nuovo Cim.* **17** 132

[9] Fiorini E and Niinikoski T O 1984 *Nucl. Instrum. Methods* **224** 83

[10] Twerenbold D 1996 *Rep. Prog. Phys.* **59** 349

[11] Alessandrello A et al. 1998 *Nucl. Instrum. Methods* **A412** 454-464

[12] Alessandrello A et al. *Phys. Lett.* (in press)

[13] Alessandrello A et al. 1997 *Proc. LTD-7, VII Int. Work. on Low Temperature Detectors, Munich, Germany* (Munich: Max Planck Insitute of Physics) p 249

[14] Alessandrello A et al. 1998 *Proc. PASCOS-98, VI Int. Symp. on Particles, Strings and Cosmology, Boston, MA, USA* (in press)

The neutrino mass from tritium β decay

E.W. Otten† **and Ch. Weinheimer**†

† Institute of Physics, Joh. Gutenberg University, 55099 Mainz, Germany

Abstract. The present status of the currently two running tritium β decay experiments at Mainz and at Troitsk is given. The resolved and remaining anomalies are reported. The recent improvement of the Mainz setup together with first data and results are presented.

1. Introduction

Whether neutrinos have a non-vanishing mass or not is still one of the most interesting questions of nowadays particle physics and cosmology. The most recent results of the Super-Kamiokande experiment claiming "evidence for neutrino oscillation" for the atmospheric neutrinos [1] seem to have answered this question in principle.[1] Also the deficit of solar neutrinos comes out more and more clearly from the different experiments and as well the hypothesis of neutrino oscillation seems to be the only possible explanation [2]. These neutrino oscillation experiments measure flavour mixing angles and differences of squares of the neutrino mass eigenstates (Δm^2), but no absolute neutrino mass values.

For this purpose a direct determination of the neutrino mass eigenstates is needed. Although the search for neutrinoless double β decay seems to be the most sensitive method, it has the limitation to require Majorana-type neutrinos and some assumptions about the mixing matrix. Therefore the determination of the mass of the electron antineutrino from tritium β decay remains very important. In spite of facing some problems in understanding the recently measured spectra fully, the sensitivity of this method is currently reaching a few eV.

[1] This result was disclosed shortly after presenting this talk.

This eV mass range might be very interesting since in both oscillation scenarios mentioned above the differences between the square values of the neutrino mass eigenstates are very small ($2 \cdot 10^{-3}$eV^2/c^4 or less than 10^{-5}eV^2/c^4, respectively). If the neutrino mass eigenstates would obey a similar generation hierarchy as the charged leptons or the quarks even the largest mass would be only of the order of 0.05 eV/c^2, much too small to contribute significantly to the dark matter in the universe. Therefore it is also very attractive to assume a degenerate neutrino mass scheme with masses of about $1 - 2.5$ eV/c^2 to fulfil the requirements of the missing hot dark matter component in the universe (*e.g.* reference 3).

The development of the electron neutrino mass measurements from tritium β decay within the last decade is plotted in figure 1. Although it seems that the obtained experimental results on m_ν^2 are mostly in the unphysical region the trend goes clearly towards $m_\nu^2 \approx 0$ and to significant higher precision.

The recent improvements are due to a new type of electrostatic retarding spectrometer with magnetic adiabatic collimation independently developed at Mainz and Troitsk [4, 5]. These instruments fulfil the above mentioned requirements of high signal-to-background-ratio by large angular acceptance of 10–100 % of 2π and of very high energy resolution of ΔE = 3–6 eV.

This paper is structured as follows: The currently two running experiments at Mainz and Troitsk are described in section 2 and 3. The improved Mainz setup and the first tritium data obtained therewith are presented in section 4. The conclusion and an outlook are given in section 5.

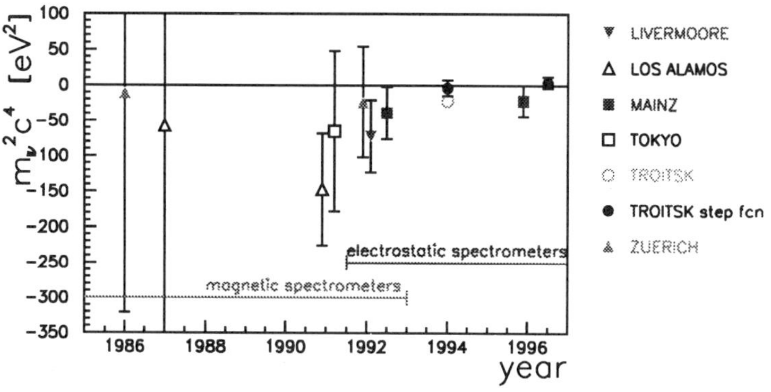

Figure 1. Results on m_ν^2 of recent tritium β decay experiments up to 1996. The time axis gives the dates, when the analysis of the data were presented in public, not when the data were taken.

2. The Mainz neutrino mass experiment

The tritium β spectrum has been investigated with the Mainz setup in 1994 for the second time [6]. As tritium source a film of about 37 monolayers of T_2 molecules quench-condensed on a graphite substrate at a temperatures of about 3 K was used.

The high signal-to-background-ratio combined with the high energy resolution allowed to determine the square value of the neutrino mass m_ν^2, which is the true observable, from a reasonably small interval of the last 140 eV below the endpoint, resulting in:

$$m_\nu^2 = -22 \pm 17_{\rm stat} \pm 14_{\rm sys} \;\; eV^2/c^4 \; , \tag{1}$$

for which a limit of $m_\nu < 5.6 \; eV/c^2$ (95% C.L.) can be deduced [6].

Looking to the experimental spectrum further below this interval a clear excess like in the 1991 run [7] was observed compared to the extrapolation of the fit over the last 140 eV. If this systematic error would be common to all experiments it could be the reason for negative values of m_ν^2 reported earlier (see figure 1). A common input to all the experiments is the final states distribution of the T_2 known only from theoretical calculations. After being under discussion in the last years the final states distribution was carefully checked by the theorists confirming the old calculations.

After checking all systematic effects one question was still remaining: Has the T_2 film undergone a transition from the quench-condensed homogeneous film into a rough inhomogeneous one [8] resulting in a harder energy loss spectrum, which could have produced the observed spectrum?

In cooperation with the group of P. Leiderer at Konstanz/Germany detailed investigations for different hydrogen isotopes were performed using conversion electron spectroscopy and light scattering techniques with the following results: The roughening transition cannot be avoided but its speed can drastically be slowed down by using lower temperatures. Extrapolating from the stable hydrogen isotopes a T_2 film at 1.8 K should have a time constant of tens of years if the β decay does not matter. This gave a clear recipe for the new generation of experiments described in section 4 on one hand, and on the other hand allowed to conclude that the earlier measurements, performed in the range 2.8 to 4.2 K, suffered most probably from (at least a partial) roughening.

3. The Troitsk neutrino mass experiment

At Troitsk a spectrometer similar to the one at Mainz was set up approximately twice in size, but using a gaseous T_2 source instead of a quench-condensed one. In their first measurements they also reported an excess count rate further below the endpoint similar to the one observed at Mainz [9]. Meanwhile the Troitsk group is able to explain their excess by electrons trapped in the source which escape into the spectrometer by large angle scattering. This disfavours the "common error" assumption and favours the "T_2 roughening transition" solution for the excess count rate reported at Mainz.

The Troitsk group reported on a second excess rate near the endpoint E_0, known as the "Troitsk anomaly". They claim that the spectral shape of this anomaly can be described by a sharp step in the experimental electron spectrum at a few eV below E_0. Since their spectrometer is integrating this sharp step in the measured spectrum corresponds to a monoenergetic line in the primary β spectrum with an amplitude of about $6 \cdot 10^{-11}$ of the total β spectrum.

Up to now the Troitsk group has performed and reported on 7 separate runs. In each measurement they found again a step-like anomaly of similar amplitude but at different positions varying from 5 eV to 15 eV below E_0. When they plot the step position versus measurement date they find a time dependence, which can be expressed by a sinus-like oscillation with a period of half a year [10].

The characteristics of the anomaly looks complicated, so is the experimental situation: The size of the step is so tiny that it can neither be checked by the Mainz 1994 data, nor is it clearly seen directly in the Troitsk raw data. It is rather extracted from the Troitsk data by the fitting procedure, which gives a significant reduction of χ^2, when a step-like anomaly is allowed.

There exist some speculative ideas about the origin of this anomaly, but there is no explanation within standard physics for the existence of a monoenergetic line within the continuous β spectrum. The Troitsk group is continuing in measuring the tritium β spectrum and in checking possible experimental reasons for this effect.

After subtracting the monoenergetic line from each run Troitsk obtains as combined result

$$\mathrm{m}_\nu^2 = -2.1 \pm 3.7_{\mathrm{stat}} \pm 2.3_{\mathrm{sys}} \quad eV^2/c^4 \tag{2}$$

from which an upper limit of $\mathrm{m}_\nu < 2.7\ eV/c^2$ (95% c.l) is deduced [10].[2]

4. The improved Mainz setup and first measurements

The main goal of the experiments to look for a non-zero neutrino mass in the range of a few eV is becoming more and more important for the reasons mentioned above. Additionally the anomaly just below the endpoint

[2] This value was presented at the Neutrino 98 conference shortly after presenting this talk.

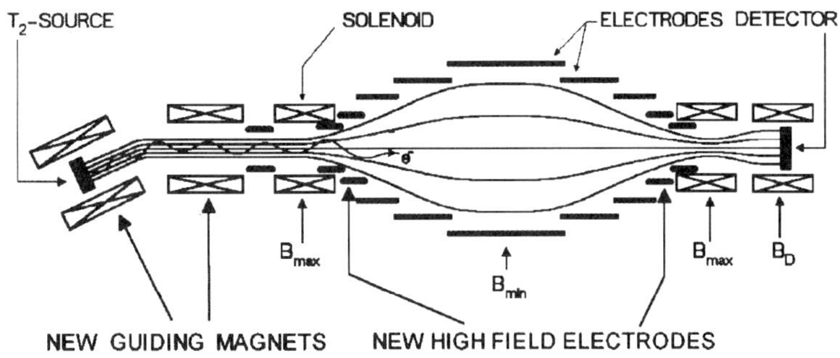

Figure 2. The improved and enhanced Mainz setup schematically, not in realistic scale. The outer diameter amounts to 1 m, the distance from source to detector is 6 m.

reported by the Troitsk experiment [10] need clearly further checks from the experimental side.

To fulfil these expectations the Mainz experiment had to solve the following problems:

- to avoid the T_2 film roughening transition
- to decrease the background rate
- to increase the signal rate
- to enable long term runs

For these reasons the Mainz experiment has improved its setup substantially (see figure 2). In two first longterm measurements of the tritium β spectrum (4 weeks in summer 1997 and 3 weeks in winter 1998), it was proven that the improved setup is working as nicely as expected:

- The new, automatically controlled source cryostat, which was installed to slow down the T_2 film roughening transition to a negligible speed by working at temperatures down to 1.7 K, was running during the two longterm measurements at a stable temperature of 1.84 K within a range of 0.03 K.
- A new doublet of superconducting solenoids, tilted by 20° to each other, was installed. Now the source is placed in the most left solenoid. β decay electrons are guided into the spectrometer without losses as before, whereas both tritium molecules evaporating from the source are prohibited from contaminating the spectrometer and residual gas molecules of the spectrometer from condensing on top of the T_2 film. These two problems, which were the biggest source of background and of systematic uncertainty for the 94 run now could been avoided:

  Although the thickness of T_2 films used for the measurements (1997:

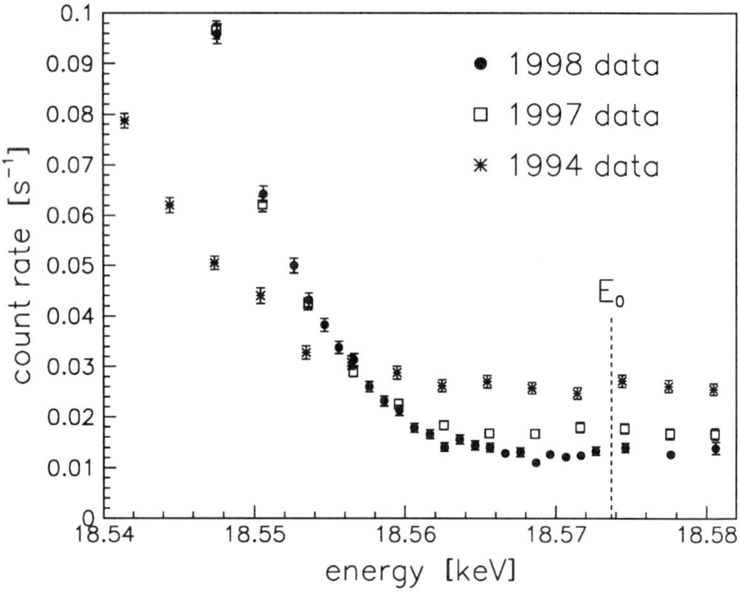

Figure 3. Mainz tritium β spectra close to the endpoint

973 ± 55 Å, 1998: 493 ± 5 Å) were much larger than the one of the measurement in 1994 (126 Å) the background was even significant lower (see figure 3). The background rates of 0.016 s^{-1} (1997) and 0.013 s^{-1} (1998) were only slightly larger than the rate without any tritium source of 0.010 s^{-1}.
Secondly the thickness of the T_2 films was measured before and after the runs with ellipsometry. Within the precision of a few percent the film thickness was the same as at the start of the measurement.
It is worth mentioning that the spatial separation of the source and the spectrometer allows a valve to be closed by a control system in the case of any problems, which is an essential feature for automatic running.

- The high field electrodes were redesigned to lower the background contribution from the spectrometer itself. Due to a better alignment of the whole system the spectrometer could run at a higher energy resolution of 4.4 eV compared to 6.3 eV in 1994 (these widths correspond to the 0 to 100 % rise of the transmission function of the integrating spectrometer).
- An experiment control system was installed in order to run the experiment automatically. Human intervention was needed only for filling of LHe and LN_2.

Figure 3 shows the endpoint region of the β spectra taken with the improved Mainz setup compared to the 1994 Mainz data. The statistical improvement of the 1997 and 1998 data due to longer data taking periods, but moreover due to a larger signal rate and a lower background rate, thus resulting in an about 10 times higher signal-to-background-ratio, is clearly visible.

Three main sources of systematic uncertainties will be discussed in the following, all other contributions are small:

1. An important systematic uncertainty is caused by the inelastic scattering of β electrons within the T_2 film. Although the film thickness was increased significantly compared to 1994 the increase of systematic uncertainty was compensated by changing the maximum starting angle for electrons accepted by the spectrometer from $\vartheta_{\max} = 78.5°$ in 1994 to $\vartheta_{\max} = 45°$. Additionally the much higher signal to background ratio allows to use much smaller energy intervals below the endpoint for the final analysis. This reduces the fraction of inelastically scattered electrons within the analysed data.
   At Mainz intensive studies have been performed within the last year on the T_2 film thickness measurements which could reduce its uncertainty significantly down to 1 % relative error for the 1998 measurement. By the comparison of our measured optical refraction index with the one for a densely packed structure we estimated the pore density of the T_2 film to be about 15 %.
   By a new series of energy loss measurements with conversion electrons from a ^{83m}Kr sub-monolayer frozen under different thick D_2 layers a better knowledge of the mean free path ($\rightarrow$ relative uncertainty: 5 %) and a better understanding of the energy loss function was obtained. It should be mentioned, that the observed shape of the energy loss function of the quench condensed hydrogen films differs quite significantly from the shape of the energy loss function of gaseous hydrogen resolving an old discrepancy between the Troitsk and the Mainz energy loss function. The different shape is due to the fact, that the excited molecular wavefunctions have a much larger overlap with the neighbours than the ground state wave function. Thus the Pauli principle causes small shifts of energy levels and population probabilities.

2. Due to the new knowledge on the existence of pores and on the shifts of the energy levels of excited states the situation for the excitation of neighbour molecules due to the fast change of nuclear charge during the β decay was reconsidered. A small energy shift of 1 eV and an amplitude reduction of 33 % was applied and fully taken into account as systematic uncertainty.

3. Whereas the first two systematic uncertainties completely vanish near

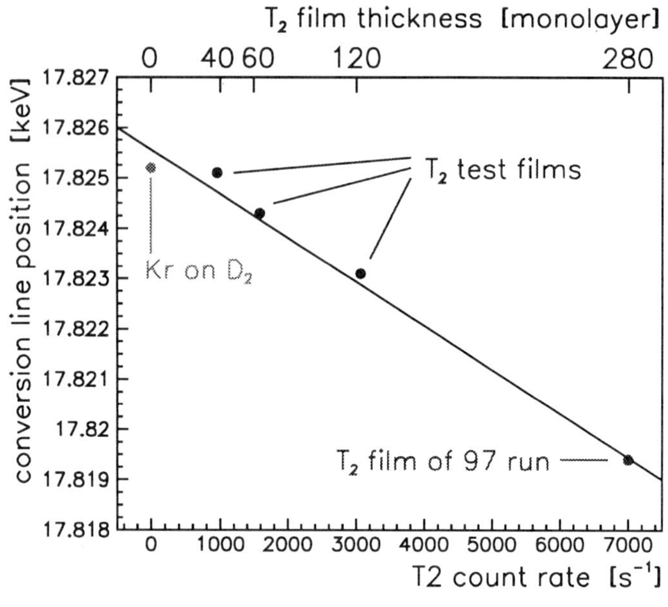

Figure 4. Position of the K conversion line from ^{83m}Kr condensed on T_2 films of different activity or thickness respectively, or on a D_2 film (reference measurement). A straight line is shown to guide the eyes. The count rate axis gives the count rate at a certain reference energy.

the endpoint (due to the absence of low lying excited states in hydrogen) the uncertainty due the charging up of the T_2 film always plays a role. This new effect has never been noticed before because it had not influenced on previous measurements. The charging of the T_2 film can be qualitatively understood by the fact that from a T_2 source of about 40 mCi activity (like the 1997 source) more than one billion β electrons per second are leaving the film, whereas the positive daughter nuclei are remaining within the T_2 film. If the charge compensation current is not large enough the film will charge up positively.

We have investigated the surface potential of the T_2 film at the end of the 4 weeks data taking by condensing a sub monolayer of ^{83m}Kr on top of the T_2 film. The shift of the position of the K conversion line with respect to a reference measurement using a 240 monolayers thick D_2 film is a measure of the surface potential due to the charging up. For several T_2 films of different thicknesses this measurement was repeated. Figure 4 shows the extracted line positions which seem

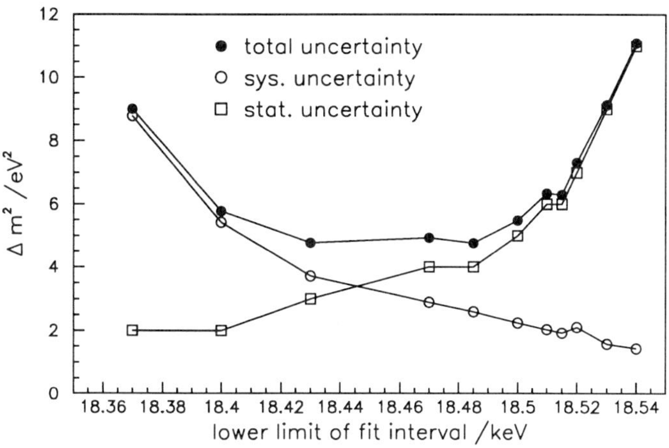

Figure 5. Statistical, systematic and total 1 σ uncertainties of m_ν^2 for the Mainz 1998 measurement in dependence on the lower limit of the data interval used for the analysis.

to depend linearly on the T_2 film thickness. Further tests with a ^{83m}Kr sub monolayer between two T_2 layers show that the shift only depend on the T_2 film thickness between the graphite substrate and the ^{83m}Kr layer, but not on the thickness of the second T_2 layer on top of the ^{83m}Kr. A more detailed description and a simple model explaining quantitatively this effect can be found in reference [11]. The charging of the T_2 film seems to be understood. This effect has the consequence of a varying potential for the different layers of the T_2 film which increases linearly with the distance to the substrate. It can be considered in the data analysis, but decreases slightly the effective energy resolution with which the β spectrum is investigated. Since this picture might simplify the reality too much 50 % of the whole effect on m_ν^2 was taken into account as systematic uncertainty.

Figure 5 shows the statistical and systematic uncertainties on m_ν^2 in dependence on the lower limit of the data interval used for the analysis at the example of the 1998 data. For small intervals the sensitivity on m_ν^2 is clearly dominated by the statistical uncertainty, whereas the systematic error dominates at larger fit intervals. The minimum total uncertainty of $\Delta m_\nu^2 \approx 5\ eV^2/c^4$ is obtained for medium fit intervals of the last 50–70 eV of the β spectrum.

Figure 6 shows the results on m_ν^2 including the total uncertainties for the measurements of 1997 and 1998, again as function of the lower limit of the data interval used for the analysis. Two things can be observed:

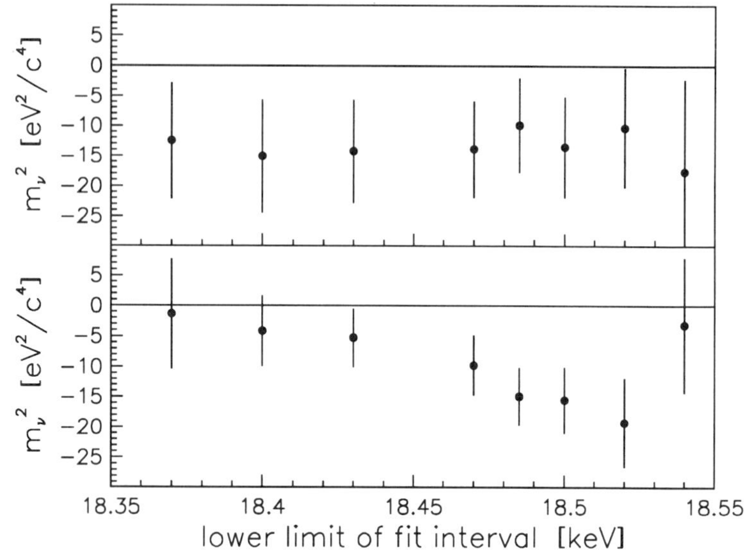

Figure 6. Fit results of m_ν^2 with total uncertainties (1 σ) in dependence on the lower limit of the data interval used for the analysis for the Mainz measurements of 1997 (top) and of 1998 (bottom).

- There is no trend anymore towards negative values of m_ν^2 for larger fit intervals as it was the case for the Mainz 1991 and 1994 data. Obviously a roughening transition of the T_2 film was the reason for this behaviour in the former measurements, which now does not appear anymore due the much lower temperature of the T_2 film.
- Still slightly but significantly negative values for m_ν^2 are obtained which are incompatible with $m_\nu^2 \geq 0$ at the 1–3 σ level.

Although the problem for the former trend towards negative values of m_ν^2 seems now to be solved the much higher sensitivity of the new data shows, that the description of the spectra still does not match completely the data. There is a small, but significant anomaly in the Mainz data. As yet the data are statistically not good enough to extract the shape of this anomaly clearly. Therefore the existence of the "Troitsk anomaly" can neither be proven nor disproven by the present Mainz data. Although the Mainz data are now competitive to the Troitsk data on a single run comparison, more data taking is necessary to check the special features of the "Troitsk anomaly" like its time dependence.

Due to the presence of an unresolved anomaly in the Mainz data it is not straightforward to give an upper limit on the neutrino mass. Therefore the Mainz group has restricted itself to the results plotted in figure 6 until more data have been taken and analysed.

5. Conclusion and outlook

Within the last years the two running experiments at Mainz and Troitsk have improved the precision of measuring the endpoint region of tritium β decay significantly. The former problems of a trend towards negative values of m_ν^2 for larger data intervals has been solved for the two experiments. The effects are now avoided at Mainz (roughening transition of the T_2 film) or corrected for in the analysis at Troitsk (electrons trapped in the T_2 source).

The Troitsk group reported on a tiny step-like anomaly at a position varying between 5 eV and 15 eV below E_0 with a period of half a year. Since there is no explanation within standard physics for such an effect, this clearly demands for an independent experimental check.

The Mainz experiment has completed the improvement of its setup in 1997 and has demonstrated with two runs that its sensitivity has become now competitive to the one of the Troitsk experiment. There is also a small anomaly in the Mainz data, which is still unresolved. Whether it corresponds to the "Troitsk anomly" or not can neither be proven nor disproven at the moment.

Because of these anomalies and moreover due to the importance to check directly neutrino masses in the cosmological relevant range down to 1 eV more data from the present experiments are needed. But it seems that the decisive answer to these questions could come only from a new experiment with superior sensitivity. For this purpose the Mainz group has suggested to build a large version of the present spectrometer, up-scaled by a factor of 5 in outer dimensions. An additional time-of-flight method could turn such an integrating spectrometer into a narrow band filter to clearly check the "Troitsk anomaly" [12].

References

[1] T. Kajita, "Atmospheric neutrino results from Super-Kamiokande and Kamiokande - Evidence for ν_μ oscillations", talk given at the International Conference on Neutrino, Physics and Astrophysics Neutrino 98 at Takayama/Japan, 4–9 June 1998, Proceedings will appear in Nucl. Phys. B Proc. Ser., Elsevier, and Y.Fukuda *et al.* hep-ex/9807003, submitted to Phys. Rev. Lett.

[2] V.N. Gavrin, T. Kirsten, Y. Suzuki, talks given at the International Conference on Neutrino, Physics and Astrophysics Neutrino 98 at Takayama/Japan, 4–9 June 1998, Proceedings will appear in Nucl. Phys. B Proc. Ser., Elsevier

[3] E. Garwiser, J. Silk, Science **280** (1998) 1405

[4] A. Picard *et al.*, Nucl. Inst. Meth. **B63** (1992) 345

[5] V.M. Lobashev *et al.*, Nucl. Instr. Meth. **A240** (1985) 305

[6] H. Backe *et al.*, "Status and perspectives of the Mainz Neutrino Mass Experiment",Proc. XVII Conference on Neutrino Physics and Astrophysics, Neutrino 96, Helsinki/Finnland, June 1996, World Scientific/Singapure

[7] Ch. Weinheimer *et al.*, Phys. Lett. **B300** (1993) 210

[8] P. Leiderer *et al.*, Jour. Low Temp. Phys **89** (1992) 219

[9] A.I. Belesev *et al.*, Phys. Lett. **B350** (1995) 263

[10] V.M. Lobashev,"Neutrino mass and anomaly in the tritium β spectrum", talk given at the International Conference on Neutrino Physics and Astrophysics Neutrino 98 at Takayama/Japan, 4–9 June 1998, Proceedings will appear in Nucl. Phys. B Proc. Ser., Elsevier

[11] H. Barth *et al.*, Prog. Part. Nucl. Phys. **40** (1998) 353

[12] J. Bonn *et al.*, *submitted to* Nucl. Instr. and Meth.

ICARUS - Proton Decay and the CERN - Gran Sasso Long Base Line Experiment

F.Arneodo[a], A.Badertscher[b], E.Barbieri[c], P.Benetti[d], A.Bettini[e], A.Borio di Tigliole[f], A.Bueno[b], E.Calligarich[d], M.Campanelli[b], F.Carli[c], C.Carpanese[e], D.Cavalli[f], F.Cavanna[g], P.Cennini[h], S.Centro[e], A.Cesana[f], C.Chen[i], Y.Chen[i], C.Cinquini[c], D.Cline[j], I.De Mitri[a], R.Dolfini[d], D.Favaretto[e], A.Ferrari[f], A.Gigli Berzolari[d], P.Goudsmit[b], K.He[i], X.Huang[i], Z.Li[i], F.Lu[i], J.Ma[i], G.Mannocchi[k], F.Mauri[d], D.Mazza[g], L.Mazzone[d], C.Montanari[d], S.Otwinowski[j], O.Palamara[a], S.Parlati[g], D.Pascoli[e], A.Pepato[e], L.Periale[l], S.Petrera[g], G.Piano Mortari[g], A.Piazzoli[d], P.Picchi[k], F.Pietropaolo[e], A.Rappoldi[d], G.L.Raselli[d], S.Resconi[f], J.P.Revol[h], M.Rossella[d], C.Rossi[g], A.Rubbia[b], C.Rubbia[h], P.Sala[f], D.Scannicchio[d], F.Sergiampietri[h], S.Suzuki[k], M.Terrani[f], P.Torre[d], S.Ventura[e], M.Verdecchia[g], C.Vignoli[d], H.Wang[j], J.Woo[j], G.Xu[i], Z.Xu[i], C.Zhang[i], Q.Zhang[d], S.Zhen[i]

[a] Lab. Naz. del Gran Sasso, S.S. 17 bis, Assergi (AQ), Italy
[b] Eidgenösische Techische Hochschule (ETH),CH 8092 Zürich, Switzerland
[c] Dip. di Meccanica Strutturale e INFN, Univ. di Pavia, via Abbiategrasso 209, Pavia, Italy
[d] Dip. di Fisica Nucleare e Teorica e INFN, Univ. di Pavia, via Bassi 6, Pavia, Italy
[e] Dip. di Fisica e INFN, Univ. di Padova, via Marzolo 8, Padova, Italy
[f] Dip. di Fisica e INFN, Univ. di Milano, via Celoria, Milano, Italy
[g] Dip. di Fisica e INFN, Univ. dell'Aquila, via Vetoio, L'Aquila, Italy
[h] CERN, CH-1211, Geneva 23, Switzerland
[i] IHEP - Academia Sinica, 19 Yuquan Road, Beijing, China
[j] Dept. of Physics, UCLA, Los Angeles, CA 90024, USA
[k] Lab. Naz. di Frascati dell'INFN, via Fermi 40, Frascati (Roma), Italy
[l] ICGF del CNR di Torino, corso Fiume 4, Torino, Italy

presented by Fulvio Mauri[1]

Abstract. The ICARUS experiment is described, in particular regarding the search for proton decay and the CERN - Gran Sasso Long Baseline Neutrino Oscillations Experiment. The first 600 ton module construction is reported in some details.

1. Introduction

Bubble chambers have played a fundamental role in Particle Physics. They provided unbiased events in 3 dimensions, with the possibility to add a magnetic field. Because of the high density of the liquid medium, bubble chambers combined target and detector functions. They have been ideal in the search for rare events produced inside their volume (Neutral Current neutrino events, SU(3) spectroscopy, etc.). The main characteristic was that of claiming unambiguous discovery at single (few) event basis.

However bubble chambers are not continuously sensitive, hard and costly to build in big dimensions and dangerous if operated underground.

An ideal detector to search for rare events in an underground laboratory, such as proton decay and neutrino interactions, could be a high resolution detector with an electronic read-out, a so called electronic bubble chamber. The construction of such a detector is the aim of the ICARUS programme [1]: a novel detector technology for underground physics.

The liquid argon image chamber ICARUS has the following main properties:

- it provides electronic bubble chamber quality 3D images,
- it is fully homogeneous,
- it is continuously sensitive,
- it can be build with large sensitive mass and high granularity,
- it is self triggering,
- it provides ionization and range measurements,
- it has an energy resolution of about 7% at few MeV and 3% at 1 GeV,
- it has a space resolution at the level of 3 mm (wire pitch).

After 5 years of R&D studies on small volumes which clarified the purification methods, the read-out schemes, the diffusion coefficients and the electronics, the ICARUS experiment operated the 3 ton prototype at CERN

[1] E-mail: Fulvio.Mauri@pv.infn.it

[2] for more than 4 years, studying the recirculation technology, the pattern recognition on real events, the doping of liquid argon with TMG, the Multi Layer Printed Board read-out technique. The next step will be the construction and operation in the LNGS underground laboratory of a 600 ton module (T600). This step is intended not only for technical reasons (cryostat technology, operation underground, safety, transportability), but fulfills also an important physics programme.

The final goal of the ICARUS programme is the construction of a detector with a sensitive mass of the order of 5000 ton [3] for ultimate physics at Gran Sasso. It is now under way the preliminary design of a cluster of 8 modules of 600 ton to reach such a volume. The next step, after the operation of the first 600 ton module, will be the construction and operation of 3 other modules to have a suitable mass (of the order of 2400 ton) for the ideal exploitation of the foreseen CERN - Gran Sasso Long Baseline neutrino beam.

2. Physics programme

The ICARUS physics programme addresses many fundamental issues:

(i) the nature of neutrinos, in particular the question of the neutrino mass and oscillation. This is being investigated in ICARUS both through the study of atmospheric neutrinos and through long baseline studies with the foreseen CERN neutrino beam [4];

(ii) the stability of the nucleon, which is the only way to access phenomena at the energy scale of Grand Unification. ICARUS will be the first high-resolution imaging study of such a phenomenon;

(iii) the study of solar neutrinos. Even though ICARUS is not optimized to study this kind of events, the issue remains extremely interesting and with a suitable neutron shielding ICARUS can reach a reasonable detection threshold;

(iv) the detection of astrophysical and cosmological neutrinos from supernova.

As a first practical step the Collaboration proposed in 1996 [5] to build a 600 ton module.

A few hundred ton detector allows an important first step in the ICARUS scientific programme. What the ICARUS technique is providing is a background-free detection of proton decay and neutrino events. While a sensitive mass in excess of few thousands tons of liquid argon is clearly needed to achieve the 10^{34} years range in proton decay lifetime of interest for the SUSY GUT decay channels, many exotic channels have been only poorly investigated so far or not at all, and would be easily covered in this first phase. Atmospheric and solar neutrinos are areas which could be also investigated with the 600 ton detector. Therefore, a 3 mm wire pitch will be used instead of the 5 mm foreseen for the 5000 ton module, in order to allow for higher precision measurements.

2.1. Solar neutrinos

The ICARUS goal in the solar neutrino area is not only to confront the Standard Solar Model, but also to provide a Solar Model independent measurement, by observing several independent processes, as envisaged in our initial proposal [3]:

(1) $\nu_x + e^- \rightarrow \nu_x + e^-$ (elastic scattering event) which occurs with all types of neutrino flavours and for both charged and neutral current exchange;

(2) $\nu_e + Ar \rightarrow K^* + e^-$ (absorption event) which only occurs with the electron neutrino.

The measurement of the ratio of rates for these two processes provides directly a measurement of the ν_e oscillation probability [3].

In our detector the signatures of these two reactions are the following:

(1) for elastic scattering events one isolated electron with energy larger than 5 MeV, within 25° from the sun direction;

(2) for absorption events one electron with energy larger than 5 MeV, in association with a total energy deposited in a sphere of 50 cm radius around the vertex, larger than 2 MeV (Compton electrons).

The main source of background for solar neutrino events comes from neutrons produced by natural radioactivity in the materials surrounding the detector (mainly the rocks around the cave). It is necessary to build around the 600 ton detector a shield made of a neutron moderating medium (paraffin or another hydrogenated material) and of thermal neutron absorber (boric acid).

The computed rates of events are of the order of 230 events per year (considering the Standard Solar Model expectations) for elastic scattering events to be compared with 42 events of background (considering a suitable neutron shield) and of 1,150 events per year for absorption events to be compared with 88 events of background.

We are considering an additional and unique way to provide a solar model independent measurement, by measuring the shape of the ^8B solar neutrino energy spectrum, introducing a small fraction ($< 5\%$) of deuterated methane (CD_4) in the liquid argon. This provides quasi-free neutrons and results in a scattering kinematics which does not suffer from important Fermi motion smearing.

2.2. Proton decay

Thanks to its large sensitive mass (of the order of 5 kton for the final detector) and to its spatial and energy resolution capability, ICARUS is an ideal device for nucleon decay detection, in particular for those channels that are not accessible to Cherenkov detectors due to the complicated topology, or because the emitted particles ($K^{\pm}$) are below the Cherenkov threshold. In particular ICARUS, due to its powerful background rejection capability, can perform exclusive decay modes measurements. It can easily distinguish between atmospheric neutrino events and true nucleon decays.

We have performed a detailed event simulation based on the standard GEANT Montecarlo code [3] and the events obtained contain very long tracks with redundant information, allowing particle identification and energy measurement with great precision. Particle identification benefits greatly from the ability to measure the ionization loss (dE/dx).

In the absence of background, the limit of the proton (bound neutron) lifetime achievable in T years of observation, assuming no backgroung events, is given by (at 90% C.L.):

$$\tau = 1.2\,(1.4) \cdot M \cdot T \cdot \eta \qquad (10^{32} years)$$

where M is the detector mass (in kton), T the effective running time (in years), η the detection efficiency.

In order to reduce the neutrino-induced background we make use of the ability of ICARUS to fully reconstruct the events. Proton decay events are characterized by a definite value of the total energy and by the fact that the total momentum of the decay products must be zero. These features, which are true for a free nucleon, are also approximately verified for a nucleon bound in a nucleus, provided the decay products do not rescatter before escaping the nucleus. As a consequence, we define the detection efficiency $\eta = \epsilon_d \cdot \epsilon_n$, where ϵ_d is the probability that the decay products do not interact with the nucleus in which they were produced, and ϵ_n is the reconstruction efficiency.

In figure 1 the achievable limits for 5000 ton sensitive mass are shown both with 1 year and 5 years running time, for some of the decay channels, taking into account the calculated background.

A Monte Carlo simulation has been performed in order to compute the detection efficiencies for a number of decay channels. For each of the decay modes considered 2000 events were simulated. We have taken the conservative approach of rejecting all events with a nuclear reinteraction. The atmospheric neutrino background to nucleon decay has been studied by means of a Monte Carlo simulation.

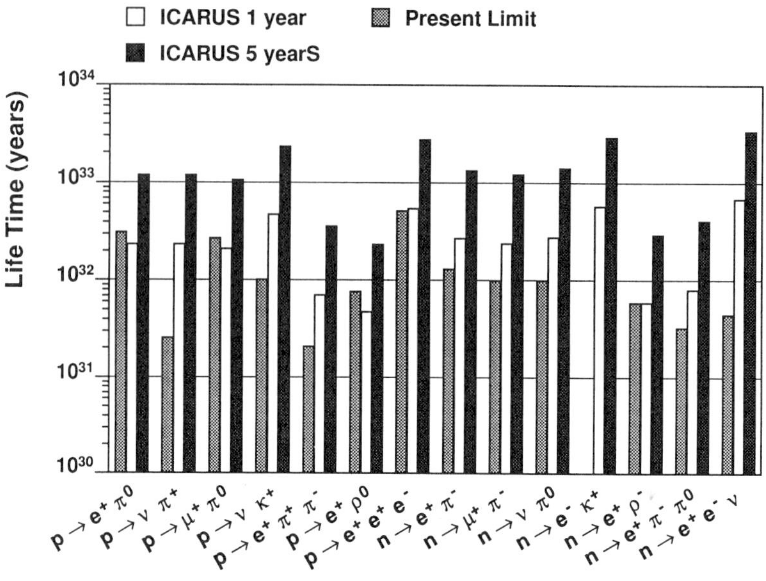

Figure 1. Proton decay achievable limits for 5000 ton sensitive mass.

In particular for the popular decay mode $p \to e^+\pi^0$ the event signature is three coplanar electromagnetic showers (one from the e^+ and two from the π^0 decay) with a total energy of $\approx$ 938 MeV and a total momentum near zero. We generated a sample of 510 $p \to e^+\pi^0$ events and a sample of 499 atmospheric neutrino events from ν_e and $\bar{\nu}_e$ charged-current interactions. All of these events were visually scanned, and proton decay candidates were selected according to the following criteria:

a) total reconstructed energy E_{tot} such that $700MeV < E_{tot} < 1000MeV$;

b) two or three showers in each one of the two views and no other track;

c) no obvious momentum imbalance.

With these criteria, the efficiency for selecting non-reinteracting proton decay events is very high (95%), while it is fairly low for reinteracting events (5%). There is no contamination by the atmospheric neutrino events.

For another interesting channel $p \to e^+\nu\nu$ the most important source of background is again charged-current interactions of atmospheric ν_e and $\bar{\nu}_e$. The event signature for the proton decay will consist of one isolated electromagnetic shower with energy between 0 and 500 MeV. We generated 519 $p \to e^+\nu\nu$ events and 1981 atmospheric neutrino charged-current

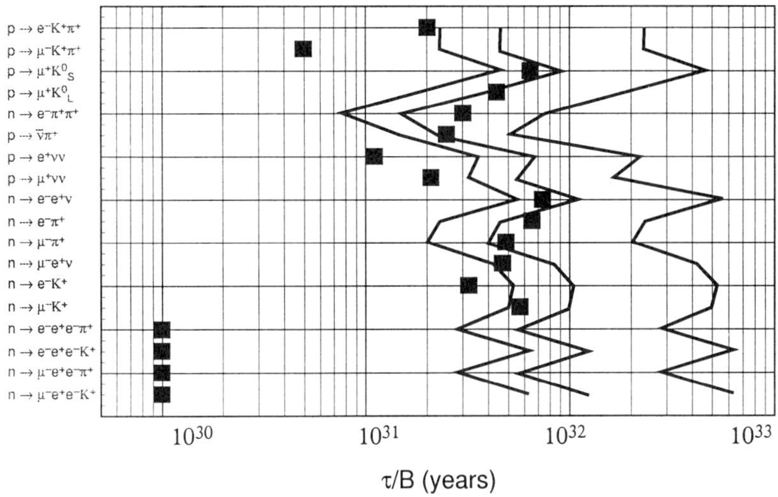

Figure 2. ICARUS 600 ton limits for exotic decay modes. The back squares show the current limits. Starting from the left the lines show the limits after 1, 2 and 10 years of run.

events. With the following selection criteria:

a) total reconstructed energy E_{tot} such that $150 MeV < E_{tot} < 450 MeV$;

b) one single electromagnetic shower in each of the two event view;

the detection efficiency for non reinteracting proton decay events is 77% and the expected background, due to the atmospheric neutrino flux, is of the order of 6 events per year (for 5000 ton of sensitive mass).

Even with a relatively low mass the 600 ton module can provide important contributions to the proton decay search, in particular for those decay modes (that we can call *exotic decay modes*) with high multiplicities (3 or 4 particles in the final state) or, more generally, with signatures particularly difficult to identify with other detector techniques.

With our 600 ton module, in a few years, we will be able to explore a lifetime region exceeding 10^{32} years for most of the decay channels, with a considerable improvement (a factor 5 to 100) of the present limits for the *exotic channels* (see figure 2). It is clear that with one single module we cannot reach the lifetime limit of 10^{34} years (see figure 3) which is needed to fully test the minimal SUSY Grand Unification Theory and which is the final goal for the 5000 ton detector.

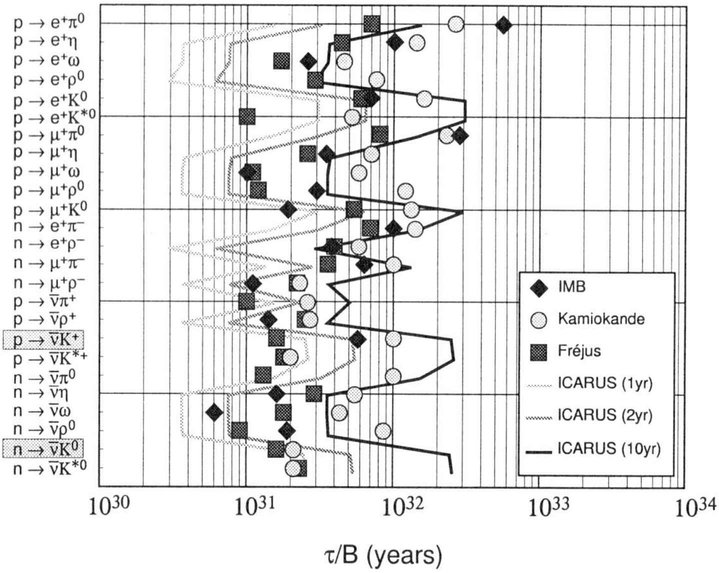

Figure 3. ICARUS 600 ton limits for most common decay modes.

2.3. Atmospheric neutrinos

The atmospheric neutrino issue is corrently considered one of the most burning and debated questions in High Energy physics. Many experiments (Kamiokande, IMB, Soudan II) found a value of the ratio of muon to electron events:

$$(\mu/e)_{data}/(\mu/e)_{MC}$$

considerebly less than 1.0.

The recent SuperKamiokande results [6] confirm the general effect, but suggest a smaller Δm^2 value ($\approx 2 \cdot 10^{-3}\ eV^2$), at the limit of consistency with the Kamiomande result. If it would be due to $\nu_\mu \to \nu_e$ oscillation, a large effect should have been seen by CHOOZ [7], which is not the case. The important, negative result from CHOOZ implies that the cosmic ray effect claimed by Kamiokande/SuperKamiokande needs to be confirmed more than ever.

ICARUS is the only new experiment which can definitely clear this problem in view of the superb quality of the recorded events. In effect the 600 ton module is ideally suited to perform the atmospheric neutrino study in a short time scale and with sufficient sensitivity to cover the present Kamiokande and SuperKamiokande allowed regions in the neutrino oscillation parameter space (see figure 4).

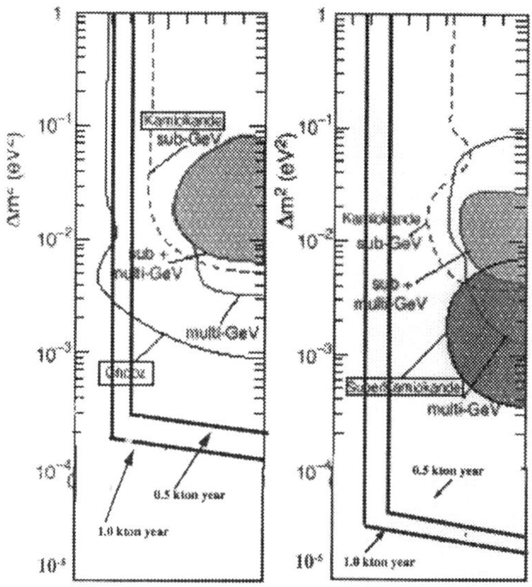

Figure 4. 90% confidence level sensitivity regions for the ICARUS 600 ton module for (a) $\nu_\mu \to \nu_e$ and (b) $\nu_\mu \to \nu_\tau$ oscillations.

The calculation of the neutrino-induced event rate expected in case of no oscillation is based on the convolution of the flux with the cross-section for neutrino interaction, in a detector with a given detection efficiency and acceptance. To calculate the rate we used the set of cosmic fluxes from the Bartol Group [8]. The total number of predicted events is 144 per year for charged-current type and 20 per year for neutral-current type.

2.4. Long Baseline Beam neutrinos

Assuming that the atmospheric neutrino anomaly is confirmed by ICARUS, the only channel left as possible is the $\nu_\mu \to \nu_\tau$ oscillation. It would be of great interest to use the long baseline neutrino beam, which gives sensibility to the same region of the oscillation parameter plane, but with a much smaller systematic error. In this case we could tune the neutrino energy spectrum of the beam in order to enhance the sensitivity in the desired oscillation parameter region.

We can make the following simple considerations:

a) the SuperKamiokande - KEK experiment will eventually see only a negative effect in the lack of ν_μ signals, since it is below threshold for ν_τ events;

b) higher energy experiments in the USA and CERN - LNGS have still the best chance of observing a positive effect as appearence of ν_τ events;

c) however only ICARUS has a demonstrated capability of identifying the ν_τ events at the single event basis, in view of its superb resolution, giving at last an unambiguous positive signal.

To fully exploit the experimental possibilities of the CERN to Gran Sasso neutrino beam, at least 4 modules of 600 ton are needed.

2.4.1. The CERN to Gran Sasso beam. Recently a Technical Committee, mandated by INFN and CERN, issued a report [9] in which a conceptual design of a new CERN Neutrino beam to Gran Sasso (NGS) is outlined.

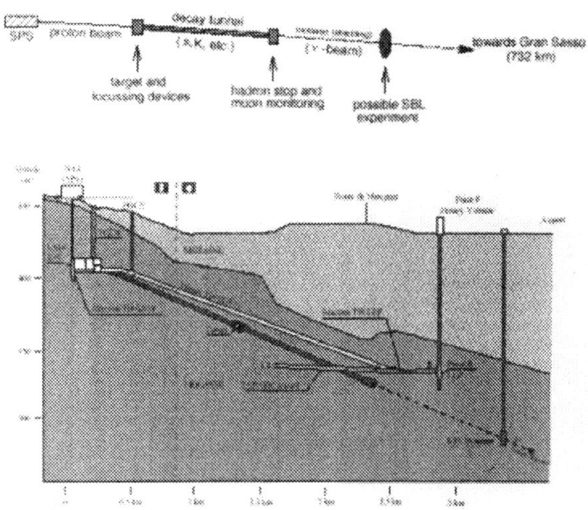

Figure 5. Schematic layout of the CERN neutrino facility and vertical cut showing the SPS and the main components of the NGS.

The main ingredients of the proposed NGS are shown schematically in figure 5:

- a proton beam with an energy up to 450 GeV extracted from the CERN SPS accelerator, transported and focused onto a small spot at the target;

- a target station consisting of a segmented graphite target, in which the secondary particles (π and K) are produced;

- a two-stage focusing system, horn and reflector, which focuses a range of chosen π and K momenta into a parallel beam, pointing toward the Gran Sasso Laboratory;

- a 1000 m long decay tunnel, which allows a fraction of the π, K to decay in fligth, producing a high-intensity ν_μ beam;

- a hadron stop, which absorb the non-interacting primary protons as well as those secondary hadrons which have not decayed;

- a muon monitoring system, permitting on-line monitoring, tuning, and control of the beam and its alignment.

The final detailed parameters of the beam need to be defined by the end of 1999, in order to assure a beam starting date of spring 2002. However, in order to proceed with the complete NGS design and to provide a basis for comparison, parameters of a *reference beam* have been defined. The energy distribution of the ν_μ is shown in figure 6. The predicted performances of the NGS *reference beam* are summarized in Table 1.

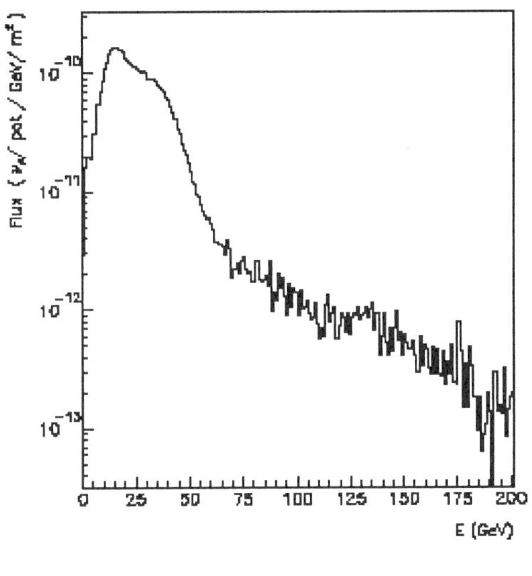

Figure 6. Energy distribution of the ν_μ flux at Gran Sasso (732 km) .

Table 1. Predicted performance of the NGS *reference beam.*

ν_μ $[m^{-2}/\text{pot}]$	$4.39 \cdot 10^{-9}$
Mean ν_μ energy [GeV]	26.7
ν_μ CC events/pot/ton	$4.73 \cdot 10^{-20}$
ν_μ CC events/year/ton	1.40
Fraction of other neutrino events:	
ν_e/ν_μ	0.8%
$\bar{\nu}_\mu/\nu_\mu$	2.2%
$\bar{\nu}_e/\nu_\mu$	0.8%

2.4.2. Search for $\nu_\mu \rightarrow \nu_x$: disappearence test. The rates of ν_μ charged current events expected at the Gran Sasso are 650 for 1 module per year, or 2,600 for 4 modules per year, considering the expected performances of the reference beam, described in the previous paragraph.

If ν_μ oscillations exist, the rate of ν_μ interactions at Gran Sasso will be depleted with respect to expectations, independently of the flavour into which the neutrinos have oscillated [4].

The oscillated events will be concentrated at low energy (see figure 7) and, considering an energy below 20 GeV, the rates of ν_μ charged current events expected at the Gran Sasso should be about 62 charged current events for 1 module per year or 248 for 4 modules per year.

2.4.3. Search for $\nu_\mu \rightarrow \nu_e$: appearence test. If, as expected, the ν_μ flux is strongly depleted at Gran Sasso, the next question to answer is *into what?*. The way to answer the question is to study the ν_e or ν_τ appearence rates. Both neutrino flavours can be explicitly and efficiently revealed by their charged current interactions given the high energy of the beam. For what concerns the $\nu_\mu \rightarrow \nu_e$ oscillation, in case of a negative result, a limit on the oscillation probability would be:

$$P(\nu_\mu \rightarrow \nu_e) < N_e{}^*/N_{\nu_\mu}$$

where $N_e{}^*$ is the upper limit on the number of signal events in presence of the beam ν_e background and N_{ν_μ} is the total number of deep inelastic events.

There is no doubt that the detector can detect unambiguously the ν_e charged current events. Since in this case the neutrino energy is measured, the background from the beam ν_e contamination, with a broad spectrum peaked at very high energy ($\approx 60GeV$) is largely suppressed.

2.4.4. Search for $\nu_\mu \to \nu_\tau$: appearence test;. Similarly to the previous case, in case of a negative result, a limit on the oscillation probability would be:

$$P(\nu_\mu \to \nu_\tau) < N_\tau / (N_{\nu_\mu} \cdot \epsilon \cdot Br \cdot (\sigma_\tau / \sigma_\mu))$$

where $N_\tau = 2.3$ (at the 90% C.L.) is the upper limit on the number of signal event (the expected ν_τ background in the neutrino beam at Gran Sasso is negligible ($\approx 10^{-6}$)), N_{ν_μ} is the total number of deep inelastic events, ϵ is the detection efficiency, Br the branching ratio and (σ_τ/σ_μ) the ratio of the cross-sections for ν_μ and ν_τ.

Considering an oscillation probability of the order of $5 \cdot 10^{-3}$, the ν_τ interactions will be of the order of 30 (for 4 modules and 4 years of run). Given a value of $\epsilon \cdot Br$ of the order of 0.1, that is very likely reachible in view of the ICARUS superb resolution, we can get an unambiguous positive signal.

3. 600 ton module

3.1. Cryogenics

The cryogenic and purification systems for the 600 ton module (T600) are being developed in co-operation with AIR LIQUIDE ITALIA industry. The basic schemes described in our proposal [5] have been revised and redesigned with some modifications and improvements.

At the beginning of 1997 the basic engineering of the T600 was completed. In agreement with our requirement of maximum useful transportable volume a rather innovative solution for the cryostat structure has been proposed with external modular insulation surrounding the liquid argon (LAr) aluminum container.

A general view of the T600 cryostat is shown in Figure 8. The LAr is contained in two mechanically independent vessels each one being transportable on the Italian highways and inside the Gran Sasso Laboratory. The containers will be realized by welding together basic structures consisting of an aluminum honeycomb panel glued to a reinforcing extruded aluminum frame. This solution, as well known in aeronautic industry, provides high mechanical stability with lightweight, with obvious advantages for what concerns trasportation: a conventional stainless steel structure with similar mechanical strength would weight about 70 ton, to be compared with the 20 ton of the proposed solution. Moreover, the high aluminum thermal and electrical conductivity will guarantee a very uniform temperature inside the LAr together with a very effective shielding against

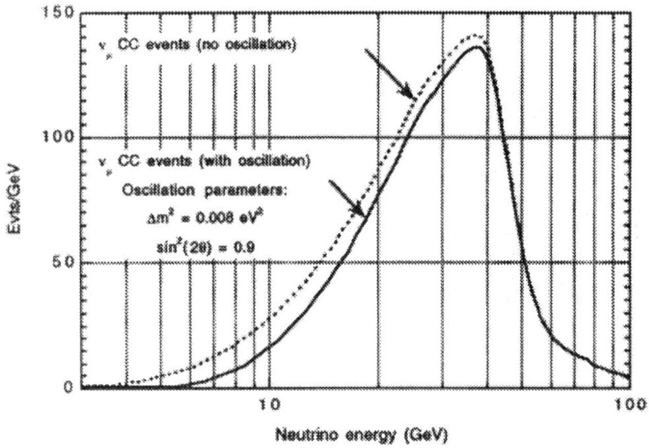

Figure 7. Comparison between the neutrino energy spectra expected at Gran Sasso in case of no oscillations (dashed) and the shown oscillation parameters (line) .

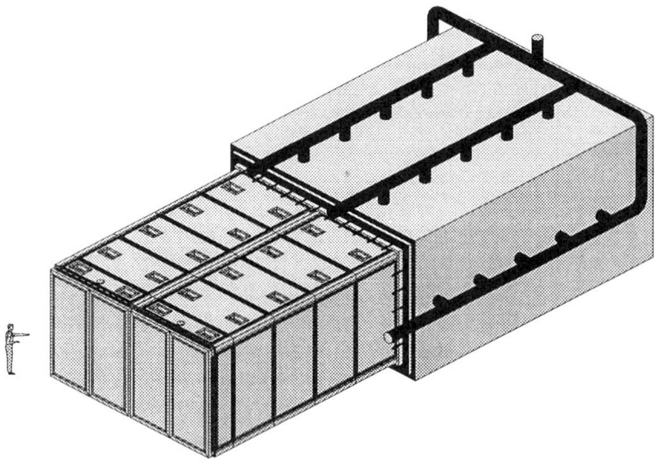

Figure 8. A 3-D sketch of the T600 cryostat.

external electronics noise. Cooling pipes are directly inserted into the honeycomb panels and cooling will be provided by a forced circulation of liquid N_2 at 3 bar abs (89 K). In order to optimize the thermal exchange between the cooling circuit and the inner wall, the honeycomb panels will be filled with He or N_2 gas. According to simulations, at regime, the LAr temperature will be uniform within 0.2 K.

The insulation is made of Aramid fiber (Nomex) honeycomb panels, about $400\,mm$ thick, refrigerated in the middle by a shield cooled with N_2 gas recuperated from the main cooling circuit. In direct contact with the LAr vessel, a $100\,mm$ layer of compressed polyester fiber foils provides the elasticity required to compensate for the thermal contraction of the aluminum LAr container. The heat losses through the insulation are estimated to be about $15\,W/m^2$, a figure quite competitive with those of other, much less handy, non-evacuated solutions (Perlite, foams, etc.).

The present solution for the cryostat offers several advantages of particular importance for our application:
a) modularity: the cryostat dimensions can be changed without major technical upgrades;
b) ease of installation: very few onsite operations are required to assemble the cryostat;
c) lightweight;
d) internal temperature uniformity.

Table 2 contains the main parameters of the T600 cryogenic system.

To test the functionality of the proposed solutions a $10\,m^3$ cryostat was built, equipped with Ar purification and recirculation units of the same size as those foreseen for the T600 module (Figure 9).

The prototype has been first operated in July 1997 by filling it with LN_2 and mechanical and thermodinamical behaviour has been tested. First (partial) filling with purified LAr was performed at the beginning of September 1997.

The tests results have been in general satisfactory, especially for what concerns the LAr purity:
1) the purification speed has been checked out and it exceeded our initial specifications ($\approx 2\,m^3h^{-1}$);
2) aluminum was demonstrated not to be a problem concerning LAr purity;
3) gas recirculation and purification system has been fully verified. Recirculation rates exceed our initial specification.

Consequently cryogenics and purification systems for the T600 have been formally commissioned to AIR LIQUIDE ITALIA in November 1997. The first half module should be transported and installed in our laboratories in Pavia at the beginning of 1999.

Figure 9. Photograph of the 10 m^3 prototype taken during the first cool-down in July 1997.

Table 2. Main parameters of the T600 cryostat

N. of containers	2
Single container:	
Internal length	19.6 m
Internal width	3.6 m
Internal height	3.9 m
External length	19.9 m
External width	3.9 m
External height	4.2 m
volume	277 m^3
weight	19.6 *ton*
Container material	*aluminum*
Design pressure	1.5 *bar abs*
Working pressure	1.25 *bar abs*
Insulation thickness	400 + 100 mm
Heat losses at regime	$\approx 12\, W/m^2$
Cooling liquid	LN_2 at 3 *bar*
Overall ext. dimensions:	
length	20.6 m
width	8.2 m
height	4.9 m
Total int. volume	**534 m^3**
Total LAr mass	**748 ton**

3.2. Wire chamber system

The wire chamber read-out complete design has been developed in co-operation with BREME TECNICA. The main concept is the elasticity of the frame in order to greatly reduce the thermal stress during the cooling phase.

In Table 3 the main characteristics of the read-out system are summarized. The wire chamber design is modular: every Peek comb holds 32 wires, with a pitch of 3 mm and every structure portion (about 2 m long) comprises 18 x 2 combs. There are 3 wire planes 3 mm apart, one horizontal and the other ones inclined $\pm 60°$. To increase the sensitive volume seen by 3 planes, 4 triangular shaped portions of wires have been added to the original design, with wire lengths ranging from 0.49 to 3.77 m, for each wire chamber.

A *Test Module* (1/12 of the complete system) was constructed to test every mechanical and cryogenic aspect of the design (see Figure 10). In particular the following aspects of the design needed a detailed study:

Table 3. Main parameters of the read-out planes of the ICARUS 600 ton module

Number of read-out chambers	4
Wires orientation respect to horizontal	0° , ±60° with 3 mm pitch
Wires length:	
horizontal wires	9.40 m
wires @ ±60°	3.77 m
wires @ ±60° (triangles)	0.49 ÷ 3.77 m
Wires diameter	0.15 mm
Wires nominal tension	12 N
Number of wires per chamber:	
horizontal wires	2,112
wires @ ±60°	9,280
wires @ ±60° (triangles)	1,920
Total	13,312
Total number of wires	53,248
Maximum drift length	1.5 m
Total sensitive LAr mass	476 ton

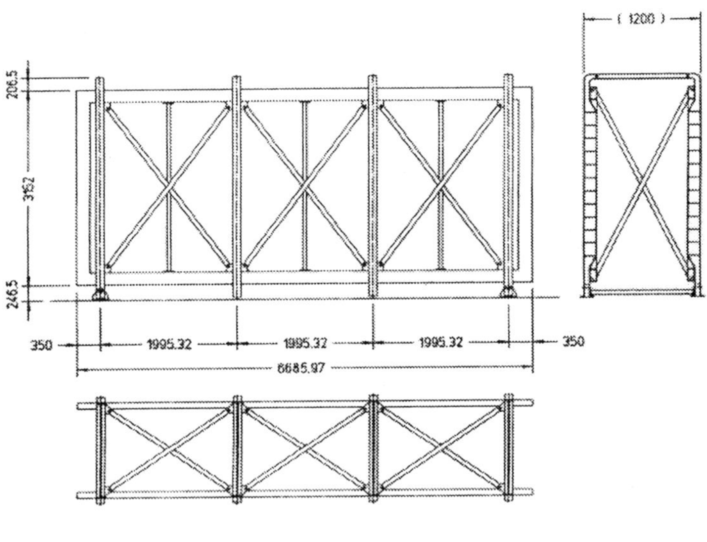

Figure 10. Test Module general view.

(i) the technical solution for the manufacturing of stainless steel (0.15 mm diameter) wires with a sleep knot around a small bush at each end, with the design and construction of a semi-automatic wiring table: there will be around 60,000 wires in the 600 ton module;

(ii) the comb (for 32 wires) structure (see Figure 11): we envisage a solution requiring the use of suitable molds for the Peek combs, together with the 32 stainless steel pins which hold the small bushes with the wires;

(iii) the elastic system, with a bucket-spring tie rod every 2 m , which keeps the wire mechanical tension constant, and minimizes the thermal stresses;

(iv) the support structure of the whole internal detector, which must be independent from the cryostat walls, in order to minimize the inconveniences due to eventual stresses in the cooling phase.

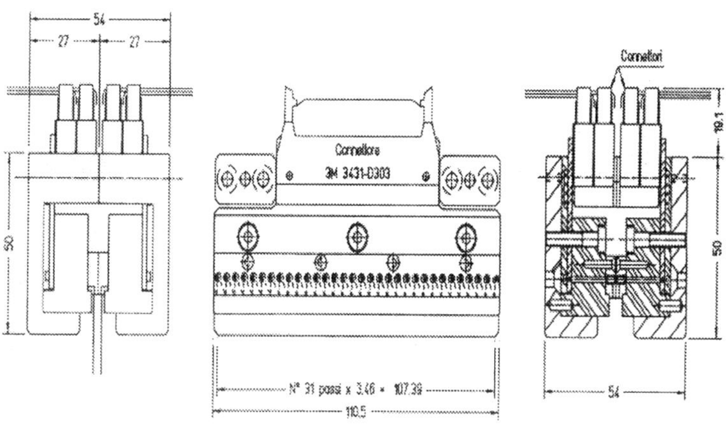

Figure 11. Peek comb structure to hold the 32 stainless steel wires.

The entire structure (about 6.0 m x 3.9 m x 1.5 m) has been mounted at room temperature to study in details every mechanical aspect: it has shown the validity of the design and persuaded the Collaboration to go ahead with the final design of the entire read-out system.

After this test, a portion of it (one module: 2.0 m x 3.9 m x 0.9 m) has been built dismounting the structure and using parts of it. This reduced module has been put in the 10 m^3 Cryogenic Test Facility to study the cryogenic behaviour of the mechanics and in particular the thermal stresses during the cooling phase, at the beginning of the year 1998. The results have been very encouraging: no wire has been broken or exceeded its elastic limit and the mechanical structure was not deformed.

4. Conclusions

The ICARUS experiment is now building its first 600 ton module. The first half of the cryostat will be delivered by industry for the end of 1998. After that the complete read-out detector will be mounted inside it. The start of data taking in the Gran Sasso underground laboratory is foreseen for mid 2000.

For fully exploit the long baseline neutrino beam nexperimental possibilities at least 3 other 600 ton module are needed.

The final goal of the experiment is to reach a sensitive mass of the order of 5000 ton to study the proton decay issue. This mass will be reached with a cluster of 600 ton modules.

References

[1] C.Rubbia, CERN-EP Internal Report 77-8 (1977)

[2] P.Benetti et al., *Nucl. Inst. and Meth* **A332** (1993) 395

[3] Cennini P et al., *ICARUS II, a second-generation proton decay experiment and neutrino observatory at the Gran Sasso Laboratory* LNGS-94/99 Vol. I&II, Proposal

[4] J.P.Revol et al., *A search programme for explicit neutrino oscillations at the long and medium baselines with ICARUS detectors* - ICARUS-TM-97/01 (1997).

[5] Cennini P et al., *A first 600 ton ICARUS detector installed at the Gran Sasso Laboratory* Addendum to Proposal LNGS-94/99 I&II

[6] A.Suzuki, This Conference

[7] M.Apollonio et al.,*Phys. Lett.* **B420** (1998) 397

[8] G.Barr, T.Gaisser and T.Stanev, *Phys. Rev.* **D39** (1988) 85; T.Gaisser and T.Stanev, *Phys. Rev. Lett.* **51** (1983) 223

[9] K.Elsener et al. *The CERN neutrino beam to Gran Sasso (NGS): conceptual technical design* - CERN-98-02; INFN-AE-98-05 (1998)

PART 3

B and L Violation, the Early Universe and Cosmology

Status of the electroweak phase transitions

Zoltán Fodor [1]

Institute for Theoretical Physics, Eötvös University, Budapest, Puskin 5-7, H-1088, Hungary

Abstract. The observed baryon asymmetry of the universe has finally been determined at the finite temperature electroweak phase transition. In order to understand the baryon asymmetry a quantitative description of the electroweak phase transition is needed. In this talk some features of this phase transition are summarized. Particular interest is paid to the analytical and lattice estimates on the sphaleron transition and to the nature of the electroweak phase transition in the standard model (SM) and in the minimal supersymmetric standard model (MSSM). On the one hand due to the large top-quark mass no SM-like Higgs boson can give strong enough phase transition. Thus, in the standard model the electroweak baryogenesis scenario can be excluded. The non-perturbative phase diagram of the electroweak model will be given and it will be shown that actually no real phase transition can happen in the standard model. On the other hand there is a small region in the parameter space of the minimal supersymmetric standard model, which might explain the observed baryon asymmetry. Since this small region is predicted perturbatively a lattice confirmation is needed.

1. Introduction

According to the standard picture of the cosmological phase transitions at high temperatures (e.g. in the early universe) the electroweak gauge symmetry has been restored. As the universe expands and supercools there is a phase transition between the high temperature "symmetric" and low

[1] E-mail: fodor@thwgs1.cern.ch

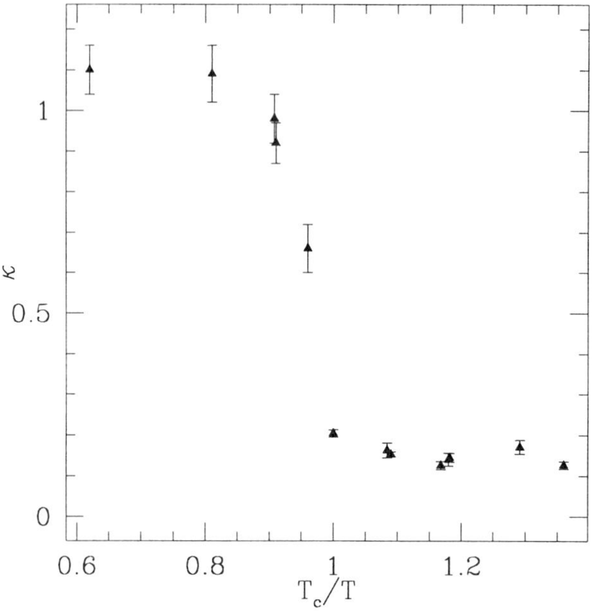

Figure 1. Transition rate normalized by T^4 as a function of the temperature.

temperature "broken" phases. The characteristics of this phase transition (critical temperature: T_c, surface tension: σ, etc.) are clearly of interest.

There are evidences that the world is made of exclusively matter, at least on the $10^{13} M_\odot$ scale. The quantitative measure of the baryon asymmetry is the dimensionless ratio of the baryon number density to the entropy density

$$\Delta_B = B/s \approx 5 \cdot 10^{-11}. \tag{1}$$

Since no known mechanism can separate on the above huge scales we have to understand the origin of this asymmetry. There are basically two possibilities. The baryon asymmetry might have been an initial condition of our universe or it could have been generated in the early universe. The second possibility (baryogenesis) is clearly more attractive.

In order produce the observed baryon asymmetry three conditions must be satisfied.

a. baryon number violating processes

b. C and CP violation

c. departure from the thermal equilibrium (the basic question is the

strength of the phase transition).
It is easy to understand the role of these conditions.
a. In the lack of baryon number violation the baryonic charge of the universe were constant. Assuming zero baryon number as an initial condition would result in no baryon asymmetry today.
b. If C or CP were conserved, then the rate of processes producing baryons would be the same as the rate of processes producing antibaryons. No net baryon number could have been generated.
c. In thermal equilibrium the universe stationary, it has no time dependence. If the initial condition were zero baryon number, it would remain zero forever.

An attractive proposal is to understand the observed baryon asymmetry as produced at the finite temperature electroweak phase transition [1]. Section 2 will discuss the first (a) Section 3 the last (c) condition in this scenario. Section 4 contains the conclusion.

2. Sphaleron rate at finite temperature

The basic source of the baryon number violating processes in the standard model is the electroweak anomaly for baryon number, which demands that

$$\Delta B \sim g^2 \int dt d^3x \mathrm{tr} F\tilde{F}, \tag{2}$$

where g is the weak coupling and F is the field strength of the weak gauge field. This anomaly equation relates the produced baryon number to topological transitions of the weak gauge fields. The vacua with different topological numbers (N_{CS}) are separated by a potential barrier.

Note, that the above anomaly equation holds not only for baryons but for leptons, too. Thus, baryon and lepton numbers are violated simultaneously ($B+L$ violation).

At zero temperature the transition between the vacua is a tunneling event with an unobservably small probability: $\approx 10^{-170}$. The transition at high temperatures, but below T_c is a thermal jump. The system jumps up to the top of the barrier (a saddle point: sphaleron) and rolls down to the neighbouring vacuum. In the meantime baryon and lepton numbers are violated. The jump to the barrier, therefore the transition rate is suppressed by the Boltzmann factor

$$\Gamma \propto \exp\left[-E_{sph}/T\right], \tag{3}$$

where the sphaleron energy is proportional to the mass of the W-boson $E_{sph} \propto m_W/\alpha_W$ [2].

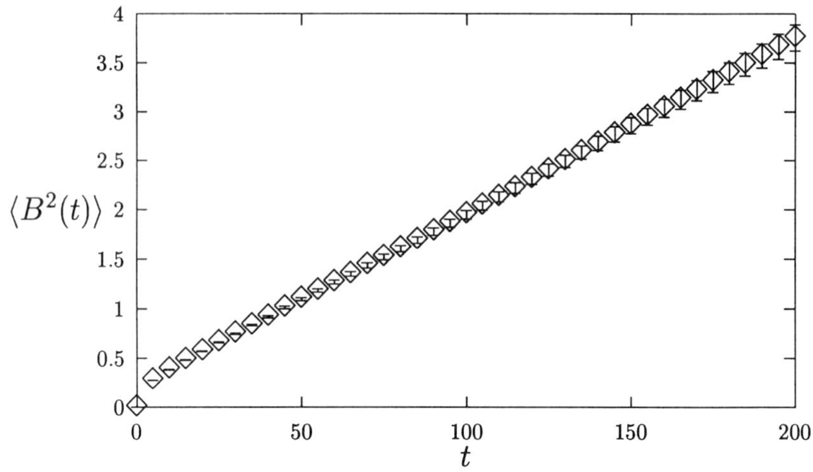

Figure 2. Diffusion of N_{CS}.

Rapid baryon violation processes in the broken phase can wash out the asymmetry generated before. Therefore a "minimal" condition for a successful baryogenesis is that these processes should be slower than the expansion rate of the universe. Comparing (3) with the Hubble constant the minimal condition can be written as $\varphi_b/T_b > 1$, where φ_b is the value of the Higgs field at the transition temperature T_b.

Above T_c the exponential suppression of the transition rate is absent. Naive power counting suggest

$$\Gamma = \kappa \cdot (\alpha_w T)^4, \tag{4}$$

where the constant κ is of order 1.

Recently it has been argued [3] that the assumptions used to derive eq. (4) are not valid. The real dependence on α_w is a different one

$$\Gamma = \kappa' \cdot \alpha_w^5 T^4. \tag{5}$$

The origin of this damping is the fact that nearly static magnetic fields can be absorbed by the system, which results in a loss of energy for the magnetic fields of interest. In the symmetric phase the baryon number violating processes are much faster than the expansion rate of the universe.

In recent years there has been considerable activity in order to determine the sphaleron transition rate at both sides of the phase transition.

No successful numerical method is known for a full Minkowskian theory; however, important results have been obtained by real time simulations in the classical approximation for the finite temperature theory.

The used procedure contains several steps. The classical theory is formulated in terms of fields and their conjugate momenta on a spatial lattice. The phase-space of the system is then sampled with the statistical weight $\exp(-H/T)$. Starting with some initial configuration from the thermal sample the classical canonical equations give the real-time evolution of the fields. As a function of time one can determine different observables and their averages (such as changes of the topological charge). There are fewer physical degrees of freedom in the theory than the number of the phase space variables used to formulate it. It is a highly non-trivial task to find effective thermalization algorithms for constrained systems. At present there are two good solutions to this problem [4].

In the broken phase numerical simulations [5] indicated an extreme difference between the lattice results and eq. (3). The observed sphaleron transition rate was a bit smaller than in the symmetric phase; however, as it can be seen on Fig. 1 no Boltzmann suppression has been observed. Going down into the broken phase the difference between the analytical and lattice results is several orders of magnitude. It is argued [3] that the reason is the problematic definition of the topological charge on a lattice, which give a systematic error in the rate.

A large scale numerical study for the sphaleron rate in the symmetric phase [6] favoured the naive law

$$\Gamma = \kappa(\alpha_w T)^4. \tag{6}$$

The classical motion of the Chern-Simons number consists of two pieces. The thermal fluctuation, which linearly diverges with the lattice cutoff, and a random walk between different vacua (see Fig. 2).

$$< B^2(t) >= c + \Gamma V t. \tag{7}$$

A detailed finite size and finite scaling analysis suggested that the above naive formula is correct with a coefficient of $\kappa = 1.09 \pm 0.05$.

According to [3] the observed rate might be only a lattice artifact. The authors reanalysed their data with an improved technique (cooling of lattice field configurations to determine the time evolution of the topological charge) [7].

They introduced, along with the real time t, a cooling time τ. The dynamical variables are functions of them: $p_i(t,\tau), q_i(t,\tau)$. The t evolution is given by the standard equation of motions, while for the evolution in the τ direction an overdamped motion is used

$$\partial_\tau q_i(t,\tau) = -\partial_{q,i} V[q(t,\tau)]. \tag{8}$$

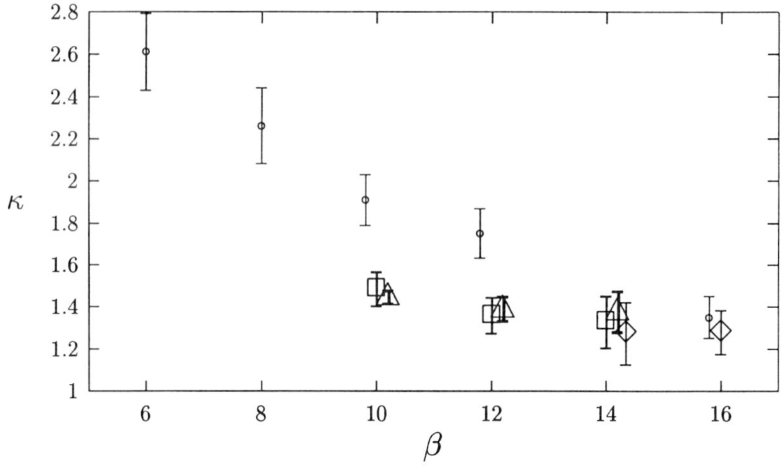

Figure 3. The dimensionless coefficient κ as a function of the inverse lattice spacing.

It is easy to see that in the vicinity of a static solution cooling leads to an exponential decay of stable eigenmodes and exponential growth of unstable ones. Moreover, the rate of decay (growth) is exponentially rapid for high frequency mode. The new results has shown some deviation from the older one; however, the $\Gamma = \kappa' \alpha_w^5 T^4$ law of the symmetric phase can not be confirmed.

Fig. 3 shows the result of this analysis (open squares and triangles) and that of [8] (open dots). According to [3] the rate should have a $1/\beta$ behaviour. Clearly, larger measurement samples and careful systematic error analysis is needed to resolve the issue of the lattice spacing dependence of the rate.

A definitely positive outcome of the cooling technique is the new simulation in the broken phase. With the cooling technique no transition between the different vacua has been observed. This result resolves the discrepancy between [5] and eq. (3); however, no definite answer to the numerical value of the transition rate could be given.

Recently the multicanonical technique has been successfully applied to determine the nonperturbative transition rate in the broken phase [9]. The reason to use the multicanonical technique is the following. Due to the large energy of the sphaleron in the broken phase the transition rate is very small,

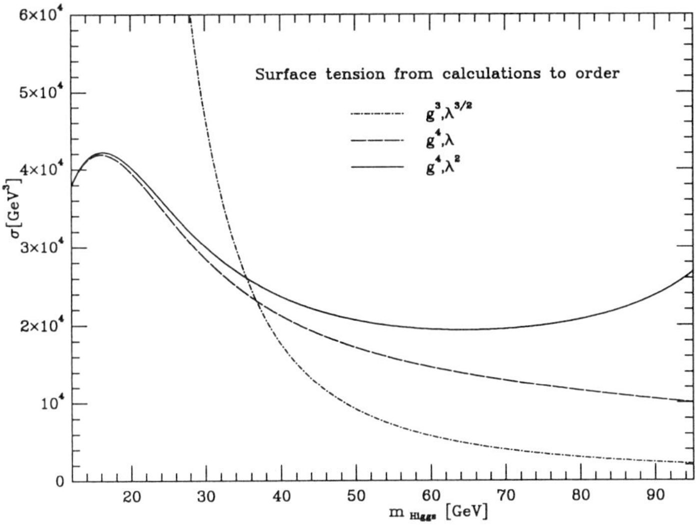

Figure 4. Surface tension from the different potentials as a function of m_H.

actually exponentially supressed. A known technique to solve this problem is the multicanonical simulation. This procedure artificially enhance the probability of regions which can not be reached by standard Monte-Carlo techniques (the top of the barrier between the vacua: sphalerons). For a given order parameter (in this case N_{CS}) one adds a term to the action $\sim \log P(N_{CS})$. Moore combined the above cooling and multicanonical techniques and obtained orders of magnitude smaller transition rate in the broken phase than in the symmetric phase.

$$\log(\Gamma \cdot T_c^4) = 24.7 \pm 0.4 \tag{9}$$

There are a few important observations. The analytic "two-loop" transition rate in the broken phase is approximately 2.5-times faster than the numerically obtained nonperturbative rate. It is a sign that not only the larger nonperturbative jump of the order parameter (Higgs-field) is responsible for the observed rate. The nonperturbative rate can give us an estimate for the bound above which the electroweak baryogenesis is excluded (wash out of the previously generated baryons). This bound is $\lambda/g^2 \approx 0.0036$.

3. The nature of the phase transition

Perturbative studies show that in the realistic Higgs mass range ($m_H >$ 70 GeV) the perturbative approach breaks down [11], it predicts $\mathcal{O}(100\%)$

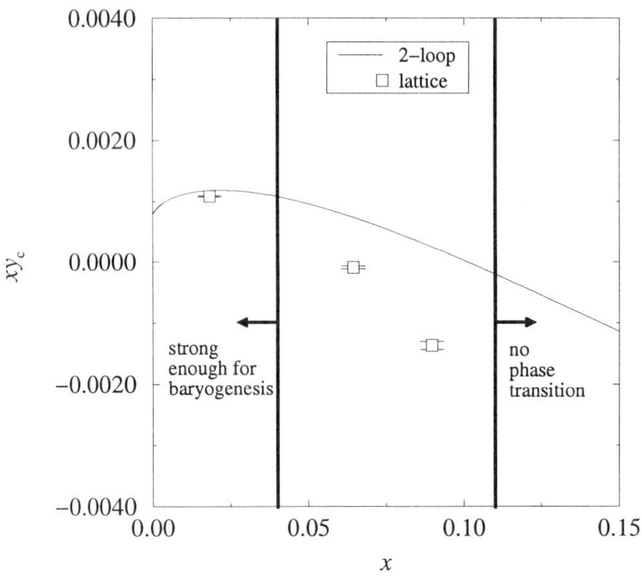

Figure 5. Phase diagram of the three-dimensional SU(2)-Higgs model.

corrections [10] for the relevant quantities (e.g. interface tension, latent heat or correction to the course of the phase transition).

A nice illustration (c.f. Fig. 4) for that is the interface tension as a function of the Higgs-boson mass in different orders of the perturbation theory. As it can be seen in the physically allowed region (according to the LEP experiments the lower bound for the standard model Higgs-boson mass is relatively large: $m_H > 85$ GeV) the perturbation theory breaks down, its predictions can not be believed.

A popular way to study this basically non-perturbative problem is first to perform a dimensional reduction in perturbation theory. One starts with the original theory (e.g. Standard Model) and integrates out the heavy degrees of freedom perturbatively. The obtained theory is a three-dimensional bosonic one. The temperature dependent parameters of this theory are determined by the matching conditions between the full theory and the reduced one.

The theory has a dimensionful coupling, which fixes the overall scale. The two dimensionless quantities are x and y. One of them is x which determines the properties of the phase transition (connected to the zero

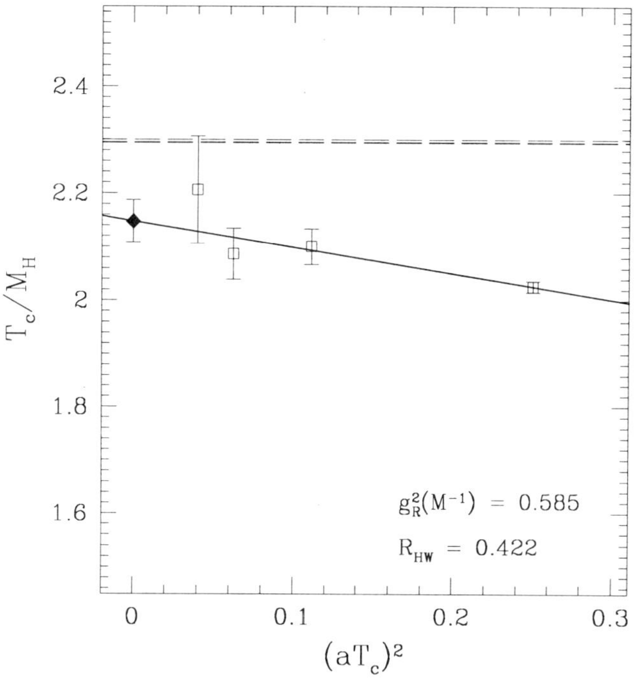

Figure 6. Critical temperature as a function of the lattice spacing.

temperature features of the original four-dimensional theory)

$$x \sim \frac{1}{8}\frac{m_H^2}{m_W^2} + c_2 \frac{m_{top}^4}{m_W^4}. \tag{10}$$

The other one is y which is used to tune the system in order to find the phase transition point (connected to the temperature of the original four-dimensional theory)

$$y \sim c_1 \frac{T - T_c^{(pert)}}{T_c^{(pert)}}. \tag{11}$$

Here c_1 and c_2 are some constants.

Static thermodynamical properties, mass spectrum and other related features have been studied both analytically and numerically for these three-dimensional models (a recent summary is [12] and see references therein).

According to these results the electroweak phase transition is of first order for small Higgs-boson masses; however, it turns out to be an analytic cross-over above $m_H \approx 67$ GeV (critical Higgs-boson mass value for the SU(2)-Higgs model) [13]. Due to the large mass of the top-quark the phase transition can not fulfill the φ_b/T_b condition for any choice of the Higgs-boson mass.

Fig. 5 shows the phase diagram of the three-dimensional theory. The solid line corresponds to the two-loop perturbation theory and the open symbols represent the result of the lattice simulations. For x values smaller than ~ 0.03 the phase transition is strong enough to satisfy the minimal condition $\varphi_b/T_b > 1$. The phase transition ends around $x \sim 0.1$ (we will discuss the determination of the endpoint). For large enough x values no phase transition can be observed, only a rapid cross-over occurs.

The determination of the endpoint of the finite temperature EWPT, thus a characteristic feature of the phase diagram, can be done by the use of the Lee-Yang zeros of the partition function $\mathcal{Z}$ [13]. One analytically continues $\mathcal{Z}$ to complex values of the couplings by reweighting the available data. Denoting κ_0 the lowest zero of $\mathcal{Z}$, i.e. the zero closest to the real axis, one expects in the vicinity of the endpoint the scaling law $\text{Im}(\kappa_0) = d_1(x)V^{-\nu} + d_2(x)$. In order to pin down the endpoint one looks for a x value for which d_2 vanishes. Again, the change in x can been done by reweighting (or by direct simulations at different x couplings).

Using this Lee-Yang technique the Bielefeld group obtained $x_{end} = 0.0951(16)$, whereas the result of the Leipzig group is $x_{end} = 0.1020(20)$ (the two groups used different lattice spacings in their simulations).

Another possibility in order to understand the non-perturbative features of the electroweak phase transition is to study the full four-dimensional theory by lattice simulations. Since fermions always have nonzero Matsubara frequencies, the perturbative treatment of these, at high temperatures very massive, modes could be satisfactory. Thus, the starting point of the lattice analyses is the $SU(2)$-Higgs model, which contains the essential features of the standard model of electroweak interactions.

In the last four years our group (DESY-Electroweak collaboration) presented a series of papers (see e.g. [14]) in order to clarify the details of the phase transition on four-dimensional lattices. Our work has been done on computers at HLRZ Jülich (CRAY-T90) and DESY-Ifh, Zeuthen (APE-Quadrics). The simulations have been performed for four set of parameters ($m_W = 80$ GeV): $m_H \approx 18$ GeV, 35 GeV, 49 GeV and 80 GeV.

For the first three masses the usual lattice formulation can be applied, thus identical lattice spacings in the spacial and temporal directions. In these cases a fairly good agreement is found between the lattice results and perturbation theory [16].

As an illustration (see Fig. 6) T_c/m_H is presented for $m_H \approx 35$ GeV. For this quantity very precise results exist and an extrapolation to zero

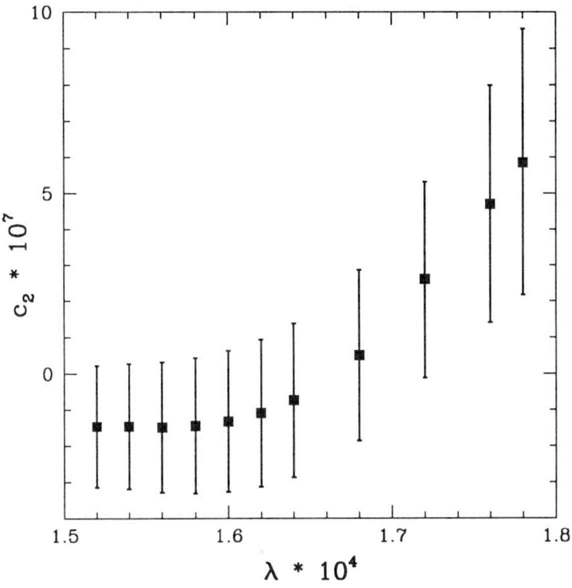

Figure 7. Dependence of c_2 on λ for $L_t = 3$.

lattice spacing is also possible. We have used $L_t = 2, 3, 4$ and 5 to extrapolate to the continuum limit. As it can be seen there is approximately three standard deviation discrepancy between the lattice and perturbative result. Note, that the result of the three-dimensional technique predicts a T_c/m_H much closer to the perturbative result. The lattice artifacts of the four-dimensional approach are expected to be proportional to a^2, whereas in the three-dimensional method they are proportional to a.

The $m_H \approx 80$ GeV case is much more difficult. The phase transition gets weaker, the lowest excitations have masses small compared to the temperature, T. From this feature one expects that a finite temperature simulation on isotropic lattice would need several hundred lattice points in the spatial directions even for $L_t = 2$ temporal extension.

In order to solve this problem we have used the simple idea that finite temperature field theory can be conveniently studied on asymmetric lattices, i.e. lattices with different spacings in temporal and spacial directions [15]. The resulting action contains anisotropies in the couplings. This action has been studied in perturbation theory and on the lattice.

One wants to tune the bare parameters in a way that the one-loop renormalized masses are finite in the continuum limit (however, their values in lattice units vanish $a_s M_{ren} = 0$ for $a_s \to 0$). At the same time the

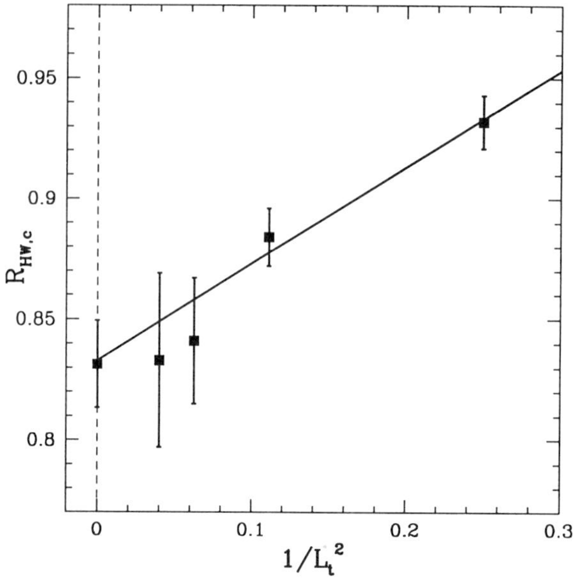

Figure 8. Dependence of $R_{HW,c}$, i.e. R_{HW} corresponding to the endpoint of first order phase transitions on $1/L_t^2$ and extrapolation to the infinite volume limit.

vacuum expectation value of the scalar field will be also zero in lattice units ($a_s v = 0$ for $a \to 0$) , i.e. we are at the phase transition point between the spontaneously broken Higgs phase and the SU(2) symmetric phase. The condition is fulfilled by an appropriate choice of the parameters (e.g. critical hopping parameter). The ratios of the other couplings are still free parameters and can be fixed by two additional conditions. We demand rotational (Lorenz) invariance for the scalar and vector propagators on the one-loop level. This ensures that the propagators with one-loop corrections have the same form in the z and t directions.

Clearly, arbitrary couplings for different directions would not lead to such rotationally invariant two-point functions. Technically the corrections to the anisotropies in the kinetic parts of the tree level propagators should be cancelled by the kinetic parts of the self-energies. This requires the knowledge of the wave function correction term in our theory. We have carried out both the perturbative and the lattice determinations of anisotropies. They are in complete agreement.

The values of the thermodynamical quantities (σ/T_c^3 and $\Delta\epsilon/T_c^4$) for $m_H \approx 80$ GeV are substantially smaller than their perturbative values. They are even consistent with a no first order phase transition scenario on the approximately 1-σ level. These results can be interpreted as a sign for the endpoint for the finite temperature electroweak phase transition.

Actually the endpoint of the phase transition is determined with a quite high accuracy. As discussed earlier the three-dimensional results show that for $m_H > 95$ GeV no first order phase transition exists and more specifically that the endpoint is $m_{H,c} \approx 67$ GeV [13]. A recent four dimensional analysis has given the value of the endpoint in the continuum limit [17], too. This work has performed the simulations on finer and finer lattices, moving along the lines of constant physics (LCP). The bare parameters are chosen in a way that the zero temperature renormalized gauge coupling g_R is held constant and the mass ratio for the Higgs- and W-bosons $R_{HW} = m_H/m_W$ corresponds to the Higgs mass at the endpoint of first order phase transitions: $R_{HW,c}$. These two conditions determine a LCP as a one-dimensional subspace in the original space of bare parameters. The position on the LCP gives the lattice spacing a. As the lattice spacing decreases $R_{HW,c} \to R_{HW,cont.}$. It is worth mentioning that the four-dimensional SU(2)-Higgs model is trivial for small gauge couplings, therefore, the $a \to 0$ limit can not be performed. Even the points on the endpoint LCP do not define continuum theories. The second order phase transitions on it merely reflect a finite temperature phenomenon, the corresponding zero temperature SU(2)-Higgs theory is still trivial.

Since our theory is a bosonic one we assumed that the corrections are quadratic in the lattice spacings; therefore an a^2 fit has been performed for $R_{HW,c}$ in order to determine its continuum value.

$V = L_t \cdot L_s^3$ type four-dimensional lattices are used. For each L_t 8 different lattices have been used, each of them had approximately twice as large lattice-volume as the previous one. The smallest lattice was $V = 2 \cdot 5^3$ and the largest one was $V = 5 \cdot 50^3$. Quite a large statistics was used and the Ferrenberg-Swendsen reweighting [18] was used to obtain information in the vicinity of a simulation point. The determination of the endpoint of the finite temperature EWPT, thus a characteristic feature of the phase diagram, is done by the use of the Lee-Yang zeros of the partition function $\mathcal{Z}$ [19]. Near the first order phase transition point the partition function reads

$$\mathcal{Z} = \mathcal{Z}_s + \mathcal{Z}_b \propto \exp(-V f_s) + \exp(-V f_b)\,, \tag{12}$$

where the indices s(b) refer to the symmetric (broken) phase and f stands for the free-energy densities. Near the phase transition point we also have

$$f_b = f_s + \alpha(\kappa - \kappa_c)\,, \tag{13}$$

since the free-energy density is continuous. It follows that

$$\mathcal{Z} = 2\exp[-V(f_s + f_b)/2]\cosh[-V\alpha(\kappa - \kappa_c)] \tag{14}$$

which shows that for complex κ $\mathcal{Z}$ = vanishes at

$$\mathrm{Im}(\kappa) = \pi \cdot (n - 1/2)/(V\alpha) \tag{15}$$

for integer n. In case a first order phase transition is present, these Lee-Yang zeros move to the real axis as the volume goes to infinity. In case a transition is absent the Lee-Yang zeros stay away from the real κ axis. Thus the way the Lee-Yang zeros move in this limit is a good indicator for the presence or absence of a first order phase transition [19] Denoting κ_0 the lowest zero of $\mathcal{Z}$, i.e. the position of the zero closest zero to the real axis, one expects in the vicinity of the endpoint the scaling law $\mathrm{Im}(\kappa_0) = c_1(L_t, \lambda)V^{\nu} + c_2(L_t, \lambda)$. In order to pin down the endpoint we are looking for a λ value for which c_2 vanish. In practice we analytically continue $\mathcal{Z}$ to complex values of κ by reweighting the available data. As an example, the dependence of c_2 on λ for $L_t = 3$ is shown in Fig. 7. To determine the critical value of λ i.e. the largest value, where $c_2 = 0$, we have performed fits linear in λ to the nonnegative c_2 values.

Having determined the endpoint $\lambda_{crit.}(L_t)$ for each L_t one calculates the $T = 0$ quantities (R_{HW}, g_R^2) on $V = (32L_t) \cdot (8L_t) \cdot (6L_t)^3$ lattices, where $32L_t$ belongs to the temporal extension, and extrapolate to the continuum limit. Having established the correspondence between $\lambda_{crit.}(L_t)$ and R_{HW}, the L_t dependence of the critical R_{HW} is easily obtained. Fig. 8 shows the dependence of the endpoint R_{HW} values on $1/L_t^2$. A linear extrapolation in $1/L_t^2$ yields the infinite volume (i.e. continuum limit) value of the endpoint R_{HW}. We obtain 66.5 ± 1.4 GeV, which is the final result.

Comparing the four-dimesional results to those of the 3d analyses [13] one observes complete agreement. Since the error bars on the endpoint determinations are on the few percent level, the uncertainty of the dimensional reduction procedure is also in this range. This indicates that the analogous perturbative inclusion of the fermionic sector results also in few percent error on M_H.

Based on the published data [14] and the endpoint result we are now able to draw the precise phase diagram of the SU(2)-Higgs model in the $(T_c/m_H - R_{HW})$ plane. This is shown in Fig. 9. The continuous line – representing the phase-boundary – is a quadratic fit to the data points.

Finally, it is possible to determine what is the endpoint value in the full SM. The nonperturbative analysis shows that the perturbative integration of the heavy modes is correct within the error bars. Therefore we can use perturbation theory [20] to transform the SU (2)-Higgs model endpoint value to the full SM. One obtains 72.4 ± 1.7 GeV, where the error includes the measured error of $R_{HW,cont.}$, g_R^2 and the estimated uncertainty [21] due to the different definitions of the gauge couplings between this paper and [20]. The dominant error comes from the uncertainty on the position of the endpoint. The present experimental lower limit of the SM Higgs-boson

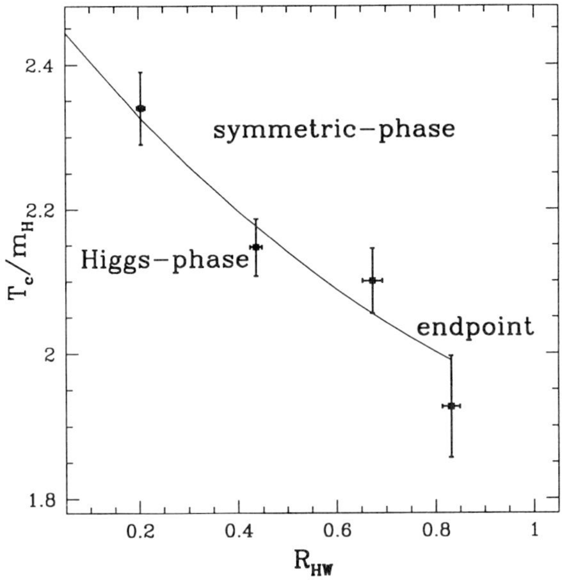

Figure 9. *Phase diagram of the SU(2)-Higgs model in the (T_c/m_H – R_{HW}) plane. The continuous line – representing the phase-boundary – is a quadratic fit to the data points.*

mass is 89.8 GeV [22]. Taking into account all errors (in particular those coming from integrating out the heavy fermionic modes), our endpoint value excludes the possibility of any EWPT in the SM. This also means that the SM baryogenesis in the early Universe is ruled out.

Similarly to the standard model case perturbative two-loop results exist for the minimal supersymmetric standard model phase transition [23]. An interesting feature of the result that it opens the baryogenesis window for light stop. Setting sun QCD diagrams (stop-gluon) can give large logarithmic contributions which increase the strength of the phase transition. Typically, setting sun diagrams give contribution proportional to $\varphi^2 \log \varphi$.

In the case of stop-gluon graph the prefactor is proportional to the strong coupling, which resulted in an enhancement effect on φ_b/T_b. In order not to have infrared divergencies for the stop sector the authors restrict themself to $m_U^2 > 0$ sector (negative m_U^2 are related to the existence of colour breaking minimum [24]). Fig. 10 shows that the ratio of the jump in the Higgs field divided by the temperature can be larger than one for low values of $\tan\beta$ (solid lines) and large enough m_A. Very low $\tan\beta$ corresponds to a light neutral Higgs (lines of constant Higgs masses are dashed). The LEP limit tells us that this mass must be larger than ≈ 70 GeV. Using this experimental bound the window for baryogenesis can be summarized as: $m_H < 85$ GeV; $2.25 < \tan\beta < 3.60$ and $m_A > 120$ GeV.

The allowed parameter space is very constrained. In particular $m_H <$ 85 GeV will be soon tested at LEP. The authors claim that the large correction compared to the one-loop results (approximately 100%) is not a serious problem here and it does not mean the breakdown of the perturbation theory. The reason for that is the fact that the QCD corrections are leading order on the two-loop level; thus the large correction is not a symptom of the perturbative expansion, it is merely the appearance of a new interaction term.

Nevertheless, the setting-sun diagrams have bad infrared behaviour in the standard model. Direct lattice Monte-Carlo approach is probably needed in order to answer the question in the MSSM case, too.

In order to answer the question for the MSSM the three-dimensional approach (dimensional reduction plus lattice Monte-Carlo simulations in three-dimensions) is used, too (see e.g. [12]). The results are qualitatively the same as summarized above. However, these analyses are less reliable precisely in the small m_U^2 region, which is the interesting one for electroweak baryogenesis.

4. Conclusions

Analytical and real time numerical result for sphaleron processes are still not in complete agreement. Above T_c new analytical estimates suggest a $\Gamma \propto \alpha_w^5 T^4$ behaviour for the transition rate, which is not seen unambigously in numerical studies. The present numerical techniques are not sensitive enough to determine the sphaleron transition rate below T_c. Perturbative studies show, that the electroweak phase transition can not be described above $m_H = 70$ GeV by perturbative methods. Lattice results exclude successful baryogenesis in the minimal standard model. The present bounds on the Higgs-boson mass most probably results in a cross-over (no phase transition scenario). The minimal supersymmetric standard model still has some constrained parameter region, which perturbatively predicts a phase transition, strong enough for baryogenesis.

Acknowledgements

The author acknowledges the organizers for the kind invitation, and the partial support of Hung. Sci. Foundation Grants T016240/T022929 and FKFP-0128/1997.

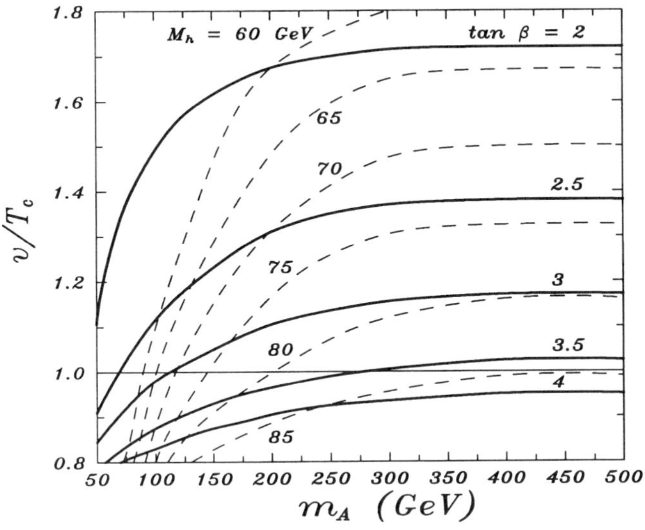

Figure 10. The normalized jump of the Higgs field in the MSSM.

References

[1] V. A. Kuzmin, V. A. Rubakov and M. E. Shaposhnikov, Phys. Lett. B155 (1985) 36.

[2] P. Arnold, L. McLerran, Phys. Rev. D36 (1987) 581.

[3] P. Arnold, D. Son, L. Yaffe, Phys. Rev. D55 (1997) 6264.

[4] A. Krasnitz, Nucl. Phys. B455 (1995) 320; G.D. Moore Nucl. Phys. B480 (1996) 657.

[5] J. Smit, W.H. Tang, Nucl. Phys. B482 (1996) 265.

[6] J. Ambjorn, A. Krasnitz, Phys. Lett. B362 (1995) 97.

[7] J. Ambjorn, A. Krasnitz, Nucl. Phys. B506 (1997) 387.

[8] . G.D. Moore, N. Turok, hep-ph/9703266.

[9] . G.D. Moore, hep-ph/9801204.

[10] Z. Fodor, A. Hebecker, Nucl. Phys. B432 (1994) 127.

[11] W. Buchmüller, Z. Fodor, T. Helbig and D. Walliser, Ann. Phys. (NY) 234 (1994) 260.

[12] M. Laine, hep-ph/9707415.

[13] K. Kajantie at al., Phys. Rev. Lett. 77 2887 (1996); F. Karsch, T. Neuhaus, A. Patkós, J. Rank, Nucl. Phys. Proc.Suppl. 53 (1997) 623; M. Gurtler, E.M. Ilgenfritz, A. Schiller, Phys. Rev. D56 (1997) 3888

[14] Z. Fodor et al. Phys. Lett. B334 (1994) 405; Nucl. Phys. B439 (1995) 147; Csikor et al. Nucl. Phys. B474 (1996) 421.

[15] F. Csikor, Z. Fodor, Phys. Lett. B380 (1996) 113.

[16] W. Buchmüller, Z. Fodor and A. Hebecker, Nucl. Phys. B447 (1995) 317.

[17] F. Csikor, Z. Fodor and J. Heitger, hep-ph/9809291,hep-ph/9809293.

[18] A. Ferrenberg and R. Swendsen, Phys. Rev. Lett. 61 2635 (1988); ibid 63 1195 (1989).

[19] C.N. Yang and T. D. Lee, Phys. Rev. 87 404 (1952).

[20] K. Kajantie at al., Nucl. Phys. B466 189 (1996).

[21] M. Laine, Phys. Lett. B385 249 (1996).

[22] P. McNamara, talk given at the ICHEP'98, Vancouver, Canada, 23-29 July, 1998.

[23] B. de Carlos, J.R. Espinosa, Phys. Lett. B407 (1997) 12.

[24] M. Carena, M. Quiros and C.E.M. Carlos, Phys. Lett. B380 (1996) 81.

Electroweak Baryon Number Violation and Constraints on Left-handed Majorana Neutrino Masses

Utpal Sarkar†[1]

†Physical Research Laboratory, Ahmedabad - 380 009, INDIA

Abstract.
During a large period of time, the anomalous baryon number violating interactions are in equilibrium, when the $(B+L)$ asymmetry is washed out. If there is any lepton number violation during this period, that will also erase the $(B-L)$ asymmetry. As a result, survival of the baryon asymmetry of the universe pose strong constraints on lepton number violating interactions. We review here the constraints on the left-handed Majorana neutrino masses arising from this survival requirement of the baryon asymmetry of the universe. We then briefly review models of leptogenesis, where lepton number violation is used to generate a baryon asymmetry of the universe and hence the constraints on the Majorana neutrino mass is relaxed.

1. Introduction

The generation of the baryon asymmetry of the universe starting from a symmetric universe is one of the very challenging question in cosmology [1]. Sakharov [2] pointed out that for this purpose we need three conditions (A) *Baryon number violation*, (B) *C and CP violation*, and (C) *Departure from thermal equilibrium*. It was then realised that grand unified theories (GUTs) satisfies all these criterion [3, 4]. Because of the quark-lepton unification baryon number is violated in GUTs. Since fermions belong to chiral representation, C is maximally violated in GUTs. Violation of CP

[1] E-mail:utpal@prl.ernet.in

was the only crucial point, which had to be incorporated in these theories. However, it was not difficult to consider some of the couplings to be complex so that there exist tree level and one loop diagrams which could interfere to give us enough baryon asymmetry in the decays of the heavy gauge and higgs bosons. It was noticed that the masses of these heavy bosons are constrained by the out-of-equilibrium condition to be very high. However, from the gauge coupling unification also the masses of these particles were found to be around the same value. In fact, the proton decay constrains also require the masses of these particles to be around the same scale [4].

This was considered to be one of the major successes of GUTs that it can explain the baryon asymmetry of the universe. After several years it was realised that the chiral nature of the weak interaction also breaks the global baryon and lepton numbers in the standard model [5]. Since these classical global baryon and lepton number symmetries are broken by quantum effects due to the presence of the anomaly, these processes were found to be very weak at the zero temperature. But at finite temperature these baryon and lepton number violating interactions were found to be very strong in the presence of some static topological field configuration - sphalerons [6]. In fact, these interactions are so strong that in no time the particles and anti-particles attain their equilibrium distributions. As a result, since CPT is conserved and hence the masses of the particles and anti-particles are same, the number density of baryons becomes same as that of the anti-baryons and that wash out any primordial baryon asymmetry of the universe, which are generated during the GUT phase transition. This started renewed interest in the subject of the baryon asymmetry of the universe.

Now this problem takes two directions. First, how to generate a baryon asymmetry of the universe, and second, which are the interactions that can wash out the baryon asymmetry of the universe and what constrains they give us on the various parameters of the particle physics models. Since the electroweak anomalous processes breaks both the baryon and the lepton numbers, still conserving the $(B - L)$ quantum number, the baryon asymmetry of the universe is no longer independent of the lepton number violation of the universe [6, 7, 8, 9]. If there is very fast lepton number violation before the electroweak phase transition, then that can erase the $(B - L)$ asymmetry of the universe and hence the baryon asymmetry of the universe. On the other hand, if any lepton asymmetry is generated at some high temperature, that can get converted to a baryon asymmetry of the universe before and during the electroweak phase transition.

The first thing then comes to mind to make use of the baryon number violation of the standard model to generate a baryon asymmetry of the universe. There were several attempts towards this direction [10, 11]. However, in these models one needs to protect the generated baryon asymmetry after the phase transition, which requires the mass of the standard model doublet higgs boson to be substantially light [12]. The present experimental limit of 95 GeV on the mass of the higgs boson *almost* rule out all these possibilities. Then the most interesting scenario remains for the understanding of the baryon number of the universe is through lepton number violation [7, 13, 14, 15], which is also referred to as leptogenesis. In this scenario one generates a lepton asymmetry of the universe, which is the same as the $(B-L)$ asymmetry of the universe. This $(B-L)$ asymmetry of the universe then get converted to the baryon asymmetry of the universe during the period when the sphaleron fields maintain the baryon number violating interactions in equilibrium. On the other hand, if there is vary fast lepton number violation in the universe during this period, that can also wash out the baryon asymmetry of the universe [8, 16, 17, 18].

In this article I shall discuss the constraints on the left-handed Majorana neutrino mass, which comes from the survival of the baryon asymmetry of the universe. Since, in models of leptogenesis these constraints are weakened, we shall also discuss models of leptogenesis briefly. The limitations of these constraints will be mentioned.

2. Sphaleron processes in thermal equilibrium and relation between baryon and lepton numbers

Anomaly breaks any classical symmetry of the lagrangian at the classical level. So, all local gauge theories should be free of anomalies. However, there may be anomalies corresponding to any global current, *i.e.*, in the triangle loop while two gauge bosons would couple to two vertices, a global current will be associated to the third vertex. That will simply mean that such global symmetries of the classical lagrangian are broken through quantum effects.

In the standard model the chiral nature of the weak interaction makes the baryon and lepton number anomalous. If we associate the $SU(2)_L$ or the $U(1)_Y$ gauge bosons at the two vertices of a triangle diagram and associate a global current corresponding to baryon or lepton numbers at the third vertex, then sum over all the fermions in the standard model will give us non vanishing axial current [5]

$$\delta_\mu j^{\mu 5}_{(B+L)} = 6[\frac{\alpha_2}{8\pi} W^{\mu\nu}_a \tilde{W}_{a\mu\nu} + \frac{\alpha_1}{8\pi} Y^{\mu\nu} \tilde{Y}_{\mu\nu}]$$

which will break the $(B+L)$ symmetry. However, the anomaly corresponding to the baryon and lepton number are same and as a result there is no anomaly corresponding to the $(B-L)$ charge.

Because of the anomaly [5], baryon and lepton numbers are broken during the electroweak phase transition,

$$\Delta(B+L) = 2N_g \frac{\alpha_2}{8\pi} \int d^4x W^{\mu\nu}_a \tilde{W}_{a\mu\nu} = 2N_g \nu,$$

but their rate is very small at zero temperature, since they are suppressed by quantum tunnelling probability, $\exp[-\frac{2\pi}{\alpha_2}\nu]$, where ν is the Chern-Simmons number.

At finite temperature, however, it has been shown that there exists non-trivial static topological soliton configuration, called the sphalerons, which enhances the baryon number violating transition rate [6]. As a result, at finite temperature these baryon and lepton number violating processes are no longer suppressed by quantum tunneling factor, rather the suppression factor is now replaced by the Boltzmann factor

$$\exp[-\frac{V_0}{T}\nu]$$

where the potential or the free energy V_0 is related to the mass of the sphaleron field, which is about TeV. As a result, at temperatures between

$$10^{12}GeV > T > 10^2GeV \tag{1}$$

the sphaleron mediated baryon and lepton number violating processes are in equilibrium. For the simplest scenario of $\nu = 1$, the sphaleron induced processes are $\Delta B = \Delta L = 3$, given by,

$$|vac> \longrightarrow [u_L u_L d_L e_L^- + c_L c_L s_L \mu_L^- + t_L t_L b_L \tau_L^-]. \tag{2}$$

These baryon and lepton number violating fast processes can, in general, wash out any pre-existing baryon or lepton number asymmetry of the universe. However, if there are any $(B-L)$ asymmetry of the universe, that will not be washed out. In fact, any $(B-L)$ asymmetry before the electroweak phase transition will get converted to a baryon and lepton asymmetry of the universe during this phase transition, which can be seen from an analysis of the chemical potential [9].

We consider all the particles to be ultrarelativistic, which is the case above the electroweak scale. At lower energies, a careful analysis has to include the mass corrections, but since they are anyway small we ignore them for our present discussion. The particle asymmetry, *i.e.* the difference between the number of particles (n_+) and the number of antiparticles (n_-) can be given in terms of the chemical potential of the particle species μ (for antiparticles the chemical potential is $-\mu$) as

$$n_+ - n_- = n_d \frac{gT^3}{6}\left(\frac{\mu}{T}\right), \tag{3}$$

where $n_d = 2$ for bosons and $n_d = 1$ for fermions.

In the standard model the quarks and leptons transform under $SU(3)_C \times SU(2)_L \times U(1)_Y$ as,

$$\begin{pmatrix} u_i \\ d_i \end{pmatrix}_L \sim (3, 2, 1/6), \quad u_{iR} \sim (3, 1, 2/3), \quad d_{iR} \sim (3, 1, -1/3); \tag{4}$$

$$\begin{pmatrix} \nu_i \\ e_i \end{pmatrix}_L \sim (1, 2, -1/2), \quad e_{iR} \sim (1, 1, -1). \tag{5}$$

where, $i = 1, 2, 3$ corresponds to three generations. In addition, the scalar sector consists of the usual Higgs doublet,

$$\begin{pmatrix} \phi^+ \\ \phi^0 \end{pmatrix} \sim (1, 2, 1/2) \tag{6}$$

which breaks the electroweak gauge symmetry $SU(2)_L \times U(1)_Y$ down to $U(1)_{em}$. When these leptons interact with other particles in equilibrium, the chemical potentials get related by simple additive relations, and that will allow us to relate the lepton asymmetry n_L to the baryon asymmetry during the electroweak phase transition.

During the period $10^2 GeV < T < 10^{12} GeV$, the sphaleron [6] induced electroweak $B + L$ violating interaction arising due to the nonperturbative axial-vector anomaly [5] will be in equilibrium along with the other interactions. In Table 1, we present all other interactions and the corresponding relations between the chemical potentials. In the third column we give the chemical potential which we eliminate using the given relation. We start with chemical potentials of all the quarks ($\mu_{uL}, \mu_{dL}, \mu_{uR}, \mu_{dR}$); leptons ($\mu_{aL}, \mu_{\nu aL}, \mu_{aR}$, where $a = e, \mu, \tau$); gauge bosons (μ_W for W^-, and 0 for all others); and the Higgs scalars (μ_-^ϕ, μ_0^ϕ).

Table 1. Relations among the chemical potentials

Interactions	μ relations	μ eliminated
$D_\mu \phi^\dagger D_\mu \phi$	$\mu_W = \mu_-^\phi + \mu_0^\phi$	μ_-^ϕ
$\overline{q_L} \gamma_\mu q_L W^\mu$	$\mu_{dL} = \mu_{uL} + \mu_W$	μ_{dL}
$\overline{l_L} \gamma_\mu l_L W^\mu$	$\mu_{iL} = \mu_{\nu iL} + \mu_W$	μ_{iL}
$\overline{q_L} u_R \phi^\dagger$	$\mu_{uR} = \mu_0 + \mu_{uL}$	μ_{uR}
$\overline{q_L} d_R \phi$	$\mu_{dR} = -\mu_0 + \mu_{dL}$	μ_{dR}
$\overline{l_{iL}} e_{iR} \phi$	$\mu_{iR} = -\mu_0 + \mu_{iL}$	μ_{iR}

The chemical potentials of the neutrinos always enter as a sum and for that reason we can consider it as one parameter. We can then express all the chemical potentials in terms of the following independent chemical potentials only,

$$\mu_0 = \mu_0^\phi; \quad \mu_W; \quad \mu_u = \mu_{uL}; \quad \mu = \sum_i \mu_i = \sum_i \mu_{\nu iL}. \tag{7}$$

We can further eliminate one of these four potentials by making use of the relation given by the sphaleron processes (2). Since the sphaleron interactions are in equilibrium, we can write down the following $B+L$ violating relation among the chemical potentials for three generations,

$$3\mu_u + 2\mu_W + \mu = 0. \tag{8}$$

We then express the baryon number, lepton numbers and the electric charge and the hypercharge number densities in terms of these independent chemical potentials,

$$\begin{aligned} B &= 12\mu_u + 6\mu_W & (9)\\ L_i &= 3\mu + 2\mu_W - \mu_0 & (10)\\ Q &= 24\mu_u + (12+2m)\mu_0 - (4+2m)\mu_W & (11)\\ Q_3 &= -(10+m)\mu_W & (12) \end{aligned}$$

where m is the number of Higgs doublets ϕ.

At temperatures above the electroweak phase transition, $T > T_c$, both Q and Q_3 must vanish. These conditions and the sphaleron induced $B-L$ conserving, $B+L$ violating condition can be expressed as

$$\begin{aligned} <Q>=0 &\implies \mu_0 = \frac{-12}{6+m}\mu_u & (13)\\ <Q_3>=0 &\implies \mu_W = 0 & (14)\\ \text{Sphaleron transition} &\implies \mu_a = -5\mu_u & (15) \end{aligned}$$

Using these relations we can now write down the baryon number, lepton number, and their combinations in terms of the $B-L$ number density as,

$$\begin{aligned} B &= \frac{24+4m}{66+13m}(B-L) & (16)\\ L &= \frac{-42-9m}{66+13m}(B-L) & (17)\\ B+L &= \frac{-18-5m}{66+13m}(B-L) & (18) \end{aligned}$$

Below the critical temperature, Q should vanish since the universe is neutral with respect to all conserved charges. However, since $SU(2)_L$ is now broken we can consider $\mu_0^\phi = 0$ and $Q_3 \neq 0$. This gives us,

$$\begin{aligned} <Q>=0 &\implies \mu_W = \frac{12}{2+m}\mu_u & (19)\\ \langle\phi\rangle \neq 0 &\implies \mu_0 = 0 & (20)\\ \text{Sphaleron transition} &\implies \mu_a = -3\mu_W - \frac{9}{2}\mu_u & (21) \end{aligned}$$

which then let us write the baryon and lepton numbers as some combinations of $B-L$ as

$$B = \frac{32+4m}{98+13m}(B-L) \tag{22}$$

$$L = \frac{-66 - 9m}{98 + 13m} (B - L) \quad (23)$$

$$B + L = \frac{-34 - 5m}{98 + 13m} (B - L) \quad (24)$$

Thus the baryon and lepton number asymmetry of the universe after the electroweak phase transition will depend on the primordial $(B - L)$ asymmetry of the universe. If there are very fast lepton number violation during the period (1), that would erase the $(B - L)$ asymmetry of the universe, and hence we will be left with a baryon symmetric universe at the end [8, 9]. On the other hand, if there is enough lepton asymmetry of the universe at some high temperature, then that will get converted to a baryon asymmetry of the universe [7, 9].

Before proceeding further, we shall briefly discuss what do we mean when we say that some interaction is fast and that will erase some asymmetry [1, 2, 20]. In equilibrium the number density of particles with non-zero charge Q would be same as the antiparticle number density since the expectation value of the conserved charge vanishes. A mathematical formulation of this statement reads that the expectation value of any conserved charge Q is given by,

$$< Q >= \frac{\mathrm{Tr}\left[Qe^{-\beta H}\right]}{\mathrm{Tr}\left[e^{-\beta H}\right]}$$

and since any conserved charge Q is odd while H is even under CPT transformation this expectation value vanishes. So for the generation of the baryon asymmetry of the universe we have to circumvent this theorem either by including nonzero chemical potential, or go away from equilibrium or violate CPT. In most of the popular models CPT conservation is assumed and one starts with vanishing chemical potential for all the fields which ensures that the entropy is maximum in chemical equilibrium. Then to generate the baryon asymmetry of the universe one needs to satisfy the out-of-equilibrium condition [1, 2, 20].

The requirement for the out-of-equilibrium condition may also be stated in a different way [1]. If we assume that the chemical potential associated with B is zero and CPT is conserved, then in thermal equilibrium the phase space density of baryons and antibaryons, given by $[1 + exp(\sqrt{p^2 + m^2}/kT)]^{-1}$ are identical and hence there cannot be any baryon asymmetry.

Whether a system is is equilibrium or not can be understood by solving the Boltzmann equations. But a crude way to put the out-of-equilibrium condition is to say that the universe expands faster than some interaction rate. For example, if some B-violating interaction is slower than the expansion rate of the universe, this interaction may not bring the distribution of baryons and antibaryons of the universe in equilibrium. In other words, before the chemical potentials of the two states gets equal, they move apart from each other. Thus we may state the out-of-equilibrium condition as

$$\Gamma < \sqrt{1.7 g_*} \frac{T^2}{M_P} \tag{25}$$

where, Γ is the interaction rate under discussion, g_* is the effective number of degrees of freedom available at that temperature T, and M_P is the Planck scale.

3. Bounds on Left-handed neutrino mass

In this section we shall discuss the constraints on the left-handed neutrino mass arising from the constraints of baryogenesis. If there is Majorana type interactions, which explains the masses of the neutrinos, then that gives a measure of lepton number violation. If this lepton number violation is too large before the electroweak phase transition is over, then that can erase all L asymmetry and hence B asymmetry of the universe.

The first attempt to constraint the neutrino mass was made in a fairly general framework [8]. In the standard model there is no lepton number violation. However, one can consider a higher dimensional effective operator which violates $(B - L)$, given by

$$L = \frac{2}{M} l_L l_L \phi\phi + h.c. \tag{26}$$

There is no origin of such interactions within the standard model. So one expects that some new interaction at some high energy will give us this effective interaction at low energy. The scale of the new interaction M, which gives us this interaction, is also the scale of lepton number violation in this scenario.

If this interaction is strong enough, it can bring the neutrinos in thermal equilibrium with the physical higgs scalars, which can wash out any lepton asymmetry of the universe. The survival of the baryon asymmetry of the universe will then require this interaction to be slower than the expansion rate of the universe,

$$\Gamma_{L\neq 0} \sim \frac{0.122}{\pi} \frac{T^3}{M^2} < 1.7 \sqrt{g_*} \frac{T^2}{M_P} \qquad \text{at } T \sim 100 GeV \tag{27}$$

which gives a bound on the mass of the heavy scale to be, $M > 10^9 GeV$. When the higgs doublets ϕ acquires a vev, the higher dimensional operator will induce a Majorana mass of the left-handed neutrinos. This bound on the heavy scale M will then imply,

$$m_\nu < 50 keV.$$

In specific models one may give stronger bounds on the mass of the neutrinos [16]. In models with right handed neutrinos ($N_{Ri}; i = e, \mu, \tau$), the neutrino masses comes from the see-saw mechanism [21]. The lagrangian for the lepton sector containing the mass terms of the singlet right handed neutrinos N_i and the Yukawa couplings of these fields with the light leptons is given by,

$$\begin{aligned}
\mathcal{L}_{int} &= \sum_i M_i[\overline{(N_{Ri})^c} N_{Ri} + \overline{N_{Ri}}(N_{Ri})^c] \\
&+ \sum_{\alpha,i} h^*_{\alpha i}\, \overline{N_{Ri}}\, \phi^\dagger\, \ell_{L\,\alpha} + \sum_{\alpha,i} h_{\alpha i}\, \overline{\ell_{L\,\alpha}} \phi\, N_{Ri} \qquad (28) \\
&+ \sum_{\alpha,i} h^*_{\alpha i}\, \overline{(\ell_{L\,\alpha})^c}\, \phi^\dagger\, (N_{Ri})^c + \sum_{\alpha,i} h_{\alpha i} \overline{(N_{Ri})^c}\, \phi\, (\ell_{L\,\alpha})^c
\end{aligned}$$

where $\phi^T = (-\overline{\phi^o}, \phi^-) \equiv (1, 2, -1/2)$ is the usual higgs doublet of the standard mode; $l_{L\alpha}$ are the light leptons, $h_{\alpha i}$ are the complex Yukawa couplings and α is the generation index. Without loss of generality we work in a basis in which the Majorana mass matrix of the right handed neutrinos is real and diagonal with eigenvalues M_i.

Once the higgs doublet ϕ acquires a *vev*, the masses of the neutrinos in the basis $[\nu_{L\alpha} \quad N_{Ri}]$ is given by,

$$\mathcal{M}_\nu = \begin{pmatrix} 0 & m \\ m & M \end{pmatrix} \qquad (29)$$

where, $m \equiv h_{\alpha i} < \phi >$ and $M \equiv M_{ij}$ are 3×3 matrices. In the limit when all eigenvalues of M are much heavier than those of m, and the matrix M is not singular, this matrix may be block diagonalised. It then gives three heavy right handed Majorana neutrinos with masses $\sim M$ and the Majorana mass matrix of the left-handed neutrinos will be given by,

$$m_\nu = m \frac{1}{M} m^T . \qquad (30)$$

In this scenario the see-saw masses of the left-handed neutrinos explain naturally why they are so light. In addition, the decay of the heavy right handed Majorana neutrinos can generate enough lepton asymmetry of the universe, which can then get converted to a baryon asymmetry of the universe, which we shall discuss in the next section.

The decay of N_{Ri} into a lepton and an antilepton,

$$\begin{aligned}
N_{Ri} &\rightarrow \ell_{jL} + \bar{\phi}, \\
&\rightarrow {\ell_{jL}}^c + \phi. \qquad (31)
\end{aligned}$$

breaks lepton number. Since the lightest of the right handed neutrinos (say N_1) will decay at the end, this interaction (N_1 decay) should be slow enough so as not to erase the baryon asymmetry of the universe, which now implies

$$\frac{|h_{\alpha 1}|^2}{16\pi} M_1 < 1.7\sqrt{g_*}\frac{T^2}{M_P} \qquad \text{at } T = M_1 \tag{32}$$

which can then give a very strong bound [16] on the mass of the lightgest of the left-handed neutrinos to be

$$m_\nu < 4 \times 10^{-3} eV.$$

In models [24, 25, 26], where the left-handed neutrino mass is not related to any heavy neutrinos through see-saw mechanism, the abovementioned bounds may not be valid. In addition, there are several specific cases even within the framework of see-saw models, where these bounds are not applicable. These bounds are also not valid if some global U(1) symmetry is exactly conserved up to an electroweak anomaly [17]. Furthermore, in some very specific models where a baryon asymmetry of the universe is generated after the electroweak phase transition [22], or there are some extra baryon number carrying singlets which decays after the electroweak phase transition [23], it is possible to avoid all the bounds from constraints of survival of the baryon asymmetry of the universe. This issue will be discussed in another talk in this meeting [23] in details.

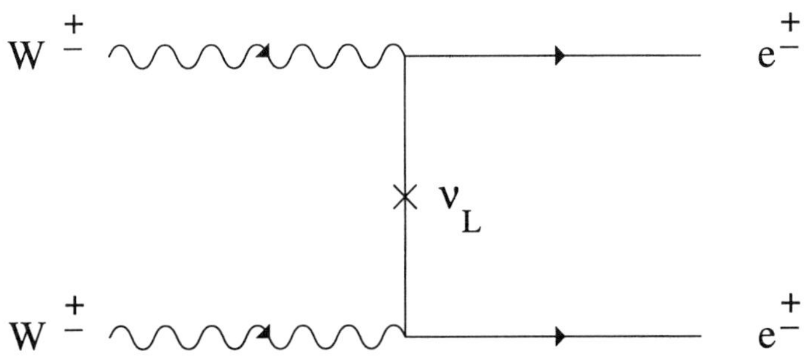

Figure 1. Lepton number violating processes $W^\pm + W^\pm \to e^\pm + e^\pm$ mediated by the left handed Majorana neutrinos.

Since some of the bounds derived indirectly from an bound on the scale of lepton number violation may be circumvented in some models, we now try to discuss the question if one can constrain the Majorana mass of the left-handed neutrinos directly. This is possible since during the electroweak phase transition, after the higgs doublets acquires a vacuum expectation value (*vev*) and the $SU(2)_L$ group is broken, there is no symmetry which can prevent the mass of the left-handed neutrinos. So if lepton number is broken before the electroweak symmetry, then as soon as ϕ acquires a *vev* the left-handed neutrinos will get a mass (m_ν). This can, in principle, induce very fast lepton number violating processes, which can wash out any primordial $(B-L)$ asymmetry and hence the baryon asymmetry of the universe.

There are several lepton number violating processes, which are active at the time of electroweak phase transition. The process,

$$W^+ + W^+ \to e_i^+ + e_j^+ \qquad \text{and} \qquad W^- + W^- \to e_i^- + e_j^- \tag{33}$$

mediated by a virtual left-handed neutrino exchange as shown in figure 1 is a lepton number violating interaction active during the electroweak phase transition. Here i and j are the generation indices. Depending on the physical mass (and also on the elements of the mass matrix) of the left-handed Majorana neutrinos these processes can wash out any baryon asymmetry between the time when the higgs acquires a *vev* and the $W^\pm$ freeze out, *i.e.*, between the energy scales 250 GeV and 80 GeV. The condition that these processes will be slower than the expansion rate of the universe,

$$\Gamma(WW \to e_i e_j) = \frac{\alpha_W^2 (m_\nu)_{ij}^2 T^3}{m_W^4} < 1.7\sqrt{g_*}\frac{T^2}{M_p} \qquad \text{at } T = M_W \tag{34}$$

gives a bound on the Majorana mass of the left-handed neutrinos to be,

$$(m_\nu)_{ij} < 20keV. \tag{35}$$

This bound is on each and every element of the mass matrix and not on the physical states and independent on the existence of any right handed neutrinos.

In general, a Majorana particle can be described by a four component real field,

$$\Psi_M = \sqrt{\frac{m_\nu}{E_\nu}}[u_\nu(b_\nu + d_\nu^*)\mathrm{e}^{(-ip.x)} + v_\nu(b_\nu^* + d_\nu)\mathrm{e}^{(ip.x)}] \tag{36}$$

and hence the charged current containing a Majorana field,

$$j_\mu = \overline{\Psi}_l \gamma_\mu (1-\gamma_5)\Psi_M$$

will have a lepton number violating part. However, this lepton number violating contribution will always be suppressed by a factor (m_ν/E_ν) and hence the rate of such processes will be suppressed by a factor $(m_\nu/E_\nu)^2$. Thus even the decay of the $W^\pm$ to e and ν will have lepton number violation at a rate,

$$\Gamma(W \to e\nu) = \frac{\alpha_W}{4} \frac{m_\nu^2 M_W^2}{T^2(T^2 + M_W^2)^{1/2}}.$$

The survival of baryon asymmetry of the universe after the electroweak phase transition again requires this process to be slow enough,

$$\Gamma(W \to e\nu) < H.$$

This translates to a bound on the Majorana mass of the left handed neutrino,

$$m_\nu < 30 \quad \text{keV}. \tag{37}$$

Similarly, the decay of the higgs doublet to an electron and an anti-neutrino will also have lepton number violating contribution, but they will be suppressed by the Yukawa coupling constants and cannot give stronger bounds. Similarly scattering processes involving the higgs, like $\phi + \phi \to l_i + l_j$ (mediated by a virtual left-handed neutrino) will contribute to the evolution of the lepton number asymmtry of the universe, but it will be much suppressed compared to the charged current interactions and hence cannot give stronger bound to the Majorana mass of the left-handed neutrinos.

4. Bounds on neutrino mass in models of Leptogenesis

In this section we shall discuss two scenarios of leptogenesis. In the first, one starts with the standard model and add to it three right handed neutrinos [7, 13, 14, 15]. In total this will add 12 degrees of freedom. In the other scenario [26] one adds two complex $SU(2)_L$ triplet higgs scalars ($\xi_a \equiv (1, 3, -1); a = 1, 2$), which also adds in total 12 degrees of freedom. At present these two scenarios are indistinguishable, except that the particle contents are different. In the left-right symmetric models [27, 28] both these scenarios are present, which we shall not discuss here. The CP violation required to generate a lepton asymmetry of the universe are different for the two scenarios.

In the standard model neutrinos are massless. In models with right handed neutrinos the left-handed neutrinos acquire a Majorana mass through see-saw mechanism as we mentioned earlier. In the triplet higgs scenario [26, 29] the *vevs* of the triplet higgses can give small Majorana masses to the neutrinos through the interaction

$$f_{ij}[\xi^0 \nu_i \nu_j + \xi^+(\nu_i l_j + l_i \nu_j)/\sqrt{2} + \xi^{++} l_i l_j] + h.c. \tag{38}$$

If the triplet higgs acquires a *vev* and break lepton number spontaneously [30], then there will be Majorons in the problem. Then from the precision measurements of the $Z-$width that model is ruled out [31]. However, in a variant of this model [26] lepton number is broken explicitly through an interaction of the triplet with the higgs doublet

$$V = \mu(\bar{\xi}^0 \phi^0 \phi^0 + \sqrt{2}\xi^- \phi^+ \phi^0 + \xi^{--} \phi^+ \phi^+) + h.c. \tag{39}$$

Let $\langle \phi^0 \rangle = v$ and $\langle \xi^0 \rangle = u$, then the conditions for the minimum of the potential relates the *vev* of the two scalars by

$$u \simeq \frac{-\mu v^2}{M^2}, \tag{40}$$

where M is the mass of the triplet higgs scalar. This is analogous to the usual seesaw mechanism for obtaining small Majorana neutrino masses, except that here we do not have any right-handed neutrinos.

Another way of handling the heavy Higgs triplet is to integrate it out. From the couplings of the triplet scalar, we obtain the effective nonrenormalizable term

$$\frac{-f_{ij}\mu}{M^2}[\phi^0\phi^0\nu_i\nu_j - \phi^+\phi^0(\nu_i l_j + l_i\nu_j) + \phi^+\phi^+ l_i l_j] + h.c. \tag{41}$$

The reduced Higgs potential involving only the doublet higgs scalar is

$$V = m^2\Phi^\dagger\Phi + \frac{1}{2}\left(\lambda_1 - \frac{2\mu^2}{M^2}\right)(\Phi^\dagger\Phi)^2, \tag{42}$$

where m is the mass of the higgs doublet. The last term comes from the exchange of ξ. As ϕ^0 acquires a nonzero vacuum expectation value v, we obtain Eq. (40) as we should, and the neutrino mass matrix becomes

$$-2f_{ij}\mu v^2/M^2 = 2f_{ij}u$$

In models with right handed neutrinos lepton number is violated when the right handed Majorana neutrinos decay as we discussed earlier [equation (31)]. The out-of-equilibrium condition is satisfied if the masses of the lightest of the right handed neutrinos satisfy $M_1 > 10^7$ GeV, which is obtained by solving the Boltzmann equations. There are two sources of CP violation in this scenario :

(*i*) vertex type diagrams which interferes with the tree level diagram given by figure 2.

(*ii*) self energy diagrams could interfere with the tree level diagrams to produce CP violation analogous to CP violation in the $K^\circ\overline{K^\circ}$ oscillation as shown in figure 3. This type of CP violation has several interesting features, which will be discussed in another talk at this meeting [32].

In the case of self energy type CP violation the amount of lepton asymmetry becomes large for a small mass difference between the two right-handed heavy neutrinos (M_1 and M_2), and is given by [15]:

$$\delta = \frac{1}{8\pi}\mathcal{C}\frac{M_1 M_2}{M_2^2 - M_1^2} \tag{43}$$

where

$$\mathcal{C} = -\mathrm{Im}\left[\sum_\alpha (h^*_{\alpha 1} h_{\alpha 2}) \sum_\beta h^*_{\beta 1} h_{\beta 2})\right] \left(\frac{1}{\sum_\alpha |h_{\alpha 1}|^2} + \frac{1}{\sum_\alpha |h_{\alpha 2}|^2}\right) \tag{44}$$

This contribution becomes significant when the two mass eigenvalues are close to each other. It indicates a resonance like behaviour of the asymmetry if the two mass eigenvalues are nearly degenerate. For very large values of the mass difference this contribution becomes similar to the vertex correction. These two contributions add up to produce the final lepton asymmetry of the universe.

In the triplet higgs scenario lepton number violation comes from the decays of the triplet higgs ξ_a. Consider the decays of ξ_a^{++},

$$\xi_a^{++} \to \left\{ \begin{array}{ll} l_i^+ l_j^+ & (L = -2) \\ \phi^+\phi^+ & (L = 0) \end{array} \right. \tag{45}$$

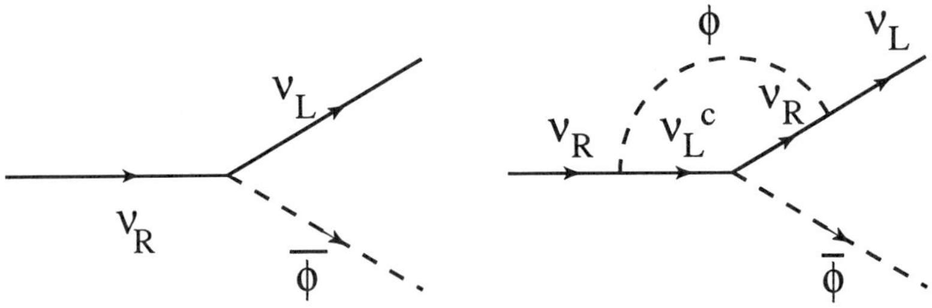

Figure 2. Tree and one loop vertex correction diagrams contributing to the generation of lepton asymmetry in models with right handed neutrinos

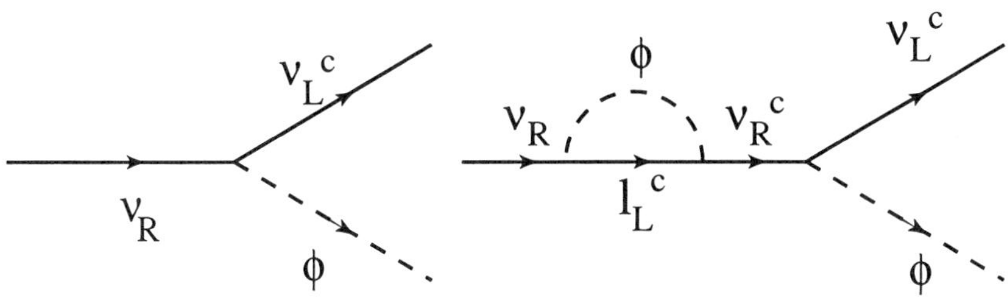

Figure 3. Tree and one loop self energy diagrams contributing to the generation of lepton asymmetry in models with right handed neutrinos

The coexistence of the above two types of final states indicates the non-conservation of lepton number. On the other hand, any lepton asymmetry generated by ξ_a^{++} would be neutralized by the decays of ξ_a^{--}, unless CP conservation is also violated and the decays are out of thermal equilibrium in the early universe. In this case there are no vertex corrections which can introduce CP violation. The only source of CP violation is the self energy diagrams of figure 4.

If there is only one ξ, then the relative phase between any f_{ij} and μ can be chosen real. Hence a lepton asymmetry cannot be generated. With two ξ's, even if there is only one lepton family, one relative phase must remain. As for the possible relative phases among the f_{ij}'s, they cannot generate a lepton asymmetry because they all refer to final states of the same lepton number.

In the presence of the one loop diagram, the mass matrix ${M_a}^2$ and M_a^{*2} becomes different. This implies that the rate of $\xi_b \to \xi_a$ no longer remains to be same as $\xi_b^* \to \xi_a^*$. Since by CPT theorem $\xi_b^* \to \xi_a^* \equiv \xi_a \to \xi_b$, what it means is that now

$$\Gamma[\xi_a \to \xi_b] \neq \Gamma[\xi_b \to \xi_a].$$

This is a different kind of CP violation compared to the CP violation in models with right handed neutrinos. If we consider that the ξ_2 is heavier than ξ_1, then after ξ_2 decayed out the decay of ξ_1 will generate an lepton

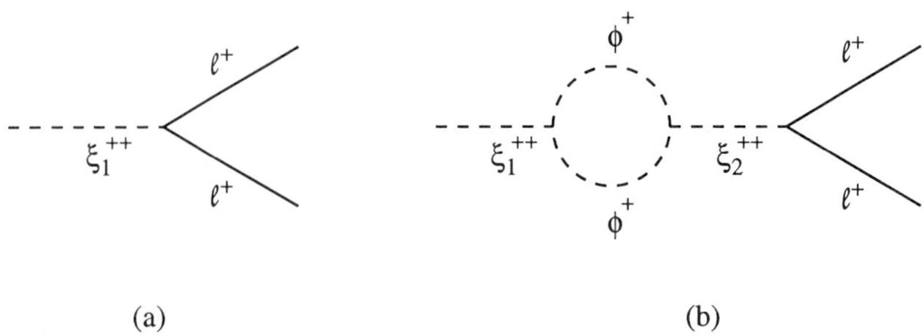

Figure 4. The decay of $\xi_1^{++} \to l^+ l^+$ at tree level (a) and in one-loop order (b). A lepton asymmetry is generated by their interference in the triplet higgs model for neutrino masses.

asymmetry given by,

$$\delta \simeq \frac{Im\left[\mu_1 \mu_2^* \sum_{k,l} f_{1kl} f_{2kl}^*\right]}{8\pi^2 (M_1^2 - M_2^2)} \left[\frac{M_1}{\Gamma_1}\right]. \tag{46}$$

In any of the above two scenarios with right handed neutrinos or with triplet higgs, the lepton number evolves with time following the Boltzmann equation in similar way. The lepton asymmetry n_L thus generated through the CP violation η (where η is the amount of CP violation in the model under consideration), would evolve with time following the Boltzmann equation,

$$\frac{\mathrm{d}n_l}{\mathrm{d}t} + 3Hn_l = \eta\Gamma_h[n_h - n_h^{eq}] - \left(\frac{n_l}{n_\gamma}\right) n_h^{eq}\Gamma_h - 2n_\gamma n_l \langle\sigma|v|\rangle \tag{47}$$

where Γ_h is the thermally averaged decay rate of the heavy particle (whose number density is n_h). The second term on the left side comes from the expansion of the universe, where $H = 1.66 g_*^{1/2}(T^2/M_{Pl})$ is the Hubble constant. n_γ is the photon density and the term $\langle\sigma|v|\rangle$ describes the thermally averaged lepton number violating scattering cross section. The density of the heavy particle satisfies the Boltzmann equation,

$$\frac{\mathrm{d}n_h}{\mathrm{d}t} + 3Hn_h = -\Gamma_h(n_h - n_h^{eq}) \tag{48}$$

It is now convenient to use the dimensionless variable $x = M_h/T$ as well as the particle density per entropy density $Y_i = n_i/s$, and the relation $t = x^2/2H(x=1)$. We also define the parameter $K \equiv \Gamma_h(x=1)/H(x=1)$ as a measure of the deviation from equilibrium. For $K << 1$ at $T \sim M_h$, the system is far from equilibrium; hence the last two terms responsible for the depletion of n_l would be negligible. With these simplifications and the above redefinitions, the Boltzmann equations effectively read:

$$\frac{\mathrm{d}Y_l}{\mathrm{d}x} = (Y_h - Y_h^{eq})\eta K x; \quad \frac{\mathrm{d}Y_h}{\mathrm{d}x} = -(Y_h - Y_h^{eq})Kx. \tag{49}$$

In this limit $K << 1$, it is not difficult to obtain an asymptotic solution for n_l. Although the decay rate of ψ_h is not fast enough to bring the number density n_h to its equilibrium density, it is a good approximation to assume that the universe never goes far away from equilibrium. In other words, we can assume $\mathrm{d}(Y_h - Y_h^{eq})/\mathrm{d}x = 0$ to get an asymptotic value for Y_l, given by $Y_l = n_l/s = \eta/g_*$. However, if $K > 1$, the terms which deplete n_l dominate for some time and the lepton number density reaches its new asymptotic value, which is lower than the value it reaches in the out-of-equilibrium case. In this case although it is difficult to get an analytic solution of the Boltzmann equations, it is possible to get an approximate suppression factor, which is proportional to K.

In our earlier discussions we have considered the out-of-equilibrium condition to be

$$K < 1,$$

which gave us all the bounds on the Majorana masses of the left-handed neutrinos. But as we can see from the above discussions, in models of leptogenesis, where a lepton number violation is associated with a CP violation, the lepton number is not washed out too fast. While in other models fast lepton number wash out any preexisting asymmetry exponentially, in this case the depletion is only linearly. As a result, in models of leptogenesis fast lepton number violation may not wash out the primordial lepton asymmetry of the universe completely and the baryon asymmetry of the universe may still be present after the lepton number violating interaction goes out of equilibrium.

5. Summary

Any fast lepton number violation in the universe can, in principle, wash out baryon asymmetry of the universe. The survival of the baryon asymmetry of the universe thus gives constraints on the left-handed Majorana neutrino masses. There are constraints which are dependent on models of the neutrino masses, but it is also possible to give constraints from the lepton number violating interactions of $W^{\pm}$ due to the Majorana masses of the left-handed neutrinos directly. Although these constraints are independent of any models, these constraints also gets weakened in models of leptogenesis.

Acknowledgement

I would like to thank the Abdus Salam International Center for Theoretical Physics, Trieste, Italy for providing financial assistance to attend this meeting, during my visit to the Center as an Associate.

References

[1] Kolb E W and Turner M S 1989, *The Early Universe* (Addison-Wesley, Reading, MA).

[2] Sakharov A D 1967, *Pis'ma Zh. Eksp. Teor. Fiz.* **5** 32.

[3] Yoshimura M 1978, *Phys. Rev. Lett.* **41** 281; E 1979: *Phys. Rev. Lett.* **42** 7461;

[4] Mohapatra R N 1992, *Unification and Supersymmetry* (Springer-Verlag);
Masiero A 1984, in *Grand Unification with and without Supersymmetry and Cosmological Implications*, International School for Advanced Studies Lecture Series No. 2, (World Scientific) 1;
Zee A 1982, (ed.) *Unity of Forces in the Universe* **1** (World Scientific);
Langacker P 1981, Phys. Rep. **72** 185.

[5] 't Hooft G 1976, *Phys. Rev. Lett.* **37** 8.

[6] Kuzmin V A, Rubakov V A and Shaposhnikov M E 1985, *Phys. Lett.* **B 155** 36.

[7] Fukugita M and Yanagida T 1986, *Phys. Lett.* **B 174** 45.

[8] Fukugita M and Yanagida T 1990, *Phys. Rev.* **D 42** 1285;
Barr S M and Nelson A E 1991, *Phys. Lett.* **B 246** 141.

[9] Khlebnikov S Yu and Shaposhnikov M E 1988, *Nucl. Phys.* **B 308** 885;
Harvey J A and Turner M S 1990, *Phys. Rev.* **D 42** 3344.

[10] Shaposhnikov M E 1987, *Nucl. Phys.* **B 287** 757;
Shaposhnikov M E 1988, *Nucl. Phys.* **B 299** 797;
Dolgov A D 1992, *Phys. Rep.* **222** 309;
Rubakov V A and M.E. Shaposhnikov 1996, *Usp. Fiz. Nauk* **166** 493; 1996 *Phys. Usp.* **39** 461.

[11] Shaposnikov M E 1987, *Nucl. Phys.* **B 287** 757;
Shaposhnikov M E 1988, **B 299** 797;
Turok N and Zadrozny J 1990, *Phys. Rev. Lett.* **65** 2331;
Kazarian A, Kuzmin S and Shaposnikov M E 1992, *Phys. Lett.* **B 276** 131.
McLerran L, Shaposnikov M E, Turok N and Voloshin M 1991, *Phys. Lett.* **B 256** 451.

[12] Bochkarev A I, Kuzmin S V and Shaposnikov M E 1990, *Phys. Lett.* **B 244** 275;
Dine M, Huet P and Singleton Jr R 1992, *Nucl. Phys.* **B 375** 625;
Shaposnikov M E 1993, *Phys. Lett.* **B 316** 112.

[13] Langacker P, Peccei R D and Yanagida T 1986, *Mod. Phys. Lett.* **A 1** 541;
Luty M A 1992, *Phys. Rev.* **D 45** 445;
Mohapatra R N and Zhang X 1992, *Phys. Rev.* **D 45** 5331;
Enqvist K and Vilja I 1993, *Phys. Lett.* **B 299** 281;
Murayama H, Suzuki H, Yanagida T and Yokoyama J 1993, *Phys. Rev. Lett.* **70** 1912;
Acker A, Kikuchi H, Ma E and Sarkar U 1993, *Phys. Rev.* **D 48** 5006;
Murayama H, Suzuki H, Yanagida T and Yokoyama J 1993, *Phys. Rev. Lett.* **70** 1912;
O'Donnell P J and Sarkar U 1994, *Phys. Rev.* **D 49** 2118;
Buchmüller W and Plümacher M 1996, *Phys. Lett.* **B 389** 73;
Ganguly A, Parikh J C and Sarkar U 1996, *Phys. Lett.* **B 385** 175;
Plümacher M 1997, *Z. Phys.* **C 74** 549;
Faridani J, Lola S, O'Donnell P J and Sarkar U 1998, hep-ph/9804261.

[14] Ignatev A Yu, Kuzmin V A and Shaposhnikov M E 1979, *JETP Lett.* **30** 688;
Botella F J and Roldan J 1991, *Phys. Rev.* **D 44** 966.
Liu J and Segre G 1993, *Phys. Rev.* **D 48** 4609;
Flanz M, Paschos E A, and Sarkar U 1995, *Phys. Lett.* **B 345** 248;
Covi L, Roulet E and Vissani F 1996, *Phys. Lett.* **B 384** 169;
Pilaftsis A 1997, *Phys. Rev.* **D 56** 5431;
Covi L and Roulet E 1997, *Phys. Lett.* **B 399** 113;
Covi L, Roulet E and Vissani F 1998, *Phys. Lett.* **B 424** 101;
Buchmüller W and Plümacher M 1997, hep-ph/9710460 (revised);
Flanz M and Paschos E A 1998, hep-ph/9805427.

[15] Flanz M, Paschos E A, Sarkar U and Weiss J 1996, *Phys. Lett.* **B 389** 693.

[16] Fischler W, Giudice G F, Leigh R G and Paban S 1991, *Phys. Lett.* **B 258** 45;
Buchmüller W and Yanagida T 1993, *Phys. Lett.* **B 302** 240.

[17] Dreiner H and Ross G G 1993, *Nucl. Phys.* **B 410** 188;
Ilakovac A and Pilaftsis A 1995, *Nucl. Phys.* **B 437** 491.

[18] Sarkar U 1997, *Phys. Lett.* **B 390** 97.

[19] Campbell B A, Davidson S, Ellis J and Olive K A 1991, *Phys. Lett.* **B 256** 457.

[20] Fry J N, Olive K A and Turner M S 1980, *Phys. Rev. Lett.* **45** 2074; 1980 *Phys. Rev.* **D 22** 2953; 1980 *Phys. Rev.* **D 22** 2977;
Kolb E W and Wolfram S 1980, *Nucl. Phys.* **B 172** 224.

[21] Gell-Mann M, Ramond P and Slansky R 1979, in *Supergravity*, Proceedings of the Workshop, Stony Brook, New York, 1979, ed. by P. van Nieuwenhuizen and D. Freedman (North-Holland, Amsterdam);
Yanagida T 1979, in *Proc of the Workshop on Unified Theories and Baryon*

Number in the Universe, Tsukuba, Japan, edited by A. Sawada and A. Sugamoto (KEK Report No. 79-18, Tsukuba);
Mohapatra R N and Senjanović G 1980, *Phys. Rev. Lett.* **44** 912.

[22] Masiero A and Mohapatra R N 1981, *Phys. Lett.* **B 103** 343.

[23] Davidson S 1998, *Proc. Int. Symp. on Lepton and Baryon number Violation, Trento*; 1997 *Phys. Lett.* **B 391** 287.

[24] Nandi S and Sarkar U 1986, *Phys. Rev. Lett.* **56** 564;
Joshipura A S and Sarkar U 1986, *Phys. Rev. Lett.* **57** 33;
Masiero A, Nanopoulos D V and Sanda A I 1986, *Phys. Rev. Lett.* **57** 663;
Mann R B and Sarkar U 1988, *Int. Jour. Mod. Phys.* **A 3** 2165;

[25] Farhi E and Susskind L 1979, *Phys. Rev.* **D 20** 3404;
Dimopoulos S 1980, *Nucl. Phys.* **B 168** 69;
Zee A 1980, *Phys. Lett.* **B 93** 389;
Wolfenstein L 1980, *Nucl. Phys.* **B 175** 93;
Nussinov S 1985, *Phys. Lett.* **B 165** 55.

[26] Ma E and Sarkar U 1998, *Phys. Rev. Lett.* **80** 5716.

[27] Pati J C and Salam A 1974, *Phys. Rev.* **D 10** 275;
Mohapatra R N and Pati J C 1975, *Phys. Rev.* **D 11** 566;
Mohapatra R N and Senjanovic G 1975, *Phys. Rev.* **D 12** 1502;
Marshak R E and Mohapatra R N 1980, *Phys. Rev. Lett.* **44** 1316.

[28] Pati J C, Salam A and Sarkar U 1983, *Phys. Lett.* **B 133** 330.

[29] Wetterich C 1981, *Nucl. Phys.* **B 187** 343;
Lazarides G, Shafi Q and Wetterich C 1981, *Nucl. Phys.* **B 181** 287;
Mohapatra R N and Senjanovic G 1981, *Phys. Rev.* **D 23** 165;
Holman R, Lazarides G and Shafi Q 1983, *Phys. Rev.* **D 27** 995;
Lazarides G and Shafi Q 1998, report no hep-ph/9803397.

[30] Gelmini G B and Roncadelli M 1981, *Phys. Lett.* **B 99** 411.

[31] Renton P B 1997, *Int. J. Mod. Phys.* **A 12** 4109.

[32] Covi L 1998, *Proc. Int. Symp. on Lepton and Baryon number Violation, Trento*

CP VIOLATION IN LEPTOGENESIS

L Covi†and E Roulet‡

† Department of Physics, Lancaster University
Lancaster LA1 4YB, U.K.

‡ Departamento de Fisica, UNLP
CC67, 1900, La Plata, Argentina

Abstract. We review the mechanism of dynamical production of baryon (lepton) number and we consider the case of baryogenesis through production of a net L number by out of equilibrium decay of heavy electroweak singlet neutrinos into light leptons and Higgs bosons. Sphaleronic processes, still active during and after the decay, reprocess such lepton number into a baryon asymmetry. We compute the CP violation in the heavy particle decay, arising from both the one-loop vertex correction and the wave function mixing. We extend the computation to the supersymmetric version of the model and then consider specifically the decay of quasi–degenerate sneutrinos; a considerable enhancement of the asymmetry arises in this case for mass-splittings of the order of the decaying particle's width. Finally we include also the finite temperature effects and find that they are negligible due to interesting cancellations.

1. Introduction

A very attractive scenario, proposed by Fukugita and Yanagida [1], for the generation of the baryon asymmetry of the Universe, is based on the production of a lepton asymmetry by the out of equilibrium decays of heavy electroweak singlet neutrinos, like those present in see-saw models and naturally predicted by any $SO(10)$ Grand Unification. Their decays into light leptons and Higgs bosons violate CP if the Yukawa couplings involved have unremovable phases, and can then lead to the production of an excess of antileptons over leptons in the final state. The lepton asymmetry so produced is then partially converted into a baryon asymmetry by anomalous electroweak processes, which are in equilibrium at temperatures larger than the electroweak phase transition one. As a result, the amount of baryons present in the Universe, $n_B/n_\gamma \simeq 5 \times 10^{-10}$, can be naturally accounted for [2, 3, 4].

2. Baryogenesis through leptogenesis

As already pointed out by Sakharov more than 30 years ago [5] three conditions are to be satisfied in order to generate any baryon number (B) dynamically:

- B violation (or better $B-L$ violation as we will see in the following);
- C and CP violation, due to the fact that B is odd under these symmetries;
- departure from thermal equilibrium, since otherwise any non conserved quantum number tend to be washed out to give the maximal entropy state.

At the beginning therefore baryogenesis was usually considered in the context of extensions of the Standard Model, like GUT's, that predict proton decay and B violating processes at high scales. Later on, the discovery of non perturbative $B+L$ violating processes in the Standard Model [6], highly suppressed at zero temperature, but active and in thermal equilibrium with respect to the universe expansion for temperatures greater than 100 GeV (sphalerons) [7], gave rise to other possibilities. For example the generation of the baryon number at the electroweak phase transition if this is strongly of the first order or otherwise through generations of a net $B-L$ number while the sphaleronic processes are still active.

In the last case, since sphalerons conserve $B-L$, but violate $B+L$, they reprocess any $B-L$ asymmetry into a final B asymmetry according to the expression [8]

$$B = \frac{8N_f + 4N_H}{22N_f + 13N_H}(B-L) \tag{1}$$

where N_f is the number of flavours and N_H the number of Higgs doublets present in the thermal bath. So any lepton number would be transformed into a baryon asymmetry by sphaleronic processes, giving

$$n_B = \left\{ \begin{array}{lc} \frac{28}{79} n_{B-L} & \text{(non SUSY)} \\ & \\ \frac{32}{92} n_{B-L} & \text{(SUSY)} \end{array} \right\} \simeq \frac{1}{3} n_{B-L}. \tag{2}$$

We will consider in the following this last case, where baryogenesis is a direct consequence of leptogenesis. For what concerns the EW baryogenesis model, such mechanism is already ruled out in the case of the Standard Model since it would require a Higgs mass lower than the present experimental bound in order that $B - L$ violation be sufficiently suppressed in the broken phase bubbles so to preserve the baryon number generated, but it is still viable in the Minimal Supersymmetric Extension of the SM if the lightest Higgs mass is $\lesssim 105$ GeV [9]. We can therefore conclude that the baryon asymmetry of the universe is at the moment, with the recent Super–Kamiocande measurement of atmosferic neutrino oscillations, one of the two experimental data that requires physics beyond the Standard Model.

3. Out of equilibrium decay

The last key ingredient for the dynamical generation of baryon or lepton number is departure from thermal equilibrium. We have in fact that in case of thermal equilibrium, unitarity itself imposes the vanishing of any CP violating asymmetry [10, 11]. Let us consider for example the 2-body L violating scatterings, which include also the case of s-channel intermediate real heavy singlet neutrino; from unitarity we have that the transition probabilities to or from a state i must add to unity, giving the relation:

$$\sum_j |M(i \to j)|^2 = \sum_j |M(j \to i)|^2 \tag{3}$$

so for our 2-body scatterings we have that the L violating CP asymmetry results to be

$$\begin{aligned} \epsilon \quad &\propto \quad \sum_{i=1}^{3}\sum_{j=1}^{3}\left[\sigma(l_i^c H^* \to l_j H) - \sigma(l_i H \to l_j^c H^*)\right] \qquad (4)\\ &= \quad \sum_{j=1}^{3}\sum_{i=1}^{3}\left[\sigma(l_i^c H^* \to l_j H) + \sigma(l_i H \to l_j H)\right] \\ &\quad -\sum_{i=1}^{3}\sum_{j=1}^{3}\left[\sigma(l_i H \to l_j H) + \sigma(l_i H \to l_j^c H^*)\right] = 0 \qquad (5)\end{aligned}$$

where we have used eq. (3).

This result brings to a vanishing collision integral in the Boltzmann equation for the lepton number density in the case of thermal equilibrium, since the number densities for the initial states depend only on the centre of mass energy and factorize out. In the case of departure from equilibrium this is no more the case and the exact cancellation is spoiled.

The crucial factor that brings out of thermal equilibrium in the early universe is the universe expansion. When the interaction rates are slow comparing to the Hubble parameter $H = \dot{R}/R$ and the intermediate real heavy particles live for a long time compared to $1/H$, the 2-body scatterings are no more the proper processes to consider: between production and decay of the intermediate heavy states the temperature changes sufficiently to allow first the compensation of the initial L asymmetry by L violating scatterings and later on the generation of a final lepton number by the heavy particle decay, when the L violating processes are no more able to cancel it [11].

We can therefore consider the CP asymmetry in the decay of the lightest singlet neutrino. If all the other L violating interactions are way out of equilibrium and can be safely ignored, the Boltzmann equation for the lepton number density is simply:

$$\dot{n}_L + 3Hn_L = \epsilon_\ell^N \langle \Gamma_{N\ell} \rangle (n_N - n_N^{eq}) \tag{6}$$

where the second term on the l.h.s. takes into account the dilution due to the universe expansion and the r.h.s. is the contribution to n_L coming from the L and CP violating decays of the singlet neutrinos N; ϵ_ℓ^N is the net lepton number produced per heavy particle decay, i.e.

$$\epsilon_\ell^N \equiv \frac{\Gamma_{N\ell} - \Gamma_{N\bar{\ell}}}{\Gamma_{N\ell} + \Gamma_{N\bar{\ell}}}, \tag{7}$$

where $\Gamma_{N\ell}(\Gamma_{N\bar{\ell}})$ is the N decay rate into leptons (antileptons) in the N rest frame. $\langle \Gamma_{N\ell} \rangle$ indicates the thermally averaged decay rate and n_N, n_N^{eq} are respectively the actual number densities of the heavy singlet states and the equilibrium one. Notice that in equilibrium no lepton number would be generated.

Equation (6) has a simple solution in the case of way out of equilibrium decay, i.e. for

$$\Gamma_N \ll H, \tag{8}$$

and we have for the ratio of the lepton number density over the entropy density, constant during the expansion,

$$\frac{n_L}{s} = \frac{45}{\pi^4} \frac{\epsilon_\ell^N}{g_\star} \tag{9}$$

where $g_\star$ is the effective number of relativistic bosonic and fermionic degrees of freedom that were in equilibrium at the time of the decay. The quantity in eq. (9) is the maximal lepton number that can be generated in the decay

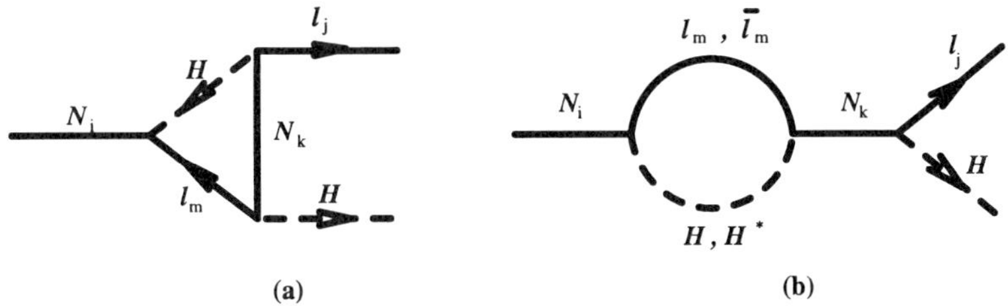

Figure 1. Vertex (a) and wave mixing (b) one loop diagrams inducing a decay asymmetry.

and is simply proportional to ϵ_ℓ^N. Using then the relation (1) it is possible to compute the baryon number obtained after sphaleron reprocessing:

$$\frac{n_B}{s} \simeq -\frac{1}{3}\frac{n_L}{s} = -\frac{15}{\pi^4}\frac{\epsilon_\ell^N}{g_\star}. \tag{10}$$

4. CP violation in the decay

Let us consider now in more detail the CP violation in the decay of the heavy Majorana neutrinos. Such quantity is vanishing at tree level due to the hermiticity of the Lagrangian; the lowest order contribution is the one–loop as can be seen using unitarity. Writing the scattering matrix $S = 1 - iT$, unitarity gives

$$T = T^\dagger - iT^\dagger T \tag{11}$$

using the CPT relation $T_{if} = T_{\bar{f}\bar{i}}$ one can obtain

$$|T_{fi}|^2 - |T_{\bar{f}\bar{i}}|^2 = 2\mathrm{Im}\left[(T^\dagger T)_{fi} T_{if}\right] + |(T^\dagger T)_{fi}|^2 \tag{12}$$

where the first term on the r.h.s. contains the one–loop contribution, while the second is already a two–loop effect. We see from this expression that the CP asymmetry is related to the interference between one–loop diagrams with real intermediate states (i.e. the absorptive parts of the one loop diagrams) and the tree level coupling; moreover non removable phases have to be present in the couplings in order to have a non vanishing imaginary part of the product.

In our case the one loop contributions are of two types, depending if the intermediate heavy state is exchanged in the s– or t–channel; they are represented in Fig 1.

We will now compute in detail the CP asymmetry in the decay of the heavy singlet neutrinos both in the non-supersymmetric and the supersymmetric case.

4.1. The non–supersymmetric model

Let us consider the non–supersymmetric Lagrangian:

$$\mathcal{L} = -\lambda_{ij}\epsilon_{\alpha\beta}\overline{N_j}P_L\ell_i^\alpha H^\beta + h.c. \ , \tag{13}$$

where $\ell_i^T = (\nu_i \ l_i^-)$ and $H^T = (H^+ \ H^0)$ are the lepton and Higgs doublets and we are taking the N_j to be the Majorana mass eigenstate fields, of mass M_j. Since the scale of the heavy neutrinos is so much larger than the electroweak scale, it is reasonable to work directly in the symmetric phase where all particles except the heavy Majorana neutrinos N_j are massless, and all components of the neutral and charged complex Higgs fields are physical.

The basic quantity we are interested in is the net lepton number produced per heavy particle decay, defined in (7). This asymmetry will survive if L violating scatterings are out of equilibrium at the decay time, and $\epsilon \simeq -10^{-8}$ is required to explain the baryon asymmetry assuming $g_\star \simeq 100$ in eq. (9).

The asymmetry ϵ_ℓ^N arises as we have seen from the interference of the one–loop diagrams depicted in Fig. 1 with the tree level coupling. The vertex correction in Fig. 1a was usually considered [1, 2], but it was pointed out [12, 13, 14] that the mixing piece in Fig. 1b also contributes to the asymmetry, in an amount which is typically comparable to the vertex contribution. We have computed these CP violating asymmetries [3] and we obtain for the vertex contribution to ϵ_ℓ^N

$$\epsilon_\ell^{N_i}(v) = \frac{1}{8\pi}\sum_{k\neq i} f(y_k)\,\frac{\mathrm{Im}\left[\left(\lambda^\dagger\lambda\right)_{ki}^2\right]}{\left(\lambda^\dagger\lambda\right)_{ii}} \equiv \frac{1}{8\pi}\sum_{k\neq i} f(y_k)\,\mathcal{I}_{ki} \ , \tag{14}$$

where $y_k \equiv M_k^2/M_i^2$ and $f(x) = \sqrt{x}\,(1-(1+x)\ln[(1+x)/x])$. For the mixing piece, assuming that $|M_k - M_i| \gg \Gamma_{N_k}$, we find

$$\epsilon_\ell^{N_i}(m) = -\frac{1}{8\pi}\sum_{k\neq i}\frac{M_iM_k}{M_k^2 - M_i^2}\,\mathcal{I}_{ki} \tag{15}$$

in agreement with the last result in [15].

In the case of a strong hierarchy among the right–handed neutrino masses, i.e. for $y_k \gg 1$, the total asymmetry produced in the decay of the lightest heavy neutrino N_1 (the asymmetry generated by the decays of the heavier ones, which take place at higher temperatures, is usually erased rapidly by L violating scattering processes), becomes

$$\epsilon_\ell^{N_1} \simeq -\frac{3}{16\pi}\sum_{k\neq 1}\frac{1}{\sqrt{y_k}}\,\mathcal{I}_{k1}. \tag{16}$$

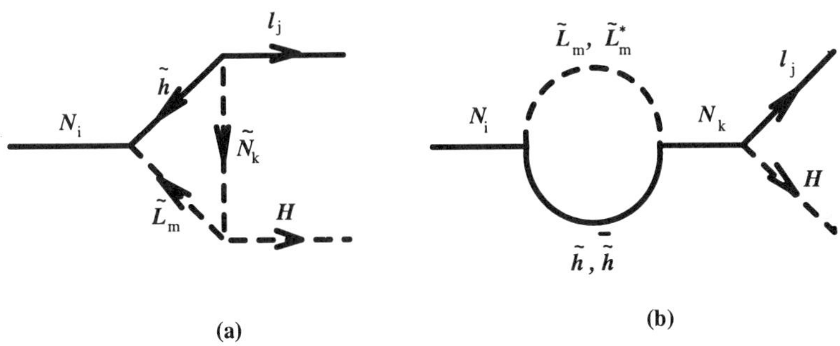

Figure 2. Supersymmetric one–loop diagrams contributing to the CP asymmetry in the heavy neutrino decay into leptons.

We see then that the inclusion of the wave function piece triplicates the total CP asymmetry in the limit $y_k \gg 1$ and that for accounting for the observed baryonic asymmetry the Yukawa couplings must satisfy

$$\sum_{k\neq 1} \frac{M_1}{M_k} \mathcal{I}_{k1} \geq 0.9 \times 10^{-6}; \tag{17}$$

the equality applies to the case of way out of equilibrium decay, when we have also

$$\Gamma_{N_1} = (\lambda^\dagger \lambda)_{11} \frac{M_1}{8\pi} \ll H \simeq g_*^{1/2} \frac{M_1^2}{M_{Pl}} \tag{18}$$

i.e. $(\lambda^\dagger \lambda)_{11}/M_1 \ll 8\pi g_*^{1/2}/M_{Pl}$.

4.2. The supersymmetric model

In the supersymmetric scenario, considered also in [16], the interactions of the heavy (s)neutrino field are given, in addition to eq. (13), by the following trilinear couplings, in terms of four component spinors,

$$\mathcal{L} = -\lambda_{ij}\epsilon_{\alpha\beta} \left\{ M_j \tilde{N}_j^* \tilde{L}_i^\alpha H^\beta + \overline{(\tilde{h}^\beta)^c} P_L \ell_i^\alpha \tilde{N}_j + \overline{(\tilde{h}^\beta)^c} P_L N_j \tilde{L}_i^\alpha \right\} + h.c.\,. \tag{19}$$

We neglect supersymmetry breaking terms since they do not contribute to the lepton number generation.

It is clear that additional one–loop diagrams are present, like those in Fig. 2, with superpartners instead of ordinary particles in the intermediate and final states and, moreover, that also the CP asymmetry in the decay of the heavy sneutrino has to be considered.

Defining $\tilde{\epsilon}_\ell^N, \tilde{\epsilon}_{\tilde{L}}^N, \tilde{\epsilon}_\ell^{\tilde{N}^*}$ and $\tilde{\epsilon}_{\tilde{L}}^{\tilde{N}^*}$ in analogy with eq. (7), we obtain that all the vertex contributions are equal among theirselves and given by

$$\tilde{\epsilon}_\ell^{N_i}(v) = -\frac{1}{16\pi} \sum_k g\,(y_k)\, \mathcal{I}_{ki} \;, \tag{20}$$

where $g(x) = \sqrt{x}\,\ln[(1+x)/x]$, while all the mixing contributions are just twice the non–supersymmetric result, e.g.

$$\tilde{\epsilon}_\ell^N(m) = 2\epsilon_\ell^N(m). \qquad (21)$$

For large hierarchy we have then the expression

$$\tilde{\epsilon}_1 = \tilde{\epsilon}^{N_1} + \tilde{\epsilon}^{\tilde{N}_1} \qquad (22)$$

$$\simeq -\frac{3}{4\pi}\sum_{k\neq 1}\frac{1}{\sqrt{y_k}}\,\mathcal{I}_{k1}. \qquad (23)$$

analogous to (16). We see therefore that in the supersymmetric case a larger CP asymmetry can be obtained with the same couplings and hierarchy, but to compute the lepton number generated we must also take into account the dilution factor $g_\star$ that in the supersymmetric case is twice as large due to the presence of the superpartners in the thermal bath. We have therefore that to account for the present baryon number a similar pattern of couplings and hierarchy is needed also in the supersymmetric cases

$$\sum_{k\neq 1}\frac{M_1}{M_k}\,\mathcal{I}_{k1} \geq 0.7\times 10^{-6}. \qquad (24)$$

5. Leptogenesis and neutrino masses

The same terms in the Lagrangian that give rise to a net lepton number in the decay of the heavy EW singlet neutrinos also generate a light Majorana mass for the SM neutrinos when the Higgs acquires a vacuum expectation value v. The Yukawa coupling λ gives rise in fact to a Dirac type mass term that mixes the light and heavy neutrinos; by diagonalizing the Majorana– Dirac mass matrix for all the neutrinos, we obtain by the see–saw mechanism [17] very light Majorana masses for the (mainly) LH neutrinos[1]. Their mass matrix is given by

$$m_L \simeq \lambda^T M^{-1}\lambda\, v^2; \qquad (25)$$

we see therefore that leptogenesis and neutrino masses can be linked together in this scenario [2], especially in the context of an $SO(10)$ unification [18, 4, 19]. In this case the singlet neutrino masses are usually at an intermediate scale corresponding to the breaking of $U(1)_{B-L}$ in the chain of symmetry breaking from $SO(10)$ to the SM.

[1] The mass eigenstates are a mixture of the SM and singlet neutrinos, but the SM component is predominant in the light ones.

6. Quasi–degenerate singlet sneutrinos

We have seen that the wave part of the CP asymmetry in the decay of a pair of heavy sneutrinos is given by twice eq. (15) and is proportional to $(M_k^2 - M_i^2)^{-1}$, due to the propagator of the intermediate state $\tilde{N}_k$. Hence, this quantity is expected to be enhanced in the limit $M_k \to M_i$, but there our result is no more valid since the particle widths can not be neglected anymore. Using the phenomenological approach in [20], we have computed $\tilde{\epsilon}^{\tilde{N}}(m)$ for arbitrary mass splittings in the case of two generations. The effect of the particle mixing can be taken into account by diagonalizing the complex mass matrix

$$H^2 = M^2 - i\Gamma^2; \tag{26}$$

the hermitian renormalised matrix M^2 includes the dispersive part of the loops, while the other hermitian matrix Γ^2 arises from the absorptive parts. H^2 is a general complex matrix whose eigenstates coincide with the propagation eigenstates; if it is non degenerate it can be diagonalized by a non unitary[2] transformation V so that the transition amplitudes from an initial state i and its conjugate are given by

$$T_{fa}(t) = \sum_b T_{fb}(V^{-1}e^{-i\omega t}V)_{ba} \tag{27}$$

$$\bar{T}_{fa}(t) = \sum_b T^*_{fb}(V^{-1}e^{-i\omega t}V)^T_{ba} \tag{28}$$

where $\omega = \mathrm{diag}(\omega_1, \omega_2)$ is related to the H^2 eigenvalues $\omega_i^2 = (m_i - i\gamma_i/2)^2$.

Then the CP asymmetry due to the sneutrino mixing is given by

$$\tilde{\epsilon}^{\tilde{N}_1}(m) = \frac{M_1}{16\pi}\int_0^\infty dt\ |T_{\ell 1}(t)|^2 - |\bar{T}_{\ell 1}(t)|^2 \tag{29}$$

$$= -\frac{M_1^2 M_2}{16\pi^2}\mathrm{Im}\left[\left(\lambda^\dagger\lambda\right)^2_{21}\right] F_1 \tag{30}$$

where the subscript ℓ denotes the leptonic and sleptonic final states, and we have assumed a diagonal M^2 matrix ($\mathrm{diag}(M_1^2, M_2^2)$); F_1 is given by:

$$\begin{aligned} F_1 &= \frac{1}{|\omega_1^2 - \omega_2^2|^2}\left\{\mathrm{Re}\left[\omega_1^2 - \omega_2^2\right]\left(\frac{1}{\gamma_2} - \frac{1}{\gamma_1}\right)\right. \\ &+ 2\mathrm{Im}\left[\omega_1^2 - \omega_2^2\right]\frac{m_1 - m_2}{|\omega_1^* - \omega_2|^2} \\ &- \left.(M_1^2 - M_2^2)\left(\frac{1}{\gamma_2} + \frac{1}{\gamma_1} - \frac{\gamma_1 + \gamma_2}{|\omega_1^* - \omega_2|^2}\right)\right\} \end{aligned} \tag{31}$$

[2] H^2 can be diagonalized by a unitary matrix only in the case when M^2 and Γ^2 commute with each other.

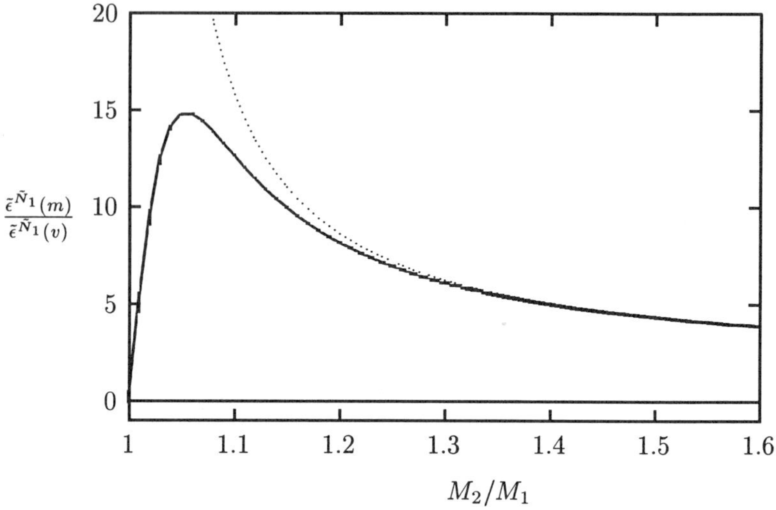

Figure 3. $\tilde{\epsilon}^{\tilde{N}_1}(m)/\tilde{\epsilon}^{\tilde{N}_1}(v)$ vs. M_2/M_1, assuming for the largest width $M/\Gamma \simeq 10$.

The detailed computations are given in [21], but in Fig. 3 it is possible to see the behaviour of $\tilde{\epsilon}^{\tilde{N}_1}(m)/\tilde{\epsilon}^{\tilde{N}_1}(v)$: at the maximum, which corresponds to a mass difference of the order of the largest particle width, the enhancement is of order M/Γ, and can then be in principle of many orders of magnitude (i.e., the asymmetry associated to the wave mixing can be of order one even if the vertex part is quite small). At large values of the mass difference, the expected behaviour $\tilde{\epsilon}^{\tilde{N}_1}(m)/\tilde{\epsilon}^{\tilde{N}_1}(v) \to 2$ is recovered.

A similar behaviour has been found in the case of quasi–degenerate fermions in [22, 23].

Such enhancement of the CP asymmetry could be fundamental in order to generate the observed baryon asymmetry in the case, for example, $\tilde{N}_1$'s are produced out of equilibrium during preheating by the resonant decay of the inflaton field. On the other hand, in the usual picture of out of equilibrium decay, one should also take into account the presence and decay in the thermal bath of the heavier quasi–degenerate species $\tilde{N}_2$, whose CP violation due to the mixing ($\tilde{\epsilon}^{\tilde{N}_2}(m)$) has opposite sign with respect to the light one.

7. Finite temperature corrections

We also took into account the effects due to the thermal background of light particles present in the universe at the decay epoch, introducing finite temperature propagators for the light states in the loop and modifying the final state phase space with the appropriate statistical factors for fermionic or bosonic states. We assumed that, since the heavy states are out of equilibrium or already decayed away, their propagation in the loop is not modified.

The effect of the bath on the decay rate is simple: due to the thermal background, the fermion's exchange and production are subject to a Pauli blocking, while the boson's exchange and production are stimulated.

We notice however that in the case of the neutrino decay, since the two light particles in the loop (ℓ and H) have opposite statistics, there is an almost perfect cancellation of the two competing effects [24]. In this case, the final phase space factors cancel in the expression for ϵ, which is a ratio of widths. From the loop integration at finite temperature we have that the main correction is

$$\epsilon^T/\epsilon^0 \simeq 1 - \bar{n}_F + \bar{n}_B - 2\bar{n}_F\bar{n}_B \tag{32}$$

where $\bar{n}_{B,F}$ are respectively the Bose or Fermi distributions evaluated at the energy equal to half the decaying particle mass M [3]; since the distribution functions satisfy the relation

$$n_B(E) - n_F(E) = 2n_F(E)n_B(E), \tag{33}$$

we see that the thermal correction to the CP violation vanishes and we can safely use the $T = 0$ result.

For the sneutrino decay, the final state phase space factor do not cancel directly and also the loop integral involve either two bosons or two fermions. However, a cancellation happens also in this case, thanks to the combined effects of the loop and final state corrections. We have in fact the bosonic loop participates to the CP violation for the fermionic final state and viceversa. We have then that, while at $T = 0$ the CP violation in the decay channel to particles and their superpartners are equal, at finite temperature the two channel acquire different corrections:

$$\tilde{\epsilon}^T/\tilde{\epsilon}^0(F) = (1 + 2\bar{n}_B + 2\bar{n}_B^2)\frac{(1-\bar{n}_F)^2}{(1-\bar{n}_F)^2 + (1+\bar{n}_B)^2} \tag{34}$$

$$\tilde{\epsilon}^T/\tilde{\epsilon}^0(B) = (1 - 2\bar{n}_F + 2\bar{n}_F^2)\frac{(1+\bar{n}_B)^2}{(1-\bar{n}_F)^2 + (1+\bar{n}_B)^2} \tag{35}$$

[3] We are considering the case in which the tree level masses of the light particles (Higgs bosons or superpartners) are negligible compared to M.

where the letter F or B indicates the different final states. But playing around with the expressions on the r.h.s. we can recast them in the form

$$\tilde{\epsilon}^T/\tilde{\epsilon}^0(F) = \frac{1}{2} + \frac{\bar{n}_B^2(1-\bar{n}_F)^2 - \bar{n}_F^2(1+\bar{n}_B)^2}{2[(1-\bar{n}_F)^2 + (1+\bar{n}_B)^2]} = \frac{1}{2} \quad (36)$$

$$\tilde{\epsilon}^T/\tilde{\epsilon}^0(B) = \frac{1}{2} + \frac{\bar{n}_F^2(1-\bar{n}_B)^2 - \bar{n}_B^2(1+\bar{n}_F)^2}{2[(1-\bar{n}_F)^2 + (1+\bar{n}_B)^2]} = \frac{1}{2} \quad (37)$$

where the last equality follows again from the relation (33). Therefore also in the singlet sneutrino case, due to the contemporaneous presence of both fermionic and bosonic channels, the finite temperature corrections to the CP asymmetry vanish at leading order. As a consequence, the results given in the previous sections are practically unmodified by thermal effects[4].

8. Conclusions

We have seen that leptogenesis is a viable mechanism for the generation of a net baryon number at intermediate scale; in this scenario it could be possible not only to generate a lepton number (reprocessed into a B by sphalerons), but also to give rise to small Majorana masses for the EW doublet neutrinos through the see-saw mechanism and link the production of lepton to CP violating phases in the neutrino mass matrix.

We have identified and studied in detail new contributions to the CP decay rate asymmetries relevant for leptogenesis. We found that the absorptive part of the mixing diagrams usually provide the dominant contribution to the asymmetry and can then help to make these scenarios even more successful.

Finally we have discussed the leading finite temperature corrections to the CP asymmetry and we have found that they vanish and that the zero temperature result is unaffected.

Acknowledgements

We would like to thank N. Rius and F. Vissani for their collaborations in part of this work.

[4] For a more detail discussion and second order effects, see [24].

References

[1] Fukugita M and Yanagida T 1986 *Phys. Lett.* **B174** 45

[2] Luty M A 1992 *Phys. Rev.* D **45** 455

[3] Covi L, Roulet E and Vissani F 1996 *Phys. Lett.* **B384** 169

[4] Plümacher M 1997 *Z. Phys.* C **74** 549

[5] Sakharov A D 1967 *JETP Lett.* **5** 24

[6] Manton N S 1983 *Phys. Rev.* D **28** 2019
Klinkhammer F R and Manton N S 1984 *Phys. Rev.* D **30** 2212

[7] Kuzmin V A, Rubakov V A and Shaposhnikov M E 1985 *Phys. Lett.* **B155** 36
Khlebnikov S and Shaposhnikov M E 1988 *Nucl. Phys.* B **308** 883

[8] Harvey J A and Turner M S 1990 *Phys. Rev.* **D42** 3344

[9] Carena M, Quiros M and Wagner C E M 1998 *Nucl. Phys.* B **524** 3

[10] Weinberg S 1979 *Phys. Rev. Lett.* **42** 850
Kolb E W and Wolfram S 1980 *Nucl. Phys.* **B172** 224 ; 1982 *Nucl. Phys.* B **195** 542 (E)

[11] Roulet E, Covi L and Vissani F 1998 *Phys. Lett.* **B424** 101

[12] Ignatiev A Y, Kuzmin V A and Shaposhnikov M E 1979 *JETP Lett.* **30** 688

[13] Botella F J and Roldan J 1991 *Phys. Rev.* **D44** 966

[14] Liu J and Segrè G 1993 *Phys. Rev.* **D48** 4609

[15] Flanz M, Paschos E A, and Sarkar U 1995 *Phys. Lett.* **B345** 248 ; 1996 *Phys. Lett.* **B382** 447 (E); 1996 *Phys. Lett.* **B384** 487

[16] Campbell B A, Davidson S and Olive K A 1993 *Nucl. Phys.* B **399** 111

[17] Gell-Mann M, Ramond P and Slansky R 1979 *Complex Spinors and Unified Theories* in *Supergravity* ed van Nieuwenhuizen P and Freedman D Z (Amsterdam: North Holland)

[18] Buchmüller W and Plümacher M 1996 *Phys. Lett.* **B389** 73

[19] Plümacher 1997 hep-ph/9704231

[20] Liu J and Segrè G 1994 *Phys. Rev.* **D49** 1342

[21] Covi L and Roulet E 1997 *Phys. Lett.* **B399** 113

[22] Flanz M, Paschos E A, Sarkar U and Weiss J 1996 *Phys. Lett.* **B389** 693

[23] Pilaftsis A 1997 *Phys. Rev.* D **56** 5431

[24] Covi L, Rius N, Roulet E and Vissani F 1998 *Phys. Rev.* D **57** 93

Cosmological Constraints on B-L Violation

Sacha Davidson

CERN Theory Division, CH-1211, Genève 23, Switzerland

Abstract. This is a review of the cosmological bounds on $B-L$ violating interactions, and the loopholes in the argument that gives these constraints. If one assumes that the baryon asymmetry we observe today was present above the electroweak phase transition in equilibrium with the non-perturbative $B+L$ violating processes, then interactions that violate all three of $\{B/3 - L_i\}$ cannot simultaneously be in equilibrium. Otherwise the baryon asymmetry would be washed out. Therefore violation of at least one of the $B/3 - L_i$ must be small. This argument can be evaded by not having the observed baryon asymmetry present in the thermal bath (for instance, make it at the electroweak phase transition), by using a non-standard cosmological model, or possibly by some mass effects in models where the difference between lepton flavour asymmetries is conserved.

1. Introduction

It is observed that at least up to the scale of our local galaxy cluster, the Universe we see is made of matter rather than anti-matter. It is difficult to build a cosmological model where the Universe contains equal numbers of baryons and anti-baryons separated on scales larger than galaxy clusters (There is a problem with causality [1]), so one usually assumes that the Universe as a whole contains a net excess of baryons over anti-baryons, and that at some time in the early history of the Universe, this asymmetry was generated. This is the Baryon Asymmetry of the Universe, or BAU (for reviews, see, *e.g.* [2, 3, 4].)

There are three ingredients required to create a baryon asymmetry [5]: baryon number violation, C and CP violation, and some out of equilibrium dynamics. The Standard Model (SM) does contain all these ingredients, but not in the right quantities to generate the BAU.

The Standard Model violates C maximally (because it is chiral), and contains small amounts of CP violation in the CKM matrix (probably not enough to generate the BAU though). It has no perturbative baryon number violation. The lowest dimensional baryon number violating operator allowed by standard model gauge symmetries and involving SM fields is $qqq\ell$ [6], which is of dimension six. However the Standard Model does contain non-perturbative baryon number violation [7]: the current associated with the quantum number $B+L$ (baryon + lepton number) is anomalous, so certain gauge field configurations can induce the $B+L$ violating vertex $(qqq\ell)^3$ (see, for instance, [3] for a review and references). This operator allows processes such as

$$u_g d_b d_r c_g s_b s_r t_g b_b b_r \longrightarrow \bar{\nu}_e \bar{\nu}_\mu \bar{\nu}_\tau \tag{1}$$

where g, r, b are SU(3) colour labels, not summed, and the quarks are SU(2) doublet members. At zero temperature the rate for such $B+L$ non-conserving processes is exponentially suppressed by the action of the weak gauge field configuration, so one does not expect to see low energy $B+L$ violation. At finite temperature the rate is much less suppressed; the rate above the electroweak phase transition is

$$\Gamma_{B\not{+}L} = \kappa \alpha^n T \tag{2}$$

where κ is a constant of order 1, and n is 4 or 5 [9]. This is clearly in thermal equilibrium in the early Universe ($\Gamma > H \equiv$ the expansion rate of the Universe; if $\Gamma > H$ the timescale for the interaction is less than the age of the Universe.). After the phase transition, the rate is Boltzmann suppressed ($\Gamma \sim \exp\{-m_W/(\alpha T)\}$), so rapidly drops out of equlbrium. This is potentially very interesting for baryogenesis (as was noted by [8]): a source of baryon number violation that is efficient in the early Universe when the asymmetry needs to be generated, and then turns off at low energy in perturbative processes where we see no baryon number violation.

Unfortunately the Standard Model lacks the third ingredient required to generate the baryon asymmetry. There is not a big enough departure from equilibrium at the electroweak phase transition; the $B+L$ violating processes turn off gradually and no asymmetry in generated [10]. The observation of the baryon asymmetry can therefore be interpreted as evidence for new physics, since it does not appear possible to generate an asymmetry within the Standard Model.

The new physics can have two effects on the generation of the baryon asymmetry. It may simply add enough CP violation and make the phase transition sufficiently strongly first order that the baryon asymmetry is generated at the phase transition using the SM $B+L$ violation. Alternatively, the beyond-the-Standard Model physics may involve new sources of B and/or L violation which generate the observed asymmetry before the electroweak phase transition. In the first case there are no cosmological bounds on $B-L$ violation. So for most of this review, I will assume the second possibility: suppose that the baryon asymmetry we observe today was generated early in the history of the Universe and is present as an asymmetry in the number density of quarks and antiquarks in the thermal bath just above the electroweak phase transition. The anomalous $B+L$ violating processes are also operating in this plasma, so if the asymmetry generated previously wishes to survive in their presence, it must carry at least one of $B/3-L_e$, $B/3-L_\mu$ or $B/3-L_\tau$. These are the three global quantum numbers exactly conserved in the Standard Model. If there are simultaneously interactions in equilibrium that violate all three of these quantum numbers, all asymmetries will be washed out. Since we are assuming that the baryon asymmetry we observe today was generated before the electroweak phase transition, this is undesirable. One can avoid this outcome by requiring that processes violating one of the $\{B/3-L_i\}$ be out of equilibrium:

$$\Gamma_{B/3-L_i} < H \qquad \text{for one } i \tag{3}$$

where H is the expansion rate of the Universe, and i is one of e, μ or τ. This condition should hold for all temperatures during which the baryon asymmetry is at risk: 100 GeV $\lesssim T \lesssim$ 100 TeV. The reasons for this range of temperatures will be discussed in section 3.

For a yukawa-type $B-L$ violating interaction $\lambda\phi\psi\psi$ (eg R-parity violating trilinears in Supersymmetry), equation (3) becomes

$$10^{-2}\lambda^2 T < \frac{25T^2}{m_{pl}} \tag{4}$$

which gives a bound of order $\lambda < 10^{-7}$. I use $H = 10T^2/m_{pl}$ in the SM, and $H = 25T^2/m_{pl}$ for SUSY particle content. One can estimate the rate associated with a majorana neutrino mass to be of order

$$\Gamma_{m_\nu} \sim \frac{10^{-3}m^2}{T} < H \tag{5}$$

which gives $m_\nu < 30$ keV in one generation. The numerical factors in these bounds will be discussed more carefully in section 2. Bounds can also be set on non-renormalisable $B/3-L_i$ violating operators of dimension $D = 4+n > 4$; if all the coupling constants are absorbed into the mass scale M, the rate can be estimated as

$$\Gamma \sim 10^{-3}\frac{T^{2n+1}}{M^{2n}} < \frac{10T^2}{m_{pl}} \tag{6}$$

For $D = n + 4 = 6$, this gives $M > 10^4 T_{\text{GeV}}^{3/4}$ GeV. It is clear that nonrenormalisable interaction rates increase faster with temperature than the expansion rate, so are more likely to be in equilibrium at higher temperatures. The maximum temperature at which the bounds can be applied, which gives the best bounds ($T \sim 100$ TeV), will be discussed in section 3.

These constraints are comparatively stringent, so it us useful to know exactly where they apply and how they can be avoided. These bounds apply to only one of the $\{B/3 - L_i\}$, so lepton number violation in one family and baryon number violation must be small. In section 2 I will review estimates of interaction rates for various coupling constants, and the bounds one can derive from these estimates. The most obvious loophole is to generate the asymmetry at the electroweak phase transition, or more precisely, to not have the baryon asymmetry present in the plasma just above the EPT. In section 3 I will show that the BAU only needs to be absent from the thermal soup for temperatures between ~ 100 TeV and the EPT, and discuss various ways of accomplishing this. Section 4 contains various other mechanisms for avoiding the bounds such as low T_{reheat} inflationary models. There is a summary of the constraints and loopholes in section 5. The remaining part of the introduction is a short overview of previous work on this topic.

It has been known for a long time that if the baryon asymmetry is generated by the out-of-equilibrium decay of heavy GUT particles, then baryon number violating processes mediated by these particles must be out of equilibrium after the asymmetry is generated [11]. Otherwise they will wipe out the asymmetry generated in the decay. For the baryon asymmetry to survive in the presence of the $B + L$ violating electroweak effects, it must be an asymmetry in $B - L$, which makes it difficult (but not impossible, see section 4.1) to preserve the baryon asymmetry in $B - L$ conserving GUTs. An attractive alternative, introduced by Fukugita and Yanagida [12], is to generate a lepton asymmetry in the decay of heavy singlet neutrinos, and use the electroweak $B + L$ violation to transform this into a baryon asymmetry. Lepton number violating interactions mediated by the heavy singlet neutrinos must be out of equilibrium after the lepton asymmetry is generated, to avoid washing out the asymmetry [13]. This bound on lepton number violating processes is not limited to leptogenesis scenarios; Barr and Nelson [14], and [15], observed that the baryon asymmetry, the $B+L$ violating processes, and interactions that violate the three $\{B/3 - L_i\}$ cannot simultaneously coexist in equilibrium, and used this to set bounds on majorana neutrino masses in one generation. This argument was then extended to R parity violating interactions in the Supersymmetric version of the Standard Model [15, 16] and to generic $B - L$ nonconserving interactions [16]. The constraints and loopholes were then discussed by many people [17, 18, 19, 20, 21].

2. Rate estimates and constraints

The argument that gives cosmological bounds of $B - L$ violation goes as follows: assume that the observed baryon asymmetry was generated before the EPT. To survive in the thermal soup, it must carry a quantum number that is "effectively conserved"—*i.e.* conserved by the interactions that are in chemical equilibrium. The SM interactions are all in chemical equilibrium just above the EPT, and the three global quantum numbers that they conserve are the $\{B/3 - L_i\}$. To protect the BAU, any additional interactions present must also "effectively conserve" at least one of the $\{B/3 - L_i\}$, or equivalently, interactions violating at least one of the $\{B/3 - L_i\}$ must be out of equilibrium.

In the first part of this section, I will estimate interaction rates associated with coupling constants of different mass dimensions. It follows from these estimates that all the SM interactions come into equilibrium before the EPT. In the second part, I will discuss the equations of chemical equilibrium and charge conservation for the SM interactions. These are interesting because there are small "second order" terms proportional to lepton masses that can be used to protect an asymmetry. Finally in the third subsection I will use the rate estimates from section 2.1 to set bounds on neutrino masses, R-parity violating coupling constants, etc.

As one can see from the estimated bounds in equations (4), (5) and (6), the best constraints are on renormalisable couplings. With the SM particle content, there are no $B-L$ non-conserving operators of dimension less than 5. However, in the supersymmetric extension of the SM, when R-parity is not imposed, baryon and lepton number violating masses and trilinear terms are possible. In this proceedings, I will discuss constraints on $B - L$ violation with SM particle content at the electroweak scale, and also with SUSY particle content.

The MSSM superpotential is

$$W = \mu H_1 H_2 + h_u^{pq} H_2 Q_p U_q^c + h_d^{pq} H_1 Q_p D_q^c + h_e^{ij} H_1 L_i E_j^c \quad . \qquad (7)$$

The Lagrangian also contains kinetic terms, gauge interactions, D-terms and soft SUSY breaking terms of the form

$$\text{soft masses} + B_H H_1 H_2 + A_u^{pq} H_2 Q_p U_q^c + A_d^{pq} H_1 Q_p D_q^c + A_e^{ij} H_1 L_i E_j^c \quad . \; (8)$$

I am abusively using capital letters for both superfields (as in eqn 7) and scalar component fields (as in eqn 8). Quark generation indices are $p, q, r, s...$ and lepton indices are $i, j, k....$ Whether indices are up or down makes no difference.

The above SUSY Lagrangian is with the symmetry R-parity imposed. Other renormalisable interactions can be constructed with the MSSM particle content, but these new interactions violate either B or L, and are often removed by imposing a symmetry like R-parity [22]. See [23] for other possibilties. These are examples of the interactions I am interested in setting bounds on. There are possible new superpotential terms

$$W_{\not R} = \epsilon^i H_2 L_i + \lambda^{ijk} L_i L_j E_k^c + \lambda^{'ipq} L_i Q_p D_q^c + \lambda^{''pqr} U_p^c D_q^c D_r^c \tag{9}$$

and new B or L violating soft terms

$$\begin{array}{l} \text{soft masses mixing } L^\dagger \text{ and } H_1 \\ +B^i H_2 L_i + A^{ijk} L_i L_j E_k^c + A^{'ipq} L_i Q_p D_q^c + A^{''pqr} U_p^c D_q^c D_r^c \quad . \end{array} \tag{10}$$

I assume that SUSY masses are of order 100 GeV $\sim T_c$. The bounds on trilinears are not substantially weakened by larger soft masses—if one sets the constraints at $T \sim m_{SUSY} \sim 1$ TeV, they are weakened by a factor of about 3. The effects of larger SUSY masses were discussed in [18, 20].

2.1. Rate estimates

In this section I will roughly estimate interaction rates for a selection of operators. See for instance [17, 18, 20] for more thorough discussions that disagree in details but all get approximately the same constraints. For a detailed calculation of the interaction rate associated with a specific coupling constant see, *e.g.* [24].

The rate for a boson to scatter off a fermion via the yukawa interaction $y\phi\bar{\psi}_1\psi_2$ can be estimated using kinetic theory (see *e.g.* [4] for an introduction) to be [17]

$$\Gamma(\phi\psi_1 \to \phi\psi_1) \simeq < \sigma v n_{\psi_1} > \sim 4 \times 10^{-4} y^4 T \tag{11}$$

where σ is the $T = 0$ cross section, n_{ψ_1} is the equilibrium thermal distribution of fermions ψ_1, v is the relative velocity of the boson and the fermion ($\simeq c = 1$), and the brackets represent averaging the cross section over thermal distributions in momentum space.

For a gauge boson, the matrix element, and therefore the numerical coefficient, will be different, but if one substitutes the gauge coupling g for the yukawa h in equation (11), it is clear that gauge interactions are in equilibrium up to temperatures of order $\lesssim m_{GUT}$.

One can similarly estimate the rate for $\phi\psi_1 \to \psi_1\gamma$ where γ is an arbitrary gauge boson, to be of order

$$\Gamma(\phi\psi_1 \to \gamma\psi_1) \sim 5 \times 10^{-3} y^2 \alpha T \quad . \tag{12}$$

The rate for a scalar (Higgs or spartner) to decay to two fermions via the interaction $y\phi\bar{\psi}\psi$ can be estimated as [17]

$$\Gamma \sim 1.4 \times 10^{-2} \frac{y^2 m_\phi^2}{T} \quad . \tag{13}$$

m_ϕ is the decaying particles mass in the finite temperature effective potential. Equation (13) neglects final state phases space suppresion due to fermion masses, which can be significant if the mass of the scalar is of order gT. (The final state fermions can also have thermal masses of order gT.)

To estimate a rate associated with a yukawa, one therefore has a choice between scattering and decay. The scattering rate has an extra power of α, but the decay may be suppressed by the small amount of final state phase space, depending on the (thermal) masses of the participating particles. I will use the decay rate estimate to set bounds on $B-L$ violating couplings, because the decaying particle is a SUSY spartner with a zero temperature mass of order 100 GeV.

In the SM case, $y \simeq \sqrt{2}m_f/v$ where $v = 246$ GeV is the Higgs vev, and m_f is a fermion mass. (This is not entirely accurate for the quarks, because the finite temperature mass eigenstates will not be the same as the zero temperature ones, but this should be a reasonable estimate). Taking $H \sim 10T^2/m_{pl}$, one finds that the electron yukawa is in thermal equilibrium below $T_{e_R} \sim 10-100$ TeV [19], the u_R quark will be in chemical equilibrium below $T_{u_R} \sim 100$ TeV, and so on. (See [24] for a careful discussion of when the electron yukawa comes into equilibrium, taking into account the thermal masses in the decay rate.). This means that all SM interactions are in equilibrium just above the EPT.

To estimate a rate associated with a majorana neutrino mass, I need to consider how it is generated. If the zero temperature mass arises from the dimension 5 operator $y^2(\ell\ell HH)/M$, where M is a heavy right-handed singlet neutrino mass, then at temperatures $T_c < T \ll M$ this operator generates two-fermion-two-Higgs scattering. The rate can be estimated [14, 15] as in equation (11) to be

$$\Gamma(\ell H \to \ell H) \simeq 4 \times 10^{-4}\frac{y^4 T^3}{M^2} \quad . \tag{14}$$

This is a conservative estimate of the lepton number violating rate, because the latter should also include the processes $\ell\ell \to HH$, $HH \to \ell\ell$ and so on.

Alternatively, if the majorana mass is present above the EPT as a mass, then it generates lepton number violating processes such as $\nu W \to \bar{\nu} W$. Majorana mass insertions on the external neutrino legs contribute the lepton number violation. The mass corrections to the scattering process can be estimated from (11) as

$$\Gamma(\nu W \to \bar{\nu} W) \simeq 10^{-3} g^4 T \left(\frac{m_\nu}{T}\right)^2 \tag{15}$$

and to a decay as

$$\Gamma(W \to \nu\nu) \sim 3 \times 10^{-2} g^2 T \left(\frac{m_\nu^2}{T^2}\right) \tag{16}$$

(To make this estimate, I took $m_W^2(T) \sim g^2T^2/3$ and $m_\nu^2(T) \sim g^2T^2/8$. For these masses, the decay $W \to \nu\nu$ is kinematically not allowed, so I ought not to use this estimate. It is nonetheless in the literature.)

2.2. Chemical equilibrium equations

There are two approaches in the literature to the equations of chemical equilibrium as applied to cosmological contraints on $B - L$ violation. The more common one uses kinetic theory [25, 18], and is reliable insofar as it agrees with the more esotheric calculations that start from the free energy [26]. I will use something in between, that I think gives the right answers as long as one is not to near the EPT (See [26] for an analysis that is valid for all temperatures.)

If an interaction is in chemical equilibrium, for example the decay $\phi \to \psi_1\psi_2$, then the sum of the chemical potentials of the participating particles is zero:

$$\mu_\phi - \mu_{\psi_1} - \mu_{\psi_2} = 0 \quad . \tag{17}$$

Above T_c, all the gauge bosons have zero chemical potential [25, 18, 28]. From equation (11) it is clear that SM gauge interactions are in equiliibrum up to very high temperatures, so all members of a gauge multiplet have the same chemical potential.

At $T \simeq T_c$, all the SM processes are in equilibrium. If the chemical potential for a gauge multiplet is written as the name of the multiplet (so the chemical potential for e_R^i is e_R^i), then the Yukawas imply

$$\begin{aligned} -e_R^i + \ell^i + H &= 0 \quad , \\ -u_R^i + q^j - H &= 0 \quad , \\ -d_R^i + q^j + H &= 0 \quad . \end{aligned} \tag{18}$$

The rate estimates in section 2.1 say that the electron yukawa is in equilibrium below $T_{e_R} \sim 10 - 100$ TeV. So equations (18) apply for $T_c < T < 10 - 100$ TeV. Above this temperature, the e_R have a thermal distribution (they have gauge interactions) but any asymmetry in the e_R is decoupled from the rest of the plasma. In the absence of an interaction that can transfer the e_R saymmetry to other particles, asymmetries must remain amoung the other particles (in particular, a baryon asymmetry) to cancel the hypercharge carried by the e_R asymmetry.

The non-perturbative $B + L$ violating processes give

$$9q_L + \sum_i \ell^i = 0 \tag{19}$$

Note that these interactions eat three units of baryon number and three units of lepton number, so they violate $B+L$ but not $B-L$. However, when they are in equilibrium with SM interactions in the early Universe, they imply $n_B = -\frac{28}{51}n_L$ and not $n_B = -n_L$. This is because the interactions in equilibrium try to minimise the free energy, and the minimum is not neccessarily where all the quantum numbers are zero. The "sphalerons" of equation (19) do not take n_{B+L} to zero because they only eat SU(2) doublets, and the SU(2) singlets (who also carry B or L) have different hypercharge. In a plasma with $Y = 0$ but non-zero n_{B-L}, the free energy minimum is at $n_B = -(28/51)n_L$ rather than at $n_{B+L} = 0$.

In the case of the MSSM, there are many new particles. The gauginos have majorana masses (sufficiently large to be in chemical equilibrium) so can carry no asymmetry and must have zero chemical potential. This means that the sfermion and fermion chemical potentials are equal and opposite to each other, because the gaugino interaction $\tilde{g}\phi\psi$ imply

$$\phi + \psi + \tilde{g} = \phi + \psi = 0 \tag{20}$$

The two Higgs have opposite chemical potentials due to the mass term $\mu H_1 H_2$, so one gets the same chemical potentials as in the SM, subject to the same constraints due to yukawa couplings.

If the flavour off-diagonal slepton masses are large enough for lepton flavour violation to be in equilibrium, then only $B - L$ is conserved, and not the three $B/3 - L_i$ separately. I will assume for the moment that the off-diagonal masses are "small enough" ($m_{ij}^2/m_{ii}^2 < 10^{-2}$), and come back to this constraint when I am discussing bounds on $B-L$ violating operators in supersymmetric theories.

This gives six equations for the 10 unknown chemical potentials. The four free parameters correspond to the four conserved charges: electric charge (or hypercharge) and the three $\{B/3 - L_i\}$. The electric charge must be zero (imposing $Y = 0$ gives an equivalent constraint):

$$\sum_{particles} Q_{em}(n - \bar{n}) = 0 \tag{21}$$

where ($\bar{n}$) n is the (anti-) particle number density. $n - \bar{n} = \partial F/\partial\mu$ where F is the free energy and μ the particle chemical potential. One finds

$$n - \bar{n} = \frac{2g}{\pi^2}\mu T^2 \begin{cases} 1 + O(m^2/T^2) & \text{fermions} \\ 2 + O(m^2/T^2) & \text{bosons} \end{cases} \tag{22}$$

for $m \ll T$. g here is the number of degrees of freedom of the field—2 for a chiral fermion and a charged scalar. One can correctly get the first term of equation (22) using kinetic theory, by expanding the equilibrium particle distributions in μ and m^2 (see section 4). However, this expansion does not always give the right coefficient for the $O(\mu m^2)$ terms. Note that I am working above the EPT, so the masses in (22) are thermal masses or soft SUSY masses, but are not due to the Higgs vev. Neglecting for the moment the mass effects, one finds in the SM (with n_H Higgs doublets)

$$Q_{em} \propto 3q_L + 6u_R - 3d_R - \sum_i [\ell^i + e^i_R] - 2n_H H = 0 \qquad (23)$$

In the MSSM, the numerical coefficients change slightly, assuming $m_{SUSY} \lesssim T_c$, because there are now two Higges and spartners as well as partners in the soup. The net effect is to change the number of Higges n_H from one to two in (23).

There are now seven homogeneous equations for the 10 unknown chemical potentials. If one fixes the asymetries in the three $\{B/3 - L_i\}$s, all the chemical potentials are determined.

The baryon asymmetry carried by the quarks is

$$\begin{aligned} n_B &= \sum_{col} \sum_{gen} (2 \times \tfrac{1}{3}(n_{q^i_L} - n_{\bar{q}^i_L}) + \tfrac{1}{3}(n_{u^i_R} - n_{\bar{u}^i_R}) + \tfrac{1}{3}(n_{d^i_R} - n_{\bar{d}^i_R})) \\ &\propto 12 q_L T^2 \end{aligned} \qquad (24)$$

where I have used equations (18) and neglected $O(\mu m^2)$ corrections. Quarks of different families have the same chemical potential so quark mass effects in equation (22) cannot be used to preserve an asymmetry. Equation (24) can be written using equations (18), (19) and (23) as

$$n_B = \frac{24 + 4n_H}{66 + 13n_H} \sum_i n_{B/3-L_i} + \frac{47 y_\tau^2}{1896\pi^2}(n_{L_e - L_\tau} + n_{L_\mu - L_\tau}) \qquad (25)$$

where n_H is the number of Higgs doublets. I have included here the lepton mass effects—specifically the τ yukawa $y_\tau = m_\tau/(175 \text{ GeV})$, because it is largest. These small corrections, which are present below [27] and above [28] the EPT, will be discussed in section four. It is clear that if $B - L = 0$, one needs very large lepton flavour asymmetries ($n_{L_i - L_j}/s \sim 10^{-3}$, where s is the entropy density) to preserve the baryon asymmetry via lepton mass differences. (As noted in [18], slepton mass differences could be more effective.)

2.3. Bounds on $B - L$ violating coupling constants

Equation (25) says, as expected, that a baryon asymmetry will remain in the plasma if an asymmetry remains in any one of the $\{B/3 - L_i\}$. Neglecting the lepton mass effects, it is clear that we need an asymmetry in at least one of the $\{B/3 - L_i\}$ to preserve the asymmetry in the quarks. This means that interactions violating one of the $\{B/3 - L_i\}$ need to be out of equilibrium. Suppose, for instance that it is $B/3 - L_e$ that is not washed out. Then one requires

$$\Gamma_{B/3 - L_e} < H \qquad (26)$$

for $T_c < T < T_{e_R} \sim 100$ TeV.

I can translate (26) into a constraint on coupling constants using the rate estimates from section 2.1.

For majorana neutrino masses that are present in the thermal soup above the EPT as a mass, requiring Γ from equations (16) or (15) to be less than H at $T = 100$ GeV gives

$$m_\nu < 10 - 100 \text{ keV} \tag{27}$$

(The weaker bound corresponds to the scattering interaction rate).

In the case where $m_\nu \sim m_D^2/M$, above the EPT there is a lepton number violating scattering process mediated by the dimension 5 operator $(\ell\ell HH)/M$. As discussed in the introduction, $\Gamma_{\not L}/H$ increases with the temperature for $D > 4$ operators, so the best bound is at the highest possible temperature. This is $T_{e_R} \sim 100$ TeV [19]. Requiring equation (14) to be out of equilibrium at $T \sim 100$ TeV gives the bound

$$m_\nu \lesssim \text{keV} \quad . \tag{28}$$

These are bounds on a neutrino mass that violates electron lepton number. So in the basis where the charged lepton mass matrix is diagonal the majorana neutrino mass matrix entries $m_\nu^{ei} < 10$ keV for $i = e, \mu, \tau$.

The R-parity violating masses $\epsilon_i \tilde{h}_2 \ell_i$ are slightly trickier [29, 30], because one can set them to zero by redefining the higgsino $\tilde{h}_1$. The "problem" here is that the H_1 superfield has the same gauge quantum numbers as the lepton superfields L_i. In a lepton number conserving theory, L_i are distinct from H_1 because they carry lepton number. However if the ith lepton number is not conserved, then there is no unique definition of which linear combination of L_i and H_1 is "the Higgs".

For setting bounds on lepton number violation, it should not matter which basis we calculate in. Whether an asymmetry in, for instance L_e, survives in the plasma is a physical question that should not be basis dependent. It is possible to construct basis independent "invariants" that parametrise the amount of R parity violation between different coupling constants (analogous to Jarlskog invariants for CP violation), and to set bounds on the invariants [29, 30, 31]. However, one can get the right answer with a less formal approach.

If we rotate the superpotential term $\epsilon_i H_2 L_i$ into $\mu H_1 H_2$ by redefining H_1:

$$H_1' = \frac{1}{\sqrt{\mu^2 + \epsilon^2}} (\mu H_1 + \epsilon_i L_i) \tag{29}$$

then we generate new trilinears:

$$\frac{h_d^{pq} \epsilon_i}{\sqrt{\mu^2 + \epsilon^2}} L_i Q_p D_q^c \qquad \frac{h_e^{jk} \epsilon_i}{\sqrt{\mu^2 + \epsilon^2}} L_i L_j E_k^c \tag{30}$$

If the soft masses $B_H H_1 H_2 + B_i H_2 L_i$ are parrallel (in H_1, L_i space) with (μ, ϵ_i), (*i.e.* $B_i/B_H = \epsilon_i/\mu$), then they are simultaneously rotated away. If in this basis the soft masses mixing H_1 and the L_i are also zero, then there is no lepton number violation in the masses. In this case, there are two possible bounds one can set on ϵ_i: one could require that the rate associated with the mass ϵ_i be out of equilibrium, or that the trilinear interaction $(h_d^{33}\epsilon_i/\sqrt{\mu^2+\epsilon^2})^2 T < H$. The first bound would imply $\epsilon_i < 10$ keV, the second $\epsilon_i < 10^{-5}\mu \simeq$ MeV. So it matters which is the right one to impose. It turns out that the second, weaker constraint is the correct one. One can see this in a one generation model; the stronger rate due to the mass term chooses the Higgs to be the direction of equation (29), so the lepton is the orthogonal direction, and lepton number is violated by the $\lambda' LQD^c$ term in this basis. One can check that in the basis (29), the trilinear $\lambda^{'i33}$ can be written in terms of the coupling constants of the original basis as $(\epsilon_i h_d^{33} - \mu\lambda^{'i33})/\sqrt{\mu^2+\epsilon^2}$. Equation (37) gives the bound

$$(\epsilon_i h_d^{33} - \mu\lambda^{'i33})/\sqrt{\mu^2+\epsilon^2} < 10^{-7} \quad \text{for one } i \tag{31}$$

or $\epsilon_i <$ MeV for $\mu \sim 100$ GeV.

It is unlikely in realistic models that (μ, ϵ_i) should be exactly parrallel to (B_H, B_i), or that the soft masses $m^2_{H_1 L_i}$ should be zero in this basis. Small lepton number violating masses parametrising the misalignment should remain. For simplicity, let me neglect the $m^2_{H_1 L_i}$, and just consider ϵ_i and B_i. The rate for these masses can be estimated as a mass correction times a gauge interaction rate, like the estimate for the neutrino majorana mass, that is

$$\Gamma \sim \min\left\{\frac{\epsilon_i^2}{\mu^2+g^2T^2}, \frac{(B_i)^2}{(B_H+g^2T^2)^2}\right\} \times 10^{-2} g^2 T \tag{32}$$

where in this formula ϵ_i (B_i) is taken in the basis where B_i (ϵ_i) is zero. The denominator contains a crude attempt to include thermal mass effects. To see why equation (32) is the right bound, imagine sitting in the early Universe above the EPT. As the temperature drops, more and more renormalisable interactions come into equilibrium. At some point the larger of the two mass terms $\sqrt{(B_H)^2+(B_i)^2}$ and $\mu^2+\epsilon_i^2$ comes into equilibrium, and thereby chooses the Higgs direction—*i.e.* if $\sqrt{(B_H)^2+(B_i)^2} > \mu^2+\epsilon_i^2$ the Higgs will be combination $B_H H + B_i L_i$. then in this basis chosen by the interactions, one needs the lepton number violating rate associated with the mass ϵ_i to be out of equilibrium, which is equation (32).

The bound (32) can be expressed in a more basis independent way as:

$$\frac{B_H\epsilon_i - \mu B_i}{\sqrt{(B_H)^2+(B_i)^2}\sqrt{\mu^2+\epsilon_i^2}} < 2\times 10^{-7} \quad , \tag{33}$$

or, in the basis where $B\epsilon_i$ is zero, $\epsilon_i/\mu < 2\times 10^{-7}$. So the cosmological bounds require that (μ, ϵ_i) be more aligned with (B_H, B_i) than with the yukawas.

In the supersymmetric Standard Model, lepton flavours are unlikely to be separately conserved. The slepton mass matrix is unlikely to be exactly diagonal in the lepton mass eigenstate basis. If lepton flavour violating processes due to flavour off-diagonal slepton masses are in thermal equilibrium, then all $B - L$ violating processes need to be out of equilibrium, not just interactions violating on of the $B/3 - L_i$, *i.e.* the bounds (31), (37) and (33) would apply to all $B - L$ violating coupling constants, and not just those involving one lepton generation. So if one wants to use the "flavour loophole", and only impose the bounds on one generation, the flavour changing soft masses must be out of equilibrium.

The rate associated with a flavour off-diagonal soft mass can be estimated analogously to the previous discussion about $\epsilon_i H_2 L_i$. There are three interactions that choose a basis in lepton flavour space: the lepton yukawa h_e^{ij}, the A term A_e^{ij}, and the soft masses m_{ij}^2. If I want to separately conserve lepton flavours in the thermal soup, then all the interactions in equilibrium must agree on the basis in lepton flavour space. Suppose I start in the basis where h_e^{ij} is diagonal, because this is a familiar basis from phenomenology, and I put a hat on the coupling constants in this basis. There will be off-diagonal terms $[\hat{m}^2]_{ij}$,$\hat{A}_e^{ij}, i \neq j$ in this basis. Since the interaction rate associated with m^2 is larger at $T \sim 100$ GeV than the rate for h_e, this is probably not a sensible basis. I should rotate to the basis where m^2 is diagonal, because the stronger interaction will choose the lepton flavour basis. In this new basis, there will be flavour off-diagonal yukawas $h_e^{ij} \sim \hat{h}_e^{ii} \hat{m}_{ij}^2 / \hat{m}_{ii}^2$ ($i \neq j$, and no sums on repeated indices). These yukawas need to be out of equilibrium:

$$\Gamma \sim 10^{-2} \left(\hat{h}_e^{ii} \frac{\hat{m}_{ij}^2}{\hat{m}_{ij}^2} \right)^2 T < H \tag{34}$$

If one wants to preserve $B/3 - L_e$, this gives the bound $\hat{m}_{ej}^2 / \hat{m}_{ee}^2 < 5 \times 10^{-2}$. Preserving an different lepton flavour would give a stronger bound (the yukawa is larger) on different elements of $\hat{m}^2$.

Now consider the A term. Suppose I estimate the associated rate from equation (13) to be

$$\Gamma_A \sim 10^{-2} \frac{A_e^2}{T} \quad . \tag{35}$$

This is equivalent to assuming that the A term mediates a decay of one scalar into the other two. It is not clear that this is kinematically allowed, but I use this estimate anyway (I estimate the scattering rate $\Gamma(\phi\phi \rightarrow \phi\gamma) \simeq 10^{-3} g^2 A^2 / T$, which gives a bound on A that is about an order of magnitude weaker.). In the basis where m^2 is diagonal, I need

$$\Gamma_A \sim 10^{-2} \frac{(A_e^{ij})^2}{T} < H \qquad (i \neq j) \tag{36}$$

for whichever lepton flavour i one wishes to conserve. This gives $A^{ij} < 10^{-5}$ GeV. This is not a particularily strict bound; recall that I have absorbed the yukawa into A (see equation (8)). Similar bounds apply to the R-violating A terms of equaltion (10).

Finally I need to calculate a bound on $B - L$ violating trilinear interactions. There are no basis confusions for these interactions, so to conserve, for instance, $B/3 - L_e$, I need all the trilinears that violate L_e and B to satisfy

$$10^{-2}\lambda^2 T < H \simeq \frac{25T^2}{m_{pl}} \tag{37}$$

This means that $\lambda''_{pqr} < 10^{-7}$ for all p, q, r, $\lambda'_{epq} < 10^{-7}$ for all p, q, and $\lambda_{ejk}, \lambda_{jke} < 10^{-7}$ for $j, k \neq e$.

2.4. Summary of bounds on R violating interactions

Suppose that I define the Higgs such that the three B_i are zero. Then to conserve $B/3 - L_e$, I need

$$\epsilon_e < 10 - 100 \text{ keV}, \qquad m^2_{H_1 L_e} < (60 - 90 \text{ MeV})^2 \tag{38}$$

for $\mu^2 \simeq m^2 \simeq B \simeq 100$ GeV.

I need all the trilinears and A terms that violate $B/3 - L_e$ to respectively satisfy

$$\lambda < 10^{-7} \quad , \qquad A < 10 - 100 \text{ keV} \tag{39}$$

I also need L_e flavour violation mediated by the R conserving soft masses to be out of equilibrium, which will be the case if

$$\frac{m^2_{ei}}{m^2_{ii}} \lesssim 5 \times 10^{-2} \quad , \qquad A^{ei}_e \lesssim 10 - 100 \text{ keV} \tag{40}$$

3. If the BAU was not there...

I wrote in the introduction that I would assume that the BAU was present in the thermal soup above the EPT. This is of course one of the loopholes in the constraints that I am discussing; if the BAU is *not* there, then there are no constraints on $B - L$ violation. So in this section I will relax this assumption.

First I would like to work out over what temperature range the BAU needs to be "not there". The lower end of this range is straightforward to find. The $B + L$ violating electroweak effects drop out of equilibrium at or shortly after the electroweak phase transition. Once they are gone, the

BAU is no longer at risk. So the baryon asymmetry can safely be present in the plasma at $T \ll T_c$.

It can also be present at high temperatures, before the electron yukawa comes into equilibrium. The interaction rate associated with a dimensionless couplings constant λ scales linearly with the temperature: $\Gamma \sim \lambda^2 T$. Since $H \sim T^2/m_{pl}$, this means that yukawa interactions are out of thermal equilibrium at sufficiently high temperatures, and come into equilibrium as the temperature drops. Since the electron yukawa is the smallest coupling constant in the SM, it is the last to come into equilibrium at $T_{e_R} \sim 100$ TeV. The e_R carries hypercharge, and the Universe ought to be hypercharge neutral, so if there is an asymmetry in the e_R, and no interaction in equilibrium that can transfer this asymmetry to other particle species, then asymmetries must remain in other particles to ensure that $Y = 0$. It is easy to check from the relevant equations of chemical equilibrium that this means a baryon asymmetry will remain in the plasma. At $T \gg T_{e_R}$, there will be other yukawa interactions out of equilibrium, and additional symmetries [32] that can protect the baryon asymmetry, but the singlet e_R is the last particle species to come into chemical equilibrium, so can protect the baryon asymmetry for the longest [19]. Any $B - L$ violating operator not involving the e_R can be in equiliibrium above T_{e_R}, without being able to eat the BAU. This was not fully realised in the earlier papers discussing bounds on non-renormalisable $B - L$ violating operators, where strong bounds were set on $D > 4$ operators by requiring that they be out of equilibrium up to T_{reheat}, or the temperature when the BAU was generated, or the temperature when the $B + L$ violation came into equilibrium.

Note that this argument assumes that there is an asymmetry stored in the e_R. This may not be easy to arrange, but in setting generic bounds, one must allow for the possibility that some baryogenesis mechanism does generate an e_R asymmetry, which can then protect the baryon asymmetry down to T_{e_R}. If there is not an asymmetry in the e_R, one must explicitly check the equations of chemical equilibrium for the interactions in equilibrium to see if a baryon asymmetry will survive.

To approximately determine T_{e_R}, one can assume that the e_R comes into chemical equilibrium when the decay of the Higgs into a left handed and a right-handed lepton is of order H [19]:

$$\Gamma \simeq 10^{-2} h_e^2 T \simeq H \tag{41}$$

which gives $T_{e_R} \simeq 100$ TeV. See [24] for a careful determination of T_{e_R}, that includes the various decay and scattering interactions mediated by h_e, and the thermal masses of the participating particles.

So the baryon asymmetry is really only "at risk" over three decades in temperature: $T_c < T < 100$ TeV. If the asymmetry is present in the thermal soup during this period, it needs to be an asymmetry in $B - L$, and interactions violating at least one of the $B/3 - L_i$ need to be out of equilibrium during this period. This gives the bounds of section 2 on renormalisable interactions.

The best generic bounds on non-renormalisable operators can be computed at $T \sim 10$ TeV. I assume that

$$\Gamma \sim 10^{-3} \frac{T^{2n+1}}{M^{2n}} < \frac{10T^2}{m_{pl}} \tag{42}$$

is the interaction rate for an operator of dimension $n + 4$. The 10^{-3} is an extrapolation from dimension 4 of the numerical factor. This gives the bound

$$\frac{M}{T} > \left(\frac{m_{pl}}{10^3 T}\right)^{1/2n} \sim 10^{6/n} \tag{43}$$

where I used $T \sim 10$ TeV in the last "$\sim$".

Alternatively, the BAU can be not in the soup between $T_c < T < 100$ TeV. The most obvious way to do this is to generate the baryon asymmetry at the electroweak phase transition (see, *e.g* [2, 3] for reviews of electroweak baryogenesis). But this is not the only possibility; if we turn out to be living in a corner of SUSY parameter space that does not allow electroweak baryogenesis, there are models where the baryon asymmetry is generated in the late decay of some particle, after the EPT [33].

Another possibility is to generate the baryon asymmetry early, and then "hide it" between T_{e_R} and T_c. One way to do this [34], is to add additional particles that carry baryon number to the SM or the SSM, decouple them from the rest of the SM particles at $T > T_{e_R}$ and have them decay after the EPT, returning the baryon asymmetry that they carry to the plasma of SM particles. Another possibilityy is to store some hypercharge in a coherent hypercharge magnetic field for $T_c < T < T_{e_R}$ [35]. If there is a sufficiently large asymmetry in the e_R, it can act as a source for the hypercharge magnetic field before T_{e_R}, and this field may survive until the EPT.

Another somewhat related loophole is to take the up-quark yukawa to be considerably smaller than is usually assumed [36]. If $h_u \lesssim 10^{-7}$ ($m_u \lesssim 20$ keV !), then an asymmetry in the u_Rs could play the same role as the e_R asymmetry does, but all the way down to the EPT. If h_u did not come into equilibrium until after the EPT, and if the u_Rs carried a primordial asymmetry, then this asymmetry will be present after the EPT, irrespective of how much $B - L$ violation is in equilibrium above the EPT (providing of course that the $B - L$ violating operators do not involve u_R).

4. Other loopholes

In this section, I list the remaining loopholes.

4.1. Lepton mass effects

In models where $n_{B-L} = 0$, lepton mass effects can protect the baryon asymmetry from SM $B+L$ violation, provided that there are lepton flavour asymmetries [27]. A small baryon asymmetry will remain below [27] and above [28] the EPT, proportional to lepton mass differences and flavour asymmetries. Since I am here discussing what happens above the EPT, the numerical factors in equation (25) are for this case, where the lepton masses are "thermal". In this subsection, I neglect most of the numerical factors.

The number density for particles in thermal equilibrium is

$$n = \frac{2g}{\pi^2} \int \frac{p^2 dp}{e^{(E-\mu)/T} \pm 1} \tag{44}$$

where the $\pm$ is for fermions or bosons, and the number density $\bar{n}$ for anti-particles has the sign in front of the chemical potential μ flipped. Expanding $n - \bar{n}$ in μ/T and m^2/T^2, assuming that both are small, gives equation (22). As previously noted, to correctly determine the $O(\mu m^2)$ term, one must calculate the free energy, and take the derivative with respect to μ. However, I am not here interested in the exact coefficient; see [26, 28] for numerical factors.

The reason that these (μm^2) contributions are interesting is that they can preserve a baryon asymmetry in the presence of the $B + L$ violating interactions, when $B - L = 0$, and there are flavour asymmetries amoung leptons (also when $B + L$ and B violation are in equilibrium, and there are flavour asymmetries). They do not protect the BAU if there are interactions in equilibrium that take all three lepton chemical potentials to zero.

So suppose that the $B + L$ violating interactions are in equilibrium (equation 19), and that $n_{B-L} = 0$:

$$12q_L - 3\sum_i \ell_i \left[1 + O\left(\frac{m_i^2}{T^2}\right)\right] - 3H = 0 \tag{45}$$

This could be the case if the BAU was generated in the out-of-equilibrium decay of heavy particles from a $B-L$ conserving GUT. There are two types of constraint imposed on the plasma: the sum of the chemical potentials participating in each interaction must be zero, and certain charge densities $(Y, B - L)$ must be zero. If one includes the $O(\mu m^2)$ terms in the charge densities, one finds that there is a small BAU in equilibrium. This is due to the non-zero chemical potentials and not equal masses for the different lepton generations.

From the condition of charge neutrality, equation (23), I can solve for the Higgs chemical potential in terms q_L and $\sum_i \ell_i$:

$$H \simeq \frac{3q_L - \sum_i \ell_i[1 + O(m_i^2/T^2)]}{7} \simeq \frac{12q_L - \sum_i \ell_i \times O(m_i^2/T^2)]}{7} \tag{46}$$

(This assumes SM particle content, and I have dropped Hm_i^2 terms). Substituted into (45), and using (19) one finds

$$n_B \propto q_L T^2 \sim \sum_i \ell_i m_i^2 \tag{47}$$

$\sum_i \ell_i m_i^2 \neq 0$, if there are lepton flavour asymmetries ($n_{L_i - L_j} \neq 0$), so the remaining baryon asymmetry (47) is proportional to lepton mass differences, and flavour asymmetries, see equation (25). It is clear from (25), where all the numerical factors are included, that one needs very large lepton flavour asymmetries ($\sim 10^{-2} - 10^{-3}$) to preserve a sufficiently large baryon asymmetry. An interesting possibility suggested in [18] is to use the slepton mass differences in a SUSY model, because these could be much larger.

4.2. Modified cosmologies

In this subsection I briefly review some modifications of cosmology that allow one to escape the bounds discussed in section 2.

One possibility is low T_{reheat} models; if the Universe after inflation reheats to a temperature $T_{reheat} < T_c$, then one never has to worry about having $B + L$ and $B - L$ violating operators simultaneously present. However, one does have to worry about how to generate the baryon asymmetry in such models. This is not neccesarily a problem; one can, for instance, generate an asymmetry from the decay of the inflaton, or in some cases via the Affleck-Dine mechanism [37]. In fact, the reheat temperature after Affleck-Dine baryogensis is often low.

Another possibility is to modify the expansion rate of the Universe. If the Universe expanded somewhat faster, the e_R could protect the BAU all the way down to T_c. This can be accomplished by, for instance, adding a component to the energy density of the Universe that scales as $1/a^6$ [38]. (This is not as far-fetched as it might sound; a scalar field in the right shaped potential can behave this way.)

It is also possible to "get rid of" the $B + L$ violating electroweak interactions. For instance, if there is a SU(2) doublet with a large vev above the EPT, it would give a mass to the W, so the rate for $B+L$ violation would be exponentially suppressed. This could occur in an an Affleck-Dine baryogenesis scenario, when the condensate carries a large baryon asymmetry [39], and therefore does not decay until $T \sim 100$ GeV.

5. Summary

The argument leading to cosmological bounds on $B - L$ violating interactions goes as follows. Assume that the baryon asymmetry was created before the electroweak phase transition as an asymmetry in at least one of the three $B/3 - L_i$, and that it is present in the thermal soup above the electroweak phase transition. The electroweak $B + L$ violating interactions are operating in this plasma. If there are also interactions in equilibrium that violate all of the $\{B/3 - L_i\}$, the baryon asymmetry will be washed out. Since it is supposed to be around today, this is undesirable. Therefore one requires that interactions violating one of the $B/3 - L_i$ be out of thermal equilibrium, so they cannot eat the baryon asymmetry.

This generically gives bounds of order $\lambda < 10^{-7}$ on "yukawa-type" coupling constants, and $m \lesssim$ keV-100 MeV for different masses. The mass bounds can be slightly tricky to calculate. These constraints apply to interactions that violate B and one particular lepton flavour. A summary of the bounds on renormalisable R parity violating couplings is at the end of section 2. Bounds on neutrino masses are in section 2, and bounds on non-renormalisable operators are towards the end of section 3.

There are various loopholes to this argument. One can avoid them completely by not having the BAU present in the plasma for 100 TeV $> T >$ 100 GeV. This is approximately when it is at risk from the $B + L$ and $B - L$ violating interactions. The most obvious way to not have the baryon asymmetry present is to generate it at the electroweak phase transition. There are other models where the BAU is generated after the phase transition, or where it is generated at very high temperatures and hidden for 100 TeV $> T >$ 100 GeV.

It is also possible to modify the cosmological model to avoid these bounds. If the reheat temperature after inflation is below that of the electroweak phase transition, there is no opportunity to wash out the BAU. (One has fewer ways of making it though). If there electroweak $B + L$ violating interactions are turned off above the phase transition, then an asymmetry in $B + L$ can survive. If the expansion rate of the Universe is increased, the slowest Standard Model interaction, which is the electron yukawa, might not come into equilibrium until after the phase transition. If there is an asymmetry in the e_R, this means that an asymmetry must remain in the baryons, because the Universe should be hypercharge neutral.

It is possible to avoid the strong bounds on B violating operators, or the conclusion that the asymmetry must carry $B - L$, via lepton mass effects. If there are large lepton flavour asymmetries, and lepton flavours are conserved, then a small baryon asymmetry proportional to $m_\tau^2/T^2 \times n_{L_\tau - L_i}$ will remain, even if $n_{B-L} = 0$ or if there are B violating interactions in equilibrium.

References

[1] see, *e.g.*, (and references therein) A.G. Cohen, A. De Rujula, S.L.

Glashow, *Astrophys.J.* **495** (1998) 539. astro-ph/9707087

[2] see e.g.
A .D. Dolgov, *Phys.Rept.* **222** (1992) 309.
A. Riotto, talk at *ICTP Summer School in High-Energy Physics and Cosmology*, Trieste, Italy, July 1998. hep-ph/9807454.

[3] V.A. Rubakov, M.E. Shaposhnikov, *Usp.Fiz.Nauk* **166** (1996) 493-537. (Phys.Usp.39 (1996) 461. hep-ph/9603208

[4] E.R. Kolb, M. Turner " *The Early Universe*", Addison-Wesley, 1990.

[5] A.D. Sakharov. 1967. *Pisma Zh.Eksp.Teor.Fiz.* **5** (1967) 32-35. (JETP Lett.5 (1967) 24.

[6] for a discussion of B and L violating operators on the Standard Model and in supersymmetry, see, *e.g.* G.G.Ross, *Grand Unified Theories*, Benjamin-Cummings, Menlo Park, CA, 1985, chapters 7 and 11.

[7] G. 't Hooft, *Phys. Rev. Lett.* **37** (1976) 8; G. 't Hooft, *Phys. Rev.* **D14** (1976) 3432.

[8] V.A. Kuzmin, V.A. Rubakov, M.E. Shaposhnikov, *Phys.Lett.* **155B** (1985) 36.

[9] P. Arnold, D. Son, L. Yaffe, *Phys. Rev.* **D 55** (1997) 6264.

[10] K. Kajantie, M. Laine, K. Rummukainen, M. Shaposhnikov, *Nucl.Phys.* **B466** (1996) 189. hep-lat/9510020

[11] see, *e.g.* N. Fry, K. A. Olive, M. S. Turner, *Phys.Rev.* **D22** (1980) 2977, *Phys.Rev.* **D22** (1980) 2953.
E. W. Kolb, S. Wolfram, *Nucl.Phys.* **B172** (1980) 224, Erratum-*ibid.* **B195** (1982) 542.

[12] M. Fukugita, T. Yanagida, *Phys.Lett.* **174B** (1986) 45.

[13] M. Fukugita, T. Yanagida, *Phys.Rev.* **D42** (1990) 1285.

[14] A. E. Nelson, S. M. Barr, *Phys.Lett.* **B246** (1990) 141.

[15] W. Fischler, G.F. Giudice, R.G. Leigh, S. Paban, *Phys.Lett.* **B258** (1991) 45.

[16] B. A. Campbell, S. Davidson, J. Ellis, K.A. Olive *Phys.Lett.* **B256** (1991) 457.

[17] B.A. Campbell, S. Davidson, J. Ellis, K. A. Olive *Astropart.Phys.* **1** (1992) 77.

[18] H. Dreiner, G.G. Ross, *Nucl.Phys.* **B410** (1993) 188.

[19] B. A. Campbell, S. Davidson, J. Ellis, K.A. Olive, *Phys.Lett.* **B297** (1992) 118.

[20] T. Inui, T. Ichihara, Y. Mimura, N. Sakai *Phys.Lett.* **B325** (1994) 392. hep-ph/9310268

[21] G. Gelmini, T. Yanagida, *Phys. Lett.* **B 294** (1992) 53; W.Buchmuller, T. Yanagida, *Phys.Lett.* **B302** (1993) 240; J.M. Cline, K. Kainulainen, K. A. Olive, *Astropart.Phys.* **1** (1993) 387. hep-ph/9304229;

C.E. Vayonakis, *Phys.Lett.* **B286** (1992) 92; A. Antaramian, L.J. Hall, A. Rasin, *Phys.Rev.* **D49** (1994) 3881. hep-ph/9311279; A. Ganguly, J.C. Parikh, U. Sarkar, *Phys.Lett.* **B385** (1996) 175. hep-ph/9408271

[22] H. Dreiner, to be published in '*Perspectives on Supersymmetry* ', Ed. by G.L. Kane, World Scientific; hep-ph/9707435.

[23] L.E. Ibanez, G. G. Ross, *Nucl.Phys.* **B368** (1992) 3.

[24] J.M. Cline, K Kainulainen, K. A. Olive, *Phys.Rev.Lett.* **71** (1993) 2372. hep-ph/9304321; *Phys.Rev.* **D49** (1994) 6394. hep-ph/9401208

[25] J.A. Harvey, M. S. Turner *Phys.Rev.* **D42** (1990) 3344.

[26] S.Yu. Khlebnikov, M.E. Shaposhnikov, *Phys.Lett.* **B387** (1996) 817. S.Yu. Khlebnikov, M.E. Shaposhnikov, *Nucl.Phys.* **B308** (1988) 885.

[27] V.A. Kuzmin, V.A. Rubakov, M.E. Shaposhnikov, *Phys.Lett.* **191B** (1987) 171.

[28] S. Davidson, K. Kainulainen, K.A. Olive, *Phys.Lett.* **B335** (1994) 339. hep-ph/9405215

[29] S. Davidson, J. Ellis, *Phys.Lett.* **B390** (1997) 210-220. hep-ph/9609451

[30] S. Davidson, J. Ellis, *Phys.Rev.* **D56** (1997) 4182. hep-ph/9702247

[31] S. Davidson, to be published in *Phys. Lett.* **B**. CERN-TH-98-161.

[32] L. E. Ibanez, F. Quevedo, *Phys.Lett.* **B283** (1992) 261. hep-ph/9204205

[33] S.Dimopoulos, L.J. Hall, *Phys. Lett.* **B 196** (1987) 135; J. Cline, S. Raby, *Phys. Rev.* **D 43** (1991) 1781; R. J. Scherrer, J. Cline, S. Raby, D. Seckel, *Phys.Rev.* **D44** (1991) 3760; A. Masiero, A. Riotto, *Phys.Lett.* **B289** (1992) 73.

[34] S. Davidson, R. Hempfling, *Phys.Lett.* **B391** (1997) 287. hep-ph/9609497

[35] M. Joyce, M. Shaposhnikov, *Phys.Rev.Lett.* **79** (1997) 1193. astro-ph/9703005

[36] for a discussion, see for instance,
J. E. Kim, *Phys.Rept.* **150** (1987) 1.
T. Banks, Y. Nir, N. Seiberg, *Gainesville 1994, Yukawa couplings and the origin of mass*, Gainesville, FL, Feb 1994. hep-ph/9403203

[37] see, for instance, I. Affleck, M.Dine, Nucl. Phys. **B 249** (1985) 361 ; J. Ellis , K. Enqvist, D.V. Nanopoulos, K.A. Olive, *Phys. Lett.* **B191** (1987) 343; M.Dine, L. Randall, S. Thomas, *Phys. Rev. Lett.* **75** (1995) 398; *ibid. Nucl. Phys.* **B458** (1996) 291; M. Gaillard, H. Murayama, K.A. Olive, *Phys.Lett.* **B355** (1995) 71.

[38] M. Joyce, T. Prokopec, *Phys.Rev.* **D57** (1998) 6022. hep-ph/9709320

[39] S. Davidson, H. Murayama, K. A. Olive, *Phys.Lett.* **B328** (1994) 354.

PART 4

B and L Violation and Astrophysics

Cosmic Antimatter Search with the Alpha Magnetic Spectrometer

W. Wallraff, 1. Physikalisches Institut, RWTH-Aachen, Germany *$

The apparent absence of antimatter (anti-helium, anti-carbon, etc.) in the universe is a major puzzle of particle physics. The ALPHA Magnetic Spectrometer AMS is a high resolution tracking detector for long-term operation in space, that recently has finished a precursor flight (STS91) with full success. It will improve the sensitivity for antimatter search by a factor of more than 10^4. For more than 3 years AMS will sample cosmic rays in space while orbiting around earth on the international space station ISSA.

1.0 Introduction - physics goals

There is not a single sufficiently sensitive experiment with data that could be used for answering the fundamental question of a matter - anti-matter (a-)symmetry of the universe. The cosmic ray flux integrated so far by balloon borne instruments is too small for detecting anti-matter from galaxies outside our local group. The AMS collaboration has built [1,2] (fig. 1) and flown (fig. 2)

AMS detector in the laboratory at **ETH Zürich** (11/1997)

FIGURE 1. AMS assembled, components visible in event displays are marked

in a precursor flight (STS 91, June 2nd - June 12th 1998) a high resolution magnetic spectrometer with large acceptance that aims for a long duration antimatter search based on the international space station ALPHA (fig. 3).

* on behalf of the AMS collaboration, CERN, Geneva, Switzerland

$ supported by DLR, Köln, Germany

The AMS detector has been shown capable of identifying nuclei of various masses. In the future the AMS detector will be able to identify antimatter nuclei ($\overline{He}$, $\overline{C}$) in the presence of a matter background 10^8 times larger. This level of sensitivity is required for confirming (excluding) the presence of astrophysical antimatter objects in a sizeable fraction of the present universe.

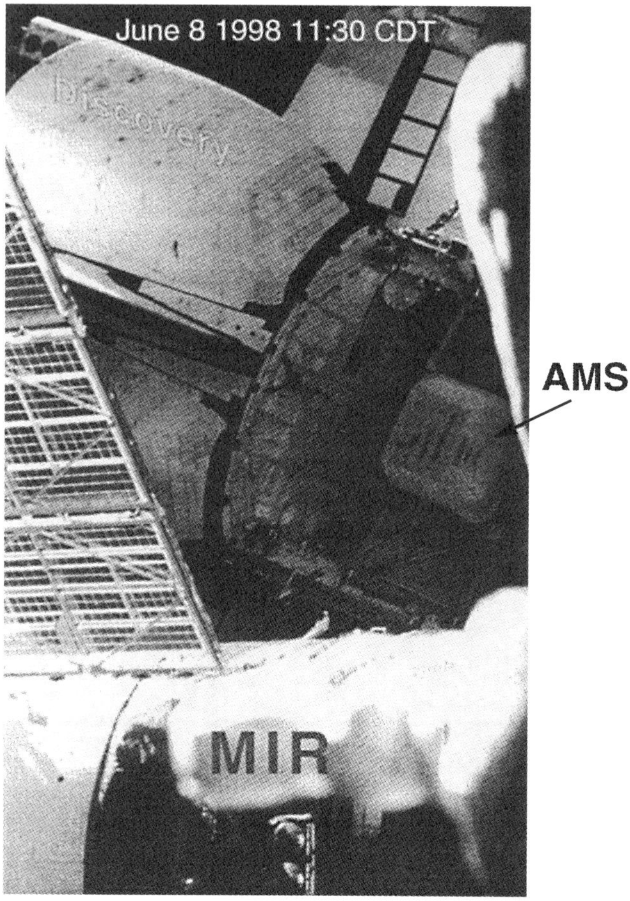

FIGURE 2. AMS in the Discovery payload bay as seen by the cosmonauts from the Russian space station MIR (photo NASA-JSC)

In addition the long observation time planned will result in a wealth of information about the spectral composition of the cosmic radiation.

From the anti-proton spectrum observed with AMS first direct hints on the contribution of supersymmetric dark matter to the mass of our galaxy may be derived [3,4,5].

With the AMS positron data a far more precise test of the particle propagation model (leaky box) for cosmic ray particle transport in galaxies will be provided. This investigation may establish the presence of a genuine positron source component [6].

The small amount of data on D/p [7] and ^3He / ^4He [8] available today will be vastly increased, such that only systematic inaccuracies will limit the precision of these isotope ratio spectra.

With the excellent charge resolving capabilities of AMS accurate measurements of the B(oron)/C(arbon) ratio will be possible over a wide range of rigidities. This will lead to a more precise determination of the galactic wind velocity and the apparent thickness of the halo.

Coupling the high resolution AMS spectrometer with specific additional detectors (transition radiation particle identifier, gamma - converter, RICH, ...) will allow to conduct investigations into the composition of the cosmic ray spectrum with a 100-fold improved precision. Important issues like the galactic confinement time derived from the ^{10}Be / ^9Be ratio can be resolved using a ring image Cherenkov counter for particle identification.

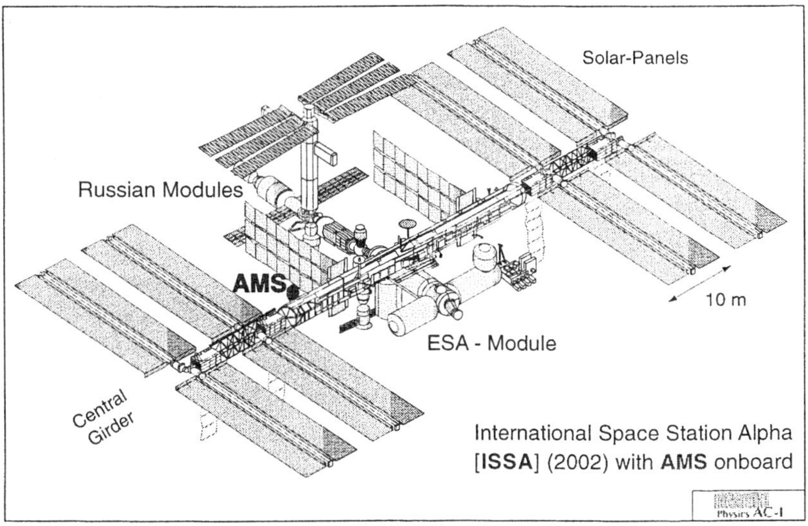

FIGURE 3. AMS placed on ISSA for a long-term search of cosmic antimatter

2.0 Detection Requirements

For reasons of abundance and nuclear stability obvious target nuclei in an antimatter search are $^4\overline{He}$ and $^{12}\overline{C}$. $^4\overline{He}$ will be mostly (≈75%) from primordial nucleosynthesis, whereas $^{12}\overline{C}$ is expected to come exclusively from thermonuclear burning of antistars. Antimatter cosmic rays leaking from antimatter galaxies will propagate almost unhindered through the extragalactic space very much in the same fashion as standard extragalactic cosmic rays [9]. Their spectrum however should be degraded in comparison to that of cosmic rays from the Milky Way as the extragalactic rays have to pass the interstellar material of at least 2 galaxies.

The probability for producing anti-nuclei in secondary processes (on the interstellar and extragalactic medium) is extremely small [10]. It has been argued [11,12] that a sensitivity of at least 10^{-6} is required to probe for extragalactic antimatter in a symmetric (on the scale of > 10Mpc) universe.

The magnetic particle spectrometer for the search of extragalactic antimatter has to provide the best momentum resolution achievable with as little as possible material in the detectors. It has to be demonstrated that mismeasurements of the sign of the track candidates can be controlled to a level lower than the search sensitivity intended. In this manner a single reliably identified anti-carbon event alone can already be taken as compelling evidence for anti-stars in the universe [13].

In the Milky Way the cosmic ray flux of nuclei with masses between those of He and C is dominated to a very large extent by spallation from collisions of heavier nuclei. Hence any antimatter signal in this mass range will likely remain undetectable due to annihilation in the spallation.

It will be impossible to identify primary extragalactic antiprotons as such. Secondary antiproton production by collisions of matter cosmic rays with nuclei from the interstellar material of the Milky Way is rather frequent ($\approx 10^{-4}$ / proton). The number of anti-protons, that have been unambiguously identified, is roughly in agreement with our knowledge about the interstellar matter density in the Milky Way, the energy spectrum and the cross-sections.

This rather small number of antiprotons observed over the past 3 decades is going to increase considerably with the long duration flights of larger acceptance experiments available now. Taken during balloon flights all these measurements have to be corrected for the interactions in the remaining atmosphere. The mass density of the atmospheric overburden is not at all small in comparison to the total amount of matter seen by a cosmic ray during its propagation in the Milky Way. At the height of the AMS orbit (350 - 400 km) however the atmospheric overburden is completely negligible.

The conceptual design of the AMS experiment aims for a large acceptance, high momentum resolution apparatus with a vastly improved sign of charge determination using an unbiased trigger for observing more than 1000 cosmic ray tracks per second.

3.0 The AMS detector

3.1 Tasks

With the goals outlined in the introduction the AMS detector was designed and built for the specific tasks given in table 1:

TABLE 1. AMS principal tasks

1	**search for cosmic antimatter in space**
2	search for darkmatter from the galactic halo
3	study of astrophysics through charged cosmic ray particles

In the following we will describe in some detail the specific techniques used in the AMS detector. These techniques have been chosen after a careful examination of the respective merits of other approaches [1].

The AMS group has greatly benefited from its earlier experience with complex particle detection equipment in accelerator environments [14, 15]. Furthermore a deliberate effort in using proven space techniques has been made [16, 17].

3.2 The general design parameters I

AMS (fig. 4) is the first large acceptance magnetic charged particle spectrometer for long-term operation in space. It has been built after studying the experience from earlier long-duration experiments - mostly high energy photon detectors - in space. We are especially grateful for the access to space environment test facilities in China, Germany, France, Italy and Switzerland [18,19,20].

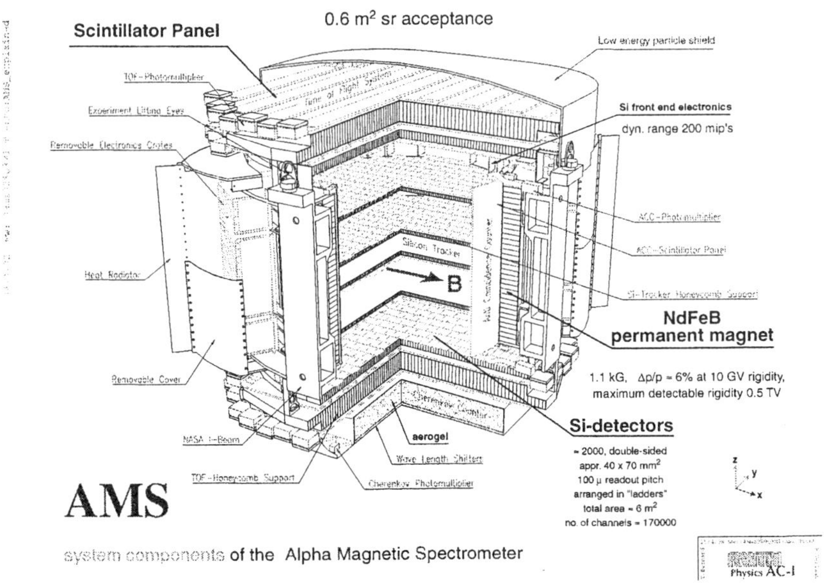

FIGURE 4. AMS system components

The basic intensity expectation for a cosmic ray experiment is shown in fig. 5. The rate (i.e. the intensity integrated over the acceptance) at low energies (T_{kin} $\approx$1Gev) exceeds 2000/m^2 s sr. It drops rather fast with a spectral index of about 2.8. There is a wide range of physical processes responsible for the typical chemical composition of the cosmic rays (p 75%, He 25%, CNO, heavy):

1. fundamental processes at the beginning of the universe
2. thermonuclear processes in the stellar evolution of the galactic matter
3. interaction of the cosmic rays with the interstellar material
4. activity of the sun (for energies < 10 GeV)

The AMS detection range spans kinetic energies between 0.2 and 500 GeV/ amu. The resolution is tailored such that at the maximum detectable momentum (MDM) of 500 GV/c amu several particles per year will be observed in the AMS acceptance of 0.6 m^2sr (MDM being defined as the measured momentum up to which the odds for assigning the wrong sign are less than 1/3).

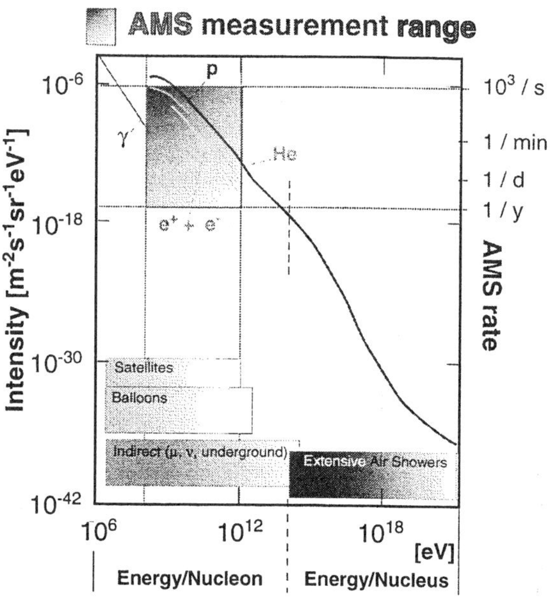

FIGURE 5. spectrum of cosmic rays; AMS measurement range and rates. The strips in the lower half of the figures indicate the energy ranges of current experimental techniques.

The high precision tracking and particle identification capabilities of AMS are needed for single particles traversing the detector only, hence AMS represents a one particle spectrometer as those used in the early days of accelerator-based experimental particle physics.

The basic features of the detector components are given in Table 2.

TABLE 2. particle detection with AMS

1			**momentum measurement** **[p] = GeV/c]**	**|Q |= 1** **$m \leq m_p$**
1.1	$p \leq 2$	GeV/c	$.06\ (1+(m/p)^2)^{1/2}$	$\Delta p/p$
1.2	$2 \leq p \leq 10$	GeV/c	.06	$\Delta p/p$
1.3	$10 \leq p \leq 500$	GeV/c	$((.06)^2 + (.0035\ p)^2)^{1/2}$	$\Delta p/p$
2			**time of flight**	
	Δdt	ns	.10	$dt \approx 5ns$
3			**particle identification**	
3.1	Cerenkov	E_{th}	3.8	GeV
3.2	dE/dx		.15/Q	$\Delta E/E$
3.3	mass $(p,\Delta t)$	$\beta > .85$	$(\gamma^4(\Delta dt/5ns)^2 + (1/E)^2\ \Delta p^2)^{1/2}$	$\Delta m/m$

In the following the basic design choices will be described in more detail. For some of the parameters preliminary performance data from the STS 91 precursor flight will also be given.

3.3 The general design parameters II

For trigger and timing the AMS spectrometer (fig. 6) is put in between 4 layers of plastic scintillation counters.

The bending of charged particles in a large acceptance high field permanent magnet is observed in 6 layers of double-sided (x,y) Si strip detectors. The accurate background-free determination of the sign of charge has been the first principal design consideration. The second goal has been to ensure stable, uncompromising long term operation.

In this spirit a high quality permanent magnet has been chosen over a superconducting one with its unavoidable limitations due to the consumption of refrigerants for maintaining the superconductive state during a measurement period of several years [1,2]. Similarly instead of a TPC type drift detector, where the operation medium has to circulated and replenished, a solid state track detector has been chosen with its intrinsically stable detection behaviour and superior position resolution [1,2].

These design considerations have been brought to fruition during the STS 91 (2.6. - 12.6.1998) precursor flight. All AMS detection systems have been made operational just 3 hours after launch. In flight AMS has accumulated more than 400 10^6 triggers out of which 100 10^6 have been stored on disks for further analysis.

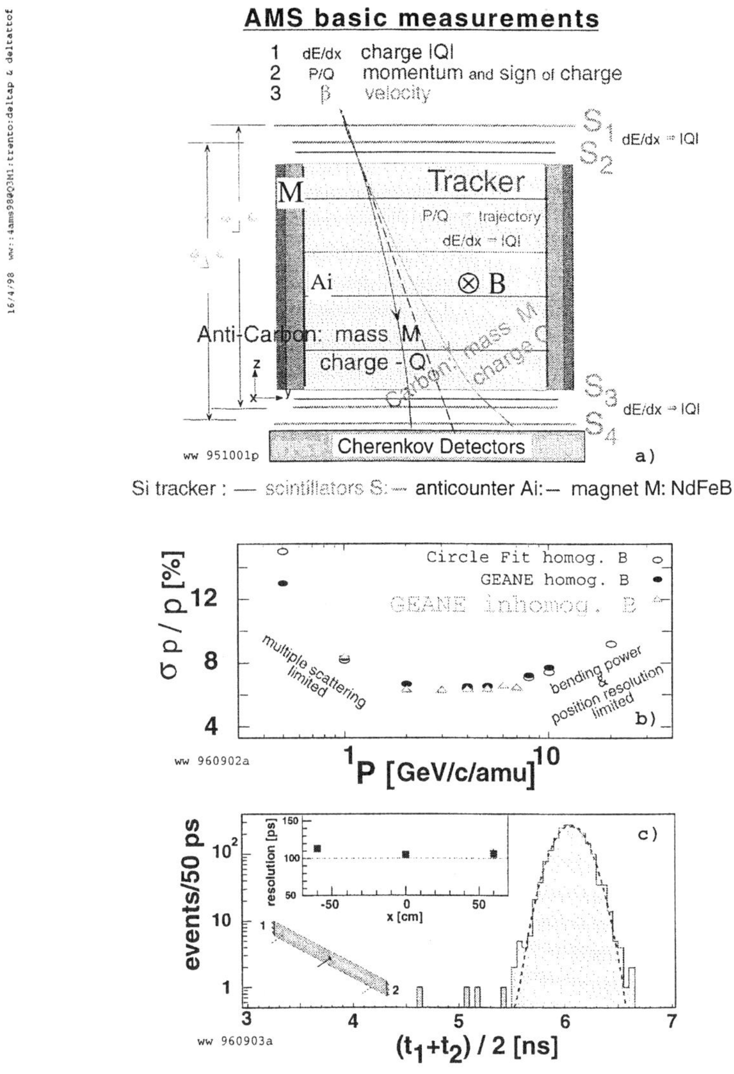

FIGURE 6. a) AMS detector basics, b) momentum resolution, c) time of flight resolution

Several antiproton events have already been identified in the ≈ 3% subsample of events that had been transferred to ground during flight (see fig. 7).

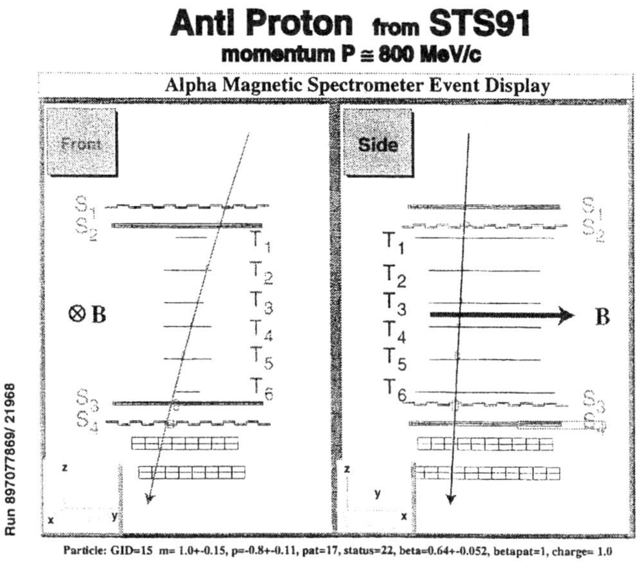

FIGURE 7. anti-proton event from the STS 91 flight, front(side) view: B perpendicular (parallel) to plane of figure. Designation of components as in figs.1, 6a.

3.4 The magnet

The centrepiece of the AMS experiment is its 1.9 to permanent magnet (fig. 8a) with a bore of 1.114 m diameter and a height of 0.8 m providing a dipole field with a bending power of $\int Bzdz = .15\ Tm^2$

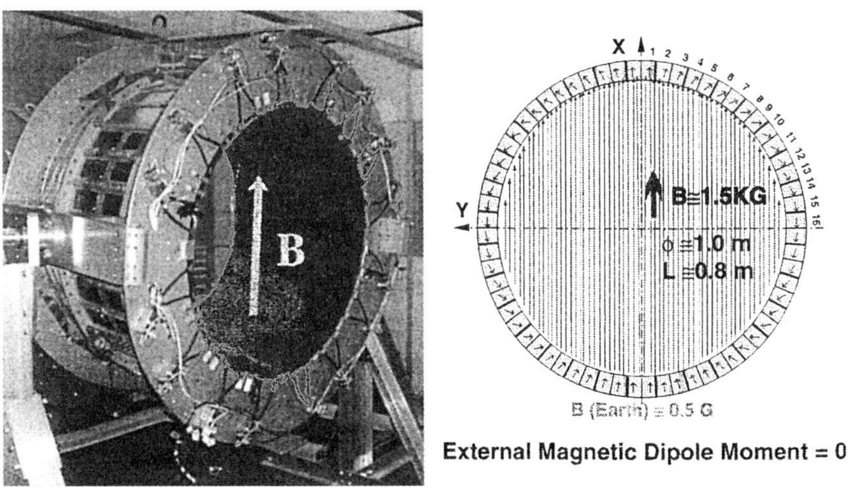

FIGURE 8. a) AMS magnet in an early phase of detector assembly; the inner cylindrical surface is covered with anti coincidence counters (ACC), see text, b) orientations of magnetic columns.

The AMS magnet has been built up from blocks of NdFeB alloy [22]. The blocks are arranged in a circle made out of 64 columns such that the vector sum of their magnetizations produces a highly homogeneous magnetic field (fig. 8b) with $\{B_x,B_y,B_z\}(x=0,y=0,z=0) = \{.15,0,0\}$T [23] (for the coordinate system see figs. 4,6,7).

This arrangement has no outside magnetic dipole moment by design, such that there is no torque due to the coupling of the AMS magnetic field with the earth field while being flown in orbit.

The columns of the magnetic material are firmly mounted into a double walled Al cylinder. This cylinder is covered at both ends by solid Al flanges. In addition the flanges are linked outside the magnetic volume by 4 I-beams (figs. 4,8) for ensuring structural integrity up to accelerations of 13 g . Both the flight and the prototype version of the magnet have undergone a full test program of flight worthiness at CALT [18] with the support of NASA. In addition a specifically built slice of the magnet (see fig.9) has been tested to destruction for the verification of the assumptions of the parameters used to assess the ultimate mechanical limits derived from finite element (FE) calculations

FIGURE 9. AMS magnet (slice) under mechanical tests for space flight

3.5 The tracker

3.5.1 overview

The AMS tracker is designed such as to provide maximum momentum resolution inside the acceptance of the AMS magnet. In the tracker the particle runs along a helix through the magnetic field. This track is observed at 6 space points with high precision $\{\Delta x, \Delta y, \Delta z\} = \{25, 8, 35\}$ μm (for the coordinate system see figs. 6,7). The tracking is realized with 6 layers of Si particle detectors *. Individual Si-sensors (41.4 x 73 mm^2) of AMS are bonded together forming ladders (see section 3.5.3) of up to 70 cm length.

* Reverse biased Si detectors work like ionization chambers. Charged particles create electron - hole pairs (typically 8000 / 100 μm) in the depletion zone through ionization.

4 of the layers are mounted inside the magnet proper. The 2 outer layers are mounted at (≈ 10 cm) above (below) the AMS magnet. The geometry of the AMS tracker is evident from figs. 4, 6, 7. Central parameters are given in the following table:

TABLE 3. AMS tracker, **STS 91** configuration in ()

plane	ø[m]	t [cm]	t[X0]	#ladders	#Si- sensors	z [mm]
1	1.6	4	.024	2*15+2 (2*4)	15	+510
2	1.0	1.4	.008	2*12+2 (2*6)	12	+214
3	1.0	1.4	.008	2*12+2 (2*6)	12	+ 78
4	1.0	1.4	.008	2*12+2 (2*6)	12	- 78
5	1.0	1.4	.008	2*12+2 (2*6)	12	- 214
6	1.6	4	.024	2*15+2 (2*3)	15	-510

The supports of the outer planes are located outside the tracking volume proper (see figs. 4, 14) Hence the scattering in these supports doesn't contribute to the resolution.

The Si sensors are glued on rails made from carbon fibre clad foam. With 5x5 mm^2 cross section rails only 6 to 10 support points are needed to keep ladders sufficiently firm in position during the vibrations of lift-off and landing. These ladders form the basic mechanical and electrical units of the AMS tracker. Up to 15 Si-sensors are assembled into ladders (table 3).

The active detector area amounts to ≈ 85% of the inner magnet bore providing for an acceptance of $0.6m^2sr$ with 5 hits out of 6. For the STS 91 precursor flight about 45% of the total Si area has been equipped with sensors. This arrangement (which has more Si than the largest collider Si detectors currently in operation) gives an acceptance of $.32m^2sr$ with 4 hits out of 6.

At the momenta AMS is designed for the track is well approximated by a segment of a parabola(circle). The minimal curvature of this track segment (hence the maximum detectable momentum (MDM)) is given by its smallest unambiguously measurable sagitta. A MDM of 500 GeV/c amu requires a sagitta resolution of 20 micron with the AMS magnetic field.

3.5.2 tracker support structure

It is evident that a position resolution of 8 μ on the individual Si sensor level is only useful if the tracker support structure is highly stable over the full acceptance. Furthermore it has to be linear in its elastic behaviour such that the vibrations during lift-off and landing don't result in permanent displacements of the Si sensors. Finally deformations due to thermal expansion and outgassing (release/uptake of water) have to be well controlled in the range of operating/ storage conditions (-30/-10°C $\leq$ T $\leq$ 30/50°C; 10^{-7}mb $\leq$ p $\leq$ 1000mb) during the space missions as well as the test phases on ground.

Ladders are mounted closely spaced on circular 0-CTE carbon fibre plates (CFC, Al honeycomb spacers) of low mass density (see fig. 10). Ladders and plates have eigenfrequencies well above the range of frequencies that are excited with high power during lift-off.

FIGURE 10. AMS tracker plane 2, Si ladders with shielding mounted on CFC plate (université de Genève), positions of laser alignment beams marked (see text).

Inner plates (2 - 5) are hold in place with rigid CFC-shells of cylindrical shape. The outer planes (1,6) are supported by 2 CFC-crowns that attach to the cylindrical inner tracker structure after insertion into the magnet. All tracker/ladder services (signal, control, power lines; heat removal bars) are run through rectangular openings in the crowns (see figs. 11, 14). The tracker support structure itself is hinged to the magnet via 8 titanium feet. They are designed to keep the tracker elastically in position but not to overstress the shell of its support. The behaviour of the complete structure has been investigated with FE methods. The lowermost resonance frequency is at 87 Hz.

It has been verified during assembly that this arrangement provides an interlock between the individual planes stable to at least 5 μ in all 3 dimensions. Furthermore the reference marks on the Si-sensors were measured after the ladders were placed on the CFC-plates. Including this data and a survey of the plane control marks accessible on the outer shell it has been found that the cumulated manufacturing (shell, plate, ladder, sensor) misalignments (deviation between actual and design position) are always less than 50μ. These misalignments are known to better than 10μ. On one central set of ladders the stability of the alignment is checked at regular intervals with the tracker alignment system (see below). Furthermore the alignment of the ladders can be controlled with low momentum (p ≤1.5 Gev/c) protons that are observed at high rate during flight [23].

FIGURE 11. AMS tracker support structure (CFC) during trial assembly at ETH Zürich

3.5.3 Si sensors and read-out

The Si sensors have a doublesided read-out structure. Their design and fabrication closely resembles those Si sensors used in the L3 detector [24]. Their basic geometry is shown in fig. 12

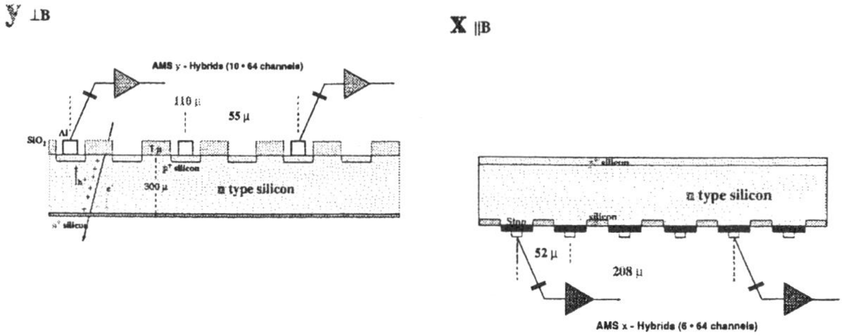

FIGURE 12. AMS doublesided Si - sensors (INFN Perugia), read-out strip geometry

and the following table:

TABLE 4. AMS Si sensor geometry (all dimensions in μm)

	# of strips	strip spacing		
width (y)		in read-out	aluminized	implanted
73000	640	110	55	27.5
length (x)				
41400	192	208	52	26.0
thickness (z)				
300				

Read-out-strips are closely spaced in the y direction (perpendicular to the main component B_x of the magnetic field for providing a position resolution of better than 8 μm [25]). This precision is possible by measuring the charge sharing between adjacent strips with a pulse height recording read-out system for each individual strip.

The orthogonal x-coordinate is measured simultaneously. To this end the ohmic side of the sensors is segmented. Recording the charge sharing between x-strips provides an x-resolution of 25 μm.

Particle rates in space are well below those encountered at accelerators (fig. 5). In consequence the noise of the frontend tracker electronics can easily be kept at low (S/N (mip) ≈ 15) levels with integration time constants of 4 μs while having the y-strips of up to 15 sensors ganged together (C ≈ 70 pF). In the x-direction the strips of alternate sensors are linked together via read-out lines printed on Kapton foils. This read-out scheme is well adapted to the task of single track localization that is characteristic for the observation of cosmic rays. The total number of channels (STS 91: 64k) to be read is fairly low for this large area Si detection system. Thus the stringent limits of power consumption (200W) and waste-heat production - especially inside the magnet (80W) - can easily be met.

3.5.4 The tracker electronics

The main design considerations of the tracker electronics were:

1. **dynamic range**
   (measuring charge (dE/dx) of nuclei from hydrogen to carbon exactly)
2. **power consumption**
   (limited thermal transfer between frontend electronics mounted on ladders and heat dumps outside the magnet)
3. **long-term stability**
4. **very high MTBF**
   (graceful system degradation in case of subcomponent failures)

The fundamental aspects are given in table 5 and fig. 13.

TABLE 5. Basic parameters of AMS Si tracker electronics; **STS 91** configuration in ().

1	no. of channels		172032	**(63488)**
1.1	y (rectifying "s"-side)		107520	**(39680)**
1.2	x (ohmic "k"-side)		64512	**(23808)**
2	no. of strips/hybrid	64		
3	no. of hybrids/ladder	16	y: 10	x: 6
4	max no. of strips connected in series		y: 14	x: 7
5	amplifier input capacitive load [pf]		y: 70	x: 130
6	equivalent noise charge [mip]		y: 1/15	x: 1/8
7	digitizers (self calibrating)		y: 16	x: 8
7.1	dynamic range [bits{mip}]		y: 12{80}	x: 12{90}

The frontend amplifiers (64 channels/chip) are mounted on hybrids (ca. 2W dissipation) . These are fixed on CFC rails of very high thermal conductivity (1000W/m K) (see figs. 10, 14) and connected (bonded) through Kapton cables to the Si sensors. The amplifier response curve is nonlinear beyond ≈ 120 mip's, allowing to observe energy deposit fluctuations of CNO group nuclei fully.

The output of the frontend amplifiers is multiplexed at 5 MHz into 12 bit flash-ADC's. There are (STS91) 24 digitization boards with 4 ADC's each, that work all in parallel (fig. 13).

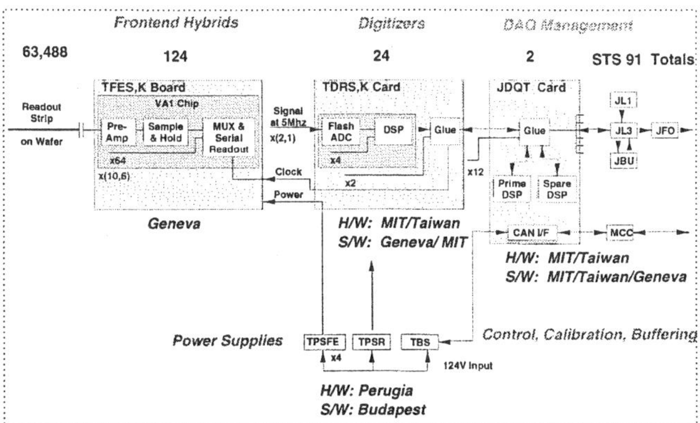

FIGURE 13. AMS tracker electronics

A digital signal processor (DSP) is used for sparsifying the data on the fly. Cluster information only (start channel-no., width, pedestal and gain corrected strip signals from width-no. of channels) is transmitted in normal runs. Basic system performance is checked typically 2 times per orbit in special calibration runs under DSP control. This self-calibration can adjust read-out thresholds used in the sparsification. In addition it manages the table of badly behaved channels that are suppressed in read-out (< 2%, mostly for electrical malfunctioning). By this method the amount of tracker information read out can be kept at a level of 600 bytes per event.

The output in clusterform is assembled into tracker bank information by the JDQT tracker data management module.

3.5.5 tracker alignment system (TAS)

The alignment of the AMS tracker is checked with 4 infrared (IR, lambda = 1082nm) rays that pass through all 6 Si layers (fig. 14).

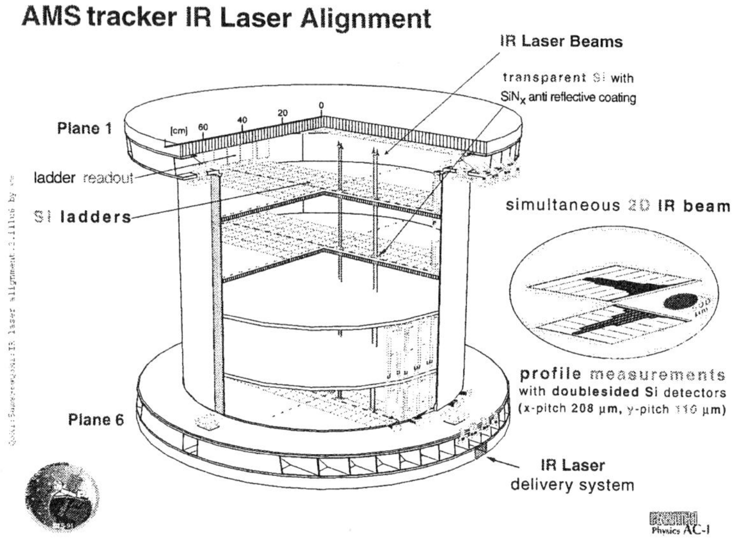

FIGURE 14. AMS infrared laser system, generating artificial straight tracks for alignment control

Si is transparent in the infrared (lambda > 1.1 μm). Nevertheless in the vicinity of the bandgap (E_g = 1.15 eV) an easily detectable amount of electron-hole pairs can be generated *throughout* the depletion zone of standard particle detectors with IR light beams.

In the AMS-TAS IR projection optics ("beamports") are installed in a channel running across (inside) the lower tracker support plane (#6). The beams are arranged in pairs, slightly offset in x and y (by about 2 beam diameters) in order to illuminate different read-out channels.

The projection optics are fed through monomode fibres via splitters from 2 individually controlled laser diodes (SDL 6702). Powerlevels and beam parameters are given in fig. 15.

The TAS system has been extensively checked for the stability of the optical parameters (see fig. 16) under the conditions of space transport (vibration) and space operation (temperature, vacuum).

The laserdiodes are pulsed (pulselength: 0.2 - 4μs) with repetition rates of 100 Hz typically. During the STS 91 flight they were operated for runs of 8 seconds duration at 4 intensity settings every 4 hours.

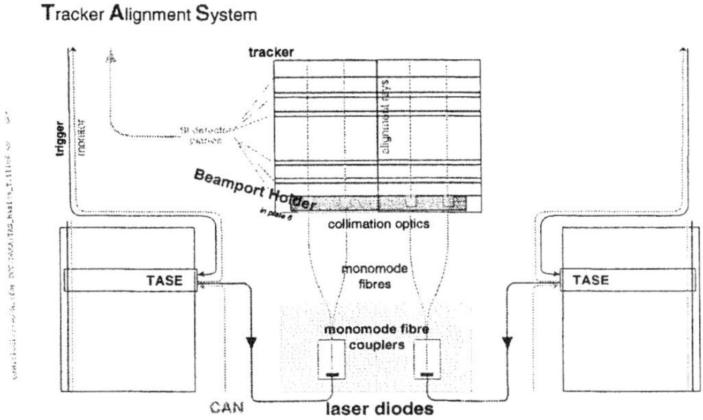

beam ø	1 mm	divergence	0.8 mrad	wavelength	1083 nm	energy per pulse	10 nJ
rep. rate	2000 Hz	pulse duration	200 - 4000 ns	attenuation per Si layer	~ 2.5		

FIGURE 15. AMS tracker alignment system, system block diagram and operating parameters

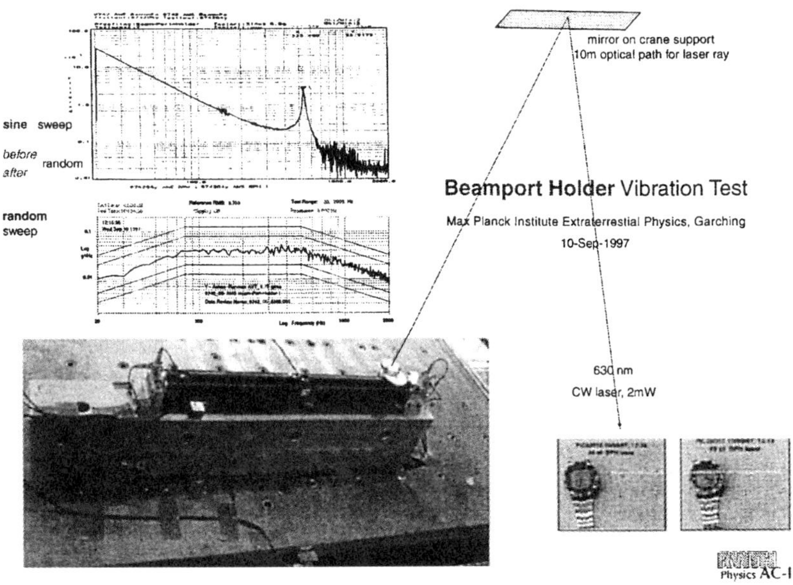

FIGURE 16. TAS beamport holder in vibration test (using 620 nm diode laser for a ≈ 10m stability control light path)

The effective transparency of the Si - sensors is determined by the transmissivity (≈ 1- reflectivity; absorption at 1082 nm in 0.3mm Si: 10 - 15%) and the fraction of the detector surface not covered with aluminized electrodes (shadowing). The contribution of the reflectivity has been considerably reduced (see fig 17a) on part of the sensors in the alignment beams by applying an antireflective SiN_x coating. With the AMS tracker flown by STS 91 in total beams are typically attenuated by a factor of ≈ 2.5 from to layer to layer. This factor is dominated by the shadowing due to the aluminization. At the present read-out pitch (table 4, fig. 12) the amount of aluminization can considerably be reduced in future AMS sensor production runs.

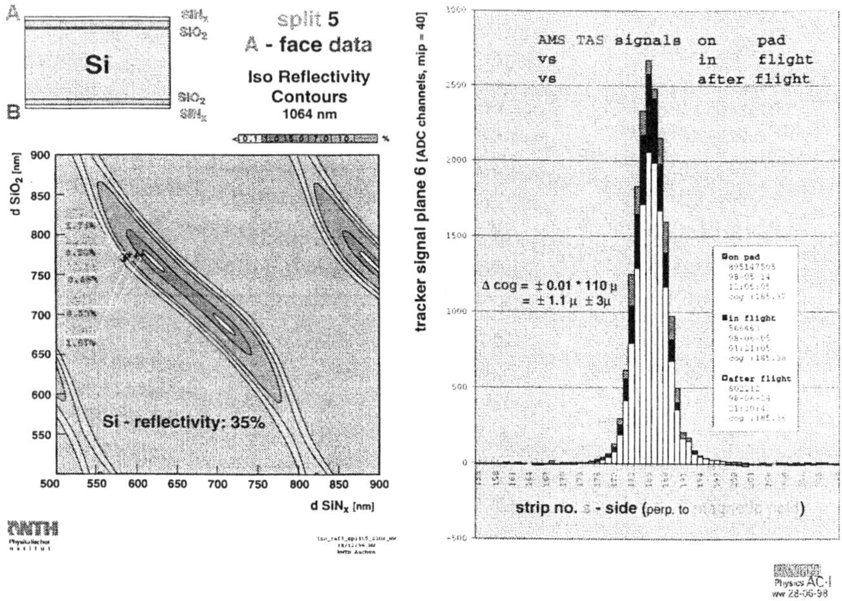

FIGURE 17. a) SiN_x on SiO_2 antireflective coating, isoreflectivity curves and sample wafer measurements, b) u1 plane 6 IR beam profiles on pad, in flight, after flight; centre of gravity (cog).

In alignment runs during flight the intensities of the IR beams are chosen such that at least 2 consecutive tracker planes receive optimal signals at 90 (40%) full scale. At full scale (saturation of the digitizer) an IR signal equivalent of approximately 300 mips is recorded summing over 1.2 mm (12/6 y/x strips). At these intensities full precision beam localization (limited by our present understanding of systematical errors to ≈ 3μ) can be done with approximately 25 IR pulses.

The first results from the STS 91 flight show (fig. 17b), that both the laser rays and the tracker planes were stable within our expectations. In fact as the example of plane 6 shows the positions of the sensors in the laser beams on plane 6 didn't move by more than ± 1.1 μ between the check-out on the launching pad, the flight and the verification run after landing.

3.6 time of flight system (ToF)

3.6.1 ToF system overview

There are 4 planes of 14 plastic scintillation counters (10 mm Bicron BC 408). One pair of planes is mounted above, another below the magnet (figs. 4, 6a). Counters do overlap by 5 mm. They are running in orthogonal directions in the 2 planes of each pair. Counters are viewed by 3 miniature photomultipliers (Hamamatsu RS5900) from each end.

Using the combination of all signals a time resolution of better than 110ps has been achieved (fig 6c) [26]. If exactly one particle is traversing the counter plane the point of impact can be determined to 1.5 cm precision from the time difference of the signals from both ends.

3.6.2 ToF electronics

For large dynamic range and high precision there are 2 timing channels and 2 amplitude measuring channels for each end of the scintillation counters (fig. 18). All digitizations are done by a single pipelined time digitizer chip [27]. The amplitude is inferred from the time over threshold.

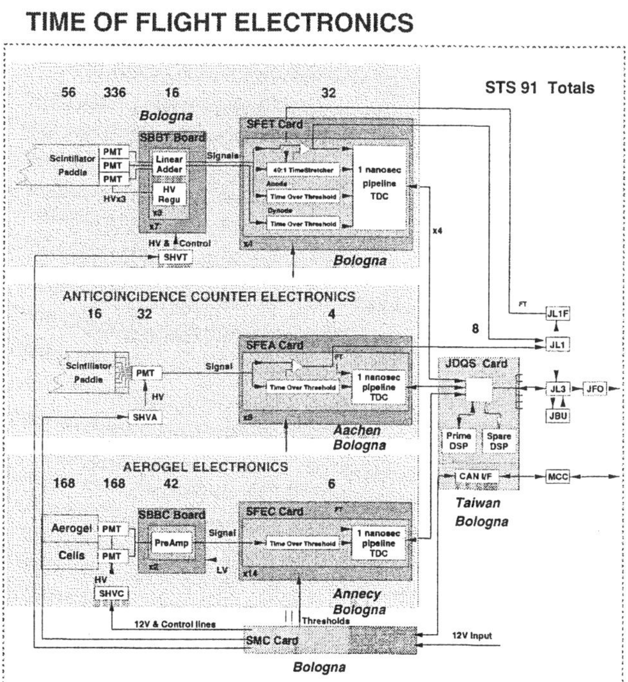

FIGURE 18. Time of Flight, anticounter and Cerenkov electronics

The low resolution time channel serves to record charged particles from interactions with the structure of AMS or the spacecraft within a time window of -6 μs to +10 μs with respect to the trigger (see below). The high precision timing is derived from a stretcher (factor 40) circuit.

One amplitude channel is connected to the photomultiplier anode, another to the last dynode (≈ 25% of the anode full amplitude). The response function of the amplitude channels have logarithmic behaviour. The 5% accuracy of the amplitude channels is matched to the dE/dx resolution for single charged particles obtainable with the AMS scintillators.

In addition to the ToF system there are 2 more photomultiplier based detection systems used in AMS (the anticounters and the aerogel Cerenkov counter, see below). The electronics chains hooked up to the AMS photomultipliers are based on identical functional units for HV generation and control (see fig. 18). For timing and amplitude measurements the same approach has been taken on the component level, but the boards have been adapted to the specific task for saving power, space and weight.

Calibration monitoring of the AMS counters is performed with high energy (E > 10 GeV) protons (in orbit) and muons (on ground).

3.7 anticounters (ACC)

The inner tracker volume is surrounded by 16 anticounters (figs. 4, 6a, 8). They allow for the suppression of unwanted events (on the trigger level) from cosmic ray (mostly several GeV protons) interactions in the magnet material. These events can lead to a loss of the important timing information such that a discrimination between downward and upward going particles is made impossible.

About 65% of all triggers in orbit have more than 1 counter fired. These triggers are reduced by a factor 4 with a fast amplitude weighted multiplicity logic [28].

The anticounters are viewed through fibre light guides by photomultipliers on both ends.

Their inefficiency has been measured to be less than $5\ 10^{-4}$.

3.8 Aerogel threshold Cerenkov counter (ATC)

High beta particles are flagged by the silica aerogel (n=1.04) threshold Cerenkov counter underneath the AMS spectrometer (fig. 6a). Its primary function (STS91) is to suppress electrons in the sample of antiproton candidates for momenta below 3.8 GeV. It consists of 2 layers of 20 (24) boxes of silica aerogel that are displaced with respect to each other such that cracks don't coincide. Each box contains 4 piles of square slabs of aerogel that are individually viewed by one photomultiplier each. A light output in mean of 4 photoelectrons has been observed with a sample box in an electron beam [29].

3.9 STS 91 AMS detector results

AMS has been thoroughly tested in the preparation phase of the STS 91 spaceflight. These tests took place in the shuttle simulation stand CITE (March/April 1998) as well as on the launch pad after AMS had been mated to the orbiter (April/May 1998). Tests included detector functions (fig. 19) and commanding through the all of the transmission chain, that is involved for reaching experimental equipment in the shuttle payload bay.

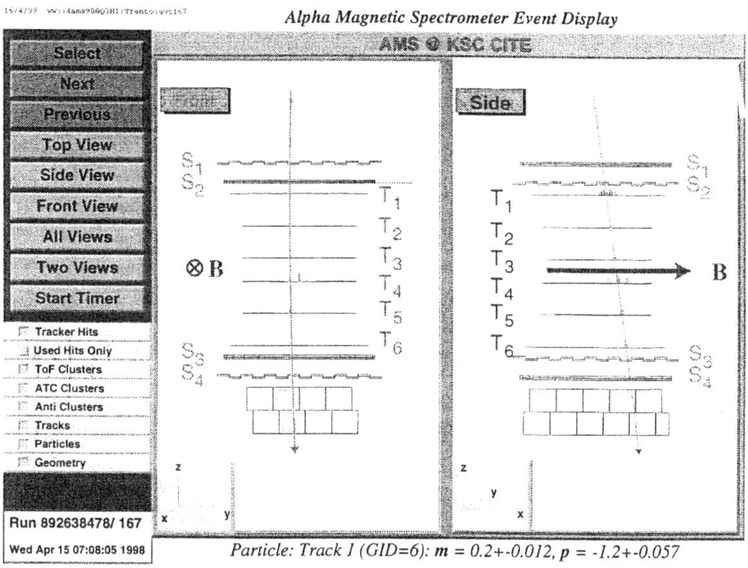

FIGURE 19. negative muon from cosmic ray testing at the NASA KSC CITE test stand

AMS has successfully taken science data for 103 hours during the STS 91 flight before and after docking of the space shuttle Discovery to the space station MIR. In addition a number of technical runs for calibration, trigger and background studies were taken while Discovery was docked to MIR.

Only a small fraction (≈ 3%) of the STS 91 data accumulated has been transferred to ground during flight due to a failure of one of the communication channels of the shuttle. The plots on detector performance shown in this report are based on this small fraction exclusively.

An overview of the composition of the raw data (while docked to MIR) is given in fig. 20. High energy particles pass through the detector both downwards and upwards. In standard data taking attitude (AMS pointing towards space) downward running particles exceed upward ones (reflected by the magnetic field of the earth) by more than a factor of 100.

Protons - by far the dominant component of the cosmic rays - can easily produce pions in the structures surrounding the AMS detector. In fact pions of both signs are observed in the upward (reverse) and downward (normal) streams

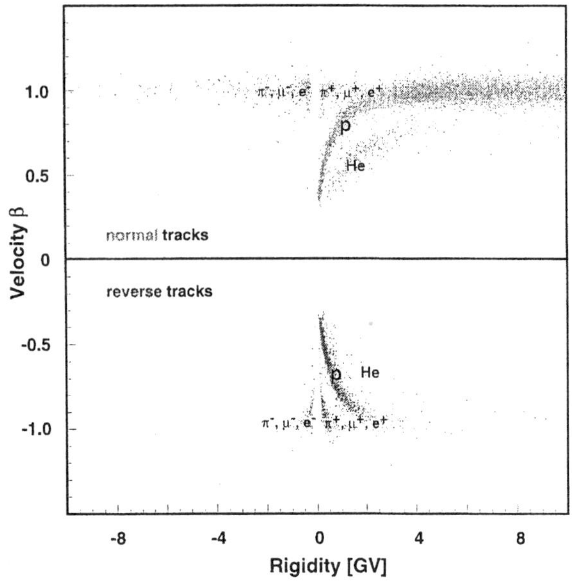

FIGURE 20. velocity vs. rigidity of tracks observed with AMS (docked with the shuttle Discovery to the space station MIR)

He nuclei - the second most important component - are identified both by their velocity (fig. 20) and their energy loss (fig. 21).

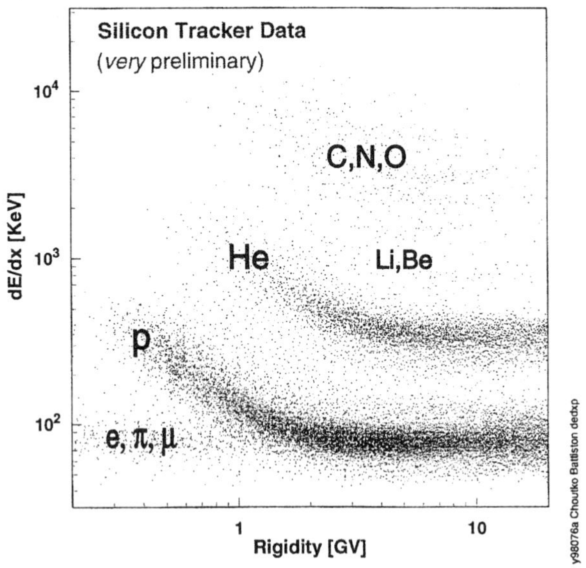

FIGURE 21. energy loss vs rigidity as measured on a sample of tracks with the AMS Si tracker

The observed particle mass spectrum as derived from the measured momenta and velocities of raw events clearly shows (fig. 22) contributions of protons, deuterons, Helium and nuclei of the group C,N,O in the proportions expected (cit.). At low mass a sizeable contribution of pions and muons is seen. These are produced by the fast protons in the structures surrounding AMS (data partly taken while the shuttle was docked to MIR).

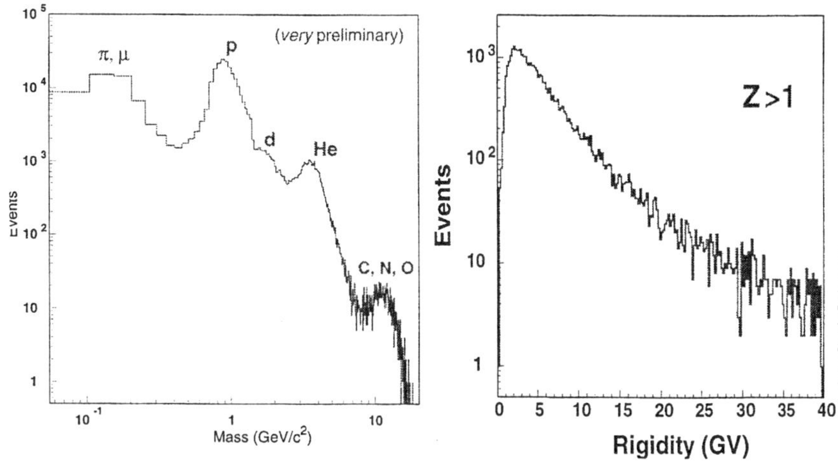

FIGURE 22. a) particle mass reconstructed from momentum and time-of-flight measurements; b) rigidity (R= p/Q) distribution for nuclei with charge > 1

The momentum spectrum of the nuclei with Z>1 observed extends well beyond 40 GeV (fig 22b). It can be expected that more than 300 particles with momenta exceeding 100 GV have been registered.

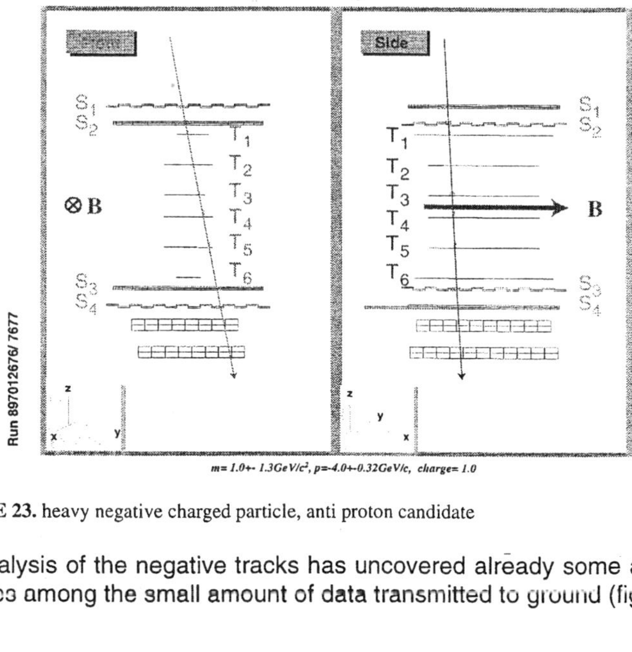

FIGURE 23. heavy negative charged particle, anti proton candidate

A fast analysis of the negative tracks has uncovered already some antiproton candidates among the small amount of data transmitted to ground (figs. 7, 23).

4.0 Physics potential of the AMS detector

4.1 AMS data taking in space

In order to illustrate the physics capabilities of the AMS experiment we will shortly discuss major items of the AMS physics program. The long-term data taking on the spacestation will start in 2002. Early this summer about 10^8 events have been recorded during the STS 91 flight. This data is now being evaluated. It is already clear that from an engineering point of view the AMS STS91 flight was a full success (figs. 2, 7, 17b). The first publications of the physics data are expected by beginning of 1999.

It can be expected that AMS in its present form will have more data taking to do over the next 2 years: There will be test runs -mostly for calibration of the particle identification systems- in high energy beams both with light particles (e, μ, π) and nuclei (p, He, Be, B, C,..). Furthermore another 10 day datataking in space is discussed for 1999.

4.2 Matter background rejection

In order to provide for the very high level of matter signal suppression required in the high sensitivity search for cosmic anti-matter with AMS on the spacestation a set of specific hard- and software procedures have been designed[1,2]:

Upward going tracks which - while missing the Si tracker - can fake in random coincidences with ordinary matter tracks downward antimatter tracks. Such events are suppressed to a high degree by rejecting events with hits in the anticounter ACC shielding the tracker volume (sections 3.7) or in panels of the ToF system (sections 3.6), that are recognized as not being associated with the track measured.

With the fine resolutions (bending 8 μ and non bending 25 μ) in each of the six Si-detector layers quality criteria based on fits to the tracks have been set up, that assign a wrong charge in 1 out of 10^7 candidate tracks only (wrong curvature matter tracks can arise from nuclear interactions in the Si layers and its supports). The internal consistency of the charge assignment of a candidate track is checked by comparing the fit results (directions, momenta) from two (or more) subsets of the measured coordinates. Rather tight cuts are possible ($\Delta\theta$ < 0.5 mrad for 8 GeV/c/amu) leading to a combined probability of charge confusion of less than 1 in 10^9.

If furthermore it is required that the signs of the total track as well as the track segments coincide and in addition the time-of-flight is matching with the momentum and absolute charge assignment of the candidate track then a charge confusion probability as low as shown in fig. 24a can be reached, while maintaining good detection efficiency for momenta $0.6 \leq p \leq 10$ GeV/c/amu.

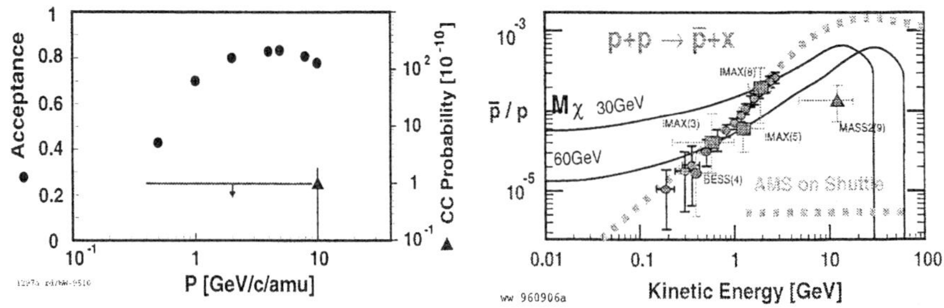

FIGURE 24. a) acceptance (righthand scale) and charge confusion probability (lefthand scale) expected for AMS on the ISSA; b) antiproton kinetic energy distribution, MC simulation of AMS detection behaviour, present anti proton data from balloonflights [31, 32, 33], expectations for neutralinos.

4.3 Anti-proton detection capabilities

Singly charged negative particles that don't fire the aerogel Cerenkov counter are almost exclusively antiprotons with $p \leq 3.8$ GeV/c. The small background from negative pions due to nuclear interaction in the counters and the radiation shield can easily be suppressed with the high segmentation of the AMS detector, its good dE/dx resolution and its accurate particle tracking. The recent 103 h data taking during the precursor flight STS91 should increase the present cosmic anti-proton data record considerably (fig. 24b). This data might be able to discern for the first time a contribution of anti-protons from the Cold Dark Matter Halo of our galaxy, that is possibly of supersymmetric nature [4, 5].

4.4 Positron detection capabilities

Only 10% of the fairly small electron fraction ($\approx$1%) of the cosmic rays appears as positrons. There is considerable interest in more and precise positron data, especially at higher (> 20 GeV) energies [4, 34]. For the spacestation run it is foreseen to equip AMS with a transition radiation detector (TRD). This device will allow to reject protons at a level of 10^4 or better for momenta between 3 and close to 100 GeV. The positron spectrum that can be observed with this combination of detectors will contain more than 10^8 positrons compared to 10^3 presently.

4.5 ^{10}Be / ^{9}Be isotope ratio

The AMS group is currently designing a NaF ring image Cerenkov counter (RICH), that is going to supplement the particle identification functions of the Aerogel counter. With this instrument a mass resolution of 0.25 amu will be possible for momenta in the best momentum resolution range. Making standard assumptions about the relative abundance of ^{10}Be/^{9}Be the spectra observed with AMS would be measured as shown in fig. 25a. The high data rate and the very clean separation will result in very accurate determination of this isotope ratio. It is well known that unstable nuclei with half-lives in the vicinity of the indirectly determined confinement time ($\approx 10^7$ y) can be used for actually measuring that time.

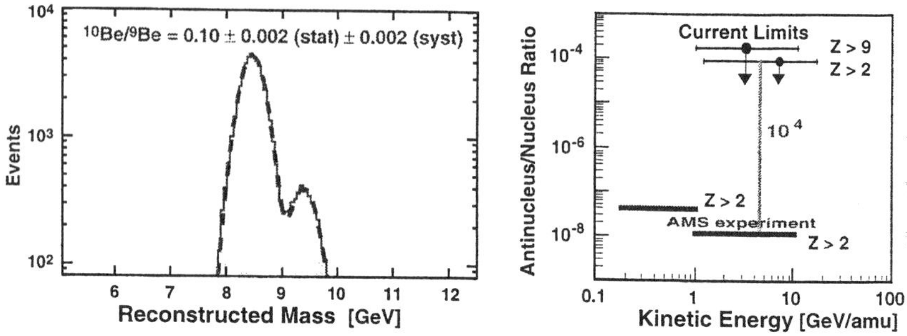

FIGURE 25. a) AMS rate measurement of Be isotopes (simulation), b) 95% CL sensitivity for antimatter search with AMS (3 years on ISSA)

5.0 Conclusion

The AMS spectrometer is designed to find cosmic antimatter with a very much improved sensitivity (fig. 25b). For the search of cosmic anti-matter AMS will be definitely installed on the spacestation ISSA in the year 2002.

Furthermore it has accumulated 10^8 events been during the STS 91 mission to MIR in June 1998. Preliminary results show that the AMS detector was fully functional. These events will serve for the technical aspects of calibration and background studies necessary in order to accomplish the full program. In addition the STS 91 data will be used for understanding in more details (antiprotons!) the composition of cosmic rays without the overburden of the atmosphere.

Acknowledgements

The AMS spectrometer construction is a collaborative effort of institutions from Asia, Europe and the US (fig. 2). We are grateful for the support given by the national research agencies involved and especially by NASA. This extends in particular to the courageous crew of astronauts (fig. 26) that took care of the AMS experiment as payload in flight STS 91.

This report is based on results from many physicists working enthusiastically for making AMS a reality. For helpful discussions the author is especially indebted to Prof. S.C.C. Ting - the AMS spokesperson -, Prof. K. Lübelsmeyer, Dr. G. Ambrosi, L. Baldini, Dr. B. Burger, Dr. M. Capell, Dr. V. Choutko, Dr. T.S. Dai, Dr. A. Lebedev, Dr. W. Lustermann, Dr. N. Produit, Dr. F. Raupach and G. Schwering.

I enjoyed the competent help of Mrs. Y. Bernard for much of the graphics of this report.

It was a pleasure to share the excitement of the AMS team with the participants of the INTERNATIONAL SYMPOSIUM on LEPTON AND BARYON NUMBER VIOLATION in Trento, 20-25 April 1998. I like to thank in particular the chairman Prof. Klapdor-Kleingrothaus for his hospitality and the careful steering of the discussions.

FIGURE 26. The STS 91 crew of astronauts (from left to right: D. Gorie, F. Chang-Diaz, W.B. Lawrence, J. Kavandi, Ch. Precourt, V. Ryumin) infront of the orbiter Discovery while the AMS team continued in testing the detector mounted in the payloadbay (May 5, 1998, photo NASA KSC).

References

1. S.C.C. Ting ed., AMS proposal, MIT-LNS, March 1995
2. S. Ahlen et al. , Nucl. Instr. and Meth. A350 (1994) 351
3. G. Jungman and M. Kamionkowski, Phys. Rev. D49 (1994) 2316
4. J. Ellis et al., Phys. Lett. B214 (1988) 403
5. R. Rudaz and F.W. Stecker , Ap. J. 325 (1988) 16
6. D. Müller and K.K. Tang, Proc. 21st Intern.Cosmic Ray Conf. 3 (1990) 249
7. W.R. Webber et al., Ap. J. 380 (1991) 230
8. J.J. Beatty et al., Ap. J. 413 (1993) 268
9. R. Silberberg and M.M. Shapiro, Proc. 17th Intern.Cosmic Ray Conf. 2 (1981) 356
10. R. Hagedorn, Cargèse Lectures in Physics, 6 (1973) 643
11. S.P. Ahlen et al., Ap. J. 260 (1982) 20
12. F.W. Stecker, NASA Tech. Memo. 83083 (1981)
13. G. Steigman, Ann. Rev. Astron. Astrophys. 14 (1976) 339
14. L3 collaboration, B. Adeva et al., Nucl. Instr. and Meth. A289 (1990) 35,
15. D.E. Dorfan et al., Phys. Rev. Lett. 14 (1965) 1003
16. Lockheed Martin, Space Systems and Services, Seabrook, MD 20706, USA
17. Max Planck Institut für Extraterrestrische Physik, Garching, Germany
18. CALT, Beijing, China
19. IABG, 85521 Ottobrunn, Germany
20. Contraves-Space, Zürich, Switzerland
21. Vakuumschmelze, Hanau, Germany
22. Y.M. Du et al. Institute of Electrical Engineering, Chinese Academy of Science, Beijing, 100080 China
23. V. Choutko, AMS note, 1997
24. M. Acciari et al. , Nucl. Instr. and Meth. A351 (1994) 300
25. B. Burger et al., université de Genève, submitted to Nucl. Instr. and Meth.
26. A. Cantin et al., INFN Bologna, private communication
27. LeCroy Corp., New York, USA
28. Han-Chang, National Central University, Taiwan, private communication
29. A. Gougas, Johns Hopkins University, private communication
30. S. Ahlen, priv. communication, ref. 8 and M.H. Salamon et al., Ap. J. 344 (1990) 78
31. J.W. Mitchell et al. [IMAX], Phys. Rev. Lett. 76 (1996) 3057
32. M. Hof et al. [MASS2], Ap. J. 467 (1996)L33
33. S. Orito et al. [BESS], Proc. 24th Intern.Cosmic Ray Conf. 3 (1995)75
34. J. Protheroe, Ap. J. 313 (1982) 391

Bounds on R-parity violation from cosmic ray antiprotons[1]

Edward A. Baltz† and Paolo Gondolo‡

† Department of Physics, University of California, Berkeley, CA 94720-3411, USA

‡ Max-Planck-Institut für Physik, Werner-Heisenberg-Institut, Föhringer Ring 6, 80805 München, Germany

Abstract. The lightest superpartner (LSP) in the minimal supersymmetric standard model is stable, and thus exists as a cold relic of the big bang. The addition of R-parity violating operators allows the LSP to decay into standard model particles. If the LSP lifetime is longer than the age of the universe, we can use the observed flux of cosmic ray protons to bound the hadronic R-parity violating couplings. We can exclude $10^{-18} \lesssim |\lambda''| \lesssim 10^{-15}$ and $2 \times 10^{-18} \lesssim |\lambda'| \lesssim 2 \times 10^{-15}$ except when there is a top quark in the final state, where we can only exclude $4 \times 10^{-19} \lesssim |\lambda'| \lesssim 4 \times 10^{-16}$.

1. Introduction

Violation of lepton- and baryon-number is allowed in supersymmetric models. A discrete symmetry called R-parity is frequently imposed to avoid this difficulty. The R-parity of a particle is $R = (-1)^{2S+L+3B}$, where S is its spin, and L and B are its lepton and baryon numbers. R is +1 for all standard model particles and −1 for all superparticles. It is also interesting exercise to allow R-parity violation. The couplings can then be constrained by experimental observations or cosmological considerations. With R-parity violation, the lightest superpartner decays into standard model particles. We will assume that the lightest superpartner is the neutralino.

[1] Presented by P. Gondolo. Based on Ref. [1]

Antiprotons in cosmic rays are rare, and are most probably produced in spallation reactions in the interstellar medium. Measurements of the antiproton flux limit additional sources of antiprotons in the galaxy. We find that the observed antiproton flux provides a powerful bound on the lifetime of the lightest superpartner (LSP). This bound permits us to bound the R-parity violating couplings in extensions to the minimal supersymmetric standard model.

We calculate the relic density of the LSP in the hot big bang model. These relic particles are then allowed to decay via the hadronic R-parity violating interactions. We use simple models for the production and propagation of antiprotons in the galaxy. We can then limit the R-parity violating couplings with observations of the cosmic ray antiproton spectrum.

Though our results strictly apply when the lifetime of the neutralino is longer than the age of the universe, there are other cosmological considerations based on nucleosynthesis in the early universe, recent structure formation, diffuse photon and neutrino fluxes, and spectral distortions in the cosmic microwave background that exclude a large fraction of neutralino lifetimes shorter than the age of the universe. These may well close the gap between our results and accelerator constraints.

2. Supersymmetric model

We consider a minimal supersymmetric extension of the standard model with explicit R-parity breaking, defined by the superpotential

$$
\begin{aligned}
W = & -h_e^{ij} E_i^c L_j H_1 - h_d^{ij} D_i^c Q_j H_1 + h_u^{ij} U_i^c Q_j H_2 \\
& -\mu H_1 H_2 + \lambda'_{ijk} L_i Q_j D_k^c + \lambda''_{ijk} U_i^c D_j^c D_k^c .
\end{aligned} \tag{1}
$$

i, j, k are generation indices; L_i and Q_i are the SU(2)-doublet lepton and quark superfields; E_i^c, U_i^c and D_i^c are the singlet superfields; and H_1 and H_2 are the doublet Higgs superfields. h_e^{ij}, h_d^{ij} and h_u^{ij} are Yukawa matrices, and μ is a free parameter [2].

The last two terms in the superpotential violate R-parity, and they also violate lepton- and baryon-number conservation explicitly. For three generations, there are 27 λ' couplings and 9 λ'' couplings, since λ''_{ijk} is antisymmetric under exchange of the final indices.

Supersymmetry is broken softly by a scalar potential that in general contains many unmeasured parameters. For our purposes, a simple parameterization suffices. We assign a common mass scale m_0 to the sleptons and squarks at the electroweak scale, we relate the gaugino mass parameters M_1, M_2, M_3 by means of GUT relations, and we keep the trilinear soft terms A_t and A_b only for stop and sbottom. Electroweak symmetry is broken at tree level by assigning vacuum expectation values v_1 and v_2 to the neutral scalar components of H_1 and H_2. The ratio $\tan\beta = v_2/v_1$ and the pseudoscalar Higgs boson mass m_A remain as free parameters. The total number of parameters is therefore seven. Further details on the class of models we consider are given by Bergström and Gondolo [3].

We use the scans of supersymmetric parameter space in Bergström, Edsjö, and Gondolo [4]. They consist of $\sim 10^4$ models that satisfy the experimental constraints reported by the Particle Data Group [5], plus the bounds on the Higgs sector and chargino masses from LEP (we update to LEP2W [6]) and the constraints on the $b \to s\gamma$ rate from CLEO [7]. The scans span values of $|\mu|$ and $|M_2|$ up to 5 TeV, of m_0 up to 3 TeV, of $|A_t|$ and $|A_b|$ up to $3m_0$, of m_A up to 1 TeV, and of $\tan\beta$ between 1 and 50. This range extends even beyond the parameter region usually considered as plausible.

3. Composition of the Galactic Halo

We do not assume that the dark matter consists of relic LSPs. Given a supersymmetric model, the relic density of neutralinos is fixed. We use the relic density computed by Edsjö and Gondolo [8], who include all neutralino self-annihilation diagrams, resonance and threshold effects, and chargino-neutralino coannihilations. In every model that is not cosmologically excluded, the lightest neutralino is a cold relic from the Big Bang. We expect that these particles are attracted into the gravitational potentials of galaxies, and are at present components of the cold dark halos.

We use the standard estimate that the dark matter density at the solar system is 0.3 GeV cm^{-3}. Neutralinos may make up the entire halo or only some fraction f. This fraction $f \geq \Omega_\chi/\Omega_{\rm DM}$, where Ω_A is the cosmological density of species A measured in units of the critical density Ω_0.

We now need an estimate of $\Omega_{\rm DM}$. A common prescription [9] is to use $\Omega_{\rm DM} = \Omega_\chi$ if $\Omega_\chi h^2 \geq 0.025$, which is the indicated by dynamical mass measurements of the mass of dark matter in galactic halos. If $\Omega_\chi h^2 \leq 0.025$, this prescription sets $\Omega_{\rm DM}h^2 = 0.025$. This prescription is not conservative enough for our purposes because there may be multiple types of cold dark matter in the galactic halo. Therefore, based on the age of the universe, we use the constraint that $\Omega_{\rm DM}h^2 < 1$ [9], to give a lower bound on $f = \Omega_\chi h^2$. In this way we obtain the conservative bound on the R-parity violating couplings. Our final results would improve by approximately a factor of five if we used the less conservative fraction.

4. Decay rate

We consider neutralino decay through both the lepton-number violating term $\lambda' LQD^c$ and the baryon-number violating term $\lambda'' U^c D^c D^c$ in the superpotential. The coupling λ'_{ijk} gives rise to the decays $\chi \to \nu_i d_j \bar{d}_k$, $\chi \to e_i^- u_j \bar{d}_k$ and their charge conjugates. The coupling λ''_{ijk} gives rise to the decays $\chi \to u_i d_j d_k$ $(j \neq k)$ and their charge conjugates.

The full decay rates for these processes are calculated in Ref. [10]. We quote the results in the Appendix. Here we give the rate for $\chi \to uds$ in the limit where the sfermions are heavy and degenerate,

$$\Gamma(\chi \to uds) = \frac{3}{32\pi^3} m_\chi |\lambda''_{112}|^2 g^2 \left(\frac{m_\chi}{m_{\tilde{q}}}\right)^4 \times \tag{2}$$
$$\left[2\tan^2\theta_W |N_{\chi 1}|^2 + \frac{m_u^2 |N_{\chi 4}|^2}{2m_W^2 \sin^2\beta} + \frac{(m_d^2 + m_s^2)|N_{\chi 3}|^2}{2m_W^2 \cos^2\beta}\right],$$

where g is the $SU(2)_L$ coupling constant, $m_{\tilde{q}}$ is the degenerate squark mass, and $N_{\chi i}$ is the projection of the lightest neutralino on the interaction eigenstates $\tilde{b}$, $\tilde{w}^0$, $\tilde{h}_1$, and $\tilde{h}_2$. A pure gaugino has the decay rate

$$\Gamma = \frac{3}{16\pi^3} m_\chi |\lambda''_{112}|^2 g^2 \tan^2\theta_W \left(\frac{m_\chi}{m_{\tilde{q}}}\right)^4. \tag{3}$$

while a pure Higgsino has

$$\Gamma = \frac{3}{128\pi^3} m_\chi |\lambda''_{112}|^2 g^2 \left(\frac{m_\chi}{m_{\tilde{q}}}\right)^4 \left[\frac{m_u^2}{m_W^2 \sin^2\beta} + \frac{m_d^2 + m_s^2}{m_W^2 \cos^2\beta}\right]. \tag{4}$$

5. Antiproton production

Antiprotons are produced in the fragmentation of the final state quarks. We are interested in the energy spectrum of antiprotons $dN_{\bar{p}}/dE$ at energies of a few GeV.

It is difficult to calculate the spectrum $dN_{\bar{p}}/dE$, especially at low antiproton energies. There are two theoretical frameworks available: a perturbative QCD and the Lund Monte-Carlo. In figure 1 we compare the theoretical curves with the measured proton spectra in e^+e^- annihilations at various jet energies (14.5 GeV [11], 29 GeV [12], and 45.6 GeV [13]). We also show the theoretical predictions at 1 TeV, one of the highest energies we are interested in.

In the perturbative QCD approach, Dokshitzer et al. [14] provide a formalism for calculating hadron spectra from parton spectra in the modified double-logarithmic approximation (MLLA). The full MLLA formula

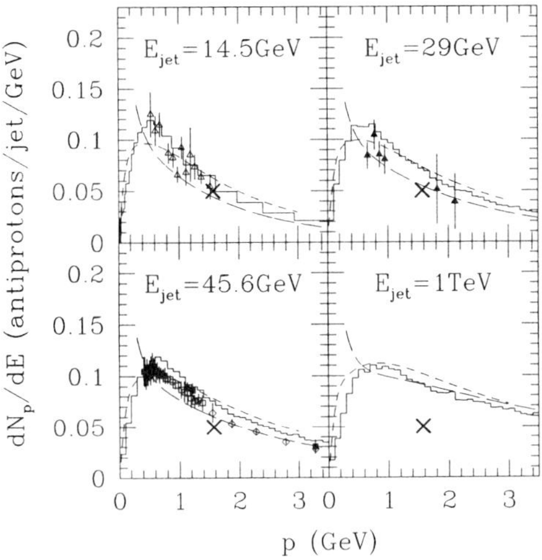

Figure 1. Antiproton spectra in quark jets from theoretical predictions and e^+e^- annihilation data. Short- and long-dashed curves are the low-energy and the limiting MLLA spectra, histograms are obtained with the Lund Monte-Carlo. Data points are from TPC (14.5 GeV), TOPAZ (29 GeV), ALEPH (41.6 GeV, full squares), DELPHI (41.6 GeV, diamonds) and OPAL (41.6 GeV, open squares). The cross indicates our choice of $dN_{\overline{p}}/dE$ at $p = 1.573$ GeV.

for inclusive particle production is difficult to evaluate, and no numerical formula is available. It has been shown that the simpler limiting spectrum reproduces the total spectrum of charged particles quite well [15]. However, as can be seen in fig. 1, neither the limiting spectrum (long-dashed curves [14]) nor the low energy limit of the full MLLA spectrum (short-dashed curves [15]) fit the proton data well. As an alternative to the MLLA formalism, the Lund Monte Carlo [16] has been used by several authors, most recently by Bottino et al. [17]. However, it also shows unsatisfactory agreement with data (histograms in fig. 1).

Inspecting both the experimental data and the results of the Lund Monte Carlo, we observe that the (anti)proton production rate in the high-energy tail of the spectrum increases with increasing jet energy. This is verified by the data for charged particle production and is a generic prediction of the MLLA formalism [14, 15].

Based on these arguments, we take the value $dN_{\overline{p}}/dE = 0.05$ antiprotons/jet/GeV at the momentum $p = 1.573$ GeV in which we want the antiproton spectra (see sect. 7), independently of jet energy. This underestimates the production rate of antiprotons, and will thus give conservative bounds on the R-parity violating couplings.

The number of jets per decay $N_{\rm jets}$ is equal to the number of final state quarks, except when there is a top quark. The top decays before it can hadronize. We approximate the process $t \to W^+ b$ using the fact that the low energy proton spectrum is insensitive to the jet energy. The bottom quark forms a jet. In addition, we take the hadronic branching fraction of the W, $B(W \to \text{hadrons}) = 0.679$, as the probability to form two additional jets in the decay $W^+ \to u\bar{d}$ and its charge conjugate. The final values for $N_{\rm jets}$ are 3 for a $U^c D^c D^c$ decay without top, 4.358 for a $U^c D^c D^c$ with top, 2 for an LQD^c without top and 3.358 for an LQD^c with top.

We now obtain the volume production rate of antiprotons in the galaxy by multiplying the flat antiproton spectrum by the number density of neutralinos ρ_χ / m_χ and the number of jets per decay $N_{\rm jets}$, and dividing by the lifetime $\tau_\chi = 1/\Gamma$,

$$\frac{dQ_{\bar{p}}}{dE} = \frac{dN_{\bar{p}}}{dE} \frac{\rho_\chi}{m_\chi} \frac{N_{\rm jets}}{\tau_\chi}. \tag{5}$$

6. Antiproton propagation in the galaxy

We use a standard diffusion model for antiproton propagation, that of Webber *et al.* [18] as implemented by Chardonnet *et al.* [19] to describe propagation of antiprotons through the galaxy. The interstellar flux of antiprotons is approximated by multiplying the source spectrum by the antiproton velocity and an appropriate diffusion time. The diffusion constant in the galaxy is energy dependent and can be approximated by the following expression,

$$K = 6 \times 10^{27} \left(1 + \frac{p}{3Z\mathrm{GV}}\right)^{0.6} \mathrm{cm}^2\ \mathrm{sec}^{-1}, \tag{6}$$

where p is the antiproton rigidity. The region of tangled magnetic fields responsible for diffusion can be approximated as a cylinder about 20 kpc in radius and 2 kpc high. This gives a diffusion time for antiprotons of

$$t_d = 6 \times 10^{15} \left(1 + \frac{p}{3\mathrm{GV}}\right)^{-0.6} \mathrm{sec}. \tag{7}$$

The interstellar flux is now given by

$$\Phi_{\bar{p}} = \frac{1}{4\pi} \frac{dQ_{\bar{p}}}{dE} v_{\bar{p}} t_d. \tag{8}$$

This flux is in units of antiprotons $\mathrm{s}^{-1}\ \mathrm{cm}^{-2}\ \mathrm{GeV}^{-1}\ \mathrm{sr}^{-1}$.

We now consider solar modulation, which is the effect that the solar wind and magnetic field have on the low energy spectrum. We use a simple model to account for this [20]. The observed antiproton energy E and momentum k are given in terms of the interstellar energy $E_{\overline{p}}$.

$$\begin{aligned} E_{\overline{p}} &= k_c \ln \frac{k+E}{k_c+E_c} + E_c + \Delta E, \qquad k < k_c; \\ E_{\overline{p}} &= E + \Delta E, \qquad k \geq k_c. \end{aligned} \tag{9}$$

We adopt values of the critical momentum k_c and energy shift ΔE that correspond to the period of minimum activity in the 11 year solar cycle, k_c =1.105 GeV and ΔE=495 MeV. This gives the largest flux in the inner solar system, thus the conservative bound on the couplings.

The solar modulated flux is given by

$$\Phi_{\overline{p}}(E) = \frac{E^2 - m_p^2}{E_{\overline{p}}^2 - m_p^2} \Phi_{\overline{p}}(E_{\overline{p}}). \tag{10}$$

7. Results

We proceed as follows. We first choose a data point for the $\overline{p}$ flux at the earth [21]. These data have large error bars, but will improve with the next generation of experiments to be done in space [22]. According to the BESS data, at an energy of 1.4 GeV the antiproton flux at the top of the atmosphere is 6.4 $\times 10^{-7}[\mathrm{cm}^2\,\mathrm{s\,sr\,GeV}]^{-1}$. We denote this flux as $\Phi_{\mathrm{obs}}(E)$. We then determine the unmodulated $\overline{p}$ energy $E_{\overline{p}}$ from the values of the observed energy E and momentum k. This interstellar energy is what we use to evaluate the diffusion time t_d and the velocity v.

We insist that the flux of antiprotons from neutralino decays be less than the observed flux. Including solar modulation corrections, we impose

$$\frac{v_{\overline{p}} t_d}{4\pi} \frac{\rho_\chi}{m_\chi} \frac{N_{\mathrm{jets}}}{\tau_\chi} \frac{dN_{\overline{p}}}{dE} \frac{E^2 - m_{\overline{p}}^2}{E_{\overline{p}}^2 - m_{\overline{p}}} < \Phi_{\mathrm{obs}}(E). \tag{11}$$

We do not make a correction for the spallation production of antiprotons. The data are consistent with the picture that cosmic ray antiprotons are produced in spallation reactions, but there are many theoretical and experimental uncertainties. Thus there is no reliable model for the amount we should subtract. We thus obtain a conservative bound on the couplings.

Using the data we have chosen, we obtain a model independent bound on the lifetime of a hadronically decaying relic with mass $m_\chi \gtrsim 10$ GeV

$$\tau_\chi \gtrsim \tau_\chi^{\overline{p}} = 7.9 \times 10^{28}\,\mathrm{s}\, \frac{N_{\mathrm{jets}} f_\chi}{m_\chi/\mathrm{GeV}}; \tag{12}$$

where we take the mass fraction of neutralinos in the galactic dark halo to be $f_\chi = \Omega_\chi h^2$. Since we assume that the neutralino is in the galactic halo at present, this bound applies for lifetimes longer than the age of the Universe t_0. In other words, we can exclude lifetimes in the range $t_0 \leq \tau_\chi \leq \tau_\chi^{\bar{p}}$ provided $t_0 \leq \tau_\chi^{\bar{p}}$.

For these large lifetimes, our bounds are in general better than those from diffuse photons [23], $\tau_\chi \gtrsim 7 \times 10^{25}\text{s}\, N_{\text{jets}} \Omega_\chi h^2$, and diffuse neutrinos [24], $\tau_\chi \gtrsim 10^{25}\text{s}\,(m_\chi/\text{GeV})^{1.7} f_\chi$, except for some supersymmetric models with neutralino masses in the TeV range.

We can limit each element of the two coupling matrices λ' and λ'' separately, by choosing the couplings such that all of the measured $\bar{p}$ flux comes from the neutralino decay via a single channel. Our excluded range of lifetimes then translates into the following *excluded* range of couplings λ

$$\lambda_{\bar{p}} \leq |\lambda| \leq \lambda_0, \tag{13}$$

where

$$\lambda_{\bar{p}} \simeq 3.48 \times 10^{-15} \sqrt{\frac{m_\chi \tilde{\tau}}{N_{\text{jets}} f_\chi}} \tag{14}$$

is the upper limit on λ coming from the antiproton flux and

$$\lambda_0 \simeq 1.8 \times 10^{-9} \sqrt{\tilde{\tau}} \tag{15}$$

is the value of λ for which the neutralino lifetime equals the age of the universe, which bounds the validity of our analysis. In the previous formulas, m_χ is in GeV, $\tilde{\tau} = |\lambda|^2 \tau_\chi$ is in seconds, and the age of the universe is taken to be 10^{10} yr.

For virtually all models we find that $\lambda_{\bar{p}}$ is smaller than λ_0 by at least three orders of magnitude. Conservatively taking $N_{\text{jets}} = 2$ we find that

$$\frac{\lambda_0}{\lambda_{\bar{p}}} = 7.3 \times 10^5 \sqrt{\frac{\Omega_\chi h^2}{m_\chi/\text{GeV}}} \gtrsim 600. \tag{16}$$

Further demanding that the neutralino provide a significant amount of dark matter, namely $\Omega_\chi h^2 > 0.025$, we expand the excluded range to $\lambda_0/\lambda_{\bar{p}} \gtrsim 4000$.

We have plotted the upper bound on λ''_{112} as a function of the neutralino mass for 10^4 supersymmetric models in figure 2. The results are similar for all generations in the R-parity violating couplings, except for λ''_{3jk} which shows the top quark threshold (see figure 3). When the neutralino is lighter than the top quark, there is no bound on λ''_{3jk} at tree level. We need to explore a one loop decay to get a bound [25].

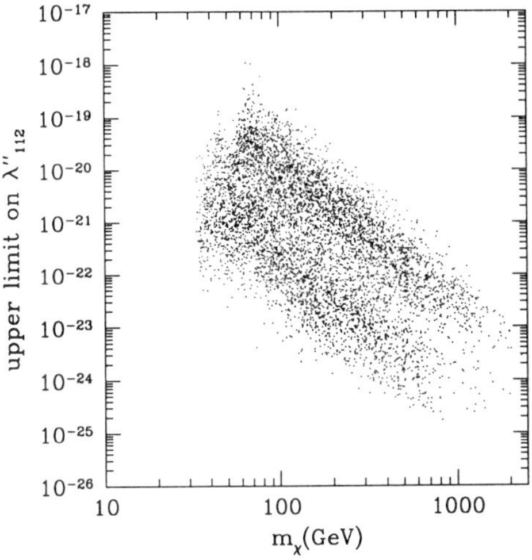

Figure 2. Bounds on λ''_{112} from cosmic ray antiprotons as a function of neutralino mass for 10^4 SUSY models. The scatter plots for λ''_{ijk} are similar when $i \neq 3$. The scatter plots for λ'_{ijk} are also similar.

In our sample of supersymmetric models, we find the absolute upper limits

$$|\lambda'_{ijk}|_{\bar{p}} \lesssim 2 \times 10^{-18} \quad (i, j, k \neq 3) \tag{17}$$
$$|\lambda'_{3jk}|_{\bar{p}} \lesssim 1 \times 10^{-18} \quad (j, k \neq 3) \tag{18}$$
$$|\lambda'_{i3k}|_{\bar{p}} \lesssim 4 \times 10^{-19} \quad (k \neq 3) \tag{19}$$
$$|\lambda'_{ij3}|_{\bar{p}} \lesssim 3 \times 10^{-19} \tag{20}$$
$$|\lambda''_{ijk}|_{\bar{p}} \lesssim 1 \times 10^{-18} \quad (i, k \neq 3) \tag{21}$$
$$|\lambda''_{ij3}|_{\bar{p}} \lesssim 1 \times 10^{-19} \quad (i \neq 3). \tag{22}$$

The apparent absence of an absolute upper bound on λ''_{3ij} is due to our neglect of one-loop diagrams. These upper limits depend on the neutralino mass, and improve considerably with higher masses. We find the following power law fits as a function of the neutralino mass

$$|\lambda'_{ijk}|_{\bar{p}} \lesssim 6 \times 10^{-22}(m_\chi/\mathrm{TeV})^{-3.1} \quad (i, j, k \neq 3) \tag{23}$$
$$|\lambda'_{3jk}|_{\bar{p}} \lesssim 3 \times 10^{-22}(m_\chi/\mathrm{TeV})^{-3.1} \quad (j, k \neq 3) \tag{24}$$
$$|\lambda'_{i3k}|_{\bar{p}} \lesssim 8 \times 10^{-23}(m_\chi/\mathrm{TeV})^{-3.2} \quad (k \neq 3) \tag{25}$$

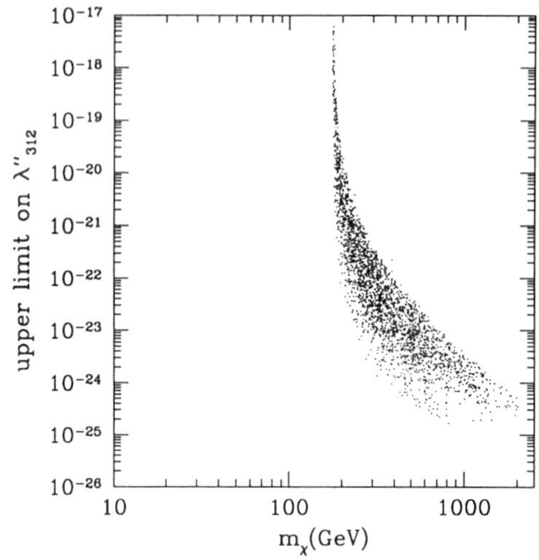

Figure 3. Bounds on λ''_{312} from cosmic ray antiprotons as a function of the neutralino mass for 10^4 SUSY models. The threshold effect is clearly seen.

$$|\lambda'_{ij3}|_{\bar{p}} \quad \lesssim 1 \times 10^{-22}(m_\chi/\text{TeV})^{-3.0} \tag{26}$$

$$|\lambda''_{ijk}|_{\bar{p}} \quad \lesssim 3 \times 10^{-22}(m_\chi/\text{TeV})^{-2.7} \qquad (i, k \neq 3) \tag{27}$$

$$|\lambda''_{ij3}|_{\bar{p}} \quad \lesssim 3 \times 10^{-23}(m_\chi/\text{TeV})^{-2.7} \qquad (i \neq 3). \tag{28}$$

In comparison, the most stringent laboratory constraints, those from proton decay [26], bound only the product $|\lambda'\lambda''| \lesssim 10^{-24}(m_0/\text{TeV})^2$, where m_0 is the sfermion mass scale. These bounds are six orders of magnitude weaker than ours, and do not apply to a single coupling. On the other hand, because our analysis only applies to lifetimes longer than the age of the Universe, our excluded region spans only 3 orders of magnitude. As mentioned in the introduction, the remaining gap of 3 orders of magnitude may be filled by other astrophysical considerations.

Finally, we comment on the dependence of our bounds on model parameters. In figure 4 we plot the dependence on the squark mass scale m_0. As expected, the bound is stricter for lighter squarks. In general, the bounds are better in the gaugino region. In the higgsino region, the bounds improve with heavier generations as the Yukawa couplings become larger. The dependence on other model parameters is unimportant in comparison to the sfermion mass scale and to the neutralino composition.

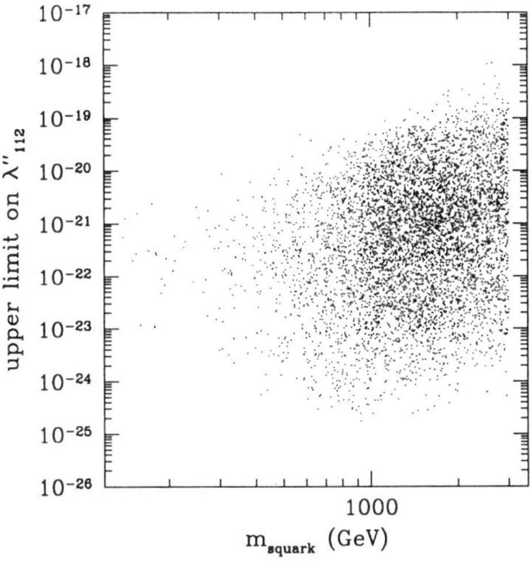

Figure 4. Bounds on λ''_{112} from cosmic ray antiprotons as a function of the squark mass scale for 10^4 SUSY models.

8. Conclusions

We obtain strong constraints on the hadronic R-parity violating couplings from observations of the cosmic ray antiproton flux. Our analysis applies in models where the neutralino lifetime is longer than the age of the universe. We are able to exclude a range $\lambda_{\bar{p}} \leq |\lambda| \leq \lambda_0$, where $\lambda_{\bar{p}}$ is the upper limit from the antiproton flux and λ_0 is the validity boundary coming from the age of the universe. Using a large selection of model parameters, we find absolute upper limits $|\lambda'_{ijk}|_{\bar{p}} \lesssim 2 \times 10^{-18}$ and $|\lambda''_{ijk}|_{\bar{p}} \lesssim 10^{-18}$ when there is no top quark in the final state, and $|\lambda'_{i3k}|_{\bar{p}} \lesssim 4 \times 10^{-19}$ when there is. In virtually all models, we find that $\lambda_0 \gtrsim 10^3 \lambda_{\bar{p}}$. Our bounds lie many orders of magnitude below the laboratory constraints coming from proton decay, and apply to each R-parity violating coupling separately.

Acknowledgments

P. G. thanks Prof. H. V. Klapdor-Kleingrothaus for the invitation to this conference.

References

[1] E. A. Baltz and P. Gondolo, Phys. Rev. D **57**, 7601 (1998).

[2] For a review see H. E. Haber and G. L. Kane, Phys. Rep. **117**, 75 (1985).

[3] L. Bergström and P. Gondolo, Astropart. Phys. **5**, 263 (1996).

[4] L. Bergström, J. Edsjö, and P. Gondolo, Phys. Rev. D **55**, 1765 (1997).

[5] Particle Data Group, R. M. Barnett *et al.*, Phys. Rev. D **54**, 1 (1996).

[6] G. Cowan (ALEPH Collaboration), talk presented at the special CERN particle physics seminar on physics results from the LEP run at 172 GeV, 25 February, 1997, http://alephwww.cern.ch/ALPUB/seminar/Cowan-172-jam/cowan.html.

[7] CLEO Coll., M. S. Alam *et al.*, Phys. Rev. Lett. **71**, 674 (1993); **74**, 2885 (1995).

[8] J. Edsjö and P. Gondolo, Phys. Rev. D **56**, 1879 (1997).

[9] G. Jungman, M. Kamionkowski, and K. Griest, Phys. Rep. **267**, 195 (1996) and references therein.

[10] E. A. Baltz and P. Gondolo, Phys. Rev. D **57**, 2969 (1998).

[11] TPC Coll., H. Aihara *et al.*, Phys. Rev. Lett. **61** 1263 (1988).

[12] TOPAZ Coll., R. Itoh *et al.*, Phys. Lett. B **345**, 335 (1995).

[13] ALEPH Coll., D. Buskulic *et al.*, Z. Phys. C **55**, 209 (1992); OPAL Coll., R. Akers *et al.*, *ibid.* **63**, 181 (1994); DELPHI Coll., P. Abreu *et al.*, Nucl. Phys. B **444**, 3 (1995).

[14] Yu. L. Dokshitzer, V. A. Khoze, and S. I. Troyan, Int. J. Mod. Phys. A **7**, 1875 (1996); Yu. L. Dokshitzer, V. A. Khoze, A. H. Mueller, and S. I. Troyan, *Basics of Perturbative QCD* (Editions Frontières, 1991).

[15] S. Lupia and W. Ochs, Phys. Lett. B **365**, 339 (1996); V. A. Khoze, S. Lupia, and W. Ochs, *ibid.* **394** 179 (1997); W. Ochs, in *ICHEP 96: Proceedings, July 1996, Warsaw, Poland*, edited by Z. Ajduk and A. K. Wroblewski (World Scientific, 1997), p. 25-31; S. Lupia, to appear in *Proceedings of the 33rd Eloiastron Workshop "Universality features in multihadron production and the leading effect"*, October 1996, Erice, Italy, preprint hep-ph/9702215 (1997).

[16] T. Sjöstrand, Computer Physics Commun. **82**, 74 (1994).

[17] A. Bottino, C. Favero, N. Fornengo, and G. Mignola, Astropart. Phys. **3**, 77 (1995); G. Mignola, preprint astro-ph/9611138 (1996).

[18] W. R. Webber, M. A. Lee, and M. Gupta, Astrophys. J. **390**, 96 (1992).

[19] P. Chardonnet, G. Mignola, P. Salati, and R. Taillet, Phys. Lett. B **384**, 161 (1996).

[20] J. S. Perko, Astron. Astrophys. **184**, 119 (1987).

[21] BESS Coll., A. Moiseev *et al.*, Astrophys. J. **474**, 479 (1997); IMAX Coll., J. W. Mitchell *et al.*, Phys. Rev. Lett. **76**, 3057 (1996).

[22] AMS Coll., S. Ahlen *et al.*, Nuc. Instrum. Methods A **350**, 351 (1994); PAMELA Coll., O. Adriani *et al.*, in *XXIV International Cosmic Ray Conference*, (Roma, 1995), p.591.

[23] G. D. Kribs and I. Z. Rothstein, Phys. Rev. D **55**, 4435 (1997).

[24] P. Gondolo, Phys. Lett. B **295**, 104 (1992).

[25] E. A. Baltz, P. Gondolo, and M. Perelstein (work in progress).

[26] I. Hinchliffe and T. Kaeding, Phys. Rev. D **47**, 279 (1993).

Appendix: Decay rate

The differential decay rate of the neutralino is given by a standard three-body phase space factor multiplied by a spin-summed squared amplitude. We average over the initial neutralino spin and sum over the final fermion spins, denoting the sum by a primed summation symbol.

$$\frac{\partial^2 \Gamma}{\partial x_1 \partial x_2} = \frac{m_\chi}{256\pi^3} \sum_{\text{spins}}{}' |\mathcal{M}|^2 . \tag{29}$$

Here $x_{1,2} = 2E_{1,2}/m_\chi$ and $E_{1,2}$ are two final state fermion energies.

We will now present the spin-summed squared amplitudes for decays via the $U^c D^c D^c$ and LQD^c operators. The sfermion mass eigenstates $\tilde{f}_\kappa$, with $\kappa = 1, \ldots, 6$, are related to left- and right-handed sfermions through the 6×3 mixing matrices $\Gamma^f_{R\kappa i}$ and $\Gamma^f_{L\kappa i}$, where $i = 1, 2, 3$ is a generation index. With no mixing, $\kappa = i$ is a left-handed sfermion with generation i and $\kappa = i + 3$ is a right handed sfermion with generation i. In general, the mixings are given by

$$\tilde{f}_\kappa = \Gamma^f_{R\kappa i} \tilde{f}_{Ri} + \Gamma^f_{L\kappa i} \tilde{f}_{Li}, \tag{30}$$

with inverse

$$\tilde{f}_{Ri} = \Gamma^{f*}_{R\kappa i} \tilde{f}_\kappa, \qquad \tilde{f}_{Li} = \Gamma^{f*}_{L\kappa i} \tilde{f}_\kappa . \tag{31}$$

There is no a right-handed sneutrino, so there is only a left handed sneutrino mixing matrix and $\kappa = 1, 2, 3$. In the case of left-right mixing without flavor mixing, the non-zero elements are

$$\begin{aligned} \Gamma^f_{Lii} &= \Gamma^f_{R(i+3)i} &= \cos\theta_f , \\ \Gamma^f_{Rii} &= -\Gamma^f_{L(i+3)i} &= \sin\theta_f , \end{aligned} \tag{32}$$

where θ_f is the left-right mixing angle for sfermion $\tilde{f}$. We will need the following coupling constants,

$$g^A_{\chi f_i \kappa} = \Gamma^f_{R\kappa i} g^{RA}_{\chi f_i} + \Gamma^f_{L\kappa i} g^{LA}_{\chi f_i}, \tag{33}$$

where A can be L or R and

$$g^{LL}_{\chi f_i} = -\sqrt{2}\left[(q_f - T_3)\, g' N_{\chi 1} + T_3 g N_{\chi 2}\right], \tag{34}$$

$$g^{LR}_{\chi u_i} = g^{RL}_{\chi u_i} = -\frac{g m_{u_i} N_{\chi 4}}{\sqrt{2} m_W \sin\beta}, \tag{35}$$

$$g^{LR}_{\chi d_i} = g^{RL}_{\chi d_i} = -\frac{g m_{d_i} N_{\chi 3}}{\sqrt{2} m_W \cos\beta}, \tag{36}$$

$$g^{RR}_{\chi f_i} = +\sqrt{2} q_f g' N_{\chi 1}. \tag{37}$$

Here T_3 is the third component of weak isospin, u is an up-type quark or neutrino ($T_3 = +1/2$), and d is a down-type quark or charged lepton ($T_3 = -1/2$). N_{ij} is the 4×4 neutralino mixing matrix, with the convention that all mass eigenvalues are positive. The charges are $q_u = 2/3$, $q_d = -1/3$, $q_\nu = 0$, and $q_e = -1$. The only non-zero neutrino couplings are $g^{LL}_{\chi\nu_i}$. We also need the weighted propagators

$$G^{AB}_{f_i} = \sum_\kappa \Gamma^{f*}_{A\kappa i} \Delta^*_{\tilde{f}_\kappa} g^B_{\chi f_i \kappa}, \tag{38}$$

with A and B either R or L and $\Delta_{\tilde{f}_\kappa} = (p^2 - m^2_{\tilde{f}_\kappa} + i m_{\tilde{f}_\kappa}\Gamma_{\tilde{f}_\kappa})^{-1}$ is the sfermion propagator. The squared amplitudes are now given as follows.

$$\sum_{\text{spins}}{}' \left|\mathcal{M}(\chi \to \bar{u}_i \bar{d}_j \bar{d}_k)\right|^2 = 8 c_f |\lambda''_{ijk}|^2 \times \text{Re}\Big\{ \tag{39}$$

$$\begin{aligned}
& d_j \cdot d_k \left[(|G^{RL}_u|^2 + |G^{RR}_u|^2)\chi \cdot u + 2 G^{RL}_u G^{RR*}_u m_\chi m_u\right] \\
& + u \cdot d_k \left[(|G^{RL}_{d_j}|^2 + |G^{RR}_{d_j}|^2)\chi \cdot d_j + 2 G^{RL}_{d_j} G^{RR*}_{d_j} m_\chi m_{d_j}\right] \\
& + u \cdot d_j \left[(|G^{RL}_{d_k}|^2 + |G^{RR}_{d_k}|^2)\chi \cdot d_k + 2 G^{RL}_{d_k} G^{RR*}_{d_k} m_\chi m_{d_k}\right] \\
& - G^{RR}_u G^{RR*}_{d_j} g(u, d_k, d_j, \chi) - G^{RL}_u G^{RR*}_{d_j} (d_k \cdot d_j) m_\chi m_u \\
& - G^{RR}_u G^{RL*}_{d_j} (d_k \cdot u) m_\chi m_{d_j} - G^{RL}_u G^{RL*}_{d_j} (\chi \cdot d_k) m_u m_{d_j} \\
& - G^{RR}_u G^{RR*}_{d_k} g(u, d_j, d_k, \chi) - G^{RL}_u G^{RR*}_{d_k} (d_k \cdot d_j) m_\chi m_u \\
& - G^{RR}_u G^{RL*}_{d_k} (u \cdot d_j) m_\chi m_{d_k} - G^{RL}_u G^{RL*}_{d_k} (\chi \cdot d_j) m_u m_{d_k} \\
& - G^{RR}_{d_j} G^{RR*}_{d_k} g(d_j, u, d_k, \chi) - G^{RL}_{d_j} G^{RR*}_{d_k} (d_k \cdot u) m_\chi m_{d_j} \\
& - G^{RR}_{d_j} G^{RL*}_{d_k} (d_j \cdot u) m_\chi m_{d_k} - G^{RL}_{d_j} G^{RL*}_{d_k} (\chi \cdot u) m_{d_j} m_{d_k} \Big\},
\end{aligned}$$

with the color factor $c_f = 6$, and $g(a, b, c, d) = (a \cdot b)(c \cdot d) - (a \cdot c)(b \cdot d) + (a \cdot d)(b \cdot c)$. Particle four-momenta have been denoted by the particle letter and we suppress unambiguous indices.

$$\sum_{\text{spins}}{}' \left|\mathcal{M}(\chi \to e^+_i \bar{u}_j d_k)\right|^2 = 2 c_f |\lambda'_{ijk}|^2 \times \text{Re}\Big\{ \tag{40}$$

$$
\begin{aligned}
&e\cdot d\left[(|G_u^{LL}|^2+|G_u^{LR}|^2)\chi\cdot u+2G_u^{LL}G_u^{LR*}m_\chi m_u\right]\\
&+u\cdot e\left[(|G_d^{RL*}|^2+|G_d^{RR*}|^2)\chi\cdot d+2G_d^{RL*}G_d^{RR}m_\chi m_d\right]\\
&+u\cdot d\left[(|G_e^{LL}|^2+|G_e^{LR}|^2)\chi\cdot e+2G_e^{LL}G_e^{LR*}m_\chi m_e\right]\\
&-G_u^{LL}G_d^{RR}g(u,e,d,\chi)-G_u^{LR}G_d^{RR}(e\cdot d)m_\chi m_u\\
&-G_u^{LL}G_d^{RL}(e\cdot u)m_\chi m_d-G_u^{LR}G_d^{RL}(\chi\cdot e)m_u m_d\\
&-G_u^{LL}G_e^{LL*}g(e,d,u,\chi)-G_u^{LR}G_e^{LL*}(e\cdot d)m_\chi m_u\\
&-G_u^{LL}G_e^{LR*}(u\cdot d)m_\chi m_e-G_u^{LR}G_e^{LR*}(\chi\cdot d)m_u m_e\\
&-G_d^{RR*}G_e^{LL*}g(e,u,d,\chi)-G_d^{RL*}G_e^{LL*}(e\cdot u)m_\chi m_d\\
&-G_d^{RR*}G_e^{LR*}(d\cdot u)m_\chi m_e-G_d^{RL*}G_e^{LR*}(\chi\cdot u)m_d m_e\Big\},
\end{aligned}
$$

$$
\begin{aligned}
&\sum_{\text{spins}}{}'\left|\mathcal{M}(\chi\to\overline{\nu}_i\overline{d}_j d_k)\right|^2=2c_f|\lambda'_{ijk}|^2\times\mathrm{Re}\Big\{ \qquad (41)\\
&\nu\cdot d\left[(|G_{d_j}^{LL}|^2+|G_{d_j}^{LR}|^2)\chi\cdot\overline{d}+2G_{d_j}^{LL}G_{d_j}^{LR*}m_\chi m_{d_j}\right]\\
&+\overline{d}\cdot\nu\left[(|G_{d_k}^{RL*}|^2+|G_{d_k}^{RR*}|^2)\chi\cdot d+2G_{d_k}^{RL*}G_{d_k}^{RR}m_\chi m_{d_k}\right]\\
&+\overline{d}\cdot d|G_\nu^{LL}|^2\chi\cdot\nu\\
&-G_{d_j}^{LL}G_{d_k}^{RR}g(\overline{d},\nu,d,\chi)-G_{d_j}^{LR}G_{d_k}^{RR}(\nu\cdot d)m_\chi m_{d_j}\\
&-G_{d_j}^{LL}G_{d_k}^{RL}(\nu\cdot\overline{d})m_\chi m_{d_k}-G_{d_j}^{LR}G_{d_k}^{RL}(\chi\cdot\nu)m_{d_j}m_{d_k}\\
&-G_{d_j}^{LL}G_\nu^{LL*}g(\nu,d,\overline{d},\chi)-G_{d_j}^{LR}G_\nu^{LL*}(\nu\cdot d)m_\chi m_{d_j}\\
&-G_{d_k}^{RR*}G_\nu^{LL*}g(\nu,\overline{d},d,\chi)-G_{d_k}^{RL*}G_\nu^{LL*}(\nu\cdot\overline{d})m_\chi m_{d_k}\Big\},
\end{aligned}
$$

where the color factor is given by $c_f = 3$ in both. Note that in Eqs. (40) and (41) for the LQD^c decays the factor in front is 2, not 8 as in Eq. (39) for the $U^cD^cD^c$ decay. The $U^cD^cD^c$ process is identical for λ''_{ijk} and λ''_{ikj}, whereas each non-zero element of the LQD^c matrix λ'_{ijk} allows a different decay channel.

LEP Constraints on Neutralino Relic Densities, and the Fate of Higgsino Dark Matter

Toby Falk

Department of Physics, University of Wisconsin, Madison, WI 53706, USA

Abstract. We examine the current state of neutralino dark matter and consider how the LEP constraints on the Minimal Supersymmetric Standard Model parameters are squeezing the available dark matter regions. We also show how cosmological constraints augment bounds coming from collider searches to further constrain the MSSM parameter space.

1. Introduction

If R-parity is conserved, the Lightest Supersymmetric Particle (LSP) is stable. In many supersymmetric models, the LSP tends to have a cosmologically interesting relic density and is a good dark matter candidate [1, 2], and it is a very appealing feature of the Minimal Supersymmetric Standard Model that it can naturally provide an answer to the dark matter question. However, due the characteristically large relic abundance of the LSP's, the MSSM is also naturally susceptible to cosmological constraints, either because the relic density of LSP's is so large that it is in conflict with a gross cosmological feature, e.g. the age of the universe, or because one may be able to detect the interactions of the LSP with terrestrial detectors or detect the self-interactions of the LSP in the galactic halo or in the cores of the earth or the sun [3].

The MSSM contains in principle two neutral dark matter candidates, the sneutrino $\tilde{\nu}$, and the lightest neutralino $\tilde{\chi}^0$. The sneutrino, though, has already been excluded[1] as a dark matter candidate by a combination of LEP bounds and direct detection experiments [4] . The lightest neutralino may be either gaugino-like (in particular $\widetilde{B}$-like) or Higgsino-like, and both the phenomenology and cosmology of neutralinos depend strongly on the neutralino composition. In this talk we will study the question of how viable these remaining MSSM dark matter candidates are in light of recent LEP data.

2. Neutralinos

In general, the neutralinos are linear combinations of the neutral gauginos and Higgsinos,

$$\chi_i = \beta_i \tilde{B} + \alpha_i \tilde{W} + \gamma_i \tilde{H}_1 + \delta_i \tilde{H}_2, \qquad i = 1, \ldots, 4 \tag{1}$$

In this notation, the gaugino purity of a neutralino χ_i is defined to be $\sqrt{{\alpha_i}^2 + {\beta_i}^2}$, and its Higgsino purity $\sqrt{{\gamma_i}^2 + {\delta_i}^2}$. In the $(\tilde{B}, \tilde{W}^3, \tilde{H}_1^0, \tilde{H}_2^0)$ basis, the neutralino mass matrix takes the form

$$\begin{pmatrix} M_1 & 0 & -M_Z s_\theta c_\beta & M_Z s_\theta s_\beta \\ 0 & M_2 & M_Z c_\theta c_\beta & -M_Z c_\theta s_\beta \\ -M_Z s_\theta c_\beta & M_Z c_\theta c_\beta & 0 & -\mu \\ M_Z s_\theta s_\beta & -M_Z c_\theta s_\beta & -\mu & 0 \end{pmatrix}, \tag{2}$$

where s_θ $(c_\theta) = \sin\theta_W$ $(\cos\theta_W)$, s_β $(c_\beta) = \sin\beta$ $(\cos\beta)$, and where gaugino mass unification implies $M_1 = 5/3 \tan^2\theta_W M_2$. The coefficients $\alpha_i, \ldots, \delta_i$ in (1) depend on $\tan\beta$ and on the soft SUSY-breaking mass parameters M_1, M_2 and μ, which appear in (2), and Fig. 1 displays the regions of high gaugino and Higgsino purity in the (μ, M_2) plane. In the limit $|\mu| \gg M_i$, the lightest neutralino is gaugino-like, specifically a $\widetilde{B}$, i.e. $\beta_1 \approx 1$ in (1), and this is typically the case in mSUGRA and in models with gauge mediated supersymmetry breaking (GMSB). In Fig. 1, contours of constant $\widetilde{B}$ purity are displayed as long-dashed lines. In the opposite limit, $M_i \gg |\mu|$, the lightest neutralino is Higgsino-like, and for small to moderate M_2, the lightest neutralino is the particular Higgsino combination defined by $\tilde{S}^0 \equiv \tilde{H}_1^0 \cos\beta + \tilde{H}_2^0 \sin\beta$, i.e., $\gamma = \cos\beta$ and $\delta = \sin\beta$, with $m_{\tilde{S}_0} \to \mu \sin 2\beta$ [7]. Contours of $\tilde{S}^0$ purity are displayed in Fig. 1 as dash-dotted lines. For large M_2, the lightest neutralino is the state $\tilde{H}_{12} \equiv \frac{1}{\sqrt{2}}(\tilde{H}_1^0 \pm \tilde{H}_2^0)$, i.e., $\delta = \pm\gamma = \pm 1/\sqrt{2}$ for $\text{sgn}(\mu) = \pm 1$, with $m_{\tilde{H}_{12}} \to |\mu|$ [8], and contours of $\tilde{H}_{12}$ purity are shown as short-dashed lines in Fig. 1.

[1] Outside the MSSM, these constraints may be evaded. See [5],[6].

In mSUGRA, the relationship between μ and M_2 is determined from the conditions of gaugino unification and sfermion and Higgs mass unification, along with the requirement of correct radiative electroweak symmetry breaking, and this relationship is shown in Fig. 1 as a thick solid line. As suggested above, the resulting contour lies in the gaugino region, and the heavier the neutralino, the more pure its $\widetilde{B}$ content. Also shown in Fig. 1 are contours of constant chargino mass, where 91 GeV represents the current LEP lower bound on $m_{\chi^\pm}$ [10], and constant neutralino mass. As a preview of the small size of the Higgsino dark matter region we'll be discussing, it consists of a subset of the shaded region between the $m_{\chi^\pm} = 91\,\text{GeV}$ and $m_\chi = M_W$ contours.

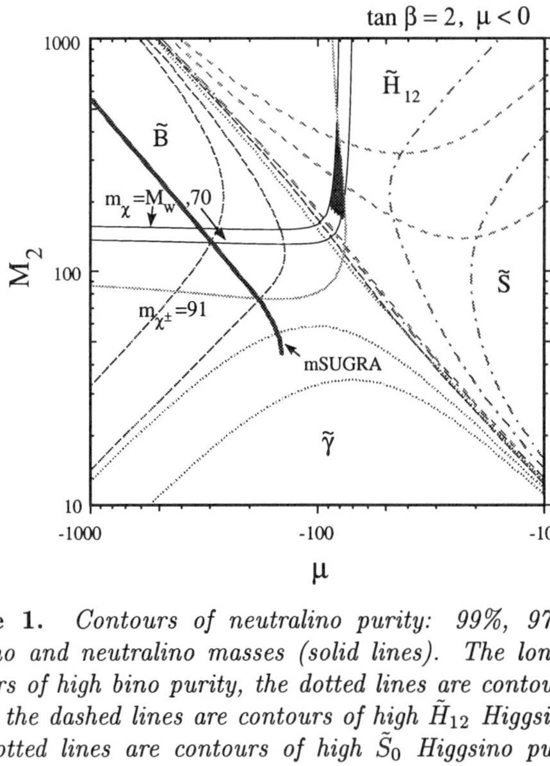

Figure 1. *Contours of neutralino purity: 99%, 97% and 75%, and chargino and neutralino masses (solid lines). The long-dashed lines are contours of high bino purity, the dotted lines are contours of high photino purity, the dashed lines are contours of high $\tilde{H}_{12}$ Higgsino purity, and the dash-dotted lines are contours of high $\tilde{S}_0$ Higgsino purity. Also shown are contours of constant m_χ and $m_{\chi^\pm}$ and the dependence of μ on M_2 in mSUGRA. We've taken $m_0 = 100\,\text{GeV}$.*

3. LEP Bounds

Recent runs at LEP at center-of-mass energies of 172 and 183 GeV have excluded large areas of MSSM parameter space, and subsequent runs at

$\sim$ 190 and 200 GeV will push the bounds even further. In this section we summarize the LEP bounds that we implement in this talk.

- Searches for chargino pair production at LEP 172 and LEP 183 have excluded chargino masses less than 86 GeV [9] and 91 GeV[10] respectively, modulo two loopholes. The first loophole is relevant when the neutralino is a gaugino and occurs when the mass of the sneutrino is close to that of the chargino. The lower limit on $m_{\chi^\pm}$ is reduced as $m_{\tilde{\nu}}$ is reduced toward $m_{\chi^\pm}$ from above, due to destructive interference with the t-channel sneutrino exchange process, and then disappears entirely for $m_{\chi^\pm} > m_{\tilde{\nu}} \gtrsim m_{\chi^\pm} - 3\,\mathrm{GeV}$, in which case $\tilde{\chi}^\pm$ decay is dominated by $\tilde{\nu}+$ soft lepton final states. Bounds on chargino production reappear when $m_{\chi^\pm} - m_{\tilde{\nu}} \gtrsim 3\,\mathrm{GeV}$ and the lepton detection efficiency picks up again. The second loophole is relevant in the Higgsino region and occurs when $\Delta M = m_{\chi^\pm} - m_\chi$ is small, and it again is due to the reduction in detection efficiency when the mass difference between the produced particle and its supersymmetric decay product is small. The effect of each of these loopholes will be discussed in more detail later on.

- Searches for associated neutralino production can provide strong bounds when the neutralino is a Higgsino. In the Higgsino region of interest for this talk, associated neutralino production is dominated by $e^+e^- \to \chi\chi_{2,3,4}$ and corresponds essentially to $m_\chi + m_{\chi'_H} = 182\,\mathrm{GeV}$, where χ'_H is the lightest mainly-Higgsino state among the $\chi_{2,3,4}$.

- Searches for Higgs production provide a lower limit of $m_h > 88\,\mathrm{GeV}$ at low $\tan\beta$ [11]. Bounds from Higgs searches are particularly constraining at low $\tan\beta$, where the experimental bounds are strongest and where the tree level Higgs mass is small. Here radiative corrections to the Higgs mass [12] must be very large, leading to strong lower bounds on the masses of the sfermions, and in particular the stops. However, the extraction of the radiatively corrected Higgs mass in the MSSM has an uncertainty of $\sim 2\,\mathrm{GeV}$, so we conservatively take $m_h > 86\,\mathrm{GeV}$ as our experimental lower limit at low $\tan\beta$.

- We also implement bounds coming from searches for sfermion production, in particular slepton [13] and stop production [14], as well as constraints on the sneutrino mass from the Z width [17].

4. Gaugino Dark Matter

As we have seen above, the lightest neutralino tends in many instances to be a $\widetilde{B}$, as is the case in mSUGRA, in particular. For our numerical examples, we will restrict our attention to mSUGRA, although the qualitative features apply to more general cases, so long as the LSP is a $\widetilde{B}$. The relic abundance of neutralinos is given by

$$\Omega_\chi h^2 \sim \frac{3 \times 10^{-10}\,\mathrm{GeV}^{-2}}{\langle \sigma_{\mathrm{ann}} v \rangle} < 0.3, \tag{3}$$

where $\langle \sigma_{\mathrm{ann}} v \rangle$ is the thermally averaged neutralino annihilation cross-section at the time the neutralinos fall out of chemical equilibrium with the thermal bath. The inequality in (3) comes from the requirements that the age of the universe t_U be greater than 12 Gyr and that the total $\Omega < 1$.

In the early universe, $\widetilde{B}$ annihilation is typically dominated by sfermion exchange into fermion pairs (Fig. (2)). Due to the Majorana nature of the neutralinos, this process generally exhibits a p-wave suppression [1] (but see [15, 16]) which decreases $\langle \sigma_{\mathrm{ann}} v \rangle$, and increases $\Omega_\chi h^2$, by roughly an order of magnitude. Now, as the sfermion masses are increased, the relic abundance of neutralinos increases, and for sufficiently heavy sfermions, $\Omega_\chi h^2$ violates the bound (3). Thus the requirement that $t_U > 12$ Gyr translates into an upper bound on the sfermion masses.

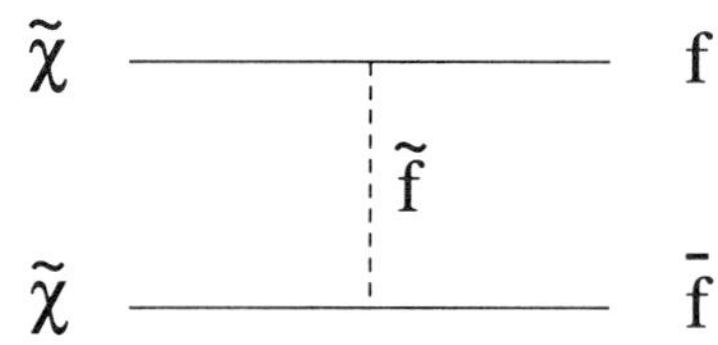

Figure 2. *The dominant contribution to $\widetilde{B}$ annihilation.*

In mSUGRA, the sfermion and gaugino masses are given by

$$m^2_{\tilde f} = m_0^2 + C_{\tilde f}\, m^2_{1/2} + O(m_Z^2), \tag{4}$$

$$(M_1, M_2) \approx (0.4, 0.8)\, m_{1/2}, \tag{5}$$

where m_0 and $m_{1/2}$ are the common scalar and gaugino mass parameters, and the coefficients $C_{\tilde f}$ are determined by the RGE evolution of the sfermion masses from the unification scale to the electroweak scale. The remaining parameters in mSUGRA are the common trilinear scalar mass parameter A_0, which appears in the sfermion mass matrices, $\tan\beta$, and the sign of μ. The magnitude of μ is fixed in mSUGRA by the conditions of correct electroweak symmetry breaking, as seen in Fig. 1. The sfermion masses are typically insensitive to A_0, so in mSUGRA, the bound on t_U translates simply into a upper bound on m_0 and $m_{1/2}$. Because the cosmological constraint provides upper limits on the soft masses, whereas particle searches typically give lower limits on the same parameters, the two types of constraints are nicely complementary. Lastly, the Higgs mass is given by

$$m_h^2 = m_Z^2 \cos^2 2\beta \; + \; \text{rad. corr.}\,(m_{\tilde{t}_i}^2, m_t, A_t), \tag{6}$$

where the dominant part of the radiative corrections depends logarithmically on the stop masses. At low $\tan\beta$, the tree level Higgs mass is small, and the Higgs mass constraint then imposes severe lower bounds on the stop masses, which translates into lower bounds on $m_{1/2}$ and m_0.

Fig. 3 summarizes [18] the LEP 172 constraints on the mSUGRA parameter space for $\tan\beta = 2$, $\mu < 0$. The solid contour at the left of the figure represents the combined chargino and slepton bounds[2]. At large m_0, the chargino bound reaches the kinematic limit, but as m_0 is decreased, the sneutrino mass falls, and the first loophole described in section 3 reduces the bound below 86 GeV. At $m_0 \sim 75$ GeV, the sneutrino mass drops below that of the chargino, and the bound retreats rapidly to the left; at lower m_0, the slepton bound comes into play and shuts the door to lower $m_{1/2}$. At this relatively low value of $\tan\beta$, both the chargino and slepton bounds are dominated by the Higgs mass constraint (~ 76 GeV), shown as the thick solid line stretching almost vertically at $m_{1/2} \sim 250$ GeV. For lower values of $\tan\beta$, the Higgs bound moves rapidly to the right, whereas for higher $\tan\beta$, the chargino bound moves to the right, and the Higgs bound retreats to the left and crosses beneath the combined chargino/slepton bound at $\tan\beta \sim 3$.

It is interesting to consider how low a neutralino mass these combined constraints admit. The Higgs bound forces one to very large $m_{1/2}$ at low m_0, giving a heavy neutralino; however, one can make the stops heavy by taking m_0 very large, rather than $m_{1/2}$, and so the Higgs contour does bend to the left at very large m_0 and strikes the chargino contour. At $\tan\beta = 2$, this occurs at $m_0 \sim 800$ GeV. Consequently, in the absence of an independent upper limit on m_0, the lower bound on m_χ is set by the intersection of the Higgs contour with the chargino contour at large m_0 and yields a minimum neutralino mass of 42 GeV. Fig. 4 displays [18] the lower bound on m_χ as a function of $\tan\beta$, given different sets of theoretical and experimental constraints. The purely experimental lower bound (i.e. not yet using cosmology) in mSUGRA is given by the solid line labelled UHM (as a reminder that it is the Unification of the Higgs Masses with the sfermion masses which leads to $|\mu|$ being an output, rather than an input, in mSUGRA).

The dark shaded area in Fig. 3 delimits the cosmologically preferred region with $0.1 < \Omega_\chi h^2 < 0.3$. The upper limit, as described above, comes

[2] All the curves in Fig. 3 are computed using tree level neutralino and chargino masses. They shift by a few GeV when 1-loop radiative corrections to the masses are included.

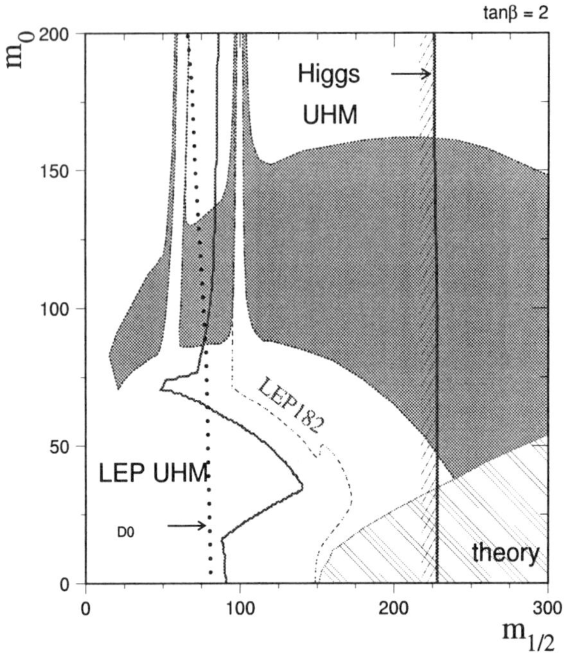

Figure 3. *We display for* $\mu < 0$ *and* $\tan\beta = 2$ *the domains of the* $(m_{1/2},\, m_0)$ *plane that are excluded by the LEP 172 chargino and selectron searches in mSUGRA, the domains that are excluded by Higgs searches, the regions that are excluded cosmologically because* $m_{\tilde{\tau}_R} < m_\chi$, *and the domains that have relic neutralino densities in the favoured range* $0.1 < \Omega_\chi h^2 < 0.3$

from an upper bound on the age of the universe. The lower limit is less a bound *per se* than a preference, stemming from the desire to have the neutralinos comprise a significant fraction of the dark matter. The two narrow vertical channels at $m_{1/2} \sim 65\,\mathrm{GeV}$ and $m_{1/2} \sim 100\,\mathrm{GeV}$ arise from s-channel neutralino annihilation on the Higgs and Z^0 poles, respectively. Note that top of the shaded area continues to fall for $m_{1/2} > 300\,\mathrm{GeV}$, and it intersects with the "theory" excluded area (where a stau is the LSP) at $m_{1/2} \sim 425\,\mathrm{GeV}$. Thus, as advertised, $\Omega_\chi h^2 < 0.3$ yields an upper bound on both m_0 and $m_{1/2}$. This independent upper bound on m_0 forbids the large m_0 solution to the Higgs mass constraint (modulo a very narrow region on the Z^0 pole) and dramatically increases the lower bound on the neutralino mass for values of $\tan\beta$ where the Higgs mass bound is significant. This is seen in Fig. 4, where the branch of the solid curve labelled "cosmo+UHM" is the result of including both the constraints from particle searches at LEP 172 and the requirement $\Omega_\chi h^2 < 0.3$.

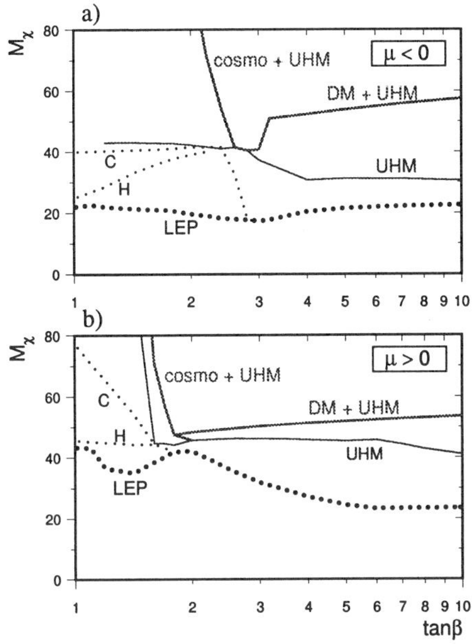

Figure 4. *Various lower limits on m_χ using different experimental and theoretical inputs are compared, as functions of* $\tan\beta$*, for both (a) $|\mu| < 0$ and (b) $|\mu| > 0$. The solid curves represent bounds in mSUGRA, including purely experimental bounds ("UHM"), separately augmented by cosmological ("cosmo+UHM") and dark matter ("DM+UHM") constraints. The dotted curves show the impact of Higgs (H) and cosmological (C) constraints in the MSSM.*

Further, for low enough $\tan\beta$, the Higgs bound moves entirely to the right of the dark-shaded region, and for $\mu < 0$, the cosmologically allowed range[3] with $\Omega_\chi h^2 < 0.3$ is actually *incompatible* with the Higgs lower limit on $m_{1/2}$ for tan$\beta \lesssim 1.7$. We conclude that in mSUGRA there is no range of $m_{1/2}$ compatible with all the constraints provided by the LEP 172 particle searches and the upper bound on the cosmological relic density, for sufficiently small tan$\beta \lesssim 1.7$. Hence, *there is a lower bound* $\tan\beta \gtrsim 1.7$, if all these constraints are applied simultaneously. Similarly, for $\mu > 0$, the bound from the LEP 172 searches is $\tan\beta \gtrsim 1.4$. These bounds have since improved. Updated chargino and slepton contours from LEP 183 are shown as dashed lines in Fig. 3. The chargino contour displayed is the kinematic limit of 91 GeV, and as discussed above, overestimates the chargino bound for low m_0. The LEP 183 Higgs bound is off the right side of the figure and in fact is very close to the intersection point of the $\Omega_\chi h^2 < 0.3$

[3] The cosmological upper bound on $m_{1/2}$ varies only weakly with tanβ for tan$\beta \lesssim 2$.

and "theory" contours. The LEP 183 Higgs bound then combines with the cosmological constraint to yield the updated [19] lower bounds

$$\tan\beta \gtrsim \begin{cases} 2.0 & \mu < 0 \\ 1.65 & \mu > 0 \end{cases} \tag{7}$$

Lastly, the branch of the solid curve labelled "DM+UHM" in Fig. 4 exhibits the effect of additionally including the constraint that there be a significant amount of neutralino dark matter, $\Omega_\chi h^2 > 0.1$. As $\tan\beta$ is increased, the chargino bound moves to the right, as does the Higgs pole, and the Higgs and Z_0 poles widen, eventually merging to form one large pole region, at $\tan\beta \sim 3$. The kink in the "DM+UHM" line emerges from the necessity to sit on the right side of the combined pole region, after the merging, in order to have sufficient dark matter.

5. Higgsino Dark Matter

5.1. Introduction

As shown in Fig. 1, when the neutralino is a Higgsino, it approaches the particular combination $\tilde{H}_{12}$. There are in principle two regions in the (μ, M_2) plane where $\tilde{H}_{12}$ can provide an interesting relic density. For large M_2, they correspond to $|\mu| < M_W$ but with $m_{\chi^\pm}$ above the LEP limit, and $|\mu| \gtrsim 1$ TeV [8, 20]. The intermediate-mass Higgsino states are not of cosmological interest, because of their rapid annihilations to W and Z pairs. We have little to add concerning the very heavy $\tilde{H}_{12}$ states, but the lighter Higgsinos lie directly in the region where the current LEP runs are eating away at the parameter plane, and thus it is of interest to examine how much of the light Higgsino parameter space remains consistent with the LEP bounds.

In the far Higgsino limit, the chargino and neutralino spectrum simplifies, and for $M_2 \gg |\mu|$,

$$\begin{aligned} m_{\chi_1^0} &\approx m_{\chi_2^0} \approx |\mu| \\ m_{\chi^\pm} &\approx |\mu|, \end{aligned} \tag{8}$$

A non-zero gaugino component provides some splitting of the spectrum, and at tree level,

$$m_{\chi_1^0} < m_{\chi^\pm} < m_{\chi_1^0}. \tag{9}$$

The LEP 182 lower bound of 91 GeV on the chargino mass then imposes a comparable lower bound on m_χ in the limit of pure Higgsinos. However, if $m_\chi > m_W$, the annihilation channel $\tilde{\chi}^0\tilde{\chi}^0 \rightarrow W^+W^-$ is available to neutralinos at their time of decoupling in the early universe. Neutralinos in fact love to annihilate into W pairs, since this process is not p-wave suppressed, in contrast to neutralino annihilation into fermion pairs, and so $\Omega_\chi h^2 \sim \langle\sigma_{\rm ann} v\rangle^{-1}$ is greatly reduced above the W threshold. Consequently, Higgsinos with masses above m_W (and below $\sim 1\,\mathrm{TeV}$) are not viable dark matter candidates. For this reason, LEP chargino searches are fatally squeezing Higgsino dark matter. In this section we examine over how much of the MSSM parameter space Higgsino dark matter remains viable, and comment on the fate of Higgsino dark matter when the last LEP runs are complete.

5.2. Loop Corrections to Ino Masses

Constraints on Higgsino dark matter are sensitive to loop corrections to the neutralino and chargino masses, and this sensitivity appears in two separate pieces of the analysis. First, as mentioned in section 3, there is a loophole which appears in the chargino experimental bounds when the mass of the chargino is close to the mass of the neutralino, and looking at eqn. 8, this is exactly the case in the Higgsino limit, where the chargino and neutralino masses are both close to $|\mu|$. The second panel of Fig. 5 displays [19] the LEP 172 lower bound on the chargino mass as a function of $\Delta M = m_{\chi^\pm} - m_\chi$, for $\tan\beta = 2$ and $m_0 = 200\,\mathrm{GeV}$. Characteristically, the bound on $m_{\chi^\pm}$ drops for small ΔM, and essentially disappears for $\Delta M < 5\,\mathrm{GeV}$. The first panel in Fig. 5 displays the chargino bound as a function of M_2, where $|\mu|$ is fixed by the chargino mass. For larger M_2 the neutralino is more pure Higgsino, the masses of the chargino and neutralino are more degenerate, and the bound on $m_{\chi^\pm}$ drops.

Since the chargino bounds are closely dependent on ΔM, it is important to consider the 1-loop radiative corrections to the neutralino and chargino masses in the Higgsino region. The 1-loop corrections to the neutralino and chargino masses have been computed [21, 22, 23], and a comparison of the dotted and solid curves in Fig. 5 shows their effect on the chargino bound. As expected, at large M_2, where ΔM is small, the effect is significant, while for $M_2 \leq 400\,\mathrm{GeV}$, there is little effect. The overlap of the two curves in the second panel of Fig. 5 demonstrates that it is really the quantity ΔM to which the chargino bound is sensitive.

The second piece of the analysis which is sensitive to the radiative corrections is the computation of the Higgsino relic abundance. Unlike in the gaugino case, the approximate degeneracy of the neutralino and chargino spectrum (8) in the Higgsino region requires the inclusion of the next-to-lightest and next-to-next-to-lightest states and their co-annihilations with the LSP (and self-annihilations) in the calculation of $\Omega_\chi h^2$ [24, 25]. That this effect is particularly important is due again to the p-wave suppression

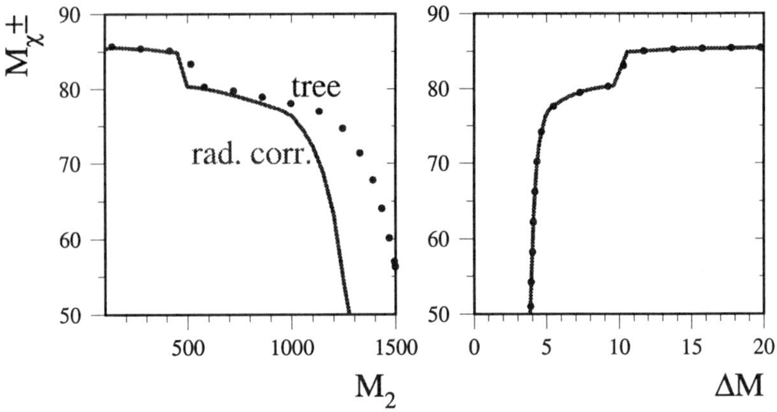

Figure 5. *The experimental limit on $m_{\chi^\pm}$ as function of M_2 and as a function of $\Delta M \equiv m_{\chi^\pm} - m_\chi$, for fixed $m_0 = 200\,\text{GeV}$ and $\tan\beta = 2$. The dotted line comes from a tree-level analysis, and the solid line is obtained using an ad hoc parameterization of the experimental efficiency, in conjunction with the radiatively-corrected mass formulae.*

of the annihilation of neutralinos into fermions. Expanding the thermally averaged annihilation cross-section in powers of (T/m_χ),

$$\langle \sigma_{\text{ann}} v \rangle_T = A + B\,(T/m_\chi) + \ldots, \tag{10}$$

one finds the zeroth order piece A suppressed by m_f^2. Neutralinos with masses of interest for dark matter fall out of chemical equilibrium with the thermal bath when the temperature is $\sim (1/20 - 1/25) \times m_\chi$, so the effect of p-wave suppression, then, is an order of magnitude reduction in the annihilation rate. By contrast, the co-annihilation processes

$$\begin{array}{rcl} \tilde{\chi}_1^0\,\tilde{\chi}_1^\pm & \longrightarrow & e^+\nu,\ \ldots \\ \tilde{\chi}_1^0\,\tilde{\chi}_2^0 & \longrightarrow & f\bar{f} \end{array}$$

do not involve the annihilation of identical particles and so do not exhibit p-wave suppression. However, the number density of the heavier scattering states is Boltzmann suppressed at low temperatures, and so the ratio of co-annihilation to annihilation rates goes as

$$R_{\tilde{\chi}_1^0 \tilde{\chi}_1^\pm} / R_{\tilde{\chi}_1^0\,\tilde{\chi}_1^0} \sim (T/m_\chi)\, e^{-\Delta M/T} \sim 25\, e^{-25(\Delta M/m_\chi)}, \tag{11}$$

and similarly for $\tilde{\chi}_1^0\,\tilde{\chi}_2^0$ co-annihilation[4]. For a degenerate spectrum, this amounts to better than an order of magnitude increase in the annihilation

[4] There are additional factors which suppress the annihilation rate for Higgsinos and increase the ratio $R_{\tilde{\chi}_1^0 \tilde{\chi}_1^\pm} / R_{\tilde{\chi}_1^0\,\tilde{\chi}_1^0}$.

rate, whereas if the mass difference $\Delta M/m_\chi$ is as much as 25 percent, this ratio of rates is less than 0.05. Clearly there is a tremendous sensitivity of $\Omega_\chi h^2$ to ΔM, and consequently to the radiative corrections to the chargino and neutralino masses.

Lastly, we note that the loophole in the chargino bound occurs for small ΔM, where co-annihilations typically make the Higgsino relic abundance tiny. By contrast, the relic density turns out to be significant only for larger ΔM, where the chargino bounds are strongest. It is therefore difficult to wriggle out of the the chargino mass constraints by appealing to this loophole in the chargino limits, while still preserving Higgsino dark matter.

5.3. The Fate of Higgsino Dark Matter

We now explore the Higgsino dark matter regions which survive the LEP bounds. In order to conservatively estimate their area in the following, we take a large sfermion mass $m_0 = 1\,\mathrm{TeV}$. This has both the effect of minimizing the impact of the Higgs mass constraint by producing large radiative corrections to the Higgs mass and of minimizing the contributions to neutralino annihilation through the neutralino's (small) gaugino component. We take a large pseudoscalar mass $m_A = 1\,\mathrm{TeV}$, similarly to give a large Higgs tree-level mass and to minimize the effect of s-channel pseudoscalar annihilation. We take A_t at the quasi-fixed point $\sim 2.25\,M_2$. This is a very good approximation at low $\tan\beta$, and the small flexibility one has to vary A_t at larger $\tan\beta$ doesn't substantially impact the allowed dark matter regions. Lastly, our working definition of a Higgsino is that $p^2 = \gamma_i^2 + \delta_i^2 > 0.81$. We note that since Higgsino and gaugino purities add in quadrature to 1, a Higgsino purity of 0.9 (our choice) corresponds to a gaugino purity of 0.44, so our choice is not too restrictive.

In Fig. 6a we display [19] for $\mu < 0$ contours of constant chargino mass $m_{\chi^\pm} = 91\,\mathrm{GeV}$, $m_\chi + m_{\chi'_H} = 182\,\mathrm{GeV}$, Higgs mass m_h, and Higgsino purity, along with a contour of constant $\Omega_\chi h^2 = 0.1$, for $\tan\beta = 2$. We see that the dashed lines in Fig. 6a representing the chargino and associated neutralino mass contours bound one away from small $|\mu|$, while the Higgs mass limit bounds one away from small M_2. The latter is particularly restrictive at low $\tan\beta$, where the tree-level Higgs mass is small, and thus where radiative corrections to m_h must be enhanced by taking large stop masses. The solid contour contains the region which leads to a significant neutralino relic density $\Omega_\chi h^2 \geq 0.1$, and its limited range in M_2 is a result of co-annihilations. For larger values of M_2, the neutralino is a purer Higgsino, and the masses of both the lightest chargino and the next-to-lightest neutralino approach the neutralino mass from above, enhancing the effect of co-annihilations that deplete the relic Higgsino abundance. For larger values of $|\mu|$, the relic density is suppressed by annihilations into W pairs. The hashed contours in Fig. 6a represent Higgsino purities. Note

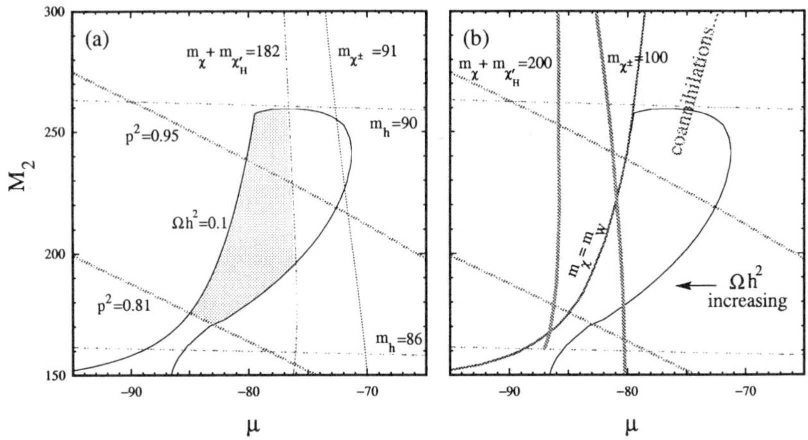

Figure 6. *Survey of experimental and cosmological constraints in the* μ, M_2 *plane, focusing on Higgsino dark matter for tan*$\beta = 2$ *and* $\mu < 0$ *In a) we plot the radiatively-corrected contours for* $m_{\chi^\pm} = 91\,\mathrm{GeV}$*, for* $m_\chi + m_{\chi'_H} = 182\,\mathrm{GeV}$*, for selected values of* m_h *and the Higgsino purity* p*, and for* $\Omega_\chi h^2 = 0.1$*. The shaded regions yield a Higgsino which satisfies the mass and relic density constraints described in the text. In b) we show the expected bounds from LEP 200.*

the limited range of μ for which the mass and relic density constraints are satisfied.

The combined effects of the above constraints, corresponding to the shaded regions of Fig. 6, are displayed for different values of tanβ in Fig. 7. We find that there are no consistent Higgsino dark matter candidates for $\tan\beta \leq 1.8$ or ≥ 2.5 for $\mu < 0$, or for any value of $\tan\beta$ for $\mu > 0$. The Higgs mass constraint cuts off the bottom of the allowed regions at low $\tan\beta$. When $\mu < 0$ it becomes a relevant constraint for $\tan\beta < 2.0$ (its effect can be seen in the flat lower edge of the $\tan\beta = 1.9$ contour) and is responsible for the complete disappearance of the allowed region when $\tan\beta \leq 1.8$. Within the allowed regions displayed, the relic densities generally increase as $|\mu|$ is increased, until the neutralino mass, whose minimum value here is $\sim 71\,\mathrm{GeV}$, becomes greater than m_W, at which point the W^+W^- annihilation channel opens, driving the relic $\Omega_\chi h^2$ below 0.1. In Fig. 6b, we plot the contour $m_\chi = m_W$, which makes evident the drop in $\Omega_\chi h^2$ above the W pair threshold[5] . In any event, $\Omega_\chi h^2$ is never greater than 0.12 anywhere in the allowed regions for $\mu < 0$.

In Fig. 6b we show the same plot as Fig. 6a, but with an estimate of the bounds that should be achieved at LEP's eventual run at $\sim 200\,\mathrm{GeV}$

[5] Sub-threshold annihilation of neutralinos, not included in this analysis, exclude a further slice of Higgsino dark matter on the right-hand side of the $m_\chi = m_W$ contour[26].

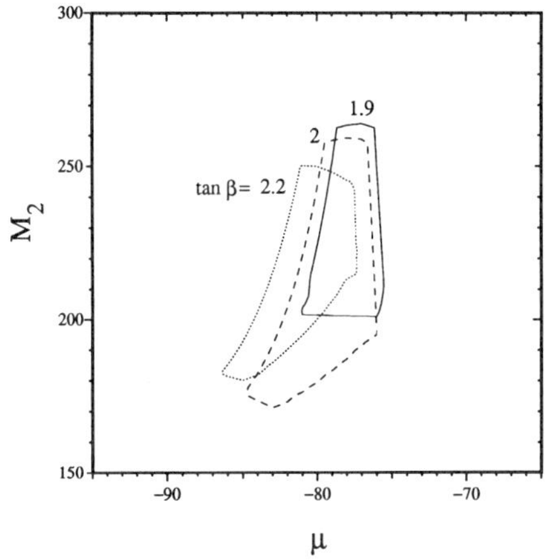

Figure 7. *The regions of the μ, M_2 plane allowed by the constraints shown in the previous figure are shown for several different values of tanβ. There are no consistent choices of Higgsino parameters for* $\tan\beta < 1.8$ *or* > 2.5 *for* $\mu < 0$*, or for any value of* $\tan\beta$ *for* $\mu > 0$.

center-of-mass energy. We see that the associated neutralino production bound of $m_\chi + m_{\chi'_H} = 200\,\text{GeV}$ excludes entirely the remaining allowed Higgsino dark matter region at $\tan\beta = 2$ for $\mu < 0$. It is also apparent that at this value of $\tan\beta$, the projected Higgs mass bound of $\sim 105\,\text{GeV}$, which lies well off the top of the plot, will play the same role. And, in fact, the entire set of allowed Higgsino dark matter regions displayed in Fig. 7 for all $\tan\beta$ are excluded by either one of the conditions $m_h > 100\,\text{GeV}$ or $m_\chi + m_{\chi'_H} > 200\,\text{GeV}$ alone. The first of these bounds should be achieved even if LEP falls somewhat shy of 200 GeV in its final run.

In Fig. 8 we show the equivalent plot to Fig. 6a, for $\mu > 0$; in this case the Higgsino purities are lower, and the entire dark matter region falls below the purity cutoff. This turns out to be the case for all $\tan\beta$; that is, for all values of $\tan\beta$, the only regions of parameter space for $\mu > 0$ which have a significant amount of neutralino dark matter $\Omega_\chi h^2 > 0.1$ have either a mixed or gaugino-like lightest neutralino. Lastly, we find the Higgsino dark matter regions are even more restricted for both $\mu < 0$ and $\mu > 0$ if the gaugino masses are related by $M_1 = M_2$ [27]. For a more detailed discussion of both the $\mu < 0$ and $\mu > 0$ cases, see reference [19].

6. Summary and Outlook

We've seen that R-parity conserving SUSY models still provide a good dark matter candidate. Neutralinos are of particular cosmological interest

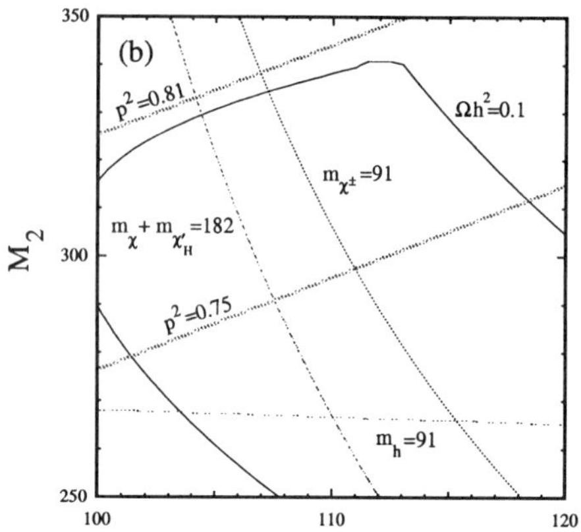

Figure 8. *Same as Fig. 6a for* $\mu > 0$

when the lightest neutralino is a gaugino, in particular a bino, and current experiments are probing into the heart of the gaugino dark matter region. Cosmological constraints combine neatly with experimental bounds, and taken together they exclude all values of $\tan\beta < 2$ (1.65) for $\mu < 0$ (> 0) in mSUGRA. By contrast, Higgsino dark matter is being fatally squeezed by LEP. Already small, restricted to an area $\sim 10\,\mathrm{GeV}$ wide in μ and for only $\tan\beta$ between 1.8 and 2.5 for $\mu < 0$, the remaining Higgsino dark matter regions (with masses $< 1\,\mathrm{TeV}$) will be finally excluded by either of the conditions $m_h > 100\,\mathrm{GeV}$ or $m_\chi + m_{\chi'_H} > 200\,\mathrm{GeV}$, bounds which should be achieved by LEP 200.

Acknowledgements

I would like to thank the organizers of the Symposium on Lepton and Baryon Number Violation for an interesting an enjoyable conference. I wish also to gratefully acknowledge John Ellis, Gerardo Ganis, Keith Olive, and Michael Schmitt, with whom this work was done in collaboration. This work was supported in part by DOE grant DE–FG02–95ER–40896 and in part by the University of Wisconsin Research Committee with funds granted by the Wisconsin Alumni Research Foundation.

References

[1] H. Goldberg, *Phys. Rev. Lett.* **50** (1983) 1419;

[2] J. Ellis, J.S. Hagelin, D.V. Nanopoulos, K.A. Olive and M. Srednicki, *Nucl. Phys.* **B238** (1984) 453.

[3] For a review of detection schemes, see G. Jungman, M. Kamionkowski, and K. Greist, *Phys. Rep.* **267** (1996) 195.

[4] T. Falk, K. A. Olive, M. Srednicki, *Phys. Lett.* **339** (1994) 248.

[5] L. J. Hall, T. Moroi, H. Murayama,*Phys. Lett.* **424** (1998) 305.

[6] T. Han, R. Hempfling, *Phys. Lett.* **415** (1997) 161.

[7] J. Ellis, J.S. Hagelin, D.V. Nanopoulos, K.A. Olive and M. Srednicki, *Nucl. Phys.* **B238** (1984) 453.

[8] K.A. Olive and M. Srednicki, *Phys. Lett.* **B230** (1989) 78; *Nucl. Phys.* **B355** (1991) 208; K. Greist, M. Kamionkowski and M.S. Turner, *Phys. Rev.* **D41** (1990) 3565.

[9] ALEPH Collaboration, D.Buskulic et al., *Phys. Lett.* **B373** (1996) 246; OPAL Collaboration, G. Alexander et al., *Phys. Lett.* **B377** (1996) 181.; L3 Collaboration, M. Acciarri et al., *Phys. Lett.* **B377** (1996) 289;; DELPHI Collaboration, P. Abreu et al., *Phys. Lett.* **B382** (1996) 323 .

[10] Presentations at the open LEPC session, Nov. 11th, 1997:
ALEPH collaboration, P. Dornan,
`http://alephwww.cern.ch/ALPUB/seminar/seminar.html`;
DELPHI collaboration, P. Charpentier,
`http://wwwinfo.cern.ch/~charpent/LEPC/`;
L3 collaboration, M. Pohl,
`http://hpl3sn02.cern.ch/conferences/talks97.html`;
OPAL collaboration, A. Honma,
`http://www.cern.ch/Opal/`.

[11] see. e.g. L3 Collaboration, CERN Preprint CERN-EP98-72 (1998).

[12] H.E. Haber, R. Hempfling and A.H. Hoang, *Z. Phys.* **C75** 539;
see also M. Carena, M. Quiros and C.E.M. Wagner, *Nucl. Phys.* **B461** (1996) 407.

[13] ALEPH Collaboration, R. Barate et al., CERN preprint EP/98-077 (1998).

[14] OPAL collaboration, K. Ackerstaff et al., CERN preprint PPE/97-046 (1997).

[15] T. Falk, R. Madden, K. A. Olive, M. Srednicki, *Phys. Lett.* **B318** (1993) 354.

[16] T. Falk, K. A. Olive, M. Srednicki , *Phys. Lett.* **B354** (1995) 99.

[17] J. Ellis, T. Falk, K.A. Olive and M. Schmitt, *Phys. Lett.* **B388** (1996) 97.

[18] J. Ellis, T. Falk, K.A. Olive and M. Schmitt, *Phys. Lett.* **B413** (1997) 355.

[19] J. Ellis, T. Falk, G. Ganis, K.A. Olive and M. Schmitt, hep-ph/9801445.

[20] P. Gondolo, J. Edsjo, hep-ph/9804459.

[21] A.B. Lahanas, K. Tamvakis and N.D. Tracas, *Phys. Lett.* **B324** (1994) 387.

[22] D. Pierce and A. Papadopoulos, *Phys. Rev.* **D50** (1994) 565 and *Nucl. Phys.* **B430** (1994) 278.

[23] D.M. Pierce, J.A. Bagger, K. Matchev and R.-J. Zhang, *Nucl. Phys.* **B491** (1997) 3.

[24] K. Griest and D. Seckel, *Phys. Rev.* **D43** (1991) 3191.

[25] S. Mizuta and M. Yamaguchi, *Phys. Lett.* **B298** (1993) 120.

[26] M. Drees, M.M. Nojiri, D.P. Roy and Y. Yamada, Phys. Rev. **D56** (1997) 276.

[27] G.L. Kane and James D. Wells, Phys. Rev. Lett. **76** (1996) 4458.

Direct Detection of Dark Matter in the Universe[1]

H.V. Klapdor-Kleingrothaus, Y. Ramachers

Max–Planck–Institut für Kernphysik
P.O.Box 10 39 80, D–69029 Heidelberg, Germany

Abstract. We present a short overview of dark matter direct detection experiments, concentrating on detectors and proposals of our group as examples. We also shortly discuss the implications of WIMP search results for accelerator experiments and vice versa.

1. Introduction

The direct detection of cold dark matter in the form of WIMPs (weakly interacting massive particles), has evolved into an intensive field of research with about 20 experiments running or starting in the near future (compare Tab. 1, for reviews see [1, 2, 3]). The search for WIMPs will automatically constrain the amount of R–parity violation in supersymmetric models [4, 5, 44]. Due to the fact that neutralinos would be required to be bound in the galaxy even today, neutralino decays via R–parity violating operators would have to be highly suppressed. The details, of course, depend on the neutralino mass and composition. For example, finding the neutralino at the lowest border of sensitivity of GENIUS (see below) would imply typical limits on R–parity violating couplings of the order of $10^{-(16-20)}$ for any of the λ_{ijk}, λ'_{ijk}, λ''_{ijk} in the superpotential. It can also give further important information on the parameter spaces of various kinds of SUSY models with R–parity conservation (MSSM, NMSSM , ...) [33–41,47–49] complementary to colliders.

[1] Talk presented by Y. Ramachers

Table 1. List of existing and planned direct detection experiments (updating [8]).

WIMP SEARCHES			
Experiment in progress	**Location**	**Detector Type**	**Ref.**
HD-Moscow	Gran Sasso, Italy	^{76}Ge ionization	[10]
Baksan	Prielbrusye, Russia	Ge ionization	[11]
Canfranc-NaI	Canfranc, Spain	NaI scintillator	[12]
COSME	Canfranc, Spain	Ge ionization	[13]
DAMA	Gran Sasso, Italy	NaI scintillator, liquid Xe scintillator, CaF_2 scintillator	[14, 15]
DEMOS	Sierra Grande, Argentina	Ge ionization	[16]
Milan	Gran Sasso, Italy	Cryogenic TeO_2 bolometer	[17]
UKDMC	Boulby mine, U.K.	NaI scintillator	[18]
Under construction			
HDMS	Gran Sasso, Italy	Ge ionization	[22]
CDMS	Stanford, U.S.	Cryogenic Ge/Si bolometer with ionization	[25]
CRESST	Gran Sasso, Italy	Cryogenic sapphire bolometer	[26]
EDELWEISS	Fréjus, France	Cryogenic Ge bolometer with ionization	[19]
ELEGANT VI	Oto cosmo obs, Japan	CaF_2 scintillator	[20]
Tokyo	Nokogiri-yama, Japan	Cryogenic LiF bolometer	[21]
LP–TPC	San Diego, USA	Direction–sensitive low–pressure TPC	[32]
Future			
PICASSO	Montreal, Canada	Superheated Freon droplets	[23]
ORPHEUS	Bern, Switzerland	Superconducting transition in tin granules	[24]
ROSEBUD	Canfranc, Spain	Cryogenic sapphire bolometer	[27]
SALOPARD	Canfranc, Spain	Superconducting transition in tin granules	[28]
SIMPLE	Paris, France	Superheated Freon droplets	[29]
CUORE	Gran Sasso, Italy	Cryogenic TeO_2 bolometer	[30]
UKDMC	Boulby mine, U.K.	Liquid Xe scintillator	[31]
GENIUS	Heidelberg, Germany	Ge ionization in liquid nitrogen	[43, 44]

The direct detection technique is defined by the observation of WIMP–nucleus elastic scattering events. These events deposit energy in the detector by the recoiling nucleus. The main uncertainties entering this technique stem from the astrophysical input data like the local WIMP halo–density, 0.3 – 0.7 GeV/cm^3 [6, 7] (note that the WIMP rates are directly proportional to this parameter) and the WIMP velocity distribution.

The mean WIMP velocity is of the order 10^{-3}c. Combined with expected masses above a few GeV/c^2, the interesting energy region for experiments results from kinematics as below $\sim$ 100 keV. The main challenge for all kinds of direct detection experiments is to reduce their background, induced for instance by radioactive impurities or neutrons.

Results of existing experiments give upper limits on the allowed WIMP–nucleon elastic scattering cross section as function of the WIMP mass. The conservative assumption underlying these limits is that the measured energy spectrum (usually in the units cpd/kg/keV [2]) in the interesting energy region consists of WIMP events. Without any further information a given energy spectrum produces a time independent exclusion curve in the cross section–mass plane (all limits in this article have a 90% confidence level) as soon as an experiment becomes background limited.

For the dark matter candidate provided by the MSSM, or the NMSSM, the lightest supersymmetric particle (LSP), the expectations for WIMP direct detection rates are at best of the order 1 cpd/kg but can be suppressed by many orders of magnitude depending on the candidate (for the neutralino as the LSP and expected rates, see [33–41,47–49]).

In order to compare theoretical expectations for WIMP-rates and experimental exclusion curves from different experiments (using different target materials), first of all, a common set of astrophysical input data has to be used. Second, to compare exclusion curves obtained with different target materials one has to separate nuclear properties due to the material from WIMP properties (both entering the elastic scattering cross section) [1, 35]. This is the reason why in recent works usually exclusion curves are given for WIMP–nucleon scattering, differentiating between the two cases of spin–independent WIMP–nucleon interactions σ^{W-N}_{scalar} and spin–dependent interactions σ^{W-N}_{axial}. Only experiments using the same target material can directly compare their results in a Rate–WIMP-mass diagram.

2. Experiments and results

According to the philosophy of using raw data, *i.e.* the measured energy spectrum without subtraction of events (with the exception of microphonic noise), to obtain a WIMP–nucleon cross section upper limit, the Heidelberg–Moscow–Experiment gives the best limit [9, 10].

The interesting experimental number which mainly determines obtainable WIMP–nucleon cross section limits is the background index near the detector threshold, typically at or below 10 keV. Background of the order 1 cpd/(kg keV) or below (Heidelberg–Moscow–Experiment: 0.042 cpd/(kg keV) between 15 keV – 40 keV) has been reached for raw data. Decrease of

[2] (cpd: counts per day)

detector thresholds improves the sensitivity for low mass WIMPs (<tens of GeV), whereas lowering the background index improves the whole WIMP exclusion curve.

The currently best limit (for spin–dependent as well as spin–independent interactions) results from the DAMA experiment [14]. This limit does not result from raw data but from a reduction of background due to the application of a highly elaborate pulse shape analysis. With the ability to use this scintillator–specific WIMP signature the DAMA group reduced their background considerably (see [14] and Fig. 1).

The first cryogenic experiment using information from the phonon and the ionization signal, the CDMS experiment, just started data taking [25]. Again, a signature of nuclear recoils is used to suppress background. Nuclear recoils in Germanium crystals deposit about 2/3 of their energy in lattice vibrations; 1/3 in ionization ([1, 42] and references therein). Energy dependent discrimination efficiencies for nuclear recoils against electron recoils are quoted as $\sim$ 99% or even higher. The background has been determined to be up to now – *after* rejection of electron and γ events – similar to the 1994 Heidelberg–Moscow–Experiment [9].

Recently our rather unusual prototype experiment started, the Heidelberg Dark Matter Search (HDMS) Experiment [22]. It is unusual in the sense that it will be a raw data experiment using the 'old–fashioned' (compare Tab. 1) Germanium ionization technique. The idea is to plug a small Ge crystal into the hole of a huge well-type Ge crystal with just a thin electric insulation plastic between the crystals.

We expect two effects to reduce the background: First, the outer well–type will act as an anticoincidence shield for the inner detector, thereby reducing Compton background from multiple scattered photons. A relative background reduction of the order of 5–6 by this anticoincidence has been measured. Second, the overall background level is expected to be reduced relative to usual Ge ionization detectors since in the immediate vicinity of the measurement crystal there is one of the radiopurest materials which we know, *i.e.* a second Ge crystal. This effect on the background level is so far undetermined and remains open until the experiment will reach a sufficient statistic (71.4 days so far) in its final location, the Gran Sasso underground laboratory. To conclude, we expect to give WIMP limits comparable to planned or started cryogenic experiments just by using raw data (microphonics subtracted).

A recent proposal, called GENIUS (Germanium Nitrogen Underground Setup) [43, 44], belongs to the category future direct detection dark matter experiment and to the most promising ones (see Fig. 1,2). It is originally designed to search for the neutrinoless double beta decay in enriched ^{76}Ge, but already with a detector mass of only 100 kg of natural Ge with its expected background reduction of a factor $10^{-3.25}$ relative to the Heidelberg–Moscow–Experiment, GENIUS would automatically give a strong WIMP limit of the order 0.003 cpd/kg from raw data [3, 43, 44]. GENIUS has the advantage of using a well known detector technology, basically HPGe detectors but immersed in liquid nitrogen as outer shielding [43, 44].

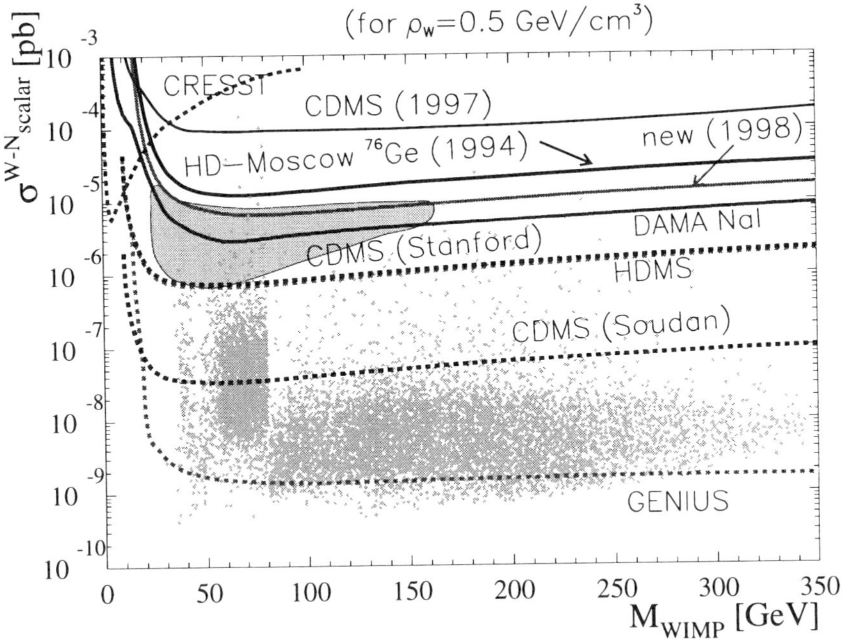

Figure 1. Comparison of already achieved WIMP–nucleon scalar cross section limits (solid lines): the Heidelberg–Moscow ^{76}Ge [9, 10], the recent CDMS nat. Ge [25] and the DAMA NaI result [14], including their evidence contour [15], in pb for scalar interactions as function of the WIMP–mass in GeV and of possible results from upcoming experiments (dashed lines for HDMS [22], CDMS (at different locations) the CRESST expectation [26] and GENIUS [44]). These experimental limits are also compared to expectations (scatter plot) for WIMP–neutralinos calculated in the MSSM framework with non–universal scalar mass unification [36] (all curves and dots scaled for $\rho = 0.5$ GeV/cm^3 for comparison of published data).

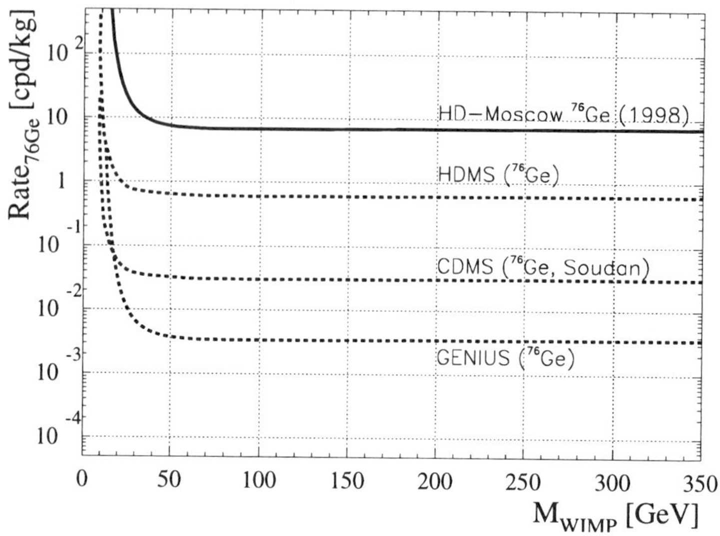

Figure 2. Comparison of rate–limits obtained (solid) or obtainable (dashed) as function of the WIMP–mass for Germanium experiments. In order to draw the picture consistently we had to assume that HDMS and CDMS would use the same isotope as the Heidelberg–Moscow–Experiment: ^{76}Ge. These possible rate–limits can roughly be compared to the expected rates from [33] since these are for ^{73}Ge. Note that the dashed lines only give approximate limits but they already show that the future experiments could indeed give strong constraints for the SUSY parameter space in the mSUGRA scheme, as discussed in [33].

3. On the relationship Neutralino Dark Matter ↔ Collider Experiments

With the assumption that WIMP dark matter consists of neutralinos as the LSP one can compare the impact of direct detection experiments on accelerator experiments looking for SUSY particles and vice versa. Since this comparison deals with regions of the MSSM parameter space one first has to specify the MSSM scheme which determines the parameter space. Well known for predictions and also comparison of dark matter experiments and accelerators is the minimal supergravity (mSUGRA) scheme for which highly developed tools exist like ISAJET (see [33, 45] and references therein). The mSUGRA scheme has a five dimensional parameter space (four numbers and the sign of the μ parameter are sufficient to fix a complete MSSM model, see for example [46]).

Unfortunately, predictions for direct detection rates are rather low (order 0.1 cpd/kg and below) [33, 34]. As soon as the unification conditions of the mSUGRA scheme are relaxed (non-universal scalar unification, non-universal gaugino mass scheme, etc.) the dimension of the SUSY parameter space grows (six or seven dimensions) and predictions for neutralino rates increase (sometimes up to already excluded rates, see Fig. 1 for a non-universal scalar mass unification scenario) [36, 37, 38, 39, 40]. With this MSSM scheme dependence in mind one can state that current WIMP limits just start to cut into the SUSY parameter space. The benefit from these marginal cuts is rather poor as long as there are no hints for a signal. Much more interesting is the possibility to reach a sensitivity for almost the whole parameter space like the GENIUS proposal. Automatically, GENIUS would be complementary to LHC for a neutralino WIMP.

As Baer and Brhlik [33] discuss for the mSUGRA scheme, complementarity between parameter regions testable by future dark matter and collider experiments results from a different SUSY parameter dependence (specifically, the $\tan\beta$ parameter) for neutralino detection. A dark matter detector will be particularly sensitive in regions of large $\tan\beta$ in the mSUGRA parameter space, where many conventional signals for supersymmetry in collider experiments are difficult to detect. If the parameter $\tan\beta$ is large, there is a significant probability that the first direct evidence for supersymmetry could come from direct dark matter detection experiments, rather than from collider searches for sparticles [33] (see also the discussion in [2]). On the other hand, extra information from collider physics about just the mass of only one SUSY particle, makes direct dark matter search very decisive. For example, a zero result with very sensitive detectors like GENIUS could be explained in the MSSM in the framework of SUSY only if the Higgs sector is very heavy [47]. It has been shown recently that the lightest neutralino could be detected in dark matter experiments also in Next–to–Minimal supersymmetric standard models with sizeable rates [48].

The detection of the neutralino at LHC would naturally have a big effect for WIMP experiments. Suddenly, one would know the kind of particle to look for but it would still be a fascinating question whether that candidate particle really constitutes the main ingredient of the universe. To answer that question directly, one needs direct detection dark matter experiments. More interesting from the point of view of a WIMP searcher is of course the case of detection of a WIMP before LHC has started. Such a scenario would have the disadvantage that one direct detection experiment can not determine the nature of the WIMP but measure its mass and the product of the local WIMP–halo density with the elastic scattering cross section, $n_W\,\sigma_{el}$. So the question to colliders in this case would be to check whether there is a neutralino or some other particle with corresponding mass and couplings to explain the WIMP search result. The maximum information about WIMPs from non–accelerator experiments can be obtained using different target nuclei [35, 49] for direct detection, find a WIMP signal in an indirect detection experiment (a neutrino telescope) [50] and combine the results.

4. Conclusion

Direct detection experiments just started to cut into the parameter space of the most favorite WIMP candidate, i.e. the supersymmetric neutralino. The present and near future experiments will be background limited to at most a factor of ten below the level which has already been reached by the Heidelberg–Moscow–Experiment for raw data. There might be a chance to improve background further using several discrimination procedures but their systematic effects are still under investigation especially for the unexplored region of extreme low background. As long as there does not appear a hint for a WIMP signal in these detectors, the benefit from cutting into the SUSY parameter space is rather poor since the lower limit for neutralino–nucleus cross section expectations is far below their sensitivity.

The GENIUS proposal from the Heidelberg group would be an outstanding future dark matter detector in the sense that it combines a high mass, extreme low background even for raw data (no systematics from discrimination procedures) and a well–known detector technology with HPGe detectors. Therefore a detector like GENIUS would be favored as a future dark matter detector and able to face the challenge of WIMP detection, in combination with future collider experiments.

References

[1] Smith P F and Lewin J D 1990 *Phys. Rep.* **187** 203
Bernabei R 1995 *Riv. Nuov. Cim.* **18** 1
Jungman G, Kamionkowski M and Griest K 1996 *Phys. Rep.* **267** 195
Lewin J D and Smith P F 1997 *Astropart. Phys.* **6** 87

[2] Klapdor–Kleingrothaus H V, Zuber K 1997 *Particle Astrophysics*, IOP (Bristol, Philadelphia)

[3] Klapdor–Kleingrothaus H V, Ramachers Y 1998 *Eur. Phys. J.* **A3** 85

[4] Gondolo P these proceedings

[5] Hirsch M and Klapdor–Kleingrothaus H V 1997 in *Proc. Workshop on Dark Matter in Astro–and Particle Physics* (Singapore: World Scientific) eds. Klapdor–Kleingrothaus H V and Ramachers Y, p. 640

[6] Flores R A 1988 *Phys. Lett.* **B215** 73

[7] Gates E I, Gyuk G and Turner M 1995 *ApJ* **449** L123

[8] Watson A 1997 *Science* **275** 1735

[9] Beck M *et al* (Heidelberg–Moscow coll.) 1994 *Phys. Lett.* **B336** 141

[10] Baudis L et al. (Heidelberg–Moscow coll.) 1998 submitted to *Phys. Rev.* **D**

[11] Osetrov S *et al* 1997 in *Proc. Workshop on Dark Matter in Astro– and Particle Physics* (Singapore: World Scientific) eds. Klapdor–Kleingrothaus H V and Ramachers Y

[12] Sarsa M L 1996 *Phys. Lett.* **B386** 458

[13] Morales E G A *et al* 1995 *Phys. Rev.* **D51** 1458

[14] Bernabei R *et al* 1997 *Phys. Lett.* **B389** 757

[15] Bernabei R *et al* 1998 *Phys. Lett.* **B424** 195

[16] Di Gregorio D E *et al* 1995 Nucl. Phys. **B38** (Proc. Suppl.) 394

[17] Alessandrello A *et al* 1997 *Phys. Lett.* **B408** 465

[18] Smith P F *et al* 1996 *Phys. Lett.* **B379** 299

[19] De Bellefon A *et al* 1996 *Nucl. Instrum. Methods* **A370** 230

[20] Hazama R *et al* 1995 in *Proceed. of the IV Int. Symposium on Weak and Electromagnetic Interactions in Nuclei (WEIN95)* (Singapore: World Scientific) eds. Ejiri H *et al* , p. 635

[21] Ootani W *et al* 1996 *Nucl. Instrum. Methods* **A372** 534

[22] Baudis L, Günther M, Hellmig J, Klapdor–Kleingrothaus H V, Müller A, Petry F, Ramachers Y and Strecker H 1997 *Nucl. Instrum. Methods* **A385** 265

[23] Hamel L A *et al* 1997 *Nucl. Instrum. Methods* **A388** 91

[24] Abplanalp A *et al* 1996 *Nucl. Instrum. Methods* **A370** 227

[25] Akerib D S et al. CDMS Collaboration 1997 in *Proc. of TAUP97* (Gran Sasso, Italy) edited by Bottino A, di Credico A and Monacelli P

[26] Cooper S 1996 in *ICHEP96* (Singapore: World Scientific); Seidel W *et al* 1993 J. Low Temp. Phys. **93** 797; 1994 *Phys. Lett.* **B323** 95; Bühler M *et al* 1996 *Nucl. Instrum. Methods* **A370** 237

[27] De Bellefon A *et al* 1996 Astrop. Phys. **6** 35

[28] Collar J I *et al* 1996 in *Proc. Workshop on Identification of Dark Matter* (Singapore: World Scientific) ed. Spooner N

[29] Collar J I *et al* 1996 in *Proc. Workshop on Identification of Dark Matter* (Singapore: World Scientific) ed. Spooner N, astro–ph/9610266

[30] Giuliani A these proceedings

[31] Cline D B 1997 in *Proc. Workshop on Dark Matter in Astro–and Particle Physics* (Singapore: World Scientific) eds. Klapdor–Kleingrothaus H V and Ramachers Y, p. 479

[32] Buckland K N, Lehner M J and Masek G E 1997 *IEEE NS* **44** 6

[33] Baer H and Brhlik M 1998 in *Proc. Beyond the Desert 97* (Bristol: IOP Publ.) 137, eds. Klapdor–Kleingrothaus H V and Päs H, hep–ph/9706509

[34] Drees M and Nojiri M M 1993 *Phys. Rev.* **D47** 376

[35] Bednyakov V, Klapdor–Kleingrothaus H V and Kovalenko S 1994 *Phys. Rev.* **D50** 7128

[36] Bednyakov V, Klapdor–Kleingrothaus H V, Kovalenko S and Ramachers Y 1997 *Z. Phys.* **A357** 339

[37] Bednyakov V, Klapdor–Kleingrothaus H V and Kovalenko S 1997 *Phys. Rev.* **D55** 503

[38] Bottino A *et al* 1997 *Phys. Lett.* **B402** 113

[39] Berezinsky V *et al* 1996 *Astropart. Phys.* **5** 1

[40] Arnowitt R and Nath P 1997 *Phys. Rev.* **D56** 2820

[41] Bergstrom L, Edsjo J and Gondolo P 1997 *Phys. Rev.* **D55** 1765

[42] Baudis L, Hellmig J, Klapdor–Kleingrothaus H V, Ramachers Y, Hammer J W and Meyer A 1998 *Nucl. Instrum. Methods* A in press

[43] Klapdor–Kleingrothaus H V 1998 in *Proc. Beyond the Desert 97* (Bristol: IOP Publ.) 485, eds. Klapdor–Kleingrothaus H V and Päs H, hep–ex/9802007
Klapdor–Kleingrothaus H V 1997 *CERN–Courier* **37,9** 16
CERN–Courier 1997 **37,10** 19
Klapdor–Kleingrothaus H V and Zdesenko Yu G 1998 *Eur. Phys. J.* **A3**

[44] Klapdor–Kleingrothaus H V, Hellmig J and Hirsch M 1998 *J. Phys.* **G24** 483–516

[45] Baer H and Brhlik M 1996 *Phys. Rev.* **D53** 597

[46] Kane G L, Kolda C, Roszkowski L and Wells J D 1994 *Phys. Rev.* **D49** 6173

[47] Bednyakov V and Klapdor–Kleingrothaus H V 1998 subm. to *Phys. Rev. Lett.*

[48] Bednyakov V and Klapdor–Kleingrothaus H V 1998 subm. to *Phys. Rev. Lett.* ; hep–ph/9802344

[49] Bednyakov V, Klapdor–Kleingrothaus H V and Kovalenko S 1994 *Phys. Lett.* **B329** 5

[50] Rich J 1996 *Astropart. Phys.* **4** 387

Neutrino Oscillations: a source of Goldstone fields and consequences for Supernovae

Luís Bento[1]

Centro de Física Nuclear da Universidade de Lisboa,
Av. Prof. Gama Pinto 2, 1699 Lisboa - *codex*, Portugal

Abstract. It is shown that true Nambu-Goldstone (NG) bosons develop coherent fields whenever the associated charges of the matter particles are not conserved in a macroscopic scale. The sources of the NG fields are the time rates of quantum number violation. If the lepton numbers are spontaneously broken at a scale below 1 TeV, the neutrino oscillation processes generate classic NG fields that are strong enough in Supernovae to modify the neutrino flavor dynamics. The oscillation patterns may change in the periods of largest ν fluxes. Two examples are given: 1. the back reaction of a NG field improves the adiabaticity of the ν_e resonant conversion; 2. $\bar{\nu}_e \leftrightarrow \bar{\nu}_\mu$ oscillations may occur even if ν_e is the lightest of the neutrinos.

1. Basic Features

The Goldstone theorem asserts that in a theory with spontaneous breaking of a global symmetry group some massless scalar bosons should exist - the so-called Nambu-Goldstone bosons (NG) - one per global symmetry broken by the vacuum. Good candidates for spontaneously broken quantum numbers are the partial lepton numbers L_e, L_μ, L_τ and total lepton number L. Indeed, they are conserved by the Standard Model (SM) interactions, but several experiments going now from solar to atmospheric neutrinos and laboratory oscillation experiments [1] indicate that the lepton flavors are not conserved.

[1] e-mail: lbento@fc.ul.pt

The NG bosons not only have zero mass but also no scalar potential terms such as ϕ^4. Futhermore, they also interact with the fermion particles however, they do not mediate long range forces. The reason lies in the very global symmetry that, broken by the vacuum still operates at the Lagrangian level. A NG boson transforms as $\phi \to \phi + \alpha$, $\alpha =$ constant, and therefore the symmetry requires that these fields have nothing but derivative couplings (this is related to the soft pion low energy theorems [2, 3]) *i.e.*, all of the NG boson interactions can be expressed with an effective Lagrangian of the form

$$\mathcal{L} = \frac{1}{2}\,\partial_\mu \phi\, \partial^\mu \phi + \mathcal{L}_{\rm int}(\partial_\mu \phi)\,. \tag{1}$$

The ϕ equation of motion is nothing but the conservation law of the total current associated with the $\phi \to \phi + \alpha$ symmetry,

$$\partial_\mu \partial^\mu\, \phi = -\partial_\mu J^\mu_\Lambda / V_\Lambda\,, \tag{2}$$

where V_Λ is essentially the scale of symmetry breaking and the current J^μ_Λ is determined in leading order by the quantum numbers of the particles. For instance, if Λ is the electron lepton number L_e,

$$J^\mu_e = \bar{e}\,\gamma^\mu\, e + \bar{\nu}_e \gamma^\mu \nu_e + \cdots\,, \tag{3}$$

where the dots stand for model dependent radiative corrections and scalar boson contributions.

This general result can be obtained using a suitable rephasing of the fields present in the theory [4, 5]. Let Λ_a and Λ_i be the quantum numbers of the fermion fields χ^a and scalars σ_i under an abelian global symmetry $\mathrm{U}(1)_\Lambda$ which is spontaneously broken by vacuum expectation values $\langle\sigma_i\rangle$. A suitable change of variables namely,

$$\chi^a = e^{-i\,\Lambda_a \phi / V_\Lambda}\, \psi^a\,, \tag{4}$$

$$\sigma_i = e^{-i\,\Lambda_i \phi / V_\Lambda}\, \left(\langle\sigma_i\rangle + \rho_i\right)\,, \tag{5}$$

makes the Lagrangian to be expressed in terms of *physical weak* eigenstates, fermions $\psi^a = e, \nu_e, \ldots$, massive bosons ρ_i and gauge bosons (assumed to be singlets of the global symmetry) all of them invariant under $\mathrm{U}(1)_\Lambda$. The only non-invariant field is the NG boson ϕ. It transforms as $\phi \to \phi + \alpha$ hence, the Lagrangian, symmetric under $\mathrm{U}(1)_\Lambda$, can only depend on ϕ through the derivatives $\partial_\mu \phi$. Its leading order couplings are derived from the kinetic Lagrangians of χ^a and σ_i by application of Eqs. (4-5). One obtains in particular, $V_\Lambda^2 = 2 \sum |\Lambda_i\, \langle\sigma_i\rangle\,|^2$.

The second member of the NG equation of motion, Eq. (2), is clearly not an ordinary scalar density. If one calculates the divergency of a vector or axial-vector current using the Dirac equation, one ends up with either scalar but flavor violating terms,

$$\partial_\mu \bar{\psi}_i \gamma^\mu \psi_j = i(m_i - m_j)\bar{\psi}_i \psi_j \ ,$$

or flavor conserving but pseudo-scalar,

$$\partial_\mu \bar{\psi}_i \gamma^\mu \gamma_5 \psi_j = i(m_i + m_j)\bar{\psi}_i \gamma_5 \psi_j \ .$$

This kind of source terms do not generate long range '$1/r$' fields, but rather spin dependent '$1/r^3$' potentials [6].

In the intent of escaping the derivative couplings law, models were made with explicit symmetry breaking terms. These might be anomaly terms [7], as in the axion case [8], or soft symmetry breaking terms [9]. In this way scalar couplings with the NG field are produced but at the cost of giving a mass to the then, so-called pseudo Nambu-Goldstone boson. The mass ultimately means a finite range for the NG bosons. I will show next that under certain conditions a true Nambu-Goldstone boson may develop a macroscopic field [5, 10].

2. Long range Nambu-Goldstone fields

First notice that the source of a NG field is different from zero if the matter current J^μ_Λ is not conserved. The volume integral

$$\int d^3x \ \partial_\mu J^\mu_\Lambda = \frac{d\Lambda^{\text{created}}}{dt}$$

represents the rate of Λ-number *created* per unity of time (in all particles except ϕ) and $\partial_\mu J^\mu_\Lambda$ is its volume density. The equation of motion (2) gives

$$\phi(t,\vec{r}) = \frac{-1}{V_\Lambda} \int d^3x \, \frac{\dot{\rho}^{\text{cr}}_\Lambda(t - |\vec{r} - \vec{x}|, \vec{x})}{4\pi \, |\vec{r} - \vec{x}|} \ . \tag{6}$$

It is a long-range Coulomb type of field but suppressed by the product of the scale V_Λ times the interval of time per Λ-violating process. To have a significant macroscopic field one needs a large rate of reactions with a net Λ number variation. That calls for astrophysical sources. There is evidence that solar neutrinos change flavor as they come out of the Sun. If ν_e oscillates into ν_μ, for instance, then L_e and L_μ are not conserved in a large scale in stars and possibly in Supernovae as well. If any of them is associated with a spontaneously broken global symmetry then, the respective NG boson acquires classic field configurations.

For definiteness suppose that L_e is spontaneously broken and that a fraction of the ν_es emitted from a star are resonantly converted into ν_μ through the Mikheyev-Smirnov-Wolfenstein (MSW) mechanism [11, 12] (also ν_μ into ν_e in a Supernova) in a certain shell of radius $\sim r_c$ inside the star. Assuming for simplicity stationary neutrino fluxes and spherical symmetry one obtains the solution for the NG boson ϕ_e:

$$\begin{aligned} \phi_e &= \text{const.}\,, \quad r < r_c\,, \\ \phi_e &= \frac{1}{V_\Lambda}\frac{1}{4\pi r}\left[\frac{dN}{dt}(\nu_e \to \nu_\mu) - \frac{dN}{dt}(\nu_\mu \to \nu_e)\right]\,, \quad r > r_c \end{aligned} \tag{7}$$

in terms of the total numbers of converted neutrinos, $N(\nu_\mu \to \nu_e)$ and $N(\nu_e \to \nu_\mu)$, per unity of time.

What are the possible consequences of such a NG field? Applying Eqs. (4) to the kinetic terms

$$\mathcal{L}_0 = \bar{e}\, i\, \partial\!\!\!/ e + \bar{\nu}_e\, i\, \partial\!\!\!/ \nu_e\,,$$

one obtains the interaction

$$\mathcal{L}_{\text{int}} = \frac{1}{V_\Lambda}\partial_\mu \phi_e\, (\bar{e}\,\gamma^\mu\, e + \bar{\nu}_e\,\gamma^\mu\,\nu_e)\,. \tag{8}$$

The fermions only care about the gradient of the NG field. In the situation described above, denoting by $\vec{j}(\nu_e \to \nu_\mu)$ the flux of e-neutrinos converted to ν_μ and $\vec{j}(\nu_\mu \to \nu_e)$ the reciprocal,

$$\vec{A}_e = -\frac{\vec{\nabla}\phi_e}{V_\Lambda} = \frac{1}{V_\Lambda^2}\left[\vec{j}(\nu_e \to \nu_\mu) - \vec{j}(\nu_\mu \to \nu_e)\right]\,. \tag{9}$$

The quantity $\vec{A}_e$ is suppressed relatively to the field ϕ_e by the scale V_Λ and radius r and that is what makes it extremely small. However, as follows from Eq. (8), it still contributes to the potential energy of an electron-neutrino with velocity $\vec{v}_\nu$ as

$$V_{\nu_e} = \vec{A}_e \cdot \vec{v}_\nu\,. \tag{10}$$

These contributions associated with breaking of partial lepton numbers are flavor dependent by nature and therefore may have a role in neutrino oscillations.

One estimates the order of magnitude of theses NG potentials as the neutrino flux divided by V_Λ^2. In the case of the Sun, evaluating the total flux at the Sun radius $R_\odot = 7 \times 10^{10}$ cm,

$$V_{\text{NG}} \sim \frac{j_\nu}{V_\Lambda^2} = \frac{2.8 \times 10^{-2}}{R_\odot}\frac{1\,\text{keV}^2}{V_\Lambda^2}\,.$$

It produces a too small variation of the neutrino phase, $\sim V_{\rm NG}\, R_\odot$, thus it does not seem significant for solar neutrinos. In a Supernova however the fluxes are many orders of magnitude larger. One has

$$\frac{j_\nu}{V_\Lambda^2} = 1.48\,\frac{G_\Lambda}{G_{\rm F}}\,\frac{L_\nu}{10^{52}\,{\rm ergs/s}}\,\frac{10\,{\rm MeV}}{\langle E_\nu\rangle}\left(\frac{r}{10^{10}\,{\rm cm}}\right)^{-2} \times 10^{-12}\ {\rm eV}\ , \qquad (11)$$

where $G_{\rm F} = 11.66\,{\rm TeV}^{-2}$ is the Fermi constant, $G_e = 1/V_\Lambda^2$, L_ν is the neutrino energy luminosity, and $\langle E_\nu\rangle$ its average energy. $L_\nu / \langle E_\nu\rangle$ gives the neutrino emission rate. This has to be compared with the Standard Model electro-weak potential V_W [11]. In the regions of a Supernova star with densities typical of the Sun the mass density goes as $1/r^3$, the constant $\tilde{M} = \rho r^3$ lying between 10^{31}g and 15×10^{31}g depending on the star [13]. Then,

$$V_W = \sqrt{2}\,G_{\rm F}\,n_e = 0.76\ Y_e\,\frac{\tilde{M}}{10^{31}{\rm g}}\left(\frac{r}{10^{10}\,{\rm cm}}\right)^{-3} \times 10^{-12}\ {\rm eV}\ , \qquad (12)$$

$Y_e \approx 1/2$ being the electron abundance and n_e the number density. Clearly, in the period of time where the neutrino luminosities are high enough say, 10^{52} ergs/s, the scale of lepton symmetry breaking can be as high as the Fermi scale and yet the NG potentials compete with V_W at the places where the resonance occurs for values of Δm^2 interesting for solar neutrino solutions ($\Delta m^2/E \sim 10^{-12}\,{\rm eV}^2$). Futhermore, the NG potentials decay as $1/r^2$ and necessarily overcome the local interactions at large enough radius.

At this point we can already draw some conclusions. First, if the lepton numbers $L_e, L_\mu, \ldots$ are spontaneously broken, the neutrino oscillations in a Supernova originate the appearance of classic Nambu-Goldstone fields. Second, if the scale of symmetry breaking lies below 1 TeV they in turn yield potential energies that are significant for the very neutrino oscillation processes. What are the possible consequences? The NG coupling are flavor dependent by nature hence, may alter the pattern of neutrino oscillations, with a fundamental difference: the NG potentials are proportional to the neutrino fluxes and that induces a time dependence of the effects in a Supernova context.

Next I show two examples. One [10] is the improvement of adiabaticity of otherwise $\nu_e \leftrightarrow \nu_X$ non-adiabatic transitions. The other [5] is the violation of the SM prediction that the anti-neutrinos do not resonantly oscillate if the neutrino masses obey the hierarchy of the charged lepton masses.

3. Improvement of adiabaticity: Majoron back reaction

Suppose that in the Sun the electron-neutrino oscillates into the muon-neutrino with the parameters of the non-adiabatic, small mixing angle solution [14, 15, 16]. Then, the $\nu_e \leftrightarrow \nu_\mu$ transitions are also non-adiabatic in a Supernova and only a fraction of each neutrino species is converted into the other. Furthermore, since the level crossing probability is an increasing function of the energy, the hotter ν_μs have larger survival probabilities than the cooler ν_es. Let us assume that the partial lepton number L_e is conserved at the Lagrangian level but the respective global symmetry, $U(1)_e$, is spontaneously broken. Then, a NG boson (Majoron), ϕ_e, exists with zero mass. Although in principle, the neutrino mass matrix violates all three lepton numbers, I will ignore for simplicity other possibly existing Majorons. Notice that other NG fields are naturally less intense if the respective scales of symmetry breaking are slightly higher.

The equations governing two flavor oscillation [17] with mixing angle θ and Δm^2 difference in the vacuum, are given by, up to an universal phase absorbed in the wave functions,

$$i\frac{\partial}{\partial r}\begin{pmatrix}\nu_e\\ \nu_\mu\end{pmatrix} = \frac{1}{2E}\begin{pmatrix}2E\,V_e & \frac{1}{2}\Delta m^2\sin 2\theta\\ \frac{1}{2}\Delta m^2\sin 2\theta & \Delta m^2\cos 2\theta\end{pmatrix}\begin{pmatrix}\nu_e\\ \nu_\mu\end{pmatrix}. \quad (13)$$

V_e is the difference between the ν_e and ν_μ potentials, which is the following in the presence of a ϕ_e field (see Eqs. (10,12):

$$V_e = V_{\nu_e} - V_{\nu_\mu} = V_W + \vec{A}_e\cdot\vec{v}_\nu\ . \quad (14)$$

Assuming spherical symmetry, $\vec{A}_e$ only has radial component and one obtains from Eq. (9)

$$V_e = \sqrt{2}\,G_F\,n_e + \frac{G_e}{4\pi\,r^2}\left[\dot{N}(\nu_e\to\nu_\mu) - \dot{N}(\nu_\mu\to\nu_e)\right]\ . \quad (15)$$

V_e would be the same if one considers alternatively a NG boson associated with the breaking of L_μ or $L_e - L_\mu$. The reason is, $\nu_e \leftrightarrow \nu_\mu$ oscillations only violate $L_e - L_\mu$ not the total number $L_e + L_\mu$.

In order to calculate the radial comp^nent $A_e(r)$ one needs to know the fluxes of ν_μ and ν_e as functions of the radius and their energy spectra. Let $f_{\nu_e}(E)$ and $f_{\nu_\mu}(E)$ be the number of neutrinos emitted per unity of time and energy, $\mathrm{d}\dot{N}_{\nu_e}/\mathrm{d}E$ and $\mathrm{d}\dot{N}_{\nu_\mu}/\mathrm{d}E$ respectively. If the transition is non-adiabatic, the number of converted neutrinos with a given energy E is only a fraction $1 - P_c$ of the radiated particles, P_c is the level crossing probability, and so

$$d\dot{N}(\nu_e\to\nu_\mu) = (1 - P_c)\,f_{\nu_e}(E)\,dE\ , \quad (16)$$
$$d\dot{N}(\nu_\mu\to\nu_e) = (1 - P_c)\,f_{\nu_\mu}(E)\,dE\ . \quad (17)$$

The problem of obtaining these rates as functions of the radius is simplified by neglecting the oscillation length of the resonantly converted neutrinos:

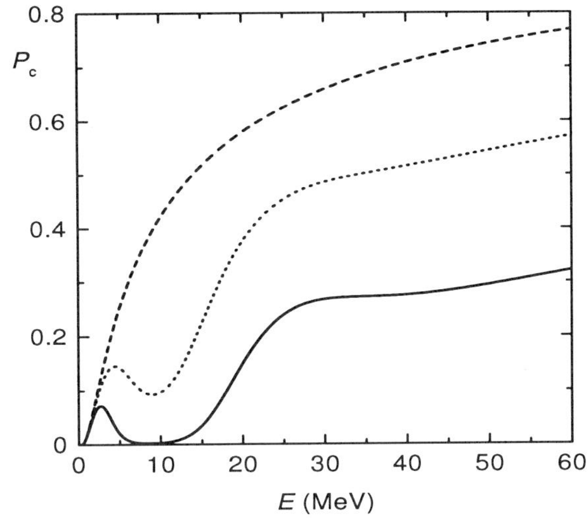

Figure 1. Level crossing probability as a function of the ν energy. Dashed curve: SM with $\tilde{M} = 4 \times 10^{31}$ g and $Y_e = 1/2$. Dotted and bold curves: a Nambu-Goldstone field exists with $G_e = G_\mathrm{F}$ and $G_e = 4\,G_\mathrm{F}$ respectively. The luminosities are 10^{52} ergs/s for ν_e and 7×10^{51} ergs/s for ν_μ.

in that limit a neutrino with energy E is converted to the other flavor at the position where the resonance condition is verified:

$$E(r) = \frac{\Delta m^2 \cos 2\theta}{2\,V_e(r)} \; . \tag{18}$$

In addition, the level crossing probability P_c can be calculated using the Landau-Zener approximation [14, 17],

$$P_c(E) = \exp\left\{ -\frac{\pi}{4} \left| \frac{\Delta m^2}{\mathrm{d}E/\mathrm{d}r} \frac{\sin^2 2\theta}{\cos 2\theta} \right| \right\}_{\mathrm{res.}} \; . \tag{19}$$

These approximations change the numbers but not, I believe, the lesson taken by comparing the results with and without a Majoron field. Eqs. (12,15 - 19) establish the differential equations for $d\dot{N}(\nu_e \to \nu_\mu)$ and $d\dot{N}(\nu_\mu \to \nu_e)$. They are non-linear because the derivative of $E(r)$ is not independent from the Majoron potential $A_e(r)$.

Let us examine the $\nu_e \leftrightarrow \nu_\mu$ oscillations with the mixing parameters of the non-adiabatic solar neutrino solution (for a recent update see [18]), choosing in particular the values $\Delta m^2 = 7 \times 10^{-6}\,\mathrm{eV}^2$, $\Delta m^2 \sin^2 2\theta = 4 \times 10^{-8}\,\mathrm{eV}^2$. In a Supernova the resonance is non-adiabatic as well and, as Eq. (19) indicates, the survival probability $1 - P_c$ increases with the ν energy. A detailed study in the framework of the SM was done in [19]. In

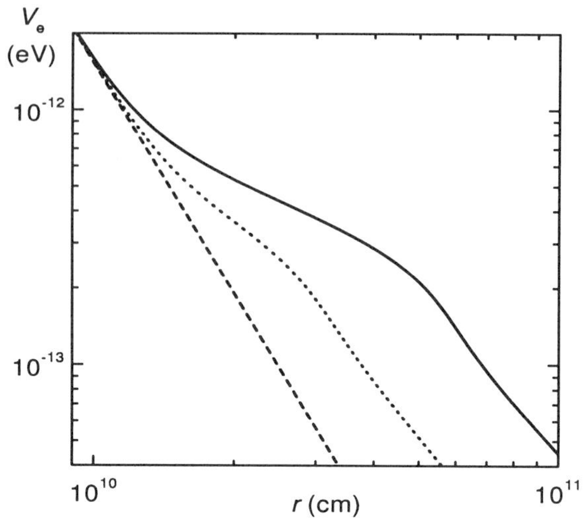

Figure 2. Total potential $V_e = V_{\nu_e} - V_{\nu_\mu}$ as a function of the radius. The dashed curve is the SM potential, the dotted and bold curves are the Majoron cases with the same parameters as in Fig. 1.

Fig. 1, P_c in the Landau-Zener approximation is plotted against the energy. The dashed curve holds for the SM potential with a constant $\tilde{M} = 4\times10^{31}$g. It is manifest the aggravation of the non-adiabaticity with the energy.

To study the Majoron case one has to specify the energy spectra and luminosities. I used Fermi-Dirac distributions with the following values of temperature and chemical potential [20]: for ν_e, $T = 2.4$ MeV and $\mu = 3.2\,T$; for ν_μ, $T = 5.1$ MeV, $\mu = 4.1\,T$. This gives average energies of 10 and 23 MeV respectively. The luminosity intensities are in turn 10^{52} ergs/s for ν_e and 7×10^{51} ergs/s for ν_μ which amount to particle emission rates of 10^{51} and 3×10^{50} ergs/s/MeV respectively. Because the e-neutrinos are more numerous, the $\nu_e \leftrightarrow \nu_\mu$ oscillations produce a net destruction of L_e-number and a positive potential A_e. The dynamics is the following: the less energetic ν_e oscillate to ν_μ at smaller radius than $\nu_\mu \to \nu_e$; the $\nu_e \to \nu_\mu$ conversion produces a positive $A_e(r)$ which attenuates the fall of the total potential $V_e = V_W + A_e$ with r; Consequently, the adiabaticity improves at larger radius and the most energetic neutrinos change flavor with higher probabilities.

Fig. 2 [10] shows the total potential V_e as a function r. The dashed curve stands for the SM potential, V_W, with $\tilde{M} = 4 \times 10^{31}$ g and $Y_e = 1/2$. The dotted and bold curves differ in the scale of L_e symmetry breaking, $G_e = G_{\rm F}$ and $G_e = 4\,G_{\rm F}$, respectively. The potential falls more slowly as the field ϕ_e grows which reflects on the level crossing probability plotted in

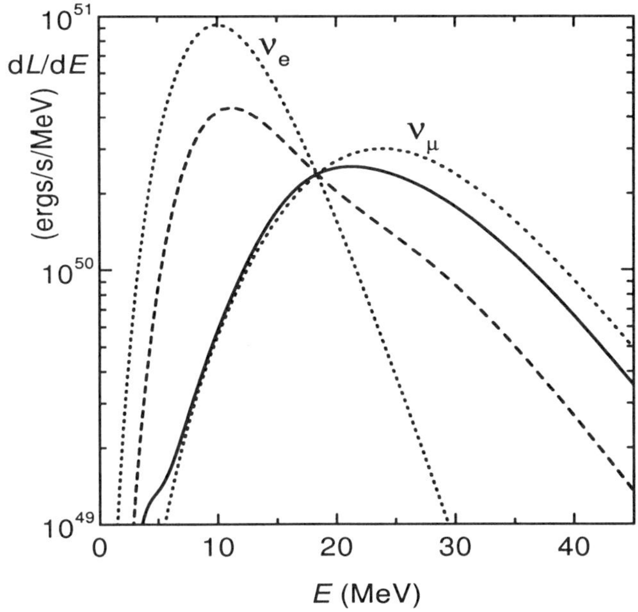

Figure 3. The dotted curves are the assumed luminosity distributions for ν_e and ν_μ as emitted from the neutrinospheres. The dashed curve represents the ν_e luminosity after electroweak neutrino oscillations whereas in the bold curve a ϕ_e field exists with $G_e = 4\,G_{\mathrm{F}}$.

Fig. 1 [10] for the same cases (dotted and bold curves). The effect is clear: the stronger the Majoron field, the more efficient is the flavor conversion. It is worth to mention that if the μ-neutrinos were more numerous than the e-neutrinos the effect would be the opposite because the Majoron potential would be negative ($\dot{L}_e > 0$). That shows up in the rise of P_c at the ν_μ energy band around 20 MeV.

Fig. 3 [10] shows the implications for the outgoing ν_e energy spectrum. The dotted curves are the assumed luminosity distributions for the emitted ν_es and ν_μs. The dashed curve represents the luminosity of the e-neutrinos that come out of the star after standard MSW oscillations and the bold curve is the same but with a Majoron field ($G_e = 4\,G_{\mathrm{F}}$). The improvement of adiabaticity makes more ν_μs convert into ν_e and less ν_e to survive, and because the μ-neutrinos are more energetic, the outgoing ν_e spectrum is harder than if there was no Majoron field. The average energy of the outgoing e-neutrinos is 13 MeV for $G_e = 0$ but reaches 17 MeV if $G_e = G_{\mathrm{F}}$ and 21 MeV if $G_e = 4\,G_{\mathrm{F}}$. These effects on the spectra can in principle be traced in those detectors such as Super-Kamiokande and SNO, capable of detecting Supernova electron-neutrinos [21].

To summarize, if the explanation of the solar neutrino deficit is the MSW non-adiabatic oscillation $\nu_e \to \nu_\mu$ ($\nu_e \to \nu_\tau$ or any other superposition of ν_μ and ν_τ) then, the Standard Model predicts that in a Supernova the $\nu_e \leftrightarrow \nu_\mu$ transitions are also non-adiabatic. It means that, to a large extent, the e-neutrinos preserve their lower energy spectrum, unless ν_e also mixes to another flavor with a too high or too low Δm^2 to show up in solar neutrinos. If however, L_e is a spontaneously broken quantum number, the associated Nambu-Goldstone boson, ϕ_e, will acquire a classic field configuration which may be strong enough to produce a back reaction that improves the adiabaticity of the $\nu_e \leftrightarrow \nu_\mu$ transitions. The outcome is a ν_e energy spectrum harder than expected. In 1987, the existing detectors were only able to detect electron anti-neutrinos but the now operating Super-Kamiokande and SNO experiments will be capable of detecting Supernova ν_e events. The analysis of the energy distribution can in principle reveal or put limits on that kind of effect.

The scenarios of neutrino mixing change considerably if one considers the evidences from atmospheric and terrestrial neutrino experiments (for a review see [1]). The atmospheric neutrino anomaly and the zenith angle dependence observed by Super-Kamiokande can be explained by $\nu_\mu \to \nu_\tau$ oscillations[22], $\nu_\mu \to \nu_e$ excluded CHOOZ [23], but that is still consistent with solar neutrinos oscillations into a linear combination of ν_μ and ν_τ. That is no longer true if one takes in consideration the LSND evidences of $\bar{\nu}_\mu \to \bar{\nu}_e$ [24] and $\nu_\mu \to \nu_e$ [25]. The much larger Δm^2 scale involved in LSND ($\Delta m^2_{e\mu} > 0.2\,\mathrm{eV}^2$), calls for a fourth flavor - a sterile neutrino - into which solar ν_es should oscillate. That changes some features concerning the effects of a NG field ϕ_e but one thing is maintained: the signature of NG fields is a *surprise i.e.*, an oscillation pattern not consistent with SM weak interactions [10].

An important point is that the Nambu-Goldstone fields are proportional to the reaction rates that violate the relevant quantum numbers. In the case of neutrino oscillations this manifests as a dependence on the neutrino luminosity magnitudes. The effects of the Majoron fields if any, will be observed in a shorter or longer interval of time depending on the actual scale of lepton number symmetry breaking. Thus, the observation of such correlation with the flux intensities, would thus provide a measurement of the scale of spontaneous symmetry breaking.

4. Resonant conversion of anti-neutrinos

In the framework of the SM weak interactions the anti-neutrinos cannot undergo a resonant oscillation if the neutrino masses follow the same hierarchy as the charged leptons. That is because the SM potential V_W is positive in an electron rich star medium and the potential reverses sign for antineutrinos. The NG boson contributions are not only non-universal in flavour but also their signs depend on the flavour content of the neutrino fluxes and dominant flavor transitions. This allows in principle other than the standard MSW type of oscillations.

If one of the partial lepton numbers is spontaneously broken one expects the same happens for all three flavours. Let σ_i be the set of scalar fields, singlets of SU(2)xU(1), with quantum numbers $\Lambda_i = (L_e, L_\mu, L_\tau)_i$ under the symmetry group $\mathrm{U}(1)_e \times \mathrm{U}(1)_\mu \times \mathrm{U}(1)_\tau$. If no residual symmetry is left after spontaneous symmetry breaking there are three NG bosons. By going to the physical basis as in Eqs. (4-5), now summing over the quantum numbers $\Lambda = L_e, L_\mu, L_\tau$,

$$\begin{aligned} \chi^a &= e^{-i\sum \xi_\Lambda \Lambda_a} \psi^a \,, & (20) \\ \sigma_i &= e^{-i\sum \xi_\Lambda \Lambda_i} \left(\langle\sigma_i\rangle + \rho_i\right) \,, & (21) \end{aligned}$$

one obtains the Lagrangian of the NG fields

$$\mathcal{L} = \frac{1}{2} V^2_{\Lambda M}\, \partial_\mu \xi_\Lambda\, \partial^\mu \xi_M + \mathcal{L}_{\rm int}(\,\partial_\mu \xi_\Lambda) \,, \tag{22}$$

the sum over Λ, M is implicit. The matrix

$$V^2_{\Lambda M} = 2 \sum_i \Lambda_i M_i \left|\langle\sigma_i\rangle\right|^2 \tag{23}$$

is in general non-diagonal, provided that some of the fields σ_i have simultaneously more than one non-trivial quantum number. Such mixing implies that the NG bosons mediate an action at a distance between two distinct currents as show the equations of motion,

$$\partial_\mu \partial^\mu \xi_\Lambda = -\sum_M G_{\Lambda M}\, \partial_\mu J^\mu_M \,, \tag{24}$$

where $G_{\Lambda M}$ is the inverse matrix of $V^2_{\Lambda M}$.

In a stationary regime, if the neutrinos oscillate causing current variations $\delta \vec{J}_M$, the expression for the gradients of the NG bosons, $\vec{A}_\Lambda = -\vec{\nabla}\xi_\Lambda$, that generalizes Eq. (9) is [5]:

$$\vec{A}_\Lambda = -\sum_M G_{\Lambda M}\, \delta \vec{J}_M \,. \tag{25}$$

Since the interaction Lagrangian of the leptons has the same form as Eq. (8) *i.e.*, $\mathcal{L}_{\rm int} = J^\mu_\Lambda\, \partial_\mu \xi^\Lambda$, the potential energy of a neutrino ν_ℓ subject to stationary NG fields is given by

$$V_{\nu_\ell} = \vec{A}_\ell \cdot \vec{v}_\nu \,, \tag{26}$$

as in Eq. (10). Plugging in Eq. (25), one gets for example,

$$V_{\nu_\mu} = -\left(G_{\mu e}\,\delta\vec{J}_e + G_{\mu\mu}\,\delta\vec{J}_\mu + G_{\mu\tau}\,\delta\vec{J}_\tau\right)\cdot\vec{v}_\nu\,.$$

This shows that the violation of a certain lepton number in one place originates fields that acts on other flavors in other places.

In the old preferred scenario (before Super-Kamikande results on atmospheric neutrinos [22]), the neutrino mixing angles are small, $\nu_e \approx \nu_1$, $\nu_\mu \approx \nu_2$, $\nu_\tau \approx \nu_3$, and the masses obey the hierarchy $m_1 < m_2 < m_3$. Then [26, 27], as the neutrinos come out of a Supernova the first resonant transitions are between ν_e and ν_τ. It is outside their resonance shell, at lower medium densities, that MSW transitions $\nu_e \leftrightarrow \nu_\mu$, with smaller Δm^2, occur.

In this meeting Atsuto Suzuki [22] reported Super-Kamiokande results on the atmospheric neutrino anomaly and zenith angle dependence that can be explained by $\nu_\mu \to \nu_\tau$ oscillations. The best values, confirmed at Neutrino 98, are $\Delta m^2 = 5\times 10^{-3}\,\mathrm{eV}^2$ and $\sin^2 2\theta = 1$. The alternative $\nu_\mu \to \nu_e$ oscillation is excluded by CHOOZ [23]. This is consistent with a scenario where the heaviest of the neutrinos is essentially a superposition, ν_H (possibly maximal), of ν_μ and ν_τ; ν_e is the lightest neutrino and oscillates in the Sun into the medium weight orthogonal superposition, ν_L, of ν_μ and ν_τ. In what follows what is said about ν_τ, ν_μ and quantum numbers L_τ, L_μ can be maintained if replaced by ν_H, ν_L and respective combinations of L_τ and L_μ.

In a Supernova the ν_e flux is larger than the ν_τ flux and consequently more ν_e transform into ν_τ than the reverse which makes a net decrease of $L_e - L_\tau$. The current densities $\vec{J}_\tau$ and $\vec{J}_e$ suffer the variations

$$\delta\vec{J}_\tau = -\delta\vec{J}_e = \vec{j}(\nu_e \to \nu_\tau) - \vec{j}(\nu_\tau \to \nu_e)\,, \tag{27}$$

that generate the NG fields ξ_e, ξ_μ, ξ_τ. Applying Eq. (25) one gets the vector potentials outside the sphere where $\nu_e \leftrightarrow \nu_\tau$ takes place. The magnitude of the constants $G_{\Lambda M}$ is directly related to the symmetry breaking scales, but the flavor structure is arbitrary and is not constrained by the neutrino mixing. The matrix $G_{\Lambda M}$ has only to be symmetric and positive definite as demands the scalar bosons kinetic Lagrangian. Consider the following hierarchy: $|G_{ee}|, |G_{e\mu}|, |G_{e\tau}| << |G_{\mu\tau}|, G_{\tau\tau}$. Then, $\vec{A}_e$ is much smaller in magnitude than $\vec{A}_\mu$ and $\vec{A}_\tau$ and, assuming spherical symmetry, the potential energies are well approximated by

$$V_{\nu_e} = V_W + A_e\,, \tag{28}$$

$$V_{\nu_\mu} = A_\mu \approx -G_{\mu\tau}\,\delta J_\tau\,, \tag{29}$$

$$V_{\nu_\tau} = A_\tau \approx -G_{\tau\tau}\,\delta J_\tau\,, \tag{30}$$

omitted the neutral current terms. The anti-neutrino potentials are the symmetric of these.

The increment of τ lepton number per unity of time, $\delta F_\tau = \dot{N}(\nu_e \to \nu_\tau) - \dot{N}(\nu_\tau \to \nu_e)$, is positive and so is the flux transfer $\delta J_\tau = \delta F_\tau / 4\pi r^2$. Defining $\delta F_\tau = \delta F_{51} \times\ 10^{51}$ergs/s/MeV, one estimates $\delta F_{51} = 2 - 5$ in the first 1/2 seconds of neutrino emission until the Supernova explosion and $\delta F_{51} = 0.2 - 0.5$ at 1 second time. See [5] for more details. The NG potential for ν_τ is negative but that is not necessarily the case of ν_μ because the element $G_{\mu\tau}$ can be negative! Using Eq. (11),

$$A_\mu = -1.48\, \frac{G^F_{\mu\tau}}{G_F} \delta F_{51} \left(\frac{r}{10^{10}\,\mathrm{cm}} \right)^{-2} \times 10^{-12}\ \mathrm{eV}\,. \tag{31}$$

The Schrödinger equation for the anti-neutrinos is the same as for neutrinos, Eq. (13), but with a symmetric potential:

$$i \frac{\partial}{\partial r} \begin{pmatrix} \bar{\nu}_e \\ \bar{\nu}_\mu \end{pmatrix} = \frac{1}{2E} \begin{pmatrix} -2E\,(V_W - A_\mu) & \frac{1}{2}\Delta m^2 \sin 2\theta \\ \frac{1}{2}\Delta m^2 \sin 2\theta & \Delta m^2 \cos 2\theta \end{pmatrix} \begin{pmatrix} \bar{\nu}_e \\ \bar{\nu}_\mu \end{pmatrix}, \tag{32}$$

where $|A_e| \ll |A_\mu|$ is implicit. Within the SM, the effective mass $-2E\,V_W$ would be negative, and the resonance condition $-2E\,V_W = \Delta m^2 \cos 2\theta$ could not be met. In presence of NG fields, the effective mass $M_e^2 = -2E\,(V_W - A_\mu)$ is negative at the high density regions, but since the NG potentials only decay with the second power of the radius their contributions inevitably overcome at large enough distances. That is a crucial point. If on top of that $G_{\mu\tau}$ is negative, the effective mass M_e^2 suddenly turns positive and may reach the value $M_\mu^2 = \Delta m^2 \cos 2\theta$. In that case there is a resonance layer where $M_e^2 = M_\mu^2$ is met. In Fig. 4 it corresponds to the interception of the "e" and "μ" curves, $M_e^2(r)$ and M_μ^2, respectively. The positive part of the 'potential' M_e^2 is plotted in Fig. 4 with values $G_{\mu\tau} = -G_{\mathrm{F}}$, $\delta F_{51} = 1$ or 3, $Y_e = 1/2$, $\tilde{M}_{31} = 4$, and a neutrino energy of 20 MeV, the same for all curves.

Take $M_\mu^2 = 6 \times 10^{-7}\,\mathrm{eV}^2$ for example. The M_μ^2 dashed line crosses the relative potential curve at two points: the first, r_1, lies in the remarkably sharp transition from the $-V_W < 0$ dominated region to the NG field dominated one and very likely a non-adiabatic level crossing occurs there *i.e.*, $\bar{\nu}_e$ continues $\bar{\nu}_e$ and $\bar{\nu}_\mu$ continues $\bar{\nu}_\mu$. In the second point, r_2, the potential is smooth enough ($M_e^2 \propto r^{-2}$) for adiabatic transitions. In general terms, the conversion probability is enhanced [12, 17] if $M_\mu^2 \tan^2 2\theta$ is larger than the rate $f(r) = |E\,\mathrm{d}V/V\mathrm{d}r|$ at resonance. In Fig. 4, $f(r)$ is plotted with squares for $V \propto r^{-2}$. One can see that small mixing angle adiabatic conversions are possible for a wide range of neutrino masses (for more details see [5]). Through this mechanism one could have resonant $\bar{\nu}_e \leftrightarrow \bar{\nu}_\mu$

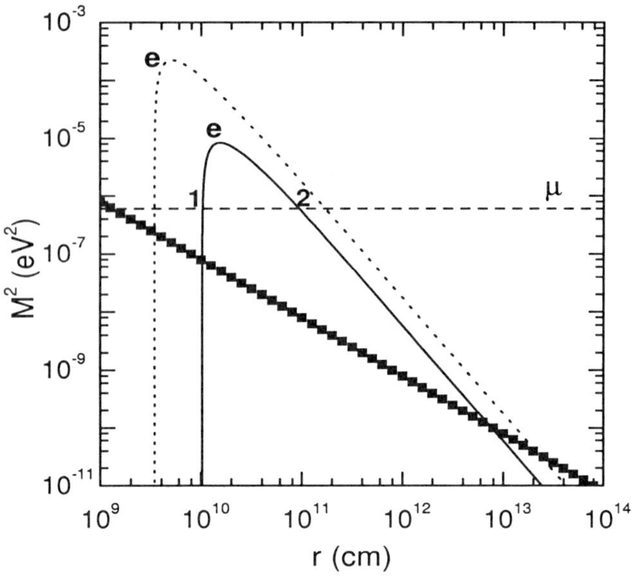

Figure 4. $2E(V_{\bar{\nu}_e} - V_{\bar{\nu}_\mu})$ is plotted in the solid and dotted curves with $\delta F_{51} = 1$ and 3 respectively. The dashed line stands for a particular value of $\Delta m^2 \cos 2\theta$ and the squares curve is the adiabatic boundary for $M_\mu^2 \tan^2 2\theta$ in resonance 2.

oscillations in a Supernova whereas $\nu_e \to \nu_\mu$ take place in the Sun. In particular, with the parameters of the non-adiabatic solution [14, 15, 16] namely, $M_\mu^2 \approx (3-12)\times 10^{-6}\,\mathrm{eV}^2$ and $M_\mu^2 \tan^2 2\theta \approx 4\times 10^{-8}\,\mathrm{eV}^2$, provided that the neutrino fluxes are high enough (compare the dotted and solid curves).

Since the predicted energy spectrum of $\bar{\nu}_\mu$ and $\bar{\nu}_\tau$ is harder than the $\bar{\nu}_e$ spectrum that kind of oscillation could be manifest in the SN1987A events [28]. Some analyses made [29, 30], but not all [31], disfavour the occurrence of $\bar{\nu}_e \leftrightarrow \bar{\nu}_\mu$ oscillations. This deserves some discussion [5], here I simply stress that the effects caused by the NG fields strongly depend on magnitude of the neutrino fluxes and on the scale of the global symmetry breaking as well. Since the fluxes vary with time in a Supernova, one could have anti-neutrino oscillations in the first instants of emission that fade away when the fluxes drop below a certain threshold. The observation in the future of such correlation between oscillation and flux magnitudes would provide a measurement of the symmetry breaking scale.

5. Conclusions

We have shown that despite the fact that true Nambu-Goldstone (NG) bosons only have derivative couplings, it is still possible to have coherent NG fields. This happens whenever the quantum numbers they are associated with are not conserved in reactions taking place in a large scale. If the lepton numbers are spontaneously broken charges then, neutrinos oscillations in stars produce classic NG fields. In addition, if the scale of symmetry breaking is below 1 TeV, these fields are strong enough in Supernovae to change the neutrino flavor dynamics.

The observable signatures are:

- Time variation of neutrino oscillation patterns in correlation with the flux magnitudes. This is a consequence of the NG fields being proportional to the time rate of reactions and therefore to the neutrino fluxes.
- Incongruence of Supernova neutrino oscillation patterns with the evidence obtained from solar, atmosferic and terrestrial neutrinos. The NG fields are not expected to be significant in these cases even if they are in Supernovae because the fluxes are much smaller. This is ultimately a result of the derivative coupling nature of the NG bosons.

Finally, if we will are lucky enough to see such kind of phenomena in the future, then, Supernova neutrinos will provide a measurement of the scale of lepton number symmetry breaking.

Acknowledgements

This work was supported in part by the project ESO/P/PRO/1127/96.

References

[1] D. O. Caldwell, in Proceedings of "COSMO 97: First International Workshop on Particle Physics and the Early Universe", Ambleside, England, 1997 (edited by L. Roszkowski, World Scientific, 1998), hep-ph/9804367; Y. Totsuka, in Proceedings of the "XVIII International Symposium on Lepton-Photon Interactions", Hamburg, 1997 (edited by A. De Roeck and A. Wagner, World Scientific, 1998).

[2] T.-P. Cheng and L.-F. Li, "Gauge Theory of Elementary Particle Physics" (Oxford University Press, New York, 1989).

[3] S. Weinberg, "The Quantum Theory of Fields: II" (Cambridge University Press, Cambridge, England, 1996).

[4] G. B. Gelmini, S. Nussinov and T. Yanagida, Nucl. Phys. B 219 (1983) 31.

[5] L. Bento, Phys. Rev. D 57 (1998) 583.

[6] Y. Chikashige, R. N. Mohapatra and R. D. Peccei, Phys. Lett. B 98 (1981) 265.

[7] F. Wilczek, Phys. Rev. Lett. 49 (1982) 1549; D. Chang, R. N. Mohapatra and S. Nussinov, Phys. Rev. Lett. 55 (1985) 2835.

[8] R. Peccei and H. Quinn, Phys. Rev. Lett. 38 (1977) 1440; S. Weinberg, Phys. Rev. Lett. 40 (1978) 223; F. Wilczek, Phys. Rev. Lett. 40 (1978) 279.

[9] C. T. Hill and G. G. Ross, Nucl. Phys. B 311 (1988) 253; J. A. Frieman, C. Hill and R. Watkins, Phys. Rev. D 46 (1992) 1226; R. N. Mohapatra and S. Nussinov, Phys. Lett. B 395 (1997) 63.

[10] L. Bento, hep-ph/9806411, Phys. Rev. D 59, 015013 (1999).

[11] L. Wolfenstein, Phys. Rev. D 17 (1978) 2369.

[12] S. P. Mikheyev and Y. Yu. Smirnov, Yad. Fiz. 42 (1985) 1441 [Sov. J. Nucl. Phys. 42 (1985) 913] and Nuovo Cimento 9C (1986) 17.

[13] J. R. Wilson *et. al.*, Ann. N. Y. Acad. Sci. 470 (1986) 267.

[14] W. C. Haxton, Phys. Rev. Lett. 57 (1986) 1271.

[15] S. J. Parke, Phys. Rev. Lett. 57 (1986) 1275.

[16] S. P. Rosen and J. M. Gelb, Phys. Rev. D 34 (1986) 969.

[17] T. K. Kuo and J. Pantaleone, Rev. Mod. Phys. 61 (1989) 937.

[18] N. Hata and P. Langacker, Phys. Rev. D 56 (1997) 6107.

[19] H. Minakata and H. Nunokawa, Phys. Rev. D 38 (1988) 3605.

[20] A. Burrows, Annu. Rev. Nucl. Sci. 40 (1990) 181.

[21] A. Burrows, D. Klein, and R. Gandhi, Phys. Rev. D 45 (1992) 3361.

[22] A. Suzuki (Super-Kamiokande collaboration), talk given at the 'International Symposium on Lepton and Baryon Number Violation', Trento, 1998.

[23] M. Appollonio *et al.*, Phys. Lett. B. 420 (1998) 397.

[24] C. Athanassopoulos *et al.*, Phys. Rev C 54 (1996) 2685; Phys. Rev. Lett. 77(1996) 3082.

[25] C. Athanassopoulos *et al.*, Phys. Rev C 58 (1998) 2489; Phys. Rev. Lett. 81 (1998) 1774.

[26] H. A. Bethe, Phys. Rev. Lett. 56 (1986) 1305.

[27] T. P. Walker and D. N. Schramm, Phys. Lett. B 195 (1987) 331.

[28] K. Hirata *et. al.*, Phys. Rev. Lett. 58 (1987) 1490; R. M. Bionta, *ibid*, 1494.

[29] B. Jegerlehner, F. Neubig and G. Raffelt, Phys. Rev. D 54 (1996) 1194.

[30] A. Yu. Smirnov, D. N. Spergel and J. N. Bahcall, Phys. Rev. D 49 (1994) 1389.

[31] P. J. Kernan and L. M. Krauss, Nucl. Phys. B 437 (1995) 243.

PART 5

Lepton Family Number Violation

R-parity violation and $\mu \to e\gamma$

Katri Huitu

Helsinki Institute of Physics
P.O.Box 9, FIN-00014 University of Helsinki

Abstract. We review the effect of R-parity violation in the muon decay to electron and photon in the framework of MSSM extended by R-parity violating couplings. We discuss all the relevant supersymmetric contributions to $\mu \to e\gamma$, including R-parity conserving ones. We consider also a related process, photonic μ–e conversion, which gives stringent bounds for the products of the couplings, and we list these bounds.

1. Introduction

It is well known that the baryon and lepton numbers, B and L, respectively, are conserved in the Standard Model (SM) of particle physics. In the minimal supersymmetric standard model (MSSM) this is not the case, and the B and L conservation are often imposed on the model by the so-called R-parity [1] conservation,

$$R_p = (-1)^{3(B-L)+2S}, \tag{1}$$

where R_p is +1 for Standard Model particles and -1 for their superpartners.

On the other hand, the supersymmetry and gauge symmetry alone allow also R_p breaking terms [2]

$$W_{\not{L}} = \lambda_{ijk} L_i L_j E_k^c + \lambda'_{ijk} L_i Q_j D_k^c - \epsilon_i L_i H_2 \tag{2}$$

and

$$W_{\not{B}} = \lambda''_{ijk} U_i^c D_j^c D_k^c \tag{3}$$

with $\lambda_{ijk} = -\lambda_{jik}$ and $\lambda''_{ijk} = -\lambda''_{ikj}$. Thus there are 45 new supersymmetric parameters in the model compared to the MSSM.

Perhaps the best motivation for imposing the R-parity conservation is that it yields a natural candidate for cold dark matter, whereas the model with R-parity violation does not contain such a candidate. Another questionable feature in the R-parity violating version of the MSSM is that simultaneous existence of λ, λ' couplings with λ'' lead to fast proton decay, unless the couplings are strictly bounded [3]. In order to prevent the proton from decaying, it is however enough to forbid a subset of the R-parity violating terms. If baryon number conservation or lepton number conservation is required instead of R_p conservation, the proton does not decay. This kind of requirement is reasonable, since there are examples of models, in which only the lepton number is broken. E.g. in supersymmetric left-right model based on the gauge group $SU(2)_L \times SU(2)_R \times U(1)_{B-L}$, the R_p is conserved by gauge symmetry, but spontaneously broken by $\langle\tilde{\nu}\rangle$ to preserve the electromagnetism [4].

The strengths of the explicitly R-parity violating couplings have been studied from a number of different sources, e.g. charged current universality, $e-\mu-\tau$ universality, forward-backward asymmetry, $\nu_\mu e$ scattering, atomic parity violation [5], neutrinoless double beta decay [6], ν masses [7, 8, 9], Z-boson partial width [10], D^+, τ-decays [11], $D^0-\overline{D}^0$ mixing, K^+, t-quark decays [12], $b\bar{b}$ production at LEP II [13] and heavy nuclei decay and $n-\overline{n}$ oscillation [14]. The bounds from these works have been collected and updated in recent reviews [15].

In this talk we'll review the muon decay to electron and photon in MSSM extended by the R_p violating couplings given in Eq.(2). When considering processes involving explicitly R_p violating couplings, special care is needed in specifying the bases in which the fermionic states are given. After discussing the dependence on the basis, we'll consider separately the contributions from R-parity preserving and violating couplings. Then we'll list the numerical results from a related process, μ–e conversion, for which the present (and foreseeable) numerical results seem to be better. However, we want to stress the use of $\mu \to e\gamma$ as a precision measurement. If the transition is observed, $\mu \to e\gamma$ can be used to severely constrain certain classes of models.

2. Basis dependence

In bounding the R-parity breaking couplings, there are ambiguities due to the basis chosen for the fermionic states. For the quark states, assumptions have to be made when choosing the basis of mass eigenstates, while in the leptonic sector, the mere identification of lepton is not well defined.

In the quark sector of the Standard Model, the rotation from gauge eigenstates to mass eigenstates is done by the Cabibbo-Kobayashi-Maskawa -matrix V_{CKM}, which appears in the Lagrangian of charged currents of quarks

$$\mathcal{L}_{cc} = \frac{g}{\sqrt{2}} \bar{u}_L \gamma_\mu V_{CKM} d_L W^{\mu +} + h.c. \tag{4}$$

with family indices suppressed and $V_{CKM} = L_u^\dagger L_d$. The matrices $L_{u,d}$ and $R_{u,d}$ rotate the quark eigenstates,

$$\psi_L' = L\psi_L \text{ and } \psi_R' = R\psi_R, \tag{5}$$

where $\psi = u$ or d and the primes indicate the gauge eigenstates. Since V_{CKM} is the only consequence of rotation, one can choose to rotate e.g. only the down quark gauge eigenstates. This also means that so far experimental information is available only for V_{CKM}.

In a supersymmetric model with explicit R-parity violation, the situation is qualitatively different [3, 12]. The lepton number violating interactions involving the quark rotations are

$$\lambda'_{ijk}(N_i D'_j D'^c_k - E_i U'_j D'^c_k), \tag{6}$$

where the gauge eigenstates of the quark superfields are denoted by primes. Thus in terms of the mass eigenstates we get

$$\lambda'_{ijk}(R_d^\dagger)_{lk}(L_u)_{js}[N_i(V_{CKM})_{sm} D_m D_l^c - E_i U_s D_l^c]. \tag{7}$$

It is evident that the bounds on the λ' couplings in the original superpotential depend on the separate rotations of the up and down type quarks, and one has to be careful in specifying the basis in which the bounds are calculated. Here we will give the bounds for $\lambda' R_d^\dagger L_u$, which will be denoted simply by λ'_{ijk}. The choice of the basis is crucial [16], since the bounds are usually obtained by assuming that only one (or one pair) of the couplings dominates. Then the different choices for bases correspond to different assumptions made in the model.

Another complication related to the basis is that in a general case the superpotential contains both bilinear and trilinear lepton number violating interactions, among which the R-parity violation can be moved by suitable field redefinitions [7]. The lepton basis independent measures of R-parity violation have been discussed in [17]. We'll discuss the contributions coming from bilinear terms in the next section, but the numerical results given in this review assume that the bilinear lepton number violating interactions vanish.

3. The process $\mu \to e\gamma$

Rare processes provide a powerful tool to test physics beyond the Standard Model. Several rare processes have been used to set bounds on the products of R-parity violating couplings. The bounds for products of couplings are typically more stringent than for individual couplings. An advantage of using $\mu \to e\gamma$ for that purpose is that it cannot occur in the SM. Thus, if the process is observed, it is always evidence for physics beyond the SM. Another useful property of $\mu \to e\gamma$ is that as a leptonic process it does not contain large hadronic uncertainties.

In supersymmetric models there are two types of graphs that contribute to lepton family number violating muon decay. First there are contributions in which separate lepton family numbers are broken, but R_p remains unbroken. Secondly, there are contributions from total lepton number and R_p violating graphs. The relative magnitudes of the different contributions depend on the model.

The $\mu \to e\gamma$ amplitude can be parametrized as

$$T(\mu \to e\gamma) = \epsilon^{\lambda}\overline{u}_e(p')(A + B\gamma_5)i\sigma_{\lambda\nu}q^{\nu}u_{\mu}(p), \tag{8}$$

where p, p' and q are the momenta of muon, electron and photon, respectively. ϵ^{λ} is the photon polarization vector and $\sigma_{\mu\nu} = \frac{i}{2}[\gamma_\mu, \gamma_\nu]$. The amplitude T is nonvanishing only when the initial and final state leptons have different helicities. Since T is a dimension five operator the decay can happen only at loop-level in any renormalizable theory. From the amplitude one finds for the width

$$\Gamma(\mu \to e\gamma) = \frac{m_\mu^3}{8\pi}(|A|^2 + |B|^2). \tag{9}$$

Using then the μ lifetime $\tau_\mu = \frac{192\pi^3}{G_F^2 m_\mu^5}$, the branching ratio for the decay is

$$BR(\mu \to e\gamma) = \frac{24\pi^2}{G_F^2 m_\mu^2}(|A|^2 + |B|^2). \tag{10}$$

Experimentally the decay is not observed and $BR(\mu \to e\gamma) < 4.9 \cdot 10^{-11}$.

4. R-parity conserving non-diagonal mass terms

In supersymmetric models one does not need to have R-parity violation to have flavour changing processes not occurring in the SM. The soft mass terms are generally non-diagonal in the generation space and provide a source for lepton family number violation,

$$-L_{soft} = m_{L\,ij}^2 \bar{L}_i L_j + m_{e_R\,ij}^2 \bar{e}_{Ri} e_{Rj} + \left[A_{ij}^e L_i H_1 e_{Rj} + h.c.\right]. \tag{11}$$

These terms change only the flavour, not L_{tot}. Thus R-parity remains unbroken.

The effects of non-diagonal mass terms have been studied assuming either universality of the soft terms at the GUT scale [18] or letting the non-diagonal scalar masses or trilinear soft couplings to be arbitrary [19, 20]. In the first case the soft mass terms and trilinear couplings are given by

$$\begin{aligned} m_{ij}^2 &= m_0^2\delta_{ij}, \\ A_{ij} &= A_0 m_{3/2} h_{ij}, \end{aligned} \tag{12}$$

where i, j are the generation indices and m_0 and A_0 are the universal soft mass and trilinear coupling at the GUT scale, and h_{ij} are the Yukawa couplings. It was found in [18] that the RGE evolution from the GUT scale to the electroweak scale is not fast enough to violate the experimental bounds.

If the assumption of universality is discarded, one can assume large family non-diagonal masses at the unification scale, e.g. $m_{ij} \sim m_{ii}$, $i \neq j$. This situation was studied in [20] assuming $A_{ij} = 0$ TeV, and a strong bound on the gaugino mass was obtained,

$$m_{1/2}^2/m_{ij} \gtrsim 34 \text{ TeV}. \tag{13}$$

When $m_{1/2}^2 \gg m_{ii}^2$, the ratio m_{ij}^2/m_{ii}^2 will in general be small at low energies due to the RGE running [20].

5. Explicit R-parity violation

The other possibility giving rise to the decay $\mu \to e\gamma$ is that R-parity is violated. The violation can be explicit or spontaneous. Here we will consider only the MSSM extended by the R-parity violating couplings. The superpotential includes both the trilinear, dimensionless, and bilinear, dimensionful, terms and both of them can separately contribute to the decay $\mu \to e\gamma$.

5.1. Trilinear terms

The relevant R-parity violating interactions can be found using the superpotential $W_{\not{R}}$ given in Eq. (2). The interactions with dimensionless couplings in the Lagrangian are the following:

$$\begin{aligned} L_{\not{R}} = {} & \lambda_{ijk}(\bar{\nu}_{Li}^c e_{Lj}\tilde{e}_{Rk}^* - e_{Li}\bar{\nu}_{Lj}^c\tilde{e}_{Rk}^* + \nu_{Li}\tilde{e}_{Lj}\bar{e}_{Rk} \\ & - e_{Li}\tilde{\nu}_{Lj}\bar{e}_{Rk} + \tilde{\nu}_{Li}e_{Lj}\bar{e}_{Rk} - \tilde{e}_{Li}\nu_{Lj}\bar{e}_{Rk}) + h.c. \\ & + \lambda'_{ijk}(\bar{\nu}_{Li}^c d_{Lj}\tilde{d}_{Rk}^* - \bar{e}_{Ri}^c u_{Lj}\tilde{d}_{Rk}^* + \nu_{Li}\tilde{d}_{Lj}\bar{d}_{Rk} \\ & - e_{Li}\tilde{u}_{Lj}\bar{d}_{Rk} + \tilde{\nu}_{Li}d_{Lj}\bar{d}_{Rk} - \tilde{e}_{Li}u_{Lj}\bar{d}_{Rk}) + h.c. \end{aligned} \tag{14}$$

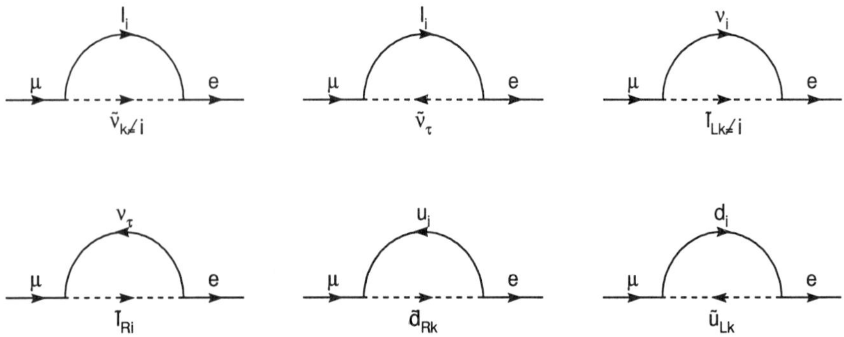

Figure 1. Contributing loop diagrams for μ–e conversion. Unless otherwise indicated $i, k = 1 \ldots 3$.

with the contributing graphs (the photon line is attached in all possible ways) depicted in Fig. 1.

All the amplitudes can be written in terms of [21]

$$A_1 \quad = \quad \frac{\lambda_1 \lambda_2 Q m_\mu}{16\pi^2 m_{\tilde{f}}^2} \frac{1}{6(\kappa-1)^3}(\kappa^2 - 5\kappa - 2 + \frac{6\kappa}{\kappa - 1}\log\kappa), \qquad (15)$$

and

$$A_2 \quad = \quad \frac{\lambda_1 \lambda_2 Q m_\mu}{16\pi^2 m_{\tilde{f}}^2} \frac{1}{6(\kappa-1)^3}(2\kappa^2 + 5\kappa - 1 - \frac{6\kappa^2}{\kappa - 1}\log\kappa), \qquad (16)$$

where $\lambda_{1,2}$ are some $\lambda-$ or $\lambda'-$type couplings. $A_{1,2}$ depend on the charge Q of the particle attached to the photon and the sfermion mass $m_{\tilde{f}}$ in the loop, as well as on the ratio κ of masses of fermion and sfermion in the loop, $\kappa = m_f^2/m_{\tilde{f}}^2$. When $\kappa \to 0$,

$$A_1,\ 2A_2 \to \frac{\lambda_1 \lambda_2 Q m_\mu}{48\pi^2 m_{\tilde{f}}^2}. \qquad (17)$$

The approximation $\kappa \to 0$ is adequate for the first two families. The proportionality to m_μ reflects the helicity flip on the external μ line. Since the external electron is approximated massless, there are no terms proportional to m_e. For the B in Eq. (8), $|B| = |A|$.

The amplitudes depend in a nontrivial way on the sfermion masses, and thus the magnitude of the contributions depend on the spectrum of the model in addition to the R_p violating couplings.

5.2. Bilinear terms

The scalar potential of the model can be found from the superpotential, and it is given by

$$\begin{aligned} V &= V_R + V_{\not{R}} \\ &= V_R + 2(m^2_{H_1 L_i} + \mu\epsilon_i) H_1 \tilde{\nu}_i + (m^2_{L_{ij}} + \epsilon_i \epsilon_j)\tilde{\nu}_i \tilde{\nu}_j \\ &\quad + 2B_i \tilde{\nu}_i H_2 + \frac{1}{4}(g_1^2 + g_2^2)\tilde{\nu}_i^2 (H_1^2 - H_2^2). \end{aligned} \tag{18}$$

Since lepton number is not conserved in the model, L_i and H_1 have the same quantum numbers and one can perform a field rotation to remove ϵ_i [7]. For small $m_{L_{ij}}$, and $\langle\tilde{\nu}\rangle/\langle H\rangle$, one then finds [22]

$$\langle\tilde{\nu}_i\rangle = -\frac{B_i v_2 + m^2_{H_1 L_i} v_1}{m^2_{L_i} + \frac{1}{2} m_Z^2 \cos 2\beta}, \tag{19}$$

which are constrained by the neutrino masses $\sim (g_1^2 + g_2^2)\langle\tilde{\nu}_i\rangle^2/(2m_{\tilde{\chi}^0})$. If one assumes that ϵ_i, $B_i = 0$ at high scale, the sneutrino VEVs are generated by loop effects.

The nonzero VEV for sneutrino resembles closely the spontaneous R-parity breaking, where the R-parity is broken exactly because the $\tilde{\nu}$ gets a VEV. The consequences are also similar, namely mixing the leptonic and Higgs sectors. Thus neutrinos mix with neutralinos, charged leptons with charginos, neutral Higgses with sneutrinos, and charged Higgses with charged sleptons.

When the field redefinitions are done so that ϵ_i disappear at some energy scale, the terms reappear in other scales due to the RGE running. The effects coming from other than R_p violating couplings are found to be small [8, 23]. The amplitude due to the effects of RGE running of R-parity violating couplings were analyzed in [22], when universality and vanishing of ϵ_i were assumed at the GUT scale. It was found that the nondiagonal soft masses at the weak scale are proportional to R_p violating couplings times the diagonal soft masses [22],

$$\frac{\Delta m_{\tilde{e}_L \tilde{\mu}_L}}{m^2_{\tilde{e}_L}}, \frac{\Delta m_{\tilde{e}_R \tilde{\mu}_R}}{m^2_{\tilde{e}_R}}, \sim \mathcal{O}(\lambda_{1ij}\lambda_{2ij}) \text{ or } \mathcal{O}(\lambda'_{1ij}\lambda'_{2ij}). \tag{20}$$

These contributions can be large.

6. Photonic coherent $\mu - e$ conversion

A process closely related to the $\mu \to e\gamma$ is the μ–e conversion. Due to the precise measuments in ongoing and planned μ–e conversion experiments, it will be one of the main probes of muon flavour conservation and thus of many extensions of the Standard Model [24, 25]. μ–e can occur at loop level via box or penguin diagrams and in the case of R_p-violation also at tree-level. The loop induced conversion is partly sensitive to different couplings than the tree-level process. Moreover, the μ–e conversion at one-loop is enhanced by large logarithms [26].

Here we will consider only the photonic mechanism, which has $\mu \to e\gamma$ as a subprocess [27], so that the relevant μ–e conversion effective Lagrangian can be expressed as

$$\mathcal{L}_{eff} = \frac{4\pi\alpha}{q^2}\, j^{\lambda}_{(ph)} J^{(ph)}_{\lambda} + \text{non} - \text{photonic}, \tag{21}$$

where q^2 denotes the photon momentum transfer. The photonic mechanism is enhanced by small q^2 but can occur only at loop level and is thus suppressed by loop factors. The non-photonic mode can happen at tree-level [28], but one may expect that both photonic and non-photonic modes are of interest, because of the enhancement of the photonic mode. The leptonic current is given by

$$j^{\lambda}_{(ph)} = \bar{e}\left[(f_{E0} + \gamma_5 f_{M0})\, \gamma_\nu \left(g^{\lambda\nu} - \frac{q^\lambda q^\nu}{q^2} \right) + (f_{M1} + \gamma_5 f_{E1})\, i\, \sigma^{\lambda\nu} \frac{q_\nu}{m_\mu} \right] \mu\,, \tag{22}$$

and the hadronic current by ($\bar{\Psi} = (\bar{p}, \bar{n})$)

$$J^{(ph)}_{\lambda} = \bar{\Psi}\gamma_\lambda \frac{1}{2}(1 + \tau_3)\Psi\,. \tag{23}$$

Contrary to the $\mu \to e\gamma$ transition, in μ–e conversion the part proportional to γ_ν is nonvanishing, since the photon needs not to be on mass-shell.

For Ti and Pb the coherent conversion rate constitutes about 91-94% of the total rate. For the photonic mechanism the coherent μ–e conversion branching ratio $R^{ph}_{\mu e}$ can be expressed as [29]

$$R^{ph}_{\mu e} = C\, \frac{8\alpha^5\, m^5_\mu\, Z^4_{eff}\, |\overline{F_p}(p_e)|^2 Z}{\Gamma_{capt}} \cdot \frac{\xi^2_0}{q^4}\,, \tag{24}$$

where Γ_{capt} is the total muon capture rate, and

$$\xi^2_0 = |f_{E0} + f_{M1}|^2 + |f_{E1} + f_{M0}|^2 \tag{25}$$

shows the process dependence on the photonic form factors and $\overline{F_p}$ is the proton nuclear form factor.

The first two diagrams in Fig. (1) give an enhanced contribution to the μ–e conversion, while those with neutrinos as an intermediate state are not enhanced, since the photon can be attached only to a charged lepton line. The diagrams involving neutrinos are neglected. When summed over the different contributions for the bounds on a pair of couplings, the form factors proportional to γ_λ, not appearing in the $\mu \to e\gamma$ amplitudes, are [27]

$$f_{E0}^{f=\ell} = \pm f_{M0}^{f=\ell} = -\frac{2\,(\lambda\lambda)}{3\,(4\pi)^2}\frac{-q^2}{m_{\tilde{f}}^2}\left(\ln\frac{-q^2}{m_{\tilde{f}}^2} + F_f(r)\right), \tag{26}$$

$$f_{E0}^{f=u} = -f_{M0}^{f=u} = -\frac{(\lambda'\lambda')N_c}{9\,(4\pi)^2}\frac{-q^2}{m_{\tilde{f}}^2}\left(\ln\frac{-q^2}{m_{\tilde{f}}^2} + F_f(r) - \frac{1}{12}\right), \tag{27}$$

$$f_{E0}^{f=d} = -f_{M0}^{f=d} = -\frac{(\lambda'\lambda')N_c}{18\,(4\pi)^2}\frac{-q^2}{m_{\tilde{f}}^2}\left(\ln\frac{-q^2}{m_{\tilde{f}}^2} + F_f(r) - \frac{1}{3}\right), \tag{28}$$

and the form factors proportional to $\sigma_{\mu\nu}$ are

$$f_{M1}^{f=\ell} = \mp f_{E1}^{f=\ell} = -\frac{(\lambda\lambda)}{3\,(4\pi)^2}\frac{m_\mu^2}{m_{\tilde{f}}^2}, \tag{29}$$

$$f_{M1}^{f=u} = f_{E1}^{f=u} = -\frac{(\lambda'\lambda')N_c}{24\,(4\pi)^2}\frac{m_\mu^2}{m_{\tilde{f}}^2}, \quad f_{M1}^{f=d} = f_{E1}^{f=d} = 0 \tag{30}$$

where

$$F_f(r) = -\frac{1}{3} + 4r + \ln r + (1-2r)\sqrt{1+4r}\,\ln\left(\frac{\sqrt{1+4r}+1}{\sqrt{1+4r}-1}\right) \tag{31}$$

with f = the fermion in the loop, and $r = m_f^2/(-q^2)$, where f stands for the virtual fermion in the loop, $N_c = 3$ is the number of colours. Some interesting limits for F_f are $F_f = -1/3$ when $r \to 0$, $F_f \approx 1.52$ when $r \approx 1$, and $F_f = \ln r + 4/3$ when $r \gg 1$. The form factors (26)–(28) contain large logarithms, while (29) and (30) do not. In calculating (26)–(28) it has been assumed that $\kappa \ll 1$. Although this is not necessarily valid for the third generation, the numerical error is not significant.

Substituting the numerical values of the parameters in Eq.(24), one has [27]

$$R_{\mu e}^{ph} = 8.6\,(12.3)\cdot 10^{-4}\ \mathrm{TeV}^4\,\frac{\xi_0^2}{q^4}, \tag{32}$$

where the first number in the expression corresponds to the conversion in Ti and the number in the brackets to the conversion in Pb and Au. The μ–e transition has not been observed so far, $R_{\mu e}^{Ti} \lesssim 4.3\cdot 10^{-12}$ [31].

	previous bounds	μ–e at loop level/$\sqrt{B}$	
	$m_{\tilde{f}} =$ 100 GeV	$m_{\tilde{f}} =$ 100 GeV	1 TeV
$\lvert\lambda_{121}\,\lambda_{122}\rvert$	$6.6 \cdot 10^{-7}$ [33]	$4.2 \cdot 10^{-6}$	$3.2 \cdot 10^{-4}$
$\lvert\lambda_{131}\,\lambda_{132}\rvert$	$6.6 \cdot 10^{-7}$ [33]	$5.3 \cdot 10^{-6}$	$3.9 \cdot 10^{-4}$
$\lvert\lambda_{231}\,\lambda_{232}\rvert$	$5.7 \cdot 10^{-5}$ [21]	$5.3 \cdot 10^{-6}$	$3.9 \cdot 10^{-4}$
$\lvert\lambda_{231}\,\lambda_{131}\rvert$	$6.6 \cdot 10^{-7}$ [33]	$8.4 \cdot 10^{-6}$	$6.4 \cdot 10^{-4}$
$\lvert\lambda_{232}\,\lambda_{132}\rvert$	$1.1 \cdot 10^{-4}$ [21]	$8.4 \cdot 10^{-6}$	$6.4 \cdot 10^{-4}$
$\lvert\lambda_{233}\,\lambda_{133}\rvert$	$1.1 \cdot 10^{-4}$ [21]	$1.7 \cdot 10^{-5}$	$1.0 \cdot 10^{-3}$

Table 1. Upper limits on the products $|\lambda\lambda|$ testable in μ–e conversion for two different scalar masses $m_{\tilde{f}}$ =100 GeV and $m_{\tilde{f}}$ =1 TeV. The previous bounds scale quadratically with the sfermion mass. The scaling factor B is defined in the text and currently $B = 1$.

7. Numerical results

The numerical bounds are based on the assumption that none of the separate contributions can be larger than the experimental limit. There could be accidental cancellations between the different contributions, but we do not consider this situation here. Bounds on the products of supersymmetric trilinear couplings are found by assuming that a pair of couplings dominates at the low scale.

For many of the testable $\lambda\lambda$ products, the μ–e conversion bounds are most stringent. In Table 1 it is also seen that they are more stringent than the bounds coming from $\mu \to e\gamma$ [21]. In μ–e conversion $\lambda\lambda$-s can be probed only at loop-level. The $\lambda'\lambda'$-type couplings give rise both to the tree-level [28] and loop-level [27] μ–e conversion. The tree-level bounds derived from this process are very strong for most products of the couplings, but not for all. The reason is that, especially if the third family quarks are involved, the contributions to the μ–e conversion can be strongly suppressed by small off-diagonal CKM matrix elements. In some of the cases ($|\lambda'_{232}\lambda'_{132}|$ and $|\lambda'_{233}\lambda'_{133}|$) the loop-induced μ–e conversion bounds are more stringent than the ones from the tree-level conversion. In μ–e conversion the $\lambda\lambda'$ -type couplings can be investigated as well. These can be probed only at tree-level, giving the most stringent bounds for some of the couplings [27].

The experimental bound on the μ–e will be improved to the level of 10^{-16} [25]. To take into account the future improvements in the experimental sensitivity, one can define a scaling factor B as [27]

$$B = \frac{R^{Ti}_{\mu e}(present\ bound)}{R^{Nu}_{\mu e}(future\ bound)} \times \begin{cases} 1.0, & \text{if } Nu = Ti, \\ 2.8, & \text{if} Nu = Au. \end{cases}$$

	previous bounds	μ–e at tree-level/ $\sqrt{B}$	μ–e at loop level/ $\sqrt{B}$	
	$m_{\tilde{f}} =$ 100 GeV	$m_{\tilde{f}} =$ 100 GeV	$m_{\tilde{f}} =$ 100 GeV	1 TeV
$\|\lambda'_{211}\,\lambda'_{111}\|$	$5.0 \cdot 10^{-6}$	$8.0 \cdot 10^{-8}$	$1.1 \cdot 10^{-5}$	$8.3 \cdot 10^{-4}$
$\|\lambda'_{21k}\,\lambda'_{11k}\|$	$1.4 \cdot 10^{-4}$	$8.5 \cdot 10^{-8}$	$1.1 \cdot 10^{-5}$	$8.3 \cdot 10^{-4}$
$\|\lambda'_{221}\,\lambda'_{111}\|$	$5.0 \cdot 10^{-6}$	$4.0 \cdot 10^{-7}$	$3.7 \cdot 10^{-5}$	$2.5 \cdot 10^{-3}$
$\|\lambda'_{22k}\,\lambda'_{11k}\|$	$4.8 \cdot 10^{-5}$ [33]	$4.3 \cdot 10^{-7}$	$3.7 \cdot 10^{-5}$	$2.5 \cdot 10^{-3}$
$\|\lambda'_{231}\,\lambda'_{111}\|$	$8.8 \cdot 10^{-5}$	$1.0 \cdot 10^{-5}$	$1.2 \cdot 10^{-3}$	$8.3 \cdot 10^{-2}$
$\|\lambda'_{23k}\,\lambda'_{11k}\|$	$4.8 \cdot 10^{-3}$	$1.1 \cdot 10^{-5}$	$1.2 \cdot 10^{-3}$	$8.3 \cdot 10^{-2}$
$\|\lambda'_{211}\,\lambda'_{121}\|$	$4.8 \cdot 10^{-5}$ [33]	$4.0 \cdot 10^{-7}$	$7.3 \cdot 10^{-5}$	$2.5 \cdot 10^{-3}$
$\|\lambda'_{21k}\,\lambda'_{12k}\|$	$4.8 \cdot 10^{-5}$ [33]	$4.3 \cdot 10^{-7}$	$7.3 \cdot 10^{-5}$	$2.5 \cdot 10^{-3}$
$\|\lambda'_{221}\,\lambda'_{121}\|$	$1.4 \cdot 10^{-4}$	$8.0 \cdot 10^{-8}$	$2.0 \cdot 10^{-5}$	$1.2 \cdot 10^{-3}$
$\|\lambda'_{22k}\,\lambda'_{12k}\|$	$1.4 \cdot 10^{-4}$	$2.1 \cdot 10^{-6}$	$2.0 \cdot 10^{-5}$	$1.2 \cdot 10^{-3}$
$\|\lambda'_{231}\,\lambda'_{121}\|$	$2.6 \cdot 10^{-3}$	$2.0 \cdot 10^{-6}$	$4.0 \cdot 10^{-4}$	$2.2 \cdot 10^{-2}$
$\|\lambda'_{23k}\,\lambda'_{12k}\|$	$4.8 \cdot 10^{-3}$	$5.3 \cdot 10^{-5}$	$4.0 \cdot 10^{-4}$	$2.2 \cdot 10^{-2}$
$\|\lambda'_{211}\,\lambda'_{131}\|$	$4.2 \cdot 10^{-4}$	$1.0 \cdot 10^{-5}$	$1.2 \cdot 10^{-3}$	$8.3 \cdot 10^{-2}$
$\|\lambda'_{212}\,\lambda'_{132}\|$	$4.0 \cdot 10^{-3}$	$1.1 \cdot 10^{-5}$	$1.2 \cdot 10^{-3}$	$8.3 \cdot 10^{-2}$
$\|\lambda'_{213}\,\lambda'_{133}\|$	$1.2 \cdot 10^{-5}$	$1.1 \cdot 10^{-5}$	$1.2 \cdot 10^{-3}$	$8.3 \cdot 10^{-2}$
$\|\lambda'_{221}\,\lambda'_{131}\|$	$4.2 \cdot 10^{-4}$	$2.0 \cdot 10^{-6}$	$4.0 \cdot 10^{-4}$	$2.2 \cdot 10^{-2}$
$\|\lambda'_{222}\,\lambda'_{132}\|$	$4.0 \cdot 10^{-3}$	$5.3 \cdot 10^{-5}$	$4.0 \cdot 10^{-4}$	$2.2 \cdot 10^{-2}$
$\|\lambda'_{223}\,\lambda'_{133}\|$	$1.2 \cdot 10^{-5}$	$5.3 \cdot 10^{-5}$	$4.0 \cdot 10^{-4}$	$2.2 \cdot 10^{-2}$
$\|\lambda'_{231}\,\lambda'_{131}\|$	$2.6 \cdot 10^{-3}$ [21]	$8.0 \cdot 10^{-8}$	$8.7 \cdot 10^{-5}$	$3.2 \cdot 10^{-3}$
$\|\lambda'_{232}\,\lambda'_{132}\|$	$2.6 \cdot 10^{-3}$ [21]	$1.3 \cdot 10^{-3}$	$8.7 \cdot 10^{-5}$	$3.2 \cdot 10^{-3}$
$\|\lambda'_{233}\,\lambda'_{133}\|$	$4.0 \cdot 10^{-4}$	$1.3 \cdot 10^{-3}$	$8.7 \cdot 10^{-5}$	$3.2 \cdot 10^{-3}$

Table 2. Upper limits on the products $|\lambda'\lambda'|$ for two different scalar masses $m_{\tilde{f}}$ =100 GeV and $m_{\tilde{f}}$ =1 TeV. The previous bounds and the tree-level bounds scale quadratically with the sfermion mass. The scaling factor B is defined in the text and currently $B = 1$. In the Table $k = 2, 3$.

To get new bounds one needs to divide the bounds in Tables 1 and 2 by the corresponding $\sqrt{B}$.

The sensitivity for $\mu \to e\gamma$ is expected to improve to 10^{-14} [32] and it can be taken into account by a corresponding B factor. With the planned improvements the bounds coming from μ–e conversion remain stronger.

8. Conclusions and discussion

The $\mu - e$ conversion leads to excellent limits on several pairs of couplings. In some of the cases, the best limits come from the loop-induced process.

On the other hand, even if the decay $\mu \to e\gamma$ does not give as good bounds on the couplings, the advantage of this process is that it has no uncertainties typical for processes involving hadronic parts and it can be accurately calculated. In a general case, the calculation includes a large number of unknown parameters, which have to be determined from the experiments.

If the transition is measured, it can be used to strongly constrain models, and in some cases also discard them. For instance, let us imagine a situation where in addition one supersymmetric partner is found. Then one can consider an interesting class of gauge mediated supersymmetry breaking models, where soft bilinear and trilinear couplings vanish at messenger scale. In these models the particle spectrum can be determined for a given messenger sector [34]. Thus the observation of $\mu \to e\gamma$ can fix the magnitude of the dominating pair of couplings.

Acknowledgement

It is a pleasure to thank professor Klapdor-Kleingrothaus and Dr. Krivosheina for creating an inspiring atmosphere during the meeting and for their warm hospitality.

References

[1] G. Farrar and P. Fayet, Phys. Lett. **B76** (1978) 575.

[2] S. Weinberg, Phys. Rev. **D26** (1982) 287; N. Sakai and T. Yanagida, Nucl. Phys. **B197** (1982) 533; C.S. Aulakh and R.N. Mohapatra, Phys. Lett. **B119** (1982) 136; L.J. Hall and M. Suzuki, Nucl. Phys. **B231** (1984) 419; J. Ellis *et al.*, Phys. Lett. **B150** (1985) 142; S. Dawson, Nucl. Phys. **B261** (1985) 297; S. Dimopoulos and L.J. Hall, Phys. Lett. **B207** (1987) 210.

[3] A. Smirnov and F. Vissani, Phys. Lett. **B380** (1996) 317.

[4] R. Kuchimanchi and R.N. Mohapatra, Phys. Rev. **D48** (1993) 4352; K. Huitu and J. Maalampi, Phys. Lett. **B344** (1995) 217.

[5] V. Barger, G.F. Giudice and T. Han, Phys. Rev. **D40** (1989) 2987.

[6] R.N. Mohapatra, Phys. Rev. **D34** (1986) 3457; M. Hirsch, H.V. Klapdor-Kleingrothaus and S.G. Kovalenko, Phys. Rev. Lett. **75** (1995) 17, and Phys. Rev. **D53** (1996) 1329.

[7] L.J. Hall and M. Suzuki, in Ref. [2].

[8] R. Hempfling, Nucl. Phys. **B478** (1996) 3.

[9] S. Dimopoulos and L.J. Hall, in Ref. [2] ; R.M. Godbole, P. Roy and X. Tata, Nucl. Phys. **B401** (1993) 67; A.S. Joshipura, V. Ravindran and S.K. Vempati, hep-ph/9706482.

[10] G. Bhattacharyya, J. Ellis and K. Sridhar, Mod.Phys.Lett. **A10** (1995) 1583; G. Bhattacharyya, D. Choudhury and K. Sridhar, Phys. Lett. **B355** (1995) 193; J. Ellis, S. Lola and K. Sridhar, Phys. Lett. **B408** (1997) 252.

[11] G. Bhattacharyya and D. Choudhury, Mod. Phys. Lett. **A10** (1995) 1699.

[12] K. Agashe and M. Graesser, Phys. Rev. **D54** (1996) 4445.

[13] J. Erler, J.L. Feng and N. Polonsky, Phys. Rev. Lett. **78** (1997) 3063.

[14] R. Barbieri and A. Masiero, Nucl. Phys. **B267** (1986) 679; S. Dimopoulos and L.J. Hall, Phys. Lett. **B196** (1987) 135; J.L Goity and M. Sher, Phys. Lett. **B346** (1995) 69, erratum Phys. Lett. **B385** (1996) 500.

[15] G. Bhattacharyya, Nucl. Phys. Proc. Suppl. **52A** (1997) 83; and IFUP-TH 43/97, hep-ph/9709395; H. Dreiner, hep-ph/9707435.

[16] K. Huitu, K. Puolamäki, D.-X. Zhang, preprint HIP-1998-50/TH (1998).

[17] S. Davidson, J. Ellis, Phys. Rev. **D56** (1997) 4182 and Phys. Lett. **B390** (1997) 210.

[18] T. Kosmas, G.K. Leontaris, J.D. Vergados, Phys. Lett. **B219** (1989) 457; F. Gabbiani, A. Masiero, Nucl. Phys. **B322** (1989) 235.

[19] D. Choudhury, F. Eberlein, A. König, J. Louis, S. Pokorski, Phys. Lett. **B342** (1995) 180.

[20] B. de Carlos, J.A. Casas, J.M. Moreno, Phys. Rev. **D53** (1996) 6398.

[21] M. Chaichian and K. Huitu, Phys. Lett. **B384** (1996) 157.

[22] B. de Carlos, P.L. White, Phys. Rev. **D54** (1996) 3427.

[23] B. Mukhopadhay, S. Roy, Phys. Rev. **D55** (1997) 7020.

[24] A. van der Schaaf (spokesman of SINDRUM II), PSI proposal R-87-03.

[25] MECO collaboration, M. Bachman *et al.*, BNL Proposal P940, 1997.

[26] M. Raidal and A. Santamaria, Phys. Lett. **B421** (1998) 250.

[27] K. Huitu, J. Maalampi, M. Raidal, A. Santamaria, Phys. Lett. **B430** (1998) 310.

[28] J.E. Kim, P. Ko and D.-G. Lee, Phys. Rev. **D56** (1997) 100.

[29] H.C. Chiang *et al.*, Nucl. Phys. **A559** (1993) 526; H.C. Chiang, E. Oset and P. Fernandez de Cordoba, Nucl. Phys. **A510** (1990) 591.

[30] D. Choudhury and P. Roy, Phys. Lett. **B378** (1996) 153; G. Bhattacharyya, A. Raychoudhury, hep-ph/9712245.

[31] SINDRUM II Collaboration, C. Dohmen *et al.*, Phys. Lett. **B317** (1993) 631.

[32] P. Wintz, these proceedings.

[33] D. Choudhury and P. Roy, Phys. Lett. **B378** (1996) 153; G. Bhattacharyya, A. Raychoudhury, Phys. Rev. **D57** (1998) 3837.

[34] H. Hamidian, K. Huitu, K. Puolamäki, D.-X. Zhang, Phys. Lett. **B428** (1998) 310.

Rare decays of kaons and muons[1]

Douglas Bryman[2]

TRIUMF, 4004 Wesbrook Mall,Vancouver, B. C. Canada V6T 2A3

Abstract. Important aspects of the generation puzzle are being addressed by experiments dealing with rare and ultra-rare kaon and muon decays. The Standard Model prediction for $K^+ \to \pi^+\nu\bar{\nu}$ is being challenged in an experiment at BNL. Progress has been made at Fermilab on the search for the unique CP-violating process $K_L^0 \to \pi^0\nu\bar{\nu}$, and new experiments are being developed to measure it. Searches for lepton flavor violation in $\mu \to e$ conversion, $K_L^0 \to \mu e$, muonium-anti-muonium conversion and other reactions are reaching unprecedented levels of sensitivity.

1. Introduction

Fifty years after the discovery of the strange quark by Rochester and Butler[1] and after the first search for $\mu \to e\gamma$ by Hincks and Pontecorvo,[2] the apparent redundancy of the generations (or families, flavors, etc.) of quarks and leptons remains a striking puzzle. In the quark sector, there are three generations of quark doublets which are related by the Standard Model (SM) Cabibbo-Kobayashi-Maskawa (CKM) mixing matrix. Unitarity of the CKM matrix with at least three generations of quarks allows for the "technical" accommodation of a complex CP-violating phase parameter, possibly providing a clue to the generation puzzle. Assuming that the product CPT is invariant, as required by general principles of relativistic quantum field theory, CP violation also implies T violation. This may be of more than passing importance since understanding CP (T)violation may

[1] A version of this paper will also appear in the Proc. WEIN Symposium, Santa Fe, eds. C. Hoffman, P. Herczeg, World Sci. (1998).

[2] E-mail: doug@triumf.ca

provide insight into phenomena as basic as the dominance of matter over anti-matter in the universe.

The lepton generation puzzle is manifested by three generations of "electrons" (e, μ, τ) and their associated neutrinos that appear to be isolated replications of each other, except for mass. Perhaps some global or accidental symmetry is responsible for the apparent conservation of lepton flavor or, as in many alternate theories, that lepton flavor-violating interactions are suppressed at very high mass scales. In addition, there seems to be an obvious, but currently unexplained, symmetry between the quarks and leptons (including the equality of the charge of the electron and proton, the matching pairs of doublets, etc.). Lepton flavor-violating (LFV) reactions are being pursued vigorously because in most extensions of the SM, these reactions occur naturally. Great progress on LFV experiments continues to be made and there is no indication that the limits of achievable sensitivity have been exhausted.

Rare and ultra-rare K decays play an important role in the search for information on the generation puzzle. Higher order K decays are potentially fertile testing grounds because nearly the entire known spectrum of quarks, leptons, and gauge bosons must interact in the internal loops which lead to the final observed states. The ultra-rare decays $K^+ \rightarrow \pi^+\nu\overline{\nu}$ and $K_L^0 \rightarrow \pi^0\nu\overline{\nu}$ may be among the best sources of new information on little known SM parameters such as top quark mixing and the CP-violating phase in the SM CKM matrix. These decays are especially attractive candidates for study because hadronic effects are well known from $K \rightarrow \pi e\nu$ measurements and the simple final states allow unusually precise calculations to be made.

In the following, I will discuss the status some rare K and μ decays which bear directly on the questions of CP violation, quark mixing and the search for new physics to solve the generation puzzle.

2. $K \rightarrow \pi\nu\overline{\nu}$ decays.

2.1. *Theory*

$K^+ \rightarrow \pi^+\nu\overline{\nu}$ presents a unique opportunity for evaluating the detailed predictions of higher order weak interactions in the SM because the second order weak effects shown in fig. 1 account for virtually the entire rate. $K^+ \rightarrow \pi^+\nu\overline{\nu}$ has comparable contributions in the SM due to top and charm quark exchange but is free of long-distance (i.e. mesonic, photonic exchange) contributions down to the 10^{-13} level.[3] Thus, short distance contributions (i.e. quark, lepton and gauge boson exchanges) dominate and SM or new physics effects, can be cleanly extracted.

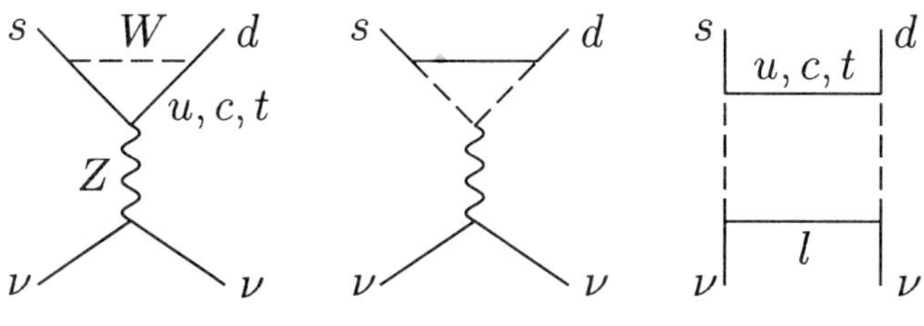

Figure 1. Second order weak Feynman diagrams leading to $K^+ \to \pi^+ \nu\overline{\nu}$ decay.

In the SM, $K^+ \to \pi^+ \nu\overline{\nu}$ is sensitive to important poorly known parameters (e.g. the magnitude of the coupling of the top quark to the down quark, V_{td}). Calculations by Buchalla, Buras and collaborators[4] beyond leading logarithms (using a two-loop renormalization group analysis for the charm contribution and $O(\alpha_s)$ for top) indicate theoretical uncertainties no greater than 5% for typical SM parameters. This result emphasizes the importance of $K^+ \to \pi^+ \nu\overline{\nu}$ as a higher order weak process which can be calculated precisely and unambiguously in the SM. The present range of branching ratios in the SM is $(0.6 - 1.5) \times 10^{-10}$.

$K_L \to \pi^0 \nu\overline{\nu}$ is also a flavor-changing neutral current (FCNC) process that is induced through the loop effects of fig. 1. $K_L \to \pi^0 \nu\overline{\nu}$ is almost entirely due to direct CP violation where only the imaginary part of V_{td} survives in the amplitude. Using present estimates for SM parameters, the branching ratio for $K_L^0 \to \pi^0 \nu\overline{\nu}$ is expected to lie in the range of $\mathrm{B}(K_L \to \pi^0 \nu\overline{\nu}) = (3 \pm 2) \cdot 10^{-11}$.[4]

Taken together $K^+ \to \pi^+ \nu\overline{\nu}$ and $K_L^0 \to \pi^0 \nu\overline{\nu}$ can be used to determine the unitarity triangle illustrated in fig. 2. A clean measure of the height of the unitarity triangle is provided by the $K_L \to \pi^0 \nu\overline{\nu}$ branching ratio. V_{td} (essentially the side from (ρ, η) to (1,0) can be accurately determined by measuring $K^+ \to \pi^+ \nu\overline{\nu}$ after a small correction is applied for the charm quark contribution as illustrated in fig 2. Precise determination of the CKM parameters in the K system would provide a basis for comparison with other observables like $B \to \pi l \nu$, x_s/x_d or V_{cb} from $b \to c$ transitions. Any significant deviation from independent determinations would point to new physics.

2.2. $K^+ \to \pi^+ \nu\overline{\nu}$ Experiment: E787

The E787 group working at the Alternating Gradient Synchrotron of Brookhaven National Laboratory has reported evidence for the reaction

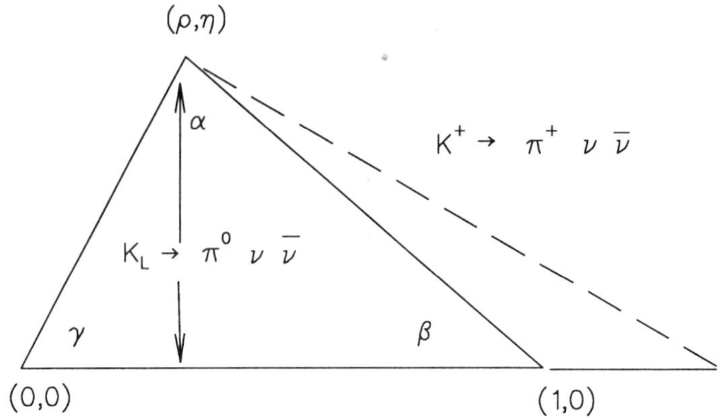

Figure 2. The unitarity triangle. ρ and η are parameters in the Wolfenstein representation of the CKM matrix.

$K^+ \to \pi^+\nu\overline{\nu}$ based on the observation of a single clean event[5]. The experiment employs an advanced design low energy kaon beam produced by the intense 30 GeV AGS and a sophisticated detection apparatus located inside a 1 T, 3 m diameter solenoidal magnet. Technically interesting features of E787 include a scintillating fiber target read out by 500 MHz GaAs CCD transient digitizers, a low mass central tracking drift chamber using inflated cathode-strip foils for the z-position measurement, and highly efficient calorimeters employing pure CsI crystals and Pb/scintillator sandwich detectors. Pions are distinguished from muons by kinematics and by observing the $\pi \to \mu \to e$ decay sequence using 500-MHz flash-ADC transient digitizers. The nearly 4π sr photon veto calorimeter is one element of an elaborate scheme to suppress backgrounds from processes which, although similar to $K^+ \to \pi^+\nu\overline{\nu}$, occur at rates of 10^8 to 10^9 higher.

Definitive observation of $K^+ \to \pi^+\nu\overline{\nu}$ requires suppression of all backgrounds to well below the sensitivity for the signal and reliable estimates of the residual background levels. Major background sources include the copious two-body decays $K^+ \to \mu^+\nu_\mu$ ($K_{\mu 2}$) with a 64% branching ratio and P = 236 MeV/c and $K^+ \to \pi^+\pi^0$ ($K_{\pi 2}$) with a 21% branching ratio and P = 205 MeV/c. The only other important background sources are scattering of pions in the beam and K^+ charge exchange (CEX) reactions resulting in decays $K_L^0 \to \pi^+ l^- \overline{\nu}$, where $l = e$ or μ. To suppress the backgrounds, techniques were employed that incorporated redundant kinematic and particle identification measurements and efficient elimination of events with additional particles.

The observation of the first $K^+ \rightarrow \pi^+\nu\overline{\nu}$ candidate event was made in data taken in 1995. Figure 3(a) shows the range in scintillator (R) vs. Energy (E) for the events surviving all other analysis cuts. Only events with measured momentum in the accepted region $211 \leq P \leq 230$ MeV/c are plotted. The rectangular box indicates the signal region specified as range $34 \leq R \leq 40$ cm of scintillator (corresponding to $214 \leq P_\pi \leq 231$ MeV/c) and energy $115 \leq E \leq 135$ MeV ($213 \leq P_\pi \leq 236$ MeV/c) which encloses the upper 16.2% of the $K^+ \rightarrow \pi^+\nu\overline{\nu}$ phase space. The residual events below the signal region clustered at $E = 108$ MeV were due to $K_{\pi 2}$ decays where both photons had been missed. A reconstruction of the candidate event is shown in Fig. 4. For this data set, the estimated background level anticipated was $b = 0.08 \pm 0.03$. If the observed event is due to $K^+ \rightarrow \pi^+\nu\overline{\nu}$, the branching ratio is $B(K^+ \rightarrow \pi^+\nu\overline{\nu}) = 4.2^{+9.7}_{-3.5} \times 10^{-10}$.

The observation of an event with the signature of $K^+ \rightarrow \pi^+\nu\overline{\nu}$ is consistent with the expectations of the SM, although the central value is well above it. Based on the result for $B(K^+ \rightarrow \pi^+\nu\overline{\nu})$, $|V_{td}|$ lies in the range $0.006 < |V_{td}| < 0.06$[5]. By the end of the 1998 run, E787 expects to have obtained a single event sensitivity of $< 7 \times 10^{-11}$, about a factor of 7 below the sensitivity of the 1995 data set which produced the observed event.

2.3. $K^0_L \rightarrow \pi^0\nu\overline{\nu}$ Experiments

The KTEV group (E799) recently reported[6] a new preliminary result on the search for $K^0_L \rightarrow \pi^0\nu\overline{\nu}$ using detection of the Dalitz decay of the π^0: $B(K^0_L \rightarrow \pi^0\nu\overline{\nu}) < 5.9 \times 10^{-7}$. Further progress on the search is anticipated by this group. Another $K^0_L \rightarrow \pi^0\nu\overline{\nu}$ experiment was recently proposed at KEK[7] seeking a single event sensitivity of 2×10^{-11} employing a highly collimated "pencil" beam. The Fermilab KTEV/KAMI group is also exploring the possibility of a new search for $K^0_L \rightarrow \pi^0\nu\overline{\nu}$ using a beam produced by the new main injector[8].

Although definitively measuring $K^0_L \rightarrow \pi^0\nu\overline{\nu}$ at levels comparable to the lowest predictions of the SM presents a formidable experimental challenge, a concept for achieving this goal was recently proposed at BNL. The

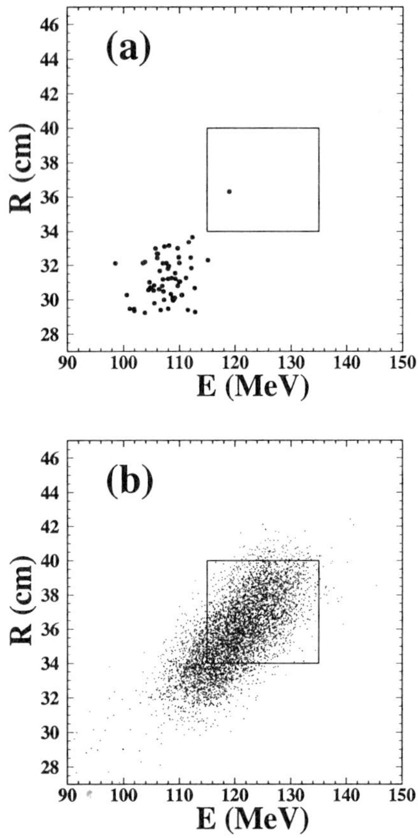

Figure 3. (a) Range (R) vs. energy (E) distribution for the E787 1995 $K^+ \to \pi^+\nu\overline{\nu}$ data set with the final cuts applied. The box enclosing the signal region contains a single candidate event. (b) The Monte Carlo simulation of $K^+ \to \pi^+\nu\overline{\nu}$ with the same cuts applied.

method to be used in BNL E926[9] is based on determining the K_L momentum by time-of-flight so that the detected π^0 can be reconstructed in the kaon center-of-mass system. Using the proposed detector sketched in fig. 5 all possible kinematic measurements on the photons from $\pi^0 \to \gamma\gamma$ decay are made: energy, time, conversion point and angle of the individual photons are accurately determined and a high efficiency veto system insures that no other particles are present in coincidence. Based on photon detection efficiency measurements made in E787 and using the extra

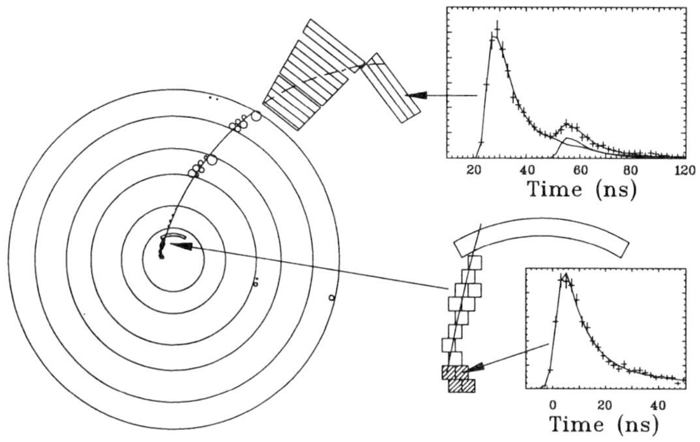

Figure 4. Reconstruction of the $K^+ \rightarrow \pi^+\nu\overline{\nu}$ candidate event[5]. On the left is the end view of the detector showing the track in the target, drift chamber (indicated by drift-time circles), and range stack (indicated by the layers that were hit). At the lower right is a blowup of the target region where the hatched boxes are kaon hits, the open boxes are pion hits, and the inner trigger counter hit is also shown. The pulse data sampled every 2 ns (crosses), in one of the target fibers hit by the stopped kaon is displayed along with a fit (curve) to the expected pulse shape. At the upper right of the figure is the $\pi \rightarrow \mu$ decay signal in the range stack scintillator layer where the pion stopped. The crosses are the pulse data sampled every 2 ns, and the curves are fits for the first, second and combined pulses.

information available from the kaon time-of-flight, it is expected that the inefficiency for detecting π^0s can be reduced to 10^{-8}, thereby suppressing the most dangerous background from $K \rightarrow \pi^0\pi^0$ decays to a level that is an order of magnitude below the expected signal. Using these and other special techniques, the goal of a $K_L^0 \rightarrow \pi^0\nu\overline{\nu}$ signal in excess of 50 events may be reached in the presence of what otherwise may be insurmountable backgrounds.

3. Searches for Non-standard model processes.

3.1. $K \rightarrow \pi$ + "nothing"

Non-SM contributions to the signature $K^+ \rightarrow \pi^+$+ 'nothing' may include $K^+ \rightarrow \pi^+\nu\overline{\nu}$ with exotic intermediate states (e.g. leptoquarks or supersymmetric particles), $K^+ \rightarrow \pi^+\nu\overline{\nu}'$, $K^+ \rightarrow \pi^+X_0$, and $K^+ \rightarrow \pi^+X_0+X_0'$,

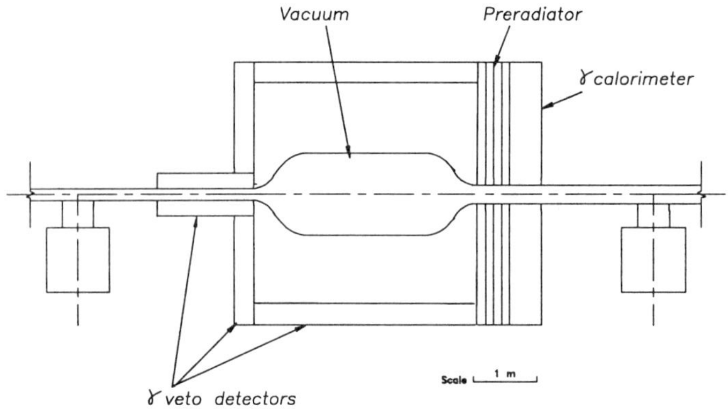

Figure 5. BNL E926 $K_L^0 \rightarrow \pi^0 \nu \overline{\nu}$ detector. The neutral beam of 1 GeV kaons enters from the left. In the forward detection region the photon detector system consists of a fine grained preradiator in which the photons are converted and the first $e^+ - e^-$ pair is tracked, followed by an 18 radiation length (X_0) calorimeter.

where X_0s represent some new weakly interacting neutral particles. One possibility for X_0 in the decay $K^+ \rightarrow \pi^+ X_0$ is Wilczek's familon[10], a Goldstone boson which arises in theories of spontaneously broken family symmetry.

Since $K \rightarrow \pi \nu \overline{\nu}$ experiments observe only the pions, an unexpected kinematic signature (e.g. a peak) or a decay rate in conflict with the SM predictions could signal the existence of new weakly interacting particles which would have eluded other investigations. It was determined from the E787 data mentioned above that no process like $K^+ \rightarrow \pi^+ X_0$ involving light ($m \approx 0$) particles occurs at a rate above 3×10^{-10} (see ref. [5]). New E787 data should eventually improve the sensitivity by another order of magnitude.

3.2. Lepton Flavor Violation

Lepton flavor violation (LFV) is absent in the SM with massless neutrinos. A minimal extension might allow LFV reactions such as $\mu \rightarrow e\gamma$ via diagrams like that shown in fig. 6. Even if the masses of the three known neutrinos were at their present experimental limits the branching ratio for $\mu \rightarrow e\gamma$ would be tiny:[11]

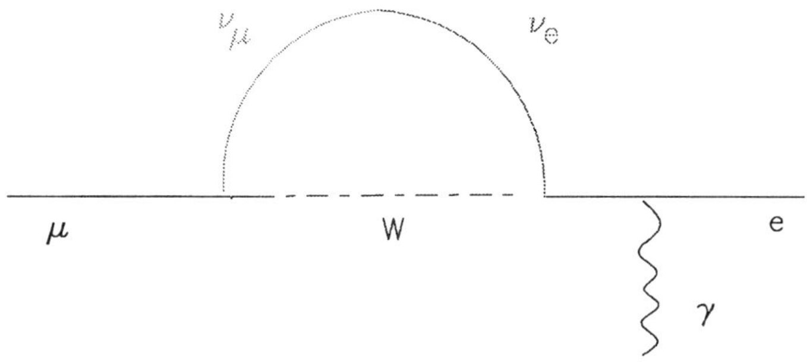

Figure 6. Possible diagram for a $\mu \to e\gamma$ transition with massive neutrinos.

$$R(\mu \to e\gamma) = \frac{\Gamma(\mu \to e\gamma)}{\Gamma(\mu \to e\nu\overline{\nu})} = 10^{-39}(\frac{m_{\nu_1}^2 - m_{\nu_2}^2}{400\text{eV}^2})^2. \tag{1}$$

Many extensions of the SM have LFV interactions naturally but heavy LFV force carriers suppress the branching ratios by factors of m_{H}^{-4}. With present limits on the LFV reactions like $\mu \to e\gamma$ and $\mu \to e$ conversion, high mass scales are being probed. As an example, LFV is expected in a supersymmetric SU(5) grand unified theory which results in large branching ratios for $\mu \to e\gamma$, $\mu \to e$ conversion and $\tau \to e\gamma$[13]. In this model, a large top-quark Yukawa coupling leads to rates of lepton flavor violating processes which are only one or two orders of magnitude below present experimental bounds. In these theories, the uncertainties for predicting the rates of LFV processes in terms of the masses of supersymmetric partner particles like "selectrons" ($e_{\tilde{R}}$) may be much less model dependent than for competing processes like proton decay (the former standard for GUTs) which would also require baryon number non-conservation. $\mu \to e\gamma$ appears to be the most sensitive process in these models for a given branching ratio, but $\mu \to e$ conversion is also competitive since the experimental levels reached so far (and perhaps, in the future) are significantly lower. It appears that improvements of several orders of magnitude in LFV experiments might have the consequence of severely restricting the attractiveness of important SUSY-GUT theories by pushing the masses to levels like $m_{e_{\tilde{R}}} \geq 1$ TeV or result in observation of a revolutionary effect.

LFV reactions have continued to be pursued vigorously on many fronts including muon, tau and kaon decays and in ep collisions.[12] Table 1 gives some current results and goals for experiments in progress.

Reaction	Present limit (90% c.l.)	Ref.	In-progress goal	Lab.
$\mu \to e\gamma$	4.9×10^{-11}	[14]	7×10^{-13}	LAMPF
$\mu \to 3e$	1×10^{-12}	[15]		
$\mu^- Ti \to e^- Ti$	6.1×10^{-13}	[16]	$few \times 10^{-14}$	PSI
$\mu^+ e^- \to \mu^- e^+$	8.2×10^{-11}	[17]		
$K_L \to \mu e$	5×10^{-12}	[21]		
$K^+ \to \pi^+ \mu^+ e^-$	2.1×10^{-10}	[20]	3×10^{-12}	AGS
$K_L \to \pi^0 \mu e$	3.2×10^{-9}	[19]	3×10^{-11}	FNAL
$K^+ \to \pi^+ X$	3×10^{-10}	[5]	10^{-11}	AGS

Table 1. Limits on flavor-violating and other exotic K and μ decays.

3.3. Muonium-anti-Muonium Conversion

The spontaneous conversion of muonium to antimuonium ($\mu^+ e^- \to \mu^- e^+$) requires a change of lepton numbers $\Delta L_e = -\Delta L_\mu = 2$. Such processes appear in models containing Majorana neutrinos which are natural consequences of the "see-saw mechanism," a candidate approach to account for the tiny (or "zero") masses of neutrinos.

A clever technique is being pursued at PSI[17] involving detection of the $\mu^- \to e^- \nu\nu$ decay and the collection of the residual positron which is accelerated electrostatically. The spectrometer is shown in fig. 7. The positron is detected by observing the 511 keV annihilation radiation. Fig. 8 shows the data in which 1 event was observed in the signal region where 1.7 ± 0.2 events were expected from backgrounds. Based on this result, the probability for muonium-antimuonium conversion was set at $P_{M\overline{M}} < 8.2 \times 10^{-11}$ and the hypothetical coupling constant for such transitions limited to $G_{M\overline{M}} < 3.3 \times 10^{-3} G_F$. This represents an improvement of three orders of magnitude in the limit on $P_{M\overline{M}}$.

3.4. $\mu \to e$ conversion

$\mu \to e$ conversion is another attractive process to pursue since the coherent nuclear capture, $\mu^- + (Z, A) \to e^- + (Z, A)$, is enhanced relative to ordinary nuclear muon capture and there are favorable experimental aspects which enhance it above other LFV candidates. The signature of $\mu - e$ conversion is a monochromatic electron energy $E_{\text{max}} = m_\mu - E_{\text{binding}} \approx 104$ MeV, which is far away from the end point of the free μ decay spectrum (53 MeV). However, the principal inherent background to $\mu - e$ conversion comes from muon decay in orbit around the nucleus where the energy can extend

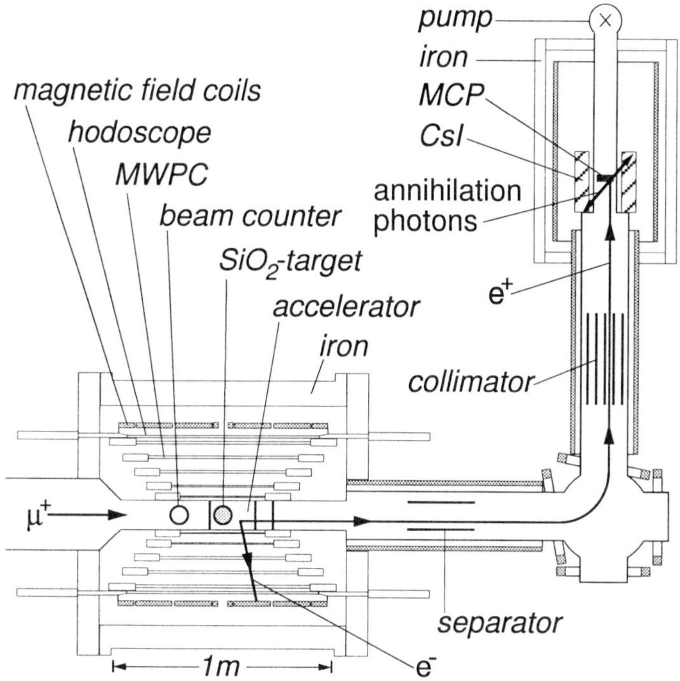

Figure 7. The spectrometer used to search for muonium-antimuonium conversion at PSI.

up to E_{max}. Additional backgrounds come from cosmic rays and radiative capture of π^- contaminating the muon beam. These backgrounds can be reduced to a manageable level by good detector resolution and excellent suppression of π^-.

The Sindrum2 group at PSI[16] recently reported a preliminary upper limit on the branching ratio $B(\mu^- Ti \to e^- Ti) < 6.1 \times 10^{-13}$ (90% c.l.) based on no observed events. The group is awaiting the development of a new high intensity muon beam to attempt to reach the level of a few $\times 10^{-14}$. A new effort to search for $\mu \to e$ conversion has been proposed for BNL[18].

3.5. LFV in Kaon decays

Also prominent among LFV reactions are $K_L^0 \to \mu e$ which is potentially sensitive to new axial-vector and pseudoscalar interactions and $K^+ \to \pi^+ \mu^+ e^-$ which can occur via vector or scalar currents. These reactions

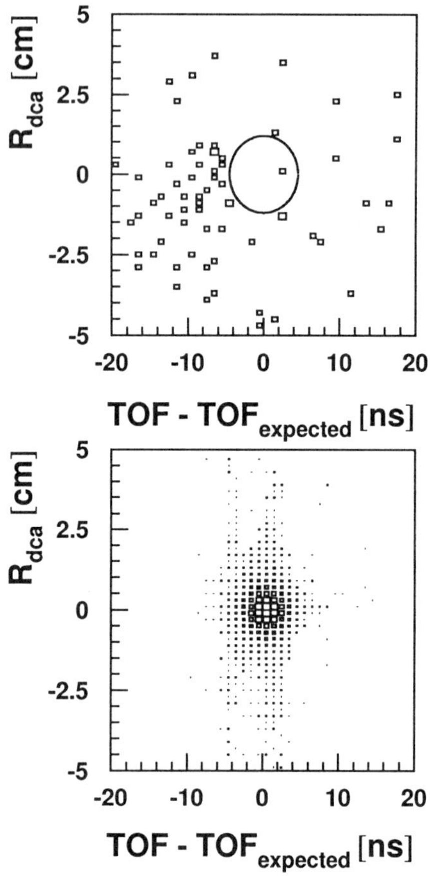

Figure 8. The difference in traced back positions (R_{dca}) for the $\mu \rightarrow e$ decay and the positron vs. the time-of-flight difference for data (upper plot) and simulation (lower plot) in the search for muonium-anti-muonium transitions.

are especially attractive in non-SM approaches which have overall "generation" numbers conserved considering both the quark and lepton sectors. They are extremely sensitive to the presence of high mass scales, possibly reaching into TeV territory where LFV processes are natural in many models such those with additional Higgs particles or leptoquarks.

Two major experiments were mounted at BNL in recent years to search for the LFV reactions $K_L^0 \rightarrow \mu e$ and $K^+ \rightarrow \pi^+\mu^+e^-$. These improved searches have been made possible in part by the substantial recent upgrading of the beam intensity at the AGS which has now exceeded 6×10^{13} protons/1.6s spill. E865 at BNL is a fine-tuned upgrade of the experiment performed by the group which obtained a limit of $B(K^+ \to \pi^+\mu^+e^-) < 2.1 \times 10^{-10}$ (see ref. [20]). The present goal is 3×10^{-12} which is being pursued using a more intense 6 GeV beam, a higher acceptance detector with improved resolutions and greater redundancy in the tracking and particle identification systems. Another search dealing with the related process $K_L^0 \to \pi^0\mu^\pm e^\mp$ is being made by the KTEV group at Fermilab[19].

Recently a preliminary limit of $B(K_L \to \mu e) < 5 \times 10^{-12}$ was reported[21] by the BNL E871 group using a straw chamber tracking apparatus and a beam plug inserted in the neutral beam upstream of the main detector to reduce accidentals.

3.6. T violation in $K_{\mu 3}$ decay

Since in the SM, transverse muon polarization (P_μ^T, out of the decay plane) in $K \to \pi\mu\nu$ decay is expected to be tiny ($\sim 10^{-6}$), measurement of a larger effect could signal the presence of a new source of T violation. For instance, in some multi-Higgs models, a diagram involving exchange of a charged Higgs particle could interfere with the SM process leading to effects as large as $P_\mu^T \sim 10^{-3}$ compared to the current limiting result of $P_\mu^T = 0.00185 \pm 0.0036$ (see ref. [22]) from a 1983 experiment which used an in-flight decay method.

An experiment is in progress at KEK (E246)[23] aiming to substantially improve the sensitivity to P_μ^T using a new technique involving stopped kaons (K^+), a toroidal spectrometer and an efficient muon polarimeter. The experiment is now operating at KEK. An extension of this effort has been proposed for BNL by the KEK group. In addition, a new in-flight approach (E923) to T violation in $K \to \pi\mu\nu$ decay has been proposed at BNL.[24]

4. Conclusion

Orders of magnitude gains in sensitivity are being made in measurements and searches for the most important rare muon and kaon decays. E787 at BNL has reached the sensitivity necessary to examine Standard Model predictions for $K^+ \to \pi^+\nu\bar{\nu}$. The KTEV group has recently improved the limit on $K_L^0 \to \pi^0\nu\bar{\nu}$ significantly. New experiments, aiming for definitive measurements of $K_L^0 \to \pi^0\nu\bar{\nu}$ are being explored. These rare kaon decays can unambiguously probe the SM origin of CP violation, potentially yielding the most accurate determinations of the CP-violating phase η and $|V_{td}|$.

Absence of $K^+ \to \pi^+\nu\overline{\nu}$ or $K_L^0 \to \pi^0\nu\overline{\nu}$ within the expected ranges or conflicts with other measures of quark mixing parameters would indicate new physics.

Considerable progress has also been made recently on searches for lepton flavor-violating reactions $K_L^0 \to \mu e$, nuclear $\mu \to e$ conversion and muonium-anti-muonium transitions. New experiments of unprecedented sensitivity are being planned. Confirmed non-zero findings would be revolutionary.

References

[1] Rochester G D and Butler C C, *Nature* **160**, 855 (1947).

[2] Hincks E P and Pontecorvo B, *Phys. Rev.* **74**, 697(L) (1948).

[3] D. Rein and L. M. Sehgal, Phys. Rev. **D39**, (1989); J. S. Hagelin and L. S. Littenberg, Prog. Part. Nucl. Phys. **23**, 1 (1989; M. Lu and M. B. Wise, Phys. Lett. **B324**, 461 (1994).

[4] Buchalla G, Buras A J *Phys. Rev.* **D54**, 6782(1996); Buchalla G, Buras A J and Lautenbacher M E, *Rev. Mod. Phys.* **68**, 1125 (1996).

[5] Adler S *et al Phys. Rev. Lett.* **79**, 2204 (1997).

[6] Barker A *et al* , *Proc. Intl. Conf. H. E. P.* , Vancouver, ed. Astbury A, *World Sci.* (1998), to be published.

[7] see Inagaki T, *Proc. Symp. on Flavor Changing Neutral Currents*, ed. Cline D, *World Sci.* (1997).

[8] Cheu E *et al* , FERMILAB-PUB-97-321-E.

[9] Chiang I-H *et al* , BNL AGS proposal E926.

[10] Wilczek F, *Phys. Rev. Lett.* **49**, 6782 (1982).

[11] Gvozdev A A *et al* , *Phys. Lett.* **B345**, 490 (1995).

[12] Sirois Y, *Proc. Symp. on Weak and E. M. Int.*, Santa Fe (1998), to be published.

[13] Barbieri R and Hall L, *Phys. Lett.* **B338**, 212 (1994).

[14] Bolton R *et al* , *Phys. Rev.* **D38** 2121 (1988).

[15] Belgardt U *et al* , *Nucl. Phys.* **B299** 1 (1988).

[16] see Wintz P, this symposium.

[17] Willmann L *et al* , hep-ex/9807011 (1998).

[18] see Molzon W, this symposium.

[19] Arisaka K *et al* , *Phys. Lett.* **B432**, 230 (1998).

[20] Lee A M *et al* , *Phys. Rev. Lett.* **64** 165 (1990).

[21] see Molzon W, this symposium, and Bachman M *et al* , *Proc. Intl. Conf. H.E.P.*, *op. cit.*.

[22] Blatt S R *et al* , *Phys. Rev.* **D27**, 1056 (1983).

[23] Imazato J *et al* , KEK E246. BNL E936 proposal (1997).

[24] Diwan M *et al* , BNL E923 proposal (1996).

Results of the SINDRUM-II Experiment

PETER WINTZ

Phys. Inst. Univ. of Zurich, CH-8057 Zurich, Switzerland

Abstract. Tests of lepton-flavor conservation (LFC) are very sensitive probes of new physics beyond the Standard Model. The most stringent upper limit on a LF-nonconserving decay mode now comes from the search for $\mu^- \rightarrow e^-$ conversion on titanium with the SINDRUM II [1] spectrometer at PSI: $\Gamma(\mu^- Ti \rightarrow e^- Ti^{g.s.})/\Gamma(\mu^- Ti\ capture) < 6.1 \times 10^{-13}$ (90% CL). This new limit improves on the best previous result by a factor of 7. A search for the charge-changing $\mu^- \rightarrow e^+$ conversion yielded new upper limits which improve on our previous results by a factor of 2.5.

1. Introduction

The strict conservation of lepton flavor (LFC), as build into the Standard Model, is primarily based on experimental evidence. In most of the proposed extensions LFC-violating decays are expected to occur at some level and an observation of these processes would be an unambiguous indication of new physics beyond the Standard Model. Until now no signal for any LF-nonconserving decay mode has been observed and the experimental tests have resulted in upper limits only. Nevertheless these very low limits put strong constraints on the allowed parameters space of the various models. See [3, 4, 5] for some reviews.

2. $\mu \rightarrow e$ Conversion in Nuclei

Neutrinoless $\mu^- \rightarrow e^-$ conversion in muonic atoms, $\mu^-(A, Z) \rightarrow e^-(A, Z)$ with A mass number and Z atomic number, gives the best constraints on

LFC-violation in a large variety of models [6]-[14]. For the conversion leaving the nucleus in its ground state the nucleons act coherently, which boosts the conversion probability relative to the rate of the dominant process of nuclear muon capture.

When negative muons stop in matter, they quickly get captured and form muonic atoms, which mostly reach their ground state before the bound muon decays $\mu^-(A,Z) \to e^-\overline{\nu}_e\nu_\mu(A,Z)$ or is captured by the nucleus $\mu^-(A,Z) \to \nu_\mu(A,Z-1)$. The capture probability increases with Z, to $f^{Ti}_{capt} = 85.3\%$ in the case of muonic titanium corresponding to a lifetime of 329 ns [15]. In the usual convention the branching ratio for $\mu \to e$ conversion is quoted relative to the rate for nuclear muon capture:

$$B_{\mu e} \equiv \frac{\Gamma_{\mu^- Ti \,\to\, e^- Ti^{g.s.}}}{\Gamma_{\mu^- Ti \; capture}}$$

The ratio $B_{\mu e}$ is predicted to increase with Z [16, 17]. The same calculations predict the coherent fraction of the $\mu^- \to e^-$ conversion to be larger than 80% for all nuclear systems. Earlier calculations estimated a maximum of $B_{\mu e}$ in the region around $Z = 30$ [18, 19, 20] so that most of the experiments have been performed on medium-heavy nuclei. A general model-independent analysis requires measurements on at least two nuclear targets with significantly different values for the normalized neutron excess $(N-Z)/(N+Z)$ [21].

The nonconservation of total lepton number may lead to the charge-changing conversion $\mu^-(A,Z) \to e^+(A,Z-2)$. Since the nucleus changes from initial to final state there is no coherent enhancement in this case.

3. The SINDRUM II Experiment

At the $\mu E1$ beamline at PSI data were taken on titanium [22, 24, 25] and lead [23]. For the search for $\mu^- \to e^-$ conversion on titanium an experimental sensitivity below 10^{-12} was achieved, which is now the highest sensitivity on any LF-nonconserving decay. In the titanium data also a search for $\mu^- \to e^+$ conversion has been performed. This article gives a summary of the results.

The SINDRUM II experiment aims at a final single event sensitivity around 10^{-14} for $B_{\mu e}$. The dedicated pion beamline $\pi E5$ with a new Pion-Muon-Converter (PMC) solenoid at the end will provide a higher muon flux with negligible pion and electron contaminations, required for the proposed sensitivity. During the last year the PMC could be brought into operation but only with limited fieldstrength. Nevertheless a search for $\mu^- \to e^-$ conversion on a gold target was carried out with a temporarily changed beam concept. The $\pi E5$ channel was tuned to select lower momentum cloud muons for which the PMC fieldstrength was sufficient. Meanwhile the PMC can be operated at full field strength. Start of the next $\mu \to e$ measurement will be in October this year.

3.1. Experimental Method

The signature of coherent $\mu^- \to e^-$ conversion is a single electron with the characteristic energy $E_{\mu e}$ given by the muon mass minus the muon binding energy E_b and the kinetic energy E_r of the recoiled atom:

$$E_{\mu e} = m_\mu c^2 - E_b(A, Z) - E_r(A, Z) \quad .$$

In case of titanium $E_{\mu e} = 104.3$ MeV with $E_b = 1.27$ MeV and $E_r = 0.12$ MeV.

The conversion energy $E_{\mu e}$ coincides with the kinematical endpoint of electrons from bound-muon decay. Their energy spectrum drops exponentially above $m_\mu c^2/2$ which is the endpoint energy of electrons from free-muon decay. By using a solenoidal spectrometer like SINDRUM II, the tracking detectors can be positioned at larger radial distances to the target to reduce their load from the vast lower momentum electrons. Low rates simplify the data readout and allow the use of high resolution drift chambers as tracking devices. Fig. 1 shows the e^- yields versus a transverse momentum threshold, both for the background of bound-muon decay and

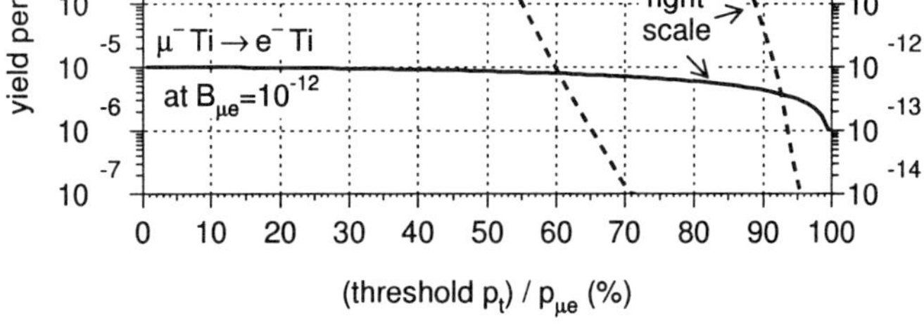

Figure 1. Total e^- yield per captured muon, both from bound-muon decay and $\mu \to e$ conversion assuming $B_{\mu e} = 10^{-12}$, as a function of the threshold on transverse momentum (p_t) normalized to the $\mu \to e$ conversion momentum ($p_{\mu e}$).

for the hypothetical $\mu \to e$ conversion. For a threshold around 80% the e^- yield from bound-muon decay is only 2×10^{-9} per captured muon whereas the acceptance for $\mu \to e$ conversion is still $\simeq$ 70%.

3.1.1. Intrinsic background The dominant intrinsic electron background comes from bound-muon decay. The energy spectrum of these electrons has been calculated for a number of muonic atoms [26, 27, 28] and the results are shown in Figure 2. Radiative corrections have not been applied and will lead to a reduction of the rate in the upper end of the spectra by 20 to 40% [26]. The probability to get an electron with energy within

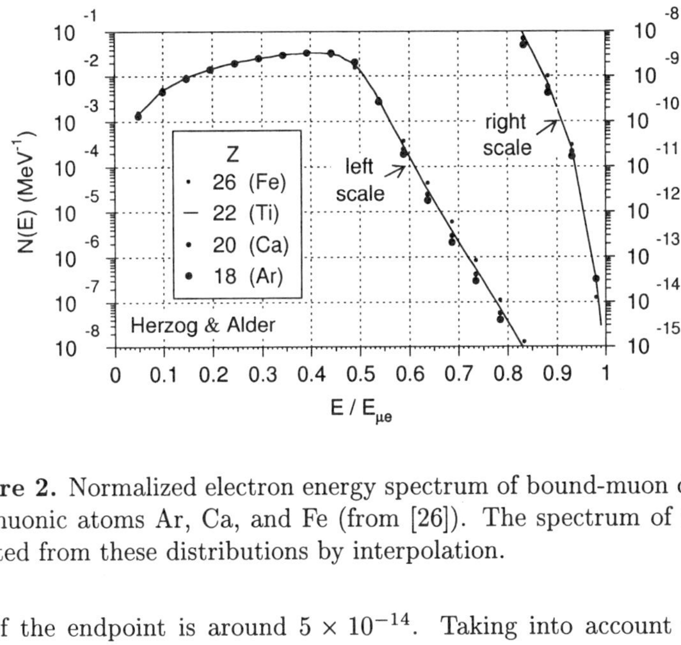

Figure 2. Normalized electron energy spectrum of bound-muon decay for the muonic atoms Ar, Ca, and Fe (from [26]). The spectrum of titanium resulted from these distributions by interpolation.

2% of the endpoint is around 5×10^{-14}. Taking into account that the probability of bound-muon decay per captured muon is around 17%, this yields a background well below 10^{-14}. Simulation confirms that a Gaussian energy resolution of around 2 MeV (FWHM) is sufficient to remain free of the background from bound-muon decay down to the proposed sensitivity.

Electron background from radiative muon capture is less significant. Due to the mass difference of initial and final nucleus the kinematical endpoint of the energy spectrum is lowered. In case of titanium with scandium as final nucleus the shift is 4 MeV so that the endpoint is well below the $\mu \to e$ conversion energy $E_{\mu e}$. For the whole measured energy spectrum the contribution of electrons from muon capture compared to the bound-muon decay background is lower than a few percent. However, radiative muon capture produces the only intrinsic positron background in the search for $\mu^- \to e^+$ conversion (see Chapt. 4.2).

3.1.2. Prompt background Prompt background may be caused by beam electrons scattering off the target and by electrons from radiative pion capture and subsequent pair production. The scattered electrons may originate from π^0 decays in the pion production target or from muon decays in flight at the end of the beamline. In the first case their momentum distribution reaches up to the selected beam momentum. Electrons from muon decay in flight may have significantly higher momenta and to avoid this background completely the beam momentum would have to be below 76 MeV/c.

Negatively charged pions at rest are immediately captured by a nucleus. With a probability of about 1% a photon is emitted with an energy above 100 MeV. These photons may convert in the target into an e^+e^- pair. If the positron remains unnoticed such events are a background source. So far, $\mu \to e$ searches had to make use of pulsed beams or a veto counter in the beam as the SINDRUM II experiment to suppress this kind of background. The $\pi E5$ beamline with the PMC solenoid will deliver the worldwide first muon beam with a pion contamination so low that the background from radiative pion capture (RPC) can be neglected. The first SINDRUM II measurement at this beamline will start in October this year.

3.1.3. Cosmic ray background There are numerous processes in which cosmic rays produce electrons in the target with an energy around $E_{\mu e}$. In almost all cases additional particles may be observed, like high-energy muons or associated positrons. This background source can be reduced by passive shielding. A cosmic background run with muon beam switched off allows to study and identify the characteristic event topologies.

3.1.4. Spectrometer The SINDRUM II spectrometer (Fig. 3) consists of a set of cylindrical detectors inside a superconducting solenoid. The muon beam enters the setup on the axis and is moderated before reaching the Ti target at the center. The incoming muons cross a CH_2 scintillation counter used to recognize prompt background events in the spectrometer and to monitor the beam intensity. The purpose of the second thinner moderator in front of the target is to absorb most of the remaining beam pions.

At the field strength of 1.27 T all electrons or positrons leaving the target with transverse momenta below 105 MeV/c are contained radially inside the magnet. The particles of interest follow helical trajectories, making at least one turn before reaching a Čerenkov hodoscope at either end of the spectrometer (Fig. 4). Two radial drift chambers are used for the spacial reconstruction of the trajectories. The average resolution of the inner drift chamber is 150 μm (σ_{xy}) in the $x-y$ plane and 2 mm (σ_z) in z direction. The main purpose of the larger outer chamber is to recognize cosmic ray induced events by coming from outside into the spectrometer

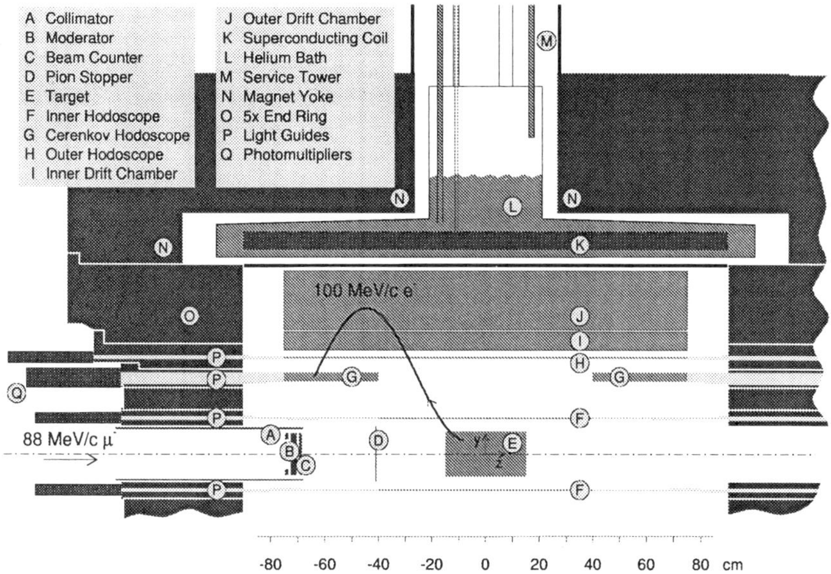

Figure 3. Vertical cross-section through the SINDRUM II spectrometer.

or by correlated signals from additional particles. The spacial resolution of this chamber is restricted to the $x - y$ plane and of the order of a few millimeters.

The trigger for data readout is mainly based on the expected hit pattern in the x-y plane from the outer hodoscope and inner drift chamber for simulated trajectories of μe conversion electrons. In addition, signals in the inner hodoscope and a Čerenkov hodoscope are required.

The time signals of the scintillator and Čerenkov hodoscopes belonging to a trajectory are used to determine the particle charge and direction. The first signal in the outer plastic-scintillator hodoscope (H in Fig. 3) defines the spectrometer reference time which is used to start the drift time measurements and to identify prompt background with the help of the beam counter. The second inner plastic-scintillator hodoscope helps to recognize background in particular from $\gamma \to e^+e^-$ conversions in the target.

In the event reconstruction analysis helical trajectories are searched and the particle momentum is fitted to the first two track elements in the inner drift chamber (see Fig. 4). Inhomogeneities of the magnetic field, energy

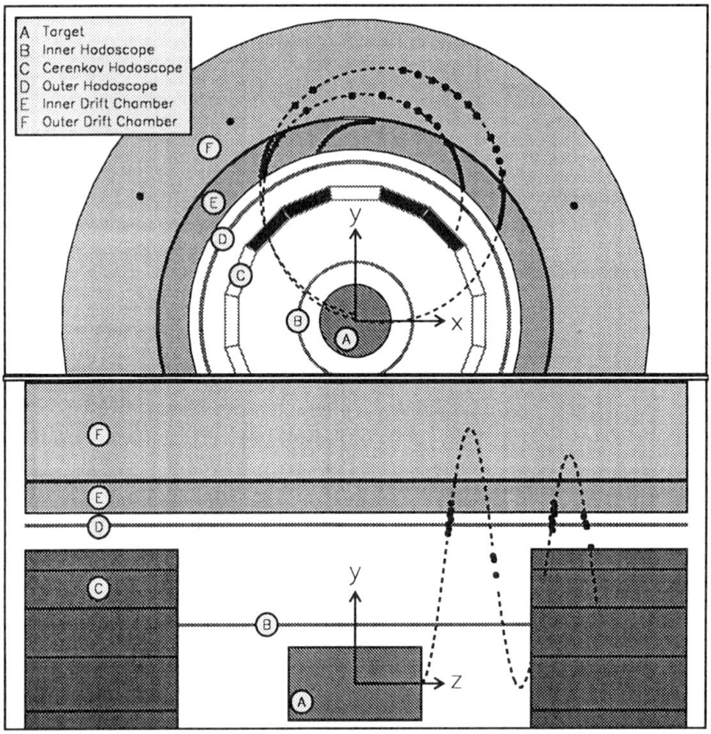

Figure 4. Example of a reconstructed electron trajectory.

loss, and multiple scattering along the particle trajectory outside the target are taken into account. The trajectory is followed backward to the point of closest approach to the spectrometer axis. This point is taken as the particle origin.

3.1.5. Momentum resolution To check the momentum calibration and resolution the beamline was tuned to select positive pions which were stopped in a low-mass polystyrene target inside the spectrometer. Fig. 5 shows the measured energy distribution of the positrons from $\pi^+ \to e^+\nu_e$ decay. Compared to the $\mu \to e$ measurement the magnetic field of the spectrometer was reversed and scaled down to the lower momentum. As may be seen the observed width is slightly larger than expected from simulation. Since the energy resolution for the μe measurement is dominated by the spread in the energy loss inside the heavier titanium target the agreement is much

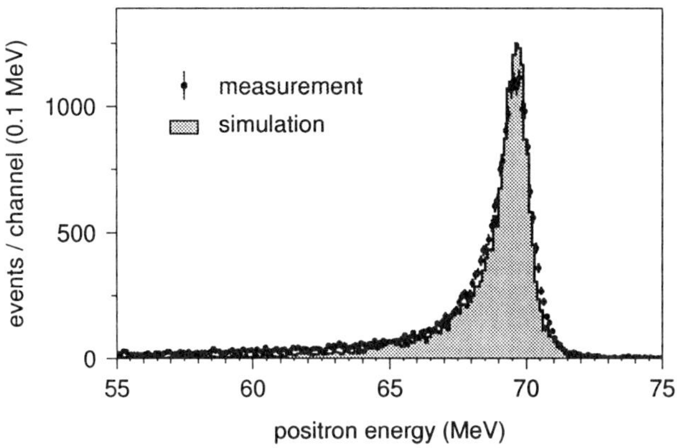

Figure 5. Energy spectrum of positrons from $\pi^+ \to e^+\nu_e$ decay in the low-mass polystyrene target. The width of the measured distribution is 1.5 MeV (FWHM).

better there. For this target a simulation of μe conversion electrons yields an energy resolution of 2.3 MeV which is sufficient to remain free from bound-muon decay background (see Chapt. 3.1.1).

4. Results of the Experiment

4.1. Search for $\mu^- \to e^-$ Conversion on Titanium

The beam intensity at the $\mu E1$ beamline rises as a function of the momentum and reached a value of 1.2×10^7 $\mu^- s^{-1}$ at the selected momentum of 88 ± 5 MeV/c. This was the highest rate for the operation of the beam counter scintillator and the momentum was low enough regarding the prompt background from electron scattering. During a measurement live time of 50.4 days with beam switched on $(3.09 \pm 0.14) \times 10^{13}$ muons stopped in the target with a pion contamination around 10^{-7}.

In total 4.0×10^6 events fulfilled the trigger conditions and were recorded from which $\simeq 170000$ electrons originating in the target could be reconstructed. Events caused by radiative pion capture and scattered beam electrons were removed by a 20 ns wide prompt veto on signals in the beam counter. The relative large width was caused by the flight-time spread of low-energy pions between second moderator and target.

The criteria to recognize cosmic background was checked and optimised by studying the data measured with muon beam switched off. To identify cosmic ray background several tests were developed looking for additional signals in the various detectors. If at least three of these tests were positive an event was declared as cosmic ray background. A specific background source survived these tests. A cosmic photon could enter the spectrometer through the unshielded cryogenic service tower and make asymmetric e^+e^- pair production in the target. If the e^+ had not enough momentum to reach the inner hodoscope only signals from the electron were seen. The only way to remove this background was a 3% cut in the geometrical acceptance.

The final electron sample consisted of $\simeq$ 21000 events, mainly from the intrinsic background of bound-muon decay. Radiative muon capture and unrecognized scattered beam electrons contributed at the level of 1% and 5%, respectively. Fig. 6 shows the resulting electron energy distributions. In the region of interest the prompt and cosmic ray background is

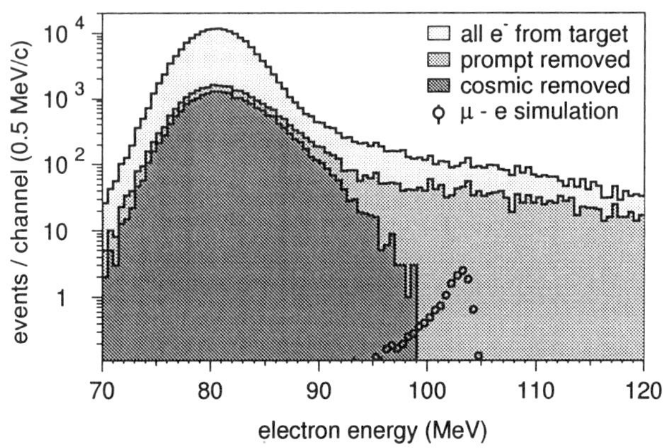

Figure 6. Energy distribution of the reconstructed electrons from the target and of the remaining events after suppression of prompt and cosmic ray background. A GEANT simulation of μe conversion electrons shows the expected amount assuming $B_{\mu e} = 4 \times 10^{-12}$.

completely removed. The endpoint of the final energy distribution is just below 100 MeV. The clear separation of this distribution from the simulated signal of μe electrons shows that no μe candidate event has been found.

In Fig. 7 various kinematical distributions of selected events from bound-muon decay in the final sample are compared with a GEANT simulation of this event type. The shift of the r_{ca} distribution to a mean value around 2.5 cm is caused by the steeply falling energy spectrum (see Fig. 2) and the vanishing acceptance towards lower values around 75 MeV. The z_{ca}

distribution reflects the range of the stopped μ^- which has been shifted to the front of the target to reduce the π^- contamination. Electrons emitted at $\Theta \simeq 90^o$ have a low probability to reach one of the Čerenkov hodoscopes which explains the dip in the Θ distribution. The simulated distributions are normalised to the observed number of events.

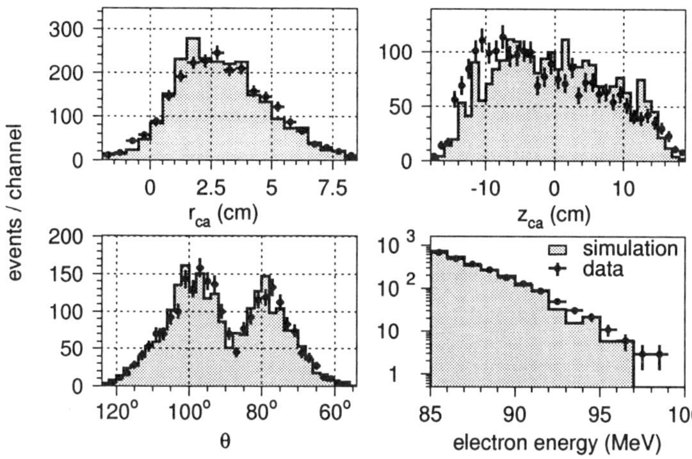

Figure 7. Kinematical distributions of electrons from bound-muon decay for data and simulation. r_{ca} and z_{ca} are the radial distance and corresponding z-coordinate of the closest point of the trajectory to the spectrometer axis. Θ is the angle of the trajectory relative to the z axis.

Since no μe conversion candidate has been found the measurement results in a new upper limit on the branching ratio

$$B_{\mu e} < \frac{n^{\mu e}_{max}\,(1 + n^{\mu e}_{max} \cdot \sigma_r^2/2)}{N^{\mu}_{stop} \cdot f^{Ti}_{capt} \cdot \epsilon^{\mu e}} \; .$$

$n^{\mu e}_{max} = 2.3$ is the upper limit on the expected number of events with 90% confidence, $N^{\mu}_{stop} = (3.09 \pm 0.14) \times 10^{13}$ is the number of stopped μ^-, and $f^{Ti}_{capt} = (0.853 \pm 0.003)$ [15] is the μ^- capture probability for titanium. $\epsilon^{\mu e}$ is the probability that a μe conversion in the target leads to the observation of an e^- with an energy above 99 MeV and σ_r [29] is the relative uncertainty of the denominator.

The experimental probability $\epsilon^{\mu e}$ is the product of the geometrical μe acceptance ϵ_ω and the efficiencies of the trigger ϵ_{trig}, track reconstruction ϵ_{track}, and selection cuts ϵ_{select} to suppress background. The acceptance $\epsilon_\omega = (0.418 \pm 0.005)$ of the spectrometer was obtained by a μe simulation whereas the efficiency factors and their energy dependencies were estimated from the events of bound-muon decay. Extrapolation to the μe energy region yielded $\epsilon_{trig} = (0.751 \pm 0.005)$, $\epsilon_{track} = (0.864 \pm 0.062)$, and $\epsilon_{select} = (0.538 \pm 0.025)$ including the energy cut of 99 MeV to suppress the bound-μ decay background (see Fig. 6). Using $\epsilon^{\mu e} = (0.146 \pm 0.012)$ and $\sigma_r = 0.11$ the new upper limit (with 90% CL) on the μe branching ratio is

$$B_{\mu e} < 6.1 \times 10^{-13}$$

which improves on the previous result [22] by a factor of seven.

4.2. Search for $\mu^- \to e^+$ Conversion on Titanium

A detailed description of this measurement can be found in [30]. Since there is no charge decision in the μe trigger also positrons coming from the target region were recorded which allows to search for $\mu^- Ti \to e^+ Ca$ conversion. The positron energy depends on the mass difference of the nuclei and varies only slightly for the different Ti isotopes if Ca is in ground state. But unlike μe conversion, one can expect an excitation of the final nucleus described by a giant dipole resonance with mean energy and width of 20 MeV in case of calcium. This leads to a broad energy spectrum of the positrons and has severe consequences for the background discrimination. Fortunately, the intrinsic background is much lower for e^+ than e^-.

Fig. 8 shows the energy distribution of the final data sample which consists mainly of positrons from radiative μ capture and some misidentified scattered beam electrons unrecognized by the beam counter. The endpoint of the positron spectrum is around 92.3 MeV with an isolated candidate event at 95.7 MeV. Also shown are the distributions resulting from $\mu^- \to e^+$ simulations for ground state and giant resonance final states. The measured distribution with beam switched off shows also one event in the interesting energy region so that the candidate event at 95.7 MeV can be explained by cosmic ray background. Therefor, above 92.3 MeV no evidence is found for $\mu^- \to e^+$ conversion and the new upper limits (with 90% C.L.) are

$$B^{gs}_{\mu^- e^+} < 1.7 \cdot 10^{-12} \quad \textit{and} \quad B^{gr}_{\mu^- e^+} < 3.6 \cdot 10^{-11}$$

which improve on the previous results [22] by a factor of 2.5.

Acknowledgments

The experiment is supported in part by the BMFT, Germany, under contract number 06AC651 and by the Swiss National Science Foundation.

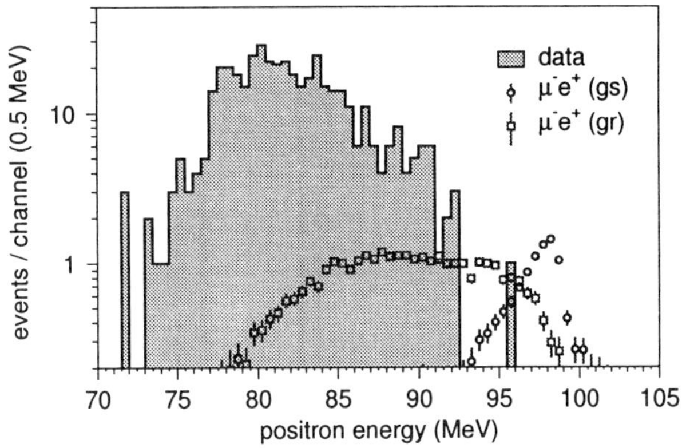

Figure 8. Energy distribution of the final positron data sample. A simulation of $\mu^- \to e^+$ conversion with Ca in ground state (gs) and with a giant resonance excitation (gr) shows the expected signals using $B^{gs}_{\mu e} = 4.3 \times 10^{-12}$ and $B^{gr}_{\mu e} = 8.9 \times 10^{-11}$.

References

[1] PSI proposal R-87-03 (1987).
SINDRUM II collaboration: W. Bertl, T. Buchsteiner, G. Cahsor, V. Djordjadze, C. Dohmen, J. Egger, S. Eggli, R. Engfer, C. Findeisen, H. Haan, E. Hermes, J. Hofmann, W. Honecker, J. Kaulard, T. Kozlowski, B. Krause, G. Kurz, M. Maas, F. Muheim, C. Niebuhr, G. Otter, S. Playfer, H. Pruys, F. Riepenhausen, F. Rosenbaum, M. Rutsche, A. van der Schaaf (spokesman), A. Schuengel, M. Starlinger, and P. Wintz.
III. Phys. Inst. B RWTH Aachen (Germany) – Paul Scherrer Institute, Villigen (Switzerland) – Phys. Inst. University of Zurich (Switzerland).

[2] Y. Ohama, *Phys. Rev.* **D57**, 6594 (1998).

[3] J.D. Vergados, *Phys. Rev.* **133**, 1 (1986).

[4] A. van der Schaaf, in Progress in Particle and Nuclear Physics, edited by A. Faessler, Vol. **31**, p. 1 (Pergamon, Oxford, 1993).

[5] P. Depommier and C. Leroy, *Rep. Prog. Phys.* **58**, 61 (1995).

[6] G. Altarelli *et al.*, *Nucl. Phys.* **B125**, 285 (1997).

[7] O. Shanker, *Nucl. Phys.* **B206**, 253 (1982).

[8] R.N. Cahn and H. Harari, *Nucl. Phys.* **B176**, 135 (1980).

[9] J. Bernabeu *et al.*, *Nucl. Phys.* **B409**, 69 (1993).

[10] L.N. Chang *et al.*, *Phys. Rev.* **D50**, 4598 (1994).

[11] T.S. Kosmas and J.D. Vergados, *Phys. Rep.* **264**, 251 (1996).

[12] G. Barenboim and M. Raidal, *Nucl. Phys.* **B484**, 63 (1996).

[13] J.E. Kim *et al.*, *Phys. Rev.* **D56**, 100 (1997).

[14] M. Raidal and A. Santamaria, *Phys. Lett.* **B421**, 250 (1998).

[15] T. Suzuki *et al.*, *Phys. Rev.* **C35**, 2212 (1987).

[16] H. C. Chiang *et al.*, *Nucl. Phys.* **A559**, 526 (1993).

[17] T. S. Kosmas *et al.*, *Phys. Rev.* **C56**, 526 (1997).

[18] S. Weinberg and G. Feinberg, *Phys. Rev. Lett.* **3**, 111, 244 (Erratum) (1959).

[19] T.S. Kosmas and J.D. Vergados, *Phys. Lett.* **B217**, 19 (1989).

[20] T.S. Kosmas and J.D. Vergados, *Nucl. Phys.* **A510**, 641 (1990).

[21] O. Shanker, *Phys. Rev.* **D20**, 1608 (1979).

[22] SINDRUM II Collab., C. Dohmen *et al.*, *Phys. Lett.* **B317**, 631 (1993).

[23] SINDRUM II Collab., W. Honecker *et al.*, *Phys. Rev. Lett.* **76**, 200 (1996).

[24] P. Wintz, *Ph.D. thesis*, RWTH Aachen (1995);

[25] S. Eggli, *Ph.D. thesis*, University of Zurich (1995)

[26] F. Herzog and K. Alder, *Helv. Phys. Acta* **53**, 53 (1980).

[27] R. Watanabe *et al.*, Atomic Data and Nuclear Data Tables **54**, 165 (1993).

[28] O. Shanker and R. Roy, *Phys. Rev.* **D55**, 7307 (1997).

[29] R. D. Cousins and V. L. Highland, Nucl. Inst. Meth. **A320**, 331 (1992).

[30] SINDRUM II Collab., J. Kaulard *et al.*, *Phys. Lett.* **B422**, 334 (1998).

The future of $\mu^+ \to e^+\gamma$ at PSI

Andries van der Schaaf

Physik-Institut der Universität Zürich

Abstract. In April 1998 a letter of intent for an experiment at PSI (R-98-05.0) was submitted by a JINR-KEK-Pavia-Pisa-PSI-Roma-Swierk-Tbilisi-Tokyo-Waseda-Zurich collaboration [1] [1]. It is proposed to perform an experimental search for the decay $\mu^+ \to e^+\gamma$ with a single-event sensitivity below 10^{-14}. This measurement would improve on the previous result by 2–3 orders of magnitude. In this talk I will discuss the motivation for such measurement, describe the various backgrounds and present two options for the experimental setup, both for the positron spectrometer and for the photon calorimeter.

1. Introduction

The non-observation of the decay $\mu \to e\gamma$ is one of the corner stones of the Standard Model. This test of lepton-flavor conservation is of similar importance as the limits on neutrino masses and mixings, proton stability, neutron electric dipole moment, and $(g\text{-}2)_{\mu,e}$. Each of them gives the best constraint to some varieties of extensions to the model. Each of them thus may give the first signal for physics beyond the Standard Model.

One of the great mysteries in present-day particle physics is the origin of the generation pattern of quark and lepton states. Transitions between the generations, as observed in charged weak interactions among quarks, can be attributed to mixing of the quark eigenstates. Within the framework of the minimal Standard Model the mass degeneracy of the neutrino states

[1] represented by Alessandro Baldini, Carlo Bemporad, Giacomo Bressi, Tadayoshi Doke, Fereydoun Foroughi, Jun Kikuchi, Spartak Korenchenko, Tadek Kozlowski, Yoshitaka Kuno, Akihiro Maki, Tetsuro Mashimo, Toshinori Mori, Marek Moszynski, David Mzavia, Shuji Orito, Dieter Renker, Andries van der Schaaf, H.Christian Walter and Dino Zanello

precludes this mechanism in the leptonic sector, in accordance with the apparent conservation of leptonic generation number (lepton-flavor conservation, LFC).

Although various anomalies in the fluxes of solar, atmospheric and beam dump neutrinos may be interpreted in terms of neutrino oscillations, no convincing signal for the violation of LFC has been found so far. Searches for LFC-violating decay modes of charged leptons, mesons and Z^0 have resulted in upper limits only.

Most extensions of the Standard Model allow neutral interactions between the generations. Whereas higher order charged interactions among quarks lead to effective flavor-changing neutral currents (FCNC), as observed in the decay $b \to s\gamma$, FCNC's among leptons would yield an unambiguous signal for new physics: neutrino mixing may contribute to the $\mu \to e\gamma$ branching ratio at a level which is at least ten orders of magnitude below the present limit. See [2, 3, 4] for some recent reviews.

Theoretical work on physics beyond the Standard Model concentrates on supersymmetry and grand-unification. Supersymmetry, which eventually unifies fermions and bosons, may be discovered at LHC. Grand unification, i.e. the unification of quarks and leptons, is presently constrained best by the limits on proton stability and double beta decay. Recently it was pointed out [5] that supersymmetric unification, the combination of these two very attractive ideas, may be observed in some low-energy signals. Since all sleptons are known to be massive and should mix just like quarks do, $\mu \to e\gamma$ would proceed with the help of a sparticle propagator with a branching ratio in the region $10^{-12\pm 2}$, i.e. within reach of this search.

Triggered by these exciting prospects, more sensitive searches are presently being designed. Whereas $\mu - e$ conversion ranks among the two top-priority experiments for the future AGS program at the Brookhaven National Laboratory [6], the feasibility of a $\mu \to e\gamma$ search at PSI is being studied by various research groups. Both initiatives were discussed at recent workshops and the improvements over previous experiments by 2-3 orders of magnitude seem feasible.

At the moment the decay is searched for by the MEGA collaboration at LAMPF. This experiment stopped data-taking in 1995. The analysis of 16% of the data resulted in an upper limit of $4.2 \cdot 10^{-11}$ (90% C.L.) [7].

Presently, only PSI delivers the high-intensity, low-momentum, DC beams required to reach the 10^{-14} sensitivity. At two workshops at PSI [2] various designs for the detection system were discussed. Two options, both for the positron and the photon detector, were selected for further study.

[2] Workshop on *a new $\mu \to e\gamma$ Experiment at PSI*, March 5–7, 1997.
Working Meeting on *A $\mu \to e\gamma$ Experiment at PSI?*, March 16–18, 1998.

1.1. Tests of lepton-flavor conservation

Searches for LFC-violating decay modes of charged leptons, mesons and Z^0 have resulted in upper limits only, as summarized in Table 1. Since

Table 1. Branching ratios of some LFC-violating decay modes. All values are upper limits at 90% confidence, with the exception of the Z^o results, which are given at the 95% confidence level. The results are taken from Ref. [8] unless quoted explicitly.

Decay mode	Branch. Ratio	Reference
$\mu^+ \rightarrow e^+\gamma$	$< 4.2 \times 10^{-11}$	[7]
$\rightarrow e^+e^+e^-$	$< 1.0 \times 10^{-12}$	
$\mu^- S \rightarrow e^- S$	$< 7 \times 10^{-11}$	
$\mu^- Ti \rightarrow e^- Ti$	$< 6.1 \times 10^{-13}$	[9]
$\rightarrow e^+ Ca^*$	$< 3.6 \times 10^{-11}$	
$\mu^- Pb \rightarrow e^- Pb$	$< 4.6 \times 10^{-11}$	
$\mu^+ e^- \leftrightarrow \mu^- e^+$	$< 3.5 \times 10^{-10}$	[10]
$\tau \rightarrow e\gamma$	$< 2.7 \times 10^{-6}$	
$\tau \rightarrow \mu\gamma$	$< 3.0 \times 10^{-6}$	
$\tau \rightarrow$ 3 charged leptons	$< 4 \times 10^{-6}$	
$K^+ \rightarrow \pi^+ \mu e$	$< 4 \times 10^{-11}$	
$K^o_L \rightarrow \mu e$	$< 1 \times 10^{-12}$	
$B^o \rightarrow \mu e$	$< 5.9 \times 10^{-6}$	
$\rightarrow \tau e$	$< 5.3 \times 10^{-4}$	
$\rightarrow \tau \mu$	$< 8.3 \times 10^{-4}$	
$Z^0 \rightarrow e\mu$	$< 1.7 \times 10^{-6}$	
$\rightarrow e\tau$	$< 9.8 \times 10^{-6}$	
$\rightarrow \mu\tau$	$< 1.7 \times 10^{-5}$	

the experimental sensitivities are mostly limited by the available beam intensities or luminosities the most stringent limits come from μ and K decays, i.e. for transitions between the lighter two generations. In models, in which the couplings involving the third generation are strongly enhanced, the constraints from Z, τ and B decays start to become interesting too.

1.2. Theoretical motivation

LFV processes appear naturally in supersymmetric (SUSY) extensions to the Standard Model, in particular in supersymmetric, grand-unified theories (SUSY-GUT). Although the soft-breaking scalar mass matrices (such as the slepton matrix) are assumed diagonal and universal at the Plank scale, radiative corrections in the renormalization group evolution from the GUT to the weak energy scale unavoidably lead to slepton mixing [11]. As a result $\mu^+ \to e^+\gamma$ occurs through loop diagrams as shown in Fig. 1.

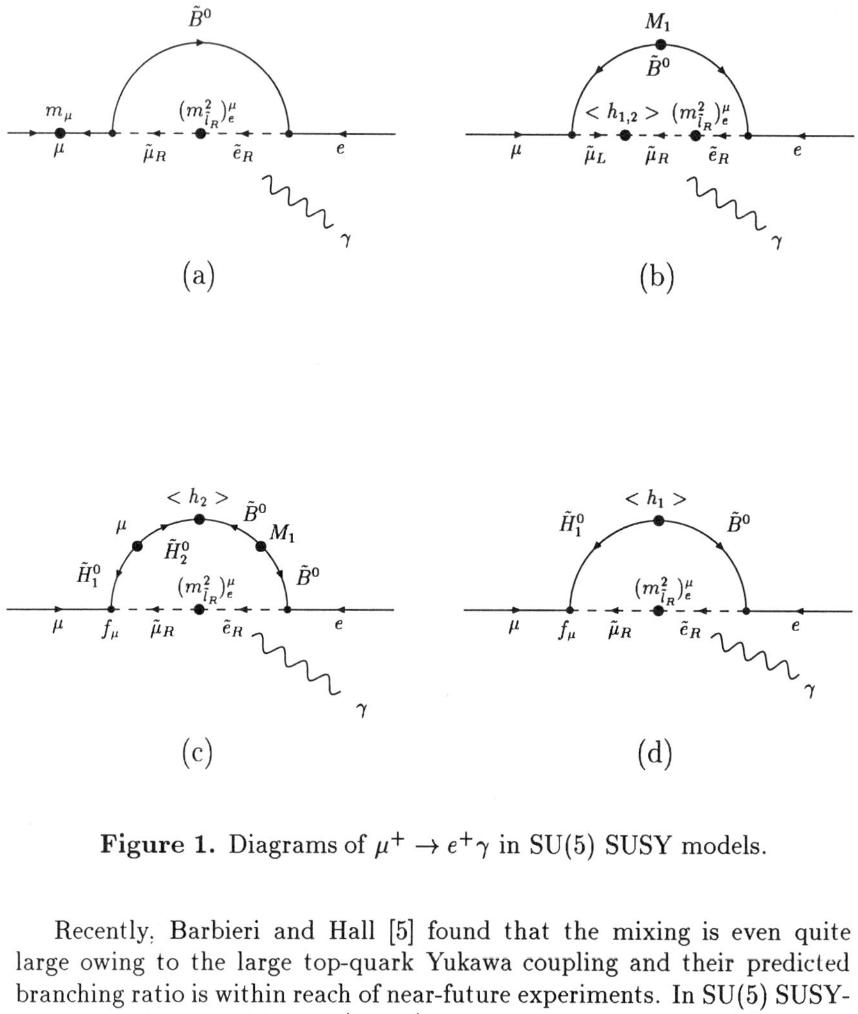

Figure 1. Diagrams of $\mu^+ \to e^+\gamma$ in SU(5) SUSY models.

Recently, Barbieri and Hall [5] found that the mixing is even quite large owing to the large top-quark Yukawa coupling and their predicted branching ratio is within reach of near-future experiments. In SU(5) SUSY-GUT models the width for $\mu^+ \to e^+\gamma$ can be approximated by [5]:

$$\Gamma(\mu \to e\gamma) \approx \frac{\alpha}{4}\left|\frac{\alpha M_Z}{24\pi\cos^2\theta_W}\frac{V_{32}^* V_{31} I(y_t)}{m_{\tilde{\mu}}^2}\right|^2 \frac{m_\mu^5}{m_{\tilde{\mu}}^4} \tag{1}$$

where $I(y_t)$ is a function of the top-quark Yukawa coupling y_t; V_{ij} is the Cabibbo-Kobayashi-Maskawa quark mixing matrix at some GUT energy

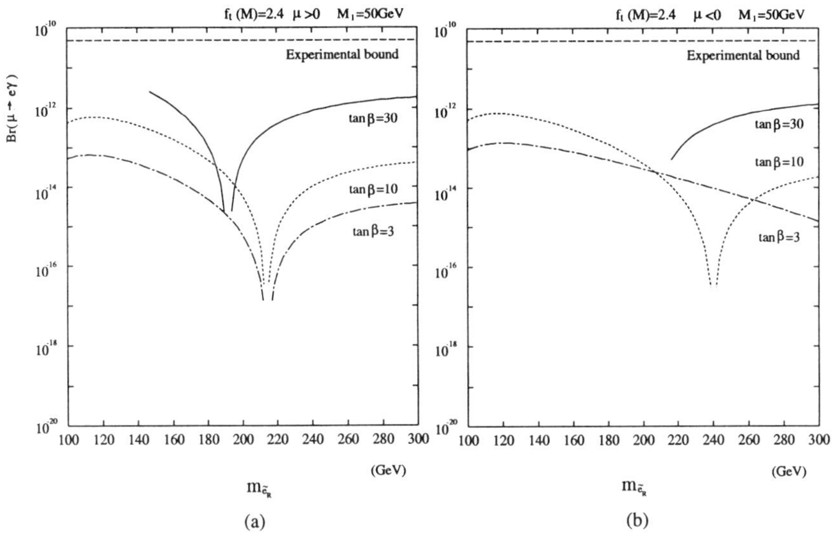

Figure 2. Predictions of $\mu^+ \to e^+\gamma$ branching ratio in SU(5) SUSY models.

scale, and m_μ and $m_{\tilde{\mu}}$ are muon and smuon masses, respectively. The branching ratio of $\mu^+ \to e^+\gamma$ predicted in SUSY SU(5) models [12] is shown in Figure 2. The prediction varies from 10^{-15} to 10^{-13} for the singlet smuon mass of $m_{\tilde{\mu}R}$ of 100 to 300 GeV. SO(10) models give even values which are enhanced by a factor $(m_\tau/m_\mu)^2$ [5] resulting from loop diagrams which contribute in proportion to the lepton mass.

2. General considerations

2.1. Sensitivity

The signature for a $\mu^+ \to e^+\gamma$ decay at rest is an $e^+\gamma$ pair, coincident in time, moving back-to-back with momenta equal to half the muon mass (52.8 MeV/c). The single event sensitivity is given by

$$S = (N_\mu \cdot t \cdot \Omega/4\pi \cdot \varepsilon_e \cdot \varepsilon_\gamma \cdot \varepsilon_{cut})^{-1}, \tag{2}$$

where N_μ is the muon stop rate, t is the measuring time, Ω is the detector solid angle, ε_e and ε_γ are the positron and photon detection efficiencies, and ε_{cut} is the efficiency of the selection cuts. Take, for example, $N_\mu = 3 \times 10^8 s^{-1}$, $t = 10^7$ s, $\Omega/4\pi = 0.1$, $\varepsilon_e = 0.9$, $\varepsilon_\gamma = 0.8$, $\varepsilon_{cut} = 0.6$, then $S = 0.8 \times 10^{-14}$, corresponding to a 90 % C.L. upper limit of 2×10^{-14} if no event candidates are found.

2.2. Backgrounds

In the following the normalized e^+ and photon energies are used, defined as $x \equiv 2E_e/m_\mu c^2$ and $y \equiv 2E_\gamma/m_\mu c^2$. Furthermore, $z \equiv \pi - \theta_{e\gamma}$ and δA (ΔA) is defined as half width (full width) of the cut on the observable A.

2.2.1. Prompt background Prompt (physics) background originates from radiative muon decay $\mu^+ \rightarrow e^+\nu\bar{\nu}\gamma$. For unpolarized muons:

$$B_{rad} = \frac{\alpha}{8\pi}[J_1 + J_2]. \qquad (3)$$

Without constraint on opening angle (CASE1) J_1 and J_2 are given by:

$$J_1 = (\delta x)^4(\delta y)^2 \quad \text{and} \quad J_2 = (8/3)(\delta x)^3(\delta y)^3 \, . \qquad (4)$$

Figure 3 shows B_{rad} as a function of δx and δy for CASE1. One sees,

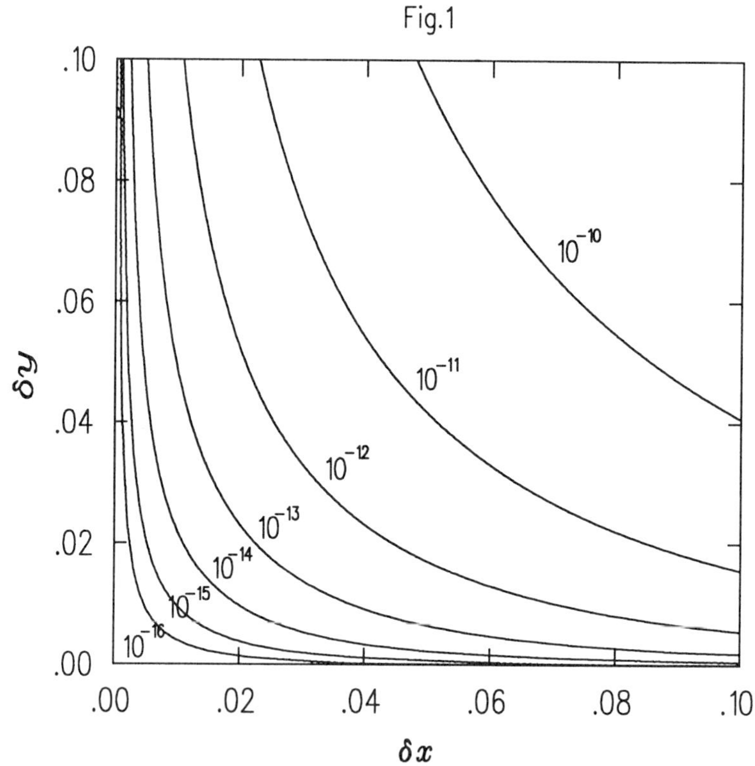

Figure 3. Prompt background contribution from $\mu^+ \rightarrow e^+\nu\bar{\nu}\gamma$ as a function of δx and δy for CASE1.

for instance, that $\Delta x = 2(\delta x) = 0.01$ and $\Delta y = 2(\delta y) = 0.05$ lead to $B_{rad} = 2 \times 10^{-15}$.

A cut on δz affects dB_{rad} only if $\delta z < 2\sqrt{\delta x \delta y}$ (CASE2), in which case:

$$\begin{aligned} J_1 &= (8/3)(\delta x)^3(\delta y)(\delta z/2)^2 - 2(\delta x)^2(\delta z/2)^4 + (1/3)(1/\delta y)^2(\delta z/2)^8 \\ J_2 &= 8(\delta x)^2(\delta y)^2(\delta z/2)^2 - 8(\delta x)(\delta y)(\delta z/2)^4 + (8/3)(\delta z/2)^6. \end{aligned} \quad (5)$$

2.2.2. Accidental background Accidental $e^+\gamma$ coincidences involve an e^+ from normal muon decay, $\mu^+ \to e^+\nu\overline{\nu}$, and a high energy photon which may originate in the decay $\mu^+ \to e^+\nu\overline{\nu}\gamma$ or from an e^+ through external bremsstrahlung or annihilation in flight. The effective branching ratio for accidental background is obtained from:

$$\begin{aligned} B_{acc} &= N_\mu \cdot f_e^0 \cdot f_\gamma^0 \cdot \frac{\delta\omega}{4\pi} \cdot 2\delta t \\ \frac{\delta\omega}{4\pi} &= \frac{\pi(\delta z)^2}{4\pi} = (\frac{\delta z}{2})^2 \end{aligned} \quad (6)$$

where f_e^0 and f_γ^0 are the e^+ and γ yields per stopped muon, respectively. The e^+ yield f_e^0 is obtained by integrating the Michel spectrum, $2x^2(3-2x)$, above $1 - \delta x$:

$$f_e^0 = 2(\delta x) \quad (7)$$

In estimating f_γ^0 first the radiative muon decay is considered [15]:

$$f_\gamma^0 = \frac{\alpha}{2\pi}(\delta y)^2\left[\ln(\delta y) + 7.33\right] \quad (8)$$

The result is shown in Fig.4 [16]. Since the muon stopping target may be very thin external bremsstrahlung of decay positrons in the target is small compared to the internal process. By taking care that positrons can not reach any material in direct sight of the photon detector this process can be ignored altogether. The contribution from annihilation in flight for a 50 mg/cm^2 target, is shown in Fig.4 as well. In this case annihilation in flight dominates for photon energies above 52 MeV.

Table 2 lists values of B_{acc} for different sets of cut values.

Table 2. Accidental background for typical examples of resolutions.

case	Δx	Δy	$\Delta\theta_{e\gamma}$	Δt	B_{acc}
	%	%	rad	ns	
1	1	2	0.01	0.3	2×10^{-15}
2	0.2	5	0.02	0.1	2×10^{-15}

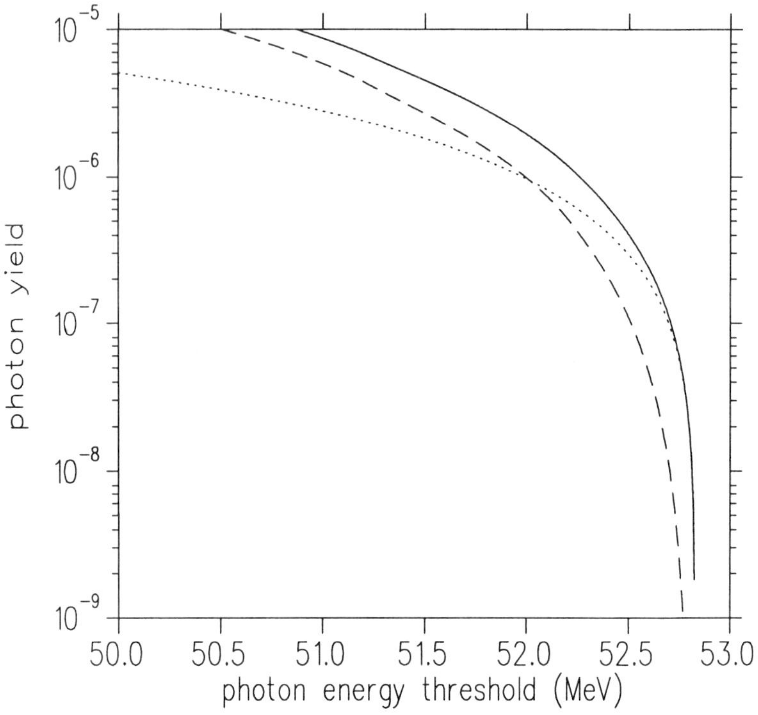

Figure 4. Photon yield v.s. energy threshold: dotted, from annihilation in flight in a 50 mg target; dashed, from $\mu^+ \rightarrow e^+\nu\overline{\nu}\gamma$; solid, total.

3. πE5 surface muon beam

The main characteristics of the beam line are listed in Table 3. The channel would be tuned to muons originating in the decays of pions which stopped close to the surface of the production target. An intensity of 4–8$\times 10^8 \mu^+$/s.

Table 3. Main properties of πE5.

momentum range	20–120 MeV/c
momentum band (FWHM)	10%
momentum resolution (FWHM)	2%
horizontal emittance	15.3 cm.rad
vertical emittance	3.6 cm.rad
spot size	2$\times$2 cm^2

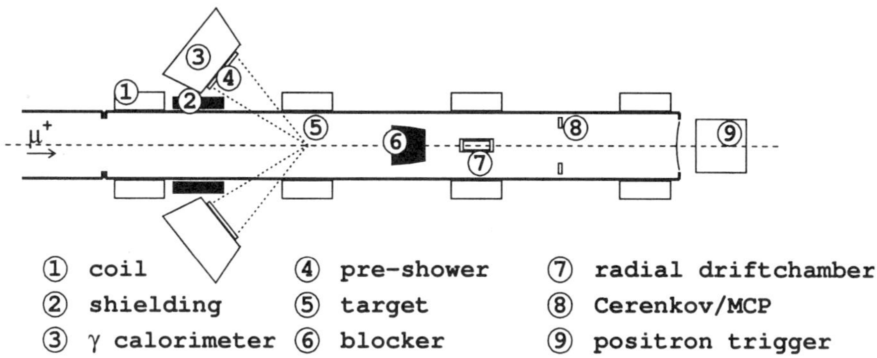

Figure 5. Intermediate-image spectrometer

4. Options for the detection system

Various options are under active study, both for the positron spectrometer and for the photon detector. Two possibilities will be discussed here (see Fig. 5 and Fig. 9). As positron spectrometer we consider:

- A two-stage intermediate image spectrometer with tracking detectors in the second stage.
- A solenoidal spectrometer with a thin coil, which allows the use of a photon detector outside the coil.

As photon detector we consider:

- A liquid Xenon detector, equipped with photo-detectors on all sides for optimal position, timing and energy resolutions.
- A segmented photon calorimeter preceeded by a pre-shower detector for optimal position and timing resolutions.

4.1. Positron spectrometer

4.1.1. Intermediate focusing spectrometer Intermediate image spectrometers were developed some 50 years ago for β-ray spectrometry. Momentum resolution was obtained by momentum selection, using variable field strengths to scan a certain momentum interval. Momentum resolution was dependent on focus quality, which gave certain limitations to source size and opening angle acceptance. With the help of tracking detectors many of these limitations disappear and dramatic improvements in resolution and/or acceptance can be achieved.

The simplest φ-symmetric spectrometer field is the constant field generated by a solenoid. In such a field particles emitted at larger angle with respect to the symmetry axis return to the axis at shorter distance and at best one may observe a sort of ring focus. A reduction of the field strength towards larger radii will lengthen the focusing distance for larger angles and may thus improve the focusing properties. The simplest configuration which generates the required field shape is a series of coils. In the following we describe the focusing properties of a system of three identical coils, with inner radius 20 cm, outer radius 30 cm and length 40 cm, placed at a distance of 60 cm from each other. The currents are chosen such that the field strength in the center of each coils is 1.877 T, optimized for particles of 50 MeV/c, emitted from a point source at $\theta = 40^o$. Figure 6 shows distributions of the distance to the axis as a function of momentum, emission angle and source size. The point source distributions show almost perfect focusing over an angular range of $30^o - 50^o$, corresponding to a solid angle of 11%.

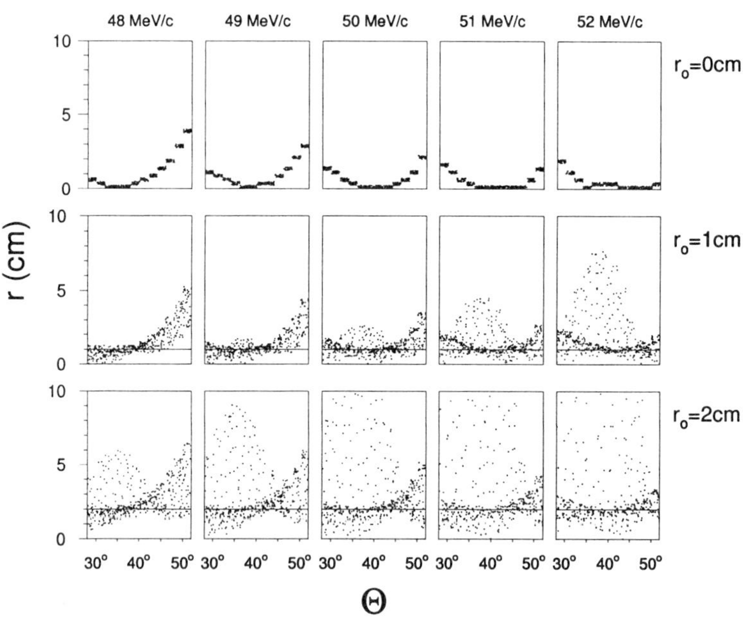

Figure 6. Distributions of r, the distance to the symmetry axis, after one spectrometer stage as a function of θ for various values of the particle momentum and ring-source radius r_0.

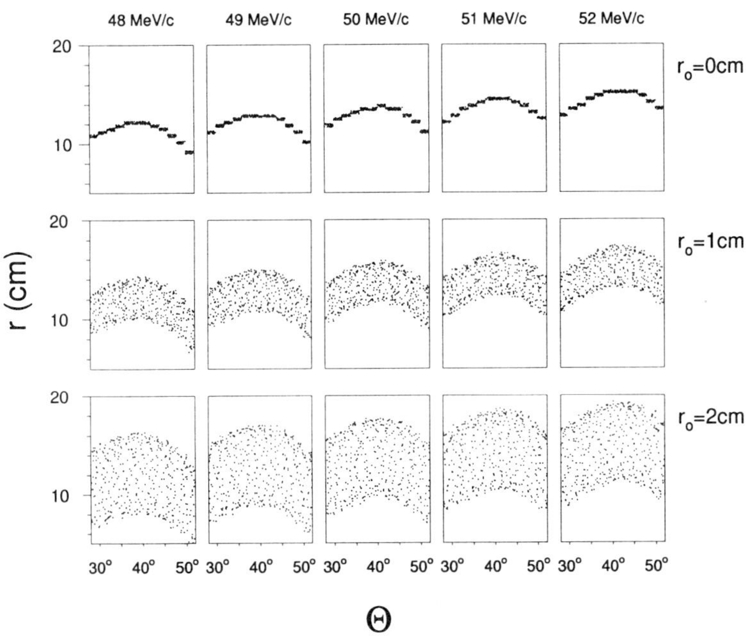

Figure 7. As in the previous figure but halfway one spectrometer stage.

The symmetry of the field with respect to the plane halfway between the coils leads to a second, intermediate, focus as is demonstrated in Fig. 7. Contrary to the "real" focus, this intermediate focus shows a very large momentum dispersion: a 10% change in momentum leads to a 50% change in radial coordinate. As has been known for intermediate image spectrometers there is a large broadening due to the source size. By measuring the trajectory in both focal planes the source contribution can be corrected. Assume three position detectors, one in the focal plane and two close to the intermediate plane. Figure 8 shows the resulting momentum resolution.

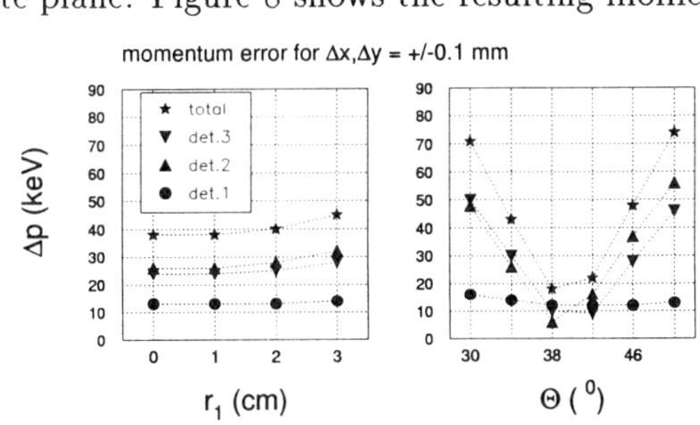

Figure 8. Momentum error, caused by finite position resolution, as a function of the distance to the origin and the emission angle.

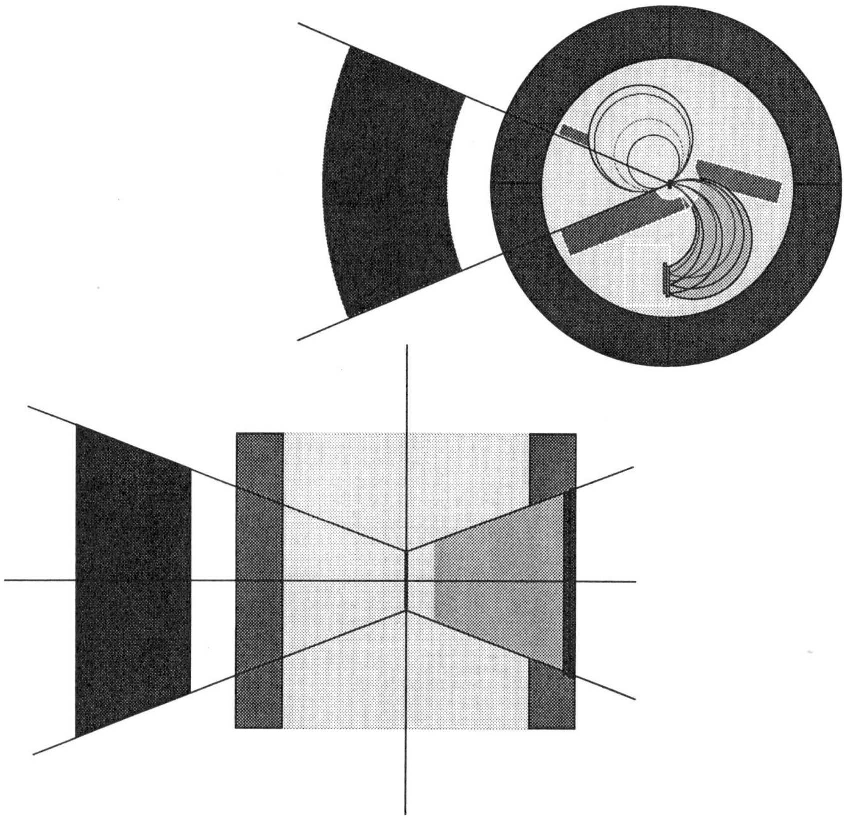

Figure 9. Solenoidal spectrometer. Top: axial view of setup with solenoidal spectrometer and photon detector on the left. Multiple e^+ crossings are suppressed by shielding walls. Bottom: side view.

4.1.2. Solenoid This option aims at the best resolution for the opening angle between the positron and the photon. This is achieved by using a very thin extended target, slanted at a small angle relative to the beam direction.

A photon detector is placed outside the solenoid magnet while a positron tracking system is installed inside. The solenoid field sweeps charged particles away from the photon calorimeter with a minimal stray field at the location of the phototubes of the photon calorimeter.

After the successful experience with a superconducting, thin-wall (0.2 radiation lengths) solenoid magnet a new magnet with the total thickness of 0.1 radiation lengths, which leads to a loss of only 5% of 50 MeV photons, is currently under design. The operating magnetic field is assumed to be 1 Tesla, which confines 52.8 MeV positrons within the inner diameter of the solenoid ($\approx$75 cm).

In a solenoid-field spectrometer (as in MEGA), curling tracks of the positrons emitted at $\approx 90^o$ might cause multiple hits in the chambers. Two different solutions are currently under discussion. In the first design (see Fig. 9), positrons are stopped after their 180° turns in a heavy metal shielding. In the second design the tracking wires are kept away from the target so that they see only particles with transverse momentum above 35 MeV/c. The central part of the sense wires is made insensitive by special coating. In addition, a slight gradient in the magnetic field will sweep all positrons out of the active area quickly.

4.2. Gamma detector

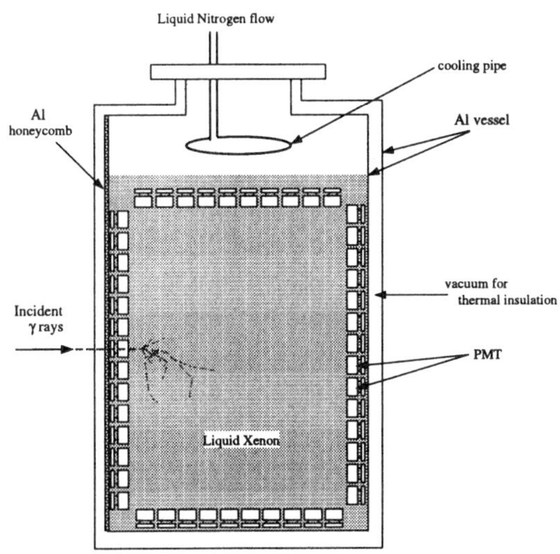

Figure 10. Schematic view of the "mini-Kamiokande" calorimeter. A simulated 52.8 MeV γ shower is superimposed.

4.2.1. Liquid xenon scintillation detector With its high light yield and short decay time (see Table 4), liquid xenon scintillator potentially achieves the energy and time resolution required for this experiment.

In the "mini-Kamiokande" scheme adopted here (see Fig. 10), a single tank (0.5–0.7 m^3) filled with liquid xenon is viewed from all sides by arrays of photo-multiplier tubes (PMTs). PMTs that are sensitive to the 175nm scintillation light and work at liquid Xe temperature have been developed in collaboration with Hamamatsu Photonics [19]. They were immersed in

Table 4. Properties of liquid xenon

Density	3.00 g/cm^3
Boiling and melting points	165 K, 161 K
Radiation length	2.77 cm
Scintillation wave length	175 nm
Energy per scintillation photon	24 eV [17]
Decay time	45 ns (75%) [18]
	22 ns, 4.2 ns
Scintillation absorption length	> 100 cm
Attenuation length (through Rayleigh scattering)	30 cm

the liquid Xe and the scintillation light from ^{241}Am α particles has been measured.

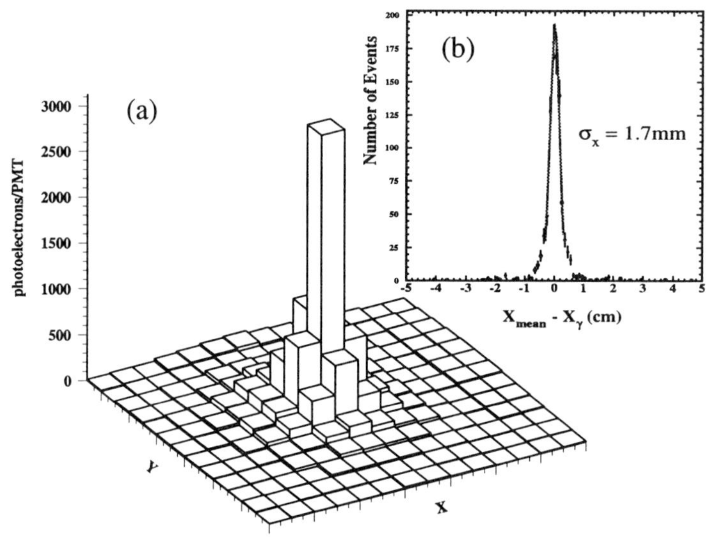

Figure 11. (a) Spatial distribution of light observed for the event in Fig. 10; (b) Distribution of the position error using weighted-mean.

A 50 MeV γ ray entering the liquid xenon loses most of its energy near the entry point, typically within 2–10 cm from the surface, resulting in a distribution of the PMT outputs as shown in Fig. 11(a). In the simulation

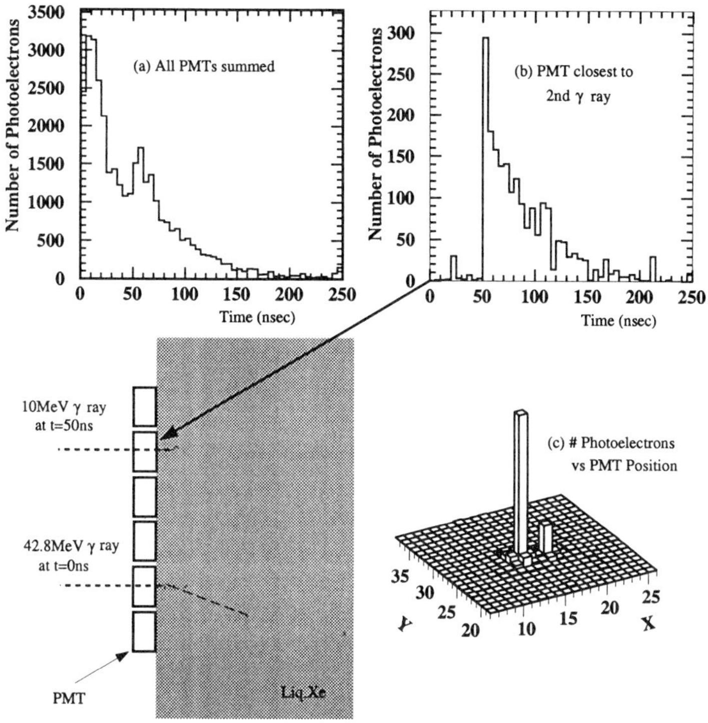

Figure 12. A simulated event in which a 43 MeV photon is followed after 50 ns by a 10 MeV photon. Top left: wave form (all PMTs summed); top right: wave form of the PMT closest to the 10 MeV γ; bottom right: spatial distribution of the scintillation light.

the PMTs are spaced at 6 cm. The γ position can be determined to a precision of $\sigma \approx 1.7$ mm, using the weighted mean of this distribution, (Fig. 11(b)). The depth of the γ interaction point can be determined with a precision of $\sigma \approx 7$ mm.

Fig. 12 shows a simulation of a potential pile-up event. Wave-form digitizers will be used to help identify the event. For example, two γ rays, each with an energy greater than 5 MeV, can be reconstructed if their time difference is more than 10 ns.

A small prototype detector with 32 PMTs is under construction to be used to study the detection principle with γ sources.

4.2.2. Crystal calorimeter with pre-shower In this scheme the photon detector consists of an electromagnetic calorimeter of conventional geometry, preceeded by a pre-shower detector. The latter not only determines the photon conversion point with a precision of a few millimeter's, but allows for extremely high timing resolution, which is important, since the potential background in this experiment is dominated by accidental coincidences.

Table 5. Properties of scintillating crystals.

	Ce concentration	light yield	fall time	density	peak wave length	energy resolution[a]	time resolution[b]	mass attenuation[c]	refractive index
YAP	0.5	18	25	5.3	365	5.7	0.16	2.63	1.94
LuAP	0.10	11	16	8.4	365	9.3	0.16	1.05	1.94
LSO	0.22	27	47	7.4	420	10	0.16	1.05	1.82
GSO	1.0	8	41	6.7	420	8.5	0.70	1.38	1.85
BGO	–	9	300	7.1	490	10	1.0	1.11	2.15
NaI(Tl)	–	40	230	3.7	415	6.5	0.35	3.05	1.85
	%	ph/keV	ns	g/cm^3	nm	%	ns	cm	

[a]at 660 keV [b]at 1 MeV [c]at 511 keV

The pre-shower detector would have an area of about 0.5 m^2 and a thickness of 2 cm and consist of about 50 single crystals. The readout would be configured as in an Anger camera with the help of some 450 small photodetectors (photomultipliers or photodiodes) mounted on the front face of the detector. Figure 13 illustrates the principle. As scintillating medium we consider one of a small number of isostructural rare-earth orthoaluminates (perovskites) whose most prominent members are $YAlO_3$ (YAP) and $LuAlO_3$ (LuAP), but also $YbAlO_3$ appears to be a very promising candidate. When activated with Ce^{3+}, these crystals have very similar spectroscopic and scintillation properties. Compared to a commonly used scintillator, such as NaI(Tl), the light yield is up to 50 percent and the decay time constant is shorter by an order of magnitude. Table 5 lists the relevant properties for a number of scintillating crystals. One of the most important parameters is the mass attenuation length which is, thanks to the higher atomic number and density, 2.5 times shorter for LuAP and YbAP, than for YAP and NaI(Tl). There are no real alternative scintillating materials for this detector. BGO is the only crystal with a similar

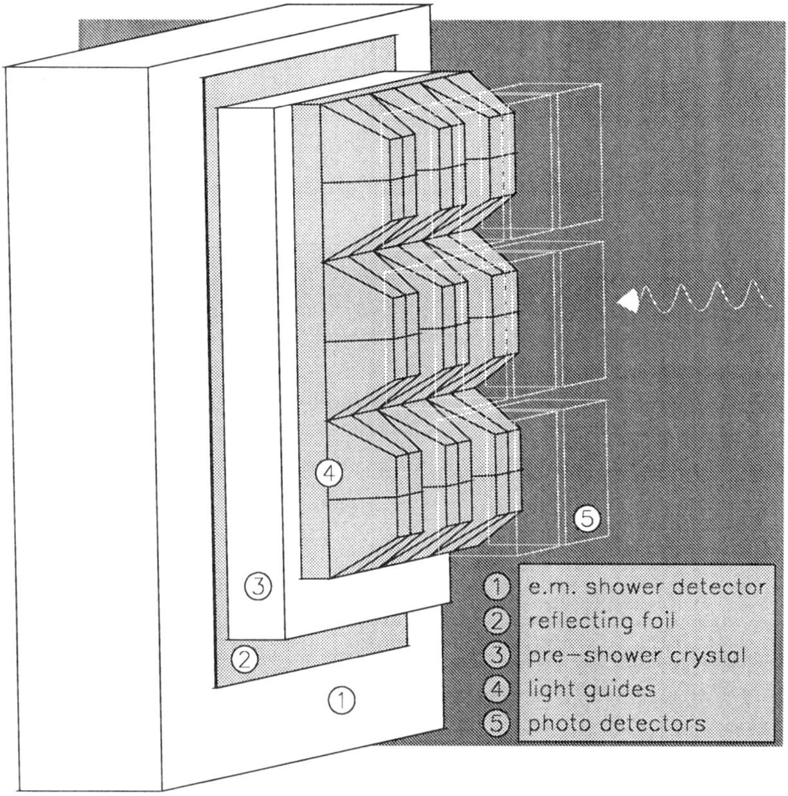

Figure 13. Schematics of photon detector

mass attenuation, but suffers from poor energy and timing resolution.

This very high performance has been obtained despite deficiencies in the Cerium concentration and optical transparency and it is commonly believed that large improvements should still be possible. Whereas it has not yet been possible to grow large (in excess of 1 cm^3) crystals of LuAP:Ce, 100 cm^3 crystals of YAP:Ce are produced in significant quantities. It is also known that the Ce content in present crystals is well below the optimal value from the point of view of light yield. A second problem is the absorptive loss of scintillation light. The source of the absorption is not clear, but is expected to be caused by unknown impurities in the crystal.

At present it is possible to produce mixed crystals of YAP and LuAP, which fulfill all our requirements with respect to size, homogeneity, light output and photon efficiency. The reason for this lucky circumstance are their very similar excitation and absorption spectra. Still we believe that we should use the time remaining until we would start the production of the required crystals to improve the crystal quality as much as possible, in particular into the direction of:

- higher Ce concentration
- reduced parasitic absorption
- higher LuAP fraction
- larger crystal sizes

As an alternative we consider $YbAlO_3$:Ce, a crystal not yet documented in the literature, which we have started to produce just recently. At present we are investigating the detection properties of this crystal which is cheaper and easier to produce than $LuAlO_3$:Ce.

For the electromagnetic calorimeter we consider good old NaI(Tl) or GSO, i.e. another new crystal type. GSO has as main advantages over NaI(Tl) a much higher density (6.7 g/cm^2 compared to 3.7 g/cm^2) and shorter decay time (41 ns compared to 230 ns) at the cost of a fourfold reduction in light yield. This crystal can be produced in the required quantities. Alternatively we might choose for YAP or CsI(Na), an extremely bright crystal which became interesting after we succeeded to reduce the decay time from $\approx 1\mu$s to ≈ 400 ns.

5. Time-table, costs and manpower

A decision about the detection system will have to await further simulations and R&D, which may take one year at least. The submission of a proposal could thus take place by the end of 1999 at the earliest. Detector construction would take another three years at least, such that data-taking could take place in the years 2003-2005.

Reliable cost estimates can not be given until the detection system is better defined and tenders have been put in. Rough guesses indicate that the experiment will not cost more than 10M CHF. A high transmission beam extraction system including DC separator costs about 0.5M CHF. The magnetic spectrometer may cost around 2M CHF. The calorimeter would cost roughly 3M CHF. The readout electronics could be quite expensive when high-quality digitizers have to be foreseen for the calorimeter. Positron tracking detectors will probably not contribute significantly.

A price tag of 10M CHF seems justified if compared with the preliminary MECO budget, which is three times larger. The MECO collaboration presently consists of seven institutions and 30 physicists, but may have to be enlarged considerably.

The sizes of the groups are listed in Table 6. We feel confident that the available manpower should be enough to finish the experiment within the time period discussed above.

Table 6. Manpower

Institution		physicists		
PSI		5		
Zurich		5		
	Switzerland		10	
Pavia		3		
Pisa		7		
Roma		2		
	Italy		12	
KEK		5		
Tokyo		5		
Waseda		4		
	Japan		14	
JINR		5		
Swierk		4		
Tbilisi		4		
	others		13	
total				49

References

[1] Search for the decay $\mu^+ \rightarrow e^+\gamma$ with a branching-ratio sensitivity of 10^{-14}, PSI-LOI: R-98-05.0, contactperson A. van der Schaaf

[2] J.D. Vergados, Phys. Rep. **133**, 1 (1986).

[3] A. van der Schaaf, in *Progress in Particle and Nuclear Physics*, edited by A. Faessler, Vol. 31, p. 1, (Pergamon, Oxford, 1993).

[4] P. Depommier and C. Leroy, *Rep. Prog. Phys.* **58**, 61 (1995).

[5] R. Barbieri and L.J. Hall, *Phys. Lett.* **B338** (1994) 212;
R. Barbieri, L.J. Hall, and A. Strumia, Nucl. Phys. **B445** (1995) 219.

[6] MECO Collaboration,
W. Molzon spokesman, *A Search for $\mu^- N \rightarrow E^- N$ with Sensitivity Below* 10^{-16}, AGS proposal P940, 1997.

[7] R. Mischke, presented at the *International Workshop on JHF Science*, Tsukuba, Japan, March 3–7, 1998.

[8] R.M. Barnett *et al.*, *Phys. Rev.* **D54**, 1 (1996) and 1997 off-year partial update for the 1998 edition.

[9] SINDRUM II Collaboration, submitted to *Phys. Rev.* C, 1998.

[10] K. Jungmann, presented at the *International Workshop on JHF Science*, Tsukuba, Japan, March 3–7, 1998.

[11] L.J. Hall, V.A. Kostelechy and S. Raby, *Nucl. Phys.* **B267** (1986) 415.

[12] J. Hisano, T. Moroi, K. Tobe and M. Yamaguchi, *Phys. Lett.* **B391** (1997) 341.

[13] M. Chaichian, K.Huitu, *Phys. Lett.* **B384** (1996) 157.

[14] Y. Kuno and Y. Okada, *Phys. Rev. Lett.* **77** (1996) 434.

[15] Y. Kuno, A. Maki, and Y. Okada, *Phys. Rev.* D55 (1997) 2517.

[16] Y. Kuno, Internal report.

[17] T. Doke, *Nucl. Instrum. Methods* **A327** (1993) 113–118.

[18] A. Hitachi *et al.*, *Phys. Rev.* **B27** (1983) 5279.

[19] R6041Q, Hamamatsu Photonics K.K.

Possible Search for Lepton Flavor Violation at Japan Hadron Facility

Yoshitaka Kuno

Institute of Particle and Nuclear Studies (IPNS),
High Energy Accelerator Research Organization (KEK)
Tsukuba, Ibaraki 305, Japan

Abstract. Future prospect on possible experimental searches for lepton flavor violation at the Japan Hadron Facility (JHF) is described.

1. What is JHF ?

Japan Hadron Facility (JHF) is a new accelerator project in Japan to build a complex consisting of a 200-MeV proton linac, a 3-GeV Booster proton ring (3-GeV PS), and a 50-GeV main proton synchrotron ring (50-GeV PS) [1]. Fig.1 shows a schematic layout of JHF at the campus of the High Energy Accelerator Research Organization (KEK). It aims to provide the highest beam intensity among the accelerators of such energies in the world. For instance, the 3-GeV PS and the 50-GeV PS will provide high-intensity beams of 200 μA and 10 μA (in average), respectively. The 3-GeV PS is a rapid-cycle machine with 25 Hz repetition rate, producing a beam power of 0.6 MW. The main 50-GeV PS has a cycle time of about 3.4 seconds with 0.7 second slow extraction time. Fig.2 shows the comparison of beam intensities of the existing and planned accelerators in the world. To reduce the total cost, the infrastructure available at KEK is intended to utilize as much as possible. For instance, the 3-GeV PS will be accommodated in the present tunnel of the KEK 12-GeV proton synchrotron, and the beamline for the neutrino oscillation experiment K2K (from KEK to Super-Kamiokande) will be used as the transport line from the 3-GeV PS to the 50-GeV PS. Recently, future energy upgrade of the linac from 200 MeV to

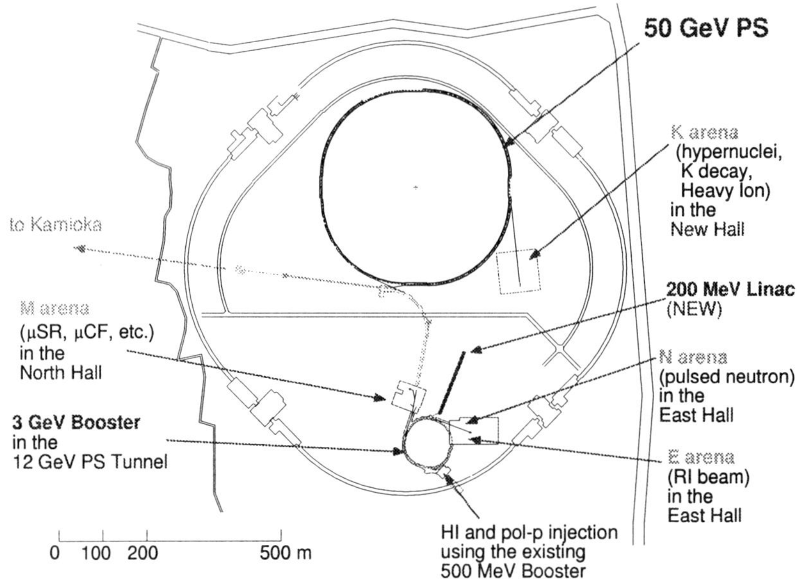

Figure 1. Schematic layout of JHF at the KEK campus.

400 MeV is also being discussed. The design report of the JHF accelerators is now available [2].

The physics programs at JHF cover a broad range of science from high-energy physics to material science by using the primary and various secondary beams from the accelerators. The experimental areas are grouped into four arenas called K-arena, M-arena, N-arena, and E-arena. They are respectively the physics programs using kaon, pion and neutrino beams (for particle and nuclear physics), those with muons (for μSR, muon-catalyzed fusion), those with neutrons (for material science), and those with exotic unstable nuclei beams (for nuclear physics). The K arena uses a proton beam from the 50-GeV PS, whereas the others use it from the 3-GeV PS. The former has both fast and slow proton extraction modes, although the latter has only fast extraction. Once the proposal is approved by the Japanese government, the construction will be completed in about 5 or 6 years.

In this talk, possible searches for lepton flavor violation (LFV) at JHF will be discussed (from my personal view). In Section 2, general overview of the muon lepton flavor violation is presented. In Section 3, a new idea of search for $\mu^+ \to e^+\gamma$ with the use of polarized muons is given. Finally,

Main Ring	Energy (GeV)	Particles per pulse	Repetition (Hz)	Current (μA)
JHF 50-GeV PS	50	20×10^{13}	0.3	10.0
BNL AGS	30	6×10^{13}	0.3	3.0
FNAL MI	120	3×10^{13}	0.3	1.3
CERN PS	26	2×10^{13}	0.5	1.6
KEK 12GeV PS	12	0.4×10^{13}	0.3	0.16
Serpukhov	70	1.7×10^{13}	0.1	0.27

Booster	Energy (GeV)	Particles per pulse	Repetition (Hz)	Current (μA)
JHF 3-GeV BS	3.0	5.0×10^{13}	25	200
Rutherford ISIS	0.8	2.5×10^{13}	50	200
LAMPF PSR	0.8			100
PSI	0.6			1500
KEK Booster	0.5	0.2×10^{13}	20	6

Figure 2. Comparison of existing and planned accelerators.

a plan of a new muon source of high intensity at JHF is briefly described in Section 4.

2. Overview of Muon Lepton Flavor Violation

The Standard Model has accomplished great success in explaining all the experimental measurements made so far. However, there are still many fundamental questions which the Standard Model can not address. Many experimental efforts therefore focus on discovery of new phenomena to understand the underlying fundamental truth. In the minimal Standard Model, lepton flavor conservation is built in by hand with the assumption of vanishing neutrino masses. In fact, virtually any new physics or interaction beyond the Standard Model would predict lepton flavor violation (LFV) at some level. Historical progress in various LFV searches is shown in Fig.3, in which the experimental upper limits have been continuously improved at a rate of about two orders of magnitude per decade for about 50 years since the first $\mu^+ \to e^+\gamma$ experiment by Pontecorvo et al. in 1947 [3]. In general,

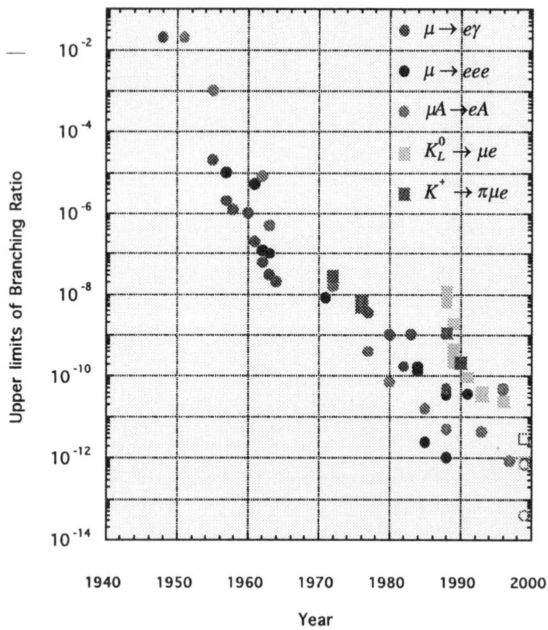

Figure 3. History of lepton flavor violation searches in various processes of kaons and muons.

a search for $\mu^+ \to e^+\gamma$ with a sensitivity of about 10^{-12} in a branching ratio could explore a mass scale of several 100 TeV, which is not accessible directly by present accelerators.

LFV processes of major interest are those with muons, which are for example $\mu^+ \to e^+\gamma$, $\mu^- + (A, Z) \to e^- + (A, Z)$ conversion in a muonic atom, $\mu^+ \to e^+e^+e^-$, and muonium-antimuonium conversion. It should be noted that the recently-claimed neutrino mixing [4] introduces only negligible effects on $\mu^+ \to e^+\gamma$. The rate from the neutrino-intermediate diagram was calculated [5] and is given by

$$R(\mu \to e\gamma) = \left(\frac{3\alpha}{64\pi}\right) \sin^2(2\theta)\left(\frac{m_{\nu_1}^2 - m_{\nu_2}^2}{m_W^2}\right)^2, \tag{1}$$

where $m_{\nu_i} (i = 1, 2)$ are masses of two neutrinos, and θ is a mixing angle between them. m_W is the mass of the W boson. This rate is very small, of the order of 10^{-52} for the neutrino mass of 0.07 eV. Therefore, LFV searches such as $\mu^+ \to e^+\gamma$ would be sensitive to *absolutely-new* physics other than the neutrino mixing.

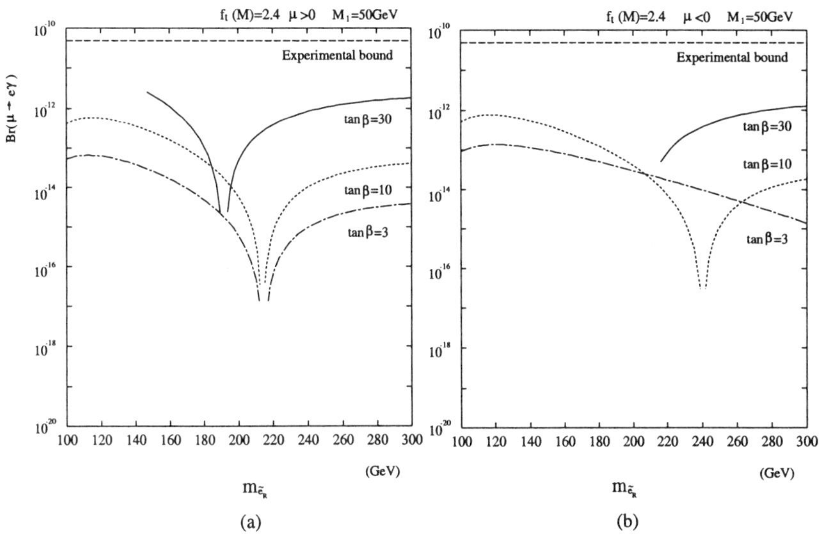

Figure 4. Predictions of $\mu^+ \rightarrow e^+\gamma$ branching ratio in SU(5) SUSY models in Ref.[9].

Recently considerable interest in LFV processes has arisen based on supersymmetric (SUSY) extensions to the Standard Model, in particular supersymmetric grand unified theories (SUSY-GUT) [6]. In many models of SUSY-GUT, LFV can be naturally introduced. For instance, in supergravity-mediated SUSY-GUT models, radiative corrections in the renormalization group evolution from the Plank scale to the weak energy scale lead to finite mixing in the slepton mass matrix even when it is assumed to be diagonal at the Plank scale [7]. Recently, Barbieri and Hall found that the slepton mixing thus generated is very large owing to the large top-quark Yukawa coupling [6, 8]. Then, $\mu^+ \rightarrow e^+\gamma$ occurs due to this slepton mixing through for loop diagrams. The predicted branching ratio is at a level which near-future experiments can reach. For instance, in SU(5) SUSY-GUT models, the prediction for $\mu^+ \rightarrow e^+\gamma$ can be given approximately by [6],

$$\Gamma(\mu \rightarrow e\gamma) \approx \frac{\alpha}{4}\left|\frac{\alpha(M_Z)}{24\pi\cos^2\theta_W}\frac{(\Delta m_{21}^2)}{m_{\tilde{\mu}}^2}\right|^2 \frac{m_\mu^5}{m_{\tilde{\mu}}^4} \quad (2)$$

$$\Delta m_{21}^2 = -V_{32}^* V_{31} I(y_t) \quad (3)$$

where $I(y_t)$ is a function of the top-quark Yukawa coupling (y_t) and V_{ij} is the Cabbibo-Kobayashi-Maskawa quark mixing matrix (at the GUT energy scale). m_μ and $m_{\tilde{\mu}}$ are respectively the mass of the muon and smuon. The branching ratio of $\mu^+ \to e^+\gamma$ predicted in SUSY SU(5) models [9] is shown in Figure 4. It ranges from 10^{-15} to 10^{-13} for the singlet selectron mass of $m_{\tilde{e}_R}$ of 100 to 300 GeV/c^2. The effect of higher dimensional operators in SU(5) SUSY-GUT was also considered, and the branching ratio was found to be enhanced for large $\tan\beta$ [10]. The SO(10) SUSY GUT models give an even larger value of 10^{-13} to 10^{-11} by an enhancement of $(m_\tau^2/m_\mu^2) \sim 100$ [8]. This is because of the existence of loop diagrams whose magnitude is proportional to the tau-lepton mass in such models.

Furthermore, the recent experimental hints of non-vanishing neutrino masses and mixing suggested by the atmospheric neutrino anomaly might invoke additional LFV contributions in SU(5) SUSY-GUT models [11]. This model includes a heavy right-handed Majorana neutrino of $10^{14} - 10^{15}$ GeV/c^2 with the ν_μ-ν_τ mixing of $\sin^2(2\theta_{\nu_\mu\nu_\tau}) \sim 1$ (as suggested by the Super Kamiokande). This contribution enhances LFV by a factor of about $(m_\tau/m_\mu)^2$ over the minimal SU(5) SUSY-GUT model, as shown in Fig.5.

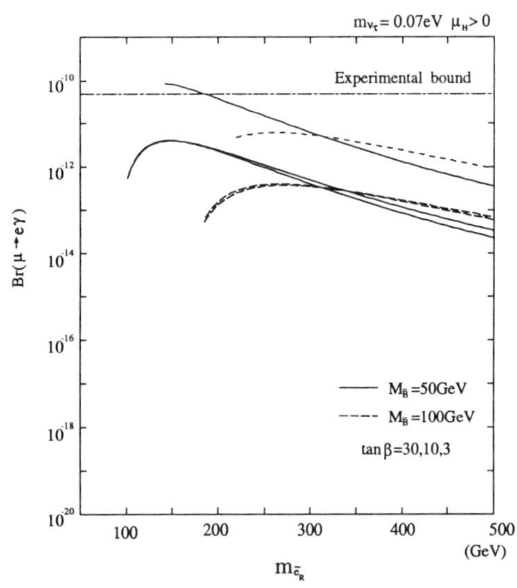

Figure 5. Predictions of $\mu^+ \to e^+\gamma$ branching ratio in SU(5) SUSY models with right-handed Majorana neutrino as a function of $m_{\tilde{e}_R}$ when the Yukawa coupling constant of ν_τ is as large as that of top quark [11].

Another class of SUSY models is gauge-mediated SUSY breaking, which in general do not lead to LFV since the messenger-matter interaction is flavor-blind. However, there have been discussions of the mechanisms to generate LFV such as messenger-matter mixing. In such models, a large branching ratio for $\mu^+ \to e^+\gamma$ is possible [12]. SUSY models with R-parity breaking also tend to predict a large $\mu^+ \to e^+\gamma$ branching ratio, which is sensitive to those couplings of R-parity breaking (λ_{ijk} and λ'_{ijk}) which are not constrained strongly by the limits on proton decay [13].

In the SUSY-GUT models, the $\mu^- - e^-$ conversion process in a muonic atom, for instance in Ti, has about 1/250 the branching ratio of $\mu^+ \to e^+\gamma$. Therefore, a search for $\mu^+ \to e^+\gamma$ at the sensitivity level of 10^{-14} is comparable to that for $\mu^- - e^-$ conversion at the level of 10^{-16}. The nuclear effects of $\mu^- - e^-$ conversion have also been calculated [14]. With taking account of relativistic atomic effects, Coulomb distortion, finite nuclear size and nucleon distribution, it was found that the ratio of $\mu - e$ conversion to $\mu^+ \to e^+\gamma$ varies from 1/389 for ^{27}Al to 1/238 for ^{48}Ti, and decreases again to 1/342 for ^{208}Pb [15]. Furthermore the extra logarithmic enhancement of the photonic loop diagrams for $\mu^- - e^-$ conversion (and also $\mu^+ \to e^+e^-e^+$) over $\mu^+ \to e^+\gamma$ has also been discussed [16]. It happens only when light charged fermions, to which a photon is attached, are involved in the loop diagrams. Therefore, it could occur for SUSY models with R-parity breaking, but not for R-parity conserving SUSY models or SUSY-GUT models.

3. $\mu^+ \to e^+\gamma$ with Polarized Muons

One of the popular processes of muon lepton flavor violation is $\mu^+ \to e^+\gamma$ decay. In general, the Lagrangian for $\mu^+ \to e^+\gamma$ can be given [17, 20] with an explicit helicity expression by

$$\mathcal{L} = A_R\ \overline{\mu}_R\sigma^{\mu\nu}e_L F_{\mu\nu} + A_L\ \overline{\mu}_L\sigma^{\mu\nu}e_R F_{\mu\nu} + h.c. \tag{4}$$

where $\mu_L(\mu_R)$ is a left-handed (right-handed) muon and $e_L(e_R)$ is a left-handed (right-handed) electron. A_R and A_L are coupling constants. $F_{\mu\nu}$ is $\partial_\mu A_\nu - \partial_\nu A_\mu$ and A_μ is an electromagnetic field. For polarized muons, from Eq.(4), the angular distribution of e^+ in $\mu^+ \to e^+\gamma$ with respect to the direction of μ^+ spin direction can be given by

$$\begin{aligned} d\Gamma(\mu^+ \to e^+\gamma) &= \frac{1}{8\pi}(1 - \frac{m_e^2}{m_\mu^2})^3 m_\mu^3 d(\cos\theta_e)\times \\ &\Big[|A_R|^2(1 - P_\mu\cos\theta_e) + |A_L|^2(1 + P_\mu\cos\theta_e)\Big] \end{aligned} \tag{5}$$

where θ_e is the angle between the muon spin and the e^+ direction. P_μ is the muon spin polarization. m_μ and m_e are the muon mass and the positron mass, respectively. For polarized muons, a seen in Fig.6, a left-handed e^+ (e^+_L) in $\mu^+ \to e^+_L\gamma$ follows a $(1 + P_\mu \cos\theta_e)$ distribution, whereas a right-handed e^+ (e^+_R) in $\mu^+ \to e^+_R\gamma$ decay does a $(1 - P_\mu \cos\theta_e)$ distribution.

This angular distribution is known to be useful to discriminate different models since the helicity of e^+ in $\mu^+ \to e^+\gamma$ is sensitive to the mechanism of lepton flavor mixing. For instance, the minimal SU(5) SUSY-GUT model predicts a non-zero A_L and a vanishing A_R, which yields a $(1 + P_\mu \cos\theta)$ distribution. On the other hand, the simplest version of the SO(10) SUSY-GUT models predict approximately-equal helicity amplitudes for right-handed and left-handed e^+s ($A_L \approx A_R$) [8], which result in an almost uniform angular distribution. For some non-unified supersymmetric models, A_L is vanishing but A_R is non-zero, which gives a $(1 - P_\mu \cos\theta)$ distribution. More importantly, it is shown in the following that the experimental backgrounds could be potentially further reduced by polarized muons.

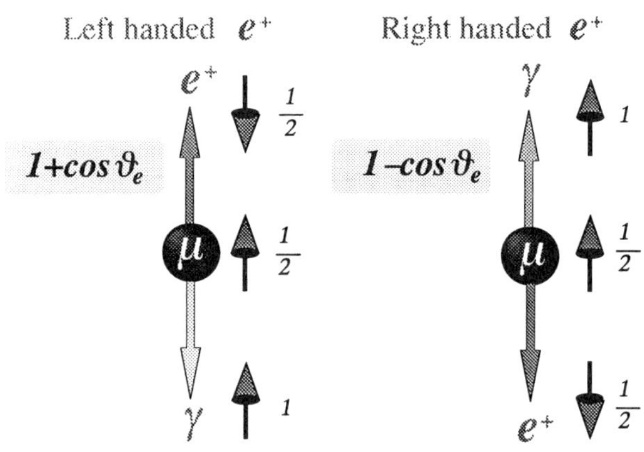

Figure 6. Angular distribution of $\mu^+ \to e^+\gamma$ with polarized muons.

The event signature of $\mu^+ \to e^+\gamma$ decay at rest is that the energies of both e^+ and photon are equal to a half of the muon mass ($m_\mu/2 = 52.8$ MeV) and they are collinear back to back. To utilize this clean kinematics, the $\mu^+ \to e^+\gamma$ searches have been done so far with stopped muons. μ^+s have been used exclusively since μ^-s are trapped by atoms in material and some of them are lost by muon capture. There are two major backgrounds to the search for $\mu^+ \to e^+\gamma$. One is a *physics (or prompt) background* from a radiative muon decay $\mu^+ \to e^+\nu\overline{\nu}\gamma$ when e^+ and photon are emitted back-to-back with two neutrinos carrying off little energy. The other background is an *accidental background* which comes from accidental coincidence of a e^+ in a normal muon decay ($\mu^+ \to e^+\nu\overline{\nu}$) accompanied by a high energy photon. The sources of a high energy photon might be either that in $\mu^+ \to e^+\nu\bar{\nu}\gamma$ decay, or external bremsstrahlung or annihilation-in-flight of e^+s in the normal muon decay. In the following, it is discussed how these backgrounds could be suppressed by using polarized muons to aim at a sensitivity of an order of 10^{-14}.

3.1. Physics Background for $\mu^+ \to e^+\gamma$

The radiative muon decay, $\mu^+ \to e^+\nu\bar{\nu}\gamma$ (branching ratio = 1.4 % for $E_\gamma > 10$ MeV), would become a background when the e^+ and photon are emitted back-to-back with two neutrinos carrying off little energy. The differential decay width of this radiative decay was calculated as a function of e^+ energy (E_e) and photon energy (E_γ) normalized to their maximum energies (of $m_\mu/2$), namely $x = 2E_e/m_\mu$ and $y = 2E_\gamma/m_\mu$, where x and y range from 0 to 1 [18, 19]. As a physics background to $\mu^+ \to e^+\gamma$, the kinematic case when $x \approx 1$ and $y \approx 1$ is important. In an approximation of the limit of $x \approx 1$ and $y \approx 1$ with an angle between e^+ and photon ($\theta_{e\gamma}$) of almost 180°, the differential decay width of $\mu^+ \to e^+\nu\bar{\nu}\gamma$ decay is given by

$$d\Gamma(\mu \to e\nu\bar{\nu}\gamma) \cong \frac{G_F^2 m_\mu^5 \alpha}{3 \times 2^8 \pi^4} \times \Big[(1-x)^2(1 - P_\mu \cos\theta) + \Big(4(1-x)(1-y) - \frac{1}{2}z^2\Big)(1 + P_\mu \cos\theta)\Big] dx dy z dz d(\cos\theta) \quad (6)$$

where G_F is the Fermi coupling constant, α is the fine-structure constant, $z = \pi - \theta_{e\gamma}$. The first term in Eq.(6) represents the e^+ being emitted preferentially opposite to the muon spin direction, whereas in the second term the e^+ is emitted along the muon-spin direction. When $x = 1$ and $y = 1$ exactly, this differential decay width vanishes. But in a real experiment, finite detector resolutions introduce background events which could limit the sensitivity of a search for $\mu^+ \to e^+\gamma$. The sensitivity limitation from $\mu^+ \to e^+\nu\bar{\nu}\gamma$ decay can be evaluated by integrating the differential decay width in Eq.(6) over the kinematic region of the signal which is determined by the detector resolutions. Take δx, δy and δz to be the kinematic range of the signal region for e^+ energy ($1 - \delta x \leq x \leq 1$), that for photon energy ($1 - \delta y \leq y \leq 1$) and that for the angle of $z = \pi - \theta_{e\gamma}$ ($0 \leq z \leq \delta z$), respectively. The integration was done with consideration of the kinematics constraints among x, y and z. Namely, if δx and δy are small, the allowed range of z is determined to be $0 \leq z \leq 2\sqrt{(1-x)(1-y)}$, instead of δz. The partial branching ratio after the integration is given by

$$dB(\mu \to e\nu\bar{\nu}\gamma) = \frac{\alpha}{16\pi}\Big[J_1(1 - P_\mu \cos\theta) + J_2(1 + P_\mu \cos\theta)\Big] d(\cos\theta), \quad (7)$$

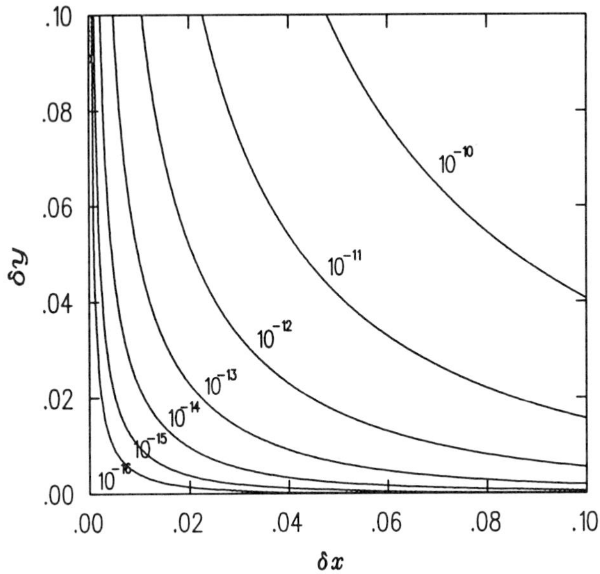

Figure 7. Sensitivity limitation imposed by physics background for unpolarized muons as functions of δx and δy.

For the case of $\delta z > 2\sqrt{\delta x \delta y}$, J_1 and J_2 are given by

$$J_1 = (\delta x)^4(\delta y)^2 \quad \text{and} \quad J_2 = \frac{8}{3}(\delta x)^3(\delta y)^3, \tag{8}$$

and when $\delta z \leq 2\sqrt{\delta x \delta y}$, they are

$$\begin{aligned} J_1 &= \frac{8}{3}(\delta x)^3(\delta y)(\frac{\delta z}{2})^2 - 2(\delta x)^2(\frac{\delta z}{2})^4 + \frac{1}{3}\frac{1}{(\delta y)^2}(\frac{\delta z}{2})^8 \quad \text{and} \\ J_2 &= 8(\delta x)^2(\delta y)^2(\frac{\delta z}{2})^2 - 8(\delta x)(\delta y)(\frac{\delta z}{2})^4 + \frac{8}{3}(\frac{\delta z}{2})^6. \end{aligned} \tag{9}$$

They can be reduced as the sixth power of a combination of δx, δy and δz. In Fig.7, the sensitivity limit imposed by the $\mu^+ \to e^+\nu\bar{\nu}\gamma$ decay with unpolarized muons for $\delta z > 2\sqrt{\delta x \delta y}$ is shown. From Fig.7, it can be seen that to achieve a sensitivity limit of a level of 10^{-15}, both δx and δy of an order of 0.01 are needed.

Experimentally the photon-energy resolution is more difficult to improve than that of the e^+ energy resolution. Assuming that δy is worse than δx by a factor of several, J_2 could be an order magnitude larger than J_1. Therefore, the distribution follows mostly $(1+P_\mu \cos\theta)$ as long as $\delta y > \delta x$. Fig.8 shows the angular distribution of $\mu^+ \to e^+\nu\bar{\nu}\gamma$ with

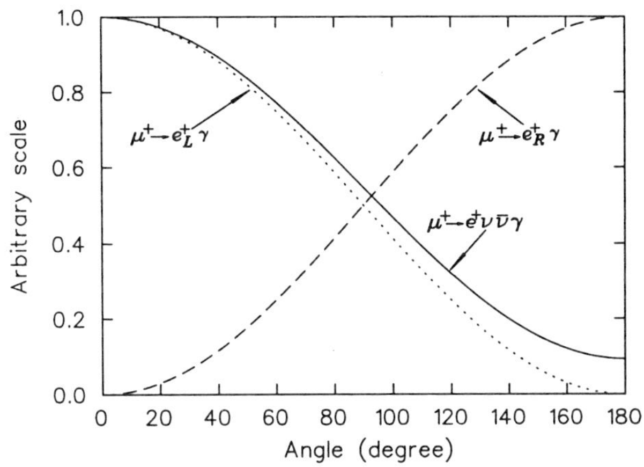

Figure 8. Angular distribution of e^+ in $\mu^+ \to e^+\nu\bar{\nu}\gamma$ decay with $\delta y/\delta x = 4$ as a function of angle from the muon-polarization direction in a solid line. A dot line and dash line show that of $\mu^+ \to e_L^+\gamma$ and $\mu^+ \to e_R^+\gamma$ decay, respectively. A vertical scale is arbitrary.

for instance $\delta y/\delta x = 4$ for $\delta z > 2\sqrt{\delta x \delta y}$. It implies that if we measure selectively e^+s in $\mu^+ \to e^+\gamma$ going opposite to the muon-polarization direction ($\mu^+ \to e_R^+\gamma$), the background from $\mu^+ \to e^+\nu\bar{\nu}\gamma$ will be reduced significantly. Furthermore, by varying δx and δy, the angular distribution of the $\mu^+ \to e^+\nu\bar{\nu}\gamma$ background can be changed according to Eq.(7), thus providing another means to discriminate the signal from the physics background.

3.2. Accidental Background for $\mu^+ \to e^+\gamma$

It turns out that the accidental background becomes more important than the physics background in a new generation experiment with a very high stopping rate of muons. The accidental background can be reduced in principle by an improved detector resolution. However, it is desirable to suppress this background by some additional means. It has been found that polarized muons are also effective to reduce the accidental background. The angular distribution of the sources of accidental background are examined for polarized muons [21]. It is known that e^+s in the normal Michel muon decay follows a $(1 + P_\mu \cos\theta_e)$ distribution. We have studied the inclusive angular distribution of a high energy photon (*e.g.* ≥ 50 MeV) from $\mu^+ \to e^+\nu\bar{\nu}\gamma$. It is found to be emitted preferentially along the muon spin direction; namely follows a $(1 + P_\mu \cos\theta_\gamma)$ distribution, where θ_γ is an angle of the photon direction with respect to the muon spin direction. It is schematically shown in Fig.9. It should be noted that this inclusive angular distribution was obtained after the integration of energy and direction of e^+s, which is in contrast to the case of physics background where only the extreme kinematics of e^+ and photon being back-to-back in $\mu^+ \to e^+\nu\overline{\nu}\gamma$ decay is relevant. It is further noted that the other sources of high energy photons such as external bremsstrahlung and annihilation-in-flight of e^+s in the normal muon decay also follow a $(1 + P_\mu \cos\theta_\gamma)$ distribution.

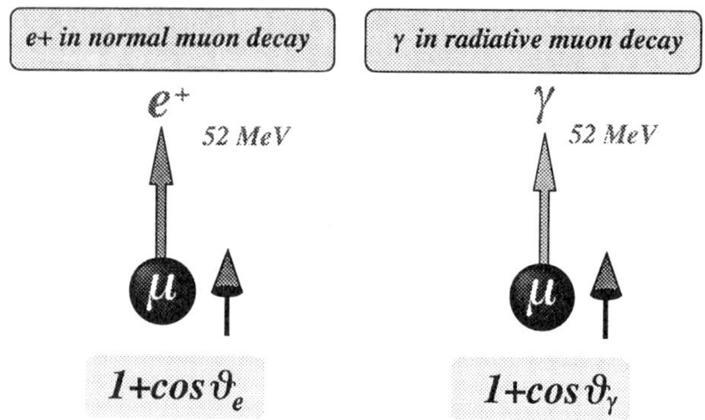

Figure 9. Schematic drawing of direction of a 52.8 MeV e^+ in Michel muon decay (left) and that of a 52.8 MeV photon in $\mu^+ \to e^+\nu\bar{\nu}\gamma$ (right). Both are preferentially emitted along the muon spin direction.

This inclusive angular distribution of a high energy photon in $\mu^+ \to e^+\nu\overline{\nu}\gamma$ implies that the accidental background could be suppressed for $\mu^+ \to e^+_L\gamma$ where high energy photons must be detected at the opposite direction to the muon polarization. A similar suppression mechanism of accidental background can be seen for $\mu^+ \to e^+_R\gamma$ when high energy e^+s are detected at the opposite direction to the muon polarization. As a result, the selective measurements of either e^+s or photons antiparallel to the muon spin direction would give the same accidental background suppression for $\mu^+ \to e^+_R\gamma$ and $\mu^+ \to e^+_L\gamma$ decays respectively. This favorable situation comes from the fact that the inclusive distributions of both high energy e^+s and photons, respectively in the normal and radiative muon decays, follow a $(1 + P_\mu \cos\theta)$ distribution, where θ is either θ_e or θ_γ.

The rate of the accidental background ($B_{\rm acc}$) normalized to the total decay rate can be estimated by

$$B_{\rm acc} = R_\mu \cdot f_e^0 \cdot f_\gamma^0 \cdot (\Delta t) \cdot (\frac{\Delta\omega}{4\pi}) \cdot \eta \qquad (10)$$

where R_μ is an instantaneous muon intensity. f_e^0 and f_γ^0 are an integrated fraction of the spectrum (with their branching ratios) of e^+ in the normal

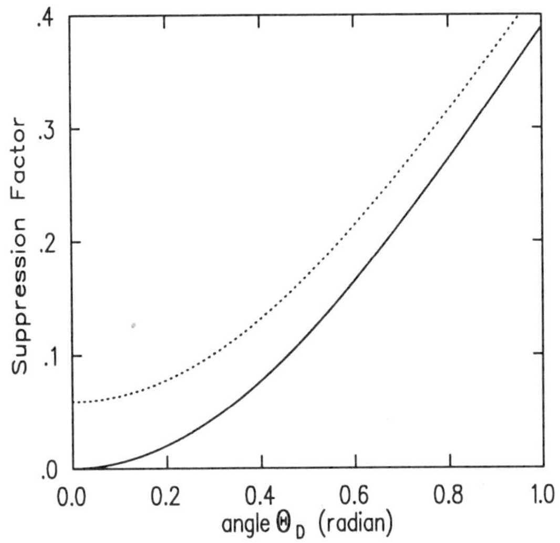

Figure 10. Accidental background suppression as a function of (half) detector opening angle.

muon decay and that of photon in $\mu^+ \to e^+\nu\overline{\nu}\gamma$ decay within the signal region for unpolarized muons, respectively. Δt and $\Delta\omega$ are respectively a full width of the signal regions for timing coincidence and angular constraint of the back-to-back kinematics. η is a suppression factor of the accidental background. $\eta = 1$ for the case of unpolarized muons. For instance, take some reference numbers such as the e^+ energy resolution (FWHM) of 1%, the photon energy resolution of 6%, $\Delta\omega = 3 \times 10^{-4}$ steradian, $\Delta t = 1$ nsec, and $R_\mu = 3 \times 10^8$ μ^+/sec, $B_{\rm acc}$ is 3×10^{-13}. Unless there are significant improvements made on the detector resolution, the accidental background might appear at a level of 10^{-13}.

As explained before, the selective measurement of either e^+s or photons antiparallel to the muon spin direction would provide a large background suppression factor. By taking account of the angular distributions of the $\mu^+ \to e^+\gamma$ signal, the suppression factor η is calculated for polarized muons by

$$\begin{aligned}\eta &\equiv \int_{\cos\theta_D}^{1} d(\cos\theta)(1 + P_\mu\cos\theta)(1 - P_\mu\cos\theta)/\int_{\cos\theta_D}^{1} d(cos\theta) \\ &= (1 - P_\mu^2) + \frac{1}{3}P_\mu^2(1 - \cos\theta_D)(2 + \cos\theta_D) \qquad (11)\end{aligned}$$

where θ_D is a half opening angle of detection with respect to the muon polarization direction. For instance, for $\theta_D = 300$ mrad, an accidental background can be suppressed down to about 1/20 (1/10) when P_μ is 100 (97)%. Note that an opening angle of $\theta_D = 300$ mrad give a geometrical solid angle coverage of 2.3%. Therefore, with a reasonable detector acceptance, a suppression of an order of magnitude could be achieved.

3.3. Experimental Status

The current experimental upper limit of $\mu^+ \to e^+\gamma$ in the Particle Data Group is 4.9×10^{-11} at 90 % confidence level done by the Crystal Box Collaboration in 1988 [22]. Since then, another experiment was done by the MEGA collaboration at Los Alamos National Laboratory (LANL). The MEGA detector consisted of dwarf chambers for positron tracking and pair spectrometers for photon detection. They were placed inside a superconducting solenoid magnet. MEGA completed its data-taking in 1995. A preliminary new limit of 4.2×10^{-11} with 16 % of the data analyzed was reported [23]. A direct extrapolation to all the MEGA data of $(3-6)\times10^{-12}$ was also expected. A new experiment of $\mu^+ \to e^+\gamma$ with a sensitivity of 10^{-14} is planned at PSI [24]. For a future JHF experiment, the use of polarized muons can be considered for further improvement on the background rejection.

4. A High-Intensity Muon Source at JHF

A high intensity source of secondary particles such as pions, muons and neutrinos is being considered at JHF to carry out various experiments including LFV searches. For conventionally-designed muon beam lines, it is expected to have an only slightly higher beam intensity for μ^+ and μ^- than those at PSI where a surface μ^+ beam of 3×10^8/sec is currently available. However, new ideas on a muon source as intense as $10^{11} - 10^{12}$ particles/sec appeared recently in conjunction with the $\mu^+\mu^-$ collider R&D [25]. The idea consists of (a) high magnetic-field capture of pions at the production target, (b) phase rotation to make a longitudinal momentum compaction, and (c) ionization cooling of muons. Such an approach could be adopted to construct a unique muon source with high intensity and high brightness and low-momentum spread. It would be crucial to carry out LFV searches with stopped muons, since they allow higher muon stopping efficiency with less background and a thinner muon-stopping target (or even a gas target). It should be noted that this kind of a new muon source would be useful as pion and neutrino sources with high intensity as well. Similar plans of intense muon sources are discussed at the RIKEN muon facility at RAL/ISIS [26] and at TRIUMF [27]. Furthermore the MECO

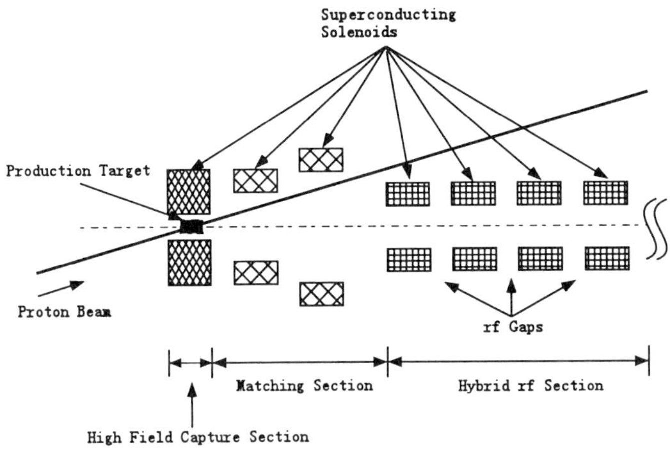

Figure 11. Schematic layout of a new high-intensity muon source at JHF. It consists of the high field capture section, and the rf section for phase rotation, which might be followed by the muon ionization cooling section (not shown).

experiment at BNL/AGS aims to use a high-field solenoid capture system for their $\mu^- - e^-$ conversion experiment of a sensitivity below 10^{-16} [28].

At this moment, the use of a proton beam from the 50-GeV PS is primarily considered at JHF for the new muon source since it has a slow-extraction mode (see the details below). A schematic layout of the preliminary muon beam design is shown in Fig.11. Protons from the 50-GeV PS hit a production target where secondary particles are produced. Those secondary particles (such as pions) are collected by a high solenoid magnetic field of 16 T or more. A magnetic field of 16 T gradually changes into a lower field of 4 T to reduce the transverse component of momentum with adiabatic condition of P_T^2/B = constant. In the case from 16 T to 4T, the average transverse momentum becomes about a half at the end of the matching section of 4 T. And then, they come into an rf phase-rotation section, where an rf field is applied to make longitudinal momentum spread narrower by phase-rotation. In this phase-rotation section, particles with higher energy than the reference energy are decelerated, whereas those with lower energy than the reference are accelerated, yielding a high-intensity beam with narrow energy spread. A solenoid field of about 5 T is given at the phase-rotation section to contain secondary particles inside the rf cavities. For muons, the ionization-cooling section follows. The requirements of muon ionization cooling as secondary beams are modest in comparison with the colliding-beam requirements.

In contrast to the case of a $\mu^+\mu^-$ collider which will run at a few Hz, the time structure of a muon beam at JHF has to be continuous or semi-continuous to carry out coincidence experiments. To do this, special approaches such as (1) micro-bunching slow extraction of protons from the 50-GeV PS and (2) superconducting rf cavity for phase rotation/ionization cooling are being considered. The time structure of micro-bunching extraction will depend on experimental requirements. For example, a search for $\mu^+ \to e^+\gamma$ needs a bunching repetition much faster than the muon lifetime (like a few 10 MHz to 100 MHz), whereas $\mu^- - e^-$ conversion and muonium-antimuonium conversion require a repetition rate slower than the muon lifetime. This slow repetition is a key issue for such experiments to eliminate backgrounds [28] and improve a signal sensitivity [29].

A superconducting rf cavity might be needed to reduce a required power consumption when rf runs in a DC operation. Furthermore, superconducting rf cavity allows a higher rf field of about 10–20 MV/m with a few 100 MHz, although a leakage of a magnetic field has to be reduced significantly at the cavity locations [30]. To achieve this, an option of alternating solenoids is studied.

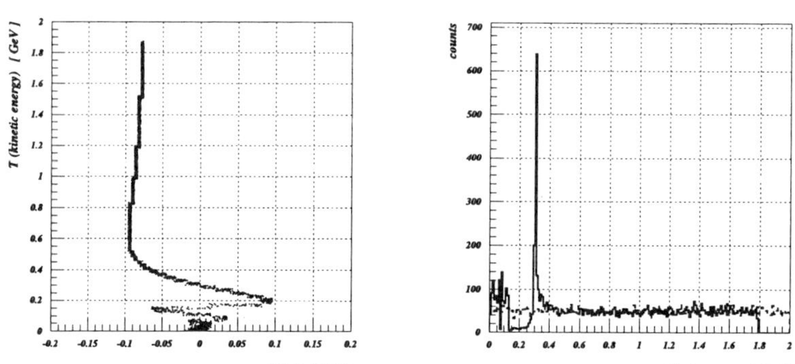

Figure 12. Energy gained or lost by phase rotation (delta-T) as a function of initial kinetic energy (T) (left). Longitudinal kinetic energy distribution after phase rotation (right).

A simple simulation of phase rotation at JHF was done. One example of the reference kinetic energy of 300 MeV is shown below. Fig.12(left) shows a kinetic energy gained or lost by rf phase-rotation (delta-T) as a function of initial kinetic energy (T) at entrance. It is clearly seen that particles of energy higher than 300 MeV lost their energy and those of energy lower gained after the phase-rotation section. The resultant longitudinal kinetic-energy distribution after the phase rotation is shown in Fig.12(right), where a strong peak at 300 MeV can be seen. Here it is assumed for simplicity that secondary particles have a flat initial energy distribution, shown in a dashed line in Fig.12(right). The intensity increase of a factor of about 10 in an energy range of about 10 MeV can be achieved by phase rotation. The yield of secondary particles is roughly estimated to be about 10^{12}/sec in 300 MeV $\pm$ 5 MeV for 6×10^{13} protons/sec of 50 GeV. To improve the beam emittance further, muon ionization cooling will be mandatory. Monte Carlo simulations have just began.

Potential LFV experimental programs which can be explored with such a new muon source are those with stopped muons. A muon beam of high-intensity and high-brightness with narrow momentum width would allow to use a very thin (or gas) target material. A thin target is crucial for a $\mu^+ \rightarrow e^+\gamma$ search to achieve a better angular resolution between the e^+ and photon directions for further background rejection. It is also important for a search for $\mu^- - e^-$ conversion to improve e^- energy resolution as well as background suppression. A search for $\mu^+ \rightarrow e^+e^-e^+$ with high sensitivity (like 10^{-16}) could be potentially carried out with a long gas target surrounded by an extended tracking detector of several meter long, where the momentum valance and vertex reconstruction of three $e^{\pm}$s can be improved and possible backgrounds is reduced owing to less materials.

5. Conclusion

Future prospect of possible searches for muon lepton flavor violation at JHF is discussed. A new idea to suppress the physics and accidental backgrounds for $\mu^+ \rightarrow e^+\gamma$ by using polarized muons is presented. Finally, a new muon source with high intensity and high brightness, which is being considered at JHF, is mentioned. Although more works and efforts should be made, it is aimed that searches for muon LFV with high sensitivity are carried out at JHF.

Acknowledgments

We would like to thank Drs. D. Bryman, L.S. Littenberg, A. Maki, W. Molzon, Y. Mori, S. Nagamiya, K. Nishikawa, Y. Okada, S. Orito, N. Sasao, S. Sawada, H.C. Walter, and T. Yokoi for discussions. This work was supported by Grant-in-Aid by the Ministry of Education, Science, Sports and Culture of Japan.

References

[1] JHF Project Office 1997 *Proposal for Japan Hadron Facility* KEK Report 97-3, JHF-97-1

[2] JHF Project Office 1998 *JHF Accelerator Design Study Report* KEK Report 97-16, JHF-97-10

[3] Hinks E P and Pontecorvo B 1947 *Phys. Rev. Lett.* **73** 257–259

[4] Fukuda Y *et al* (The Super-Kamiokande Collaboration) 1998 *Evidence for Oscillation of Atmospheric Neutrinos* Preprint hep-ex/9807003

[5] Bilenky S M, Petcov S T, and Potecorvo B 1977 *Phys. Lett.* **B67** 309–312

[6] Barbieri R and Hall L 1994 *Phys. Lett.* **B338** 212–218

[7] Hall L J and Kostelecky V A and Raby S 1986 *Nucl. Phys.* B **267** 415–432

[8] Barbieri R, Hall L and Strumia A *Nucl. Phys.* B **445** 219–251

[9] Hisano J, Moroi T, Tobe K, and Yamaguchi M 1997 *Phys. Lett.* **B391** 341–350; Erratum 1997 *Phys. Lett.* **B397** 357

[10] Hisano J, Nomura D, Okada Y, Shimizu Y and Tanaka M 1998 *Enhancement of* $\mu^+ \to e^+\gamma$ *in the Supersymmetric SU(5) GUT at Large* $\tan\beta$ Preprint KEK-TH-575, UT-815, OU-HET-294, hep-ph/9805396

[11] Hisano J, Nomura D and Yanagida T 1997 *Atmospheric Neutrino Oscillation and Large Lepton Flavor Violation in the SUSY SU(5) GUT* Preprint KEK-TH-548, hep-ph/9711348

[12] Dubovsky S L and Gorbunov D S 1997 *Messenger-matter Mixing and Lepton Flavor Violation* Preprint hep-ph/9706272

[13] Chaichian M and Huitu K 1996 *Phys. Lett.* **B384** 157–160.

[14] Kosmas T S, Faessler A and Vergados J D 1997 *J. Phys. G: Nucl. Part. Phys.* **23** 693–703
Kosmas T S and Vargados J D 1996 *Phys. Rep.* **264** 251–266
Shanker O 1979 *Phys. Rev.* D **20** 1608–1615

[15] Czarnechi A, Marciano W J and Melnikov K 1998 *Coherent Muon-Electron Conversion in Muonic Atoms* Preprint BNL-HET-98/2, TTP 97-19, hep-ph/9801218

[16] Raidal M and Santamaria A 1998 *Phys. Lett.* **B421** 250–258

[17] Weinberg S and Feinberg G 1959 *Phys. Rev. Lett.* **3** 111-114; Erratum 1959 *Phys. Rev. Lett.* **3** 244–245

[18] Fronsdal C and Überall H 1959 *Phys. Rev.* **118** 654–657

[19] Eckstein S G and Pratt R H 1959 *Ann. Phys., Lpz.* **8** 297–309

[20] Kuno Y and Okada Y 1996 *Phys. Rev. Lett.* **77** 434–438

[21] Kuno Y, Maki A, and Okada Y 1997 *Phys. Rev.* D **55** 2517–2520

[22] Bolton R D *et al* 1988 *Phys. Rev.* D **38** 2077–2101

[23] Mischke R 1998 *Proc. Int. Workshop on JHF Science* (KEK Publication); 1998 Private communications

[24] Van der Schaaf A 1998 *The Future of $\mu^+ \to e^+\gamma$ at PSI* in this workshop

[25] Muon-Collider Collaboration 1996 *$\mu^+\mu^-$ Collider – A Feasibility Study* Preprint BNL-52503, Fermi Lab-Conf.-96/092, LBNL-38946

[26] Nagamine K 1998 private communications

[27] Blackmore E *et al* 1998 *MUONS at TRIUMF* (Proposal of the 5-year Plan at TRIUMF) unpublished

[28] Bachman M *et al* 1997 *A Search for $\mu^- N \to e^- N$ with Sensitivity Below* 10^{-16} *MECO* Proposal to BNL AGS (E940)
Molzon W 1998 *Experimental Prospects for Muon-Electron Conversion Search below* 10^{-16}, *MECO at BNL* in this workshop

[29] Jungman K 1998 Private communications

[30] Noguchi S 1998 Private communications

Search for muon and electron lepton number violation with MECO at BNL

William R. Molzon[1]

Department of Physics and Astronomy,
University of California, Irvine, CA, USA

Abstract. I review the motivation for and status of searches for violation of muon and electron lepton number conservation in muon initiated processes. I discuss the expected progress in muon conversion experiments, concentrating on the MECO experiment, E940 at BNL. It will improve the sensitivity for the process $\mu^- N \to e^- N$ to below 10^{-16}, roughly 4 orders of magnitude better than the current limit.

Since the discovery of the muon and the realization that all differences between muons and electrons are due only to their different masses, there has been interest in understanding why more than one family of leptons exists and how they are related. An important question is whether or not muon, electron and tau lepton number are conserved. Non-conservation of these numbers is often referred to as lepton flavor violation (LFV). Experiments to search for LFV have mow been done for nearly fifty years, with no direct evidence for such processes having been found.

In principle, LFV could be induced by neutrino mixing, by much the same mechanism that effective flavor changing neutral currents are induced in the quark sector. Recently, compelling evidence for neutrino oscillations has been seen. This can be accommodated most easily in an expanded Standard Model by ascribing different masses to neutrino types, resulting in a leptonic matrix equivalent to the CKM matrix in the quark sector. However, current information on neutrino mass differences and mixing angles precludes the possibility that experimentally observable LFV effects in the charged sector can be due to neutrino mixing in processes involving loop diagrams containing neutrinos. This means that observation of LFV

[1] E-mail: wmolzon@uci.edu.

would indicate the existence of new physics processes beyond the Standard Model, other than neutrino mass and mixing. LFV searches are among the most likely places in which such evidence may be seen in the next few years.

1. Motivation for searches for LFV

Because there is no theory in which muon and electron number conservation follows from invariance under a local gauge transformation, there is strong theoretical prejudice that LFV will be seen eventually. Aside from the underlying motivation to test conservation laws with the best possible sensitivity, there is theoretical motivation derived from the many proposed extensions to the Standard Model that allow LFV. In general, these models are not devised for the purpose of predicting LFV. In many cases, the stringent limits already set restrict the allowed values of parameters within these models. Figure 1 shows Feynman diagrams for new physics processes in representative models, including technicolor and its variants, new massive electroweak gauge bosons, multiple Higgs doublet models, models with heavy neutrinos, and supersymmetry.

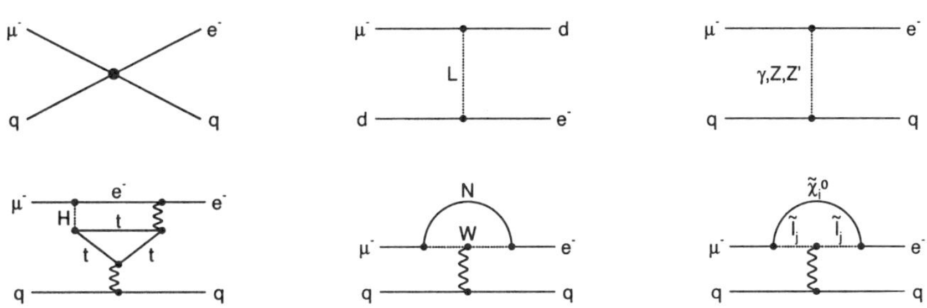

Figure 1. Feynman diagrams for the process $\mu^- N \to e^- N$ in different scenarios for non Standard Model physics.

A particularly interesting possibility is the existence of LFV processes in grand unified supersymmetric models. It was realized, first by Barbieri and Hall [1], that in some SU(5) and SO(10) models, LFV occurs naturally, and the degree of mixing is related to mixing in the quark sector. Hisano et. al [2] showed that for values of the masses of the supersymmetric partners of the leptons and of $\tan(\beta)$ within their expected range, LFV occurs at experimentally accessible levels. Figure 2 shows the expected branching fraction for the process $\mu^- N \to e^- N$ and $\mu^+ \to e^+\gamma$ in their model.

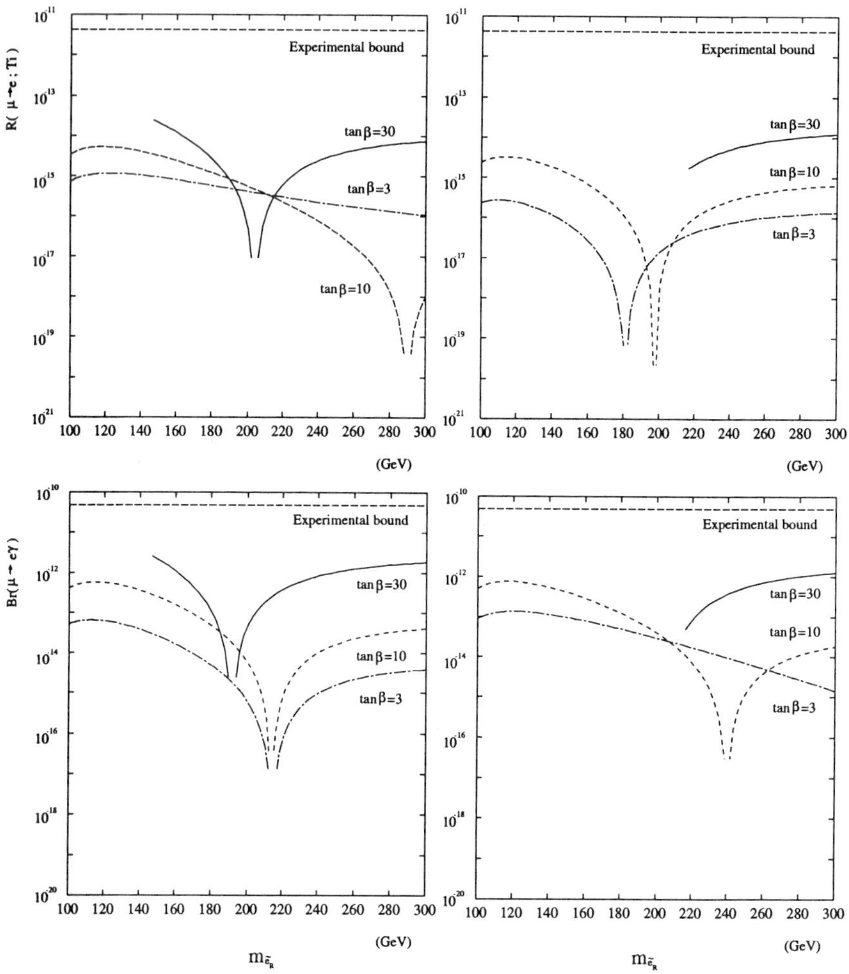

Figure 2. Rates for $\mu^- N \to e^- N$ and $\mu^+ \to e^+ \gamma$ in the model of Hisano et al. [2]

The two sets of curves are for the parameter $\mu > 0$ and $\mu < 0$. The experimental limits on the former have improved to 6.1×10^{-13} since this paper appeared.

2. Status of experimental searches for LFV

Experimental programs are searching actively for LFV in both muon and kaon induced processes. Table 2 summarizes the current limits on LFV processes and the limit which are expected to be achieved by completed or currently running experiments if no events are seen.

Table 1. The table gives current limits on LFV processes and the limits expected from currently running experiments if no signal is found.

Process	Current limit	Expected limit
$K_L^0 \to \mu^\pm e^\mp$ [3, 4, 5]	2.4×10^{-11}	$\sim 5 \times 10^{-12}$
$K_L^0 \to \pi^0 \mu^\pm e^\mp$ [6]	3.2×10^{-10}	$\sim 3 \times 10^{-11}$
$K^+ \to \pi^+ \mu^+ e^-$ [7, 8]	2.1×10^{-10}	$\sim 2 \times 10^{-11}$
$\mu \to e^+ e^+ e^-$ [9]	1.0×10^{-12}	
$\mu^+ \to e^+ \gamma$ [10]	3.8×10^{-11}	$\sim 3 \times 10^{-12}$
$\mu^- N \to e^- N$ [11]	6.1×10^{-13}	$\sim 5 \times 10^{-14}$

In the kaon system, the most sensitive search mode is $K_L^0 \to \mu^\pm e^\mp$, where the current limit is 2.4×10^{-11}, and the experimental sensitivity recorded in a recently completed experiment [5] is one event for a branching fraction of 2×10^{-12}. Experiments are either taking data or will soon take data for the modes $K_L^0 \to \pi^0 \mu^\pm e^\mp$ and $K^+ \to \pi^+ \mu^+ e^-$. While these experiments will not have the sensitivity of the $K_L^0 \to \mu^\pm e^\mp$ searches, they are sensitive to vector, rather than axial-vector interactions, and in that sense are complementary.

In the muon system, sensitivity of 10^{-12} and beyond has been achieved. For $\mu^+ \to e^+\gamma$, the state of the art is the MEGA experiment [12] at LAMPF; it completed data taking in 1995. The experiment stopped $\sim$1.5 $\times 10^{14}$ muons. Briefly, the detector consisted of a set of proportional wire chambers, drift chambers, and scintillation counters in a solenoid. The helical trajectory of electrons was measured in drift chambers. Photons were converted in thin radiators and the trajectories of the resulting e^+e^- pairs measured to deduce the photon momentum. Results of $\sim$16% of the data have been reported [12]; from these data an upper limit $B(\mu^+ \to e^+\gamma) < 3.8 \times 10^{-11}$ at 90% confidence level was set. Based on full analysis of the data and possible improvements in analysis efficiency, an upper limit of $3 - 6 \times 10^{-12}$ is expected to be set, assuming no events are seen. There are no proposals for more sensitive searches for this process, although a number of ideas for experiments that could reach a sensitivity close to 10^{-14} have been discussed [13, 14] previously. A letter of intent to do this experiment has been submitted to PSI. A. van der Schaaf discussed that program and Y. Kuno discussed prospects for experiments at the Japanese Hadron Facility during this Symposium.

Currently, there is no proposal for an improved search for $\mu^+ \to$

$e^+e^+e^-$. Since the current limit is from a rather old experiment, improvements could probably be made.

For $\mu^-N \to e^-N$, the state of the art is SINDRUM2 at PSI. The first phase of that experiment is completed, and a limit $R_{\mu e} < 6.1 \times 10^{-13}$ at 90% confidence level has been set [11], where $R_{\mu e} = \Gamma(\mu^-N \to e^-N)/\Gamma(\mu^-N \to \nu N')$. This program is continuing and an improvement by more than a factor of 10 is hoped for. The status of that experiment was discussed at this conference by P. Wintz. We discuss in section 4 an approved experiment with the goal of improving the sensitivity to below 10^{-16}.

3. Overview of $\mu^-N \to e^-N$ experiment

Coherent conversion of muons to electrons in the field of a nucleus is a particularly sensitive probe of LFV. Searches for $\mu^-N \to e^-N$ are conceptually simple. A large flux of μ^- is brought to rest in a thin target. There they either decay or are captured on the nucleus, at about the same rate for nuclei with Z about 10. If they convert to electrons, the signature is an electron originating in the stopping target with energy $\sim m_\mu c^2$, less small binding and nuclear recoil energies. The decay rate is conventionally normalized to that of the kinematically similar process of nuclear capture, and we define the quantity $R_{\mu e} \equiv \Gamma(\mu^-N \to e^-N)/\Gamma(\mu^-N \to \nu N')$. Since this process is coherent, the rate is enhanced with respect to nuclear capture by about a factor of Z. It is reduced by the square of the nuclear form factor at 105 MeV/c momentum transfer; this factor is near unity for small to moderate sized nuclei.

This process is closely related to $\mu^+ \to e^+\gamma$, if the process is mediated by a photon. In that case, for a given strength interaction, $B(\mu^+ \to e^+\gamma)$ would be $\sim 400 \times R_{\mu e}$[15]. Even though the branching fraction is smaller for given strength interaction, searches for $\mu^-N \to e^-N$ have a significant experimental advantage. Since only one final state particle is observed, the experiments are not limited by accidental backgrounds, as are $\mu^+ \to e^+\gamma$ experiments, and data can be collected at very high rates.

Background processes which produce 105 MeV electrons are heavily suppressed, but originate from a variety of sources: μ^- decay in a Coulomb bound orbit, radiative μ^- or π^- capture with photon conversion in the stopping target, 105 MeV beam electrons, μ^- or π^- decay in flight, $\bar{p}$ induced electrons, and cosmic ray induced electrons.

Muon decay in orbit and radiative muon capture are intrinsic to μ^- stopped in the target; they can be minimized only by improving the measurement of the electron energy. The most important background is μ^- decay in orbit. The endpoint energy is equal to that of the signal, and the rate is approximately proportional to $(E_{max} - E_e)^5$ near the endpoint [16].

The background level is extremely sensitive to both the central part of the detector response function and possible high energy tails. Eliminating this source of background places stringent requirements on the stopping target and on the spectrometer used to measure the electron energy, as we discuss in section 4.1.

Most of the other sources derive from prompt processes, with the electron detected close in time to the arrival of a beam particle at the stopping target. These can be reduced with a number of techniques, and we discuss this in section 4.1. Very slow $\overline{p}$'s have a very long transit time in the muon beam-line and arrive at the stopping target essentially continuously. They produce pions when they annihilate, and this can be a source of background. Hence, if the muon beam is produced using protons above $\overline{p}$ production threshold, the flux of $\overline{p}$'s must be controlled. Eliminating these backgrounds places stringent requirements on the muon beam design.

Finally, cosmic ray background can be reduced with appropriate active and passive shielding. This background source is unique in that the absolute number of background events depends on the integrated running time and not the muon flux or sensitivity. Hence, improving sensitivity by orders of magnitude does not require reducing cosmic ray induced background by the same factor.

Based on considerations of rates and backgrounds, we have concluded that the techniques of previous and current experiments cannot be used to extend the sensitivity by several orders of magnitude. First, a much more intense muon beam is needed. Second, to improve the background rejection by orders of magnitude at high rates, neither of the techniques used by the SINDRUM2 experiment to remove prompt backgrounds (vetoing events in time with beam particles or making a very pure beam) appear to be feasible. We conclude that a pulsed beam, as discussed below, is required. Third, to improve the sensitivity significantly will require a detector with much better resolution, to reduce backgrounds from muon decay in orbit.

4. The MECO muon to electron conversion experiment

4.1. Overview

The **M**uon to **E**lectron **Co**nversion experiment [17] is designed to extend the experimental sensitivity for $\mu^- N \to e^- N$ to below 10^{-16} using a new beam and detector operating at the Brookhaven National Laboratory (BNL) Alternating Gradient Synchrotron (AGS). E940 recently received scientific approval to do the MECO experiment.

E940 addresses the three crucial requirements to make a major improvement in this search. First, it will use a very intense μ^- beam which uses the idea from the MELC proposal of Djilkibaev and Lobashev [18, 19] to place the production target in a graded solenoidal field and collect π's over essentially 4π solid angle. MECO uses a higher energy proton beam than that proposed for MELC. The smaller current at a high energy machine like the AGS is largely offset by the increased pion production cross section. Second, MECO will use a pulsed beam, as used in an earlier experiments [20], which reduces the prompt backgrounds to acceptable levels. The idea is to stop a pulse of muons and detect conversion electrons only after all π^- and e^- in the beam have either decayed or passed through the detector region. These considerations lead to a beam pulse frequency of 0.74 MHz (twice the revolution frequency of the AGS) and the use of an aluminum stopping target, in which the μ^- lifetime is 880 ns. Third, the MECO experiment will use a high precision magnetic spectrometer capable of operating in the high rate environment resulting from the very high μ^- stopping rate.

A schematic drawing of the MECO beam-line and experimental area is shown in figure 3. Pions are produced from a tungsten target on the axis of a super-conducting solenoid. The axial component of the field is large and varies with axial position, largest at the proton beam exit and smallest at the muon beam exit. This results in a very high probability for containing π's and μ's in the evacuated bore. Particles with sufficiently small transverse momentum are captured, and particles emitted away from the μ^- beam direction are reflected in the graded field and captured in the beam with high efficiency.

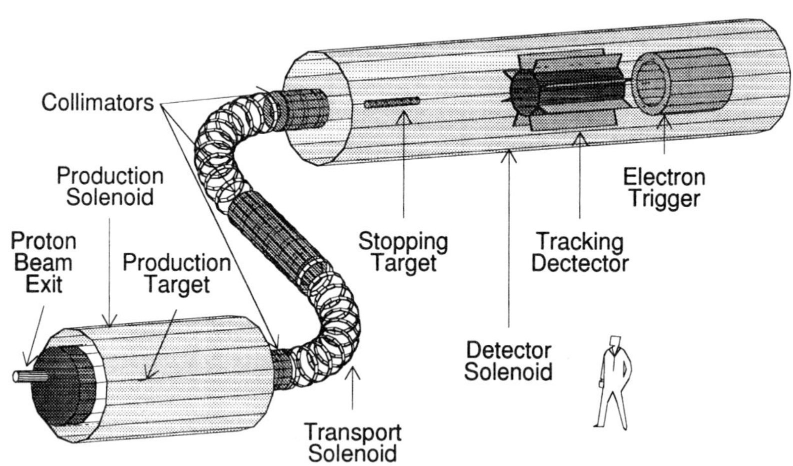

Figure 3. Schematic drawing of the MECO beam and apparatus. The cosmic ray shield surrounding the detector region is not shown.

The μ^- resulting from π^- decay are guided to the stopping target in a curved transport solenoid. Particles propagating in helical trajectories in the curved solenoid drift vertically and are sign and momentum selected with appropriate collimators. The μ^- stopping target and the electron detector are located in a solenoid in which the axial component of the field is also graded. The use of an axially graded field results in very good acceptance and reduces backgrounds from electrons produced in the upstream pole piece of the detector solenoid. The heart of the detector is a low mass magnetic spectrometer with an intrinsic resolution of ~100 keV. An electron trigger detector's is used primarily to select potential signal events and secondarily to check the e^- energy, help distinguish e^- from other particles, and help reject cosmic ray induced backgrounds. The detector region is surrounded by a cosmic ray shield.

A New Pulsed, High Intensity Muon Beam

An appropriately pulsed, high intensity muon beam is critical to MECO, both to achieve the desired sensitivity and to reduce background. We have developed a conceptual design for and tested some aspects of this beam.

It will be pulsed by using the RF structure of the AGS, operating with two filled RF bunches in the 2.7 μs revolution time, and extracting the beam bunched. The AGS will be operated at an energy of 7.5-8.0 GeV, below transition energy ($\gamma_T = 8.6$). This will allow higher bunch intensity than has been previously achieved, since the full aperture of the machine can be used if the beam does not have to be accelerated through transition. It is anticipated that a bunch(total) intensity of 2(4) $\times$ 10^{13} protons can be achieved. Further, since only two transfers from the booster to the AGS are required, and the energy has to be ramped only to 8 GeV, the cycle time with 50% duty factor will be 1 second. Hence, an average intensity of 6.4 μA can be achieved.

An extinction (ratio of total beam between bunches to beam in the bunches) of below 10^{-9} is required. First tests of a bunched beam extraction have been done, and the beam properties measured. One RF bunch was filled, accelerated to 24 GeV, and extracted. The pulses were measured to be ~ 15 ns wide and the extinction between bunches and in unfilled bunches is below 10^{-6} and 10^{-4}, respectively. Figure 4 shows the measured time structure of the beam. These measurements were made with only minimal tuning of the AGS, and substantially improved performance is expected. In particular, a means of removing particles at the wrong phase in the AGS has been devised. Recently, tests of high bunch intensity and bunched beam extraction at 8 GeV have begun. It may not be possible to reach an extinction below 10^{-9} in the beam extracted from the AGS, and other means of improving it have been studied. The preferred solution is a pulsed electric or magnetic kicker in the proton transport line, which could achieve another factor of $10^{-2} - 10^{-3}$ in extinction.

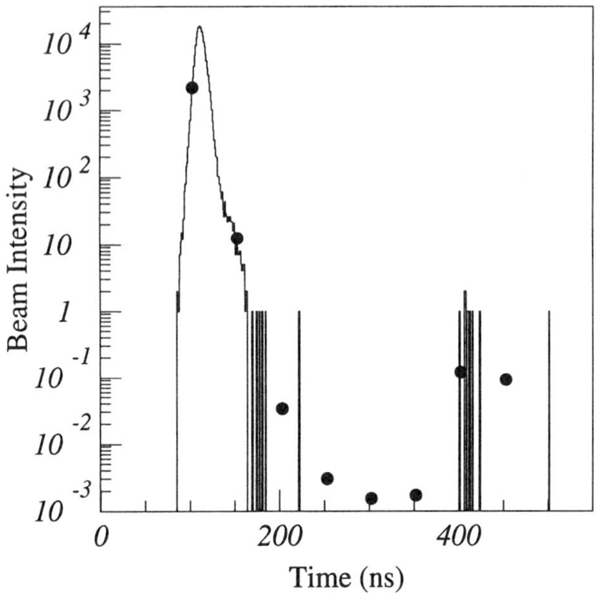

Figure 4. The measured time structure in a beam extracted from the BNL AGS. The main pulse is less than 50 ns. wide. Extinction between pulses is about 10^{-7}, while the extinction in an unfilled bunch is about 10^{-6}.

The design of the μ^- production area uses some of the ideas of the MELC experiment [18, 19]. Pions are produced in a radiation cooled, 0.8 cm diameter, 16 cm long tungsten target in a high field super-conducting solenoid. The field is graded, with the axial component varying from 3.3 T at the upstream end to 2 T at the muon beam channel entrance. The super-conducting coil is protected with a heat and radiation shield with inner and outer radii of 30 and 60 cm. At the intensity proposed, the maximum target temperature is near the limit imposed by the rate of evaporation of the target. Various means of reducing the target temperature by increasing its surface area are being studied. Heat load from the particle spray on the super-conducting solenoid was studied with a GEANT simulation of the target region. Less than 50 W is deposited in a 6 cm thick coil pack outside the shield, the maximum instantaneous local heat load is below 0.2 mW/gm and the total radiation load in a 10^7 s run is below 50 MRad, all well within the demonstrated range of operating parameters for super-conducting magnets.

The expected muon yield is based on data from an experiment [21] which measured pion yields over the full kinematic range relevant to the MECO beam. The experiment used a 10 GeV proton beam incident on a bubble chamber containing thin sheets of tantalum. Pions were identified and their energy and production angle measured. The yield is about a factor of 2.5 smaller than that predicted by the GHEISHA code in the GEANT package. We have used the GHEISHA code to model pion production and decay and muon transport, and scaled the yield by the ratio of the measured production cross section to that used by GHEISHA. Additional information on pion yields may be forthcoming from E910 [22] at BNL, which has recorded data over a fraction of the relevant kinematic range.

Charged particles are transported to the detector solenoid using a curved solenoid, one purpose of which is to decrease the transmission of both high momentum and positive particles. Charged particles of sufficiently low momentum follow helical trajectories centered on magnetic field lines. In a torus, they drift in a direction perpendicular to the plane of the torus by a distance given by $D = 1/0.3B \times s/R \times (p_s^2 + \frac{1}{2}p_t^2)/p_s$, where D is the drift distance, B is the magnetic field in Tesla, s/R is the bend angle of the solenoid, and p_t and p_s are the perpendicular and parallel momentum components in GeV/c. For $s/R = \pi/2, p_t = 0.09$ GeV/c, $p_s = 0.12$ GeV/c, and $B = 2$ T, the drift of the center of the helix is 49 cm. The drift direction depends on the charge. Hence, by putting appropriate collimators in the straight sections of the transport solenoid, positive particles and high momentum negative particles can be absorbed. The drift effect is illustrated in figure 5. It is exploited to remove $> 90\%$ of the positive particles and essentially all negative particles above 100 MeV/c.

This beam-line has been simulated fully by the MECO collaboration. The simulation included a model of the graded magnetic field in the production solenoid and a calculation of the field in the curved transport solenoid based on a realistic coil configuration. The effects of pion decay, absorption in the production target, and the scattering of all particles in materials in the production and transport region were included. The design was approximately optimized for yield and cost. For example, the yield could be increased with higher field strength in the production solenoid, but that would increase substantially the cost. Parameters that were optimized included the production solenoid inner bore and length, the target length and radius, target position, precise form of the graded field, sizes of collimators, and variation of the field in the transport solenoid.

Many of the design choices represented a compromise between muon yield and background suppression. For example, the flux could be increased by opening the collimator sizes, but this would increase the momentum acceptance of the transport and increase backgrounds. The default configuration yields about 0.006 muons entering the detector solenoid per incident proton, 37% of which stop in the thin aluminum stopping target.

Particular care was taken to ensure that the time structure of the beam is maintained, with sufficiently low flux of pions and muons arriving at

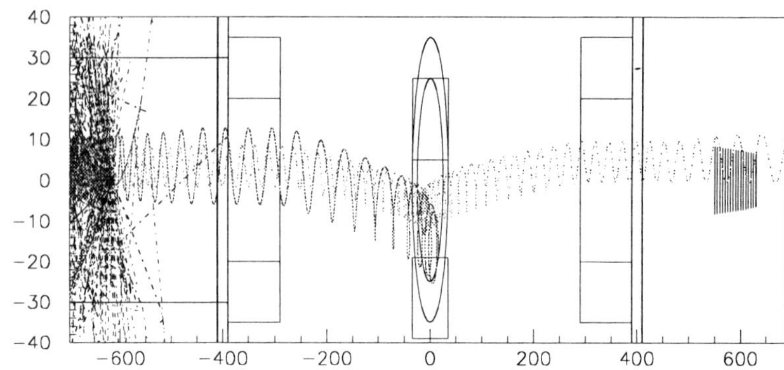

Figure 5. An example of a GEANT simulation of proton interactions in the MECO beam-line. The view shown is in the plane of the curved transport solenoid. The production target is at -730 cm, and the muon stopping target is centered at 600 cm. Collimators in the transport solenoid are shown centered at -350, 0, and 350 cm. Two negative particles enter the transport solenoid; the higher momentum one is absorbed in a collimator and the lower momentum one is transported to the detector solenoid at the right end of the figure. The scale is given in cm.

the stopping target during the detection time. The calculations involved a combination of Monte Carlo and analytic techniques. The late arrival of pions could be simulated by MC technique at full sensitivity by turning off pion decays and then weighting events by the survival probability, given the arrival time distribution. The calculation of the time spectrum of late arriving muons or electrons required a combination of analytic and MC techniques. For example, muons could spend a long time in uniform field regions of the transport, and, if they were sufficiently high momentum, could subsequently decay to electrons which could simulate background. This potential source could not be simulated at full sensitivity due to limited computing power. By considering analytically the maximum transit time in various segments of the transport, the expected time distribution was calculated. The choice of collimator apertures was based in part on reducing the transport for high energy particles. Further, a small gradient was introduced in the straight parts of the transport solenoid to reduce maximum transit times.

Another potential source of background is $\overline{p}$ annihilation in the muon stopping target or in the transport line. This is particularly dangerous for low energy $\overline{p}$'s, for which the transit time is very long, and hence the arrival time is essentially uniform. The flux of $\overline{p}$'s produced in the proton interactions was taken from measured cross sections, scaled slightly for the target material dependence. The data were at 6.3 GeV incident energy, and a phase space model for the energy dependence, including the effects of Fermi motion, was used to determine the dependence of the flux on the incident energy. Those $\overline{p}$'s which can be transported through the limited aperture of the transport solenoid are sufficiently low energy that they are absorbed fully in a 100 μm thick beryllium window in the middle collimator. The energy and number of annihilation products were taken from tabulated cross sections, and then transported using GEANT. Based on these calculations, the background from $\overline{p}$ annihilation was found to be negligible.

The backgrounds which are sensitive to the beam-line properties are summarized in section 4.1.

The Stopping Target and Experimental Apparatus

The stopping target and detectors are located in a solenoid of radius ~ 0.9 m, with a graded axial field, varying from 2T at the entrance to 1 T in the region of the detectors. The use of a graded field has two consequences. First, the quantity p_T^2/B_Z is constant, and hence 105 MeV e^-, either in the beam or produced at the upstream end of the solenoid, will have $p_T < 74$ MeV/c at the detector and can be eliminated as background by requiring $p_T > 75$ MeV/c. Second, conversion electrons emitted at angles of $90° \pm 30°$ with respect to the solenoid axis will have trajectories which intercept the tracking detector and which have a restricted range of p_T. Those initially moving away from the detector bounce in the graded field. Sample trajectories illustrating this are shown in figure 6.

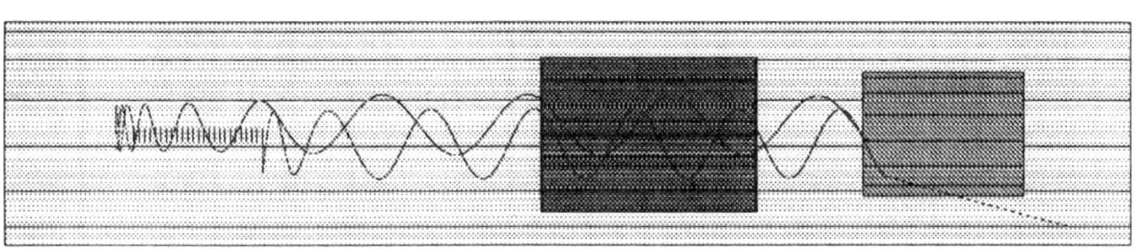

Figure 6. A schematic drawing of the MECO detector region with two typical conversion electron trajectories produced by the GEANT simulation superimposed.

The stopping target consists of 17 disks of Aluminum, each 0.02 cm thick and ~6 cm radius. The thickness chosen results from a tradeoff between stopping efficiency and energy loss straggling of conversion electrons. Approximately 40% of the μ^- in the beam are stopped in the target, and the energy straggling contributes ~200 keV to the energy resolution.

The high stopping rates in MECO will result in significant fluxes of low energy photons, protons and neutrons exiting the stopping target. The MECO design has a geometry designed to minimize the effect of these particle fluxes on the detector rates. The detector gains will be reduced during the stopping pulse when rates are very high. During the detection interval, rates are dominated by p's and γ's from μ capture processes and e^- from μ^- decay in orbit; a full GEANT simulation was used to calculate them. Most electrons from muon decay are contained at radii smaller than the inner radius of the detectors. To reduce the flux of p's and γ's, the detector is located 2 m from the stopping target, reducing the subtended solid angle. Thin absorbers before the spectrometer chambers reduce the flux of low energy protons hitting them. The absorbers are sufficiently thin that the electron energy resolution is not significantly degraded. The detector is very low mass to minimize photon conversions.

The goal for the tracking detector is to measure with good efficiency the parameters of the helical trajectory of electrons. The accuracy of the measurement is dominated by multiple scattering. The detector consists of a cylinder and 8 equally spaced *vanes* of tracking detectors, as shown in figure 3. All individual detector elements are oriented in the axial direction. The baseline design of the cylinder and vanes uses three layers of 5 mm diameter, 2 m long straw tubes. The axial coordinate is measured using pads external to the straws, in which signals are induced. The straws will be made of carbon loaded kapton. This technique has been demonstrated in prototype detectors [23]. The total rate per detector element was calculated to be $<$ 300 kHz; it results primarily from protons and γ's. The typical detector channel occupancy is under 1%.

The performance of the detectors was calculated using a full GEANT simulation of the stopping target and detectors. Using the resolution function determined from Monte Carlo simulated events, the level of muon decay in orbit background was calculated by convolving the resolution function and the background electron energy spectrum. Figure 7 shows the spectrometer lineshape for 105 MeV electrons and the expected measured energy distribution of electrons from muon decay in orbit. This and other background contributions are summarized in section 4.1.

The purpose of the electron trigger calorimeter is to detect 105 MeV electrons that have passed through the tracking system. The proposed detector is a scintillator cylinder with inner and outer radii of $\sim$ 45 and $\sim$80 cm, respectively, and 1.2 m long, segmented in azimuthal and axial directions. The performance was studied with a GEANT simulation, including effects of extra particles in the detector. The trigger rate will be dominated by muon decay in orbit electrons in which pile-up from the high rate environment contributes enough localized energy to exceed a trigger threshold. The rate has been estimated by convolving the electron energy

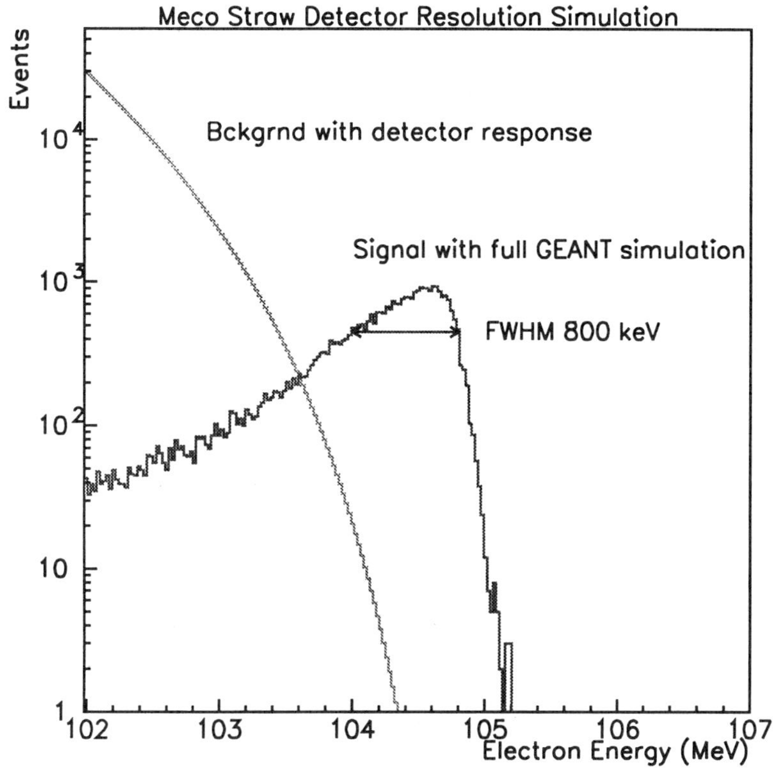

Figure 7. Histogram of the full energy response of the MECO spectrometer to 105 MeV electrons, including all effects of detector resolution and energy loss in the stopping target. Also shown is this response function convoluted with the energy distribution of background from μ^- decay in orbit. The relative normalizations correspond to a value of $R_{\mu e} = 10^{-16}$.

distribution with the response function of the calorimeter (including the pile-up energy) and imposing a minimum energy requirement of 65 MeV. Including the effects of pile-up of photons and neutrons at twice the expected level, the rate is calculated to be below 10 kHz, well within the low level trigger rates of current experiments.

The whole detector solenoid is surrounded by a nearly hermetic cosmic ray shield. containing active and passive elements. It consists of 0.5 m of steel surrounding the detector solenoid (some of which is provided by the return yoke of the magnet), two layers of plastic scintillator, and 2 m of concrete shielding. It is assumed that the probability of not detecting a penetrating charged particle in either layer of scintillator is 10^{-4}. The cosmic ray induced background was calculated using measured cosmic μ^- fluxes and a GEANT simulation of the shielding and detector. From this simulation, the cosmic ray background is predicted to be 0.0035 events in 10^7 seconds of exposure.

Expected performance and sensitivity of MECO

The sensitivity that will be achieved by MECO is summarized in table 2. In 10^7 s running time, a few events can be detected at a value of $R_{\mu e} = 10^{-16}$. Dead-time losses are expected to be small; losses in pattern recognition in the tracking detector are also expected to be small but have not yet been estimated. In one year (10^7 s) running time, a few events can be detected at a value of $R_{\mu e} = 10^{-16}$.

Table 2. A summary of the factors affecting the expected MECO sensitivity for a one year (10^7 s) run.

Running time (sec)	10^7
Proton flux (sec^{-1})	4×10^{13}
μ/p entering solenoid	0.006
Stopping probability	0.370
μ capture probability	0.600
Fraction of μ that capture in time window	0.480
Electron trigger efficiency	0.900
Fitting and selection criteria	0.250
Detected events for $R_{\mu e} = 10^{-16}$	5.800

As discussed above, the muon yield used in table 2 is calculated using the GHEISHA hadronic interaction package, and scaling the yield to agree with measured pion production cross sections. The numbers given are updated from those in the MECO proposal to account for this effect and for the increased repetition rate of the AGS. We note that the experiment is not limited by backgrounds, and increased running time will improve the sensitivity proportionally.

Table 3 shows the expected background rates for the sensitivity quoted above. It is dominated by the μ^- decay in orbit contribution. Substantial improvement in discrimination against this source of background can be had with modest loss in acceptance. A proton beam extinction of 10^{-10} is assumed.

Table 3. A summary of the level of background from various sources calculated for a sensitivity of ~5 events for $R_{\mu e} = 10^{-16}$.

Source	Events	Comment
μ decay in orbit	0.290	S/N = 20 for $R_{\mu e} = 10^{-16}$
Radiative μ capture	$\ll 0.050$	
* μ decay in flight	< 0.003	without scatter in target
* μ decay in flight	0.004	with scatter in target
* Radiative π capture	0.007	from out of time protons
Radiative π capture	0.004	from late arriving π
* π decay in flight	$\ll 0.001$	
* Beam electrons	< 0.002	
$\overline{p}$ induced	0.004	mostly from π^-
Cosmic ray induced	0.004	10^{-4} CR veto inefficiency
Total background	0.370	* assumes 10^{-10} extinction

5. Summary

Experiments to search for LFV have now been done for about 50 years, with ever increasing sensitivity. Current limits are at the level of 10^{-11} to below 10^{-12} for the processes $\mu^+ \to e^+\gamma$, $\mu^+ \to e^+e^+e^-$, and $\mu^-N \to e^-N$. These limits place stringent constraints on many scenarios for physics beyond the Standard Model.

Improvements in muon beams and detector technology hold promise for making further significant improvements. The SINDRUM2 experiment is expected to improve the sensitivity to $\mu^-N \to e^-N$ to below 10^{-13} in the next year or two. Further improvement, to a sensitivity below 10^{-16}, is promised by the MECO experiment, now approved at BNL. These very substantial improvements in experimental sensitivity allow some optimism that the first evidence for muon and electron number violation may be found. If these proposed experiments are successfully executed, they will be sensitive to the level of lepton flavor violating signals suggested in many scenarios for physics beyond the Standard Model. In particular, predictions of a class of grand unified supersymmetric models will be confronted directly by experimental measurements.

Acknowledgements

I thank my collaborators on MECO who have done much of the work discussed in this paper. I acknowledge the help of W. Marciano, with whom

I have had many useful conversations about muon conversion experiments. I also thank the organizers of the workshop for arranging a stimulating program and the staff of the European Center for Theoretical Physics forr providing a pleasant working environment.

References

[1] R. Barbieri *et al.*, IFUP-TH 72/94 (1995).

[2] J. Hisano *et al.*, LBL-38653 (1996).

[3] K. Arisaka *et al.*, Phys. Rev. Lett. **70**, 1049 (1993).

[4] T. Akagi *et al.*, Phys. Rev. Lett. **67**, 2614 (1991).

[5] A. Heinson *et al.*, *A New Search for Very Rare K_L Decays*, BNL Proposal 871, 1990.

[6] K. Arisaka *et al.*, , 1998.

[7] A. M. Lee *et al.*, Phys. Rev. Lett. **64**, 165 (1990).

[8] M. Zeller *et al.*, BNL Proposal 865, 1990.

[9] U. Bellgardt *et al.*, Nucl. Phys. **B299**, 1 (1988).

[10] R. Bolton *et al.*, Phys. Rev. D **38**, 2077 (1988).

[11] F. Riepenhausen, talk at the Sixth Conference on the Intersections of Particle and Nuclear Physics, 1997.

[12] D. Barlow *et al.*, LAMPF Proposal 969, 1985.

[13] Y. Kuno and Y. Okada, LANL hep-ph/9604296, 1996.

[14] Y. Kuno, A. Maki, and Y. Okada, LANL hep-ph/9609307, 1996.

[15] A. Czarnecki, W. Marciano, and K. Melnikov, in *AIP Conference Proceedings of Workshop on Physics at the First Muon Collider and at the Front End of the Muon Collider*, edited by S. H. Geer and R. Raja (American Institute of Physics, Woodbury, New York, 1998).

[16] O. Shanker, Phys. Rev. D **25**, 1847 (1982).

[17] M. Bachman *et al.*, *A Search for $m^-N \to e^-N$ with Sensitivity Below* 10^{-16}, BNL Proposal 940, 1997.

[18] R. Djilkibaev and V. Lobashev, Sov. J. Nucl. Phys. **49(2)**, 384 (1989).

[19] V. S. Abadjev *et al.*, *MELC Experiment to Search for the $\mu^-A \to e^-A$ Process*, INR preprint 786/92, 1992.

[20] Badert, Nucl. Phys. **A377**, 406 (1979).

[21] D. Artmutliski *et al.*, Sov. J. Nucl. Phys. **48**, 161 (1988).

[22] I. Chemakin *et al.*, BNL Proposal 910, 1995.

[23] S. Majewski *et al.*, Nucl. Instrum. Meth. **348**, 307 (1994).

PART 6

B and L Violation and Collider Physics

Inverse neutrinoless double beta decay and $\Delta L = 2$ processes at linear colliders

Geneviève Bélanger

LAPTH,[1] Chemin de Bellevue, B.P. 110, F-74941 Annecy-le-Vieux, Cedex, France.

Abstract. Prospects for observation of the $\Delta L = 2$ lepton-number-violating process $e^-e^- \to W^-W^-$ at linear colliders are reviewed. In the standard model with additional Majorana neutrinos, constraints from neutrinoless double beta decay render the process unobservable at a linear collider of $\sqrt{s} < 2$ TeV except in the most contrived scenarios. In that respect, some models beyond the standard model fare better.

1. Introduction

The problem of the mass of the neutrino has been at the forefront of particle physics recently. What is even more intriguing than the value of the neutrino mass would be to know whether it has a Dirac or a Majorana ma If the neutrino has a Majorana mass, then it will contribute to ΔL 2 lepton-number-violating processes such as neutrinoless double beta decay ($\beta\beta_{0\nu}$) through the subprocess $W^-W^- \to e^-e^-$. The neutrinoless double beta decay data constrains both the mass of a Majorana neutrino and its mixing with the electron neutrino. In the design of a future high energy linear e^+e^- collider, one option being considered seriously is to replace the positron beam by an electron beam in order to look at e^-e^- collisions. The inverse neutrinoless beta decay process $e^-e^- \to W^-W^-$ would then be accessible.

The process $e^-e^- \to W^-W^-$ has been looked at several times, by different authors, over the last decade or so [1]-[6]. In one recent analysis,

[1] URA 14-36 du CNRS, associée à l'Université de Savoie.

it was found that within the context of the standard model with additional Majorana neutrinos, once the constraint from $\beta\beta_{0\nu}$ is taken into account, the cross section for $e^-e^- \to W^-W^-$ is in general too small for it to be seen at a .5-1.5 TeV linear collider[7]. Here I will review the current situation taking into account the latest data[8] and considering the high luminosity version of a linear collider being discussed now[9]. Recently, some efforts were made to explore ways to evade the constraint from double beta decay [10]. While it was found that under specific conditions this could be done, in general, $e^-e^- \to W^-W^-$ is not the most suitable process to test the Majorana nature of the neutrino. Processes involving the single production of a Majorana neutrino have a much larger cross-section[10, 11]. I will as much as possible keep to a model independent approach not relying on any assumptions on neutrino masses and mixings.

When going beyond the standard model, the constraint from neutrinoless double beta decay still plays a role although it is possible to find models where the inverse neutrinoless double beta decay process can occur at an observable rate, such as composite neutrinos models. Furthermore, in extensions of the standard model, such as left-right symmetric models or supersymmetric models, there are other $\Delta L = 2$ processes involving the production of new particles that could be interesting to probe the Majorana nature of the neutrino. Of course these entail that new particles such as W_R or charginos are kinematically accessible[12, 13].

At a linear collider many options are under consideration. It is possible to backscatter laser light off one or both of the beams, creating an $e\gamma$ or $\gamma\gamma$ collider [14]. There are a number of $\Delta L = 2$ processes mediated by a Majorana electron neutrino that could in principle take place et these colliders, for example $e^-\gamma \to e^+W^+W^-$. Unfortunately these higher order processes are also subject to the neutrinoless double beta decay constraint. One would need energies in the multi TeV range to observe them, certainly a very remote possibility[7].

Clearly the limits from $\beta\beta_{0\nu}$ apply only to ν_e. Should the ν_μ have a Majorana mass, it will contribute to the processes $\mu^-\mu^- \to W^-W^-$ (and similarly for the ν_τ), with no constraints from low-energy processes. The best way to find evidence of the Majorana nature of the ν_μ would be with a $\mu^-\mu^-$ collider [15]. Lacking such a collider such lepton-number-violating processes cannot take place directly but they can occur as subprocesses in any of the proposed options of the linear collider. For example, the observation of $\gamma\gamma \to \mu^+\mu^+W^-W^-$ would be evidence for a Majorana ν_μ. However again a linear collider with a centre-of-mass energy of at least 4 TeV is necessary[7].

2. Inverse neutrinoless double beta decay, $e^-e^- \to W^-W^-$

2.1. Neutrino Mixing

Suppose that the ν_e mixes with an unspecified number of other neutrinos, after diagonalization of the mass matrix, ν_e can be expressed in terms of the mass eigenstates N_i:

$$\nu_e = \sum_i U_{ei} N_i \ , \tag{1}$$

where the mixing matrix U is unitary. Phenomenologically, we have observed two things. First, the ν_e does not mix much with other neutrinos [16]:

$$\sum_{i \neq 1} |U_{ei}|^2 < 6.6 \times 10^{-3} \quad (90\% \ c.l.) \ . \tag{2}$$

This limit is essentially independent of the $SU(2)_L$ transformation properties of the neutrino(s) with which the ν_e mixes. Also, the limit is quite conservative – it allows for the possibility that the other charged fermions also mix with new, exotic charged particles [17]. If one assumes that the only new particles are neutrinos, then the above limit improves somewhat to 5.0×10^{-3}. Thus, the ν_e is mainly N_1, furthermore, from muon decay, we know that the N_1 is very light: $M_1 < 7$ eV [18].

2.2. Cross Section for $e^-e^- \to W^-W^-$

Assuming that the N_i are Majorana neutrinos, they will contribute to the process $e^-e^- \to W^-W^-$ through the diagrams of Fig. 1.

The cross section for unpolarized electrons is rather simple in the limit that $s \gg M_W^2$. This is a reasonable approximation for the linear collider,

$$\frac{d\sigma}{d\cos\theta} = \frac{g^4}{1024\pi M_W^4} \left(\sum_i M_i \left(U_{ei}\right)^2 \left[\frac{t}{(t - M_i^2)} + \frac{u}{(u - M_i^2)} \right] \right)^2 \ . \tag{3}$$

(The full expression given for example in [7] is used when presenting numerical results.)

In this energy regime the production of longitudinal W's dominates the process $e^-e^- \to W^-W^-$, in fact Eq. 3 is precisely what is obtained if one calculates the cross section for the process $e^-e^- \to W^-_{long} W^-_{long}$, where W^-_{long} is the longitudinal component of the W^-. The full helicity amplitudes are given for example in [7].

There are two limiting cases of Eq. 3 which will be useful in what follows. First, if $s \gg M_i^2$, the cross section becomes

$$\frac{d\sigma}{d\cos\theta} = \frac{g^4}{256\pi M_W^4} \left(\sum_i M_i \left(U_{ei}\right)^2 \right)^2 \ . \tag{4}$$

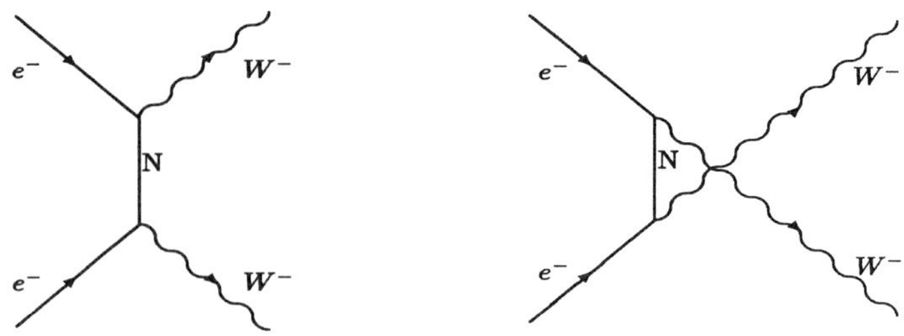

Figure 1. *Diagrams contributing to $e^-e^- \to W^-W^-$.*

Second, in the other limit, $M_i^2 \gg s$, the cross-section grows like s^2

$$\frac{d\sigma}{d\cos\theta} = \frac{g^4}{1024\pi M_W^4} s^2 \left(\sum_i \frac{(U_{ei})^2}{M_i} \right)^2 . \qquad (5)$$

2.3. Unitarity Considerations

From Eq. 4, we see that, in the high-energy limit ($s \to \infty$), the cross section tends towards a constant:

$$\sigma_{s\to\infty} = \frac{g^4}{128\pi M_W^4} \left(\sum_i M_i (U_{ei})^2 \right)^2 . \qquad (6)$$

In this particular case this indicates a violation of unitarity, since the amplitude (which is a pure s-wave) grows as $\sqrt{s}$.

There are basically two ways in which this unitarity violation can be cured. The first through the inclusion of a Higgs triplet. If the neutrinos with which the ν_e mixes are $SU(2)_L$ doublets, then they can acquire Majorana masses by giving the Higgs triplet a vacuum expectation value (v.e.v.). This Higgs triplet includes a doubly-charged Higgs, H^{--}. In this case unitarity is restored through the inclusion of a diagram in which the H^{--} is exchanged in the s-channel. However this solution is not favoured phenomenologically and this direction will not be pursued here.

In the absence of Higgs triplets, the only way to restore unitarity is to require that the neutrinos' masses and mixing angles satisfy

$$\sum_i (U_{ei})^2 M_i = 0 . \tag{7}$$

Although this relation may appear arbitrary at first sight, it is in fact automatically satisfied. It is straightforward to show that

$$\sum_i (U_{ei})^2 M_i = M_{ee}^* , \tag{8}$$

where M_{ee} is the Majorana mass of the ν_e. However, because there are no Higgs triplets, this mass is equal to zero, so that Eq. 7 holds. This is exactly what happens in the famous seesaw mechanism where one adds a right-handed neutrino N_R to the spectrum. This neutrino acquires a large Majorana mass M through the v.e.v. of a Higgs singlet, and the combination $\bar{N}_R \nu_{eL} + h.c.$ obtains a Dirac mass m once the ordinary Higgs doublet gets a v.e.v. The mass matrix looks like

$$\begin{pmatrix} 0 & m \\ m & M \end{pmatrix} . \tag{9}$$

The two mass eigenstates are N_1 and N_2, with masses $-m^2/M$ and M, respectively (the minus sign in front of M_1 can be removed by a γ_5 rotation). For m of the order of the electron mass and M about 1 TeV, one obtains a mass of about 1 eV for the lightest neutrino. (Thus, in such models, the large range of neutrino masses is explained in a natural way, unlike the Higgs triplet models.) The ν_e is a linear combination of these two physical neutrinos:

$$\nu_e = \cos\theta N_1 + \sin\theta N_2 , \qquad \sin\theta = \frac{m}{M} . \tag{10}$$

It is clear that, with these masses and mixing angles, the relation in Eq. 7 is automatically satisfied.

The downside of this particular solution is that the mixing of the ν_e with the N is tiny: for $m \sim m_e$ and $M \sim 1$ TeV, $\sin\theta \sim 10^{-6}$! This would make the cross section for $e^-e^- \to W^-W^-$ invisible, since $\sigma(e^-e^- \to W^-W^-) \propto \sin^4\theta$, and is typical of what happens in left-right symmetric models [13]. However, if the ν_e mixes with more neutrinos, it is, in principle, possible to satisfy Eq. 7 without having such small values of U_{ei}. (This is the assumed solution in Ref. [3].) This is perhaps a bit artificial, and probably requires some fine-tuning, but it is possible. If this is how unitarity restoration comes about, then Eq. 3 contains all the contributions to $e^-e^- \to W^-W^-$.

Although it is interesting to understand how unitarity is restored in different models, the above discussion demonstrates that the cross section for $e^-e^- \to W^-W^-$ is essentially unconstrained by such considerations –

the U_{ei} and M_i can take any values consistent with the phenomenological limits in Sec. 2.1. This is not the case when the experimental limits on neutrinoless double beta decay are taken into account.

2.4. *Limits from $\beta\beta_{0\nu}$*

As mentioned in the introduction, $e^-e^- \to W^-W^-$ is essentially the inverse of neutrinoless double beta decay. We might therefore expect that the limits on the latter process could constrain the former.

If some of the neutrinos have masses $M_i \ll 1$ GeV, then, for these neutrinos, the quantity which contributes to $\beta\beta_{0\nu}$ is

$$\langle m_\nu \rangle = \sum_i{}' \left(U_{ei}\right)^2 M_i \ , \tag{11}$$

where the sum is over the light neutrinos. (For simplicity, we have ignored factors corresponding to complications from the nuclear matrix elements – their inclusion does not change our conclusions. For more details we refer the reader to Ref. [19],[20].) At present, the best experimental limit on $\langle m_\nu \rangle$ is obtained by the Heidelberg-Moscow group[8],

$$\langle m_\nu \rangle \lesssim .45 \text{ eV} \ . \tag{12}$$

As for the neutrinos which are heavy, $M_i \gg 1$ GeV, they can still mediate $\beta\beta_{0\nu}$ decay. In this case the relevant quantity is

$$\langle m_\nu^{-1} \rangle_H = \sum_i{}'' \left(U_{ei}\right)^2 \frac{1}{M_i} \ , \tag{13}$$

where the sum is over the heavy neutrinos. Now the experimental limit on $\beta\beta_{0\nu}$ implies the following [21]:

$$\langle q^2 \rangle \sum_i{}'' \left(U_{ei}\right)^2 \frac{1}{M_i} \lesssim .45 \text{ eV} \ , \tag{14}$$

where q is an average nuclear momentum transfer. If one takes q to be roughly about 100 MeV, one obtains the right order-of-magnitude constraint. However, a more careful calculation, including all the nuclear effects, gives

$$\sum_i{}'' \left(U_{ei}\right)^2 \frac{1}{M_i} < 5 \times 10^{-5} \text{ TeV}^{-1} \ . \tag{15}$$

This limit has been arrived at by combining the nuclear matrix element calculation of Ref. [22] with the latest lower bound of 1.2×10^{24} yr on the half-life for ${}^{76}Ge \to {}^{76}Se + 2e^-$ [8]. While there are some variations in the matrix elements calculations, it is now generally accepted [20] that the uncertainty is at most a factor of 2 or 3. In fact the $\beta\beta_{0\nu}$ transition matrix elements are expected to be much less sensitive than the $\beta\beta_{2\nu}$ to the details of the nuclear structure. In particular they are much less affected by the particle-particle interactions.

Assuming no cancellations between contributions of different heavy Majorana neutrinos, the $\beta\beta_{0\nu}$ constraint implies a generic lower bound on the mass of the heavy neutrino:

$$M_i > 1.4 \times 10^4 \left(U_{ei}\right)^2 \ \text{TeV} \ . \tag{16}$$

For $\left(U_{ei}\right)^2 \sim 5 \times 10^{-3}$, this gives $M_i > 100$ TeV. However, it is possible to evade this order-of-magnitude bound if one allows cancellations among the various terms, I will come back to that point below, for the moment I will assume that this constraint holds for all neutrinos.

As was reported earlier in this conference[23], the Heidelberg-Moscow experiment should improve by a factor of 5 their limit on the lifetime of neutrinoless double beta decay. There has also been a proposal, GENIUS, to probe half-life at the level of $5.8 \times 10^{27} yr$ or even $6.4 \times 10^{28} yr$ with a one ton detector[24]. The limit on neutrino mass improving only as the square root of the life-time, the proposed experiment could reach a sensitivity for the neutrino mass below the .01 eV level.

2.5. $e^-e^- \to W^-W^-$ at the linear collider

The constraint from $\beta\beta_{0\nu}$ gives us one of two conditions, depending on whether the new neutrinos are very light or very heavy relative to the energy scale of neutrinoless double beta decay. For the case of light neutrinos, the cross section for $e^-e^- \to W^-W^-$, which is independent of $\sqrt{s}$ is much too small to be observable at any future collider, $\sigma(e^-e^- \to W^-W^-) \leq 1.3 \times 10^{-17}\ fb$.

If no cancellations are allowed in Eq. 15, then $\beta\beta_{0\nu}$ constrains the neutrinos to be very massive (Eq. 16). For linear colliders with centre-of-mass energies in the TeV range, we have $M_i \gg \sqrt{s}$ and Eq. 5 can be used to calculate the cross section for $e^-e^- \to W^-W^-$. Using the limit in Eq. 15, we find that, at $\sqrt{s} = 500$ GeV,

$$\sigma(e^-e^- \to W^-W^-) < 4.9 \times 10^{-3}\ fb \ . \tag{17}$$

Even with a high luminosity collider, $\mathcal{L} = 500 fb^{-1}$, the process $e^-e^- \to W^-W^-$ would be unobservable.

In this section we assume that the cross section for $e^-e^- \to W^-W^-$ is dominated by the exchange of a single neutrino. (Of course, additional, heavier neutrinos must be present to satisfy the bound from unitarity.) Note that even if one assumes that more than one neutrino contributes to

$e^-e^- \to W^-W^-$, this will not change the cross section significantly, since the mixing angles of all the neutrinos must be correspondingly reduced in order to satisfy the constraint in Eq. 15.

In Ref.[7] it was shown that a linear collider of at least 2 TeV would be necessary to improve on the present neutrinoless double beta decay limits on massive Majorana neutrinos (see Fig. 2 in Ref. [7]). In particular it was found that with the luminosity, $80(\sqrt{s}/(1\ \mathrm{TeV}))^2\ fb^{-1}$, for $\sqrt{s} = 500$ GeV and 1 TeV, the values of M_i and $(U_{ei})^2$ which produce an observable $e^-e^- \to W^-W^-$ cross section are already ruled out by neutrinoless double beta decay. For $\sqrt{s} = 2$ TeV, the discovery limit and the limit from $\beta\beta_{0\nu}$ are roughly equal. Although the 2 TeV linear collider opens a very small region of parameter space when polarized e^- beams are used and hence does slightly better than $\beta\beta_{0\nu}$.

Since that work was published, the linear collider design team has claimed that a luminosity of $\mathcal{L} = 500 fb^{-1}$ could be achieved with the 500 and 800 GeV linear collider[9]. This could improve the prospects for inverse neutrinoless double beta decay. Unfortunately the $\beta\beta_{0\nu}$ constraint is still too good to have any chance of seing the inverse process at those energies.

In Fig. 2, the discovery limit for $e^-e^- \to W^-W^-$ at the linear collider for several centre-of-mass energies as a function of M_i and $(U_{ei})^2$ are presented. We demand 10 events for discovery, and assume polarized e^- beams and a luminosity of $500 fb^{-1}$ for 500 and 800 GeV. For higher energies, $\sqrt{s} = 2, 4$ and 10 TeV, the luminosity $80(\sqrt{s}/(1\ \mathrm{TeV}))^2\ fb^{-1}$ and unpolarized beams are assumed. Note that perfect polarization basically increases the cross-section by a factor of 4. The improvement that polarization provides at 2 TeV is also displayed in Fig. 2. Note that we have not included efficiencies for the detection of the W's, nor have we included any backgrounds. Our discovery limits are therefore as optimistic as they possibly could be. In this figure the phenomenological limit on $(U_{ei})^2$, as well as the constraints from $\beta\beta_{0\nu}$ are superimposed. By the time a linear collider is built the $\beta\beta_{0\nu}$ limits is expected to improve by a factor of 5 on the lifetime measurement, the limit achievable in this case is also shown as well as the limit from the proposed GENIUS experiments.

From Fig. 2 we conclude that in the context of the standard model with additional Majorana neutrinos, there are no chance of seing the inverse neutrinoless process at a collider of energy below 2TeV. Even a very high luminosity collider of 800 GeV with perfectly polarized beams is no match for the neutrinoless double beta decay experiment. Note that the bound on $U_{ei}^2 - M_i$ improves as the square root of the luminosity. A 2 TeV version of a linear collider with polarized beams can extend the discovery reach compare with present day low-energy experiments. However by the time such a collider is built, the low-energy constraint would have improved as displayed. In summary, prospects for 2 TeV collider to improve upon neutrinoless double beta decay are marginal at best.

Finally, for 4 TeV and 10 TeV linear colliders, there exists a sizeable region of M_i-$(U_{ei})^2$ parameter space, not ruled out by $\beta\beta_{0\nu}$, which produces an observable signal for $e^-e^- \to W^-W^-$. In Fig. 2, one can easily see that a 10 TeV collider is needed to do better than the next generation of neutrinoless double beta decay experiments for probing Majorana neutrinos.

In all this discussion the bound used for $\beta\beta_{0\nu}$ (Eq. 5) was obtained from the calculation in Ref. [22], it is clear that a factor of 2 or 3 uncertainty in the nuclear matrix element does not alter any of the conclusions. In fact, the matrix elements calculation quoted here is not the most restrictive and using the published results of various groups would only strengthen the conclusions[22].

2.6. Evading the $\beta\beta_{0\nu}$ constraint

The analysis just presented might be too restrictive and one should wonder if there is a way to evade the constraint from neutrinoless double beta decay and still obtain an observable cross-section for the inverse process. As the nuclear matrix element calculation is well under control[20], the only alternative to evade the $\beta\beta_{0\nu}$ constraint rests with particle physics. To obtain some cancellation between the contributions of different neutrinos in the bound in Eq.16 , the following conditions must be met: many Majorana neutrinos that mix with ν_e and some of the mixing matrix elements square are purely imaginary. In other words at least one neutrino must have a different intrinsic CP parity, η_{CP}. Some explicit examples of what is needed are given in Ref. [7].

First, consider the case where there are only two neutrinos. To have a cancellation the heavier neutrino must have a larger mixing with the electron neutrino than the lighter one, this is certainly possible but not very natural. Putting aside the naturality argument, it was shown by scanning over the parameter space of masses and mixing angles that when the neutrinoless double beta decay constraint is loosened, the $e^-e^- \to W^-W^-$ cross-section goes down as well and becomes unobservable at either 500 GeV or 1 TeV[10]. Perhaps what is even more interesting is that a neutrino of mass less than $\sqrt{s}$ which has a significant mixing could in principle be observed at a linear collider in the single production process $e^+e^- \to \nu_e N$. With the decay products of the heavier neutrino one could determine whether it is Majorana in nature[25]. After taking into account the neutrinoless double beta decay constraint Gluza and Zrałek showed recently that one could get an observable cross-section for $e^+e^- \to \nu N$ at 1 TeV[10, 11]. Large cross-sections ($\mathcal{O}(100 fb)$) are obtained when the two

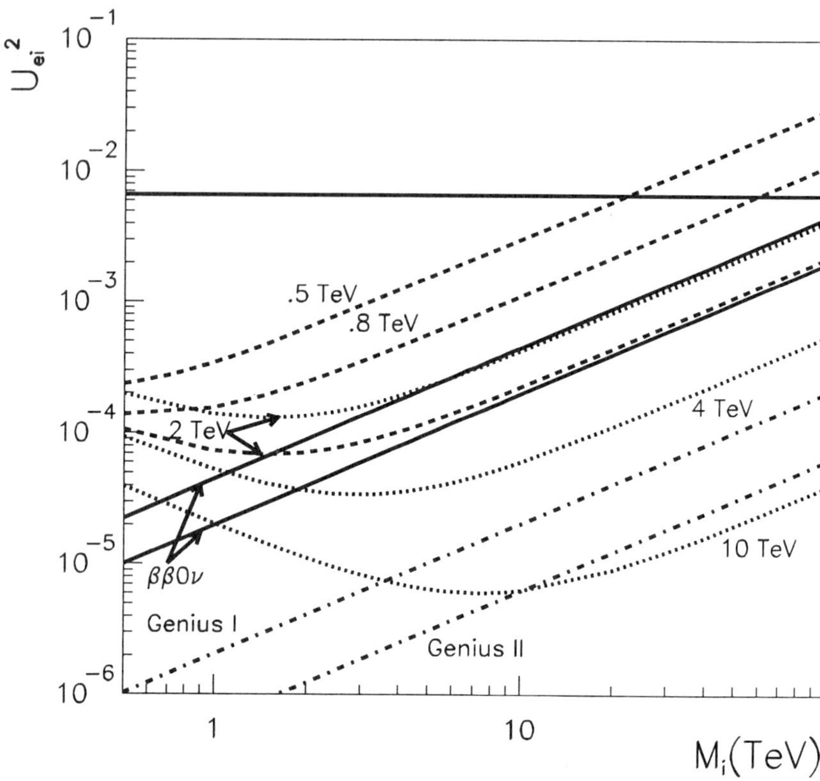

Figure 2. *Discovery limit for* $e^-e^- \rightarrow W^-W^-$ *at a linear collider as a function of* M_i *and* $(U_{ei})^2$ *for* $\sqrt{s}$ = 500 *GeV, 800 GeV, 2 TeV, 4 TeV and 10 TeV with luminosities of* $500 fb^{-1}$ *for energies below 1 TeV and* $80(\sqrt{s}/(1\ TeV))^2\ fb^{-1}$ *above. In all cases, the parameter space above the line corresponds to observable events. Dashed lines correspond to perfectly polarized beams while dotted ones to unpolarized beams. We also superimpose the present and future experimental limit from* $\beta\beta_{0\nu}$ *(diagonal solid line), from the Heidelberg Moscow experiment as well as the prospective limit from the future Genius experiment (dash-dot) The horizontal line represents the limit on neutrino mixing,* $(U_{ei})^2$. *Here, the parameter space above the line is ruled out.*

neutrinos are nearly degenerate, while neutrino masses ratio around 100 made for at most one fb.

As one adds more neutrinos, there is more freedom in the choice of parameters and one might be able to have some signal in inverse neutrinoless double beta decay. It is sufficient to look more closely at the three neutrinos case. As above one still needs one of the neutrinos with opposite intrinsic CP parity in order to evade the $\beta\beta_{0\nu}$ constraint. Assuming this being the neutrino with mass M_2, with $M_1 < M_2 < M_3$, Gluza and Zrałek [10] scanned the full parameter space for three neutrino masses. They found values for the masses and mixings where $\sigma(e^-e^- \to W^-W^-)$ exceeds one fb while the $\beta\beta_{0\nu}$ constraint is satisfied. In general, this implies that two neutrinos are nearly degenerate. However, when the Majorana neutrino mass is below the center of mass energy, the single production process is also allowed and the cross-section is quite large. This process is much favoured to discover a heavy Majorana neutrino, furthermore it takes place in the e^+e^- version of the linear collider. There exists a small region of parameter space where $e^-e^- \to W^-W^-$ is above the discovery limit at 1TeV and $\mathcal{L} = 80fb^{-1}$(see Fig. 3 in Ref.[10]). Such is also the case at 500GeV if the high luminosity collider is available.

Clearly with more neutrinos there is more freedom and the same conclusion would be reached, it might be possible to observe the inverse neutrinoless double beta decay process at a linear collider with specific choices of masses, mixings and CP parity but in the major part of parameter space the single production process is the favoured one.

3. Other $\Delta L = 2$ Processes at the linear collider

For various reasons, it is conceivable that, even if a linear collider is built, the e^-e^- option may never be used while the $e\gamma$ or $\gamma\gamma$ option might be available. In the absence of an e^-e^- collider, the only real possibility for detecting a Majorana ν_e through $\Delta L = 2$ processes is the $2 \to 3$ process $e^-\gamma \to e^+W^-W^-$ which contains $e^-e^- \to W^-W^-$ as a subprocess as shown in Fig. 3. Being subject to the $\beta\beta_{0\nu}$ constraint one does not expect this process to do better than inverse neutrinoless double beta decay. Indeed it was shown that a few events could be observed in this process only at extremely high energies, beyond $\sqrt{s} = 4$ TeV[7]. A similar conclusion is reached when considering other processes such as $\gamma\gamma \to e^+e^+W^-W^-$.

Once again for the $e\gamma$ collider there exist a more direct way of seeing a Majorana neutrino: single production in $e\gamma \to WN$. The authors of Ref. [26] have shown recently, by scanning over all the possible mixing angles and masses that up to the kinematic limit one could get large cross-sections for single neutrino production. In particular in the case of

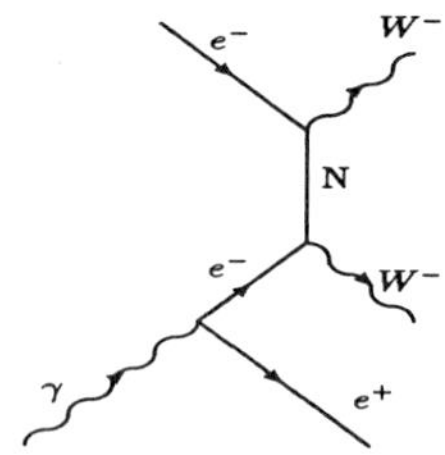

Figure 3. *The dominant diagram in* $e^-\gamma \to e^+W^-W^-$.

three neutrinos, cross sections above $100 fb$ could be reached at a collider of 460GeV. Of course this presupposes that single neutrino production is kinematically accessible.

Although the constraint from neutrinoless double beta decay has proved to be very stringent in the context of the standard model, one should not forget that $\beta\beta_{0\nu}$ constrains only the ν_e –it says nothing about the ν_μ or the ν_τ. It therefore seems reasonable to ask about the possibilities for observing other $\Delta L = 2$ processes at the linear collider, specifically those involving a Majorana ν_μ or ν_τ. For this one could use the following options currently discussed in the context of the design of a linear collider: $e\gamma, \gamma\gamma$ or of a muon collider, $\mu^+\mu^-, \mu^-\mu^-$.

The most interesting option is obviously the $\mu^-\mu^-$ collider. A Majorana ν_μ will mediate processes such as $\mu^-\mu^- \to W^-W^-$. In order to observe $\Delta L = 2$ processes involving the ν_μ, this neutrino must mix with heavy Majorana neutrinos, just as was the case for the ν_e. Indeed even if the mass of the ν_μ lies at the upper allowed limit, $m_{\nu_\mu} < 0.27$ MeV but do not mix with heavy neutrinos, then the cross section for $\mu^-\mu^- \to W^-W^-$ is still unobservable. From Eq. 4, it is at most $O(10^{-3})$ fb.

For a muon neutrino mixing with a heavy Majorana neutrino, the analysis is very similar to the one presented in the electron case, in particular the cross section formula and the numerical values for $\mu^-\mu^- \to W^-W^-$ are exactly the same. The curves in Fig. 2 then also represent the potential of a muon collider for probing the Majorana nature of the neutrino safe for the constraint from $\beta\beta_{0\nu}$ which is irrelevant here. The upper limit on the mixing angle of the ν_μ differs slightly from the ν_e case, $\sum_{i\neq 2}|U_{\mu i}|^2 < 6.0\times 10^{-3}$[16]. As with the ν_e, this conservative limit is for the case where the other fermions also mix with new particles. If one assumes that only the neutrinos mix, then the limits improve to 1.8×10^{-3}.

In the absence of a muon collider, a muon Majorana neutrino can in principle also be looked for since the process $\mu\mu\to WW$ is also a subprocess in a number of 2 to 4 processes at linear colliders, from $\gamma\gamma\to\ell^+\ell^+W^-W^-$ to $e^-\gamma\to\nu_e\ell^-\ell^-W^+$ (here $l=\mu,\tau$ since these processes can also be used for a tau neutrino). Unfortunately again one needs to go to very high energies (4-10 TeV)to get a reasonable cross-section even if one does not have to deal with the $\beta\beta_{0\nu}$ constraint[7].

4. Beyond the standard model and heavy neutrinos

Beyond the context of the standard model the picture could be quite different as far as the prospect of observing the effect of a heavy Majorana neutrino mixing with the electron neutrino. In some models the constraint from $\beta\beta_{0\nu}$ is less stringent and in other models, additional processes with lepton number violation mediated by Majorana neutrinos are available. Composite models are example of the former and left-right symmetric models and supersymmetric models of the latter.

In composite models one could have an excited neutrino ν_e^* which can mix with a number of Majorana neutrinos. The coupling of the excited neutrino to leptons has the generic form

$$\mathcal{L}_{int} = g\frac{\lambda_W}{m_\nu^*}\bar{e}\sigma^{\mu\nu}(1\pm\gamma_5)\nu_e^*\partial_\mu W_\nu^- + h.c. \tag{18}$$

where the mass dimension m_ν^* is taken to be of the order of the excited neutrino mass and the sign of the chirality operator depends on the chirality of the neutrino. The present limit on such an excited neutrino from accelerator searches is $m_{\nu^*} > 91GeV$[27]. With the $\sigma^{\mu\nu}$ type of coupling, the $\beta\beta_{0\nu}$ constraint is much less stringent in composite models [28]. In fact Kinuya and Takasagi[29] found that the constraint from neutrinoless double beta decay is basically irrelevant when considering the $e^-e^-\to W^-W^-$ cross-section, it can exceed the pb. On the other hand if one assumes that $\lambda_W\approx 1$ the cross -section is significantly reduced but remains in the observable range (i.e $>$ .1 fb) for masses $m^*\approx 20TeV$ even after taking into account the $\beta\beta_{0\nu}$ constraint. One characteristic feature of this class of models is that the W's produced are transverse rather than longitudinal.

In left-right symmetric models it is possible to choose parameters to evade both low-energy limits (including double beta decay) on Majorana neutrinos and lower bounds on right-handed W's in order to look for lepton-number violation in e^-e^-. However, the process of interest involves the production of a right-handed W either singly or in pair, see Ref. [13]. While certainly interesting, this scenario relies on the production of a new particle whose existence would first be probed in the e^+e^- version of the linear collider not to mention hadronic machines.

Other scenarios for lepton-number violation include those classes of supersymmetric models with R-parity violation, see Ref. [30]. These models have been shown to be very much constrained by double beta decay [30, 31]. In fact, in the context of the analysis presented in Section 2, the question is whether these new contributions could interfere destructively with those from a Majorana neutrino (that could also be present) with the result of evading those limits on the Majorana neutrino that we have used. Though such fine tuning seems unnatural, in this eventuality $e^-e^- \to W^-W^-$ could be used to single out the $\Delta L = 2$ lepton-number violation due to the Majorana neutrino. However, it is found from the analysis on double beta decay in Ref. [31] that these contributions add up coherently and that the limits on the Majorana neutrinos used above remain valid.

Another $\Delta L = 2$ process exists in supersymmetric models, $e^-e^- \to \chi^-\chi^-$ mediated by a Majorana sneutrino. Indeed it was shown in [32] that a Majorana sneutrino is tied to the existence of its standard superpartner, a Majorana neutrino. The cross-section for this process could be sizeable and should be easily observable [12].

5. Conclusions

I have reviewed the prospects for the observation of $e^-e^- \to W^-W^-$, mediated by a Majorana neutrino, at a high-energy e^-e^- collider . This process is essentially the inverse of neutrinoless double beta decay ($\beta\beta_{0\nu}$). Once the constraints from $\beta\beta_{0\nu}$ are taken into account, we have found that $e^-e^- \to W^-W^-$ is unobservable at an linear collider of $\sqrt{s} < 2$ TeV within the context of the standard model plus an extra Majorana neutrino . It is possible to evade the constraints, but this requires models which are extremely contrived and fine-tuned. In order to evade the constraint one needs at least three neutrinos mixing with the electron neutrino, the mixing of some heavier neutrinos larger than the one of the lighter ones, and at least one neutrino with an opposite intrinsic CP. Then for some values of the parameters $e^-e^- \to W^-W^-$ is observable. However in general the best way to probe the Majorana nature of the neutrino is through the single direct production process, if kinematically accessible.

Going beyond the standard model one can also find processes with lepton number violation mediated by Majorana neutrinos or other particles.

In most cases those involve the production of a new particle, such as a right-handed W or a chargino. Only in the case of composite Majorana neutrinos can one evade easily the constraint from $\beta\beta_{0\nu}$ and get a large cross-section in the inverse process.

Other $\Delta L = 2$ process mediated by ν_μ or ν_τ are not subject to the $\beta\beta_{0\nu}$ constraint. However even in this case subprocesses involving the μ or τ sector still require colliders of very high energies unless a $\mu^-\mu^-$ collider exists. In conclusion, the only way to find evidence of Majorana neutrinos in the standard model at a lepton collider of energies below 2 TeV without appealing to very specific models of Majorana neutrino masses would be at a $\mu^-\mu^-$ collider.

References

[1] T.G. Rizzo, *Phys. Lett.* **116B** (1982) 23.

[2] D. London, G. Bélanger and J.N. Ng, *Phys. Lett.* **188B** (1987) 155.

[3] D.A. Dicus, D.D. Karatas and P. Roy, *Phys. Rev.* **D44** (1991) 2033.

[4] J. Maalampi, A. Pietilä and J. Vuori, *Phys. Lett.* **297B** (1992) 327.

[5] T.G. Rizzo, *Phys. Rev.* **D50** (1994) 5602.

[6] C.A. Heusch and P. Minkowski, *Nucl. Phys.* **B416** (1994) 3.

[7] G. Bélanger , F. Boudjema, D. London,H. Nadeau, *Phys. Rev.* **D53** (1996) 6292.

[8] Heidelberg-Moscow Collaboration, *Phys. Lett.* **407B** (1997) 219.

[9] ECFA Study on Physics and Detectors for a linear electron-positron Collider.

[10] J. Gluza and M. Zrałek,*Phys. Lett.* **362B** (1995) 259, J. Gluza and M. Zrałek,*Phys. Lett.* **362B** (1995) 148.

[11] J. Gluza and M. Zrałek, *Phys. Rev.* **D55** (1997) 7030.

[12] M. Hirsch, H. V. Klapdor-Kleingrothaus, S. Kolb and S. G. Kovalenko, *Phys. Rev.* **D57** (1998) 2020.

[13] T. G. Rizzo, hep-ph/9510349, Oct. 95.

[14] I.F. Ginzburg, G.L. Kotkin, V.G. Serbo and V.I. Telnov, *Pis'ma ZhETF* **34** (1981) 514; *Sov. Yad. Fiz.* **38** (1983) 372; *Nucl. Instr. Methods* **205** (1983) 47; I.F. Ginzburg, G.L. Kotkin, S.L. Panfil, V.G. Serbo and V.I. Telnov, *Sov. Yad. Fiz.* **38** (1983) 1021; *Nucl. Instr. Methods* **219** (1984) 5.

[15] For a discussion of $\mu^-\mu^-$ colliders, see C.A. Heusch and F. Cuypers, hep-ph/9508230, August 1995.

[16] E. Nardi, E. Roulet and D. Tommasini, *Phys. Lett.* **344B** (1995) 225.

[17] For the formalism of fermion mixing, see P. Langacker and D. London, *Phys. Rev.* **D38** (1988) 886.

[18] *Review of Particle Properties,* Particle Data Group, *Phys. Rev.* **D50** (1994) 1173, Part I.

[19] For a review, see, for example, M. Moe and P. Vogel, *Ann. Rev. Nucl. Part. Sci.* **44**, (1994) 247.

[20] A. Staudt, K. Muto and H. V. Klapdor-Kleingrothaus, Europhys. Lett., **13** (1990) 31.

[21] For discussions of the contribution of heavy Majorana neutrinos to $\beta\beta_{0\nu}$, see, for example, J.D. Vergados, *Phys. Rep.* **C133** (1986) 1; R.N. Mohapatra and P.B Pal, *Massive Neutrinos in Physics and Astrophysics,* World Scientific Lecture Notes in Physics – Vol 41, Singapore, 1991, p. 204; G. Pantis, A. Faessler, W.A. Kaminski and J.D. Vergados, *J. Phys. G: Nucl. Part. Phys.* **18** (1992) 605.

[22] G. Pantis *et al*, Ref. [21].

[23] H. V. Klapdor-Kleingrothaus, these proceedings.

[24] H. V. Klapdor-Kleingrothaus, hep-ex/9802007.

[25] W. Buchmuller and C. Greub, *Nucl. Phys.* **B363** (1991) 345.

[26] J. Gluza, J.Maalampi, M. Raidal and M. Zrałek, *Phys. Lett.* **407B** (1997) 45.

[27] Aleph Collaboration, *Phys. Rep.* **C216** (1992) 253.

[28] O. Panella and Y. N. Srivastava, *Phys. Rev.* **D52** (1995) 5308.

[29] R. Kinyua and E. Takasugi, hep-ph/9804245.

[30] For a recent review on neutrinoless double beta decay which also discusses supersymmetric models with R-parity violation, see R. N. Mohapatra, University of Maryland Preprint, UMD-PP-95-147, hep-ph/9507234, July 95.

[31] M. Hirsch, H. V. Klapdor-Kleingrothaus and S. G. Kovalenko, *Phys. Lett.* **352B** (1995) 1, *ibid., Phys. Lett.* **372B** (1996) 181, *ibid., Phys. Rev.* **D53** (1996) 1329.

[32] M. Hirsch, H. V. Klapdor-Kleingrothaus and S. G. Kovalenko, *Phys. Lett.* **398B** (1997) 311

Sneutrino Physics with Lepton Number Violation

St Kolb, H V Klapdor-Kleingrothaus, M Hirsch†and O Panella‡

Max-Planck-Institut für Kernphysik, P.O. 10 39 80, D-69029 Heidelberg, Germany

†Present address: Instituto de Física Corpuscular – C.S.I.C., Departamento de Física Teòrica, Universitat of València, 46100 Burjassot, València, Spain

‡Istituto Nazionale di Fisica Nucleare, Sezione de Perugia, Via A. Pascoli, I-06123 Perugia, Italy

Abstract. Lepton number violation in the sneutrino sector has numerous consequences like sneutrino oscillations, neutrinoless double beta decay etc. In view of the possibilty that sneutrinos could be the cold dark matter in the universe the consequences of the sneutrino being the lightest supersymmetric particle is examined for facilities like the Next Linear Collider. Heavier to light sneutrino decays, giving rise to new signatures, and cross sections for sneutrino production with accompanying lepton are considered.

1. Introduction

Recently there has been extensive interest in sneutrinos, the scalar counterparts of the neutrinos appearing in supersymmetric (SUSY) extensions of the Standard Model (SM) [1], due to the fact that a sneutrino "Majorana" mass term present in the low energy SUSY Lagrangian [2] may have huge impacts both on sneutrino and neutrino phenomenology. The intimate connection between sneutrino and neutrino properties implies that if the sneutrino bears Majorana properties so does the neutrino, and vice versa.

Sheding light on the as yet unresolved questions concerning the neutrino mass [3] is an unexpected feature of the supersymmetric version of the SM once a sneutrino Majorana mass $\tilde{m}_M$ different from zero exists. Further potentially observable consequences of a finite $\tilde{m}_M$ include lepton number violating processes such as neutrinoless double beta $0\nu\beta\beta$ decay [5], lepton number violating processes such as *e.g.* $e^-e^- \to \chi^-\chi^-$ [6] and sneutrino oscillations [7, 6].

Particularly interesting is the impact of neutrino/sneutrino properties on Cosmology and the still unresolved questions of nature and origin of the dark matter in the universe. Whereas massive neutrinos are long known to be an interesting candidate for the hot component of dark matter (HDM) in the universe, it emerged only recently that in the presence of $\tilde{m}_M$ sneutrinos may well serve as cold component of dark matter (CDM) [8]. In view of these facts it is fair to say that neutrino/sneutrino physics may act as a key player in the quest of a consistent description of both Particle Physics and Cosmology.

The big experimental efforts that are being made to observe e.g. neutrino oscillations (LSND, SAGE, Super Kamiokande etc. [9]), $0\nu\beta\beta$ decay (Heidelberg-Moscow, GENIUS [10], Nemo [9] etc.), to establish SUSY (e.g. at the LHC, NLC [9]), to observe CDM (HDMS [11], GENIUS [10], CDMS [9] etc.) will hopefully provide interesting insight into these questions soon.

The organization of this contribution is as follows: in the next part lepton number violating sneutrino masses will be introduced and their possible consequences for $0\nu\beta\beta$ decay, $e^-e^- \to \chi^-\chi^-$, sneutrino oscillations, neutrino masses and CDM will be briefly reviewed. The motivation for the third part is the observation that, if $\tilde{m}_M \neq 0$, the lightest sneutrino state may be a CDM candidate and consequentely the LSP. In such a scenario the transitions from heavier to the lightest sneutrino give rise to new signatures at collider facilities like the NLC, or LHC. In section four processes are considered where sneutrinos are produced alongside a lepton which allows to tag the sneutrino flavour. Such processes may be of interest in the $\not{L}$MSSM in order to gain insight into the mass texture in the sneutrino sector since the dominant decay modes of the sneutrinos may be invisible. The same argument holds in the case that $\tilde{m}_M = 0$ when the sneutrinos are lighter than the charginos, so that, again, the main sneutrino decay channels are invisible and therefore not suitable for sneutrino spectroscopy.

2. Lepton number violating sneutrinos

Lepton number violation in the sneutrino sector has been introduced in a model independent way in [2]. The basic observation is that the SUSY-part

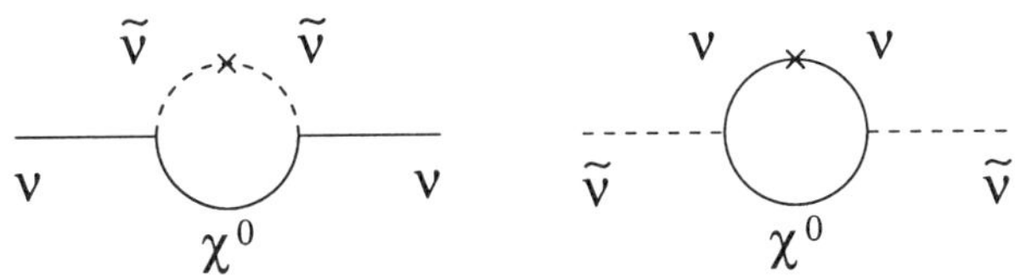

Figure 1. Radiative generation of neutrino masses via a loop containing a sneutrino with $\tilde{m}_M \neq 0$ (left), and radiative generation of a $\tilde{m}_M \neq 0$ via a loop containing a massive Majorana neutrino (right).

of the Lagrangian may contain in addition to the usual lepton number conserving "Dirac" sneutrino mass term a lepton number violating "Majorana" sneutrino mass term:

$$\mathcal{L} \supseteq \tilde{m}_D^2 \tilde{\nu}_L^* \tilde{\nu}_L + \frac{1}{2}(\tilde{m}_M^2 \tilde{\nu}_L \tilde{\nu}_L + h.c.) \tag{1}$$

In the following a scenario with such a mass term present for brevity will be called $\not{L}$MSSM. The consequence is that the weak eigenstates $\tilde{\nu}_L, \tilde{\nu}_L^*$ are no longer mass eigenstates. Using the real field representation of a complex scalar field

$$\tilde{\nu}_L = \frac{1}{2}(\tilde{\nu}_1 + i\tilde{\nu}_2) \tag{2}$$

($\tilde{\nu}_1, \tilde{\nu}_2$ real) the mass eigenstates are now $\tilde{\nu}_1, \tilde{\nu}_2$ with masses

$$\tilde{m}_{1,2}^2 = \tilde{m}_D^2 \pm |\tilde{m}_M|^2 \tag{3}$$

respectively. In the following it will be assumed that $|\tilde{m}_M|^2 < \tilde{m}_D^2$, otherwise the lighter sneutrino state aquires a vacuum expectation value which breaks spontaneously R-parity rendering the LSP unstable. The result of (3) is that the sneutrino mass states $\tilde{m}_1, \tilde{m}_2$ are no longer degenerate in mass implying potentially observable phenomenons: i) generation of neutrino majorana masses; ii) generation of a non-vanishing neutrinoless double beta ($0\nu\beta\beta$) decay amplitude without the need of broken R-parity; iii) generation of lepton number violating processes at colliders; iv) generation of sneutrino oscillations; v) sneutrino cold dark matter (CDM). These effects will be briefly summarized in the following. Note that in the $\not{L}$MSSM generally mixing in the lepton and slepton sectors will occur, *cf.* the discussion in [4]. For simplicity all constraints given below do not take into account such mixing.

i) In the $\not{L}$MSSM Majorana neutrino masses are generated radiatively by contributions of the type depicted in figure 1 (left). In turn, if a tree level Majorana neutrino mass exists a nonvanishing $\tilde{m}_M$ is generated via the mechanism shown in figure 1 (right) [2]. From the current laboratory bounds on neutrino masses the following constraints on $\tilde{m}_M$ can be extracted (for details see [2]). These bounds depend on the neutralino mixing matrix. In the case of a bino-dominated neutralino for typical SUSY mass scale of order 100 GeV the values

$$\tilde{m}_{M(e)} \leq 120\text{MeV}, \quad \tilde{m}_{M(\mu)} \leq 13\text{GeV}, \quad \tilde{m}_{M(\tau)} \leq 149\text{GeV} \tag{4}$$

are obtained. The third generation $\tilde{m}_M$ is virtually unconstrained, a fact which is important for the possibility of sneutrino CDM, see below.

ii) A first generation $\tilde{m}_M$ different from zero induces a contribution to $0\nu\beta\beta$ decay. The complete set of $\tilde{m}_M$-induced $0\nu\beta\beta$ decay diagrams and the details of the calculation can be found in [5]. Note that in this case it is not necessary to break R-parity (this case has been considered in [12]), the lepton number violation is provided by $\tilde{m}_M$. The constraints on $\tilde{m}_M$ stemming from the various contributions are in general less stringent than those from the neutrino masses, but some are independent of the neutralino composition.

iii) In the $\not{L}$MSSM lepton number violating processes like $l^- l^- \to \chi^- \chi^-$ (l: lepton) may be observable at facilities as the NLC or a muon collider. A detailed analysis of this process [6] shows that the bounds from the first generation neutrino is too restrictive to allow this process to be observable since the cross section is proportional to $(\tilde{m}_M/\tilde{m}_D)^4$. The situation is different in the case of a muon collider. With a annual integrated luminosity of $50 fb^{-1} y^{-1}$ and the bound from the second generation neutrino one may expect an event rate of around 50 per year, which may well be observable.

iv) In the $\not{L}$MSSM sneutrinos may oscillate due to the mass difference of the sneutrino states. This behaviour is in close analogy to the K^0-$\overline{K}^0$ and B^0-$\overline{B}^0$ systems and was considered in [7]. In the case that sneutrinos decay into a chargino and a lepton the transition of a initially produced sneutrino into an antisneutrino with the subsequent decay into a "wrongly" charged lepton may be observable and depends on the sum of the decay widthes $\tilde{\Gamma}$ of the sneutrino mass states, on $\tilde{m}_D$ and $\tilde{m}_M$, on the chargino parameters and the branching ratio of this decay. For the values $\tilde{\Gamma}$=$1GeV$, $\tilde{m}_D$=$100GeV$ and $\tilde{m}_M$=$13GeV$ ($1GeV$) on expects 40 % (< 1 %) of the events to show up with the wrong lepton charge (for a branching ratio one) which makes sneutrino oscillations probably the most promising phenomenon to look for a $\tilde{m}_M$ different from zero.

v) The possibiliy of one sneutrino state being the LSP and in the case of conserved R-parity being a CDM candidate has particularly interesting consequences for the sneutrino decay signatures to be expected at colliders. The sneutrino being CDM is firmly excluded in the case that $\tilde{m}_M$ vanishes: sneutrinos with masses of order $100GeV$ annihilate rapidly via s-channel Z^0-exchange so that no cosmologically interesting relic abundance is obtained. To reduce the annihilation rate it was proposed that sneutrinos

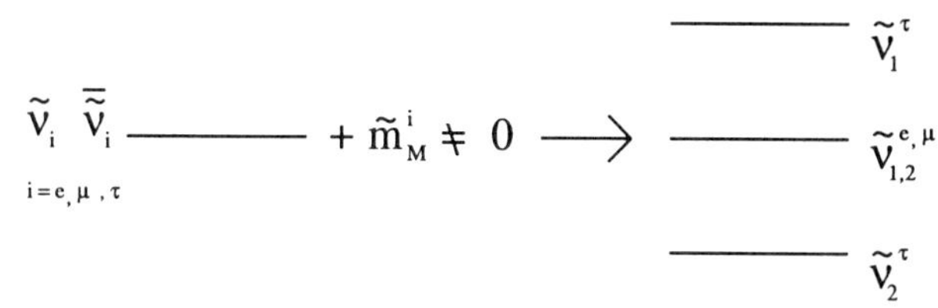

Figure 2. Possible mass relations of the $\not{L}$MSSM sneutrino sector allowed by neutrino mass bounds

should be either very light ($2GeV$ [13]) or very heavy (of order $1TeV$ [14]). The first case is excluded from Z^0-width measurements, the second case is excluded from direct nuclear recoil experiments like the Heidelberg-Moscow setup ([14] and ref. therein).

However, the situation changes in the $\not{L}$MSSM . In [8] it was shown that if the mass splitting of two sneutrino states of the same generation is bigger than $5GeV$ the sneutrino may be a viable CDM candidate due to the fact that the coupling to the Z^0 is off-diagonal, *i.e.*

$$Z^0 \tilde{\nu}_{i,1} \tilde{\nu}_{i,2} \tag{5}$$

for any generation i. This behaviour is a consequence of angular momentum conservation and Bose symmetry, thus suppressing both the annihilation of sneutrinos in the early universe and the elastic sneutrino scattering rates in counting experiments like HDMS, GENIUS or CDMS. Therefore the reasons excluding sneutrino CDM in the pure "Dirac" like case are no longer true if a lepton number violating sneutrino mass is present in the low energy SUSY Lagrangian. The lower bound on the mass splitting implies that only the lighter τ-sneutrino $\tilde{\nu}_{\tau,2}$ may be a CDM candidate since the bounds from neutrino masses result in a sneutrino mass splitting being not large enough to suppress annihilation and counting rates sufficiently. Furthermore it was argued in [8] that for the relic sneutrino density to be sufficiently high sneutrino annihilation via t-channel neutralinos must not occur too fast. It is not possible to reconcile this condition with the requirement of one sneutrino state being the LSP in a gravity mediated GUT scenario. However, in more general circumstances such a situation is possible and phenomenologically interesting.

3. Sneutrino Decays

Depending on the mass relations of the sneutrinos to the other SUSY-particles one can distinguish between three qualitatively different scenarios

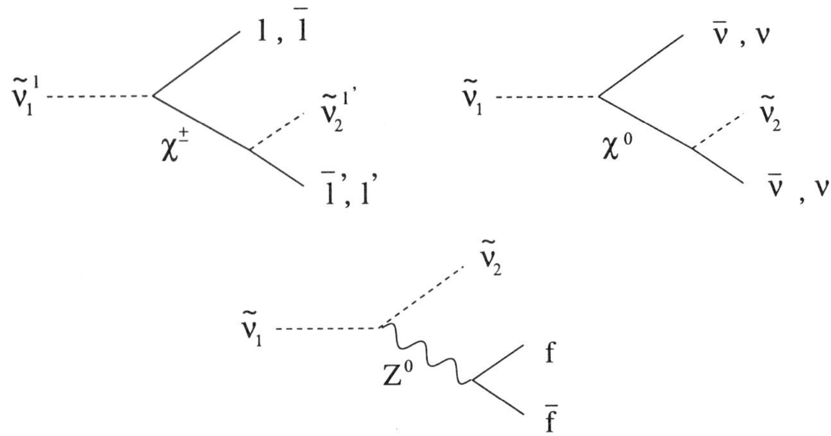

Figure 3. Graphs contributing to heavy to light sneutrino transitions in the $\not{L}$MSSM

of sneutrino decays: i) one sneutrino is the LSP; ii) no sneutrino is the LSP but sneutrinos are lighter than charginos; iii) sneutrinos are heavier than charginos.

First, consider the case that in the context of the $\not{L}$MSSM one sneutrino is the LSP. This implies that exotic sneutrino decay signatures may be observable in collider experiments. For definiteness consider the case that the lighter τ-sneutrino is the LSP, since this possibly is the cosmologically interesting one. In such a scenario the mass relations in the sneutrino sector can be qualitatively visualized as in figure 2, with the sneutrinos of the first two generations being almost degenerate in mass and a substantial mass splitting (of order $10GeV$) in the τ-sector. Sneutrinos not being the LSP decay via neutralino, chargino and Z^0 three body decays into the lighter τ-sneutrino plus fermions (note that there is no Higgs mediated channel since the Higgs couples only either to two light or two heavy sneutrinos). These decay modes are depicted in figure 3. The explicit expressions for the various contributions are listed in [15]. Provided the escaping fermions being not too soft to be detected signatures as $\tilde{\nu}_\mu \to \mu^\pm + \tau^\mp + \not{E}$ are expected. Consider as an example the decay $\tilde{\nu}_\tau^1 \to \tilde{\nu}_\tau^2$ + fermions and for the sake of definiteness assume gaugino mass unification but vary the parameters of the sneutrino sector freely. With the values

$$\mu = 150.0GeV, M_2 = 250.0GeV,\ \tan\beta = 1.23,\ \tilde{m}_D^\tau = 50GeV,\ \tilde{m}_M^\tau = 15GeV$$

one obtains

$$\tilde{m}_\tau^1 = 52.2GeV,\ \tilde{m}_\tau^2 = 47.7GeV,\ m_{\chi_0^1} = 75.7GeV,\ \Gamma_{\tilde{\nu}_\tau^1}^{tot} = 3.3eV$$

The branching ratio to $\tau\overline{\tau}$ pairs is 15%, the branching ratio to other visible decay products is 12%. If the Z^0-mediated channel (which does not allow to determine the sneutrino flavour) dominates strongly the chargino mediated one, one may study processes where a sneutrino is produced alongside with a lepton which allows to determine the sneutrino generation and use missing momentum to deduce its mass. Two examples of such processes will be presented in the next chapter.

Second, consider the case that no sneutrino and the lightest neutralino is the LSP, but that both charginos are heavier than the sneutrinos. This possibility arises both in the MSSM and the $\not{L}$MSSM . In such a scenario the dominant sneutrino decay channel will be the two body decay into a neutrino and a neutralino, *i.e.* sneutrinos decay invisibly. In this case the sneutrino decay products do not allow to measure their properties so that the dominant sneutrino production channel $e^+e^- \rightarrow \tilde{\nu}^i_L\overline{\tilde{\nu}}^i_L$ (MSSM); $\tilde{\nu}^i_1\tilde{\nu}^i_2$ ($\not{L}$MSSM) (i=e,μ,τ) is of no interest for sneutrino spectroscopy, and, as above, lepton associated sneutrino production may be a remedy.

Third, in the case that charginos are lighter than sneutrinos the latter will decay into the first with an accompanying lepton which allows to tag the sneutrino flavour. The dominant two sneutrino production channel is the most suitable place to examine their properties and the lepton associated production is of no particular interest. This case is, as mentioned before, the best one for direct observation of sneutrino oscillations.

4. Lepton associated sneutrino production

If sneutrinos decay invisibly the dominant two-sneutrino production mechanism does not allow to deduce the properties of the MSSM or $\not{L}$MSSM sneutrino sector. A possibility to do so may be the study of processes where sneutrinos are produced together with a lepton which allows to tag the flavour of the sneutrino and use missing momentum to determine sneutrino masses. Examples of processes at *e.g.* the Next Linear Collider (NLC) or a electron-photon machine where a sneutrino of any flavour is produced alongside a lepton or slepton which allows to tag the sneutrino flavour include

$$l_i^+ l_i^- \rightarrow \tilde{\nu}_j l_j^+ \chi_k^- \tag{6}$$

$$l_i^+ \gamma \rightarrow \tilde{\nu}_j \tilde{l}_j^+ \overline{\nu}_i \tag{7}$$

$$l_i^+ l_i^- \rightarrow \tilde{\nu}_j \tilde{l}_j^+ W^- \tag{8}$$

These reactions can be regarded as the leptonic SUSY analogues of single top-quark production [17] in the SM. The subject of this chapter will be

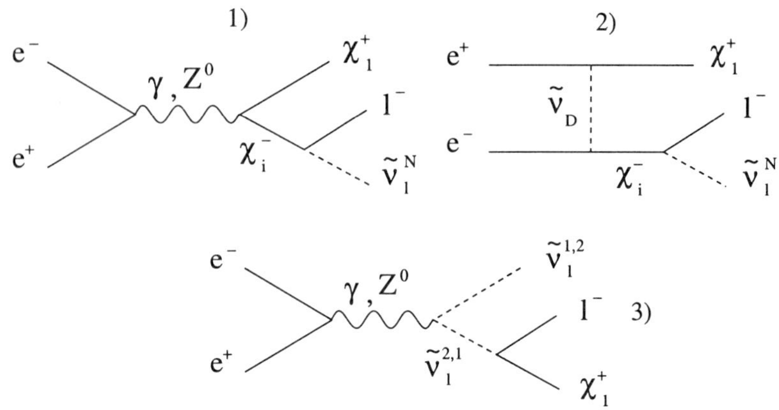

Figure 4. Contributions to $e^+e^- \to \tilde{\nu}_l^N l^- \tilde{\chi}_i^+$ where N=1,2 and $l = \mu, \tau$ (for $l = e$ more graphs contribute)

the cross-sections of the first and second of these processes (6,7) for j=μ, τ. The third one (8) produces a more difficult signature due to the additional final state W and therefore will not be analysed.

The Feynman graphs contributing to (6) in the case of production of μ- and τ-sneutrinos are depicted in figure 4, for the case of e-sneutrino production further contributions similar to those examined in [18] (replacing the proton with an electron) have to be taken into account. In the case $l_j = \mu, \tau$ and $k = 1$ (*i.e.* production of the lighter of the two charginos) the resulting cross section can be written as

$$\sigma_j^{tot} = \sum_{k,l=1;k\leq l}^{3} \sigma_j^{kl} \tag{9}$$

where *e.g.* σ_j^{22} is the contribution from t-channel sneutrino exchange (fig 4 2)) (which is proportional to the "Dirac" sneutrino mass $\tilde{m}_D$), and so on. The explicit expressions of the various contributions can be found in [15].

The cross-section (9) depends strongly on the input parameters of the chargino sector, *i.e.* on chargino masses and mixing coefficients. In figure 5 the cross-section for the lighter chargino in the final state is plotted *not* taking into account the contribution of fig. 4; 3) and the interferences with figs. 4; 1),2). The full cross section is presented in [15]. For obtaining numerical results a scenario with gaugino and scalar unification was assumed and the following input parameters were used:

$$m_0 = 70GeV,\; M_2 = -36.4GeV,\; \mu = 72.9GeV,\; \tan\beta = 1.23,\; A_0 = 0$$

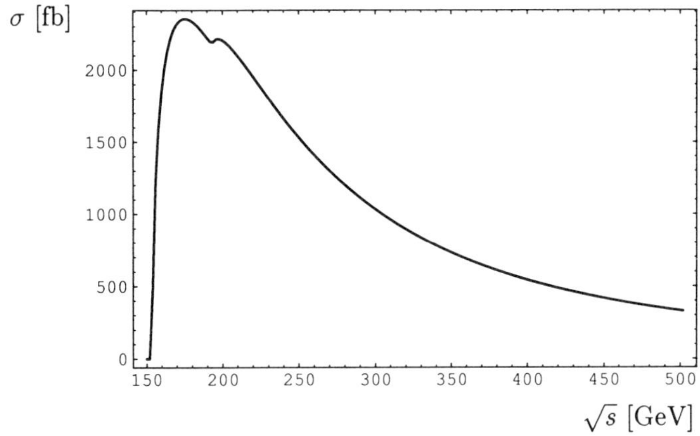

Figure 5. Cross section for the process $e^+e^- \to \tilde{\nu}_\tau \tau^+ \chi_2^-$ for the parameters specified in the text.

Diagonalization of the chargino mass matrix yields $m_{\chi_1^+} = 116.8GeV$, $m_{\chi_2^+} = 76.8GeV$ and the sneutrino mass used is $m_{\tilde{\nu}} = 71.3$. The chargino widthes have been calculated taking into account both- two and three-body decay final states (our results for the chargino decay widthes agree with the results given in [16]) and the values are $\Gamma_1 = 1.8GeV$, $\Gamma_2 = 0.003GeV$. In the case that both charginos are heavier than the sneutrinos the cross-section exhibits typically two peaks corresponding to the two poles in the chargino propagator (or equivalently to resonant on-shell production of light-light and light-heavy chargino pairs). The cross-section reaches values of order nb not only in the region of resonant chargino pair production but also at higher energies producing $\mathcal{O}(10^4)$ signal events for an integrated luminosity of order $10fb^{-1}y^{-1}$ like one expects at the NLC.

The situation differs in the case of reaction (7). The graphs contributing to (7) are depicted in figure 6. The cross section of process (7) can be written as

$$d\sigma_{tot} = d(PS)\frac{e^2g^4}{4}\sum_{i,i'=1}^{4}\sum_{spins} -g^{\mu\mu'}\frac{1}{1+\delta_{ii'}}T_\mu^{(i)}(T_{\mu'}^{(i')})^\dagger \qquad (10)$$

The explicit expressions for the individual amplitudes $T_\mu^{(i)}$ are listed in [15]. The square of the total amplitude is a lengthy function of the particle momenta's scalar products. The integration over the three particle phase space $d(PS)$ has been performed, as in [17], parametrizing (10) in the slepton rest frame $\vec{p}_{\tilde{e}} + \vec{p}_{\tilde{\nu}} = 0$ and using the Monte-Carlo integration

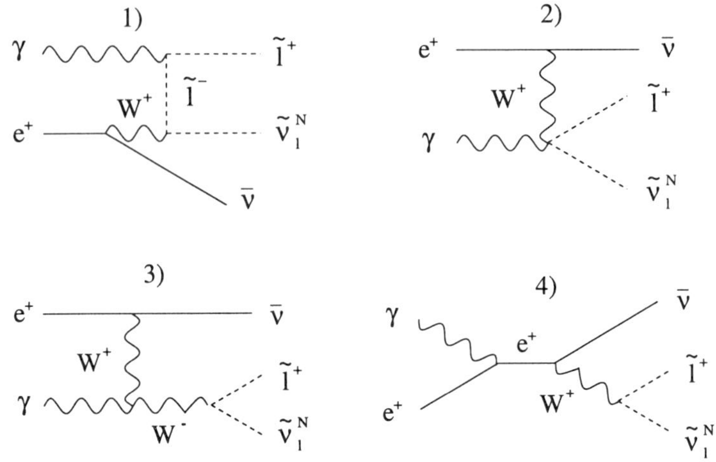

Figure 6. Contributions to $e^+\gamma \to \tilde{\nu}_l^N \tilde{l}^+ \overline{\nu}_e$

code VEGAS. The resulting cross section is plotted in figure (7) yielding values of order fb (like its SM counterpart) rendering this process less interesting to produce μ- and τ-sneutrinos. In order to estimate the cross section (7) at an e^-e^+ facility the initial photon line may be attached to an external electron line using the equivalent photon approximation ([17] and ref. therein). Doing so the results plotted in figure 7 do not change significantly.

The main SM-background to both processes stems from the production of W-bosons. Discrimination between signal and SM-background seems very difficult in the case of process (7) due to the smallness of the cross-section leading to an only small amount of signal events. Therefore the right place to search for generational independent sneutrino production seems to be reaction (6) yielding sufficient statistics to renounce the background. Certainly, the precise way of selecting the SUSY signal requires a profound analysis of *e.g.* the energy and angular distributions of the final state leptons and jets (stemming from the decaying final state chargino and slepton). This lies beyond the scope of this note.

5. Summary

In the search for windows to beyond SM physics currently the probably most promising, yet controversial experimental hints point to a nonvanishing Majorana neutrino mass manifesting itself in neutrino oscillations. If SUSY is realized in nature this immediately implies that lepton number

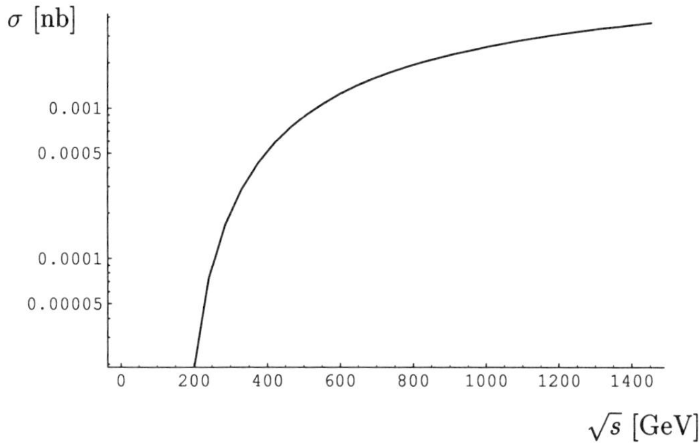

Figure 7. Cross section of the reaction $e\gamma \rightarrow \overline{\nu}_e \tilde{\nu}_l^N \tilde{l}$ for sneutrino and selectron masses of $50GeV$ and $100GeV$ respectively.

violation is present as well in the sneutrino sector extending the MSSM to the $\not{L}$MSSM . Additionally, there could be higher order operators effective at some high scale contributing to such sneutrino masses at the Fermi scale as was speculated in [8]. Possible phenomenological consequences include $0\nu\beta\beta$-decay, sneutrino oscillations, $e^- e^- \rightarrow \chi^- \chi^-$ and sneutrino CDM. These facts make it seem worthwhile to pay special attention to sneutrinos at some future collider facility, especially at the NLC, since precise measurements of superpartner masses are easier to accomplish at a lepton collider than at a hadronic collider like the LHC.

In the case that one sneutrino state is the LSP exotic fermionic final states will be observable stemming from the transition from a a non-LSP sneutrino to the LSP sneutrino. Visible final states typically make up up to 30% of the total decay width of such a transition, the rest being totally invisible LSP-neutrino final states.

In the case that sneutrinos mainly decay into invisible channels the only possibility, both in the MSSM and in the $\not{L}$MSSM , to gain insight into the sneutrino sector are processes where sneutrinos are produced alongside a lepton which allows to tag the sneutrino flavour. The processes presented in this note allow the production of a sneutrino of any flavour alongside an accompanying lepton of the same generation at a future electron collider like the NLC or a electron-photon facility. Some of the resulting cross sections seem big enough to explore the mass relations in the sneutrino sector, which may be of utmost importance not only in particle physics but also in cosmology.

References

[1] For a phenomological overview and references see *e.g.* Haber H E and Kane G L 1985 *Phys. Rep.* **117** 75

[2] Hirsch M, Klapdor-Kleingrothaus H V and Kovalenko S 1997 *Phys. Lett.* B **398** 311

[3] Mohapatra R N and Pal P B 1991 *Massive neutrinos in physics and astrophysics* (Singapore: World Scientific)
Klapdor-Kleingrothaus H V and Staudt A 1995 *Non-accelerator Particle Physics* (Bristol: Adam Hilger, IOP)
Klapdor-Kleingrothaus H V and Zuber K 1997 *Particle Astrophysics* (Bristol: IOP)

[4] Kolb St, Hirsch M, Klapdor-Kleingrothaus H V and Kovalenko S 1998 in *Proc. Int. Workshop Beyond the Desert, Castle Ringberg, Germany*, eds. Klapdor-Kleingrothaus H V and Päs H (Bristol: IOP)

[5] Hirsch M, Klapdor-Kleingrothaus H V and Kovalenko S 1998 *Phys. Rev.* D **57** 1947
Klapdor-Kleingrothaus H V and Stoica S 1995 *Proc. Int. Workshop on Double Beta Decay and Related Topics, Trento, Italy* (Singapore: World Scientific)

[6] Hirsch M, Klapdor-Kleingrothaus H V, Kolb St and Kovalenko S 1998 *Phys. Rev.* D **57** 2020

[7] Hirsch M, Klapdor-Kleingrothaus H V and Kovalenko S 1997 *Phys. Lett.* B **403** 291
Grossman Y and Haber H E 1997 *Phys. Rev. Lett.* **78** 3082

[8] Hall L J, Moroi T and Murayama H 1998 *Phys. Lett.* B **424** 305

[9] see *e.g.* corresponding contributions in Klapdor-Kleingrothaus H V and Päs H (eds.) 1998 *Proc. Int. Workshop Beyond the Desert 1997, Castle Ringberg, Germany* (Bristol: IOP) and references therein

[10] Klapdor-Kleingrothaus H V, these proceedings.

[11] Klapdor-Kleingrothaus H V and Ramachers Y, these proceedings.

[12] Hirsch M, Klapdor-Kleingrothaus H V and Kovalenko S 1996 *Phys. Rev. Lett.* **75** 17 and *Phys. Rev.* D **53** 1329

[13] Ibáñez L E 1984 *Phys. Lett.* B **137** 160
Hagelin J S, Kane G L and Raby S 1984 *Nucl. Phys.* B **241** 638

[14] Falk T, Olive K A and Srednicki M 1994 *Phys. Lett.* B **339** 248

[15] Kolb St, Hirsch M, Klapdor-Kleingrothaus H V and Panella O, in preparation.

[16] Bartl A, Fraas H and Majerotto W, 1986 *Z. Phys.* C **30** 441

[17] Panella O, Pancheri G and Srivastava Y N 1993 *Phys. Lett.* B **318** 241

[18] Wöhrmann T and Fraas H 1995 *Phys. Rev.* D **52** 78

In Search of New Physics at HERA

Gerd W. Buschhorn
(H1 Collaboration)

Max-Planck-Institut für Physik
(Werner-Heisenberg-Institut)
München

Abstract. An update is presented of the analysis of the H1 experiment at the HERA electron/positron-proton collider including the luminosity taken in 1997. Topics covered are: Status of the excess in deep-inelastic positron-proton scattering at very high Q^2 reported earlier and new results on neutral and charged current cross-sections at high Q^2, limits on the couplings of leptoquarks of the 1st and 3rd generation, limits on the production of excited fermions, observation of events with a high momentum isolated lepton and unbalanced transverse momentum. An outlook is given on the high luminosity upgrade of HERA and the H1 detector planned for 2000/2001.

1. Introduction

The HERA collider at the Deutsches Elektronensychrotron DESY in Hamburg went into operation in 1992. Electrons or positrons with an energy of 27.5 GeV are colliding with protons with an energy of 820 GeV yielding a center-of-mass energy of 300 GeV.

The unique features of HERA allow the investigation of the structure of the proton in kinematical regimes inaccessible so far. Fig. 1 shows how the two detectors H1 and ZEUS in the years 1994-1997 extended the range of Q^2 and x in deep inelastic scattering beyond that of previous fixed target experiments.

In the first years of operation, the experiments concentrated on the small x region where strong interaction properties of the proton were dominating. Meanwhile, the steady increase of the luminosity of HERA (Fig. 2)

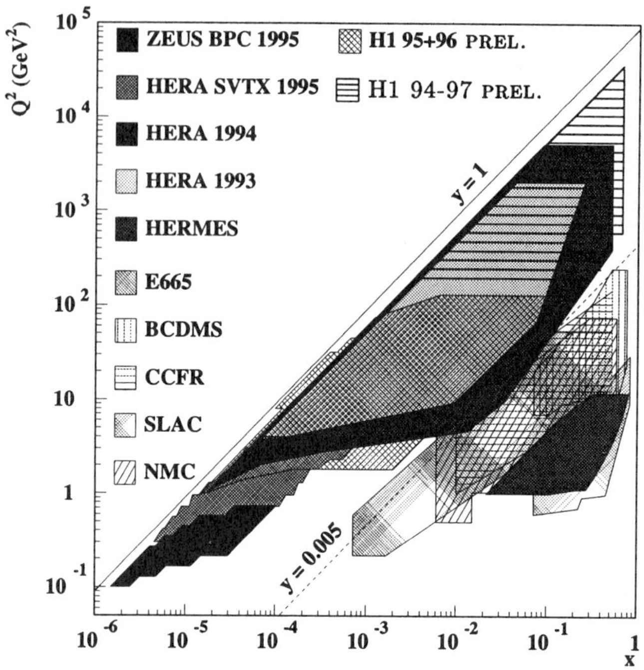

Figure 1. Regions in $x - Q^2$ covered by H1 and ZEUS in comparison with fixed target experiments.

also allows the study of weak interaction processes with increasing sensitivity.

In this paper, we present preliminary results from the H1 detector which include the full 1997 luminosity on deep-inelastic scattering (DIS) at high Q^2 (Chapter 2), limits on the couplings of leptoquarks (Chapter 3) and excited leptons (Chapter 4) and an update on events with an isolated high momentum lepton and unbalanced transverse momentum dubbed "odd lepton events" (Chapter 5). Some remarks on the planned upgrade of HERA and the H1 detector (Chapter 6) conclude the paper.

Details of HERA and the H1 detector can be found elsewhere [1], [2] and are not repeated here; the same holds for definitions and notations used in DIS [3].

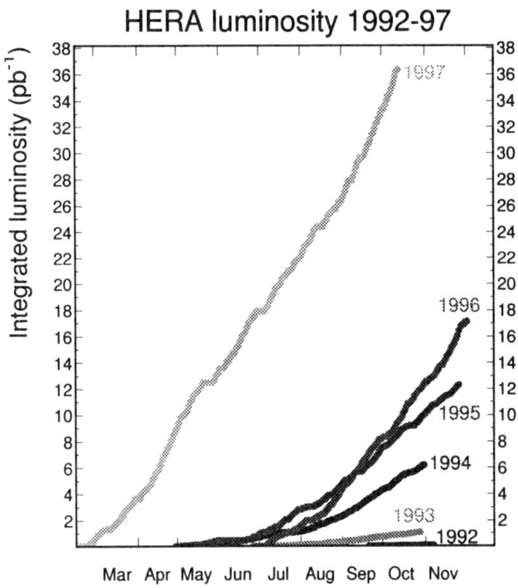

Figure 2. Yearly integrated luminosity of HERA running.

2. Deep inelastic scattering at high Q^2

2.1. The excess at very high Q^2

In February 1997, H1 and ZEUS independently and simultaneously announced the observation of an excess of events compared with the standard model-expectation for DIS of positrons on protons at $Q^2 > 15.000$ GeV^2 [4], [5]. An update of the data with increased luminosity was presented in summer 1997 [6], [7]. The results gave rise to extended theoretical discussions on their interpretation in particular as leptoquarks produced as a s-channel lepton-quark resonance [8]. While the H1 data indicate some clustering of events at a lepton-quark mass of about 200 GeV, this is not supported by the combined analysis of H1 and ZEUS [9].

After summer 1997, both experiments further increased the collected luminosity; here we present the status of the H1 analysis including the full 1997 luminosity [10], [11]. For the analysis, the liquid argon calorimeter calibration has been improved *in situ* making use of the large NC event sample; it results in an electromagnetic energy resolu-

$Q^2{}_{min}(\text{GeV})^2$	2500	5000	10000	15000	20000	30000
N_{obs}	1297	322	51	22	10	2
N_{DIS}	1276	336	55	14.8	4.39	0.58
	± 98	± 29.6	± 6.4	± 2.1	± 0.73	± 0.11
$P(N > N_{obs})$	42%	56%	60%	5.9%	1.8%	11%

Table 1. NC-events in the high Q^2-region: Observed (N_{obs}) and expected (N_{DIS}) number of events for $Q^2 > Q^2_{min}$ with the probability (P) for the observation of $N \geq N_{obs}$.

tion of $\sigma(E)/E_{em} = 12\%/\sqrt{E/GeV} \oplus 1\%$ with an error in the absolute energy calibration of $\Delta E/E_{em} = 1 - 3\%$ and a hadronic energy resolution $\sigma(E)/E_{had} = 50\%/\sqrt{E/GeV} \oplus 2\%$ with a corresponding error of $\Delta(E)/E_{had} = 4\%$; the error in the measurement of the electron angle amounts to $\Delta\Theta_e$=2-5 mrad.

While for neutral current (NC)-DIS events the scattered positron can be used in order to reconstruct the kinematic quantities Q^2_e; y_e and $M_e = \sqrt{xs_ep}$ of each event, for charged current (CC)-events one has to rely on the less well measured hadrons for obtaining Q^2_h; y_h and M_h. As has first been pointed out by Blondel and Jaquet [12], the hadronic quantity $\Sigma = \Sigma(E_h - P_{z,h})$, where E_h and $P_{z,h}$ are the energy and longitudinal momentum component (z = proton beam direction) of a hadron h and the sum is taken over all measured hadrons, is less sensitive to hadrons which are produced in forward i.e. proton direction and escape detection by staying inside the beam pipe or are badly measured. For NC-events, a combination of leptonic and hadronic information was used in order to reduce errors resulting from the absolute energy calibration of the electromagnetic calorimeter or from initial state radiative losses of the lepton; for a detailed discussion see [4]. The analysis was performed taking into account the improved calorimeter calibration.

Including the 1997 runs, a total luminosity of 37 pb^{-1} was collected by H1. Fig. 3 compares for NC events the measured Q^2-distribution with the standard DIS-Monte Carlo based on parton densities taken from the Martin-Roberts-Stirling MRSH parametrization and evolved to high Q^2 values using NLO-DGLAP equations (comp.[4]). Including the 1997 data results in a decrease of the significance of the previously reported excess, the 1997 data alone only marginally indicate an excess in the $Q^2 > 15.000$ GeV^2 region. The NC-results in the very high Q^2-region are summarized

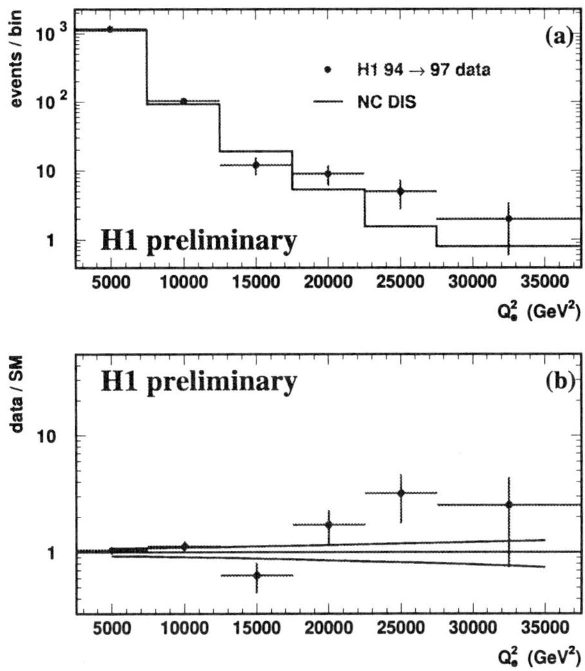

Figure 3. (a) Q_e^2 distribution of NC events (points) in comparison with standard DIS expectation (histogram). (b) Ratio of observed and expected number of NC events (points); the upper and lower line correspond to the $\pm$ 1σ-error determined from combining statistical and systematic errors of the standard model expectation.

in Table 1.

The CC-results shown in Fig. 4 are summarized in Table 2. While a slight excess in the $Q^2 > 7.500$ GeV^2 region is observed, the data are compatible with the standard model within errors.

2.2. Inclusive high Q^2 NC/CC-cross sections

The HERA data nevertheless immensely increase the knowledge of the DIS cross sections. A kinematic range of 200 GeV$^2 < Q^2 <$ 30.000 GeV^2 for $.005 < x < .65$ is covered by now enabling extended comparisons with QCD predictions.

The NC-differential cross section can be written as

$Q^2{}_{min}(\mathrm{GeV})^2$	2500	7500	15000
N_{obs}	100	41	9
N_{DIS}	95.3	27.6	5.07
	± 16.7	± 8.4	± 2.8
$P(N > N_{obs})$	40%	9%	17%

Table 2. CC-events in the high Q^2-region.

$$\frac{d^2\sigma^{NC}(e^{\pm}p)}{dxdQ^2} = \frac{2\pi\alpha^2}{xQ^4}\left[Y_+F_2(x,Q^2) - y^2F_L(x,Q^2) \mp Y_-xF_3(x,Q^2)\right] \quad (1)$$

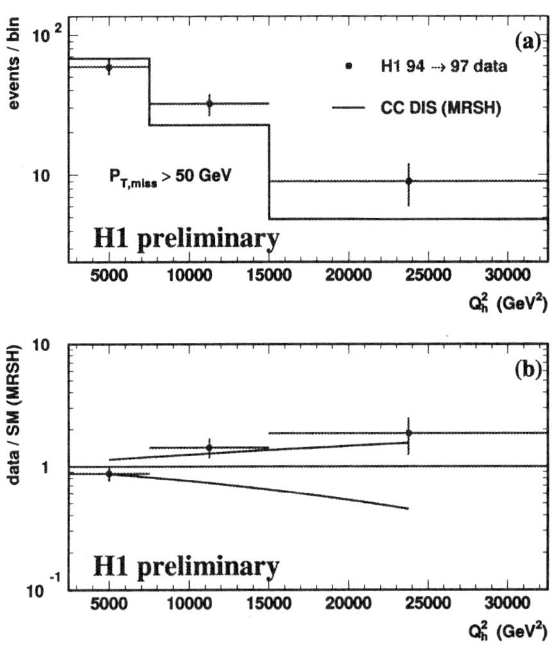

Figure 4. (a) Q_h^2 distribution of CC events. (b) Ratio of observed and expected number of CC events (for explanation see Fig. 3).

with the kinematical factors $Y_{\pm}(y) = 1 \pm (1-y)^2$ ($\pm$ for $e^{\pm}p$ - scattering). It can be approximated as

$$\frac{d^2\sigma^{NC}(e^{\pm}p)}{dxdQ^2} \cong \frac{2\pi\alpha^2}{xQ^4} Y_+ F_2(x, Q^2)$$

with corrections due to F_3 remaining $< 1\%$ for $Q^2 < 1.500$ GeV^2 and $< 10\%$ for $Q^2 = 5.000$ GeV^2 (at x=.08) and those due to F_L being negligible for $y < .5$ and $\cong 5\%$ for $y = .9$, respectively. A "reduced cross section" is defined as

$$\sigma^{NC}(e^+p) = \frac{xQ^4}{2\pi\alpha^2} \frac{1}{Y_+} \frac{d^2\sigma}{dxdQ^2}. \tag{2}$$

In Fig. 5 the measured reduced cross section for e^+p-scattering in the high x-region is compared with the QCD prediction ([10], [11]). The MRSH prediction, based on measured parton densities, as well as the H1-NLO QCD fit which includes the new H1-high Q^2 data, describe the data well; in the high x- and high Q^2-region the H1 data are lowering the fit by up to 5%. The excess of the data observed in particular in the x= .45-distribution at the highest Q^2-values reflects the high Q^2-excess discussed in 2.1.

The CC-differential cross section may be written [3] in leading order QCD and neglecting quark mixing in terms of parton densities $xq(x, Q^2)$ as

$$\frac{d^2\sigma^{CC}(e^{\pm}p)}{dxdQ^2} = \frac{G_F^2}{2\pi}(P_W)^2 \left[(\bar{u}+\bar{c}) + (1-y)^2(d+s)\right]$$

with the Fermi coupling constant $G_F = \pi\alpha/\sqrt{2} sin^2\Theta_W M_W^2$ and the W-propagator $P_W = M_W^2/(Q^2 + M_W^2)$.

Similarly to (2), a reduced cross section is defined

$$\sigma^{CC}(e^+p) = \frac{2\pi x}{G_F^2} \left(\frac{Q^2 + M_W^2}{M_W^2}\right)^2 \frac{d^2\sigma^{CC}}{dxdQ^2}$$

The measured x-integrated cross sections $d\sigma/dQ^2$ for NC- and CC-processes are compared in Fig. 6 with the standard model predictions. Good agreement within errors is found; the remaining difference between the two cross sections at high Q^2 is due to the different couplings of the neutral and charged current to the quark flavours as well as to the different propagators.

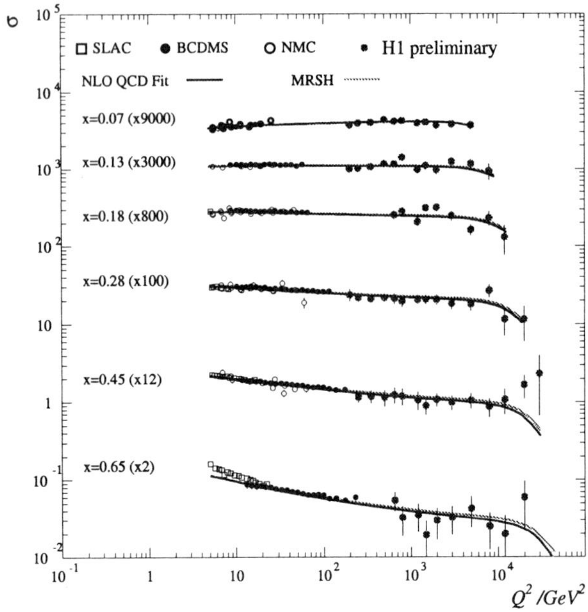

Figure 5. Reduced cross-section $\sigma^{NC}(e^+p)$ at fixed x vs. Q^2 (1994-1997 data). The NLO QCD fit includes the H1 data.

3. Limits on leptoquark-production

The high Q^2-data can be used to derive limits on the couplings of leptoquarks (LQ). Preliminary results are presented for scalar LQ couplings to 1st and 3rd generation fermions for the 1994-1997 data corresponding to a total integrated luminosity of 37 pb^{-1} [10], [11]. These results update the previous H1 analyses [14].

A scalar s-channel LQ is produced at a fixed mass $M = \sqrt{sx}$ with an intrinsic width $\Gamma = \lambda^2 M/16\pi$ where λ is the dimensionless Yukawa coupling of the LQ to lepton-quark pair and a production cross section approximately given by [15] $\sigma = (\pi/4s)\lambda^2 q(x = M^2/s)$.

A LQ decaying into a 1st generation lepton and a quark jet cannot be distinguished from NC- or CC-DIS events on an event by event basis but statistically only by searching for a resonance peak in the invariant mass distribution and by making use of the decay angular distribution of the LQ. Scalar LQs decay isotropically in their CM-frame; with $y = (1 + cos\Theta^*)/2$ where Θ^* is the decay polar angle in the LQ CM frame, this results in a

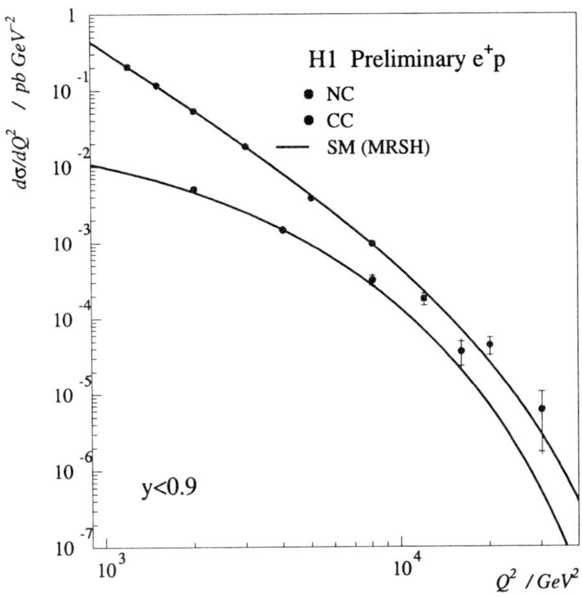

Figure 6. Differential cross-section $d^2\sigma/dxdQ^2$ for NC-events (upper data points) and CC-events (lower data points) in e^+p-scattering vs. Q^2 (1994-1997 data). Lines give the standard model prediction.

flat $d\sigma/dy$. In contrast, the dominating background from t-channel photon exchange at fixed x is characterized by $d\sigma/dy \sim 1/y^2$. 1st generation LQs are, therefore, searched for in high mass NC- or CC-DIS events at high y. 2nd and 3rd generation LQs are characterized by an "exotic" final state i.e. the μ or τ and are, therefore, expected to be essentially free of background for high transverse momentum leptons. 2nd generation LQs are not further discussed here; the μ-events discussed in Chapter 5 fail significantly the kinematic constraints for LQ production and no candidate for LQ→ μ+jet was observed.

3.1. 1st generation LQs

LQ→ eq: In order to maximize the significance of a LQ signal as a function of the LQ mass M, a mass dependent y_e-cut $y_{cut}(M)$ was applied such that only events with $y_e > y_{cut}(M)$ were accepted. The optimization of the cut was obtained from Monte Carlo studies; it ranges from $y_e \cong .5$ around M_e=45 GeV to $y_e \cong .1$ around M_e=250 GeV.

In Fig. 7a, the event distributions for NC-like candidates before and after applying the y-cut are compared with the expected distribution, taking into account the event detection efficiency. Details of this procedure and of the cut optimization are given in [15]. From the resulting limits on the LQ production cross section one can derive mass dependent limits on the coupling constant λ keeping the branching ratio β fixed e.g. due to a specific model or limits on β keeping λ fixed.

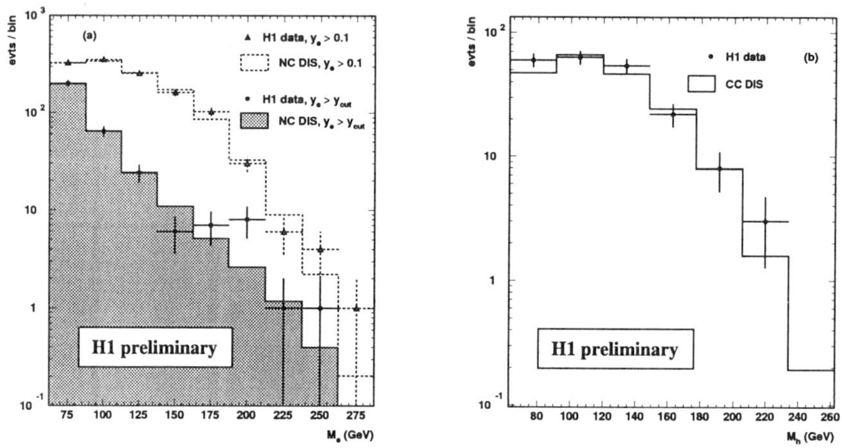

Figure 7. 1st generation on leptoquark analysis: (a) Mass distribution for NC DIS-like final states for different y_e-cuts i.e. $y_e > .1$ (fixed) and $y_e > y_{cut}$ (optimized). (b) Mass distribution for CC DIS-like final states with P_T^{miss} >30 GeV.

The limits obtained on λ_{11} (λ_{ij}: $i =$ lepton generation, $j=$ quark generation) shown in Fig. 8 are based on the LQ model of Buchmüller, Rückl and Wyler [16] in which for the 7 scalar LQs β_e is fixed to .5 and 1, respectively. The new analysis improves earlier H1-results on LQs with fermionic number $F = 0$ [14] by about a factor of 3, implying $M_{LQ} > 275$ GeV (95% c.l.) for $\lambda^2/4\pi = \alpha_{em}$ i.e. $\lambda = 0.3$; for LQs with $F = 2$ the improvement is less significant. For this coupling strength the present LEP limit of 142 GeV [17] is not competitive with the new H1 limit.

The resulting constraints on β_e for given λ_{11} are shown in Fig. 9. A drop in sensitivity around $M = 210$ GeV is observed (note that M_e underestimates M_{LQ} by about 5 GeV). The H1-results extend the excluded domain beyond the $D0$ limits [18] in particular for u quark-couplings and small λ($\lambda = .1$ corresponds to $\alpha_{em}/10$).

LQ→ νq: Candidates for this channel correspond to CC-DIS events

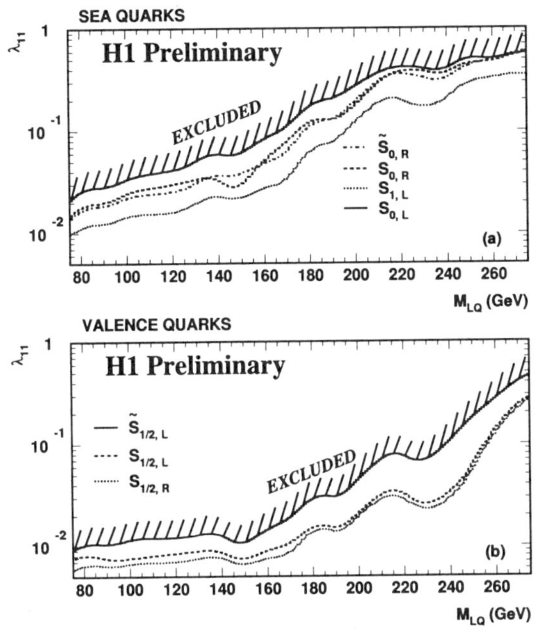

Figure 8. Exclusion limits (95% c.l.) on couplings λ_{11} for scalar leptoquarks in the BRW-model: (a) fermionic number $F = 2$, (b) $F = 0$.

and, therefore, were searched for in an inclusive CC analysis with a cut of $p_T^{miss} > 30$GeV. The mass distribution shown in Fig. 7b is well described by the DIS-prediction.

Scalar LQs decaying to eq with the branching ratio β_e and to νq with the branching ratio β_ν ($\beta_e + \beta_\nu \leq 1$) are outside the BRW model considered above. Gauge invariance implies that only the $e^+\bar{u}$-coupling contributes. The constraints on λ_{11} are shown in Fig. 10. The combination of NC- and CC-DIS like channels results in an improvement of the λ_{11} limit when compared with the pure NC-like channel ($\beta_e = 0.1$). For $\lambda = .3$ one finds $M_{LQ} > 210$ GeV (95%c.l.), far exceeding the TeVatron limit of $\cong$ 110 GeV for such small β_e [18], [19].

3.2. 3rd generation LQs

LQs of this type with couplings λ_{11} and λ_{3j} are characterized by their decay into a jet and a τ-lepton. The search was performed using the hadronic decay $\tau \to$ hadrons + ν, where the hadrons have the signature of a pencil like

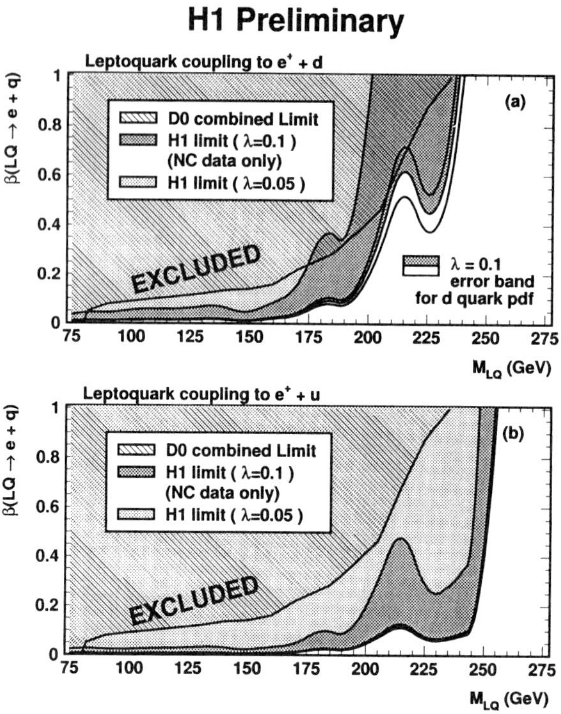

Figure 9. Limits on branching ratios β_e of scalar leptoquarks formed in e^+q-fusion for fixed λ. (a) Leptoquark coupling to e^+d. For $\lambda = .1$ the $\pm 1\sigma$-error band due to uncertainties in the parton density functions is shown. (b) Leptoquark couplings to e^+u. Shown is also the $D0$ limit.

jet. Using simulated events the cuts in the mass distribution as well as in the multiplicity distribution of the jets were optimized. The event selection required 2 hadronic jets in the liquid argon calorimeter with $P_T > 30$ GeV where one jet has τ-characteristics i.e. $M_{jet} < 7$ GeV and $1 \leq N_{tracks} \leq 3$.

Further cuts have been applied on the basis of Monte Carlo-studies of the P_T^{miss}- and $\Delta\phi$-distribution (with $\Delta\phi$ the azimuthal angle between the high multiplicity jet and the low multiplicity jet) of the LQ-decays. Setting the cuts to $P_T^{miss} > 10$ GeV and $\Delta\phi > 165^0$ leaves no event candidate where $.8 \pm .3$ events are expected.

Limits on λ_{11} for a scalar LQ coupling to e^+u for different branching ratios β_e; β_ν are derived as above and shown in Fig. 11a. For $\lambda_{11} = .3$ and $\beta_e = .1$; $\beta_\tau = .9$ a lower limit of $M > 275$ GeV (95% c.l.) is found; it is to be compared with the TeVatron limit of about 100 GeV obtained from $\tau\tau bb$- and $\nu\nu bb$- final states for $\beta_\tau = 1$ and $\beta_\nu = 1$ respectively [20], [21].

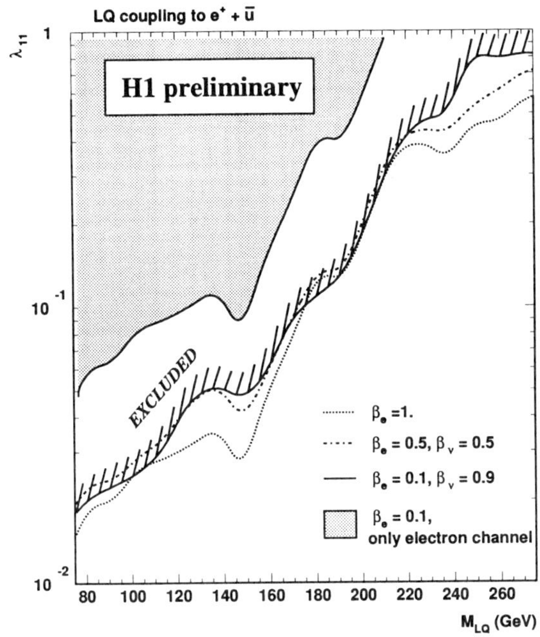

Figure 10. Limits on couplings λ_{11} (95% c.l.) for scalar leptoquarks from e^+p-scattering for given β_e and β_ν.

Alternatively, limits on λ_{3j} ($j = 1, 2$) for different λ_{11} assuming $\beta_e + \beta_\tau = 1$ are shown in Fig. 11b. For λ_{11} and λ_{3j} of electromagnetic strength, a limit of $M > 255$ GeV (95% c.l.) is obtained.

4. Excited Leptons

Excited leptons e^* and ν^* decaying into a lepton and a gauge boson were searched for using the full 1994-1997 ep-luminosity of 37pb^{-1} [13]. The results update and improve the previous H1 analysis [22].

For the 7 different final state topologies resulting from the 11 decay channels which have been considered the observed number of events is compared in Table 3 with the number predicted by the SM. The numbers of observed and expected events are compatible within errors and limits only on $\sigma \cdot \beta$ can be derived which improve the previous H1 results by about an order of magnitude.

In the specific model of Hagiwara, Komamiya and Zeppenfeld [23], the cross sections depend on coupling constants f and f' for the gauge groups SU(2) and U(1), respectively, as well as on a substructure scale Λ. Assuming specific relations between f and f', the branching ratios depend

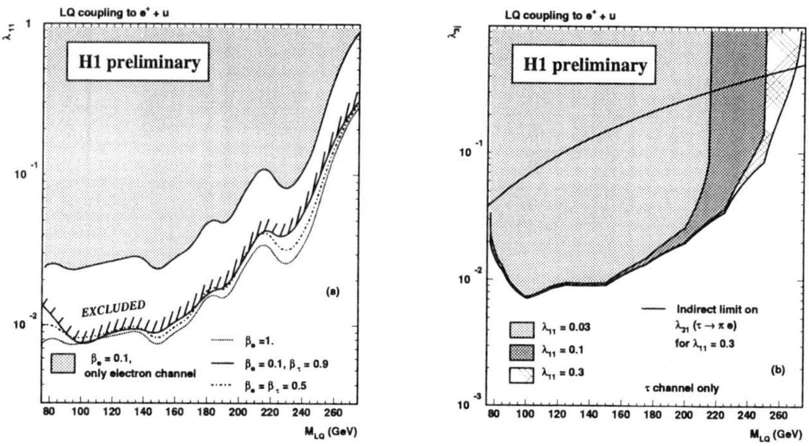

Figure 11. Exclusion limits (95% c.l.) on couplings λ_{11} (a) and $\lambda_{3j}(j = 1.2)$ (b) for 3rd generation scalar leptoquarks from e^+u-fusion for different fixed β_e, β_τ and λ_{11}, respectively.

on f/Λ and constraints on f and f' can be derived which are shown in Fig. 12. The LEP limits are more stringent but so far reach lower mass values only [24].

Channel	Selection	Data	SM exp.	Efficiency
$e^* \to e\gamma$	2 EM clusters	223	239 ± 7	85 %
$e^* \to eZ \hookrightarrow ee$	3 EM clusters	3	1.4 ± 0.3	78 %
$e^* \to eZ \hookrightarrow \nu\bar{\nu}$	1 electron + P_t^{miss}	1	3.6 ± 0.7	70 %
$e^* \to eZ \hookrightarrow q\bar{q}$	2 jets + 1 electron	38	48 ± 3	41 %
$e^* \to \nu W \hookrightarrow e\nu$	1 electron + P_t^{miss}	1	3.6 ± 0.7	70 %
$e^* \to \nu W \hookrightarrow q\bar{q}$	2 jets + P_t^{miss}	3	3.8 ± 0.5	40 %
$\nu^* \to \nu\gamma$	1 photon + P_t^{miss}	0	1.3 ± 0.8	38 %
$\nu^* \to \nu Z \hookrightarrow ee$	2 electrons + P_t^{miss}	0	0.38 ± 0.2	40 %
$\nu^* \to \nu Z \hookrightarrow q\bar{q}$	2 jets + P_t^{miss}	3	3.8 ± 0.5	40 %
$\nu^* \to eW \hookrightarrow e\nu$	2 electrons + P_t^{miss}	0	0.38 ± 0.2	40 %
$\nu^* \to eW \hookrightarrow q\bar{q}$	2 jets + 1 electron	38	48 ± 3	41 %

Table 3. Decay channels and final state topologies for excited leptons: Number of observed events vs. Standard Model expectation.

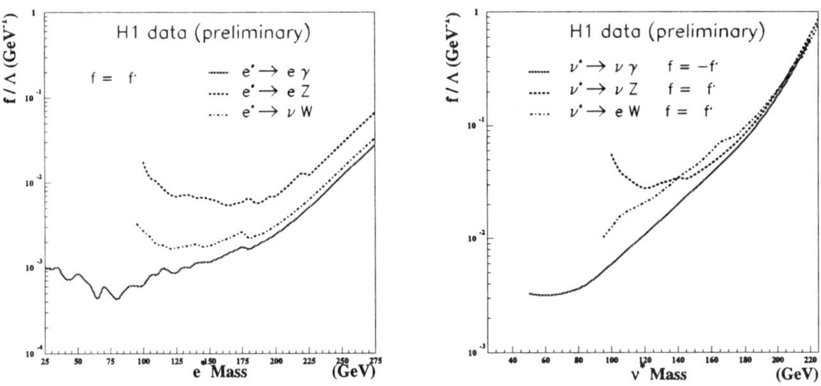

Figure 12. Exclusion limits (95% c.l.) for excited leptons as function of the lepton mass for e^* (left) and ν^* (right).

5. "Odd leptons"

In 1994 with an integrated e^+p-luminosity of 3.2 pb^{-1}, an event was observed in the H1-detector [25] which showed an isolated high $P_T - \mu^+$ recoiling back to back in azimuth against a high P_T hadronic system. The event is unbalanced in P_T indicating an undetected particle in the final state. The event kinematics is compatible with W production, the rate, however, is about a factor of 10 too high compared with the standard model-calculations of Baur, Vermaseren and Zeppenfeld [26] leaving room for other contributions. Here we present an update of the search for similar events based on the total e^+p-luminosity of 37 pb^{-1} [27].

In order to identify events with significant missing transverse momentum, the standard H1-selection criteria for CC-DIS-events requiring in the calorimeter a $P_T > 25$ GeV have been applied. From the resulting event sample events with an isolated track were selected by essentially requiring a track with $P_T > 10$ GeV. In Fig. 13, the separation of these events to the closest hadronic jet (D_{jet}) as well as to the closest single particle track (D_{track}) is shown. The correlation between D_{jet} and D_{track} exhibits 6 events well separated from the rest (Fig. 13); applying in addition electron- and μ-identification criteria shows that all the isolated tracks are due to leptons: $1e^-$, $2\mu^+$, $2\mu^-$, 1 μ of unidentified charge. Full details of the event kinematics are given in [27].

The imbalance in total transverse momentum P_T^{miss} is ascribed to unobserved particles. Additional evidence for unobserved particles is provided by a significant deficit in longitudinal momentum balance in some

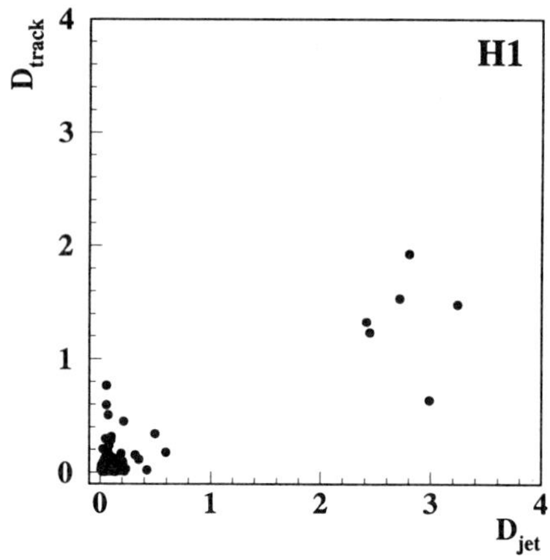

Figure 13. Correlation between the distances D_{jet} and D_{track} (see text) to the closest hadronic jet and closest track, respectively, for all high P_T tracks in the inclusive event sample.

of the events and an acoplanarity $\Delta\phi$ between the high P_T-lepton and the hadronic system X in most events. The correlation between $\Delta\phi$ and P_T^{miss} for the six events is compared with that measured in standard NC events (subject to proper cuts in order to cover a similar kinematic region) in Fig. 14. Except for one μ^+-event, the events are characterized by both high $\Delta\phi$ and high P_T^{miss}.

The rates for standard model processes which can yield lepton events of the observed type i.e. with an isolated high P_T lepton and missing transverse momentum are summarized in Table 4. The e^- event is compatible with the expected rate from W production; the 5 μ events are to be compared with an expectation of .8 ± .2 events dominated by W production and photon-photon-processes.

For the 6 events the kinematic parameters P_T^X and $M_T^{\mu\nu}$ are compared in Fig. 15 with the distributions expected for W production and photon-photon-processes (with P_T^X = total hadronic transverse momentum of the event and P_T^l = transverse momentum of the isolated lepton one has the high P_T lepton-neutrino transverse mass $M_T^{l\nu} = \sqrt{(P_T^{miss} + P_T^l)^2 - (\vec{P}_T^{miss} + \vec{P}_T^{\nu})^2}$ where the measured missing transverse momentum $\vec{P}_T^{miss}$ is attributed to a neutrino). While the e^- event and the $\mu3$ event show kinematic properties typical for W production and the $\mu5$ event is compatible with W production within 2σ, the other μ events are found in regions of the phase space where the expected

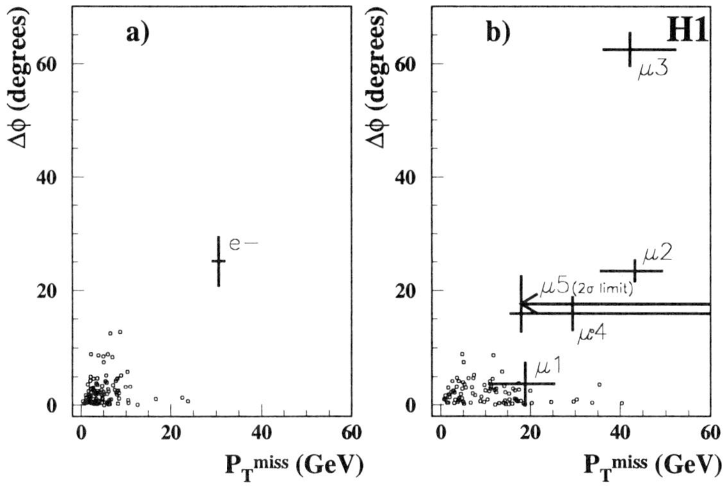

Figure 14. Distribution of events in the P_T^{miss}-$\Delta\phi$ plane. (a) electron channel: (b) muon channel. Crosses show the individual measurement errors, open circles show for comparison the NC event distribution (see text).

	Electron Channel	Muon Channel
Data	$0\ e^+, 1\ e^-$	5
W production	1.65 ± 0.47	0.53 ± 0.11
Z production	0.01 ± 0.01	0.01 ± 0.01
CC - DIS	0.02 ± 0.01	0.01 ± 0.01
NC - DIS	$0.51 \pm 0.10\ e^+, 0.02 \pm 0.01\ e^-$	0.09 ± 0.06
Photoproduction	< 0.02	< 0.02
Heavy Quarks	< 0.04	< 0.04
Photon-Photon	$0.09 \pm 0.03\ e^+, 0.04 \pm 0.01\ e^-$	$0.14^{+0.14}_{-0.07}$

Table 4. Observed and predicted number of events with isolated high P_T-leptons$^{\pm}$ and P_T^{miss}.

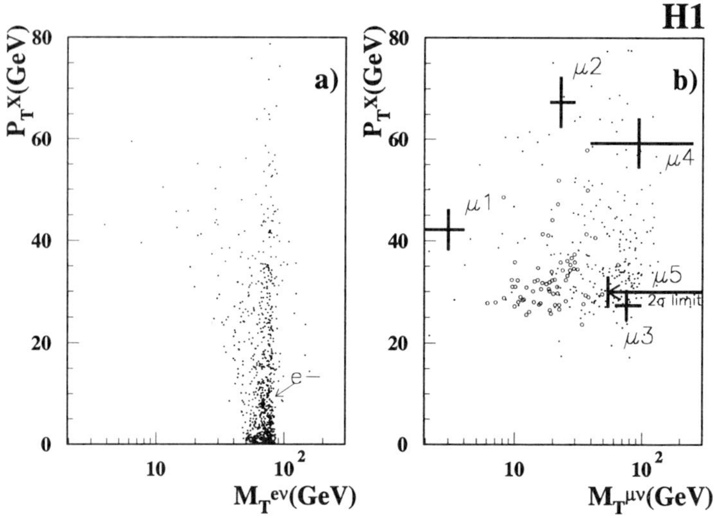

Figure 15. Distribution of events in P_T^X and $M_T^{l\nu}$. (a) electron channel. (b) muon channel. Crosses indicate the 1σ-uncertainty on the measured kinematical parameters of the observed events (for $\mu5$ the 2σ-lower limit for $M_T^{l\nu}$ is shown). Expected events from an 500 x higher luminosity are shown for W production (dots) and photon-photon-processes (open circles).

W rate is low. $\mu2$ and $\mu4$ have a large P_T^X and $\mu1$ has a small $M_T^{l\nu}$, both of which are unlikely for W production or photon-photon-interaction. These three muon events are, therefore, unlikely to be due to the SM processes considered above.

6. Future Prospects

The scientific potential of HERA very much depends on the luminosity attainable. As the maximum possible beam currents have essentially been reached, a major increase of the luminosity can be achieved only by a modification of the beam optics. The standard approach is to reduce the beam cross section at the interaction point by reducing the β function. The adopted solution which promises to lead to an increase of the luminosity at least by a factor of five foresees the installation of superconducting combined function magnets very close to the interaction point i.e. into the detectors. The luminosity increase is expected to become effective in the years 2000/2001.

The upgrade of the collider will be complemented by an upgrade of the H1 and ZEUS detector. In H1 besides improving the tracking components of the inner detector and increasing the sensivity of the liquid argon calorimeter, efforts are concentrating on improvement of the triggering including the development of advanced triggers.

A physically very interesting feature of HERA is the possibility to provide a longitudinal polarization of the $e^{\pm}$-beam. The polarized running mode of HERA foreseen after the luminosity upgrade obviously would profit from the higher luminosity and ensure a rich physics program for the first decade of the next century.

Acknowledgement: I am grateful to J. Meyer for his help and reading the manuscript.

References

[1] G.-A. Voss, B.H. Wiik, Ann. Rev. Nucl. Part. Sci. **44** (1994) 413

[2] H1 Collab., I. Abt et. al., *Nucl. Instrum. Methods* **A. 386** (1997), 310, 348

[3] comp. e.g. A.M. Cooper-Sarkar, R.C.E. Devenish, A. DeRoeck, DESY 97-226

[4] H1 Collab., C. Adloff et. al., *Z. Phys.* **C74** (1997) 191

[5] ZEUS Collab., J. Breitweg et. al., *Z. Phys.* **C74** (1997) 207

[6] B. Straub, Int. Symp. on Lepton Photon Interactions, Hamburg 1997

[7] E. Elsen, Int. Europhys. Conf. on High Energy Physics, Jerusalem (1997)

[8] comp. e.g. G. Altarelli et. al., *Nucl. Phys.* **B506** (1997); G. Altarelli, CERN-TH. 97/278

[9] U. Bassler, G. Bernardi, DESY 97-136

[10] H1 Collab., Ph. Bruel, Rencontres de Moriond 1998

[11] H1 Collab., B. Heinemann, 6th Int. Workshop on Deep-Inelastic Scattering and QCD, Bruxelles (1998)

[12] A. Blondel, F. Jacquet, Proc. of the Study of an *ep*-Facility for Europe, ed. U. Amaldi, DESY 79-48

[13] H1 Collab., M.C. Cousinou, 6th Int. Workshop on Deep-Inelastic Scattering and QCD, Bruxelles (1998)

[14] H1 Collab., I. Abt et.al., *Nucl. Phys.* **B396** (1993) 3; ibid. T. Ahmed et.al., *Z. Phys.* **C64** (1994) 545; ibid. T. Ahmed et.al., *Phys. Lett.* **B369** (1996) 173

[15] B. Schrempp, Proc. Workshop Physics at HERA (1991), eds. W. Buchmüller and G. Ingelman, 1034

[16] W. Buchmüller, R. Rückl, D. Wyler, *Phys. Lett.* **B191** (1987) 442

[17] DELPHI Collab., F. Barao et.al., Contrib. to Int. Europhys. Conf. on High Energy Physics, Jerusalem (1997)

[18] *D*0 Collab., B. Abbott et.al., *Phys. Rev. Lett.* **79** (1997) 4321, **80** (1998) 2051

[19] CDF Collab., F. Abe et.al., *Phys. Rev. Lett.* **79** (1997) 4327

[20] CDF Collab., F. Abe et.al., *Phys. Rev. Lett.* **78** (1997) 2906

[21] *D*0 Collab., B. Abbott et.al., FERMILAB-PUB-98-081-E

[22] H1 Collab., T. Ahmed et.al., *Phys. Lett.* **B340** (1994) 205; H1 Collab., S. Aid et.al., *Nucl. Phys.* **B483** (1997) 44; see also ref.[14], I. Abt et.al.

[23] K. Hagiwara, S. Komamiya, D. Zeppenfeld, *Z. Phys.* **C29** (1985) 115

[24] DELPHI Collab., P. Abren et.al., *Phys. Lett.* **B393** (1997) 245; L3 Collab., M. Acciari et.al., *Phys. Lett.* **B401** (1997) 139

[25] H1 Collab., T. Ahmed et.al., DESY 94-248; H1 Collab., S. Aid et.al., *Z. Phys.* **C71** (1996) 211

[26] U. Bauer, J.A.M. Vermaseren, D. Zeppenfeld, *Nucl. Phys.* **B375** (1992) 3

[27] H1 Collab., C. Adloff et.al., DESY 98-063

Search for Physics Beyond the Standard Model with the ZEUS Detector at HERA

Massimo Corradi,
on behalf of the ZEUS collaboration

INFN Bologna, via Irnerio 47, Bologna, Italy

Abstract. Searches for new physics based on the ZEUS 1994-'97 e^+p data are reviewed. Hig-Q^2 deep inelastic scattering cross section is compared with the Standard Model predictions. Limits on different processes are reported: contact interactions, lepton flavor violation, anomalous W production, MSSM selectrons and squarks and excited fermions.

1. Introduction

HERA is a unique place to search for signals of new physics beyond the Standard Model (SM). Electron-proton Deep Inelastic Scattering (DIS) can be studied at high-Q^2 ($Q^2_{\rm max}9 \times 90000{\rm GeV}^2$), where the effect of new physics at high-mass scale could manifest itself as a deviation from the SM prediction. Moreover, exotic processes such as lepton flavor violation, the production of excited fermions or supersymmetric particles can be searched for with small backgrounds.

In the period 1994-'97, HERA has been colliding positrons and protons with beam energies of $E_e = 27.5$ GeV and $E_p = 820$ GeV, corresponding to a center of mass energy of $\sqrt{s} = 300$ GeV. The integrated luminosity useful for data analysis collected by ZEUS in this period is $\mathcal{L} = 47$ pb^{-1}. More than one half of the integrated luminosity (27 pb^{-1}) was collected during 1997.

Figure 1 shows an example of a high-Q^2 neutral-current event in the ZEUS detector. The tracks of charged particles are reconstructed with the central tracking detector, a cylindrical drift chamber placed in a 1.43 T solenoidal magnetic field that provides a resolution of

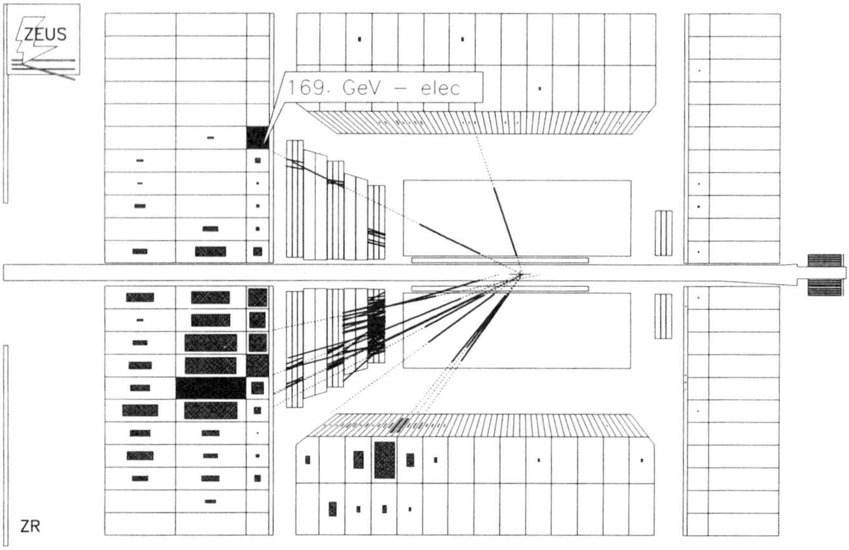

Figure 1. Display of the highest-Q^2 NC DIS event collected in 1997 (Q^2_{DA} = 19000GeV^2) in a Z-R projection of the ZEUS detector. The tracks associated to the electron and to the hadronic system are shown, the energy deposited in the calorimeter is represented by the size of the black boxes. The label indicates the scattered electron. The electron beam enters from left and the protons from right.

$\sigma_{p_t}/p_t = 0.005 p_t(\mathrm{GeV})$ on the transverse momentum of full length tracks. The Uranium-scintillator calorimeter provides an energy measurement with a resolution $\sigma_E/E = 35\%/\sqrt{E(\mathrm{GeV})}$ for hadrons and $\sigma_E/E = 18\%/\sqrt{E(\mathrm{GeV})}$ for electromagnetic showers. In the ZEUS coordinate system the Z-axis points in the direction of the proton beam and all the polar angles are measured from this direction. With this notation the electrons are scattered at large polar angle θ in low-Q^2 interactions while at high-Q^2 they are back-scattered at small θ. Event variables are reconstructed from the interaction vertex measured by tracks and from clusters of energy deposited in the calorimeter. In fully contained events, the missing transverse momentum $\not{P}_t = |\vec{\not{P}_t}| = |-\sum_i \vec{p}_t^{\,i}|$ (the sum runs over all the calorimeter clusters) is close to zero and the longitudinal momentum variable $\delta = \sum_i E^i - P_z^i$ is close to $2E_e = 55$ GeV.

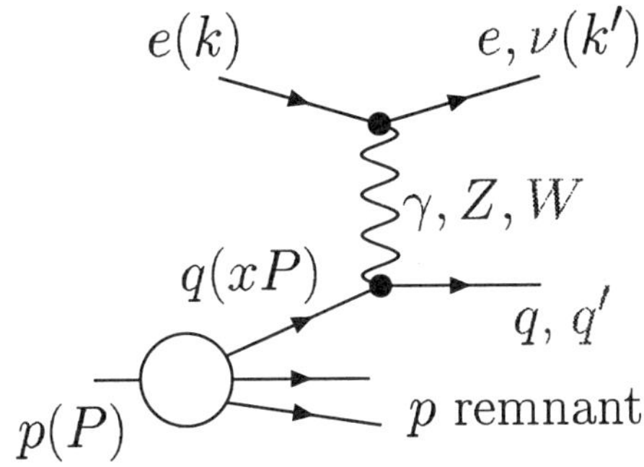

Figure 2. Diagram of a DIS process at LO.

2. High-Q^2 DIS

The Neutral Current (NC) and Charged Current (CC) inclusive reactions[1]

$$\begin{array}{lll} e^+(k)p(P) \rightarrow & e^+(k')X & \text{(NC)} \\ e^+(k)p(P) \rightarrow & \bar{\nu}(k')X & \text{(CC)} \end{array}$$

are studied as a test of the SM. They are characterized by the following variables: the squared four-momentum transfer $Q^2 = -(k-k')^2$, the Bjorken scaling variable $x = \frac{Q^2}{2P(k-k')}$ and the inelasticity $y = \frac{P(k-k')}{Pk}$. Only two of them are independent due to the relation $Q^2 = xys$. At high Q^2 and high x, i.e. in the DIS regime, the ep scattering is well described by the Leading Order (LO) ($\mathcal{O}(\alpha_{em}^2)$) diagram of figure 2: a gauge boson (γ,Z^0 in NC and W in CC) is exchanged between the lepton and a quark that carries a fraction x of the proton momentum. The DIS cross section is given by the sum of the partonic processes weighted by the Parton Density Functions (PDF's) $q(x,Q^2)$. The PDF's are known at low Q^2 from fixed target experiments and can be evolved at higher Q^2 through the QCD evolution equations.

The SM prediction of DIS cross section at high-Q^2 used in this analysis includes QCD $\mathcal{O}(\alpha_S)$ matrix elements, QED and electroweak radiative corrections (as implemented in the MC generator DJANGO [1]) and recent PDF's parametrizations (CTEQ4 [2]). The theoretical uncertainty on the SM prediction is 6-8% for NC and 9-17% for CC in the Q^2 range under study and is dominated by the uncertainty on the fixed target measurements. Older ZEUS results on high-Q^2 DIS cross section can be found in [3].

[1] The variables in parenthesis are the four-momenta of the corresponding particle.

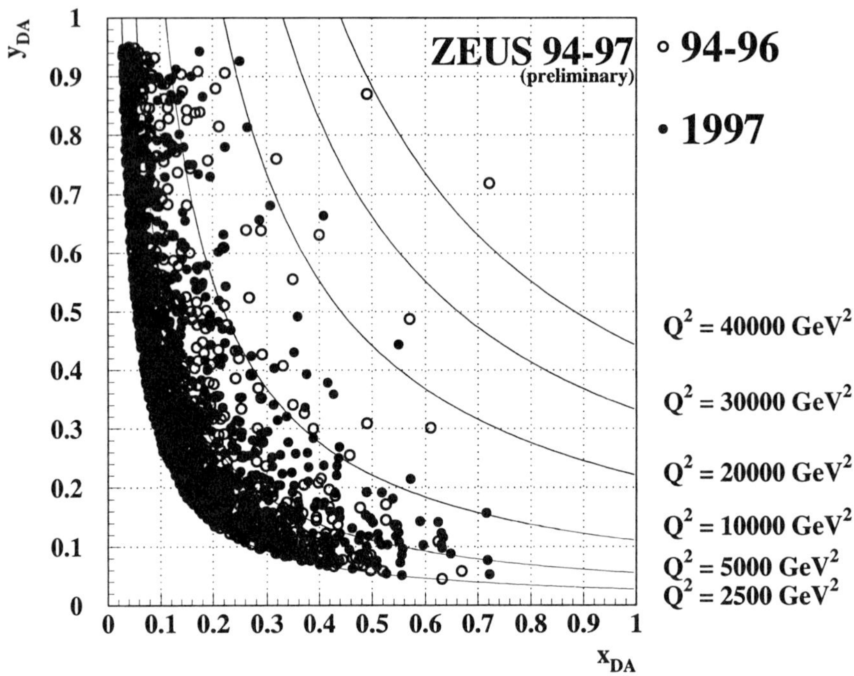

Figure 3. Scatter plot showing the position of the NC DIS events with $Q^2_{\rm DA} > 2500$ GeV^2 in the $x_{\rm DA} - y_{\rm DA}$ plane. Events from 1994-'96 data taking are shown as empty circles, those collected in 1997 as filled circles. Lines of constant $Q^2_{\rm DA}$ are also shown.

Table 1. Observed and expected number of NC DIS events with $Q^2_{\rm DA} > Q^2_{\rm min}$ from ZEUS 1994-'97 preliminary results.

$Q^2_{\rm min}$[GeV^2]	$N_{\rm obs}$	$N_{\rm exp}$	
2500	1817	1792	± 93
5000	440	396	± 24
10000	66	60	± 4
15000	20	17	± 2
35000	2	0.29	± 0.02

2.1. Neutral current DIS

NC DIS events at high-Q^2 are characterized by a high-energy electron and by $\delta \sim 2E_e$. The electron is identified by requiring an isolated high-energy electromagnetic cluster. In the region of full tracking efficiency ($\theta > 17^o$) a match with a track, electron energy $E_e > 10$ GeV and $38 < \delta < 65$ GeV are required. In the forward region ($\theta \leq 17^o$) the electron transverse momentum must be $p_t^e > 30$ GeV and $44 < \delta < 65$ GeV. For $Q^2 > 400$ GeV^2 the trigger and selection efficiency is 90% while the background from low-Q^2 interactions is estimated to be $\sim 1/1000$ of the signal.

The measurement of the DIS variables is overconstrained. Four independent measurements (energy and angle of the electron and of the final hadronic system) can be used for the determination of two DIS variables. The double angle (DA) method has been chosen because it provides a very good resolution over all the kinematic plane and is almost insensitive to errors in the energy scale. This method uses the electron scattering angle and the hadronic angle γ that corresponds at LO to the quark scattering angle:

$$Q^2_{\mathrm{DA}} = 4E_e^2 \frac{\sin\gamma(1+\cos\theta)}{\sin\gamma + \sin\theta - \sin(\gamma+\theta)},$$

$$x_{\mathrm{DA}} = \frac{E_e}{E_p}\frac{\sin\gamma + \sin\theta + \sin(\gamma+\theta)}{\sin\gamma + \sin\theta - \sin(\gamma+\theta)}.$$

The hadronic angle γ is obtained from the energy and the transverse and longitudinal momenta of the hadronic system. The resolution on the variable Q^2_{DA} is RMS$\simeq$ 5% ($\sigma \simeq 2\%$) over the whole Q^2 range.

An excess of data with respect to the SM expectation was reported in a previous publication [4] based on 1994-'96 data (20 pb^{-1}). Four events were found in the region $x_{\mathrm{DA}} > 0.5$, $y_{\mathrm{DA}} > 0.25$ where only 0.9 were expected by the SM. In 1997 data (27 pb^{-1}) only one more event has been found in this region, in agreement with the SM expectation. All the events with $Q^2 > 2500$ GeV^2 from the whole 1994-'97 data set are shown in the $x_{\mathrm{DA}} - y_{\mathrm{DA}}$ plane in figure 3. Table 1 shows the observed and the expected number of events above a given Q^2_{min}. Data and SM expectation are in good agreement up to $Q^2 = 30000$ GeV^2, while two outstanding events with $Q^2 > 35000$ GeV^2, both found in the 1994-'96 running period, are above the SM expectation of 0.29 ± 0.02.

The measurement of the differential cross section $d\sigma/dQ^2$, corrected to the Born level, is reported in figure 4 with the SM prediction obtained using the CTEQ4 PDF's. The agreement is generally good with the exception of the last point that is above expectation due to the two very high Q^2 events. For $Q^2 \ll M_z^2$, the cross section falls like $d\sigma/dQ^2 \sim 1/Q^4$ because of the photon propagator. The effect of the Z^0 exchange is to reduce the cross section at high Q^2 (by $\sim 25\%$ at $Q^2 = 10000$ GeV^2).

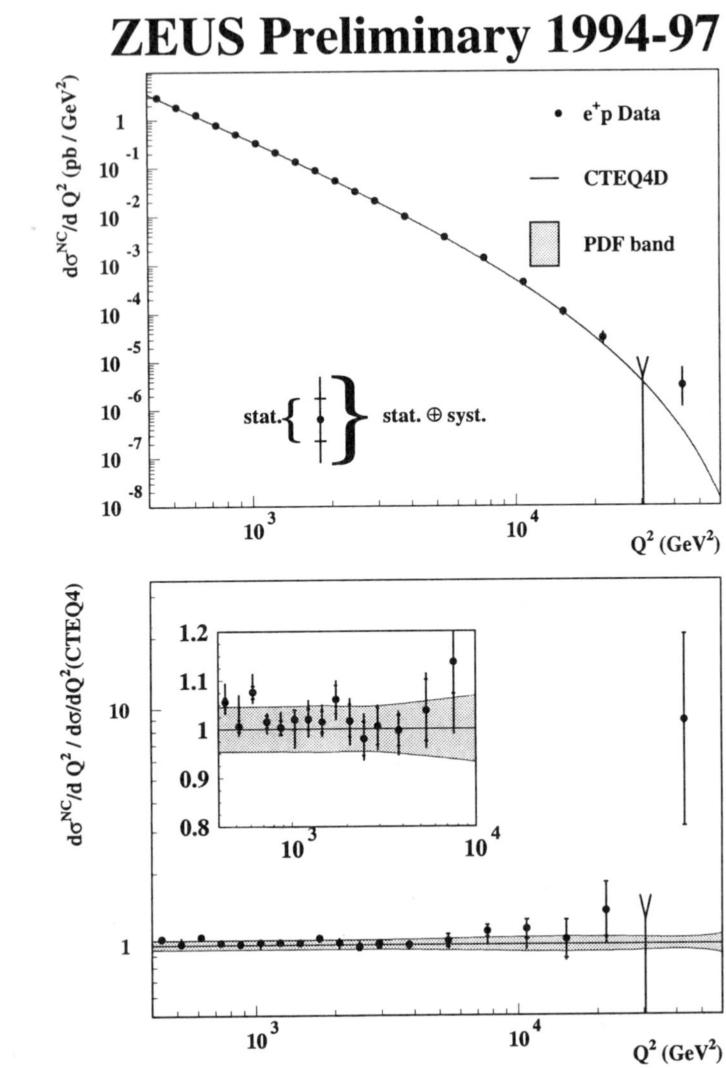

Figure 4. The NC DIS differential cross section $d\sigma/dQ^2$ for the data (points) and the SM expectation from the CTEQ4 PDF (line). The bottom plot shows the ratio of the measured cross section to SM prediction. The statistical errors are shown by the inner error bars and the full error bar is the sum in quadrature of the statistical and the systematic uncertainty. The uncertainty on the SM prediciton from the PDF is shown as shaded band.

2.2. Contact Interaction fits

A departure from the SM behavior of $d\sigma/dQ^2$ can be an indication of new physics. A wide class of hypothetical new interactions at the TeV mass scale (e.g. the exchange of a new neutral vector boson) can be parametrized by an effective four-fermion $eeqq$ contact interaction (CI). Its effect is to modify $d\sigma/dQ^2$ at high Q^2, where the interference between CI and SM can produce different behaviors, depending on the choice of the particular CI scenario.

A generic Lorenz-invariant CI lagrangian that conserves $SU(2) \times U(1)$ contains scalar, vector and tensor terms. We consider here only the vector terms that are less constrained by other experiments:

$$\mathcal{L}_{CI}^{V} = \frac{4\pi}{\Lambda^2} \sum_{f=u,d} \sum_{\alpha=L,R} \sum_{\beta=L,R} \eta^f_{\alpha\beta} (\bar{e}_\alpha \gamma_\mu e_\alpha)(\bar{q}^f_\beta \gamma^\mu q^f_\beta)$$

We assume that $\eta^f_{\alpha\beta}$ can be only zero, -1 or $+1$. Atomic parity violation measurements require that $\eta_{LL} + \eta_{LR} - \eta_{RL} - \eta_{RR} = 0$ reducing the number of possible combinations. 24 CI scenarios, summarized in table 2, are considered. For each scenario a binned log-likelihood fit to the measured Q^2 distribution is performed to obtain the best value of $1/\Lambda^2$ as in the examples of fig. 5. Figure 6 shows the $1/\Lambda^2$ value of the best fits, the 1σ error bars and the 95% C.L. upper limits. Limits from $\bar{p}p$ and e^+e^- experiments are also reported [5]. In some CI scenario the best fit is $\lambda = \infty$, i.e. the SM fits best. In the other cases the probability for observing the same or a lower value of Λ from the SM is also reported and is always larger than 20% with the exception of the scenario $X2^-$ that gives a SM probability of 4.5%. The corresponding lower limits on Λ range from 1.5 to 4.9 TeV depending on the scenario. In many cases ZEUS limits are competitive with other experiments.

2.3. Charged current DIS

The CC selection is based on the requirement of a large missing transverse momentum $\not{P}_t > 10$ GeV, due to the undetected neutrino. Special cuts are applied to reduce the backgrounds from low-Q^2 interactions, NC DIS, cosmic muons and beam-gas interactions. The sample is restricted to a region where the resolution on the kinematical variables is good and the background is small, by taking only events with a reconstructed vertex, $y < 0.9$ and $Q^2 > 400$ GeV^2. The background was estimated to be at the 1% level.

In the case of CC DIS the final lepton is undetected and therefore the kinematical variables must be obtained from the hadronic system. We used

Table 2. Classification of the CI scenarios. The value of a can be either +1 or -1.

Label	η^u_{LL}	η^u_{LR}	η^u_{RL}	η^u_{RR}	η^d_{LL}	η^d_{LR}	η^d_{RL}	η^d_{RR}
VV	$+a$	$+a$	$+a$	$+a$	$+a$	$+a$	$+a$	$+a$
AA	$+a$	$-a$	$-a$	$+a$	$+a$	$-a$	$-a$	$+a$
VA	$+a$	$-a$	$+a$	$-a$	$+a$	$-a$	$+a$	$-a$
XA	$+a$	$-a$	0	0	$+a$	$-a$	0	0
X2	$+a$	0	$+a$	0	$+a$	0	$+a$	0
X3	$+a$	0	0	$+a$	$+a$	0	0	$+a$
X4	0	$+a$	$+a$	0	0	$+a$	$+a$	0
X5	0	$+a$	0	$+a$	0	$+a$	0	$+a$
X6	0	0	$+a$	$-a$	0	0	$+a$	$-a$
U5	0	$+a$	0	$+a$	0	0	0	0
UA	$+a$	$-a$	0	0	0	0	0	0
U4	0	$+a$	$+a$	0	0	0	0	0

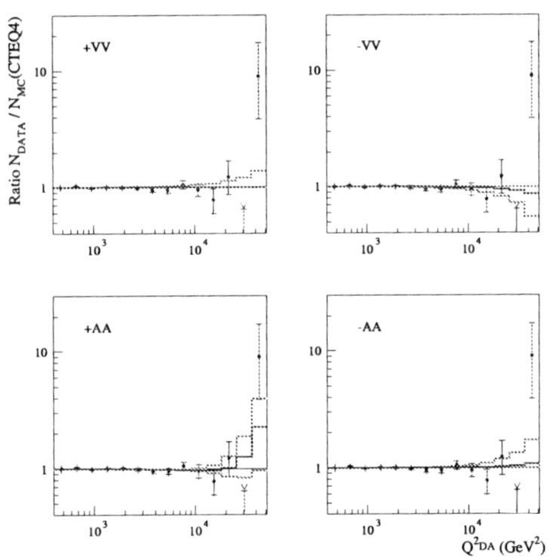

Figure 5. Examples of contact interaction fits for 4 different CI scenarios (labeled $+VV, -VV, +AA, -AA$). The points show the ratio of the measured dN/dQ^2_{DA} to the SM expectation (using CTEQ4 PDF). The central curve is the best fit to the CI scenario under study. The side curves show the $\pm 1\sigma$ limits.

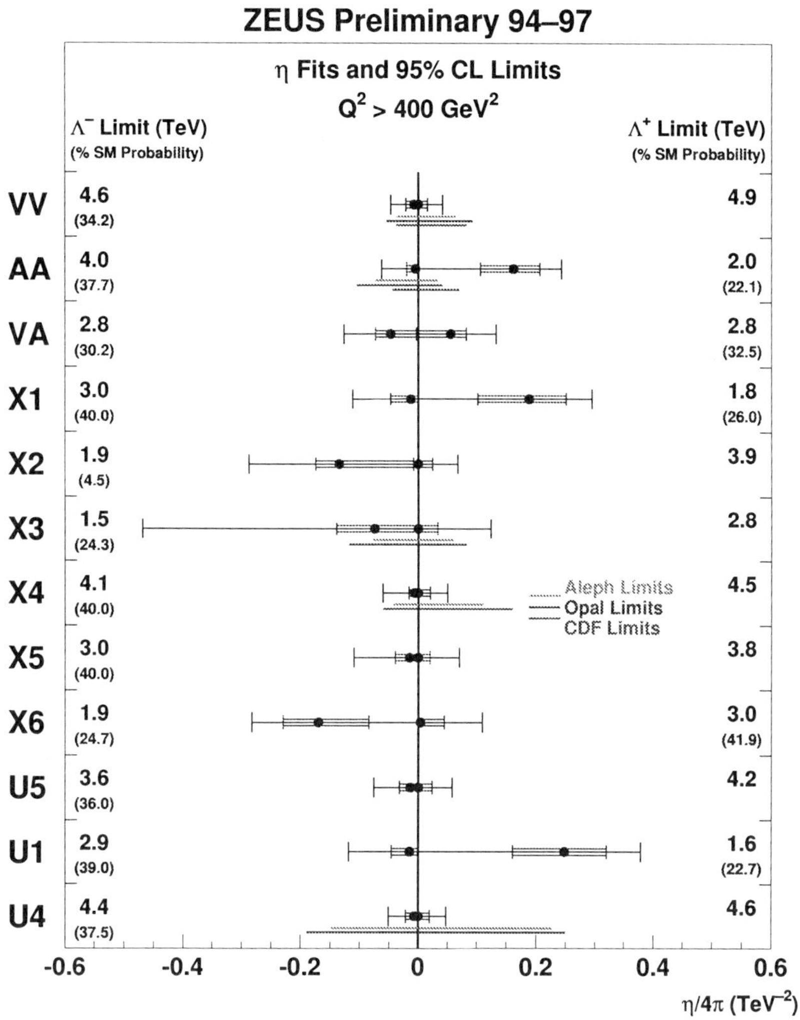

Figure 6. Results of the CI fit. Each row represents two scenarios, for $a = -1$ on the left and for $a = +1$ on the right. The dots are the most likely value of $1/\Lambda^2$ and are presented with 1σ errors represented as double error bars. Single error bars represent the 95% C.L. upper limits on $1/\Lambda^2$. The corresponding lower limit on Λ is reported on the left and right margins of the plot. If this limit is lower than ∞, the probability of obtaining a similar or smaller limit on Λ from SM is also reported in parenthesis. CI limits from CDF, ALEPH and OPAL are indicated as single lines where available.

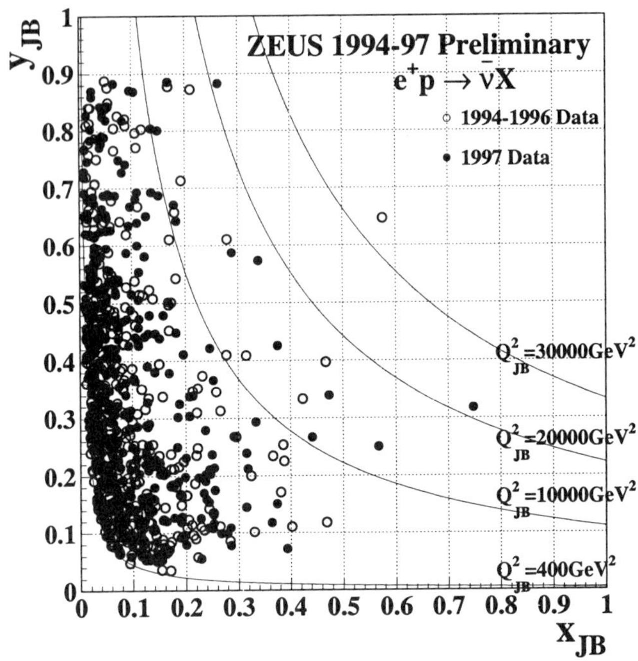

Figure 7. Distribution of the charged-current DIS events with $Q^2_{JB} > 400$ GeV^2 in the x_{JB}-y_{JB} plane. The events collected during the '94-'96 data taking are shown as empty circles, those from '97 as filled circles. Lines of constant Q^2_{JB} are also shown.

Table 3. Observed and expected number of CC DIS events with $Q^2_{\rm JB} > Q^2_{\rm min}$ from ZEUS preliminary 1994-'97 results.

$Q^2_{\rm min}$[GeV^2]	$N_{\rm Obs}$	$N_{\rm exp}$	
1000	586	600	± 52
10000	22	17	$^{+5.7}_{-5.2}$
15000	8	3.9	$^{+1.9}_{-1.6}$
20000	3	0.97	$^{+0.65}_{-0.47}$
30000	1	0.06	$^{+0.08}_{-0.04}$

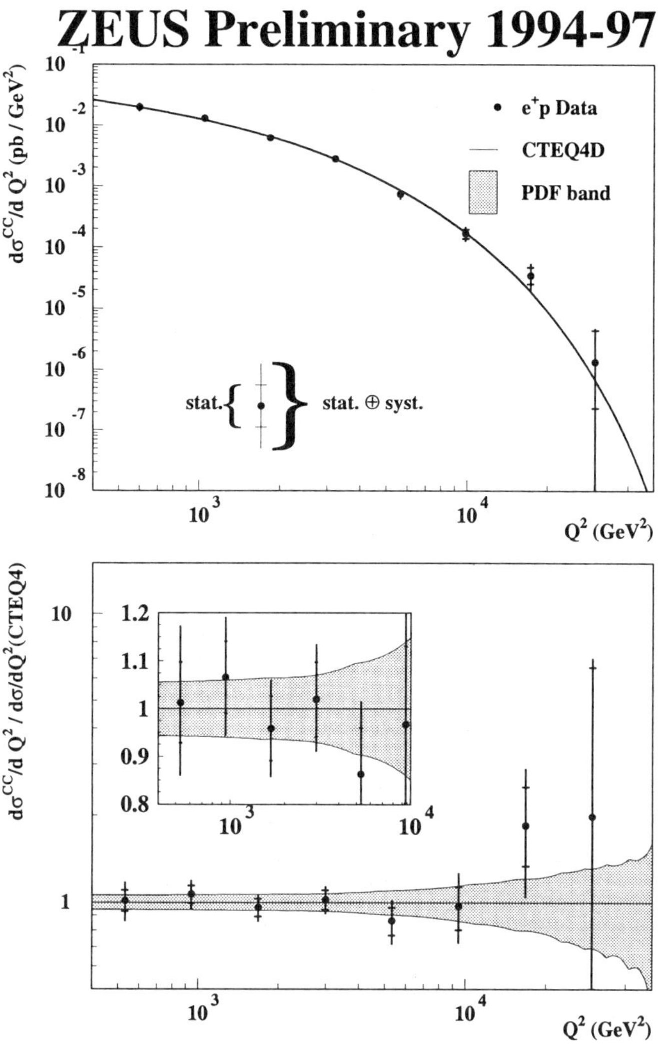

Figure 8. The CC DIS differential cross section $d\sigma/dQ^2$ for data (points) and the SM expectation using the CTEQ4 PDF (line). The bottom plot shows the ratio of the measured cross section to the SM prediction. The statistical errors are shown by the inner error bars and the full error bar is the sum in quadrature of the statistical and the systematic uncertainty. The uncertainty on the SM prediction from PDF's is shown as a shaded band.

the Jaquet-Blondel (JB) method:

$$y_{JB} = \frac{\delta}{2E_0}; \; Q^2_{JB} = \frac{\not{p}_t^{\,2}}{1 - y_{JB}},$$

corrected to obtain unbiased variables. It gives a resolution on Q^2_{JB} that ranges from RMS= 20% at low y to 50% at $y \simeq 0.9$. The distribution of the CC events in the $x_{JB} - y_{JB}$ plane is shown in fig. 7. The number of events observed above a given $Q^2_{\rm min}$ is compared with the SM expectation in table 3. The differential cross section $d\sigma/dQ^2$ is shown in figure 8 with the SM prediction obtained with CTEQ4 PDF's. The systematic uncertainty ranges from $\sim 10\%$ at low Q^2 to $\sim 100\%$ at $Q^2 \sim 30000$ GeV^2, where it is dominated by the 3% uncertainty on the calorimeter energy scale. A good agreement with SM is observed.

3. Limits on Lepton Flavor Violation

Lepton flavor is conserved in all the SM interactions. HERA is well suited to search for lepton flavor violation (LFV) because events of the type $ep \to lX$, with the high p_t lepton being a muon or a tau ($l = \mu, \tau$), are almost background free and easy to detect.

A possible scenario for LFV is the s- or u-channel exchange of an object of mass M_{LQ} with the quantum numbers of a leptoquark (LQ), which could couple to eq with strength λ_{eq} and decay into leptons and quarks of different generation with the branching ratio $B_{lq'}$. If the mass of the object is below the HERA center of mass energy, the production cross section is dominated by the resonant production:

$$\sigma_{eq \to lq'} = \frac{\pi}{4s} \lambda^2_{eq} B_{lq'} q(\frac{M^2_{LQ}}{s}, M^2_{LQ}) V$$

where V =1(2) for a scalar (vector) object. In the high-mass limit ($M_{LQ} >> \sqrt{s}$) both s- and u- channel contribute. In this case the cross section is proportional to

$$\sigma_{eq \to lq'} = \frac{s}{32\pi} \left(\frac{\lambda_{eq} \lambda_{lq'}}{M^2_{LQ}} \right)^2 F$$

where $\lambda_{lq'}$ is the coupling to the wrong flavor lepton and F is a factor that depends on the PDF's and on the u- or s- channel nature of the process. Similar processes are expected also in R-parity (R_p) violating supersymmetry, from the s-channel production of a squark through the λ' lepton-number-violating coupling and its decay into leptons and quarks of different generations:

$$eq^i \;\; \overset{\lambda'_{1ik}}{\rightarrow} \;\; \tilde{q}^k \;\; \overset{\lambda'_{ljk}}{\rightarrow} \;\; l^l q^j,$$

where generation indices are given.

A search for LFV was performed on 1993-94 data, corresponding to a total luminosity of 2.94 pb^{-1} of e^+p and 0.84 pb^{-1} of e^-p interaction. Events with high-p_t muons or taus are characterized by missing momentum in the calorimeter as muons leave only a small fraction of their energy in the calorimeter and tau decays are accompanied by neutrinos. The event selection requires large $\vec{\not{P}}_t$ aligned with an energy deposition compatible with a muon or a tau. Zero candidates were found.

In the low mass case, limits on $\lambda_{eq}^2 B_{lq'}$ as a function of M_{LQ} are derived. In the case of $e \to \mu$ processes, limits from low energy experiments are in general stronger, with the exception of the case of the production of a b quark ($ed \to \mu b$) that is better constrained by this analysis than by the $B \to \mu e$ decay for masses below $\sim$ 200 GeV. In the case of $e \to \tau$ the ZEUS limits are better than previous ones from tau and B decays for many kinds of LQ and squarks. Examples of upper limits on the production coupling are given in fig. 9 for leptoquarks (for $B_{lq'} = 0.5$) and in fig. 10 for squarks. The gauge decay of R_p-violating squarks to a quark and a photino ($\tilde{q} \to q\tilde{\gamma}$) can compete with the LFV decay $\tilde{q} \to lq'$. The squark branching ratio $B_{lq'}$ is evaluated under two extreme assumptions: no gauge decays allowed ($m_{\tilde{\gamma}} \geq m_{\tilde{q}}$) or a massless photino ($m_{\tilde{\gamma}} = 0$).

In the high-mass case limits on $\frac{\lambda_{eq}\lambda_{lq'}}{M_{LQ}^2}$ are obtained. For particular leptoquark kinds, in which couplings to heavy quarks are involved, the ZEUS limits are better than those from previous experiment. More details can be found in [6].

4. Events with leptons and missing transverse momentum

Events with high-p_t leptons in the final state and large missing transverse momentum $\not{P}_t$ are rare in the SM. They represent therefore an important tool for new physics searches. The SM process with the highest cross-section that presents this signature is the production of a W boson with its subsequent leptonic decay. The cross section for this process is expected to be $\sigma_{ep \to WX} \times B_{W \to l\nu} \sim 1 \times 0.1$ pb for each lepton, an extremely small value. This process is interesting because its cross section could be substantially increased by an anomalous triple gauge-boson coupling $WW\gamma$. Supersymmetry could be a source of events with leptons and $\not{P}_t$ as in the production of a selectron ($\tilde{e}$) and a squark ($\tilde{q}$) or in the production of an R_p violating stop $\tilde{t}$ with the following decay into $\tilde{b}W$.

The interest for the leptons and $\not{P}_t$ signal has been recently motivated also by the 5 events with a muon, a jet and $\not{P}_t$ found by the H1 collaboration [7]. Here we present two preliminary studies, one with the aim of evaluating the W production cross section and the other intended to set limits to the masses of selectrons and squarks.

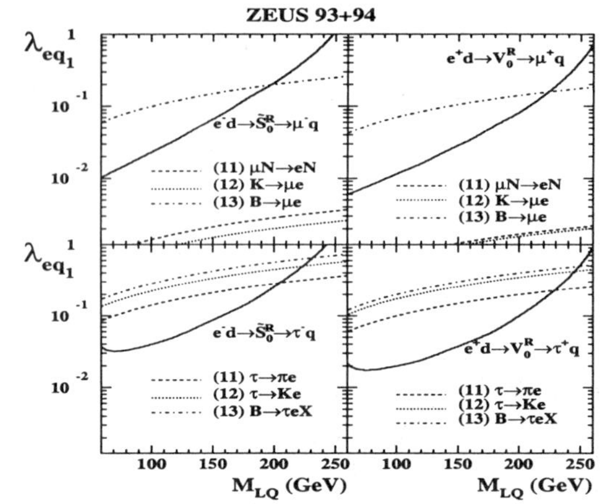

Figure 9. Upper limits at 95% CL on the coupling to eq for selected kinds of lepton flavor violating leptoquarks. The branching ratio to μq (above) and to τq (below) is assumed to be 50%. The full line is the ZEUS limit. Limits from previous experiments involving decay to quarks of the first (11), second (12) and third (13) generation are also shown.

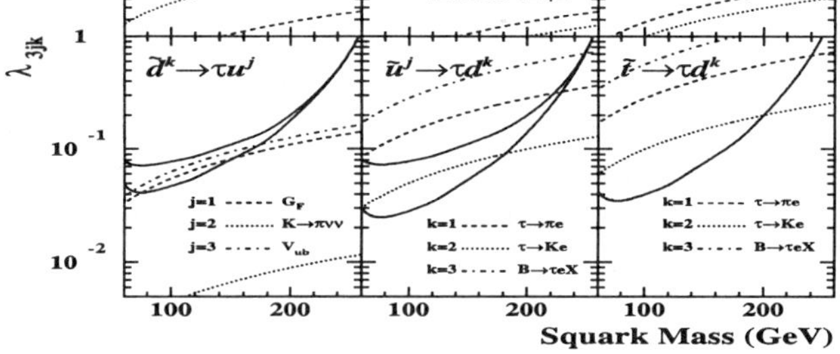

Figure 10. Examples of limits on the λ'_{ljk} coupling from R_P-violating squarks ($\tilde{q}$) decaying into muons (top) or taus (bottom). The full lines show the ZEUS limit for $m_{\tilde{\gamma}} \geq m_{\tilde{q}}$ (lower line) and for $m_{\tilde{\gamma}} = 0$. Production and decay couplings are assumed to be equal $\lambda'_{1ik} = \lambda'_{ljk}$. Limits from other experiments are also shown.

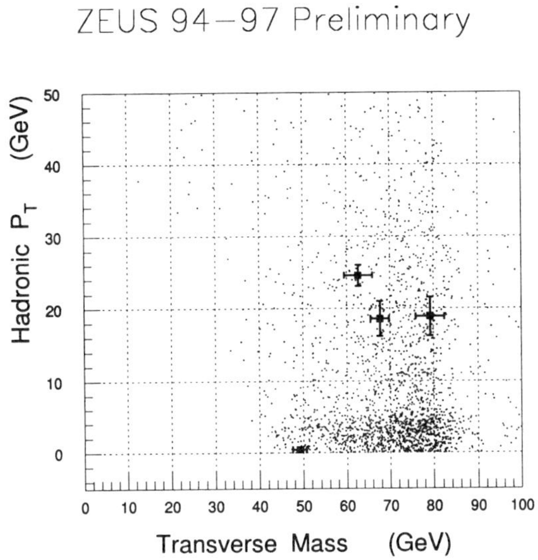

Figure 11. Distribution of the four $W \to e\nu$ candidates in the plane defined by the transverse mass M_T and by the transverse momentum of the associated hadronic system P_T^X. The error bars represent measurement uncertainties. The small points are $W \to e\nu$ events generated with the EPVEC Monte Carlo generator.

Table 4. ZEUS preliminary '94-'97 $W \to e\nu$ candidates.

candidate	1	2	3	4
Electron p_t (GeV)	24.7 ± 1.2	47.6 ± 1.9	44.4 ± 1.9	36.8 ± 1.6
Electron polar angle	37°	58°	70°	54°
Hadronic P_t (GeV)	0.4 ± 0.2	18.9 ± 2.7	24.6 ± 1.5	18.6 ± 2.4
Missing P_t (GeV)	24.3 ± 1.2	33.8 ± 2.4	22.7 ± 2.1	32.6 ± 1.4
Transverse mass (GeV)	49.0 ± 1.6	79.2 ± 3.3	62.3 ± 3.3	67.8 ± 2.2
Track p_t (GeV)	$22.3^{+6.8}_{-4.2}$	44^{+33}_{-13}	43^{+18}_{-10}	36^{+14}_{-8}
Track Charge	+1	+1	+1	+1

4.1. $W \to l\nu$

The selection for the $W \to e\nu$ decay requires an isolated electron, energy $E_e > 15$ GeV, in the angular region ($15^o < \theta < 74^o$), not back to back with the hadronic system (i.e. acoplanarity $> 17^o$) and a total missing momentum $\not{P}_{\rm t} > 19$ GeV. The selection efficiency is $\sim 40\%$. Four events survive this selection from the whole 1994-'97 data set and this number is compatible with the SM expectation of 2.2 events from W production and 1.2 from background sources (0.3 from NC DIS, 0.65 from CC DIS and 0.3 from lepton-pair production in $\gamma\gamma$ interactions). The properties of these events are shown in table 4.

The transverse mass $M_T = \sqrt{(|\vec{p_t^e}| + |\vec{\not{P}}_{\rm t}|)^2 - (\vec{p_t^e} + \vec{\not{P}}_{\rm t})^2}$ distribution is expected to show a Jacobian peak at $M_T = M_W$ for W events. The distribution of the $W \to e\nu$ candidates in the plane defined by M_T and by the transverse momentum of the hadronic system P_t^X is compared with the distribution of W events generated with the EPVEC[8] Monte Carlo in fig. 11. The distribution is compatible with W events. In three events the value of P_t^X is somewhat large if compared to the expectation for W events. An upper limit at the 95% CL on the W production cross section is derived from this channel for any P_t^X and with the condition $P_t^X > 30$ GeV:

$$\begin{aligned} \sigma_{ep\to WX} &<_{95\%CL} 3.9\text{pb}, \\ \sigma_{ep\to WX}(P_t^X > 30) &<_{95\%CL} 1.9\text{pb}. \end{aligned}$$

A similar selection was made for the muon channel by requiring a muon (identified by a track matching a MIP in the calorimeter), calorimeter missing momentum $\not{P}_{\rm t} > 18$ GeV and missing momentum corrected with the muon track momentum $\not{P}_{\rm t}^{\rm corr} > 18$ GeV.

Since the muon leaves only a small amount of energy in the calorimeter, events with small P_t^X would have small detectable transverse energy. For this reason the efficiency that is $30 - 40\%$ for $P_t^X > 20$ GeV drops to zero at $P_t^X < 20$GeV. Zero events have been observed, in agreement with the 1.3 expected from SM (0.5 from W production and 0.8 from CC and $\gamma\gamma$ backgrounds). Cross section upper limits are derived for $P_t^X > 20$ GeV and for $P_t^X > 30$ GeV:

$$\begin{aligned} \sigma_{ep\to WX}(P_t^X > 20) &<_{95\%CL} 2.0\text{pb}, \\ \sigma_{ep\to WX}(P_t^X > 30) &<_{95\%CL} 1.5\text{pb}. \end{aligned}$$

The last limit should be compared with 3 events with $P_t^X > 40$GeV in 24.7 pb^{-1} observed by H1 with an efficiency of 60%, that would lead to a cross section of $\sigma_{ep\to WX}(P_t^X > 40) \sim 2$ pb.

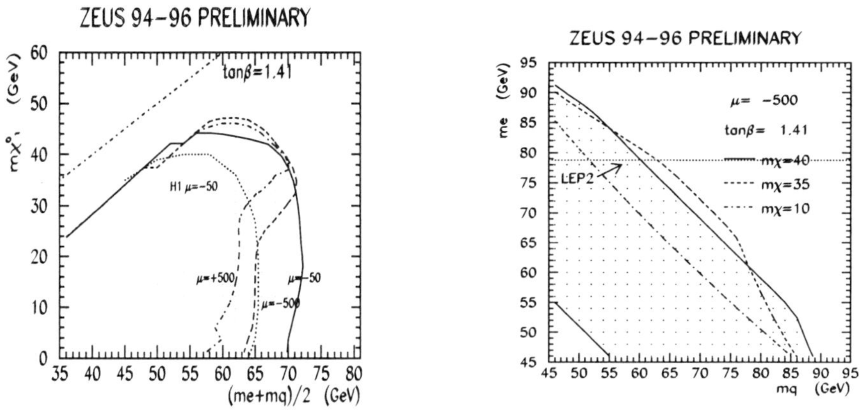

Figure 12. Excluded regions for the masses of MSSM particles. The left plot shows the excluded area in the plane $m_{\chi_1^0}$ vs. $(m_{\tilde{e}} + m_{\tilde{q}})/2$ for three different values of μ (-500, -50 and +500 GeV). The straight line at the top left corner shows where the mass of χ_1^0 is equal to $(m_{\tilde{e}} + m_{\tilde{q}})/2$. H1 results from 1994-95 data are also shown. The right plot shows the excluded area in the plane $m_{\tilde{q}}$ - $m_{\tilde{e}}$ for fixed values of $m_{\chi_1^0}$ (10, 35 and 40 GeV) for $\mu = -500$ GeV. LEP2 results from 161-172 GeV runs are also shown. The two plots are given for $\tan\beta = 1.41$ and $M_2 = 2M_1$.

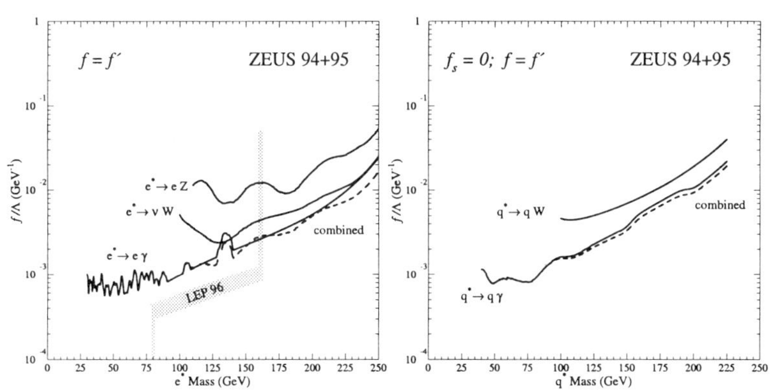

Figure 13. 95% C.L. upper limits on the excited electron (left) and the excited quark (right) coupling f/Λ as a function of the mass. The limits obtained from different decay channels (full lines) and from their combination (dashed lines) are reported. e^* LEP limits are also reported.

4.2. Selectrons and Squarks

In the Minimal Supersymmetric Standard Model (MSSM) with R-parity conservation supersymmetric particles can only be produced in pairs. The lowest order process in ep is the production of a selectron and a squark ($ep \to \tilde{e}\tilde{q}X$). This process is mediated by the t-channel exchange of a neutralino (χ_i^0). The direct decays into the lightest neutralino were sought: $\tilde{e} \to e\chi_1^0$ and $\tilde{q} \to q\chi_1^0$. The χ_1^0, which is assumed to be the Lightest Supersymmetric Particle (LSP), escapes detection giving a the typical signature of large $\not{P}_{\rm t}$ and low δ, accompanied by an electron and a jet. Events of this kind were searched for in the 1994-'96 data[2]. Two events of this kind were found, in agreement with a SM expectation of 2.7 (mainly from $W \to e\nu$, NC DIS, CC DIS and heavy flavors). Excluded areas in the MSSM parameter space were derived. In the MSSM, the $ep \to \tilde{e}\tilde{q}$ cross section depends on the sum of the masses of the produced scalars $m_{\tilde{e}} + m_{\tilde{q}}$ and on the parameters that determine the neutralino masses and couplings: the zino and bino mass terms M_2 and M_1 and the Higgs parameters μ and $\tan\beta$. For $|\mu| >> M_2 > M_1$, the cross section depends weakly on the parameters μ, M_2, M_1 and $\tan\beta$ while depends steeply on $m_{\tilde{e}} + m_{\tilde{q}}$. Figure 12 shows the excluded areas in the plane defined by $(m_{\tilde{e}} + m_{\tilde{q}})/2$ and $m_{\chi_1^0}$ and in the plane $m_{\tilde{e}} - m_{\tilde{q}}$. The following simplifying assumptions are made: $M_2 = 2M_1$, the mass of the scalar partners of the left and right-handed electrons are equal (i.e. $m_{\tilde{e}_L} = m_{\tilde{e}_R} = m_{\tilde{e}}$) and all the squarks have the same mass (except for stop) $m_{\tilde{q}}$. If the latter assumption is released and only the $\tilde{u}$ mass is assumed to be small enough to contribute to the process, the excluded areas are reduced by only ~ 2 GeV, because the u quark dominates the PDF at high x.

These limits extend the published LEP limits on the selectron mass from $\sqrt{s} = 161 - 172$ GeV runs for light squarks as shown in fig. 12. The strong limits on the squark mass obtained at Tevatron are based on the GUT relation that links the LSP and the gluino mass. No $\bar{p}p$ limits are available if this relation is relaxed, allowing the gluino to be heavy ($m_g > 500GeV$) and keeping the mass difference between the squarks and the LSP small (for example $m_{\tilde{q}} = 50$ GeV, $m_{\chi_1^0} = 40$ GeV).

5. Search for excited fermions

The existence of high mass excited fermions would provide a clear evidence for substructure in SM fermions. Excited electrons (e^*), neutrinos (ν^*) and quarks (q^*) could be produced by the t-channel exchange of a gauge boson and subsequently detected through their decay into the corresponding

[2] At the time of writing this analysis is superseded by the final '94-'97 analysis [9]

fermion and a gauge boson. A general way to model the production and the decay of excited fermions, without specifying the dynamics of the compositeness, is through an effective interaction lagrangian that describes the interaction of ordinary left-handed fermions, spin-$\frac{1}{2}$ right-handed excited states and SM gauge fields:

$$\mathcal{L}_{f\to f^*} = \frac{1}{\Lambda}\bar{f}^*_{\rm R}\sigma^{\mu\nu}\left[gf\frac{\tau^{\rm a}}{2}\mathbf{W}^{\rm a}_{\mu\nu} + g'f'\frac{Y}{2}B_{\mu\nu} + g_{\rm s}f_{\rm s}\frac{\lambda^{\rm a}}{2}\mathbf{G}^{\rm a}_{\mu\nu}\right]f_{\rm L},$$

where $\mathbf{W}^{\rm a}_{\mu\nu}$, $B_{\mu\nu}$ and $\mathbf{G}^{\rm a}_{\mu\nu}$ are the field-strength tensors, $\tau^{\rm a}$, Y and $\lambda^{\rm a}$ are the gauge structure constants and g, g', g_s are the coupling constants SM gauge fields. The coupling to the gauge fields $SU_{\rm W}(2)$, $U_{\rm Y}(1)$ and $SU_{\rm c}(3)$ is determined by the parameters f ,f' and f_s and by compositness scale Λ.

The present results are from the published analysis of 1994-'95 e^+p data [10]. Eight different decay topologies were selected:

1. $e\gamma$ from $e^* \to e\gamma$,
2. $eq\bar{q}$ from $e^* \to eZ$,
3. $\nu q\bar{q}$ from $e^* \to \nu W$ and $\nu^* \to \nu Z$,
4. $\nu\gamma$ from $\nu^* \to \nu\gamma$,
5. $e\nu\bar{\nu}$ from $e^* \to eZ$ and $e^* \to \nu W$,
6. $ee\nu$ from $\nu^* \to eW$,
7. $q\gamma$ from $q^* \to q\gamma$,
8. $qe\nu$ from $q^* \to qW$.

The excited fermion mass was reconstructed from the selected candidates, searching for a resonance signal. As no significant deviation from the SM background mass spectra was found, limits are set on the excited fermion couplings as a function of the mass.

In figure 13 the upper limits on f/Λ for excited electrons and for excited quarks, assuming $f = f'$ and $f_s = 0$, are reported. The limits on e^* from LEP at $\sqrt{s} = 161$ GeV are superimposed, showing that ZEUS limits improve LEP ones at high energy. Tevatron limits on excited quarks rely on the quark-gluon fusion production mechanism and therefore they hold only for $f_s \neq 0$ while HERA limits are valid also for $f_s = 0$ due to the purely electroweak production mechanism.

6. Conclusions and outlook

The preliminary analysis of the ZEUS 1994-97 e^+p data shows a general agreement with SM predictions. A tendency of data to be above the SM is observed in DIS at $Q^2 > 30000$ GeV^2. The limits derived for many hypothetical processes beyond the SM are competitive with those obtained in other experiments. In 1998-'99 HERA will collide e^-p (instead of e^+p) extending the sensitivity to particular processes (like particular LFV or CI scenarios). Searches for very-low cross section processes and a more precise tests of the SM will be possible after the high-luminosity upgrade foreseen for the year 2000.

References

[1] G. Schuler and H. Spiesberger, in *Physics at HERA*, vol. 3 (1991) 1419.

[2] H. Lai et al.,*Phys. Rev.* **D** (1997). 1280.

[3] ZEUS collab., M. Derrick et al. , *Phy. Rev. Lett.* **75** (1995) 1006.

[4] ZEUS collab., J. Breitweg et al., *Z. Phys.* **C74** (1997) 207.

[5] OPAL collab. K.Ackerstaff et al., *Eur. Phys. J.* **C2** (1998) 441; CDF collab. F.Abe et al., Phys. Rev. Lett **79** (1997) 2198.

[6] ZEUS collab., M. Derrick et al.,*Z. Phys.* **C73** (1997) 613.

[7] H1 collab., C. Adolff et al., DESY 98-063, submitted to *Eur. Phys. J.*.

[8] U. Baur, J.A.M. Vermaseren and D. Zeppenfeld, *Nucl. Phys.* **B 375**(1992)3.

[9] ZEUS collab., J. Breitweg et al., DESY 98-069 , submitted to *Phys. Lett. B.*

[10] ZEUS collab., J. Breitweg et al. ,*Z. Phys.* **C76** (1997) 631.

Electroweak Baryon- And Lepton-Number Violating Processes At High Energies By The Valley Method

Masatoshi SATO

Department of Physics, University of Tokyo, Tokyo 113-0033, Japan

Abstract. To obtain reliable amplitudes of the electroweak baryon- and lepton-number violating processes at high energies, new non-perturbative techniques are required. We report on a recent development of the non-perturbative techniques based on the optical theorem and the valley method. A method of the separation of the baryon- and lepton-number violating processes from the preserving ones in the optical theorem is suggested.

Electroweak baryon- and lepton-number violating processes [1] induced by collisions of highly energetic particles [2, 3, 4] remain an object of controversial discussion. In naive sphaleron [6] picture, it is expected that the processes are enhanced at the sphaleron energy $E_{\rm sph}$ ($\sim O(10)$ TeV) and observed in future colliders. On the other hand, there are arguments which show that the instanton-like transition rates are likely exponentially small at all energies [7, 8, 9]. In spite of much effort [10], the final answer has not been found yet. The computation of the process requires new non-perturbative techniques.

The purpose of this article is to report a recent development of the non-perturbative technique based on the optical theorem and the valley method. We will suggest a method of the separation of the baryon- and lepton-number violating processes from the preserving ones in the optical theorem. The method is based on the Euclidean path-integral formalism and the real configuration. (For the development based on the Minkowski path-integral formalism, see Ref.[11].)

First, we will briefly review difficulties of computing reliable amplitudes of the baryon- and lepton-number violating processes. The standard technique for computing the tunneling effects in the quantum field theories is

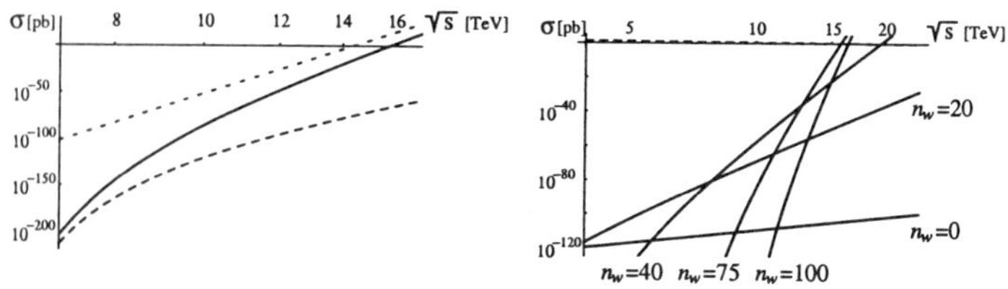

Figure 1. The cross section of the baryon- and lepton-number violating process computed by the instanton [5]. The right figure denotes the energy dependence of the cross section for $n_{\rm w} = 75$, and the left one denotes the dependence on the number of W-boson in the final state. We have set $n_{\rm f} = 3$, $n_{\rm h} = 0$, $m_{\rm w} = 80{\rm GeV}$, $g^2 = 0.42$, $m_{\rm h} = 200{\rm GeV}$ and $m_{\rm top} = 180{\rm GeV}$. The other fermion masses except top-quark have been neglected.

the perturbation about the classical solution - the instanton calculus. By use of the instanton, quantitative studies of the electroweak baryon- and lepton-number violating process were started in Refs.[3, 4]. The instanton induces the following baryon- and lepton-number violating process

$$\rm q + q \to (3n_f - 2)\bar{q} + n_f\bar{l} + n_w W + n_h H, \tag{1}$$

where W represents gauge boson, $W^{\pm}$ and Z, and H represents the Higgs particles. The number of boson, $n_{\rm w}$ and $n_{\rm h}$, is arbitrary. In Fig.1, we show the latest result [5] of the cross section computed by the instanton calculus. These lead to the picture of exponentially increasing total cross sections at a relatively low center of mass energies, $\sqrt{s} \ll E_{\rm sph}$.

The perturbation about the instanton, however, breaks down in the most interesting region, $\sqrt{s} \sim E_{\rm sph}$ [3]. For the kinematic reason, the dominant process for this region is accompanies with the huge number of the final boson, $n_{\rm w}, n_{\rm h} \sim O(1/g^2) \sim O(100)$. The interaction between the boson is too complicated to evaluate by the perturbation theory about the instanton.

There are another problem in the instanton calculus. Large anomalous quantum corrections of the initial hard particles exist in the background of the instanton [12, 13], and they also make the perturbation about the instanton meaningless. This anomalous behavior comes from a zero-mode in the background of the instanton [14]. Due to the existence of the zero-mode, the high-energy behavior of the propagator in the background of the instanton is different from the usual one.[1]

[1] Strictly speaking, there is a subtlety in this problem. The instanton in the elec-

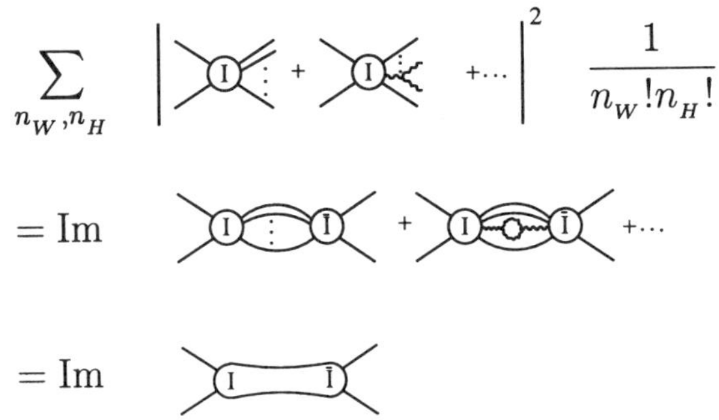

Figure 2. The optical theorem for the baryon- and lepton-number violating processes. The complicated interaction of the final particles (in the first line) is taken into account by the deformed instanton–anti-instanton pair (in the third line).

An approach to the complicated interaction between the bosonic particles in the final state is to consider the optical theorem, i.e. to calculate the total cross section for the baryon- and lepton-number violating processes via the imaginary part of the forward elastic scattering amplitude [17, 18]. In this approach, the complicated interactions between bosonic particles are automatically taken into account.

As well as the usual perturbation theory, the optical theorem holds in the instanton calculus. A schematic explanation of the optical theorem is given by Fig.2. (For more details, see Ref.[19].) It shows that the instanton–anti-instanton system is a relevant configuration of the baryon- and lepton-number violating processes, and the total cross section $\sigma_{\not{B}+\not{L}}$ can be computed by the following relation,

$$\sigma_{\not{B}+\not{L}} = \frac{1}{s} \mathrm{Im} \int_{A=A_{\mathrm{I}\bar{\mathrm{I}}}} \mathcal{D}A\mathcal{D}\psi\mathcal{D}\bar{\psi}\psi(p_1)\psi(p_2)\bar{\psi}(p_1)\bar{\psi}(p_2)e^{-S_{\mathrm{SM}}}, \tag{2}$$

where A denotes the gauge or Higgs fields, and ψ denotes the quark or lepton fields. $A_{\mathrm{I}\bar{\mathrm{I}}}$ indicates the instanton–anti-instanton system.[2] The above formula (2) reproduces the result of the standard instanton calculus in the region of $\sqrt{s} \ll E_{\mathrm{sph}}$ [18].

troweak theory is not a classical solution except a zero-radius one [15], thus the existence of the zero-mode depends on a way of the construction of the instanton. Indeed, for the valley-instanton [16], which is the relevant instanton to the high energy processes, the zero-mode does not exist in general. But at least for the small-radius one, we find that the anomalous behavior remains.

[2] For simplicity, we have neglected the multi-instanton effects discussed in Refs.[20, 21, 22]. They can be taken into account by the suitable extension of the method in Section 4 in Ref.[28].

The construction of the instanton–anti-instanton system is not trivial. Unless the distance between the instanton and the anti-instanton is infinite, the instanton–anti-instanton system is not a classical solution. The proper definition of the instanton–anti-instanton system is given by the valley method [23, 24, 25]. The instanton–anti-instanton system forms a valley in the configuration space.

Although the optical theorem was originally introduced to take into account the complicated interaction of the final boson, we would like to point out that it also resolves the problem of the initial hard particles. This is because that the zero-mode disappear in the relevant configuration for $\sqrt{s} \sim E_{\rm sph}$. When $\sqrt{s} \sim E_{\rm sph}$, the instanton–anti-instanton system dominating the path-integral (2) collapses due to the attractive force between them, and the separation between the instanton and the anti-instanton becomes comparable to the electroweak scale $1/m_{\rm W}$. The configuration is almost the fluctuation about the vacuum, thus the bound state, especially the zero-mode causing the anomalous behavior does not exist in the background.[3]

Unfortunately, a new difficulty appears in the above approach, which have prevented us from computing reliable amplitudes for the baryon- and lepton-number violating processes at $\sqrt{s} \sim E_{\rm sph}$. Since the instanton–anti-instanton system is smoothly deformed to the vacuum, the path-integral (2) contains the baryon- and lepton-number preserving processes also. The distinction of the baryon- and lepton-number violating processes from the preserving ones is not trivial. On the contrary to the standard instanton calculus, the topological charge is useless to separates these two processes.

Together with H.Aoyama, H.Kikuchi, I.Okouchi and S.Wada, the author recently examined the above problem in a simplified model, and found that the separation of the baryon- and lepton-number violating processes from the baryon- and lepton-number preserving processes is possible [27, 28].

Our idea is based on the investigation of the large order behavior of the perturbation theory [29]. In perturbation theory, the physical quantity O is expressed as

$$O = \sum_n C_n g^{2n}, \tag{3}$$

where g^2 is the coupling constant of the theory. As was first denoted by Dyson [30], the perturbative series is the asymptotic expansion in general, and it diverges as follows,

$$C_n \overset{n\to\infty}{\sim} n!. \tag{4}$$

[3] The vanishing of the bound state was pointed out in Ref.[9, 26] from the different context.

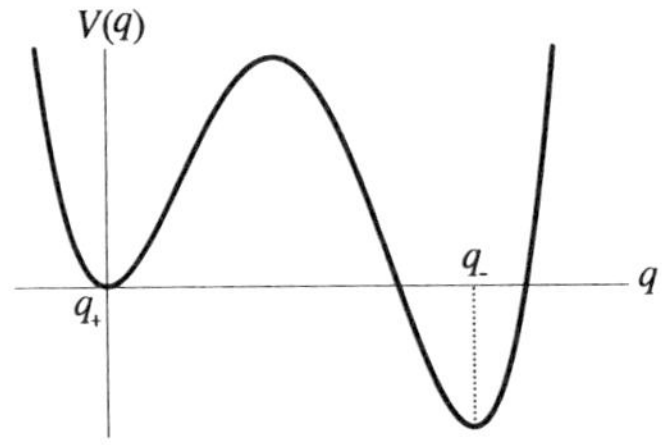

Figure 3. The asymmetric double-well potential $V(q)$ defined by Eq.(6) for $\epsilon = 5$ and $g = 0.1$.

Therefore, a straightforward summation of the perturbation series is impossible. A useful technique of the summation is the Borel-resummation: Rewrite formally the perturbative series as follows,

$$\begin{aligned} O &= \sum_n \int_0^\infty dt \frac{C_n}{n!} t^n e^{-t/g^2}, \\ &= \int_0^\infty dt \left(\sum_n \frac{C_n}{n!} t^n \right) e^{-t/g^2} \\ &\equiv \int_0^\infty dt F(t) e^{-t/g^2}. \end{aligned} \tag{5}$$

Here we have introduced the Borel transformed $F(t)$ of the perturbative series. While the summation of the perturbative series (3) is impossible for any non-zero value of g^2, the summation in the second line of the above equation is possible for small enough t [31]. In theories with tunneling effects, it is known that $F(t)$ has a singularity on the real positive axis of t. This singularity is a signal of the the tunneling effects in the perturbative series.

The model we study is a one-dimensional system with coordinate q, with the following Hamiltonian;

$$\begin{aligned} H &= -\frac{1}{2}\frac{d^2}{dq^2} + V(q), \\ V(q) &= \frac{1}{2} q^2 (1 - gq)^2 - \epsilon g q, \end{aligned} \tag{6}$$

where $g(> 0)$ is a coupling constant and $\epsilon(> 0)$ is a parameter which represents the asymmetry of the potential. For $\epsilon g^2 < \sqrt{3}/18$, this potential has two vacua $q = q_\pm$. (See Fig.3.) The tunneling effect between the two vacua corresponds to the baryon- and lepton-number violating processes in the electroweak theory. If we rewrite the above model with the fermion degree of freedom, the correspondence becomes clearer. By use of the

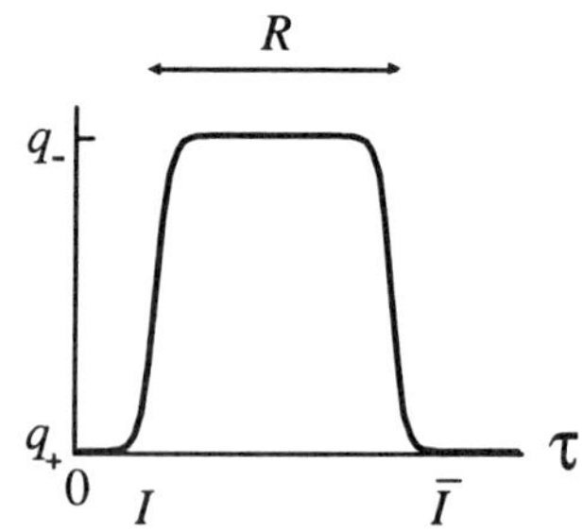

Figure 4. The instanton–anti-instanton configuration

following fermion ψ,

$$\psi = \begin{pmatrix} 0 & 0 \\ 1 & 0 \end{pmatrix}, \quad \psi^\dagger = \begin{pmatrix} 0 & 1 \\ 0 & 0 \end{pmatrix}, \tag{7}$$

the model (6) is equivalently rewritten by the following Lagrangian,

$$L = \frac{1}{2}\dot{q}^2 - \frac{1}{2}q^2(1 - gq)^2 + i\psi^\dagger\dot{\psi} + \epsilon q\{\psi^\dagger, \psi\}. \tag{8}$$

This Lagrangian describes the $0 + 1$ dimensional similitude of the electroweak theory. In addition, this model has a generalized supersymmetry when ϵ is a positive integer [28]. The non-renormalization theorem of the supersymmetry makes clear the relation between the perturbative effects and the non-perturbative effects.

When $\epsilon = 0$, the system (6) has the instanton solution,

$$q(\tau) = \frac{1}{g}\frac{1}{1 + e^{-\tau}}. \tag{9}$$

When $\epsilon \neq 0$, the configuration connecting the two different vacua does not exists as a classical solution, but it exists as a solution of the valley equation [32]. This configuration is the instanton in the case of $\epsilon \neq 0$.

Now let us evaluate the transition amplitude from the left vacuum $q = q_+$ to itself

$$Z = \lim_{T\to\infty} \langle q_+ | e^{-HT} | q_+ \rangle, \tag{10}$$

in the background of the instanton–anti-instanton system. For simplicity, we focus mainly on the case of $\epsilon = 0$. Except for the collective coordinates of the instanton–anti-instanton system, the path integral can be performed by Gaussian integrals. Then we obtain

$$Z = \lim_{T\to\infty} \text{const.} \times \frac{1}{\pi g^2} \int_0^T dR (T - R) e^{-S(R)/g^2}, \tag{11}$$

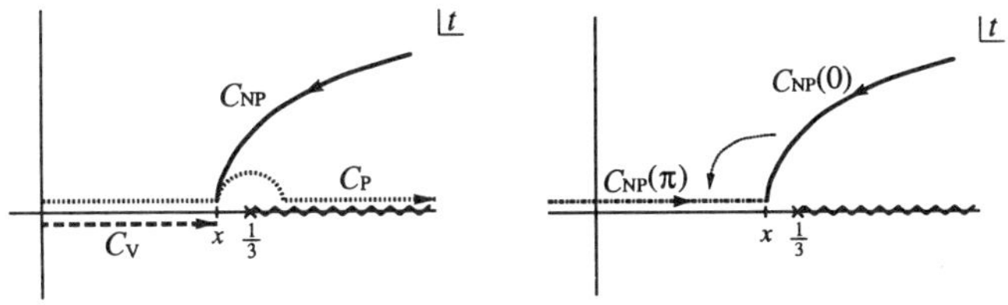

Figure 5. The left figure denotes that deformation of the contour C_V to the sum of C_P and C_{NP}. The right one denotes that the change of the contour $C_{NP}(\theta)$ of $Z_{NP}(|g^2|e^{i\theta})$ as θ is changed from zero to π.

where R is the collective coordinate corresponding to the relative distance between the valley-instanton and the anti-valley-instanton (see Fig.4), and $S(R)/g^2$ is the action of the instanton–anti-instanton system. The action $S(R)$ behaves as

$$S(R) = \begin{cases} \sigma R^2 & \text{for } R \to 0\,; \\ \dfrac{1}{3} - 2e^{-R} & \text{for } R \to \infty, \end{cases} \tag{12}$$

where σ is a constant [28]. The configuration at $R = 0$ is the left vacuum $q = q_+$, and at $R \sim \infty$ the well-separated instanton–anti-instanton pair. As well as the path-integral (2), Eq.(11) contains both the perturbative contribution at $R \sim 0$ and the non-perturbative one at $R \sim \infty$.

To separate these contributions, we change the integration variable R to $t = S(R)$. The amplitude (11) becomes

$$Z = \lim_{x \to 1/3} \text{const.} \times \frac{1}{\pi g^2} \int_{C_V} dt F(t)\, e^{-t/g^2}, \tag{13}$$

where $x = S(T)$, $C_V = [0, x]$, and $F(t)$ is the Jacobian. From Eq.(12), it is found that the Jacobian behaves as

$$F(t) = \frac{f(t)}{\sqrt{4\sigma t}\,(1/3 - t)} \ln\left(\frac{1/3 - t}{1/3 - x}\right), \tag{14}$$

where $f(t)$ is a function that satisfies

$$f(0) = -\frac{T}{3\ln(1 - 3x)}, \quad f(1/3) = \sqrt{\frac{4\sigma}{3}}. \tag{15}$$

Note that there is a significant resemblance between Eq.(13) and the formal Borel-summation of the perturbation series (5). The only difference is the contour of the integral. Deforming the contour C_V to the sum of C_P and

C_{NP} as is shown in the left-hand side of Fig.5, the amplitude becomes

$$\begin{aligned} Z &= \lim_{x\to 1/3} \text{const.} \times \frac{1}{\pi g^2} \int_{C_{\mathrm{P}}} dt F(t)\, e^{-t/g^2} \\ &+ \lim_{x\to 1/3} \text{const.} \times \frac{1}{\pi g^2} \int_{C_{\mathrm{NP}}} dt F(t)\, e^{-t/g^2}, \end{aligned} \tag{16}$$

so *we identify that the first term of Eq.(16) is the formal Borel-summation of the perturbative series and second term is the non-perturbative contribution.* The singularity of $F(t)$ at $t = 1/3$ indicates the non-Borel-summability of the perturbation series. We denote the first term of Eq.(16) as Z_{P} and the second term as Z_{NP}.

The physical meaning of Z_{NP} becomes evident if we perform the analytic continuation in the coupling constant g^2. Let us denote the contour of $Z_{\mathrm{NP}}(g^2)$ for $g^2 = |g^2|e^{i\theta}$ as $C_{\mathrm{NP}}(\theta)$. If θ is changed from zero to π, the contour $C_{\mathrm{NP}}(0)$ changes to $C_{\mathrm{NP}}(\pi)$ (see the right-hand side of Fig.5) and in the weak coupling limit, $C_{\mathrm{NP}}(\pi)$ can be safely replaced with C_{V}. Thus $Z_{\mathrm{NP}}(g^2)$ can be evaluated by the instanton–anti-instanton system, but contrary to the case of $g^2 > 0$, this instanton–anti-instanton system never collapse to the vacuum, since the force between instanton and anti-instanton is repulsive for $g^2 = |g^2|e^{i\pi} < 0$.

We have tested in Ref.[28] the above identification of the non-perturbative contribution by comparison with

1. WKB result,
2. large order behaviors of the perturbation theory,
3. numerical solutions of the Schrödinger equation,
4. and the non-renormalization theorem of the supersymmetries.

All of the above tests support our identification. A part of the results is given by Fig.6. The left figure denotes the comparison of the large order behavior against g^2 for perturbative part of the energy of the ground state. Here $A^{(-)}(\epsilon, 0)$ is defined as follows,

$$E_{\mathrm{P}}^{(-)}(\epsilon, 0) = \sum_{m=0}^{\infty} E_{\mathrm{P}}^{(-)}(\epsilon, 0, m) g^{2m},$$

$$\begin{aligned} E_{\mathrm{P}}^{(-)}(\epsilon, 0, m) &= A^{(-)}(\epsilon, 0) 3^m \Gamma(-\epsilon + m + 1) \\ &\times \left[1 + O\left(\frac{1}{m}\right)\right], \end{aligned} \tag{17}$$

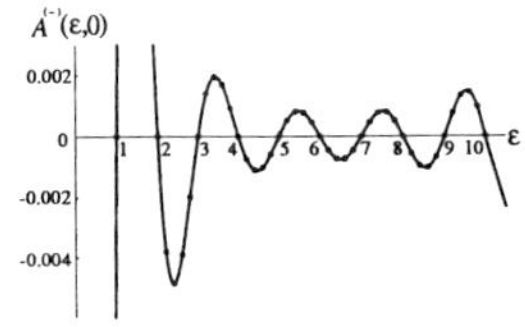

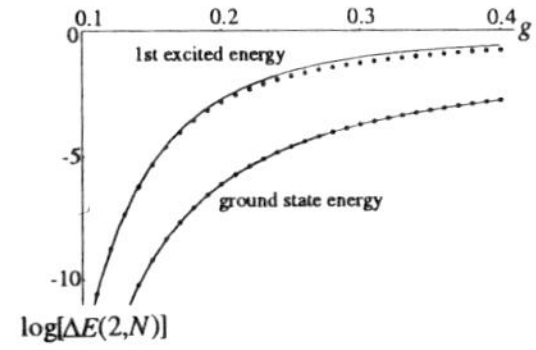

Figure 6. Tests of the identification of the non-perturbative (or tunneling) effect. The left figure denotes the comparison of the large order behavior of the perturbative part of the ground state energy. The result of our identification is indicated by the solid line, and the results of the ordinary perturbation theory is indicated by the dots. The right figure denotes the comparison between the result of the our identification for the non-perturbative part of the energy spectrum (the solid line) and that of the numerical solutions of the Schrödinger equation (the dots).

where $E_{\rm P}^{(-)}(\epsilon, 0)$ is perturbative part of the energy of the ground state. The solid line denotes the result of the identification (5), and the dots denotes the result of the ordinary perturbation theory. They agree excellently. The right figure is the comparison of the non-perturbative part of the energy spectrum in the case of $\epsilon = 2$. The solid line is the result of our identification, and the dots are the numerical solutions of the Schrödinger equation. They also agree excellently.

The extension of the separation (5) to the electroweak theory is straightforward, although additional complications arise. We expect that this make it possible to separate the baryon- and lepton-number violating processes from the the preserving ones in the path-integral (2). The basic ingredient of the instanton–anti-instanton system of the electroweak theory has been already constructed in Ref.[16], and we hope to find the final answer for the high-energy electroweak baryon- and lepton-number violating processes in the near future.

References

[1] G. 't Hooft, Phys. Rev. Lett. **37** (1976) 8; Phys. Rev. **D14** (1976) 3422.

[2] H. Aoyama and H. Goldberg, Phys. Lett. **B188** (1987) 506.

[3] A. Ringwald, Nucl. Phys. **B330** (1990) 1.

[4] O. Espinosa, Nucl. Phys. **B343** (1990) 310.

[5] T. Harano and M. Sato, hep-ph/9703457.

[6] F. R. Klinkhamer and N. S. Manton, Phys. Rev. **D30** (1984) 2212.

[7] V. I. Zakharov, Phys. Rev. Lett. **67** (1991) 3650.

[8] G. Veneziano, Mod. Phys. Lett. **A7** (1992) 1661.

[9] R. Guida, K. Konishi, and N. Magnoli, Int. Mod. Phys. **A9** (1994) 795.

[10] For a review, V. A. Rubakov and M. E. Shaposhnikov, hep-ph/9603208.

[11] C. Rebbi and R. Singleton, Jr., Phys. Rev. **D54** (1996) 1020.

[12] S. Yu. Khlebnikov, V. A.Rubakov and P. G. Tinyakov, Nucl. Phys. **B347** (1990) 783.

[13] A. H. Muller, Nucl. Phys. **B348** (1991) 310.

[14] M. B. Voloshin, Nucl. Phys. **B359** (1991) 301.

[15] I. Affleck, Nucl. Phys. **B191** (1981) 429.

[16] H. Aoyama, T. Harano, M. Sato and S. Wada, Mod. Phys. Lett. **A11** (1995) 43; Nucl. Phys. **B466** (1996) 127.

[17] M. Porrati, Nucl. Phys. **B347** (1990) 371.

[18] V. V. Khoze and A. Ringwald, Nucl. Phys. **B355** (1991) 351.

[19] A. H. Muller, Nucl. Phys. **B364** (1991) 109.

[20] H. Aoyama and H. Kikuchi, Phys. Lett. **B247** (1990) 75; Phys. Rev. **D43** (1991) 1999; Int. Mod. Phys. **A7** (1992) 2741.

[21] V. I. Zakharov, Nucl. Phys. **B353** (1991) 683.

[22] M. Maggiore and M. Shifman, Nucl. Phys. **B365** (1991) 161.

[23] I. I. Balitsky and A. V. Yung, Phys. Lett. **B168** (1986) 113.

[24] H. Aoyama and H. Kikuchi, Nucl. Phys. **B369** (1992) 219.

[25] H. Aoyama, H. Kikuchi, T. Harano, I. Okouchi, M. Sato and S. Wada, Prog. Theor. Phys. Suppl. **127** (1997) 1.

[26] H. Aoki, Prog. Theor. Phys. **91** (1994) 171.

[27] H. Aoyama, H. Kikuchi, I. Okouchi, M. Sato and S. Wada, Phys. Lett. **B424** (1998) 93.

[28] H. Aoyama, H. Kikuchi, I. Okouchi, M. Sato and S. Wada, preprint, UT-816, hep-th/9808034.

[29] For a review, see "Large-Order Behaviour of Perturbation Theory", ed. by J. C. Le Guillou and J. Zinn-Justin, (North-Holland, 1990), and references cited therein.

[30] F. J. Dyson, Phys. Rev. **85** (1952) 631.

[31] G. Auberson and G. Mennessier, Phys. Lett. **B126** (1983) 263.

[32] H. Aoyama, H. Kikuchi, T. Harano, M. Sato, S. Wada, Phys. Rev. Lett. **79** (1997) 4052.

Neutrino electron scattering and electroweak gauge structure: Probing the masses of a new Z boson

O. G. Miranda†[1], V. B. Semikoz‡ and José W. F. Valle†[2]

†Instituto de Física Corpuscular - C.S.I.C.
Departament de Física Teòrica, Universitat de València
46100 Burjassot, València, Spain
http://neutrinos.uv.es

‡Institute of the Terrestrial Magnetism, the Ionosphere and Radio Wave Propagation of the Russian Academy of Science, IZMIRAN, Troitsk, Moscow region, 142092, Russia.

Abstract.
Low-energy high-resolution neutrino-electron scattering experiments may play an important role in testing the gauge structure of the electroweak interaction. We discuss the use of strong radioactive neutrino sources (e.g. ^{51}Cr) in underground experiments such as BOREXINO, HELLAZ and LAMA. We display the sensitivity of these detectors in testing the possible existence of extra neutral E_6 gauge bosons.

1. Introduction

Despite the success of the Standard Model (SM) in describing the electroweak interaction, there have been considerable interest in extensions of the gauge structure of the theory [1]. In this talk we discuss the proposal [2] of using $\nu_e e$ and $\overline{\nu_e} e$ scattering from terrestrial neutrino sources with improved statistics as a test of the electroweak gauge structure. The

[1] On leave from Departamento de Física CINVESTAV-IPN, A. P. 14-740, México 07000, D. F., México
[2] E-mail: valle@flamenco.ific.uv.es

coupling constants governing $\nu_e e \to \nu_e e$ scattering in the SM have been well measured from $e^+e^- \to l^+l^-$ at high energies by the LEP Collaborations and have given strong constraints on additional neutral currents, specially on the mixing of the standard Z boson with other hypothetical neutral gauge bosons. This carries an important weight in global fits of the electroweak data [3]. However, we argue that low-energy experiments can give complementary information, namely, they allow a better sensitivity to the mass of the new gauge boson than available from high energies, e.g. from LEP physics. On the other hand, although the Tevatron does give relatively good limits on Z' masses, one may argue that a neutrino-electron experiment is a cleaner environment that will provide useful complementary information on the gauge structure of the electroweak interaction.

Using $\nu_e e$ and $\overline{\nu_e} e$ scattering from terrestrial neutrino sources has also been suggested as a test of non-standard neutrino electromagnetic properties, such as magnetic moments [4]. In contrast to reactor experiments such MUNU [5], a small (but intense) radioactive isotope source can be surrounded by detectors with full geometrical coverage. Here we demonstrate that a low-energy high-resolution experiment can play an important role in testing the structure of the neutral current weak interaction.

We explicitly determine the sensitivity of these radioactive neutrino source experiments as precision probes of the gauge structure of the electroweak interaction and illustrate how it works in a class of E_6-type models as well as models with left-right symmetry.

2. The νe Cross Section

In a generic electroweak gauge model the differential cross section for the process $\nu_e e \to \nu_e e$ is given by

$$\frac{d\sigma}{dT} = \frac{2m_e G_F^2}{\pi}\{(g_L+1)^2 + g_R^2(1-\frac{T}{E_\nu})^2 - \frac{m_e}{E_\nu}(g_L+1)g_R\frac{T}{E_\nu}\} \qquad (1)$$

where T is the recoil electron energy, and E_ν is the neutrino energy. For the SM case we have $g_{L,R} = \frac{1}{2}(g_V \pm g_A)$, with $g_V = \rho_{\nu e}(-1/2 + 2\kappa \sin^2\theta_W)$ and $g_A = -\rho_{\nu e}/2$ where $\rho_{\nu e}$ and κ describe the radiative corrections for low-energy $\nu_e e \to \nu_e e$ scattering [6]. For the case of $\overline{\nu_e} e \to \overline{\nu_e} e$ scattering we just need to exchange $g_L + 1$ with g_R and vice versa.

As already mentioned, the values of the coupling constants governing $\nu_e e \to \nu_e e$ scattering in the SM have been well measured through the $e^+e^- \to l^+l^-$ process at LEP. A combined LEP fit at the Z peak gives [7] $g_V = -0.03805 \pm 0.00059$ and $g_A = -0.50098 \pm 0.00033$. These results have given strong constraints on the mixing of the standard Z boson with an additional Z' ; in the framework of global fits of the electroweak data [3]. As a result we will, in what follows, focus mainly on the possibility of probing the Z' mass at low-energy $\nu_e e \to \nu_e e$ scattering experiments. For convenience we define the parameter

$$\gamma = \frac{M_Z^2}{M_{Z'}^2} \tag{2}$$

and we will neglect the mixing angle θ' between the SM boson and the extra neutral gauge boson.

For extended models, the neutral contribution to the differential cross section will be, for $\theta' = 0$,

$$\delta \frac{d\sigma}{dT} = \gamma\Delta = \gamma \frac{2m_e G_F^2}{\pi}\{D + E\frac{T}{E_\nu}(\frac{T}{E_\nu} - 2) - F\frac{m_e}{E_\nu}\frac{T}{E_\nu}\} \tag{3}$$

with Δ in obvious notation and

$$\begin{aligned} D &= 2(g_L + 1)\delta g_L + 2g_R\delta g_R & (4)\\ E &= 2g_R\delta g_R & (5)\\ F &= (g_L + 1)\delta g_R + g_R\delta g_L & (6) \end{aligned}$$

where g_L and g_R are the SM model expressions and $\delta g_{L,R}$ give the corrections due to new physics. In the particular case of the LRSM these corrections are given by [8, 9]

$$\begin{aligned} \delta g_L &= \frac{s_W^4}{r_W^2} g_L + \frac{s_W^2 c_W^2}{r_W^2} g_R & (7)\\ \delta g_R &= \frac{s_W^4}{r_W^2} g_R + \frac{s_W^2 c_W^2}{r_W^2} g_L & (8) \end{aligned}$$

while for the E_6 models we have [10, 11],

$$\begin{aligned} \delta g_L &= 4\rho s_W^2 (\frac{3\cos\beta}{2\sqrt{24}} + \frac{\sqrt{5}}{\sqrt{8}}\frac{\sin\beta}{6})(\frac{3\cos\beta}{\sqrt{24}} + \frac{1}{3}\frac{\sqrt{5}}{\sqrt{8}}\sin\beta) & (9)\\ \delta g_R &= 4\rho s_W^2 (\frac{3\cos\beta}{2\sqrt{24}} + \frac{\sqrt{5}}{\sqrt{8}}\frac{\sin\beta}{6})(\frac{\cos\beta}{\sqrt{24}} - \frac{1}{3}\frac{\sqrt{5}}{\sqrt{8}}\sin\beta) & (10) \end{aligned}$$

where, ρ includes the radiative corrections to the ratio $M_W^2/M_Z^2 cos\theta_W$ and β defines the E_6 model in which we are interested in.

The correction to the $\nu_e e$ scattering depends on the model as well as on the energy region. In order to illustrate how this corrections affect the SM prediction we define the expression

$$R = \frac{\Delta}{(\frac{d\sigma}{dT})^{SM}}. \tag{11}$$

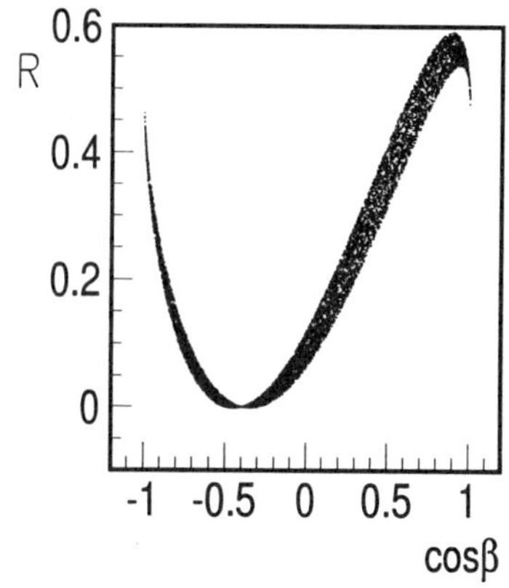

Figure 1. Plot of the ratio given in eq. 11 as a function of the model for different values of T and for $E_\nu = 746$ KeV.

This ratio depends on the specific model through the angle β and depends also on the electron recoil energy, as well as on the neutrino energy. As we are interested in artificial neutrino sources we can study what would be the value of R in Eq. (11) for the case of a neutrino coming from a ^{51} Cr source, which corresponds to a neutrino energy $E_\nu = 746$ KeV. We show this ratio in Fig. 1 as a function of β for different values of T. We can see from the plot that the sensitivity is bigger at $\cos\beta \simeq 0.8$ and it is almost zero for $\cos\beta \simeq -0.4$. Of the most popular models (χ, η and ψ models) we can say that the χ model is the most sensitive to this scattering. A similar result can be obtained for the case of anti-neutrino sources, such as ^{147} Pm, now proposed for the LAMA experiment [12, 13]. We can also see from the figure that, in order to reach a constraint on $\gamma \simeq .1$ in the χ model we need a resolution of the order of 5 %.

3. Experimental prospects

The first high-activity artificial neutrino sources have been recently developed in order to calibrate both GALLEX and SAGE solar neutrino experiments [14]. These are ^{51}Cr sources producing neutrinos by electron capture through the reaction $^{51}\ Cr + e \rightarrow ^{51}\ V + \nu_e$. The main line is at 746 KeV and represents 90 % of the neutrino flux. Besides the neutrino flux, there is also γ emission which is stopped by a tungsten shielding. The activity of the GALLEX source was 1.67 $\pm$0.03 MCi.

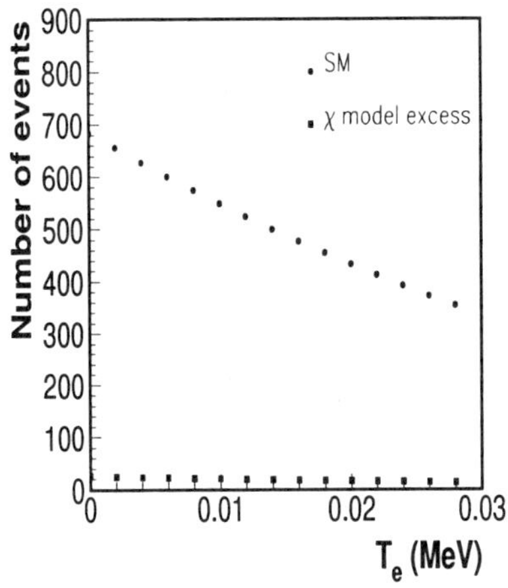

Figure 2. Expected number of events per bin for the LAMA proposal. The additional contribution of an extra neutral gauge boson with $M_{Z'} =$ 330 GeV in the χ model is also shown.

An anti-neutrino source has recently been considered by the LAMA proposal in order to probe for the neutrino magnetic moment [12]. This is a ^{147}Pm source that produces anti-neutrinos through the $^{147}Pm \rightarrow ^{147} Sm + e + \overline{\nu_e}$ beta decay. In this case we have an anti-neutrino spectrum with energies up to 235 KeV. The spectrum shape is well known and the activity of the source can be measured with an accuracy better than 1 % [15]. A tungsten shielding of 20 cm radius plus a Cu shielding of 5 cm is considered in order to stop the γ emission. In this case we can use as a good approximation for the anti-neutrino spectrum the expression

$$f(E_\nu) = \frac{1}{N} F(Z, p) E_\nu^2 (W - E_\nu) \sqrt{(W - E_\nu)^2 - m_e^2} \qquad (12)$$

where $F(Z, p)$ stands for the Coulomb correction to the spectrum, N is a normalisation factor and $W = m_e + 235$ KeV.

The possibility of surrounding the source with a NaI(Tl) detector is now considered by the LAMA team. As a first step they plan to use a 400 tones detector (approximately 2×10^{29} electrons) that will measure the electron recoil energy from 2 - 30 KeV; the source activity will be 5 MCi. A second stage with a one tone detector and 15 MCi of ^{147}Pm is under study.

We can now estimate the event rates expected both in the Standard Model as well as in extended models for the configuration discussed above. In order to do this we need to integrate over the neutrino energy spectrum and to take the average over the electron recoil energy resolution. The expected number of events per bin in the Standard Model is shown in Fig. 2. For definiteness we have considered 2 KeV width bins. In Fig. 2 we also show the excess in the number of events for the case of an extra neutral gauge boson in the χ model for a Z' mass of 330 GeV. The prospects of the experiment to be sensitive to such an excess in the shape of the electron energy distribution will depend on the error achieved in the event numbers per bin. At the moment we can only estimate the statistical error, but not for the systematic.

In order to estimate the LAMA sensitivity to the mass of a Z' in the χ model we have considered an experimental set up with 5 MCi source and a one tone detector. Assuming that the detector will measure exactly the SM prediction and taking into account only the statistical error, we obtain a sensitivity of the order of 600 GeV at 95 % C. L., comparable to the present Tevatron result. A more detailed analysis can be found in ref. [13].

In the case of the BOREXINO proposal [16], they have considered the use of a ^{51}Cr source that will be located at 10 m from the detector [17]; unfortunately the experimental set up does not allow one to surround the source and, therefore the statistics is not high enough to provide a strong sensitivity to the Z' mass. In this case, if we consider again that the experiment will measure the SM prediction, the sensitivity to the Z' mass in the χ model will be 275 GeV, if only the statistical error is considered. If we take into account the background [18] the sensitivity will decrease to 215 GeV.

Finally we now move to the HELLAZ proposal. Although this collaboration has not considered the use of an artificial source, there is room to speculate about the experimental set up and expected event rates. For definiteness we assume in our analysis a ^{51}Cr source and the originally designed HELLAZ detector [19]. Since the error will depend on the specific topology of the experiment, we have computed the sensitivity at 95 % C. L. for different values of the total error in the number of events per bin (a detailed explanation of this analysis can be found in [2]). The results, for four different models are shown in Fig. 3 where we have considered two possible energy regions for the detector. First we consider the case of an energy window from 100 KeV-250 KeV, that is the energy region that HELLAZ is considering for the study of solar neutrinos. We can see that the prospects for getting a better sensitivity than other indirect searches seems to be very realistic. On the other hand the chances of improving the Tevatron constraint seems feasible only if a high-statistics experiment is done. This can be achieved either by constructing a more intense source, by increasing the mass of the detector, by enriching the source several times (in order to increase the exposure time), or a combination of the above. Extending the energy window to 100 KeV-560 KeV is also helpful in getting a better sensitivity, as can be seen from the same Fig 3.

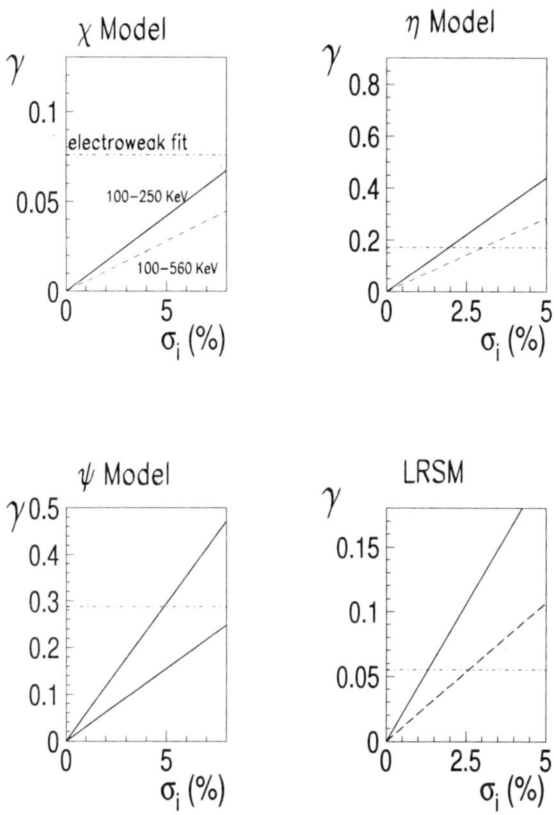

Figure 3. Sensitivity to the γ parameter for different models for the case of the HELLAZ proposal. The solid line correspond to the case of an energy region from 100 KeV - 250 KeV while the dotted line is for the region from 100 KeV - 550 KeV. We also show in the plot the present constraint from indirect searches [3].

Acknowledgements

We would like to thank useful discussions with Igor Barabanov, Marco Giammarchi, Concha González-García, and Tom Ypsilantis. This work was supported by DGICYT under grant number PB95-1077, by the TMR network grant ERBFMRXCT960090 and by INTAS grant 96-0659 of the European Union. O. G. M. was supported by a CONACYT and SNI-México, and V. S. by the sabbatical grant SAB95-506 and RFFR grants 97-02-16501, 95-02-03724.

References

[1] See, for example, J.W.F. Valle, *Prog. Part. Nucl. Phys.* **26** (1991) 91, and references therein.

[2] O. G. Miranda, V. Semikoz and José W. F. Valle, *Phys. Rev.* **D58** (1998) 013007.

[3] M. Cvetic, P. Langacker, hep-ph/9707451.

[4] P. Vogel and J. Engel, *Phys. Rev.* **D39** (1989) 3378.

[5] MUNU Collaboration, Nucl. Inst. & Meth. **A396**, 115-129 (1997).

[6] John N. Bahcall, Marc Kamionkowski, Alberto Sirlin, *Phys. Rev.* **D51** (1995) 6146.

[7] D. Stickland, Electroweak results from ee colliders talk given at the XVI International Workshop on Weak Interactions and Neutrinos, Capri 1997.

[8] J.C. Pati and A. Salam, *Phys. Rev.* **D10** (1975) 275; R.N. Mohapatra and J.C. Pati, *Phys. Rev.* **D11** (1975) 566, 2558.

[9] R.N. Mohapatra and G. Senjanović, *Phys. Rev.* **D23** (1981) 165 and references therein.

[10] M. C. González-García and J. W. F. Valle, *Nucl. Phys.* **B345** (1990) 312.

[11] M. C. González-García and J. W. F. Valle, *Phys. Lett.* **B29** (1991) 365.

[12] I. R. Barabanov *et. al.*, *Astrop. Phys.* **8** (1997) 67.

[13] I. Barabanov *et. al.*, hep-ph/9808297.

[14] GALLEX Coll., *Phys. Lett.* **B342** (1995) 440; SAGE Coll., J. N. Abdurashitov, *et. al. Phys. Rev. Lett.* **23** (1996) 4708.

[15] V.N.Kornoukhov, Preprint ITEP N90 (1994), ITEP N26 (1996), ITEP N2 (1997) and *Phys. of Atomic Nuc.* **60** (1997) 558.

[16] J. B. Benziger *et. al.*, A proposal for participating in the Borexino solar neutrino experiment, October 30, 1996.

[17] N. Ferrari, G. Fiorentini and B. Ricci *Phys. Lett.* **B387** (1996) 427.

[18] We thanks Marco Giammachi for providing us with the expected BOREXINO background.

[19] F. Arzarello *et. al.*: CERN-LAA/94-19.

Signals of R-parity breaking in MSSM and SLRM models

Jukka Maalampi[1]

Department of Physics, Theoretical Physics Division, University of Helsinki, Finland

Abstract. The R-parity breaking can be tested in high-energy electron-positron collisions through the single production of supersymmetric particles. The discussion in this talk concentrates mainly on the supersymmetric $SU(2)_L \times SU(2)_R \times U(1)_{B-L}$ theory (SLRM), where the R-parity breaking is spontaneous. The resulting single production of charginos and neutralinos is studied. In the framework of the minimal susy model (MSSM), where the R-parity violation is explicit, the production of a single sneutrino is discussed as well.

1. Introduction

In the Standard Model (SM) the baryon number B and the lepton number L are conserved automatically due to an accidental discrete symmetry of the Lagrangian arising from the choice of gauge symmetry and field representations. In the minimal supersymmetric extension of the model (MSSM) this symmetry disappears and one has to eliminate the phenomenologically dangerous B and L violating couplings, in particular those which would make the proton unstable, in some other way. A conventional method is to introduce a discrete multiplicative symmetry called R-parity, which can be written as

$$R = (-1)^{3B+L+2S}, \tag{1}$$

where S is the spin of the particle [1] . As a particle and its superpartner have the same quantum numbers except the spin which differs by 1/2 units,

[1] **E-mail: Jukka.Maalampi@Helsinki.Fi**

they have opposite R-parities: the ordinary SM particle has $R = +1$ and the superpartner has $R = -1$. The R-parity conservation allows for all the SM couplings, but at the same time it eliminates the following B and L violating part of the superpotential [2]:

$$W_R = \lambda_{ijk} L_i L_j E_k^c + \lambda'_{ijk} L_i Q_j D_k^c + \lambda''_{ijk} U_i^c D_j^c D_k^c, \tag{2}$$

where L, E, Q, U and D denote the superfields of the left-handed doublets of leptons, the right-handed singlets of charged leptons, the left-handed doublets of quarks, the right-handed singlets of up-type and down-type quarks, respectively, and i, j, k are generation indices. The first two terms are identical with the ordinary Yukawa terms except that the Higgs doublet superfield is there replaced with the lepton doublet superfield L_i which has the same quantum numbers. The last term consisting of three $SU(2)_L$ singlets is also allowed by the gauge symmetry.

The couplings λ and λ' violate lepton number and the couplings λ'' violate baryon number. If they all were non-vanishing the proton would be unstable, and this would be in conflict with observations unless the relevant combinations of the couplings are adjusted to very small values [3]. R-parity was introduced to avoid such unnatural adjustments. Nevertheless, in the framework of MSSM R-parity is quite an arbitrary concept not well motivated on theoretical grounds. It is therefore reasonable to study the phenomenological consequences of the possible R-breaking interactions, and plenty of such studies has appeared in the literature [4].

In contrast with the situation in the MSSM, in the supersymmetric left-right model (SLRM) [5] the R-parity is not introduced *ad hoc* but it is there from the beginning. The SLRM is based on the gauge group $SU(2)_L \times SU(2)_R \times U(1)_{B-L}$, where the gauge symmetry $U(1)_{B-L}$ automatically makes the Lagrangian R-conserving [6]. Indeed, for all terms in the superpotential $(-1)^{2S} = +1$ due to Lorentz invariance, and since for the known particles $(-1)^{3B+L} = (-1)^{3(B-L)}$ we see that the $U(1)_{\mathrm{B-L}}$ gauge symmetry implies the conservation of R-parity. Since $B - L$ is conserved due to gauge symmetry, all terms of the superpotential are R-even.

It turns out, however, that the minimum of the scalar potential in the SLRM corresponds to an unphysical ground state [8, 9], unless at least one of right sneutrinos ($\tilde{\nu}_R$), the scalar partners of the right-handed neutrinos, acquire a non-vanishing VEV. In this situation R-parity is spontaneously broken. Nevertheless, the proton decay does not pose any phenomenological problem in this case as $\langle \tilde{\nu}_R \rangle \neq 0$ breaks only the lepton number but not the baryon number.

An important consequence of the spontaneous breaking of the R-parity is that R-odd particles, the superpartners of the SM particles, can be produced singly and on the other hand they can decay into the SM particles. The single production is possible due to the mixing between the R-odd and

R-even particles: the neutralinos mix with neutrinos, the charginos with leptons, and sneutrinos with the Higgs bosons. The R-breaking single production mechanisms provide probes for heavier supersymmetric particles than is the mass reach in the case of conserved R-parity.

It should be noted that while in the MSSM with explicit R-parity violation many of the three point interaction couplings (2) should be fine tuned to small values so as to satisfy the severe constraints set by experiments (see e.g. [?]), in models with spontaneous R-parity violation, like SLRM, these couplings are naturally suppressed since they are proportional to small mixing angles or are generated by loops.

My main focus in this talk will be on the phenomenology of the spontaneous R-parity breaking in the framework of the SLRM. I shall discuss the signals of single chargino and neutralino production in electron positron collisions. I will start with a summary of the basic features of the supersymmetric left-right theory. Then I shall describe a model where the spontaneous R-parity breaking is realized, and discuss the phenomenology of this model. This part of the talk is based on the work done in collaboration with K. Huitu and K. Puolamäki [10].

As a separate topic I shall discuss new processes for testing the explicit R-parity breaking terms of (2) in the MSSM. It will be pointed out that the single production of sneutrinos in $e^+e^- \to e^+\mu^-\tilde{\nu}$ has clear experimental signal and can be used to constrain some couplings λ to values not probed by the present tests. This discussion is based on the study made together with I. Vilja and A. Pietilä [11].

2. Description of the left-right symmetric model

Let us first consider the conventional, non-supersymmetric left-right symmetric electroweak model (LRM) [12]. This model has features which make it a serious rival of the Standard Model. It gives a better understanding of parity violation and it maintains the lepton-quark symmetry in weak interactions. Perhaps the most important property is its ability to provide a simple and natural explanation for the smallness of neutrino masses in terms of the see-saw mechanism [13]. The anomalies measured in the solar [14] and atmospheric [15] neutrino fluxes seem to require that neutrinos indeed have a mass, manifested in these phenomena through flavour oscillations [2]. In the framework of SM these observations are difficult to deal with as there neutrinos are predicted to be massless.

[2] Soon after this Trento Workshop the SuperKamiokande experiment confirmed that at least the muon neutrino does have a mass [16]

In the LRM quark and leptons are assigned to the following $SU(2)_L \times SU(2)_R \times U(1)_{B-L}$ representations [17]:

$$L_L = \begin{pmatrix} \nu_e \\ e^- \end{pmatrix}_L = (2,1,-1), \quad L_R = \begin{pmatrix} \nu_e \\ e^- \end{pmatrix}_R = (1,2,-1),$$
$$Q_L = \begin{pmatrix} u \\ d \end{pmatrix}_L = (2,1,\tfrac{1}{3}), \quad Q_R = \begin{pmatrix} u \\ d \end{pmatrix}_R = (1,2,\frac{1}{3}), \tag{3}$$

and similarly for the other families. The minimal set of fundamental scalars, the theory to be symmetric under the $L \leftrightarrow R$ transformation, consists of the following Higgs multiplets:

$$\Phi = \begin{pmatrix} \phi_1^0 & \phi_1^+ \\ \phi_2^- & \phi_2^0 \end{pmatrix} = (2,2,0),$$
$$\Delta_L = \begin{pmatrix} \Delta_L^+ & \sqrt{2}\Delta_L^{++} \\ \sqrt{2}\Delta_L^0 & -\Delta_L^+ \end{pmatrix} = (3,1,2), \tag{4}$$
$$\Delta_R = \begin{pmatrix} \Delta_R^+ & \sqrt{2}\Delta_R^{++} \\ \sqrt{2}\Delta_R^0 & -\Delta_R^+ \end{pmatrix} = (1,3,2).$$

They transform according to $\Phi \rightarrow U_L \Phi U_R^\dagger$, $\Delta_L \rightarrow U_L \Delta_L U_L^\dagger$ and $\Delta_R \rightarrow U_R \Delta_R U_R^\dagger$, where $U_{L(R)}$ is an element of $SU(2)_{L(R)}$.

The vacuum expectation values of the scalar triplets are denoted by

$$\langle \Delta_{L,R} \rangle = \frac{1}{\sqrt{2}} \begin{pmatrix} 0 & 0 \\ v_{L,R} & 0 \end{pmatrix}. \tag{5}$$

The right-triplet Δ_R breaks the $SU(2)_R \times U(1)_{B-L}$ symmetry to $U(1)_Y$, and at the same time the discrete $L \leftrightarrow R$ symmetry, and it yields a Majorana mass to the neutrinos through the Yukawa coupling

$$\mathcal{L}_\Delta^{\text{Yukawa}} = i h_L L_L^T C \sigma_2 \Delta_L L_L + i h_R L_R^T C \sigma_2 \Delta_R L_R + h.c., \tag{6}$$

where $\Delta_{L,R} = \Delta_{L,R}^i \sigma_i$. .

The vacuum expectation value of the bidoublet Φ, given by

$$\langle \Phi \rangle = \frac{1}{\sqrt{2}} \begin{pmatrix} \kappa_1 & 0 \\ 0 & \kappa_2 \end{pmatrix}, \tag{7}$$

breaks the SM symmetry $SU(2)_L \times U(1)_Y$ further to $U(1)_{\text{em}}$, and it generates masses to fermions through the Yukawa couplings

$$\mathcal{L}_\Phi^{\text{Yukawa}} = \bar{L}_L^i (f_{ij}\Phi + \tilde{f}_{ij}\tilde{\Phi}) L_R^j + \bar{Q}_L^i (f_{ij}\Phi + \tilde{f}_{ij}\tilde{\Phi}) Q_R^j + \text{h.c.}, \tag{8}$$

where $\tilde{\Phi} = \sigma_2 \Phi^* \sigma_2$.

The most general form of the Higgs potential, given in ref. [18], contains 18 independent parameters. Allthough there are severe constraints on the parameters, the most crucial one concerning flavour changing neutral currents (FCNC) [18], there is a finite range of parameter values that realize the physically reasonable symmetry breaking chain $SU(2)_L \times SU(2)_R \times U(1)_{B-L} \rightarrow SU(2)_L \times U(1)_Y \rightarrow U(1)_{\rm em}$ corresponding to the vevs (5) and (7).

If there is a hierarchy $v_R \gg \kappa_{1,2}$, so that the LR-symmetry is broken much above the electroweak scale, then at the energies probed by experiments so far the LRM and the SM are more or less indistinguishable. The negative direct searches of heavy W and Z bosons at Tevatron [7] indicate the breaking scale of the LR-symmetry to lay above the 1 TeV scale.

3. Supersymmetric left-right model (SLRM)

The left-right symmetric model has a naturality problem similar to that in the SM, namely the masses of the fundamental Higgs scalars diverge quadratically. As in the SM, the supersymmetry can be used to stabilize the scalar masses and cure this hierarchy problem. The supersymmetric extension of the LRM has been investigated, e.g., in [5].

Apart from the existence of the superpartners of the ordinary LRM particles, the most significant difference between the ordinary and the supersymmetric left-right model is in the Higgs sector. In supersymmetrization, the cancellation of $U(1)_{B-L}$ chiral anomalies among the fermionic partners of the triplet Higgs fields requires that the Higgs triplet Δ_R is accompanied by another triplet, δ_R with opposite $U(1)_{B-L}$ quantum number, and similarly the triplet Δ_L by a triplet δ_L. Due to the conservation of the $B-L$ symmetry, δ_R and δ_L do not couple with leptons and quarks. Also, to avoid a trivial Kobayashi-Maskawa matrix for quarks, one has to add another bidoublet. This comes about because supersymmetry forbids a Yukawa coupling where the bidoublet appears as conjugated. The two bidoublets will be denoted by ϕ_u and ϕ_d.

We have chosen the vacuum expectation values for the Higgses, which break the $SU(2)_L \times SU(2)_R \times U(1)_{B-L}$ into the $U(1)_{em}$, to be as follows

$$\begin{aligned} \langle\phi_u\rangle &= \tfrac{1}{\sqrt{2}}\begin{pmatrix} \kappa_u & 0 \\ 0 & 0 \end{pmatrix}, \quad \langle\phi_d\rangle = \tfrac{1}{\sqrt{2}}\begin{pmatrix} 0 & 0 \\ 0 & \kappa_d \end{pmatrix}, \\ \langle\Delta_{L,R}\rangle &= \tfrac{1}{\sqrt{2}}\begin{pmatrix} 0 & 0 \\ v_{L,R} & 0 \end{pmatrix}, \quad \langle\delta_{L,R}\rangle = \tfrac{1}{\sqrt{2}} v^{\delta}_{L,R}. \end{aligned} \tag{9}$$

This is the breaking pattern one wants to have, so as to reproduce the results of the SM at low energies. Nevertheless, in contrast with the non-susy case, things do not work in this manner in the presence of supersymmetry. Investigation of the superpotential reveals that with the vevs (9) alone one would end up with the breaking of electric charge in the ground state, as was shown by Kuchimanchi and Mohapatra on general grounds [8]. In [9] Huitu and myself came across this problem in a specific model described below. We found that when diagonalizing the Higgs mass Lagrangian with this set of vevs some of the scalars turn out to be unphysical. A solution to the dilemma is to assume that at least one of the superpartners of right-handed neutrino ($\tilde{\nu}_R$) achieves a non-zero vacuum expectation value. (Another possibility is to add non-renormalisable terms to the Lagrangian [19], but I do not consider it in this talk.)

Let us now define our supersymmetric left-right symmetric model. The superpotential is assumed to have the following form [9]:

$$\begin{aligned} W &= h_u^Q Q_L^{cT} \phi_u Q_R + h_d^Q Q_L^{cT} \phi_d Q_R \\ &\quad + h_u^L L_L^{cT} \phi_u L_R + h_d^L L_L^{cT} \phi_d L_R + h_\Delta L_R^T i\tau_2 \Delta L_R \\ &\quad + \mu_1 \mathrm{Tr}(\tau_2 \phi_u^T \tau_2 \phi_d) + \mu_2 \mathrm{Tr}(\Delta \delta). \end{aligned} \tag{10}$$

The generation indices of the quark and lepton superfields are not shown.

In the superpotential (10) the R-parity, $R = (-1)^{3(B-L)+2S}$, is preserved. This ensures that the susy partners with $R = -1$ are produced in pairs and that the lightest supersymmetric particle (LSP) is stable. The parameters μ_i in Eq. (10) are supersymmetric mass parameters. They are assumed to be close to the scale of the soft supersymmetry breaking parameters, in order to preserve the naturalness of the theory. In supersymmetric models that have also a gauge singlet Higgs field, the μ-type terms are generated by giving a vacuum expectation value to the singlet Higgs. It is assumed here that the parameters $|\mu_i|$ are of the order of the weak scale.

In addition to the mass terms arising from the superpotential (10) as a result of spontaneous gauge symmetry breaking, there are also soft mass terms, appearing in the soft supersymmetry-breaking Lagrangian

$$\mathcal{L}_{\mathrm{soft}} = -\frac{1}{2}\sum_i m_i^2 |\varphi_i|^2 - \frac{1}{2}\sum_\alpha M_\alpha \lambda_\alpha \lambda_\alpha + B\varphi^2 + A\varphi^3 + h.c., \tag{11}$$

where φ_i denotes scalar fields and the second sum is over the gauginos. The last two terms, φ^2 and φ^3, are also allowed by gauge symmetry.

As mentioned already, the diagonalization of the ensuing mass matrix leads to unphysical scalar masses if the vev pattern of eq. (9) is assumed

[9]. An acceptable mass spectrum is achieved by allowing at least one right-handed sneutrino to generate a non-vanishing value in the vacuum, $\langle\tilde{\nu}^c\rangle = \sigma_R$.

The sneutrino vev breaks, along with the vevs of Δ_R and δ_R, the $SU(2)_R \times U(1)_{B-L}$ symmetry. All these vevs break the $U(1)_{B-L}$ symmetry due to their non-vanishing lepton number (their baryon number is $B = 0$). The triplet Higgses Δ_R lead to $|\Delta L| = 2$ Majorana mass terms $\nu^c\nu^c$ for neutrinos, which play a central role in the see-saw mass mechanism and make neutrinos Majorana particles. As L is broken by two units these terms preserve the R-parity.

In contrast, for the vev of the sneutrino $|\Delta L| = 1$, and it therefore gives rise to R-parity breaking mass terms. These are terms that mix leptons (l, ν) with gauginos (λ) and higgsinos $(\tilde{h})$. Let us note that in the MSSM theory one could add R-parity breaking $L_i H_u$ mass terms to the superpotential (2) and generate similar mixings. At tree level such terms can be rotated away and absorbed in the trilinear Higgs term $H_u H_d$ by a suitable redefinition of the Higgs superfield H_d, but after higher order effects are taken into account they may reappear. One should stress, however, the fundamental difference in the origin of the lepton-gaugino and lepton-higgsino mixing in these two models. In SLRM they appear as a concequence of the spontaneous R-parity breaking, whereas in the MSSM they are introduced on *ad hoc* basis.

In the following subsection I shall consider some phenomenological signals of R-parity breaking in the SLRM model. Before that it should be mentioned incidentally out that adding supersymmetry into the LR model brings a interesting constraint on the energy scale of the breaking of the left-right symmetry. In the SLRM with spontaneous R-parity breaking the mass of the W_R boson is at most 10 TeV [20], otherwise susy were broken at energies which were too high for susy to solve the hierarchy problem.

3.1. Signals of R-parity breaking in SLRM

As the physical leptons in the SLRM are composed of gauginos, higgsinos and leptons, their couplings to Z-boson will deviate from the standard form. However, the modifications are small in general as the components which form the dominant part of the mass eigenstates of charginos and neutralinos corresponding to the SM leptons have all approximately the SM couplings with $SU(2)_L$ gauge bosons. This is so due to the symmetry breaking structure: the states that diagonalize the Lagrangian after the $SU(2)_R \times U(1)_{B-L}$ breaking have definite $SU(2)_L \times U(1)_Y$ quantum numbers and couplings. The subsequent breakdown to $U(1)_{em}$ cause only small deviations to the composition of the physical states, characterized by the mass ratios $(M_{W_L}$ or $m_\tau)/M_{W_R}$ or m_τ/m_{susy}, where m_{susy} is a general susy scale.

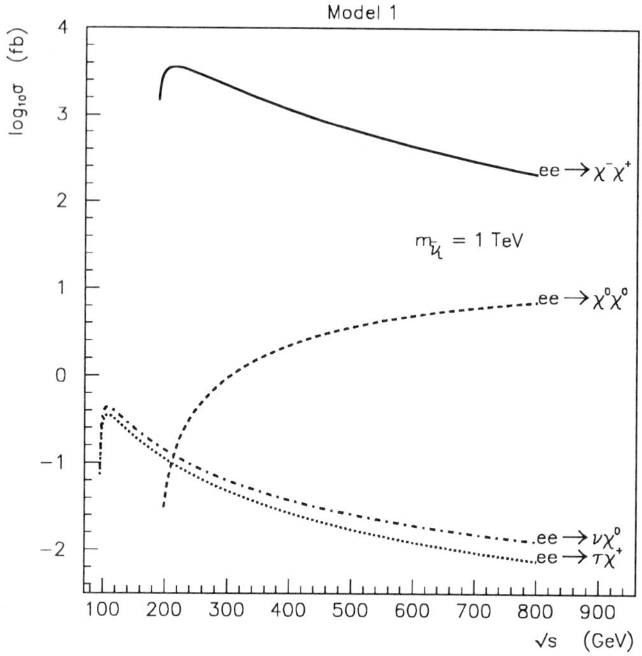

Figure 1. The production cross section of single or pair produced neutralinos and charginos as a function of center of mass energy in the Model 1 with $m_{\tilde{e}_L} = m_{\tilde{\nu}_{eL}} = 100$ GeV.

In [10] two representative models have been considered. In both models it is assumed, for simplicity, that only one sneutrino ($\tilde{\nu}_{R\tau}$) has a non-vanishing VEV. In the first model (Model 1) the soft gaugino mass terms are taken to be about 100 GeV. The diagonalization of the mass Lagrangian reveals that the lightest supersymmetric particles are almost mass-degenerate neutralino ($m_{\tilde{\chi}_2^0} \simeq 93$ GeV) and chargino ($m_{\tilde{\chi}_2^\pm} \simeq 94$ GeV), both of them almost purely gaugino λ_L states.

In Fig. 1 the cross sections have been plotted for single production of the lightest charginos and neutralinos via reactions $e^+e^- \to \tilde{\chi}_2^+\tau^-$ and $e^+e^- \to \tilde{\chi}_2^0\nu_\tau$ and for the pair production via the reactions $e^+e^- \to \tilde{\chi}_2^+\tilde{\chi}_2^-$ and $e^+e^- \to \tilde{\chi}_2^0\tilde{\chi}_2^0$ as a function of the center of mass energy $\sqrt{s}$ assuming $m_{\tilde{e}_L} = m_{\tilde{\nu}_L}$ =1 TeV and $m_{\tilde{\chi}} = 0.1$ TeV. These processes occur via a gauge boson exchange in s-channel and via a t-channel sneutrino or selectron exchange.

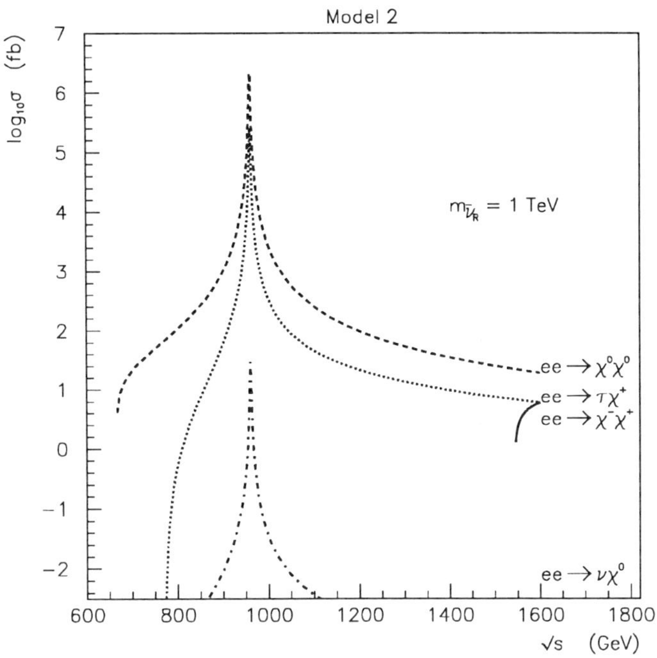

Figure 2. The production cross section of single or pair produced neutralinos and charginos as a function of center of mass energy in Model 2 with $m_{\tilde{e}_R} = m_{\tilde{\nu}_{eR}} = 1$ TeV and $m_{\tilde{\chi}} = 0.1$ TeV.

One can test the model by studying the production and decays of charginos and neutralinos in electron-positron collisions. Assuming a linear collider to operate at center of mass energy $\sqrt{s} = 0.5$ TeV with an annual luminosity of $\mathcal{L}_{year} = 500$ fb^{-1} there would be a few tens of events during a one-year run. While this is several orders of magnitude below the pair production rate (when the latter is kinematically allowed), the signal may be visible.

In Model 1 the neutralino decays at the interaction point to the tau lepton and the ordinary W-boson with the decay width of $\Gamma(\tilde{\chi}_2^0 \to \tau + W) \simeq 10$ keV. The chargino can decay either via gauge or Higgs boson, decay widths being $\Gamma(\tilde{\chi}_2^+ \to \nu_\tau W) \simeq 20$ keV and $\Gamma(\tilde{\chi}_2^+ \to \tau^+ H_1^0) \simeq 10$ keV. In the latter case the Higgs boson will further decay, mainly to a pair of bottom quarks.

In Model 2 all soft masses are of the order 1 TeV. An interesting feature

ijk	λ_{ijk}	ijk	λ'_{ijk}	ijk	λ'_{ijk}	ijk	λ'_{ijk}	ijk	λ''_{ijk}
121	0.05	111	0.001	211	0.09	311	0.16	112	10^{-6}
122	0.05	112	0.02	212	0.09	312	0.16	113	10^{-5}
123	0.05	113	0.02	213	0.09	313	0.16	123	1.25
131	0.06	121	0.035	221	0.18	321	0.20	212	1.25
132	0.06	122	0.06	222	0.18	322	0.20	213	1.25
133	0.004	123	0.20	223	0.18	323	0.20	223	1.25
231	0.06	131	0.035	231	0.22	331	0.26	312	0.43
232	0.06	132	0.33	232	0.39	332	0.26	313	0.43
233	0.06	133	0.002	233	0.39	333	0.26	323	0.43

Table 1. Present bounds on the R breaking Yukawa couplings for susy particle mass $\tilde{m} = 100$ Gev. (From ref. [2].)

of this model is the existence of a relatively light doubly charged Higgs boson H^{++} ($m_{H^{++}} = 334$ GeV). The lightest supersymmetric chargino ($m_{\tilde{\chi}_2^{\pm}} = 772$ GeV) and neutralino ($m_{\tilde{\chi}_2^0} = 333$ GeV) pair production would be clearly visible in linear collider, whenever kinematically allowed. In this particular model a relatively light extra Z-boson, Z_2, ($m_{Z_2} \simeq 1$ TeV) also contributes and allows for measureable production cross section for single chargino $\tilde{\chi}_2^+$ and $\tilde{\chi}_1^- \sim \tau^-$, as shown in Fig. 2.

The neutralino would decay mainly to a neutrino and the lightest Higgs boson $\Gamma(\tilde{\chi}_2^0 \to \nu_\tau H_1^0) \simeq 100$ keV. The chargino would decay predominantly to the tau lepton and the doubly charged Higgs boson, which would further decay to same sign lepton pair. The decay widths for the chargino are $\Gamma(\tilde{\chi}_2^+ \to \tau^- H^{++}) \simeq 90$ MeV and $\Gamma(\tilde{\chi}_2^+ \to \tau^+ Z) \simeq 3$ MeV.

4. Slepton production in the MSSM model

Let us now move to consider the minimal supersymmetric extension of the standard model (MSSM) where R-parity is broken explicitly through the superpotential terms (2). The present constraints on the various coupling constants λ and λ' are summarised in Table 1 (obtained from ref. [2]).

In particular, I will discuss the possibility of testing the LLE^c coupling via single production of sneutrinos.

The single resonant production of sneutrinos in e^+e^- collisions due to the operator $4L_1L_{2,3}E_1^c$ has been investigated in [21, 22]. This is a useful reaction as soon as one has information on the mass of the sneutrino from some other measurement. If the sneutrino mass is not known with a good precision, so that one cannot adjust the collision energy to the pole

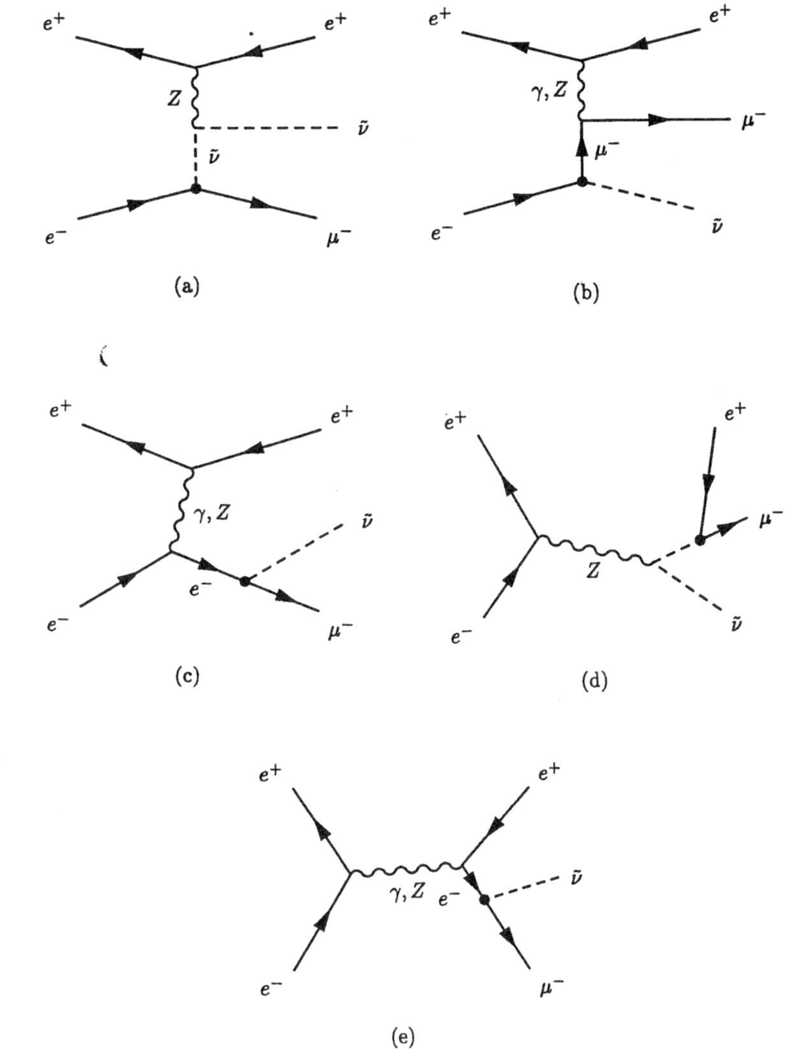

Figure 3. The lowest order Feynman diagrams for the process $e^+ + e^- \to \tilde{\nu} + l^+ + l'^-$.

value, associated production of sneutrino with gauge bosons could provide a better tool for study [23]. Effects of the single sneutrino can be studied also indirectly through their contribution to $e^+e^- \to l^+l'^-$, where the final state consists of various combinations of e, μ, τ depending on the coupling under test [24].

Here I will report a study [11] of the single sneutrino production in the reactions

$$e^+ + e^- \to \tilde{\nu} + l^+ + l'^- . \tag{12}$$

The lowest order Feynman diagrams for these processes are presented in Fig. 3.

In the reactions with a photon or Z boson exchange in s-channel l^+ is always the beam particle e^+, and due to the antisymmetry of the interaction $\lambda_{ijk} L_i L_j E_k^c$ under the exchange of the first two indices these diagrams contribute to the production of $\tilde{\nu}_\mu$ and $\tilde{\nu}_\tau$ but not to that of $\tilde{\nu}_e$. The couplings one can test in these processes are λ_{ijk} with indices 121, 122, 123, 131, 132, and 133. For the diagrams d and e these limitations do not apply and there one can test, in princible, all the couplings λ_{ijk}.

It is obvious that for a light sneutrino ($m_{\tilde{\nu}} \lesssim \sqrt{s}/2$) the t channel photon exchange gives the dominant contribution. In ref. [25] this contribution was separately studied, completely neglecting the other diagrams, and it was found that, for example, for $m_{\tilde{\nu}} = 100$ GeV, $\lambda_{ijk} = 0.05$ and $\sqrt{s} = 500$ GeV the cross section is about 1 pb, corresponding to tens of thousands of events in NLC with anticipated luminosity of 500 fb^{-1}. It was assumed in this analysis that the beam particle that emits the photon continues into the beam pipe and is not detected, thereby allowing for a large soft-photon contribution.

In our analysis we have assumed that all final state particles are detected, so as to diminish background. We have set a lower limit of xxxx for the scattering angle of the beam particle, which cuts out a substantial part of the soft-photon forward peak. Nevertheless, for light sneutrinos the photon exchange diagrams still dominate. For heavier sneutrinos also the rest of the diagrams are relevant and should be taken into account.

In Fig. 4 the cross section for the reaction $e^+e^- \to e^+\mu^-\tilde{\nu}$ is presented as a function of the center of mass energy for a presentative sneutrino mass value of 0.1 TeV and $\lambda_{i12} = 0.05$ for the coupling. The produced sneutrino is $\tilde{\nu}_\mu$ or $\tilde{\nu}_\tau$. All the diagrams of Fig. 3 contribute to this reaction. Taking the luminosity of 500 fb^{-1} one can find that NLC can improve the present limit on the R-parity breaking couplings, $\lambda_{i12} = 0.05$. Of course, the same statements apply to the process $e^+e^- \to e^+\tau^-\tilde{\nu}$.

In above it was tacitly assumed that the processes are detected with a full efficiency. The actual effeciency depends on the decay mode of the

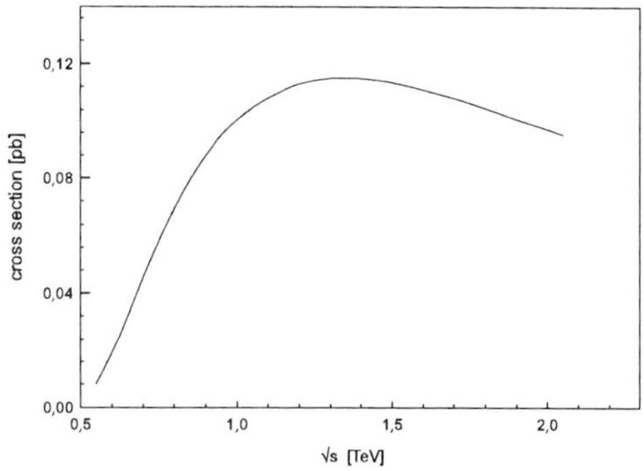

Figure 4. The cross section for the reaction $e^+e^- \to e^+\mu^-\tilde{\nu}$ as a function of the center of mass energy for $m_{\tilde{\nu}} = 0.1$ TeV and $\lambda_{i12} = 0.05$.

sneutrino, which in turn depends on the mass of sneutrino in comparison with the masses of other supersymmetric particles, in particular neutralinos and charginos. In the case the sneutrino is the lightest susy particle, it will decay via R-breaking interactions into a pair of charged leptons. The final state would then consist of four charged leptons, e.g. $e^+\mu^-\mu^+\mu^-$, with no missing energy. The signal is less striking in the case where the sneutrino is kinematically allowed to decay into a chargino or neutralino, associated with a charged lepton or a neutrino, respectively, which then decays via R-breaking interactions into three leptons, resulting in a substantial amount of missing energy.

5. Summary

In the supersymmetric $SU(2)_L \times SU(2)_R \times U(1)_{B-L}$ model (SLRM) the superpotential conserves the R-parity, but the R parity is spontaneously broken by via sneutrino vev. Consequently, supersymmetric particles can produced singly and missing energy is no longer a necessary signature of supersymmetric processes. I discussed the production and decay of neutralinos and charginos at high-energy lepton collisions in the framework of a supersymmetric $SU(2)_L \times SU(2)_R \times U(1)_{B-L}$ model. A difference in comparison with the same process in the supersymmetric extension of the SM (MSSM) is due to the fact that in the SLRM for an interesting part of the parameter space the dominant decay mode is mediated by a doubly charged Higgs.

I also discussed the single production of sneutrinos in the MSSM in the the reactions $e^+ + e^- \rightarrow \tilde{\nu} + l^+ + l'^-$. Given the present constraints on the explicit R-breaking couplings, these processes may lead to a detectable signal in the Next Linear Collider.

This Workshop nicely presented the interplay of low-energy and high-energy physics in constraining the rare R, L and B breaking interactions. In particularly the double beta decay experiments from the low-energy side compete very well with various high-energy tests [26]. I just want to add here as an example the recent observation that it gives better constraints on the Majoron couplings than what seems to be possible to achieved from the inverse double beta decay in high-energy e^-e^- collisions [27].

Acknowledgements

I thank professor H. Klapdor-Kleingrothaus and the other organizers of this Workshop for creating an enjoyable atmosphere. I thank K. Huitu, J. Laitinen, A. Pietilä, K. Puolamäki and I. Vilja for collaboration. My work and participation in the Workshop was supported by the Academy of Finland under the project no. 40677.

References

[1] G. Farrar and P. Fayet, *Phys. Lett.* **B76**(1978) 575

[2] For a review on R-arity, see *e.g.*, H Dreiner, hep-ph/9707435.

[3] A. Yu. Smirnov and F. Vissani, *Nucl. Phys.* **B 460** (1996) 37, and references therein.

[4] C. S. Aulakh and R. N. Mohapatra, *Phys. Lett.* **119B**, 136 (1983); F. Zwirner, *Phys. Lett.* **132B**, 103 (1983); L. Hall and M. Suzuki, *Nucl. Phys.* **B231**, 419 (1984); G. G. Ross and J. W. F. Valle, *Phys. Lett.* **B151**, 375 (1985).

[5] Y. Ahn, *Phys. Lett.* **149 B** (1984) 337;
R.M. Francis, M. Frank, C.S. Kalman, *Phys. Rev.* **D43** (1991) 2369;
R. M. Francis, C. S. Kalman and H.N. Saif, *Z. Phys.* **C59** (1993) 655;
K. Huitu, J. Maalampi, M. Raidal, *Nucl. Phys.* **B 420** (1994) 449; *Phys. Lett.* **B 328** (1994) 60.

[6] S. Martin, *Phys. Rev.***D54** (1996) 2340.

[7] F. Abe *et al.*, CDF Collaboration, *Phys. Rev. Lett.* **74** (1995) 2900;
S. Abachi *et al.*, D0 Collaboration, *Phys. Lett.* **B 358** (1995) 405.

[8] R. Kuchimanchi, R.N. Mohapatra, *Phys. Rev.* **D48** (1993) 4352.

[9] K. Huitu, J. Maalampi, *Phys. Lett.* **B344** (1995) 217.

[10] K. Huitu, J. Maalampi and K. Puolamäki, *Europhys. J.* in press.

[11] J. Maalampi, A. Pietilä and I. Vilja, in preparation.

[12] J.C. Pati and A. Salam, *Phys. Rev.* **D 10** (1974) 275;
R.N. Mohapatra and J. C. Pati, *Phys. Rev.* **D 11** (1975) 566, 2558;
G. Senjanovic and R. N. Mohapatra, *Phys. Rev.* **D 12** (1975) 1502;
R. N. Mohapatra and R. E. Marshak, *Phys. Lett.* **91 B** (1980) 222.

[13] M. Gell-Mann, P. Ramond and R. Slansky, in *Supergravity*, eds. P. van Niewenhuizen and D. Z. Freedman (North Holland 1979);
T. Yanagida, in Proceedings of *Workshop on Unified Theory and Baryon Number in the Universe*, eds. O. Sawada and A. Sugamoto (KEK 1979).
See also, R.N. Mohapatra and G. Senjanovic, *Phys. Rev. Lett.* **44** (1980) 912.

[14] K. Lande and P.S. Wildenhain (for the Homestake Program), in *Proc. of the 17th International Conference on Neutrino Physics and Astrophysics*, June 13-19, 1996, Helsinki, Finland, eds. K. Huitu, K. Enqvist and J. Maalampi (World Scientific, Singapore, 1997), p. 25;
Y. Suzuki (for the Kamiokande and the Super-Kamiokande collaboration), ibid. p. 73;
T. Kirsten (for the Gallex Collaboration), ibid. p. 3;
V.N. Gavrin et al. (The Sage Collaboration), ibid. p. 14.

[15] K.S. Hirata et al., *Phys. Lett.* **B280** (1992) 146;
D. Casper et al., *Phys. Rev. Lett.* **66** (1993) 2561.

[16] Y. Fukuda et al., hep-ex/9805006.

[17] R.N. Mohapatra, *Unification and Supersymmetry*, Springer-Verlag New York Inc., 1986.

[18] N.G. Deshpande, J.F. Gunion, B. Kayser, and F. Olness, *Phys. Rev.* **D 44** (1991) 837.

[19] R.N. Mohapatra and A. Rašin, *Phys. Rev. Lett.* **76** (1996) 3490 ; *Phys. Rev.***D54**, 5835 (1996); R. Kuchimanchi, *Phys. Rev. Lett.* **76**, 3486 (1996); R. N. Mohapatra, A. Rasin and G. Senjanović, *Phys. Rev. Lett.***79**, 4744 (1997);C. S. Aulakh, A. Melfo and G. Senjanović, *Phys. Rev.* **D 57** (1998).

[20] R. Kuchimanchi, R.N. Mohapatra, *Phys. Rev. Lett.* **75** (1995) 3989.

[21] S. Dimopoulos and L. Hall, *Phys. Lett.* **B207** (1987) 210.

[22] V. Barger, G. Giudice and T. Han, *Phys. Rev.* **D40** (1989) 2987.

[23] R.A. Alanakyan, hep-ph/9709501 (1997) and hep-ph/9804247 (1998).

[24] J. Kalinowski, R. Rückl, H. Spiesberger and P. Zerwas,*Phys. Lett.* **B406**(1997)314.

[25] B.C. Allanach, H. Dreiner, P. Morawitz and M.D. Williams, *Phys. Lett.* **B420** (1998) 307.

[26] H. Klapdor-Kleingrothaus, in these Proceedings.

[27] J. Laitinen and J. Maalampi, in preparation.

Experimental review of $\not{R}$-parity violation searches at LEP collider

R Nicolaidou †

† Institut des Sciences Nucléaires,
53 Avenue des Martyrs, Grenoble, France

Abstract. The current status of $\not{R}$-parity violation searches in e^+e^- collisions at LEP is reviewed. After a brief introduction on the phenomenology of the $\not{R}$-parity violation and the working framework at LEP, the most recent results on different searches of all four collaborations (ALEPH, DELPHI, OPAL, L3) are stated. Reported results (unless stated otherwise) are obtained from 1997 data at center of mass energy of 183 GeV.

1. Introduction

In the recent years, supersymmetry (SUSY) provides the framework of possible physics beyond the standard model (SM). This symmetry transforms bosons into fermions and vice-versa, predicting the existence of new particles that differ from their standard model partners by a half-integer spin. One of the simplest model available is the Minimal Supersymmetric Standard Model (MSSM) [1] which contains the minimal number of new particles and interactions that are consistent with the SM gauge group. In this model, many problems such as the hierarchy problem and the unification of couplings are solved in a very elegant way [2].

Nevertheless, this symmetric picture is modified by two facts. First the fact that no partner of standard model particles, degenerate in mass, has been found yet, implies that the symmetry is not exact but has to be broken in a way that the supersymmetric solutions to hierarchy problems are still valid. Second, the fact that many measurements on particle flavor properties such as the proton lifetime, are not affected by a possible virtual exchange of these new particles, implies that the parameters of the standard model particles have to be protected by a new quantum number which is called R-parity R_p [3]. This number assures the conservation of the $(B - L)$ symmetry, where B represents the baryonic number and L the leptonic number. For a particle of spin S, the R-parity is expressed in the form R_p=$(-1)^{3B+L+2S}$. Standard model particles have even R-parity and the corresponding superpartners have odd R-parity. The conservation of R-parity has two major phenomelogical consequences: a) supersymmetric particles are produced in pairs and b) the Lightest Supersymmetric Particle (LSP) is stable.

The mass spectrum of supersymmetric particles (sparticles), is obtained assuming a SUSY breaking mechanism which introduces many new parameters.[1] Nevertheless, out of this multiparametric space, LEP searches for sparticles are assuming simplifying suppositions on sparticle masses and their relationships. From the possible mixing of supersymmetric partners of the gauge and Higgs bosons (the so called *gauginos* and *higgsinos*) the physical mass eigenstates obtained are called *charginos* and *neutralinos.* Their masses and mixing angles are determined by the values of four parameters M_1, M_2, the U(1) and SU(2) gaugino mass parameters at the electroweak scale, μ, the mixing mass term of the Higgs doublets at the electroweak scale and $tan\beta$, the ratio of the vacuum expectation values of the two Higgs doublets. Assuming that gaugino masses unify at the GUT scale implies $M_1 = \frac{5}{3}\tan^2\theta_W M_2 \simeq \frac{1}{2}M_2$ at the electroweak scale. Finally the fifth important parameter m_0, is defined as the common sfermion mass at GUT scale.

One approach to go *beyond the MSSM* is to retain the minimal particle content of the MSSM, but remove the assumption of R-parity invariance. In this scenario, the supersymmetric Langrangian derived within the MSSM is modified by the addition of new invariant terms [6]:

$$\lambda_{ijk} L_i L_j \bar{E}_k + \lambda'_{ijk} L_i Q_j \bar{D}_k + \lambda''_{ijk} \bar{U}_i \bar{D}_j \bar{D}_k$$

where L and E denote the lepton superfields and Q, U, D the quark superfields. The λ_{ijk}, λ'_{ijk} and λ''_{ijk} are the new Yukawa couplings where i, j and k are the generation indices running from 1 to 3. The first two terms violate L explicitly whereas the last term violates B explicitly. Due to SU(2) invariance, there are only 9 allowed λ terms (i.e. $i \neq j$). There are 27 allowed λ' terms and 9 allowed λ'' (i.e. $j \neq k$) due to SU(3) invariance. But all R-parity violating ($\not{R}_p$) terms cannot be simultaneously present, otherwise the proton would rapidly decay [7, 8].

In $\not{R}_p$ models, the Lightest Supersymmetric Particle (LSP) is allowed to decay to standard fermions. Therefore the signatures of the supersymmetric particle production are modified compared to the expected signatures

[1] The details of this mechanism are not of present interest. References can be found in [4, 5].

in case of R-parity conservation. Moreover, the single sparticle production is possible.

Two kinds of $\not\!R_p$ decays exist: direct and indirect ones. In a *direct* $\not\!R_p$ decay the sparticle decays directly or via a virtual exchange to standard particles through an $\not\!R_p$ vertex. This is always the case when the sparticle is the LSP. If e.g. the $\tilde{\nu}$ is the LSP, it can decay directly to a pair of fermions through the $\not\!R_p$ operators (see fig. 1). If on the other hand the lightest neutralino $\tilde{\chi}_1^o$ is the LSP, then it can decay into a fermion virtual-sfermion pair with the subsequent decay of the sfermion to standard fermions via the R-parity terms (see fig. 2). In an *indirect* decay, the sparticle first decays through an R_p-conserving vertex to a standard particle and an on-shell sparticle which then decays through an $\not\!R_p$ vertex. This mode usually dominates when there is enough phase space between "mother" and "daughter" sparticles. As a rule of thumb, when the mass difference between these 2 sparticles is larger than 5-10 GeV, the indirect mode tends to dominate. But regions of the parameter space where one has a "dynamic" suppression of the R_p-conserving modes also exist. In this case, even if the sparticle is not the LSP, it decays through an $\not\!R_p$ mode. One has e.g. the case of the light stop decay, where the R_p decay to charm and neutralino is naturally suppressed. A typical example of indirect decay is the R_p decay $\tilde{\chi}_1^+ \to \tilde{\chi}_1^0 + W^{*+}$ and the subsequent decay of $\tilde{\chi}_1^0$ through the $\not\!R_p$ couplings. In the case of e.g. λ'' dominance, the direct $\tilde{\chi}_1^+$ decay has a signature of 3 jets, while the indirect would give either 5 jets or 3 jets + 1 lepton + missing energy (coming from the undetected neutrinos), depending on the decay modes of W^*.

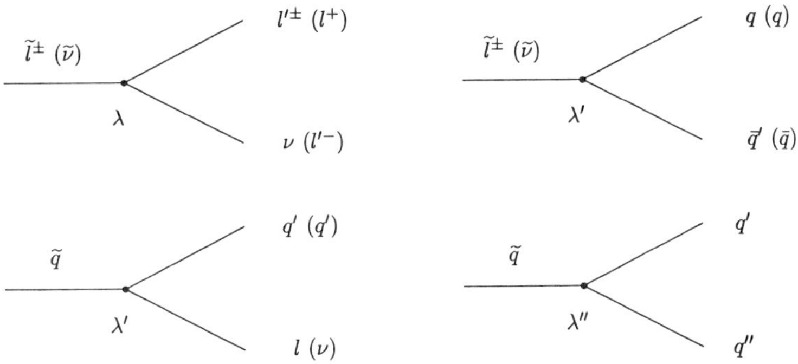

Figure 1. $\not\!R_p$ decays of sfermions.

In case of pair production of supersymmetric particles (see figure 3), R_p is conserved at the production vertex; the cross section does not depend

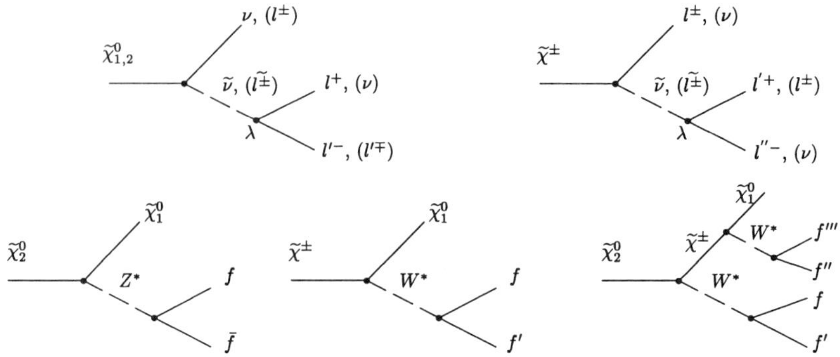

Figure 2. upper part: $\widetilde{\chi}^0$ and $\widetilde{\chi}^{\pm}$ direct decay via λ coupling, lower part: $\widetilde{\chi}^0$ and $\widetilde{\chi}^{\pm}$ indirect decay (valid for any coupling).

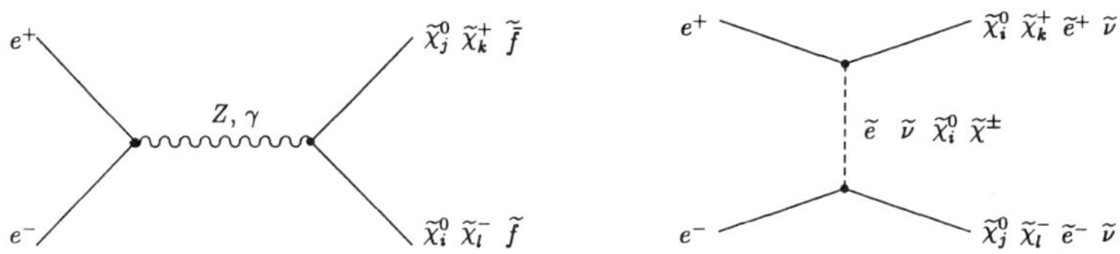

Figure 3. Neutralino, chargino and sfermion pair production diagrams $(i, j = 1, ..4; k, l = 1, 2)$. Only charged particles are produced via a γ in the s-channel.

on the $\not{R}_p$ couplings. On the contrary, the $\not{R}_p$ decay width depends on the λ coupling strength, which then determines the mean decay length of the LSP. If the LSP is a neutralino or a chargino, the mean decay length is given by [9]:

$$L(cm) = 0.3\ (\beta\gamma) \left(\frac{m_{\tilde{l}}}{100\mathrm{GeV/c^2}} \right)^4 \left(\frac{1\mathrm{GeV/c^2}}{\mathrm{m}_{\widetilde{\chi}}} \right)^5 \frac{1}{\lambda^2} \tag{1}$$

where $\lambda = \lambda_{ijk}, \sqrt{3}\lambda'_{ijk}$, or $\sqrt{3}\lambda''_{ijk}$, and with $\beta\gamma = P_{\widetilde{\chi}}/m_{\widetilde{\chi}}$. If the LSP is a sfermion, the previous formula becomes:

$$L(cm) = 10^{-12}\ (\beta\gamma) \left(\frac{1\mathrm{GeV/c^2}}{\mathrm{m}_{\tilde{\mathrm{f}}}} \right) \frac{1}{\lambda^2} \tag{2}$$

The condition that the LSP decays close to the production vertex ($L \lessapprox 1$ cm), implies a lower limit in sensitivity on the λ coupling in the order of $10^{-4} - 10^{-5}$ in case of $\widetilde{\chi}^0$, $\widetilde{\chi}^{\pm}$, and 10^{-7} in case of sfermions. Upper limits on the λ couplings can be derived from Standard Model processes [10], mainly charged-current universality, lepton universality, $\nu_\mu - e$ scattering, forward-backward asymmetry in e^+e^- collisions, and bounds on ν_e-majorana mass.

2. Pair production of neutralinos, charginos and sfermions

The pair production of neutralinos, charginos and sfermions is described by the standard MSSM processes. As previously mentioned, the new $\not{R}_p$ coupling affect only the decay and not the production of the sparticles.

Searches for pair production of supersymmetric particles in e^+e^- collisions at LEP [11] have been performed, assuming one $\not{R}_p$ coupling dominant at a time and that the strength of the couplings is such that the lifetimes can be neglected (i.e. decays near the interaction point). Under the assumption that only one operator is responsible for the $\not{R}_p$ decay, different searches are classified as ones with λ, λ' or λ'' dominance, the results of which are described in sections 2.1, 2.2, 2.3 respectively.

2.1. Dominance of λ couplings

Studies in the λ dominance case have been done by all four collaborations.

Neutralinos and charginos decay into charged leptons and neutrinos. The decay of the lightest neutralino leads to one neutrino and two charged leptons. The heavier neutralinos and the charginos, depending on their mass difference with $\widetilde{\chi}^0_1$, can either decay directly into 3 standard fermions, or decay to $\widetilde{\chi}^0_1$, via for example virtual Z or W, as illustrated in figure 2 and in table 1. The neutralino being the LSP in most of the parameter space decays directly, while the chargino decays mostly indirectly. The signatures of all these decay modes are either purely leptonic or semileptonic. The LEP experiments have searched for the above signatures. The selection efficiencies are of the order of 50-90% for the most efficient case that contain muons or electrons in the final state (e.g λ_{122}) and 30-50% for the most difficult case that contain taus in the final state (e.g λ_{133}). Searches for both cases have been performed and the most conservative limits have been derived. For low m_0 values ($m_0 < 100$ GeV/c^2) the neutralino production cross section is enhanced by the t-channel $\tilde{e}$ echange, while the chargino cross section decreases due to destructive interference between s and t-channel amplitudes. The situation is reversed for large values of m_0. Both cases were studied by DELPHI as can be seen in figure 6 which shows the exclusion area obtained by these searches for two values of $tan\beta$ and m_0. From these exclusion areas, a lower limit on neutralino mass of 27 GeV/c^2 is obtained by DELPHI independently of m_0 values and for all possible values of *tanβ* as shown in figure 7. These searches, as has been shown by all four experiments, have no difficulty in reaching the kinematic limit for chargino pair production, therefore at center of mass energy of 183 GeV a chargino with mass smaller than 91 GeV/c^2 is excluded. Similar results are obtained by R-parity conservation studies.

direct decay	indirect decay
$\tilde{\chi}_1^0 \to \nu l^+ l^-$	
$\tilde{\chi}_2^0 \to \nu l^+ l^-$	$\tilde{\chi}_2^0 \to Z^* \, \tilde{\chi}_1^0, \quad Z^* \to ff$ $\tilde{\chi}_2^0 \to \tilde{\chi}_1^0 l^+ l^-, \, \tilde{\chi}_1^0 \gamma$
$\tilde{\chi}_1^+ \to \nu \nu l^+, \, l^+ l^- l^+$	$\tilde{\chi}_1^+ \to W^{*+} \, \tilde{\chi}_1^0, \; W^{*+} \to f \bar{f}'$ $\tilde{\chi}_1^+ \to \tilde{\chi}_1^0 \nu_l l^+$

Table 1. Possible decays of neutralinos and charginos when a λ coupling is dominant.

Charged slepton pairs decay directly to 2 charged leptons and neutrinos ($\tilde{l}^\pm \to l^\pm \nu$), while the indirect decays ($\tilde{l}^\pm \to l^\pm \tilde{\chi}^0$, $\nu \tilde{\chi}^\pm$) give the same signatures as the neutralino or chargino decays but with 2 extra leptons or missing energy in the final state (see table 2). The dominance of indirect over direct decays depends on the value of λ couplings and the mass difference between the slepton and the neutralino. The most conservative mass limits come from direct decay searches ($2l + \not{E}$ in the final state), in which the remaining background from SM processes is quite high. OPAL obtains limits for $\tilde{\mu}$, $\tilde{\tau}$ production at 66 GeV/c^2 while ALEPH and DELPHI limits are of the order of 61 GeV/c^2 (figure 8). Contrary to the $\tilde{\mu}$, $\tilde{\tau}$ pair production cross section which depends only on the slepton mass, in the case of $\tilde{e}$, limits depend strongly on MSSM parameters as the cross section depends also on the neutralino mass exchanged in the t-channel.

Sneutrino pairs decay directly to 4 leptons each sneutrino giving a clear resonant signature of 2 leptons ($\tilde{\nu} \to l^+ l^-$). The indirect decay ($\tilde{\nu} \to \nu \, \tilde{\chi}_1^0 \to \nu l^+ l^- \nu$) leads also to 4 leptons but with missing energy in the final state (see table 2). Contrary to the charged slepton case, the most conservative mass limit now is given by the indirect decays. OPAL limit for $\tilde{\nu}_\mu$, $\tilde{\nu}_\tau$ production is at 66 GeV/c^2. ALEPH and DELPHI exclude a $\tilde{\nu}_\mu$, $\tilde{\nu}_\tau$ with mass lower than 62 GeV/c^2. Figure 9 shows this limit as a function of $\tilde{\chi}_1^0$ mass. As in the case of $\tilde{e}$, the $\tilde{\nu}_e$ production cross section depends on the choise of MSSM parameters, via the mass of the chargino, exchanged in the t-channel. The most conservative limit for the $\tilde{\nu}_e$ is the same as the $\tilde{\nu}_\mu$, $\tilde{\nu}_\tau$ one.

Squark indirect decays through λ couplings ($\tilde{t}, (\tilde{b}) \to c(b) \; \tilde{\chi}_1^0$), lead to semileptonic signatures. ALEPH most recent limits at $\sqrt{s}$=183 GeV is

Decay type		Pair production signature
Direct	$\tilde{\nu} \to l^+l^-$	$4l$
Indirect	$\tilde{\nu} \to \nu\ \tilde{\chi}_1^0$	$4l+ \not{E}$
	$\tilde{\nu} \to \nu\ \tilde{\chi}_2^0$	same as χ^0 analysis
	$\tilde{\nu} \to l^{\pm}\tilde{\chi}^{\mp}$	multilepton or lepton+jets
Direct	$\tilde{l}^{\pm} \to l^{\pm}\nu$	$2l + \not{E}$
Indirect	$\tilde{l}^{\pm} \to l^{\pm}\ \tilde{\chi}_1^0$ $\tilde{l}^{\pm} \to \nu\ \tilde{\chi}^{\pm}$	same as $\tilde{\chi}^{\pm}$, $\tilde{\chi}^0$ analyses with ≤ 2 extra leptons or $\not{E}$

Table 2. Final states in charged slepton and sneutrino pair production when a λ coupling is dominant. $\not{E}$ on table, states for missing energy.

61(44) GeV/c^2 for $\tilde{t}$ and $m_{\tilde{\chi}^0}$ >23 GeV/c^2 in the case of minimal(maximal) mixing and for the most difficult case (i.e. λ_{133} coupling), as can be seen in figure 10. Previous studies at $\sqrt{s}$=172 GeV, done by ALEPH also, exclude sbottoms with mass smaller than 58 GeV/c^2 in case of no mixing.

2.2. Dominance of λ' couplings

All kind of sparticles are allowed to decay either directly or indirectly through the λ' terms.
Neutralino and chargino direct and indirect decays lead to multi-jets+charged leptons with or without $\not{E}$ or multijets $+\not{E}$ signatures as can be seen in table 3. The exclusion area obtained by ALEPH and DELPHI in μ M_2 plane also reach the kinematic limit for chargino pair production in case of large sfermion masses (i.e. large values of m_0), as shown in figure 11.

Charged sleptons and sneutrinos decay to topologies with at least 4-jets as can be seen in table 4. Due to the structure of λ' operator, only left handed sleptons are allowed to decay directly through λ' terms, while there is no restriction for the indirect decays. Concerning the direct decays, a typical "4-jet" analysis is used to set limits on slepton masses. The corresponding ALEPH mass limits are $m_{\tilde{\mu}_L,\tilde{\tau}_L}$ >61 GeV/c^2 and $m_{\tilde{\nu}_\mu,\tilde{\nu}_\tau}$ >60 GeV/c^2. DELPHI has studied in particular the λ'_{333} operator controlling the decay of $\tilde{\nu}_\tau$ to a pair of b quarks, giving a very clear signature of 4b's in the final state, as a possible interpretation of small remnant deviations in b asymmetries [12]. From this search, a limit of $\tilde{\nu}_\tau$ mass lower than 76 GeV/c^2 has been set, while for couplings leading to final states with no b jets, a limit of $m_{\tilde{\tau}_L}$ >71 GeV/$c^2$and $\tilde{\nu}_\tau$ >60 GeV/c^2 has been derived. OPAL studies of left handed sleptons indirect decays, lead to $m_{\tilde{\mu}_R,(\tilde{\tau}_R)}$ >53(65) GeV/c^2,

Signature	Channel(s)
4 jets ($\not\!\!E$)	$\tilde\chi_1^0\tilde\chi_1^0$ - $\tilde\chi_1^0\tilde\chi_2^0$ (D,I) - $\tilde\chi_1^+\tilde\chi_1^-$ (D)
6 jets ($\not\!\!E$)	$\tilde\chi_1^0\tilde\chi_2^0$ (I)
8 jets ($\not\!\!E$)	$\tilde\chi_1^+\tilde\chi_1^-$ (I)
1 lepton 4 jets ($\not\!\!E$)	$\tilde\chi_1^0\tilde\chi_1^0$ - $\tilde\chi_1^0\tilde\chi_2^0$ (D,I) - $\tilde\chi_1^+\tilde\chi_1^-$ (D)
1 lepton 6 jets ($\not\!\!E$)	$\tilde\chi_1^0\tilde\chi_2^0$ (I) - $\tilde\chi_1^+\tilde\chi_1^-$ (I)
1 lepton 8 jets ($\not\!\!E$)	$\tilde\chi_1^+\tilde\chi_1^-$ (I)
2 leptons 4 jets	$\tilde\chi_1^0\tilde\chi_1^0$ - $\tilde\chi_1^0\tilde\chi_2^0$ (D) - $\tilde\chi_1^+\tilde\chi_1^-$ (I)
2 leptons 4 jets ($\not\!\!E$)	$\tilde\chi_1^0\tilde\chi_2^0$ (I) - $\tilde\chi_1^+\tilde\chi_1^-$ (D)
2 leptons 6 jets	$\tilde\chi_1^0\tilde\chi_2^0$ (I)
2 leptons 6 jets ($\not\!\!E$)	$\tilde\chi_1^+\tilde\chi_1^-$ (I)
2 leptons 8 jets	$\tilde\chi_1^+\tilde\chi_1^-$ (I)
3 leptons 4 jets ($\not\!\!E$)	$\tilde\chi_1^0\tilde\chi_2^0$ (I) - $\tilde\chi_1^+\tilde\chi_1^-$ (I)
3 leptons 6 jets ($\not\!\!E$)	$\tilde\chi_1^+\tilde\chi_1^-$ (I)
4 leptons 4 jets	$\tilde\chi_1^0\tilde\chi_2^0$ (I)
4 leptons 4 jets ($\not\!\!E$)	$\tilde\chi_1^+\tilde\chi_1^-$ (I)

Table 3. Final states in neutralino and chargino pair production when a λ' coupling is dominant; D (I) means direct (indirect) decay of $\tilde\chi_1^+$ or $\tilde\chi_2^0$.

as shown in figure 12. Finally, for $\tilde e_R$ and $\tilde\nu_e$, limits at 73 GeV/c^2 and 85 GeV/c^2 respectively are obtained by OPAL indirect searches, for low $\tilde\chi_1^0$ mass (see figure 13).

	Decay type	Pair production signature
Direct	$\tilde l_L^\pm(\tilde\nu) \to q\bar{q}'(q\bar{q}')$	4 jets
Indirect	$\tilde l_{L,R}^\pm(\tilde\nu) \to l^\pm\tilde\chi^0(\nu\tilde\chi^0)$	4 jets + $\geq$ 1 lepton
	$\tilde l_{L,R}^\pm(\tilde\nu) \to \nu\tilde\chi^\pm(l^\mp\tilde\chi^\pm)$	with or not $\not\!\!E$

Table 4. Final states of charged slepton and sneutrino pair production when a λ' coupling is dominant.

Squark semileptonic decays would give 2 charged leptons and 2 jets in the final state in the case of $\tilde t$ pair production, assuming the $\tilde t$ is the LSP. OPAL studied these signatures for all three lepton flavour (i.e. e, μ, τ in the final state) separately. The mass limits obtained (the best so far of all four collaborations), are summarized in table 5 and the cross section limit in the tau channel is shown in figure 15. The specificity of the $\tilde t$ channel is that due to the very low width of the $\tilde{t} \rightarrow c\tilde{\chi}_1^0$, this indirect decay is not dominant for λ' values down to $\lambda' > 10^{-4}$. The sensitivity on λ' value reached by these searches is shown in the figure 14, in which DELPHI results are compared with relevant searches from HERA. While the range of the accessible stop masses is modest when compared to the range of H1, λ'_{131} coupling values can be excluded down to 10^{-4} level which is about 2 orders of magnitude below H1 limits.

Limits	$\theta_{\tilde{t}}$= 0 rad	$\theta_{\tilde{t}}$=0.98 rad
$\tilde{t} \rightarrow e + q$	86 GeV	82 GeV
$\tilde{t} \rightarrow \mu + q$	86 GeV	82 GeV
$\tilde{t} \rightarrow \tau + q$	81 GeV	73 GeV

Table 5. OPAL mass limits for stop searches with λ' dominant coupling for the two extreme values of the mixing angle in the electron, muon and tau channels.

2.3. Dominance of λ'' couplings

Neutralinos charginos and squarks can decay through the λ'' terms.

Neutralino and chargino decays lead to multijet topologies as stated in table 6. ALEPH, DELPHI, OPAL studies showed that in this case also, the limits can be extended close to kinematic limit for large sfermion masses (i.e. large value of m_o). Figure 16 shows the exclusion area in the plane μ, M_2.

Squarks decay directly to 4 jet topologies through the λ'' terms. The stop can decay to 2 down d quarks ($\tilde{t} \rightarrow dd$) and the sbottom to an up u and a down d quark ($\tilde{b} \rightarrow ud$). In case of stop, the direct decay dominates over the indirect ($\tilde{t} \rightarrow c\ \tilde{\chi}_1^0$ through suppressed CKM matix elements), while for sbottoms the indirect decay ($\tilde{b} \rightarrow b\ \tilde{\chi}_1^0$) dominates whenever it is kinematically possible, leading to 8 jet topologies. DELPHI studies on the 4-jet topologies combining the data at $\sqrt{s}$=161,172 and 183 GeV, exclude stop masses lower than 76(73) GeV/c^2 for mixing angle $\theta_{\tilde{t}} = 0^0(56^0)$ and sbottom masses lower than 74(62) GeV/c^2 for mixing angle $\theta_{\tilde{b}} = 0^0(40^0)$ respectively (see figure 17).

3. Single production of squarks and sleptons

R-parity violation searches can be extended to masses close to center of mass energy, provided that at least one $\not{R}_p$ vertex is introduced at the

final states	direct decay of	indirect decay of
$6j$	$\tilde{\chi}_1^0\tilde{\chi}_1^0$, $\tilde{\chi}_2^0\tilde{\chi}_1^0$, $\tilde{\chi}_1^+\tilde{\chi}_1^-$	
$6j+\not{E}$		$\tilde{\chi}_2^0\tilde{\chi}_1^0$
$6j+$ 2 l		$\tilde{\chi}_2^0\tilde{\chi}_1^0$
$6j+$ 2 l $+\not{E}$		$\tilde{\chi}_1^+\tilde{\chi}_1^-$
$8j$		$\tilde{\chi}_2^0\tilde{\chi}_1^0$
$8j+$ 1 l $+\not{E}$		$\tilde{\chi}_1^+\tilde{\chi}_1^-$
$10j$		$\tilde{\chi}_1^+\tilde{\chi}_1^-$

Table 6. Final states in neutralino and chargino pair production with a dominant λ'' coupling.

production level. This is the case of single production processes, which are of two types, as described in sections 3.1 and 3.2.

3.1. Single squark and slepton production

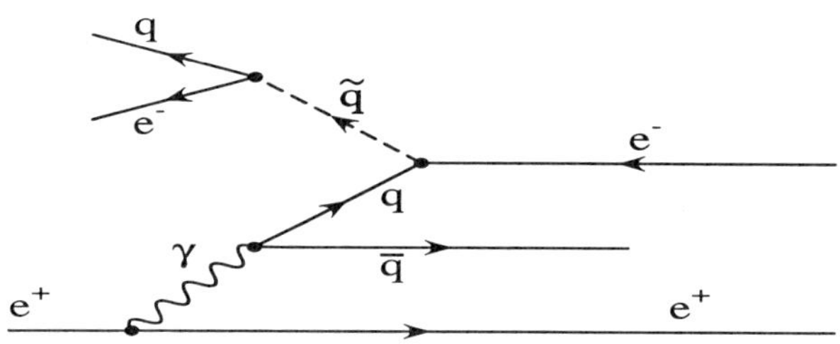

Figure 4. Single squark production via a resolved γ.

It is manifested through the interaction of a quark or a lepton contained in a radiated resolved γ from one of the incoming particles (e^+ or e^-) [13] as shown in figure 4. Direct decays of this type give a striking signature of (eq or νq) of one lepton or missing energy opposite to a jet, while indirectly, an R-parity decay of $q\tilde{\chi}_1^0$ or $q'\tilde{\chi}^\pm$ lead to a topology of one jet opposite to 2 jets and a lepton or missing energy. The "leptoquark" type analyses can be used to set limits on squark masses and couplings involved, as it is done by DELPHI (see [14]).

3.2. $\tilde{\nu}_\mu$ $\tilde{\nu}_\tau$ resonant production

The operators involved in this type are λ_{121} or λ_{131} [10]. When the neutralino or chargino mass is lower than the $\tilde{\nu}_{\mu,\tau}$ mass one $\not{R}_p$ vertex is present at the production level and one at the decay (figure 5). The signature in

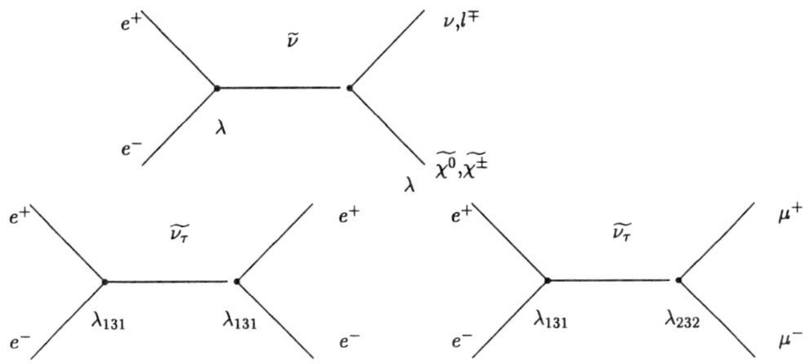

Figure 5. Sneutrino resonant production.Upper plot: $m_{\tilde{\nu}} > m_{\tilde{\chi}^0, \tilde{\chi}^\pm}$ with one $\not{R}_p$ vertex at the production. Two down plots: $m_{\tilde{\nu}} < m_{\tilde{\chi}^0, \tilde{\chi}^\pm}$ with two $\not{R}_p$ vertices on the production.

this case is 2 or 4 leptons with or without missing energy. DELPHI has set limits [15] on λ_{121} between 0.003-0.007 for a $\tilde{\nu}$ between 130-172 GeV/c^2. When the $\tilde{\chi}_1^0$ or $\tilde{\chi}^\pm$ have masses larger or equal than the $\tilde{\nu}$ mass, the $\tilde{\nu}$ can decay either through the dominant operators responsible for its production (λ_{121}, λ_{131}) or some others (figure 5). L3 [16] studied the possibility of anomalous e^+e^-, $\mu^+\mu^-$ distributions with all the data collected at LEP at $\sqrt{s}$=91,130,136,161,172,183 GeV. No evidence for deviations of the measured cross sections and forward backward asymmetries for these reactions from standard model expectations is found. Upper limits for the couplings λ_{121}, λ_{131}, λ_{232} for sneutrino masses up to 220 GeV/c^2 are determined, as shown in figure 18.

4. Conclusion

Searches for R-parity violation have been performed at LEP for all types of sparticles and all three types of Yukawa couplings λ, λ', λ''. LEP searches for supersymmetry now cover most of the signatures originating from R-parity violation with a sensitivity comparable to the one obtained in the R_p-conserving case. In most cases limits from both models follow closely; therefore we can tentatively claim that at LEP we could obtain "assumption free" limits (e.g the neutralino mass), provided that we have enough luminosity and we continue to look for all possible signatures of SUSY manifestations, leaving no unexplored gap in supersymmetric parameter space.

Acknowledgments

I would like to thank my colleagues from the four LEP Collaborations who provided me with plots and results helping me to prepare this report. I am grateful to C. Bérat for helpful discussions and for the very careful reading of this report. It is a pleasure for me to thank the organizers of the conference for providing such a pleasant atmosphere.

References

[1] For reviews, see e.g. H.P. Nilles, *Phys. Rep.***110** (1984) 1;
H.E. Haber and G.L. Kane, *Phys. Rep.* **117** (1985) 75.

[2] For a recent review see e.g S. Dawson, hep-ph/9712464 and references therein.

[3] P. Fayet, *Phys. Lett.* **B69** (1977) 489;.
G. Farrar and P. Fayet, *Phys. Lett.* **B76** (1978) 575.

[4] For a recent review see P. Nath hep-ph/9708221 and references therein.

[5] For a recent review see G.F. Guidice and R. Rattazzi hep-ph/9801271 and references therein.

[6] S. Weinberg, *Phys. Rev.* **D26** (1982) 287.

[7] I. Hinchliffe and T. Kaeding, *Phys. Rev.* **D47** (1993) 279.

[8] C.E. Carlson et al., *Phys. Lett.* **B357** (1995) 99.

[9] S. Dawson, *Nucl. Phys.* **B261** (1985) 297.
H. Dreiner and G.G. Ross, *Nucl. Phys.* **B365** (1991) 597.

[10] V. Barger, G.F. Guidice, T. Han, *Phys. Rev.* **D40** (1989) 2987.
H. Dreiner, hep-ph/9707435.
G. Bhattacharyya, hep-ph/9709395 and *Nucl. Phys.***B** *(Proc. Suppl.)* **52A** (1997) 83.

[11] ALEPH HEP'98 no 949, DELPHI HEP'98 no 210,
OPAL HEP'98 no 278,279, L3 HEP'98 no 496.
See also older publications: ALEPH *Phys. Rev. Lett.* **B349** (1995) 238,ALEPH *Phys. Rev. Lett.* **B384** (1996) 461,
ALEPH 98-017 CONF 98-007,ALEPH 98-027 CONF 98-015,
OPAL *Phys. Rev. Lett.* **B313** (1993) 333,
OPAL *Physics Note* PN-307,OPAL *Physics Note* PN-335.

[12] J. Erler, J.L. Feng and N. Polonsky, *Phys. Rev. Lett.* **78** (1997) 3063.

[13] M. Donchenski, S. Godfrey *Phys. Rev.* **D49** (1994) 6220.

[14] F. Barao et al., DELPHI 98-35 CONF 127.

[15] DELPHI HEP'97 no 589

[16] L3 coll., "Search for R-parity Breaking Sneutrino Exchange at LEP", L3 note no 2235. See also L3 *Phys. Rev. Lett.* **B414** (1997) 373.

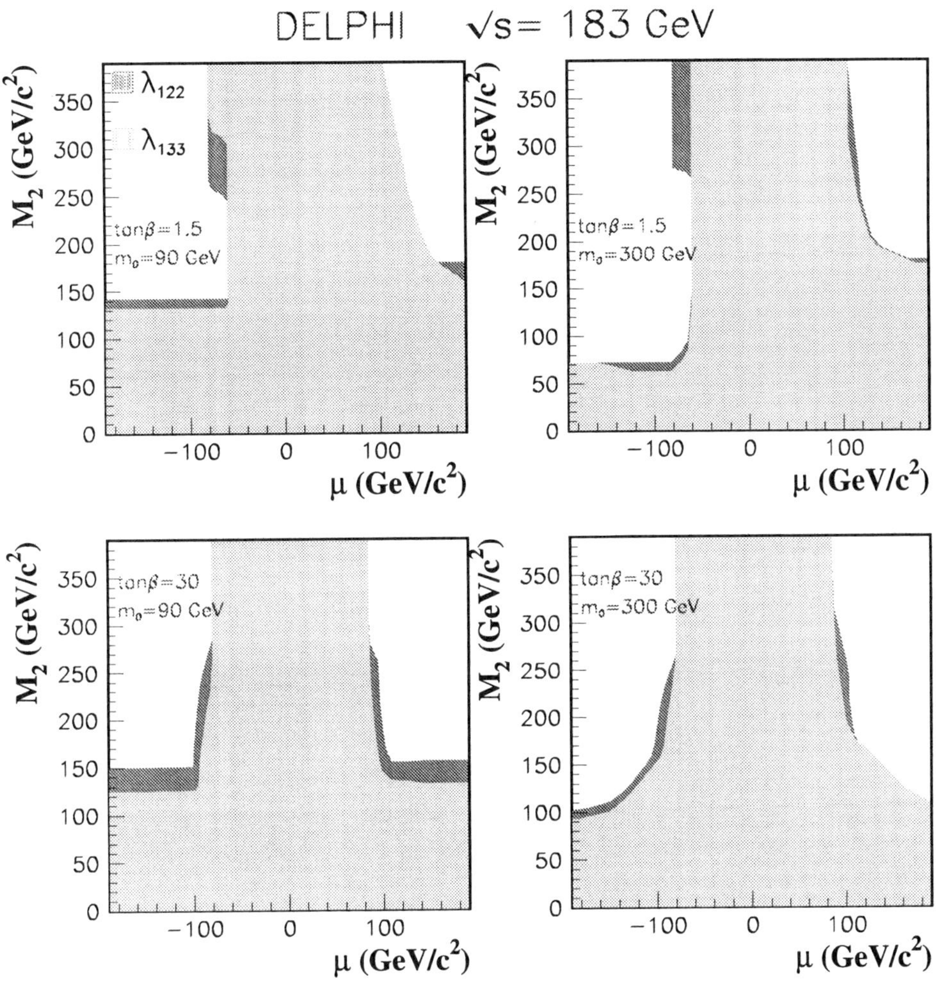

Figure 6. Neutralino and chargino searches with a dominant λ coupling: regions in μ, M_2 parameter space excluded at 95 % C.L. for two values of $tan\beta$ and two values of m_0. The exclusion area obtained from the λ_{133} search is shown in light grey and the corresponding area for the λ_{122} search is shown in dark grey. The second exclusion area includes the first. The data collected at E_{CM} = 183 GeV are used.

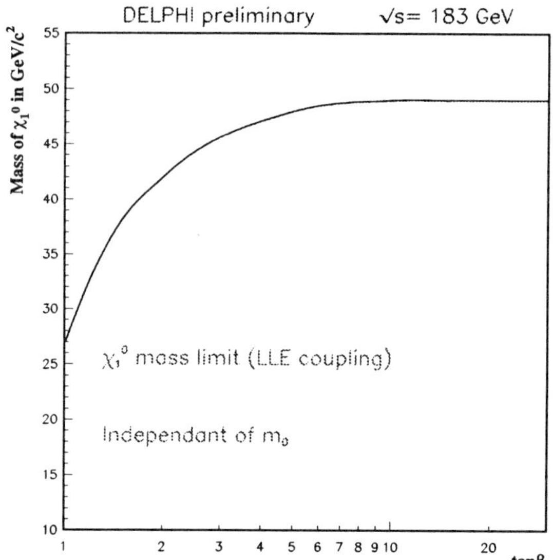

Figure 7. The lightest neutralino mass as a function of $tan\beta$ at 95 % confidence level. This limit is independent of the choice of m_0 and the generation indices i,j,k of the λ_{ijk} coupling.

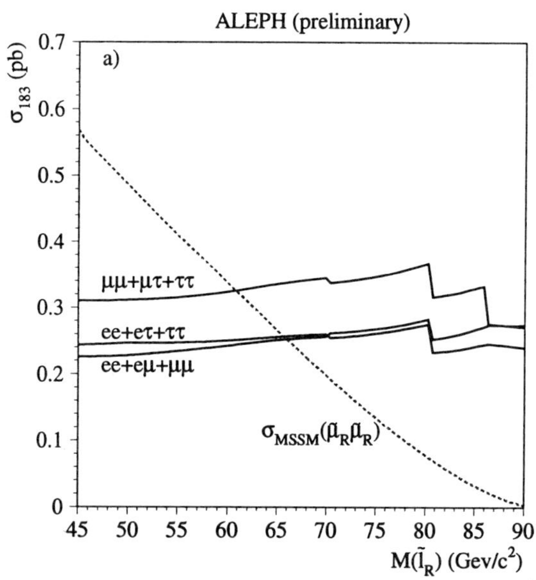

Figure 8. The 95% confidence level exclusion cross sections for sleptons decaying via λ couplings. The MSSM cross section for pair production of right-handed smuons is superimposed.

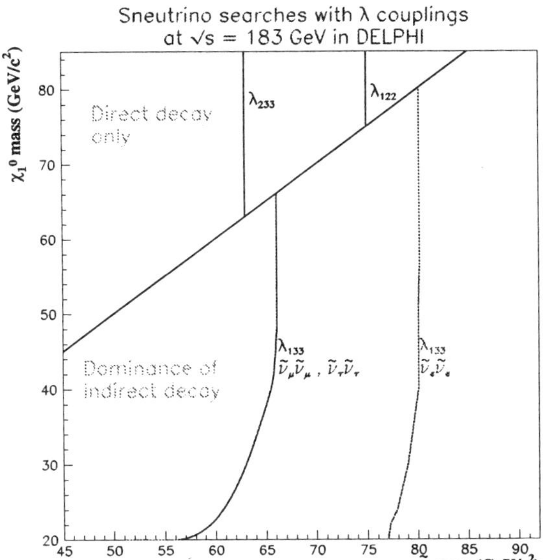

Figure 9. Sneutrino search with λ coupling: exclusion domain in $m_{\tilde{\chi}^0}$ $m_{\tilde{\nu}}$ plane for the $\tilde{\nu}$ pair production cross section; Above the diagonal, only the direct decay is allowed. In the lower part, the indirect decay dominates and the exclusion limit depends also on the $m_{\tilde{\chi}^0}$ mass. In both cases, only the most conservative limit has been reported, for the $\widetilde{\nu}_\mu$ and $\widetilde{\nu}_\tau$ production, and for the $\widetilde{\nu}_e$ in case of chargino mass close to the kinematic limit.

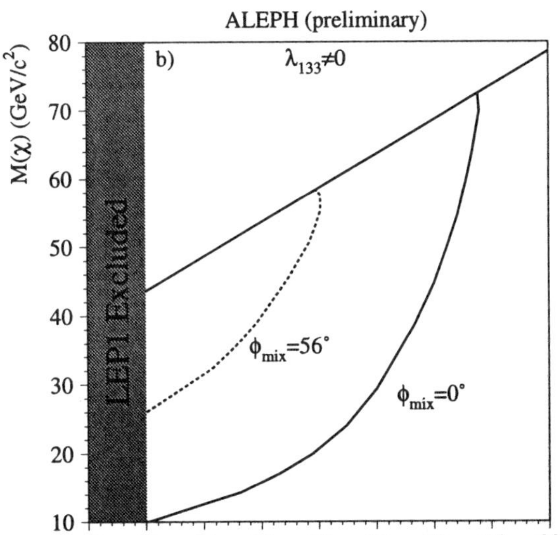

Figure 10. The 95% C.L limits in the $m_{\tilde{\chi}^0}$, $m_{\tilde{t}}$ plane for the λ_{133} coupling. The two mixing angles correspond to the minimum and maximum production cross section.

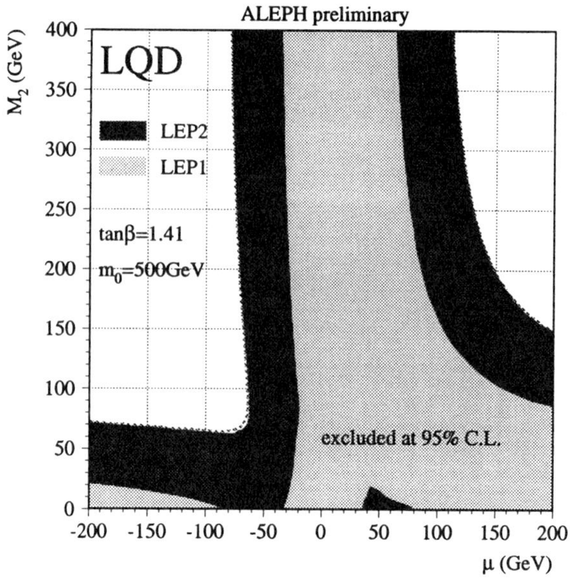

Figure 11. Exclusion region in μ, M$_2$ plane for $\tilde{\chi}_1^0\tilde{\chi}_1^0$, $\tilde{\chi}_1^+\tilde{\chi}_1^-$ production in case of λ' dominant couplings. The dotted line represents the kinematic limit.

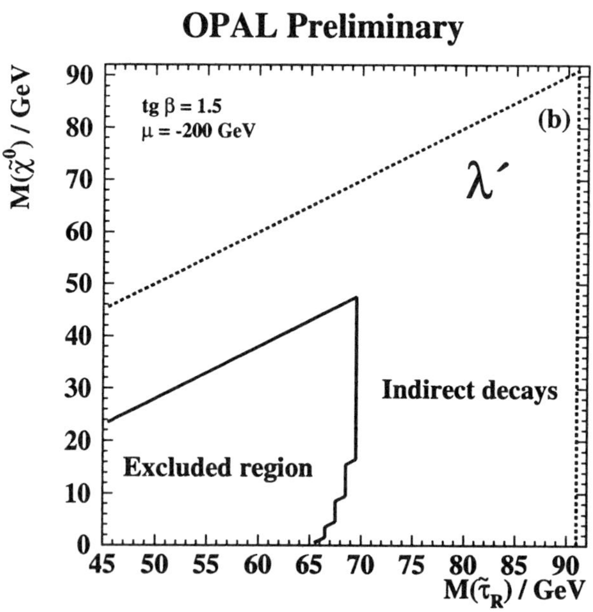

Figure 12. Slepton searches with λ' couplings: exclusion region for $\tilde{\tau}_R^+\tilde{\tau}_R^-$ production in the $(m_{\tilde{\tau}_R}, m_{\tilde{\chi}^0})$ plane at 95% confidence level. The kinematic limit is shown as the dashed line.

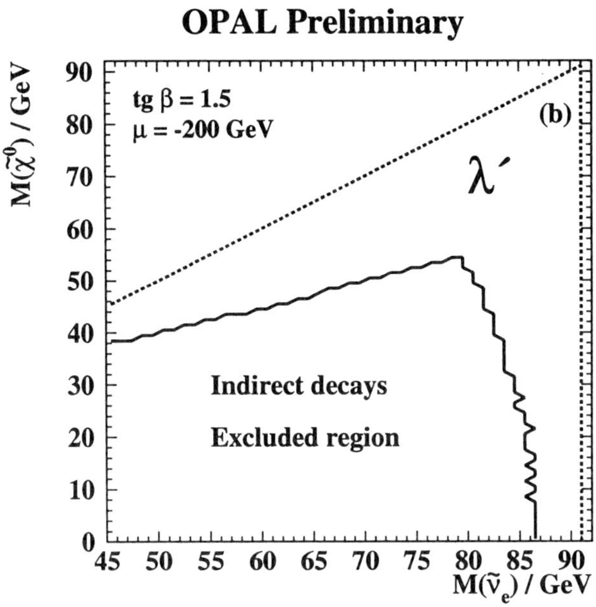

Figure 13. Sneutrino searches with λ' couplings: exclusion region for $\tilde{\nu}_e\tilde{\nu}_e$ production in the $(m_{\tilde{\nu}_e}, m_{\tilde{\chi}^0})$ plane at 95% confidence level. The kinematic limit is shown as the dashed line.

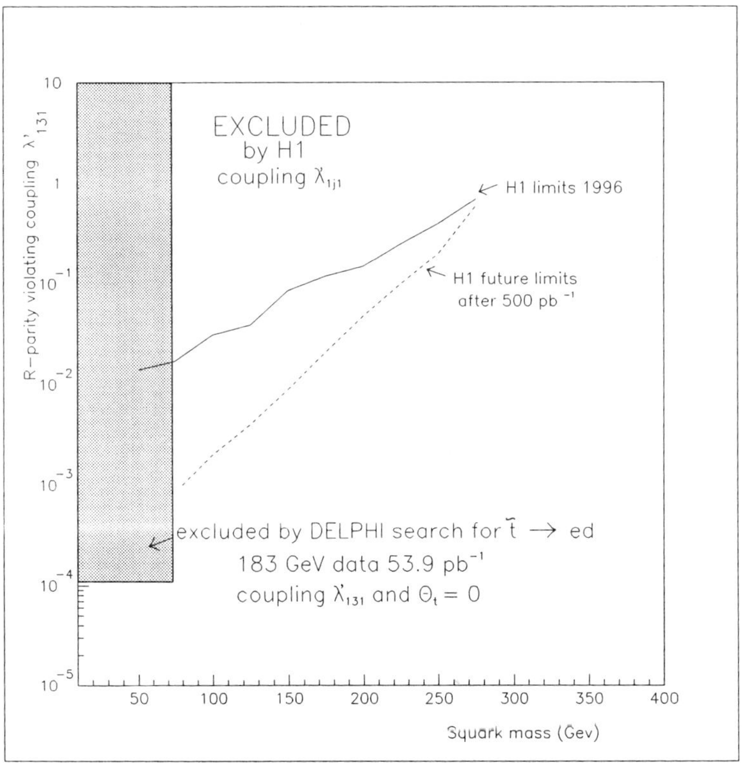

Figure 14. Stop searches with a dominant λ' coupling. Exclusion domain in the λ'_{131} $m_{\tilde{t}}$ plane in case of a pure left stop.

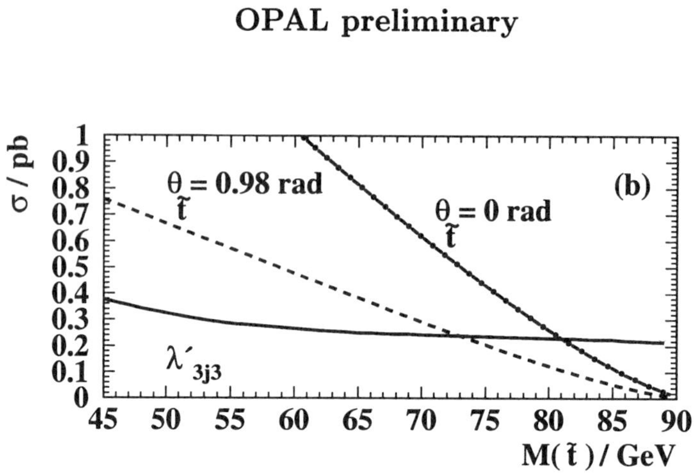

Figure 15. Stop direct decays via λ' coupling: cross section limit in the tau channel. The maximum and minimum cross sections predicted by MSSM for minimal (0 rad) and maximal (0.98 rad) mixing angles is also shown.

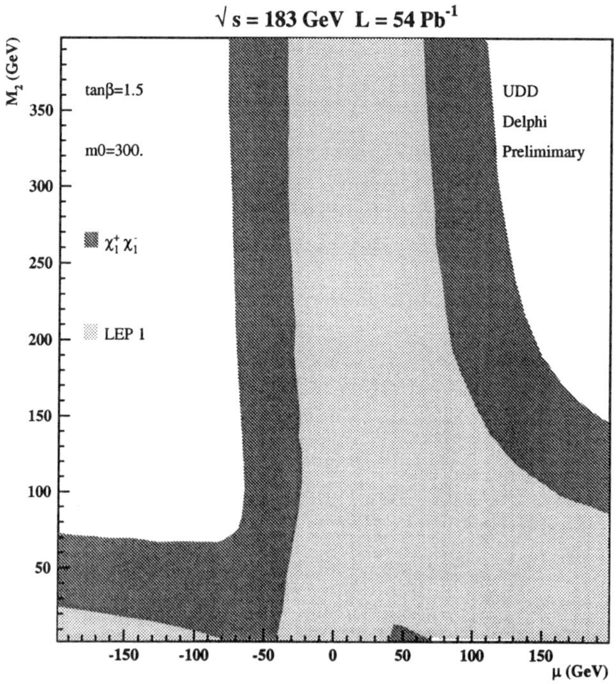

Figure 16. Exclusion region in μ, M_2 plane for $\tilde{\chi}_1^0\tilde{\chi}_1^0$, $\tilde{\chi}_1^+\tilde{\chi}_1^-$ production in case of λ'' dominant couplings.

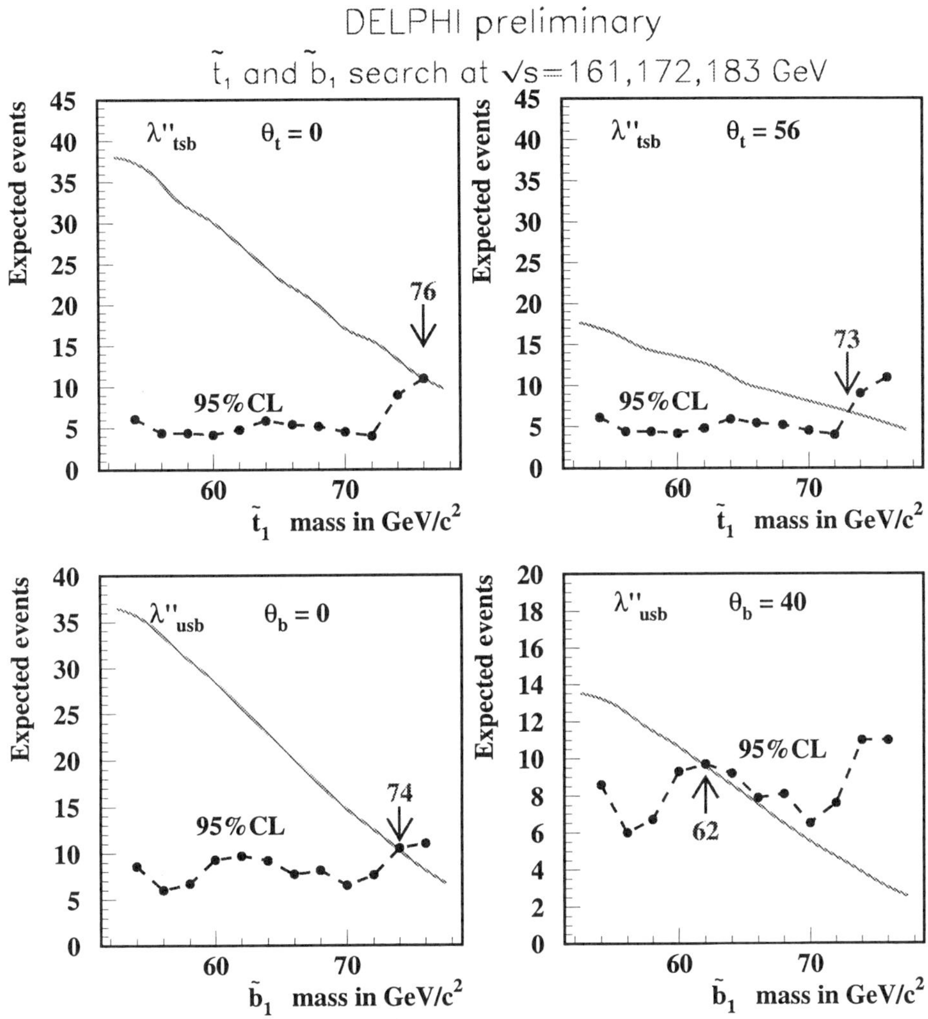

Figure 17. Stop and sbottom analysis with λ''_{UDD} couplings: expected events (dots connected with a dotted line) versus mass at 95% for the $\lambda''_{tsb}(323)$ and the $\lambda''_{usb}(123)$ terms for maximal and minimal mixing angles. The solid line represents the number of expected events from the MSSM cross section taking into account the integrated luminosity ($\mathcal{L} \sim 74$ pb^{-1}) and the efficiency at each center of mass energy.

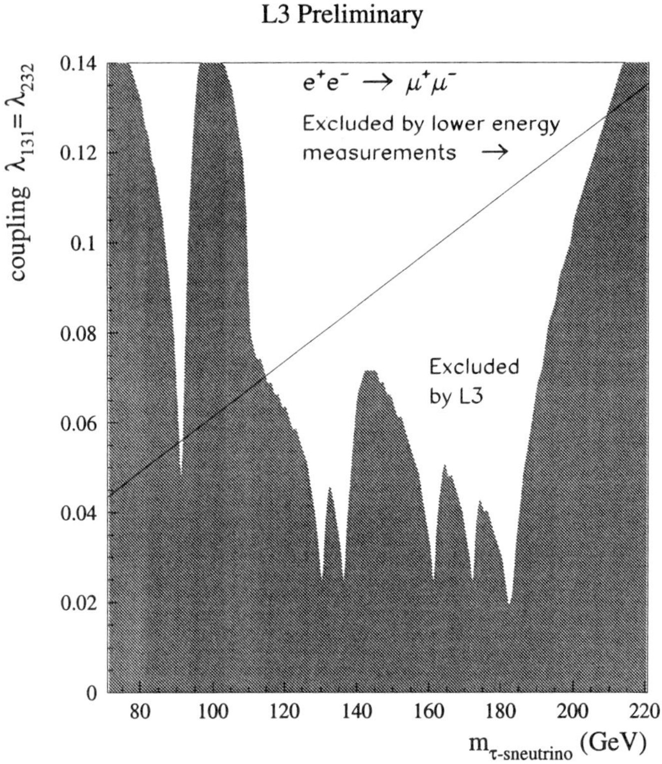

Figure 18. Upper limits on the coupling strenght λ_{232} (assumed to be equal to λ_{131}) as a function of $m_{\tilde{\nu}_\tau}$ derived from the measurements of $\mu^+\mu^-$ pair-production. The area above the straight line is excluded by lower energy measurements at 95% confidence level. The jagged curve is the 95% confidence level upper limit from this analysis.

Search for Charginos and Neutralinos with R–Parity Violation with the L3 Detector at LEPII

Silvia Costantini § [1]

§ Humboldt University Berlin

Abstract. A search for charginos and neutralinos produced in e^+e^- collisions at center-of-mass energies between 161 and 183 GeV has been performed under the assumptions that R-parity is not conserved, that the R-parity violating terms involve only leptonic fields, and that the lifetime of the LSP, assumed to be the lightest neutralino, is negligibly small. No candidate events are found in the total data sample, corresponding to an integrated luminosity of 76.42 pb^{-1}. Limits on the Minimal Supersymmetric Standard Model parameters are set at 95% confidence level.

1. Introduction

The search for supersymmetric particles has been performed in the past under the assumption that R-parity is conserved. R is a multiplicative quantum number defined as:

$$R = (-1)^{3B-L+2S}, \tag{1}$$

where B and L are the baryonic and leptonic quantum numbers and S is the spin. R takes the value +1 for all the ordinary particles, and -1 for their supersymmetric partners. As a consequence, if R-parity is conserved, supersymmetric particles can be produced only in pairs and they decay in cascade to the lightest supersymmetric particle (LSP, usually assumed to

[1] E-mail: Silvia.Costantini@cern.ch.

be the lightest neutralino because of cosmological arguments suggesting the LSP should be colourless and electromagnetically neutral), which is stable. The lowest-threshold SUSY process:

$$e^+e^- \to \tilde{\chi}_1^0\tilde{\chi}_1^0 \tag{2}$$

is thus completely invisible.
However, with the MSSM particle content, the requirement that the theory is supersymmetric, renormalizable and gauge invariant does not imply R-parity conservation, and the most general Lagrangian can contain the following additional terms [1]:

$$\lambda_{ijk}L_iL_j\overline{E_k} + \lambda'_{ijk}L_iQ_j\overline{D_k} + \lambda''_{ijk}\overline{U_iD_jD_k}, \tag{3}$$

after the bilinear term $\mu_i L_i H_2$ has been rotated away. Here i, j and k are the generation indices; L and Q are the left-handed lepton and quark-doublet superfields; $\overline{E}$, $\overline{D}$ and $\overline{U}$ are the right-handed singlet superfields for charged leptons, down and up-type quarks respectively. Antisymmetry requires $\lambda_{ijk} = -\lambda_{jik}$ and $\lambda''_{ijk} = -\lambda''_{ikj}$, so that there are in total 9 + 27 + 9 independent Yukawa couplings. A recent review can be found e.g. in [2]. The $LL\overline{E}$ and $LQ\overline{D}$ terms violate the leptonic quantum number L, while the $\overline{UDD}$ terms violate the baryonic quantum number B: their simultaneous presence would lead to a fast proton decay [3] and is thus forbidden in the MSSM by imposing R-parity conservation. However, the absence of either the B-violating or the L-violating terms is enough to prevent the proton decay at a too high rate. If only B-parity conservation is required, the lepton number violating terms $LL\overline{E}$ and $LQ\overline{D}$ can still be present, and the process (2) can be observable. The neutralino[2] can in fact decay into SM fermions via scalar fermion exchange, as listed below [4]:

$$\lambda_{ijk} \quad : \quad \tilde{\chi}_1^0 \to \ell_i^- \nu_j \ell_k^+ \quad + \text{c.c.} \tag{4}$$

$$\lambda'_{ijk} \quad : \quad \tilde{\chi}_1^0 \to \ell_i^- u_j \bar{d}_k, \ \nu_i d_j \bar{d}_k \quad + \text{c.c.} \tag{5}$$

Limits obtained under the hypothesis of R-parity conservation are not valid any more if R-parity is violated, and dedicated analyses have to be performed.

In most experimental analyses, only one of the 45 coupling constant is assumed to dominate over all the others. More complicated situations, where several λ and λ' contribute, are discussed for example in ref. [5]. The coupling is assumed to be strong enough, that the LSP lifetime can be neglected. The mean neutralino decay length can in fact be expressed by means of the lifetime calculated in [6] as:

$$\ell_{\tilde{\chi}_1^0}(\text{cm}) = 0.3\,\frac{p_{\tilde{\chi}_1^0}}{M_{\tilde{\chi}_1^0}}\left(\frac{M_{\tilde{f}}}{100\text{ GeV}}\right)^4\left(\frac{1\text{ GeV}}{M_{\tilde{\chi}_1^0}}\right)^5\frac{1}{\lambda^2}, \tag{6}$$

[2] In the following we consider only the R-parity violating decays of the lightest neutralino.

where $p_{\tilde{\chi}_1^0}/M_{\tilde{\chi}_1^0} = \beta_{\tilde{\chi}_1^0}\gamma_{\tilde{\chi}_1^0}$ is the neutralino Lorentz factor, $M_{\tilde{f}}$ is the mass of the scalar particle exchanged in the neutralino decay, λ is the relevant R-parity violating coupling considered and masses and momenta are expressed in GeV. For instance, if $M_{\tilde{\chi}_1^0} = 80$ GeV the neutralino decay products are seen in the detector if $\ell_{\tilde{\chi}_1^0} \lesssim 1$ cm, that is if $\lambda \gtrsim 10^{-4}$.

In this analysis only the $LL\bar{E}$ mediated decays are considered, in which each neutralino decays into two charged plus one neutral lepton. In particular, two extreme cases have been studied, in which the dominant R-parity violating couplings are, respectively, λ_{122} and λ_{133}. In the former case one expects in fact the highest selection efficiency, because of the presence of electrons and muons in the final state. In the latter, the efficiency will be the lowest, because each neutralino decays into at least one tau.

2. Data and Monte Carlo samples

The data used in this analysis were collected with the L3 detector during the 1996 and 1997 high energy runs. The integrated luminosity is 76.42 pb^{-1}, of which 10.91 pb^{-1} were taken at 161.3 GeV, 1.00 pb^{-1} at 170.3 GeV, 9.25 pb^{-1} at 172.3 GeV and 55.27 pb^{-1} at 182.7 GeV.

The selection procedure has been developed on Monte Carlo signal and background events at $\sqrt{s} = 182.7$ GeV. The signal events $e^+e^- \to \tilde{\chi}_1^0\tilde{\chi}_1^0$, $e^+e^- \to \tilde{\chi}_2^0\tilde{\chi}_1^0 \to \tilde{\chi}_1^0 Z^* \tilde{\chi}_1^0$ and $e^+e^- \to \tilde{\chi}_1^+\tilde{\chi}_1^- \to W^*\tilde{\chi}_1^0\, W^*\tilde{\chi}_1^0$, with subsequent R-parity violating decays of the lightest neutralino, have been generated with SUSYGEN [7] at center-of-mass energies of $\sqrt{s} = 172.3$ and 182.7 GeV, for two possible values of the λ couplings: λ_{122} and λ_{133}.

In the case of direct neutralino production, samples of 400 or 500 events have been generated with different values of the neutralino mass $M_{\tilde{\chi}_1^0}$ between 5 GeV and the kinematic limit. In the case of the process $e^+e^- \to \tilde{\chi}_2^0\tilde{\chi}_1^0$, samples of 125 or 400 events have been generated with $\Sigma M = M_{\tilde{\chi}_2^0} + M_{\tilde{\chi}_1^0} = 130$ and 170 or 180 GeV, for different values of $\Delta M = M_{\tilde{\chi}_2^0} - M_{\tilde{\chi}_1^0}$ between 5 and 75 GeV, at $\sqrt{s} = 172.3$ GeV and 182.7 GeV. Additional samples of 400 events have been generated for the reactions $e^+e^- \to \tilde{\chi}_3^0\tilde{\chi}_1^0 \to \tilde{\chi}_1^0 Z^* \tilde{\chi}_1^0$ and $e^+e^- \to \tilde{\chi}_4^0\tilde{\chi}_1^0 \to \tilde{\chi}_1^0 Z^* \tilde{\chi}_1^0$ at $\sqrt{s} = 182.7$ GeV, with $\Sigma M = 180$ GeV and $\Delta M = 75$ GeV.

In the case of the process $e^+e^- \to \tilde{\chi}_1^+\tilde{\chi}_1^-$, samples of 400 or 500 events have been generated with a chargino mass $M_{\tilde{\chi}_1^\pm}$ equal to 85 or 91 GeV, for different values of the neutralino mass, corresponding to different $\Delta M = M_{\tilde{\chi}_1^\pm} - M_{\tilde{\chi}_1^0}$ values between 5 and 55 GeV. Smaller samples have been generated with larger values of ΔM and with lower chargino masses. Since in some regions of the parameter space $M_{\tilde{\chi}_1^\pm}$ is greater than $M_{\tilde{\chi}_2^0}$, the lightest chargino can decay in cascade to $\tilde{\chi}_1^0$ through $\tilde{\chi}_2^0$ [8], via the process

$e^+e^- \to \tilde{\chi}_1^+\tilde{\chi}_1^- \to \tilde{\chi}_2^0 W^* \tilde{\chi}_2^0 W^* \to \tilde{\chi}_1^0 Z^* W^* \tilde{\chi}_1^0 Z^* W^*$. For this reaction, 150 events have been generated with $M_{\tilde{\chi}_1^\pm} = 91$ GeV and $M_{\tilde{\chi}_2^0} = 55$ GeV at $\sqrt{s} = 182.7$ GeV.

For the background studies the following Monte Carlo samples have been used, at $\sqrt{s} = 161$, 172 and 183 GeV:

- two-fermion processes: $e^+e^- \to Z \to q\bar{q}$ (PYTHIA [9]), e^+e^- (BHAGENE3 [10]), $\mu^+\mu^-$ and $\tau^+\tau^-$ (KORALZ [11]), corresponding to integrated luminosities of about 70, 5, 250 and 250 times the luminosity of the data;
- two-photon processes: $e^+e^- \to \gamma\gamma \to e^+e^-q\bar{q}$ (PHOJET [12]) and $e^+e^- \to \gamma\gamma \to e^+e^-e^+e^-, e^+e^-\mu^+\mu^-$ and $e^+e^-\tau^+\tau^-$ (DIAG36 [13]), corresponding to integrated luminosities of about 3, 7, 7 and 10 times the luminosity of the data;
- events with four fermions in the final state, coming from W and Z decays: $e^+e^- \to W^+W^-$ (KORALW [14]), $e^+e^- \to Z e^+e^-$ and $e^+e^- \to ZZ$ (PYTHIA), corresponding to integrated luminosities of about 300, 300 and 1000 times the luminosity of the data.

The detector response is simulated using the GEANT 3.15 package [15]. It takes into account effects of energy loss, multiple scattering and showering in the detector materials and in the beam pipe. Hadronic interactions are simulated with the GHEISHA program [16]. Time dependent inefficiencies of the different subdetectors are also taken into account in the simulation procedure.

3. Analysis Procedure

3.1. Signal Topologies

The processes $e^+e^- \to \tilde{\chi}_1^0\tilde{\chi}_1^0$, $e^+e^- \to \tilde{\chi}_2^0\tilde{\chi}_1^0 \to \tilde{\chi}_1^0 Z^* \tilde{\chi}_1^0$ and $e^+e^- \to \tilde{\chi}_1^+\tilde{\chi}_1^-$ are searched for, where in the latter case each chargino decays into $\tilde{\chi}_1^0 f\bar{f}'$, with f and $\bar{f}'$ belonging to the same weak isodoublet. Subsequent R-parity violating decays of the two neutralinos into $\ell^+\ell^-\nu$ take place. The possible topologies arising from the different final states are listed in Table 1.

In the following only the two couplings λ_{122} and λ_{133} are taken into account, leading to the decays of the neutralinos into two possible types of final states (with muons or with electrons and muons, and with taus or with electrons and taus, respectively). When the λ_{122} coupling is dominant, each neutralino can decay as follows:

$$\lambda_{122}: \quad \tilde{\chi}_1^0 \to \bar{\nu}_e\mu^+\mu^-,\ \nu_\mu e^+\mu^-,\ \nu_e\mu^-\mu^+ \text{ or } \bar{\nu}_\mu e^-\mu^+. \tag{7}$$

Process				Topology
$\tilde{\chi}^0_1\tilde{\chi}^0_1 \rightarrow$	$\ell\ell\ell\ell\nu\nu$			4 lept. + $\not{E}$
$\tilde{\chi}^0_i\tilde{\chi}^0_1 \rightarrow$ $(i \geq 2)$	$\tilde{\chi}^0_1\tilde{\chi}^0_1$ Z* $\rightarrow$	$\ell\ell\ell\ell\nu\nu$	qq	4 lept. + 2 jets + $\not{E}$
		$\ell\ell\ell\ell\nu\nu$	$\ell\ell$	6 lept. + $\not{E}$
		$\ell\ell\ell\ell\nu\nu$	$\nu\nu$	4 lept. + $\not{E}$
$\tilde{\chi}^+_1\tilde{\chi}^-_1 \rightarrow$	$\tilde{\chi}^0_1\tilde{\chi}^0_1$ W*W* $\rightarrow$	$\ell\ell\ell\ell\nu\nu$	$qq'qq'$	4 lept. + 4 jets + $\not{E}$
		$\ell\ell\ell\ell\nu\nu$	$qq'\ell\nu$	5 lept. + 2 jets + $\not{E}$
		$\ell\ell\ell\ell\nu\nu$	$\ell\nu\ell\nu$	6 lept. + $\not{E}$

Table 1. Processes and topologies considered in this analysis.

On the other hand, when the λ_{133} coupling is dominant, each neutralino can decay as listed below:

$$\lambda_{133}: \quad \tilde{\chi}^0_1 \rightarrow \bar{\nu}_e\tau^+\tau^-,\ \nu_\tau e^+\tau^-,\ \nu_e\tau^-\tau^+ \text{ or } \bar{\nu}_\tau e^-\tau^+. \tag{8}$$

The decay products of the neutralino pair are thus four leptons and two neutrinos. If the λ_{133} coupling dominates, the final state contains at least four neutrinos, due to the presence of at least two τ leptons. If the neutralino mass is high, the event topology consists of four charged particles, distributed in a roughly isotropic way, with missing energy. On the contrary, for low neutralino masses, the events have two back-to-back jets with missing energy. In this case, the selections are less efficient, because the topology is more similar to two jet background processes.

In the case of chargino production, the signal topologies depend not only on the value of the neutralino mass, but also on the chargino decay modes. Both charginos can in fact decay into $\tilde{\chi}^0_1$ and hadrons , or one chargino into $\tilde{\chi}^0_1$ and hadrons and the other into $\tilde{\chi}^0_1\,\ell\,\nu$ or both charginos can decay into $\tilde{\chi}^0_1$ and leptons. These final states are referred to as hadronic, mixed and leptonic respectively. Since for high neutralino masses the energy of the event is mainly carried by the neutralino decay products, almost independently on the different chargino decay channels, a common selection has been developed for all the chargino decay modes. For small ΔM in fact the signal topology is given by four energetic charged leptons and two neutrinos, which come from the decay of the two neutralinos and are isotropically distributed, plus four low energy leptons or hadronic jets. On the contrary, for medium and large ΔM, different sets of cuts have been developed for all the possible configurations. Common to all selections is the requirement of a relatively low number of clusters and tracks in the hadronic decay channel as well. This is due to the fact that the main background sources for the hadronic mode are represented by the processes $e^+e^- \rightarrow q\bar{q}$ and

$e^+e^- \to W^+W^- \to q\bar{q}q\bar{q}$: thus a relatively low multiplicity has to be required in order to be able to reject the background.

In the case of the process $e^+e^- \to \tilde{\chi}_2^0\tilde{\chi}_1^0 \to \tilde{\chi}_1^0 Z^* \tilde{\chi}_1^0$, three possible final states are present, corresponding to the Z^* decays into hadrons, charged leptons or neutrinos. As it is shown in Table 1, the topologies arising from this process are already taken into account by combining the neutralino and chargino pair selections.

3.2. Event Preselection

A common preselection has been developed for all the processes listed in Table 1. The corresponding cuts are summarised in Table 2. The events are required to have at least 3 charged tracks and 4 calorimetric clusters, in order to remove the bulk of $e^+e^- \to e^+e^-, \mu^+\mu^-$ and double-leptonic $\tau^+\tau^-$ and W^+W^- decays. Very high multiplicity events, like W^+W^- and ZZ hadronic decays, are reduced by requiring the number of tracks not to exceed 32, and the number of clusters not to exceed 75. The visible energy is required to be smaller than 90% of the center-of-mass energy $\sqrt{s}$, against $q\bar{q}$ and hadronic W^+W^- and ZZ decays. The contamination from hadronic events is also reduced by requiring the presence of at least 2 leptons. Two-fermion back-to-back events are reduced by applying the cut on the polar angle of the total momentum (θ_{tot}), and on the acollinearity and acoplanarity angles. The acollinearity and acoplanarity angles are calculated by grouping all clusters in every event into exactly two jets by using the DURHAM algorithm. The cuts on θ_{acol} and θ_{acop} are also very efficient against W^+W^- and ZZ decays.

Most preselection cuts are applied in order to reduce the contribution to the background coming from two-photon events. Tagged two-photon interactions are rejected by requiring that the sum $E_{\mathrm{lum}} + E_{\mathrm{alr}}$ of the energies measured in the lead-scintillator ring calorimeter and in the luminosity monitor, corresponding to a polar angle between 1.5° and 9.0°, is less than 10 GeV. In addition, the visible energy must be greater than 20% of $\sqrt{s}$. The total number of leptons has to be at least 2. This cut is very efficient against $e^+e^- \to e^+e^-q\bar{q}$ events, when one or both electrons are not detected. Two-photon background is further reduced by means of the three angular cuts.

The preselection efficiency is $\sim$ 5% for $\tau^+\tau^-$ and $\sim$ 1% for $q\bar{q}$ events. Less than 10^{-3} $\mu^+\mu^-$ events pass the preselection. The efficiency on Ze^+e^- is $\sim$ 2%, and it is $\sim$ 10% for W^+W^- and ZZ events. Less than 10^{-4} e^+e^- and $\gamma\gamma$ events are selected. The preselection efficiency on the signal $e^+e^- \to \tilde{\chi}_1^+\tilde{\chi}_1^-$ is $\sim$ 80% for $M_{\tilde{\chi}_1^\pm} = 91$ GeV. On $e^+e^- \to \tilde{\chi}_1^0\tilde{\chi}_1^0$ it is $\sim$ 80% for $M_{\tilde{\chi}_1^0} \geq 50$ GeV and between 20% and 40% for $M_{\tilde{\chi}_1^0} \leq 20$ GeV.

N_{tracks}	3 – 32
$N_{clusters}$	4 – 75
$N_{e,\mu,\tau}$	≥ 2
$E_{lum} + E_{alr}$	$\leq$ 10 GeV
$\frac{E_{vis}}{\sqrt{s}}$	0.2 –0.9
θ_{tot}	15° – 165°
θ_{acol}	$\leq$ 178°
θ_{acop}	$\leq$ 178°

Table 2. Summary of all the preselection cuts for both the neutralino and the chargino selections.

After the preselection is applied, 358 events are selected in the data and 354 events are expected in total from 161, 172 and 183 GeV background processes, as reported in Table 3. The distributions of two variables at this stage are shown in Fig. 1 and Fig. 2.

3.3. Analysis Optimization

For each selection, some cuts are optimized simultaneously in order to maximize the analysis sensitivity. Discrete variables, whose cut values are directly deduced from the event topologies, and variables, whose cut values remain the same as in the preselection, are not included in the optimization procedure.

The optimization procedure uses Monte Carlo signal and background events and maximizes the sensitivity $1/\sigma_{95}$, where σ_{95} is given by [17]:

$$\sigma_{95} = \frac{1}{\varepsilon} \sum_{n=0}^{\infty} c_n P_b(n). \tag{9}$$

Here ε is the signal selection efficiency, $P_b(n) = e^{-b} \frac{b^n}{n!}$ is the poissonian distribution of the background events and c_n are the upper limits at 95% C.L. on the signal expectation values, in absence of background subtraction.

Two-photon background has been further analysed in order to check whether it affects the maximization, due to the lower statistics available. It has been checked that two-photon events don't influence the cut values obtained from the optimization procedure, since the final cut positions depend only on other background sources. As a further check, the maximization has been performed on half of the two-photon event sample and the background contamination has been estimated on the other half.

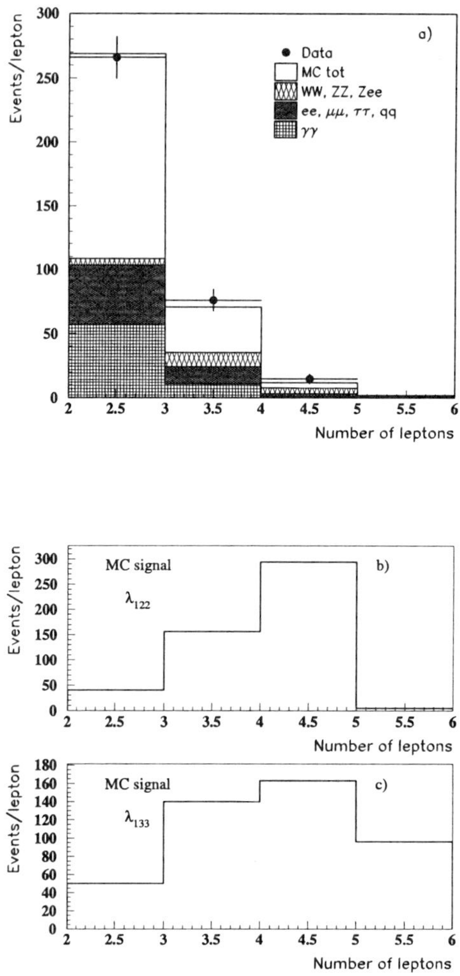

Figure 1. Lepton multiplicity distribution after the preselection of Section 3.2. Fig. a) shows data and background Monte Carlo at $\sqrt{s} = 161$, 172 and 183 GeV. Figures b) and c) show two examples of signal, with low and high multiplicity respectively, arbitrarily normalised. Fig. b): $\tilde{\chi}_1^0\tilde{\chi}_1^0$, with $M_{\tilde{\chi}_1^0} = 91$ GeV and coupling λ_{122}. Fig. c): $\tilde{\chi}_1^+\tilde{\chi}_1^-$, with $M_{\tilde{\chi}_1^\pm} = 91$ GeV, $\Delta M = 5$ GeV and coupling λ_{133}.

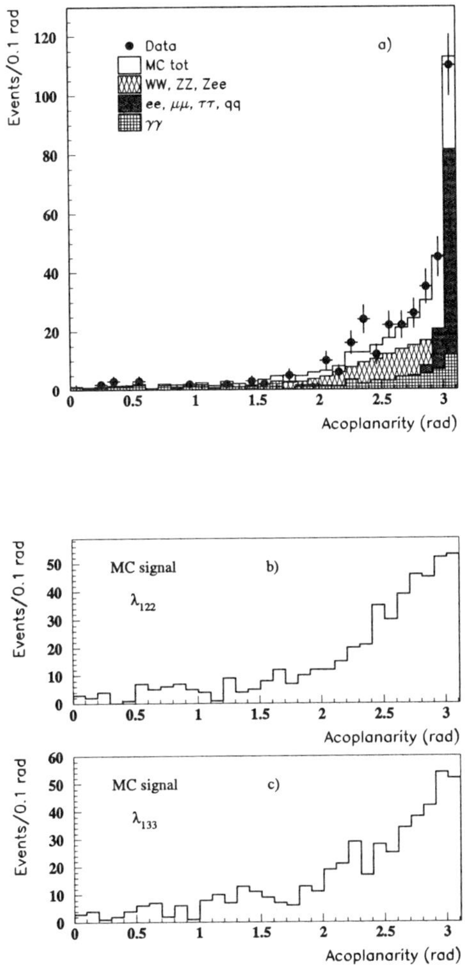

Figure 2. Acoplanarity distribution after the preselection of Section 3.2. Fig. a) shows data and background Monte Carlo at $\sqrt{s} = 161$, 172 and 183 GeV. Figures b) and c) show two examples of signal, with low and high multiplicity respectively, arbitrarily normalised. Fig. b): $\tilde{\chi}_1^0\tilde{\chi}_1^0$, with $M_{\tilde{\chi}_1^0} = 91$ GeV and coupling λ_{122}. Fig. c): $\tilde{\chi}_1^+\tilde{\chi}_1^-$, with $M_{\tilde{\chi}_1^\pm} = 91$ GeV, $\Delta M = 5$ GeV and coupling λ_{133}.

	Data	MC tot	2f	2γ	4f
161 GeV	55	54.9	42.7	8.3	3.9
172 GeV	70	54.7	25.7	6.7	22.3
183 GeV	233	243.9	63.2	53.6	127.1
Total	358	353.6	131.6	68.6	153.4

Table 3. The number of expected data and background events after the preselection of Table 2, for the different center-of-mass energies. The different background contributions are also listed.

4. Results

The selection efficiencies of the process $e^+e^- \to \tilde{\chi}_1^0\tilde{\chi}_1^0$ depend on $M_{\tilde{\chi}_1^0}$ and range from 15% to 65% (λ_{122}) and from 3% to 50% (λ_{133}). In the case of the process $e^+e^- \to \tilde{\chi}_1^+\tilde{\chi}_1^-$, the efficiencies go from 70% to 45% (λ_{122}) and from 55% to 25% (λ_{133}), depending on $\Delta M = M_{\tilde{\chi}_1^\pm} - M_{\tilde{\chi}_1^0}$. The selection efficiencies of $e^+e^- \to \tilde{\chi}_2^0\tilde{\chi}_1^0$ depend on $\Delta M = M_{\tilde{\chi}_2^0} - M_{\tilde{\chi}_1^0}$ and range from 65% to 45% (λ_{122}) and from 50% to 25% (λ_{133}). The selection efficiencies of the reactions $e^+e^- \to \tilde{\chi}_3^0\tilde{\chi}_1^0 \to \tilde{\chi}_1^0 Z^* \tilde{\chi}_1^0$ and $e^+e^- \to \tilde{\chi}_4^0\tilde{\chi}_1^0 \to \tilde{\chi}_1^0 Z^* \tilde{\chi}_1^0$ at $\sqrt{s} = 182.7$ GeV, with $\Sigma M = 180$ GeV and $\Delta M = 75$ GeV, are similar to the efficiencies of the processes $e^+e^- \to \tilde{\chi}_2^0\tilde{\chi}_1^0 \to \tilde{\chi}_1^0 Z^* \tilde{\chi}_1^0$ for the same mass point. As a consequence, the efficiencies obtained for $\tilde{\chi}_2^0\tilde{\chi}_1^0$ are used also in the case of $\tilde{\chi}_3^0\tilde{\chi}_1^0$ and $\tilde{\chi}_4^0\tilde{\chi}_1^0$ production. Since in some regions of the parameter space $M_{\tilde{\chi}_1^\pm}$ is greater than $M_{\tilde{\chi}_2^0}$, the lightest chargino can decay in cascade to $\tilde{\chi}_1^0$ through $\tilde{\chi}_2^0$. The selection efficiencies of the process $e^+e^- \to \tilde{\chi}_1^+\tilde{\chi}_1^- \to \tilde{\chi}_2^0 W^* \tilde{\chi}_2^0 W^* \to \tilde{\chi}_1^0 Z^* W^* \tilde{\chi}_1^0 Z^* W^*$, with $M_{\tilde{\chi}_1^\pm} = 91$ GeV and $M_{\tilde{\chi}_2^0} = 55$ GeV, at $\sqrt{s} = 182.7$ GeV, are higher than the efficiencies of the process $\tilde{\chi}_1^+\tilde{\chi}_1^- \to \tilde{\chi}_1^0 W^* \tilde{\chi}_1^0 W^*$, which are conservatively assumed to be valid in all the parameter space.

The number of expected background events is shown in detail in Table 4. In total, 1.52 events are expected for the Yukawa coupling λ_{122} from 161-183 GeV background processes, and no events are found in the data, corresponding to the total luminosity of 76.4 pb^{-1} considered. In the case of the coupling λ_{133}, 1.55 events are expected and no candidates are found in the data sample.

4.1. MSSM Limits

The exclusion regions at 95% C.L. in the $M_2 - \mu$ plane are shown in Figures 3 to 8, for $m_0 = 500$ GeV and three different values of tanβ: 1, 1.41

and 40. The plots show the regions excluded by $\tilde{\chi}_1^0\tilde{\chi}_1^0$, $\tilde{\chi}_2^0\tilde{\chi}_1^0$, $\tilde{\chi}_3^0\tilde{\chi}_1^0$ and $\tilde{\chi}_1^+\tilde{\chi}_1^-$ analyses at LEP II and by the L3 precision measurements of the Z lineshape at LEP I. In the latter case, a point is excluded if:

$$\frac{\sigma_{\mathrm{SUSY}}}{\sigma_{\mathrm{Z}}}\Gamma_{\mathrm{Z}} > \Gamma_{\mathrm{LIM}}, \tag{10}$$

where σ_{SUSY} is the sum of the pair production cross sections of the supersymmetric particles involved, calculated with SUSYGEN [7], σ_{Z} is the total Z cross section, Γ_{Z} is the total Z width and $\Gamma_{\mathrm{LIM}} = 22.61$ MeV [18] is the 95% upper limit on non-SM contributions to the total Z width, obtained from the L3 data. For low $\tan\beta$, some regions beyond the chargino kinematic limit are excluded by $\tilde{\chi}_2^0\tilde{\chi}_1^0$ and $\tilde{\chi}_3^0\tilde{\chi}_1^0$ analyses.

For $m_0 = 500$ GeV, $0 \leq M_2 \leq 2000$ GeV, -500 GeV $\leq \mu \leq 500$ GeV and $1 \leq \tan\beta \leq 40$, we derived the following lower limits at 95% C.L. on the neutralino and chargino masses,

- $M_{\tilde{\chi}_1^0} > 26.8$ GeV
- $M_{\tilde{\chi}_2^0} > 44.3$ GeV
- $M_{\tilde{\chi}_1^\pm} > 91.2$ GeV.

Those limits are valid for both λ_{122} and λ_{133}, since in both cases we reached the chargino kinematic limit, and can thus be considered as not dependent on the Yukawa coupling analysed.

	2f	2γ	4f	Sum	Data
λ_{122}: 161–172 GeV	0.49	0.04	0.32	0.85	0
λ_{122}: 183 GeV	0.14	0	0.52	0.66	0
λ_{122}: total	0.63	0.04	0.85	1.52	0
λ_{133}: 161–172 GeV	0.31	0	0.32	0.63	0
λ_{133}: 183 GeV	0.14	0	0.78	0.92	0
λ_{133}: total	0.45	0	1.09	1.55	0

Table 4. The number of expected background events for the different selections and the different center-of-mass energies. All the background processes at $\sqrt{s} = 161$, 172 and 183 GeV are included.

References

[1] C.S. Aulakh and R.N. Mohapatra, *Phys. Lett. B* **119** (1982) 316;
F. Zwirner, *Phys. Lett. B* **132** (1983) 103;

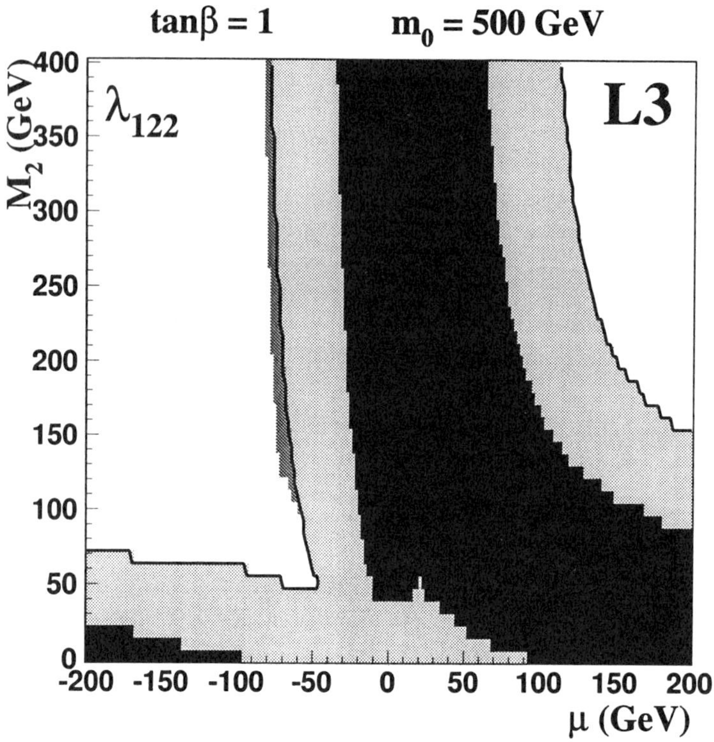

Figure 3. Exclusion region at 95% confidence level for λ_{122}, tan$\beta = 1$ and $m_0 = 500$ GeV. The black region is excluded by LEP I measurements and the yellow (light grey) one by the chargino analysis. The black solid line indicates the chargino kinematic limit. The green (grey) region beyond the kinematic limit is excluded by $\tilde{\chi}^0_2\tilde{\chi}^0_1$ and $\tilde{\chi}^0_3\tilde{\chi}^0_1$ analyses. The total data sample of 76.4 pb^{-1} is included.

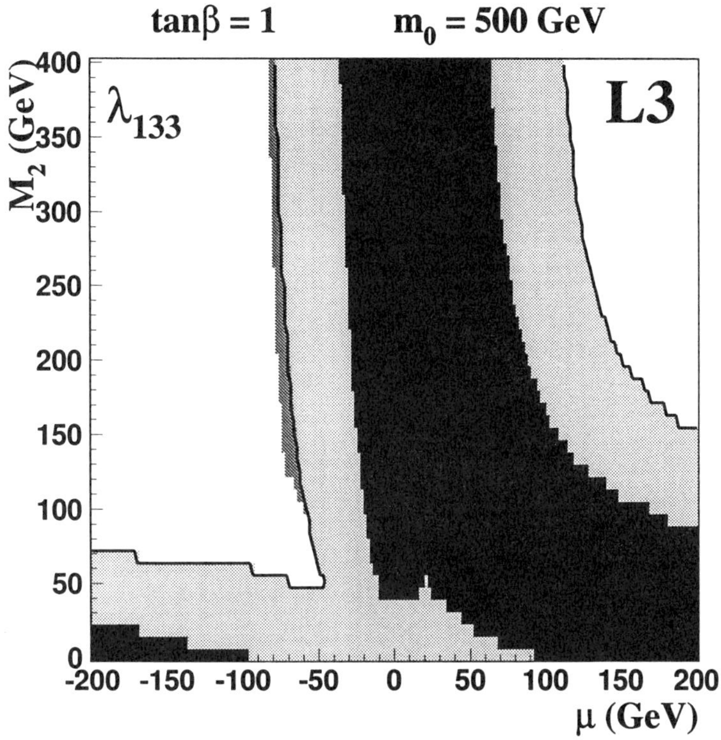

Figure 4. Exclusion region at 95% confidence level for λ_{133}, tan$\beta = 1$ and $m_0 = 500$ GeV. The black region is excluded by LEP I measurements and the yellow (light grey) one by the chargino analysis. The black solid line indicates the chargino kinematic limit. The green (grey) region beyond the kinematic limit is excluded by $\tilde{\chi}_2^0\tilde{\chi}_1^0$ and $\tilde{\chi}_3^0\tilde{\chi}_1^0$ analyses. The total data sample of 76.4 pb^{-1} is included.

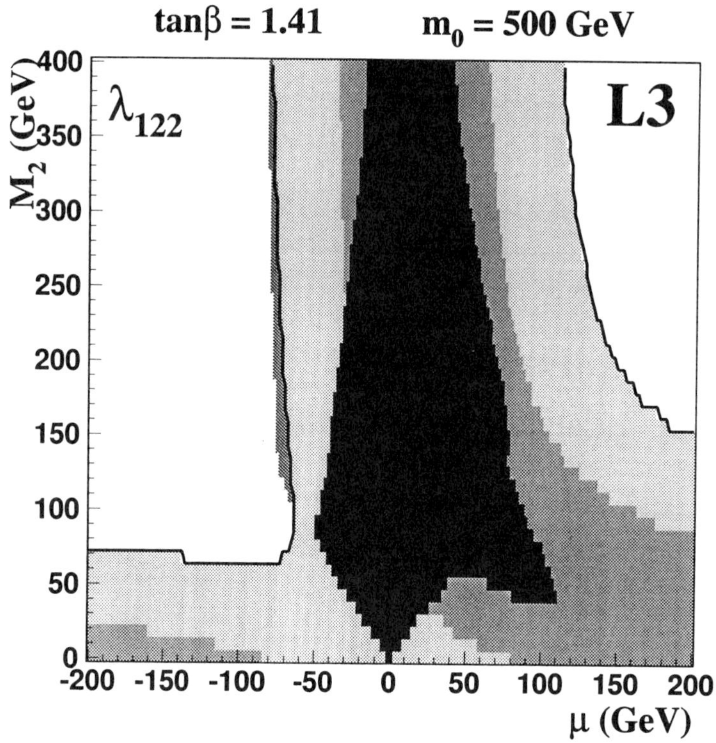

Figure 5. Exclusion region at 95% confidence level for λ_{122}, tan$\beta = 1.41$ and $m_0 = 500$ GeV. The blue (black) region is excluded by the neutralino analysis, the light blue (dark grey) region by LEP I measurements and the yellow (light grey) one is excluded by the chargino analysis. The black solid line indicates the chargino kinematic limit. The green (grey) region beyond the kinematic limit is excluded by $\tilde{\chi}^0_2\tilde{\chi}^0_1$ and $\tilde{\chi}^0_3\tilde{\chi}^0_1$ analyses. The total data sample of 76.4 pb^{-1} is included.

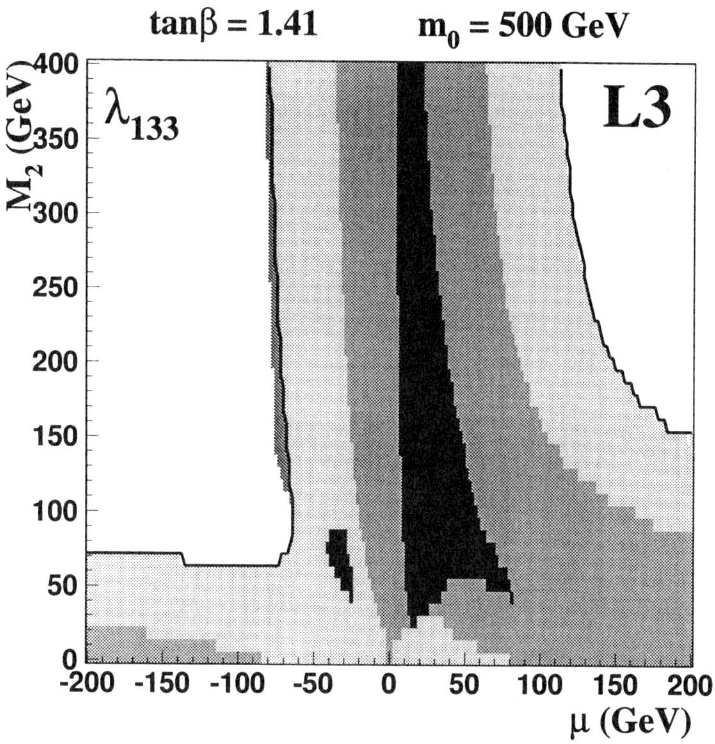

Figure 6. Exclusion region at 95% confidence level for λ_{133}, tan$\beta = 1.41$ and $m_0 = 500$ GeV. The blue (black) region is excluded by the neutralino analysis, the light blue (dark grey) region by LEP I measurements and the yellow (light grey) one is excluded by the chargino analysis. The black solid line indicates the chargino kinematic limit. The green (grey) region beyond the kinematic limit is excluded by $\tilde{\chi}^0_2\tilde{\chi}^0_1$ and $\tilde{\chi}^0_3\tilde{\chi}^0_1$ analyses. The total data sample of 76.4 pb^{-1} is included.

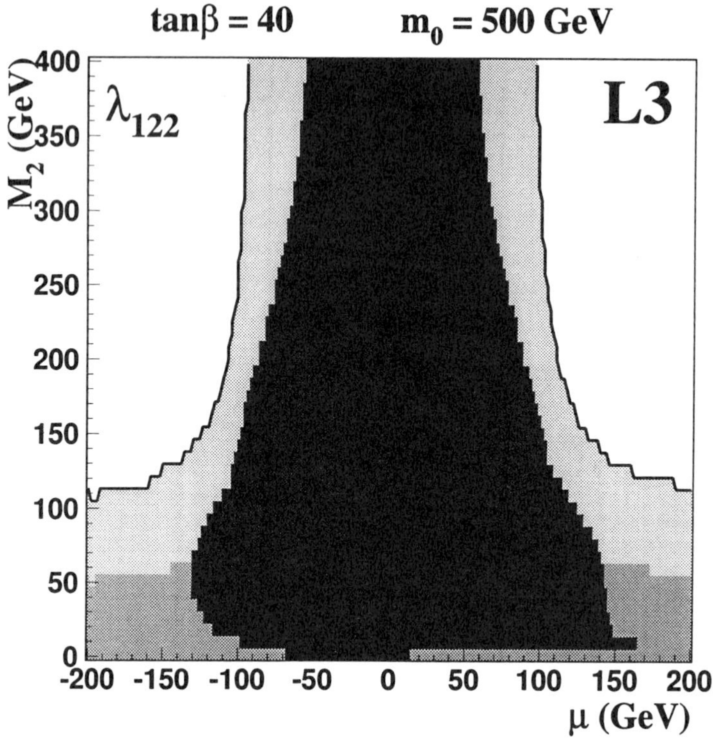

Figure 7. Exclusion region at 95% confidence level for λ_{122}, tanβ = 40 and m_0 = 500 GeV. The blue (black) region is excluded by the neutralino analysis, the light blue (dark grey) region by LEP I measurements and the yellow (light grey) one is excluded by the chargino analysis. The black solid line indicates the chargino kinematic limit. The total data sample of 76.4 pb^{-1} is included.

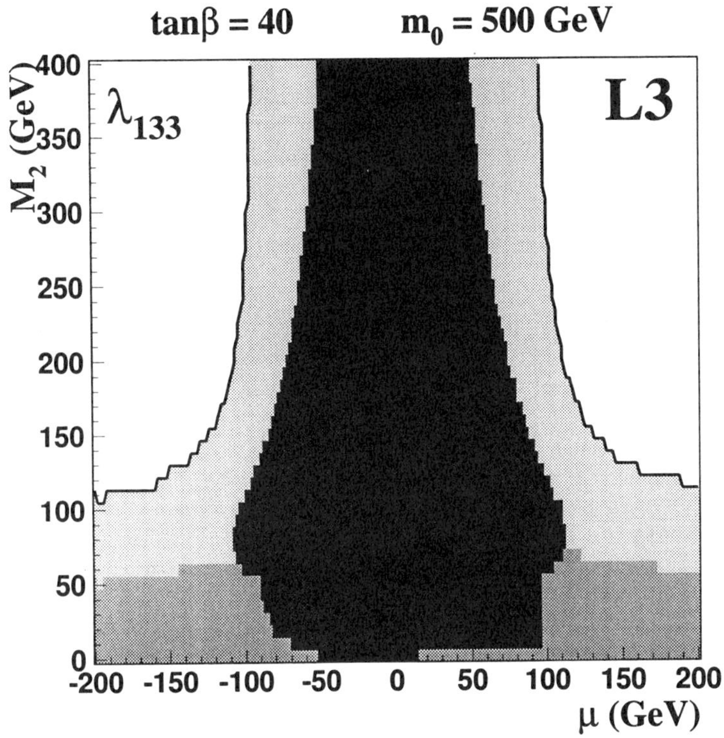

Figure 8. Exclusion region at 95% confidence level for λ_{133}, $\tan\beta = 40$ and $m_0 = 500$ GeV. The blue (black) region is excluded by the neutralino analysis, the light blue (dark grey) region by LEP I measurements and the yellow (light grey) one is excluded by the chargino analysis. The black solid line indicates the chargino kinematic limit. The total data sample of $76.4\ \mathrm{pb}^{-1}$ is included.

L.J. Hall and M. Suzuki, *Nucl. Phys. B* **231** (1984) 419;
R. Barbieri and A. Masiero, *Nucl. Phys. B* **267** (1986) 679;
S. Dimopoulos and L.J. Hall, *Phys. Lett. B* **207** (1987) 210;
V. Barger, G.F. Giudice and T. Han, *Phys. Rev. D* **40** (1989) 2987;
L. Hall, Mod. *Phys. Lett.* A **5** (1990) 467;
V. Barger, W.Y. Keung and R.J.N. Phillips, *Phys. Lett. B* **356** (1995) 546.

[2] H. Dreiner, "An introduction to explicit R-parity violation", hep–ph/9707435.

[3] S. Weinberg, *Phys. Rev. D* **26** (1982) 287.

[4] H. Dreiner and P. Morawitz, *Nucl. Phys. B* **428** (1994) 31.

[5] R. Godbole, P. Roy and X. Tata, *Nucl. Phys. B* **401** (1993) 67.

[6] S. Dawson, *Nucl. Phys. B* **261** (1985) 297.

[7] SUSYGEN:
S. Katsanevas and S. Melachroinos, Proceedings of the Workshop "Physics at LEP 2", eds. G. Altarelli, T. Sjöstrand and F. Zwirner, CERN 96-01 (1996), vol. 2, p. 328.

[8] A. Bartl, H. Fraas, W. Majerotto and B. Mösslacher, *Z. Phys. C* **55** (1992) 257.

[9] T. Sjöstrand, PYTHIA 5.7 and JETSET 7.4 Physics and Manual, CERN-TH/7112/93 (1993), revised August 1995; *Comp. Phys. Comm.* **82** (1994) 74.

[10] BHAGEN3: Version 3 was used.
J.H. Field, *Phys. Lett. B* **323** (1994) 432;
J.H. Field and T. Riemann, *Comp. Phys. Comm.* **94** (1996) 53.

[11] KORALZ: Version 4.02 was used.
S. Jardach, B.F.L. Ward, and Z. Wąs, *Comp. Phys. Comm.* **79** (1994) 503.

[12] PHOJET: Version 1.05 was used.
R. Engel, *Z. Phys. C* **66** (1995) 203;
R. Engel and J. Ranft, *Phys. Rev. D* **54** (1996) 4244.

[13] DIAG36:
F.A. Berends, P.H. Daverfeldt and R. Kleiss, *Nucl. Phys. B* **253** (1985) 441.

[14] KORALW: Version 1.21 was used.
M. Skrzypek, S. Jardach, W. Placzek, and Z. Wąs, *Comp. Phys. Comm.* **94** (1996) 216;
M. Skrzypek, S. Jardach, M. Martinez, W. Placzek, and Z. Wąs, *Phys. Lett. B* **372** (1996) 289.

[15] GEANT:
R. Brun et al, CERN DD/EE/84-1 (Revised 1987).

[16] GHEISHA:
H. Fesefeldt, RWTH Aachen Report PITHA 85/2 (1985).

[17] J.F. Grivaz and F. Le Diberder, Complementary analyses and acceptance optimization in new particle searches, LAL 92-37 (1992).

[18] J. Mnich, private communication.

List of Participants

Wanda M. Alberico
Dipartimento di Fisica Teorica
dell'Universita di Torino
and INFN, Sezione di Torino
via P.Giuria 1, 10125 Torino
ITALY
tel: 011 67 07 236
fax: 011 67 07 214
e-mail: alberico@torino.infn.it
e-mail: alberico@to.infn.it

Geneviéve Bélanger
LAPTH
Chemin de Bellevue, B.P. 110
F-74941, Annecy-le-Vieux, Cedex
FRANCE
tel: 33 4 50 09 16 76
fax: 33 4 50 27 94 95
e-mail: belanger@lapp.in2p3.fr

Luis Bento
Centro de Fisica Nuclear da
Universidade de Lisboa
Av. Prof. Gama Pinto 2
1699 Liboa -codex
PORTUGAL
tel: 351 01 79 04821
fax: 351 01 79 54288
e-mail: LBENTO@fc.ul.pt

Zurab Berezhiani
Dipartimento di Fisica
Universita di Ferrara
Via Paradiso 12
I-44100 Ferrara
ITALY
tel: 39 532 781871
fax: 39 532 762057
Institute of Physics
Georgian Academy of Sciences
Tbilisi
GEORGIA
e-mail: BEREZHIANI@fe.infn.it

Douglas Bryman
TRIUMF
4004 Wesbrook Mall
Vancouver
B. C. CANADA
V6T 2A3
tel: 604 222 7338
tel: (TRIUMF) 604 222 1047
fax: 604 222 1074
e-mail: doug@triumf.ca

Gerd W. Buschhorn
Max-Planck-Institut fuer Physik
Werner - Heisenberg Institut
Foehringer Ring 6
D - 80805, Muenchen
GERMANY
tel: 89 32 354 208
tel: 89 32 354 207
fax: 89 32 354 436
e-mail: gwb@mppmu.mpg.de

Massimo Corradi
Bologna University
and INFN, Bologna
via Irnerio, 47
40127, Bologna
ITALY
e-mail: CORRADI@vxdesy.desy.de

Silvia Costantini
Humbold University Berlin
Invalidenstr.110
10125 Berlin
GERMANY
tel:0049 30 20 93 7671
fax:0049 30 20 93 7642
e-mail: Silvia.Costantini@cern.ch

Laura Covi
School of Physics and Chemistry
Lancaster Univ. Bailrigg
Lancaster LA1 4YB
ENGLAND
tel: 44 1524 59 3930
fax: 44 1524 84 4037
e-mail: l.covi@lancaster.ac.uk

Sacha Davidson
CERN
Theory Division
CH - 1211, Geneve 23
SWITZERLAND
tel: 41 22 767 4523
fax: 41 22 767 3850
e-mail: Sacha.Davidson@cern.ch

Toby Falk
Phenomenology Inst.
Departm. of Physics
1150 Univ. Avenue
Univ. of Wisconsin
Madison, WI 53706
USA
tel: 608 262 4699
fax: 608 262 8628
e-mail: falk@pheno.physics.wisc.edu

Zoltan Fodor
Institute for Theoretical Physics
Eoetvoes University
Budapest, H - 1088
Puskin ul. 5 - 7
HUNGARY
e-mail: fodor@hal9000.elte.hu
e-mail: z.fodor@cern.ch

Katri Huitu
Helsinki Institute of Physics
P. O. Box 9
FIN - 00014
University of Helsinki
FINLAND
tel: 358 9 1918405
fax: 358 9 1918458
e-mail: katri.huitu@helsinki.fi

Gerry Garvey
LANL
Los Alamos
USA
e-mail: garvey@lanl.gov

Marco G. Giammarchi
Instituto Nazionale di Fisica Nucleare
Milano, Via Celoria 16
20133 Milano
ITALY
tel: 39 2 2392 305
fax: 39 2 2392 617
e-mail: Marco.Giammarchi@mi.infn.it

Andrea Giuliani
INFN - Sezione di Milano
via Celoria 16, Milano
I - 20133
ITALY
tel: 39 2 239 2639
fax: 39 2 706 09512
e-mail: Andrea.Giuliani@mi.infn.it

Paolo Gondolo
Max-Planck-Institut fuer Physik
Werner-Heisenberg-Institut
Foehringer Ring 6
80805 Muenchen
GERMANY
tel: 49 89 323540
fax: 49 89 3236704
e-mail: gondolo@mppmu.mpg.de

Hans Volker Klapdor-Kleingrothaus
Max-Planck-Institut fuer Kernphysik
Heidelberg
Postfach 10 39 80
D - 69029
Heidelberg
GERMANY
tel: 06221 516 262
fax: 06221 516 540
e-mail: klapdor@mickey.mpi-hd.mpg.de

Stephan Kolb
Max-Planck-Institut fuer Kernphysik
Heidelberg
Postfach 10 39 80
D - 69029
Heidelberg
GERMANY
tel: 06221 516 259
fax: 06221 516 540
e-mail: kolb@mickey.mpi-hd.mpg.de

Irina Krivosheina
Radio-Physical Research Institut (NIRFI)
Nishnij-Novgorod
Bolshaja Pesherskaya 25
603 000 Nishnij-Novgorod
RUSSIA
tel: 06221 516 262
fax: 06221 516 540
e-mail: irina@mickey.mpi-hd.mpg.de

Yoshitaka Kuno
Institute of Particle and Nuclear Studies (IPNS)
High Energy Accelerator
Research Organization (KEK)
Oho 1-1 Tsukuba
Ibaraki 305
JAPAN
tel: 81 298 64 5419
fax: 81 298 64 7831
e-mail: yoshitaka.kuno@kek.jp

Marta Losada
CERN TH
1211 Geneva 23
SWITZERLAND
tel: 767 24 39
e-mail: losada@mail.cern.ch

Jukka Maalampi
Dep. of Physics
Helsinki Istitute of Physics
P. O. Box 9
FIN - 00014
University of Helsinki
FINLAND
tel: 358 9 191 8438
fax: 358 9 191 8366
e-mail: jukka.maalampi@helsinki.fi

Fulvio Mauri
Department di Fisica Nucleare e Teorica e INFN
University di Pavia, Sezione di Pavia
via Bassi 6
27100 Pavia
ITALY
tel: 39 382 507419
fax: 39 382 423241
e-mail: mauri@pv.infn.it

Antonio Masiero
SISSA, Via Beirut 2-4
34013 Trieste
ITALY
INFN, Sezione di Trieste
Padriciano 99
34012 Trieste
ITALY
e-mail: masiero@padova.infn.it
e-mail: ANTONIO.MASIERO@CERN.CH
e-mail: MASIERO@IPDGR4.PD.INFN.IT
e-mail: antonio.masiero@pd.infn.it

Omar Miranda
Instituto de Fisica Corpuscular - C.S.I.C.
Departament de Fisica Teorica
Universitat de Valencia
C/ Dr Moliner 50
46100 Burjassot
Valencia
SPAIN
tel: 34 6 3 86 4594
fax: 34 6 3 86 4555
e-mail: omr@flamenco.ific.uv.es

William R. Molzon
Department of Physics and Astronomy
University of California
Irvine, CA 92697 - 4575
USA
tel: 714 - 824 - 5987
fax: 714 - 824 - 7478
e-mail: wmolzon@uci.edu
e-mail: molzon@ucihep.ps.uci.edu

Rosy Nicolaidou
Institut des Sciences Nucleaires
53 Avenue des Martyrs
38026 Grenoble Cedex
FRANCE
tel: 0033 - 47 62 84 179
fax: 0033 - 47 62 84 004
e-mail: Rosy.Nicolaidou@cern.ch

Pran Nath
Department of Physics
Northeastern University Boston
Mass 02115
USA
tel: 617 373 4669
fax: 617 373 2943
e-mail: nath@neu.edu
e-mail: nath@albert.physics.neu.edu

Holger Bech Nielsen
Niels Bohr Institute
Blegdamsvej 17 Copenhagen
DENMARK
tel: 45 3532 5259 (7)
e-mail: hbech@nbivms.nbi.dk

Ernst Otten
Institute fuer Physics
Johannes Gutenberg Univ. Mainz
Staudinger Weg 7
D - 55099 Mainz
GERMANY
tel: 49 613139 2518
secr.: 49 613139 2279
fax: 49 613139 3428
e-mail: exakt@dipmza.physik.uni-mainz.de
e-mail: Ernst.Otten@uni-mainz.de

Martti Raidal
DESY
Theory Group
Notkestrasse 85
22603, Hamburg
GERMANY
tel: (49-40) 89 98 2414
fax: (49-40) 89 98 2777
e-mail: raidal@mail.desy.de

Yorck Ramachers
Max-Planck-Institut fuer Kernphysik
Heidelberg
Postfach 10 39 80
D - 69029
Heidelberg
GERMANY
tel: 06221 516 259
fax: 06221 516 540
e-mail: Yorck.Ramachers@mpi-hd.mpg.de

Utpal Sarkar
Physical Research Laboratory
Ahmedabad – 380 009
INDIA
tel: 9179 46 2129 (off.)
tel: 9179 674 1951 (Residence)
fax: 9179 656 0502
e-mail: utpal@prl.ernet.in
e-mail: Utpal.Sarkar@cern.ch

Masatoshi Sato
Department of Physics
University of Tokyo
Tokyo 113 – 0033
JAPAN
tel: 81 3 3812 2111 ex.4139
fax: 81 3 5684 9642
e-mail: msato@hep-th.phys.s.u-tokyo.ac.jp
e-mail: msato@siren.phys.s.u-tokyo.ac.jp

Andries van der Schaaf
Physik - Institut der
Universitaet Zuerich
c/o PSI CH – 5234
Villigen PSI
SWITZERLAND
tel: 41 56 310 3653
fax: 41 56 310 3292
e-mail: vanderschaaf@psi.ch

Atsuto Suzuki
Aramaki, Aoba–ku
Sendai, Miyagi, 980 – 8578
JAPAN
tel: 81 22 217 6720
fax: 81 22 217 6728
e-mail: suzukia@awa.tohoku.ac.jp

Morimitsu Tanimoto
Science Education Laboratory
Ehime University
Bunkyo-cho 3
Matsuyama, 790-8577
JAPAN
and
Dep. de Fisica and CFIFUTL
Inst. Superior T.Enrico
1096 Lisboa, PORTUGAL
tel: 81 89 927 9441
fax: 81 89 927 9441
e-mail: tanimoto@edserv.ed.ehime-u.ac.jp

Jose W. F. Valle
Department de Fisica Teorica
IFIC/CSIC
University de Valencia
c/ Dr. Moliner, 50
46100 Burjassot
Valencia
SPAIN
tel: 34 6 3864555
fax: 34 6 3864555
e-mail: valle@neutrinos.uv.es
e-mail: valle@flamenco.ific.uv.es

Wolfgang Wallraff
AMS PWTH
Aachen
Germany
tel: 0241 807 187
tel:(USA) 001 407 867 1702
fax: 0241 88 88 190
fax:(USA) 001 407 867 1550
e-mail: Wallraff@Physik.RWTH-Aachen.De
e-mail: wallraff@acphyz.physik.rwth-aachen.de

Peter Wintz
Physik Institut
University of Zurich
Winterthurerstrasse 190
CH - 8057 Zuerich
SWITZERLAND
tel: 41 56 310 3648
fax: 41 56 310 3294
e-mail: Peter.Wintz@psi.ch

Author Index